MONOGRAPHS ON THE
PHYSICS AND CHEMISTRY OF
MATERIALS

MONOGRAPHS ON THE PHYSICS AND CHEMISTRY OF MATERIALS

Strong solids (Third edition) A. Kelly and N. H. Macmillan
Optical spectroscopy of inorganic solids B. Henderson and G. F. Imbusch
Quantum theory of collective phenomena G. L. Sewell
Experimental high-resolution electron microscopy (Second edition) J. C. H. Spence
Experimental techniques in low-temperature physics Guy K. White
Principles of dielectrics B. K. P. Scaife
Surface analytical techniques J. C. Rivière
Basic theory of surface states Sydney G. Davison and Maria Steślicka
Acoustic microscopy Andrew Briggs
Dynamic light scattering W. Brown
Quasicrystals: a primer (Second edition) C. Janot
Interfaces in crystalline materials A. P. Sutton and R. W. Balluffi
Rare-earth iron permanent magnets J. M. D. Coey (ed.)

Rare-earth Iron Permanent Magnets

Edited by

J. M. D. COEY
Trinity College, Dublin

CLARENDON PRESS • OXFORD
1996

Oxford University Press, Walton Street, Oxford OX2 6DP
Oxford New York
Athens Auckland Bangkok Bombay
Calcutta Cape Town Dar es Salaam Delhi
Florence Hong Kong Istanbul Karachi
Kuala Lumpur Madras Madrid Melbourne
Mexico City Nairobi Paris Singapore
Taipei Tokyo Toronto
and associated companies in
Berlin Ibadan

Oxford is a trade mark of Oxford University Press

Published in the United States
by Oxford University Press Inc., New York

A catalogue record for this book is available from the British Library

Library of Congress Cataloging in Publication Data
Rare-earth iron permanent magnets / edited by J.M.D. Coey.
(Monographs on the physics and chemistry of materials)
Includes index.
1. Permanent magnets. 2. Rare earth metals. 3. Iron–Magnetic properties. I. Coey, J. M. D. II. Series.
QC757.9.R36 1996 538'.44–dc20 95-47856

ISBN 0 19 851792 0

Typeset by EXPO Holdings, Malaysia
Printed in Great Britain by
Bookcraft (Bath) Ltd
Midsomer Norton, Avon

A wonder I experienced as a child of 4 or 5 years, when my father showed me a compass. That this needle behaved in such a determined way did not at all fit into the nature of events. I can still remember—or at least believe I can remember—that this experience made a deep and lasting impression upon me. – *Albert Einstein*

Truth is not found by proofs, but by exploration. It is always experimental. – *Simone Weil*

Preface

Permanent magnets have been objects of fascination for millenia. Lodestones offered the first palpable demonstration of 'action at a distance', providing a stable magnetic field which could be manipulated by the observer. Early magnet applications involved state ritual and medical quackery, besides the compass and the celebrated modelling of the Earth's dipole field, published by Gilbert in 1600 in *De Magnete*, perhaps the first modern scientific text.

By the Sung period (960–1279), technology for manufacturing steel wire for use as compass needles and magnetizing them with a lodestone or by cooling in a magnetic field was firmly established, both in China and in Europe, thus beginning the series of scientific advances which has culminated in the subject matter of this book. Until the turn of the nineteenth century, magnets were weak, unstable and made of carbon steel. Some improvements came with the discovery of cobalt magnet steels in Japan in 1917, and later with the development of a family of iron–aluminium–nickel–cobalt alloys — the Alnicos — the precursor of which was also discovered in Japan by Mishima in 1931. Alnicos were the first truly *permanent* magnets; they owe their properties to shape anisotropy associated with a special metallurgical microstructure.

The next step forward came in the early 1950s, with the development of synthetic hard ferrites at Philips in The Netherlands. Like lodestone, these magnets are composed of an iron oxide ($BaFe_{12}O_{19}$ or $SrFe_{12}O_{19}$). Although they have never held the record energy product $(BH)_{max}$, which is a convenient measure of the quality of a magnet, they are cheap, effective and reliable. Ferrites today account for 90% of the mass and a little over half the value of all magnets produced worldwide, an annual market worth roughly 2–3 billion US dollars.

The other half of the market is mostly for rare-earth magnets having energy products ten times greater than ferrites. Ferromagnetic alloys of 3d and 4f elements offer the best prospects for high-performance magnets. The role of the 3d partner is to provide most of the magnetization and ensure an adequate Curie temperature, whereas the 4f partner is present mainly to provide magnetocrystalline anisotropy, potentially a much stronger source than the shape anisotropy of the Alnicos. Elements in the first half of the 4f series, the light rare-earths, are preferred because their exchange coupling with the ferromagnetic 3d elements leads to parallel alignment of 3d and 4f moments. Despite their name, crustal abundances of light rare-earths are similar to those of zinc or lead.

Strnat, in the USA, is credited with drawing attention to the potential of rare-earth cobalt alloys for permanent magnets. The first successful samarium–cobalt magnet was produced in 1967 at Philips, by Velge and Buschow, who bonded oriented $SmCo_5$ powder in resin to yield an energy product of 65 kJ/m^3. Subsequent development of sintered magnets based on $SmCo_5$ or the related Sm_2Co_{17} phase led to energy products in excess of 200 kJ/m^3, already double that

(a)

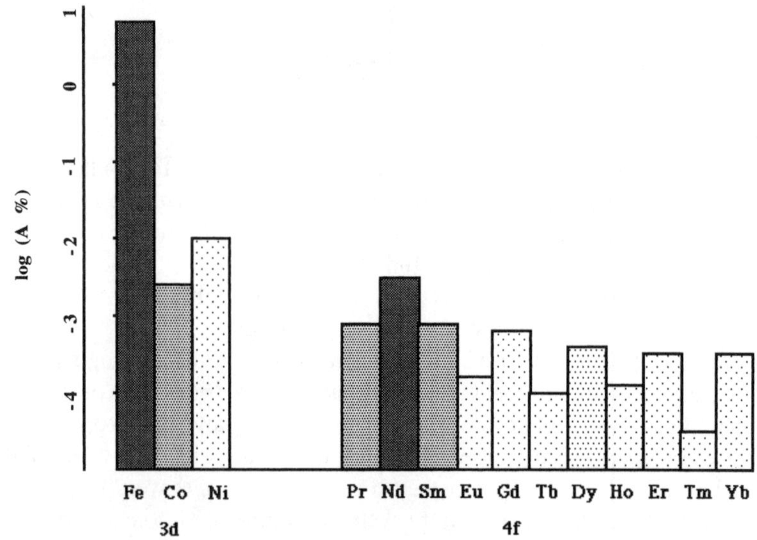

FIG. P1 (a) Relative abundances of the magnetic elements of the 3d and 4f series in the Earth's crust. There is a rough correlation between abundance and cost. Note that iron is about 40 times as abundant as all the other magnetic elements together. (b) Atomic magnetic moments in Bohr magnetons (black bars) and magnetic moment density in arbitrary units (white bars). Positive and negative magnetic moments indicate the sense of the exchange coupling with the 3d elements. ▶

of the best Alnicos. These magnets established themselves for applications where miniaturization is required (e.g. small stepper-motors, light headphones), or where cost is not an overriding consideration (e.g. synchrotron wigglers and demonstration magnetically-levitated transportation systems).

Uncertainties in the world cobalt supply in the late 1970s led to huge fluctuations in the price of this strategic raw material, thereby spurring a search for new magnetic materials containing little or no cobalt. A glance at Fig. P.1 will reveal why attention naturally focused on neodymium–iron alloys. Not only is neodymium the most abundant magnetic light rare-earth (lanthanum and cerium are more abundant, but they are usually nonmagnetic in alloys), it also has the largest 4f moment, 3.3 μ_B. Iron is by far the most abundant ferromagnetic 3d element, thanks to the 'doubly magic' ^{56}Fe nucleus, and ferromagnetic α-Fe also has the largest 3d atomic moment of the ferromagnetic metals, 2.2 μ_B. Regrettably, no suitable Nd–Fe binary compound exists.

Several routes were open in the search for a new Nd–Fe magnet material. One was to look for novel ternary phases with a suitable crystal structure; another was to seek to stabilize a metastable phase by rapid quenching. In fact, the existence of a new iron-rich compound in the Nd–Fe–B ternary diagram had been signalled by Kuz'ma and coworkers in the Ukraine as early as 1979, but it was not until 1983 that Sagawa announced that Sumitomo in Japan had succeeded in making a

(b)

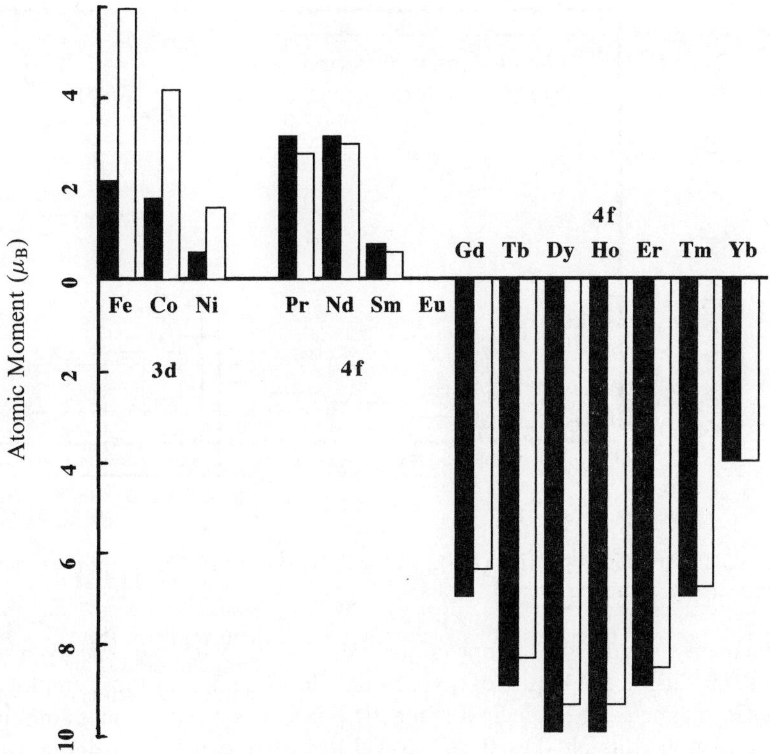

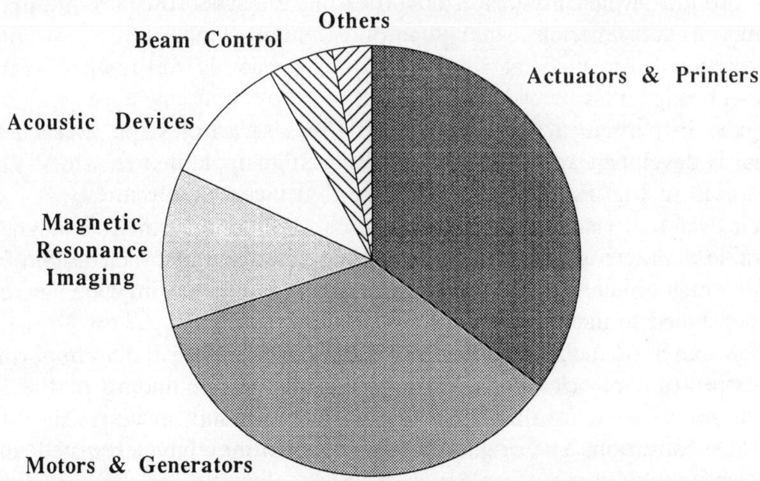

FIG. P2 The market for rare-earth permanent magnets, showing the main areas of application.

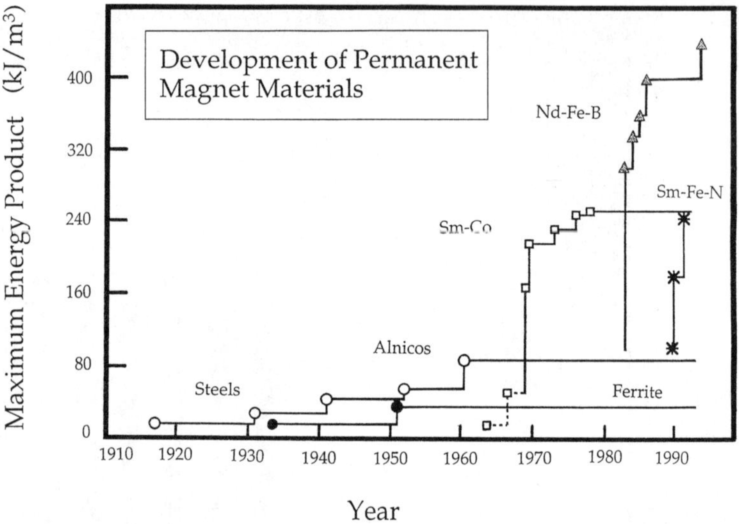

FIG. P3 Progress in improving $(BH)_{max}$ over the past century. Energy product has increased exponentially, doubling every 12 years.

permanent magnet of composition $Nd_{15}Fe_{77}B_8$ with an energy product of 290 kJ/m^3, using techniques of powder metallurgy similar to those employed for the Sm–Co magnets. The principal magnetic phase was a tetragonal compound, soon to be identified as $Nd_2Fe_{14}B$. A parallel line of research on melt-spun rare-earth iron boron in the USA involving Koon, Hadjipanayis and, most notably, Croat and Herbst at General Motors led independently to permanent magnets based on the same ternary phase. The neodymium–iron–boron magnets hold the record for energy product, which now exceeds 400 kJ/m^3, and their uses are multiplying in such areas as consumer electronics, computer components, robotics, portable tools and equipment for medical imaging. An overview of the uses of rare-earth permanent magnets is provided in Fig. P.2.

Progress in permanent magnets proceeds in a series of steps, as each type of material is developed and improved, before being supplanted by a new one. The trend traced in Fig. P.3 is illustrated somewhat more vividly in Fig. P.4. Intense research over the past decade has uncovered other rare-earth iron magnets with favourable characteristics, notably interstitial intermetallics such as $Sm_2Fe_{17}N_3$, some of which are considered for commercial development. Sm–Co is increasingly being consigned to niche applications.

New materials such as Nd–Fe–B offer both a challenge and an opportunity to a wide spectrum of scientists and engineers. An understanding of the *intrinsic magnetic properties* is invaluable for any rational attempt at optimizing them by chemical substitution. The original Sm–Co magnet may have consisted simply of particles of $SmCo_5$, but modern, optimized Sm–Co magnets may now include three or more additional elements. Intrinsic magnetic properties set the limits on the potential development of any particular material. That development is aimed at

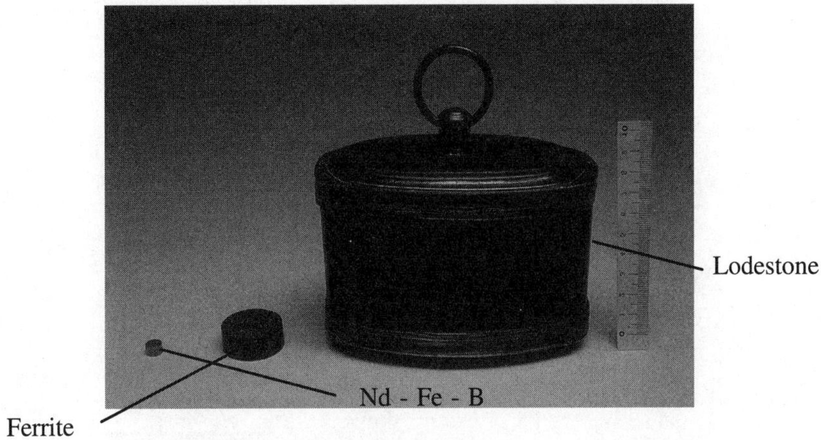

FIG. P4 260 years of progress with permanent magnets. The brass-bound lodestone (presented to Trinity College Dublin for experiments in natural philosophy in 1724), the ferrite block and the neodymium iron boron magnet each store the same magnetic energy (≈ 0.4 J), and each contain about 70% of iron by weight, yet the mass has decreased a thousand fold.

improving the practically useful quality of a magnet, its *hysteresis* and the associated *coercivity* which is a function of the metallurgical microstructure, itself a reflection of thermodynamic *phase relations* in the chemical system. Microstructure is an area where theory is as yet unable to offer a comprehensive guide to action. It remains the domain of art, and jealously guarded commercial secrecy. A *fabrication process* which yields fine magnets in one environment may not be readily transplanted to another. Ultimately, of course, the new magnets must find *applications* in products. A record energy product, strong anisotropy and high remanence offer opportunities for realizing novel static field configurations, and present new challenges for the design of motors and other electromagnetic machines that fully exploit the properties of the material. Figure P.5 illustrates the ubiquity of rare-earth magnets in a computer peripheral device.

It would be difficult for anyone to write authoritatively on all those subjects, so this book is composed of interlocking chapters written by specialists in each area. Emphasis is on Nd–Fe–B magnets, but Sm–Co magnets and other iron- or cobalt-rich intermetallic compounds are also covered. The text should be useful to physicists, metallurgists, materials scientists and electrical engineers interested in permanent magnets. The necessary background will be found in undergraduate courses on physics or electrical engineering. There is a cross-disciplinary sweep that will appeal to those who like to see beyond the cloisters of their own speciality, whether it be to look into the practical consequences of research in solid-state magnetism, or to gain an understanding of the physical basis of new engineering materials. Core chapters can be chosen according to the reader's profile. Physicists can concentrate on Chapters 1–6. Metallurgists and materials scientists may focus their attention on

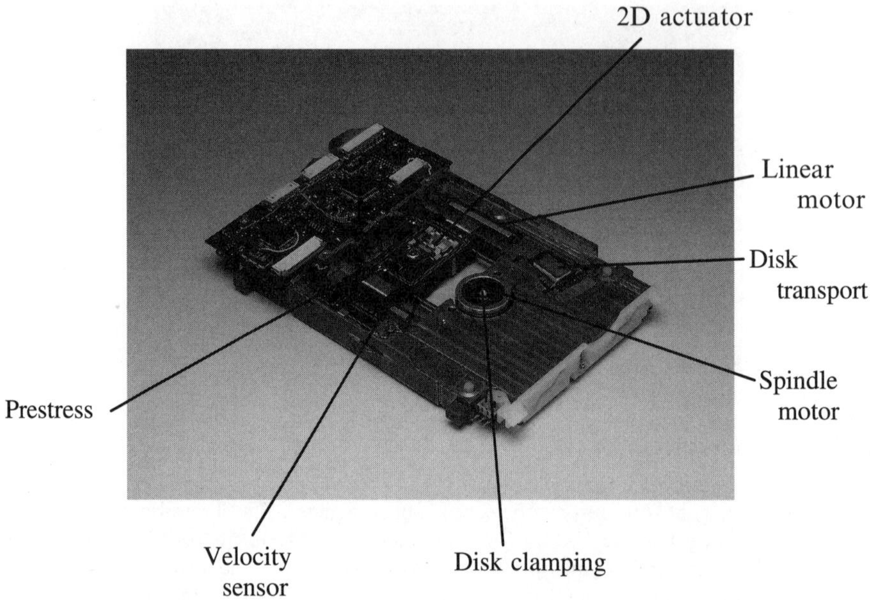

FIG. P5 A 3.5 inch magneto-optic data drive, showing the locations and uses of ten rare-earth permanent magnets. Sintered magnets are used in the 2-d lens actautor, the linear motor (2) and the velocity sensor, bonded magnets are used in the spindle motor, the disk clamp and for prestressing the bearing system (4). A ferrite magnet is used in the disk transport motor and the recording medium is an amorphous rare-earth iron film (Photo: Philips Electronics NV, by courtesy of R. Fastenau).

Chapters 1, 3, 4, 6, and 7. Engineers can follow Chapters 1 and 8–11. In each case, the remaining chapters should be scanned for background information.

This book is a fruit of the Concerted European Action on Magnets, a unique experiment in broad-based multinational scientific collaboration. During the years 1985–94, more than eighty industrial and academic laboratories cooperated in an all-embracing effort to promote materials development, magnet processing and application of new permanent magnets in Europe, with support from the European Commission. Almost all the contributors here were CEAM participants, and this book is a tribute to the remarkable spirit of collaboration and common purpose that CEAM engendered.

The book would never have been produced without the patience and enthusiasm of the twelve contributers, in the face of shifting editorial demands during the gestation process. Many members of my research group, summer students and staff at Trinity College Dublin have helped in innumerable respects. I am grateful to all of them.

Dublin J. M. D. C.
January 1995

Contents

List of Contributors

J. M. D. Coey	*Physics Department, Trinity College, Dublin, Ireland*
J. J. M. Franse	*Van der Waals-Zeeman Institute, University of Amsterdam, The Netherlands*
D. Givord	*Laboratoire Louis Néel, CNRS, Grenoble, France*
G. Hadjipanayis	*Department of Physics and Astronomy, University of Delaware, Newark, USA*
R. Hanitsch	*Institute of Electrical Machines, University of Technology, Berlin, Germany*
I. R. Harris	*School of Metallurgy and Materials, University of Birmingham, England*
D. Howe	*Department of Electronic and Electrical Engineering, University of Sheffield, England*
K. G. Knoch	*Philips Components Ltd, Southport, England*
R. J. Radwański	*Institute of Physics and Computer Sciences, Pedagogical University and Center for Solid State Physics, Cracow, Poland.*
M. F. Rossignol	*Laboratoire Louis Néel, CNRS, Grenoble, France*
R. Skomski	*Physics Department, Trinity College, Dublin, Ireland*
H. A. Leupold	*US Army Research Laboratory, Fort Monmouth, NJ, USA*
J.–P. Yonnet	*Laboratoire d'Electrotechnique de Grenoble, Institut National Polytechnique de Grenoble, France*

Units

Units are a perennial source of confusion in magnetism, mainly due to the attachment of physicists to the obsolete cgs system. A coherent modern text involving electromagnetic applications is better written using SI units, which have the virtue that the dimensions of any expression are immediately obvious. The defining relations for the flux density in a magnet, $\mathbf{B} = \mu_0(\mathbf{H} + \mathbf{M})$ and $\mathbf{B} = \mu_0\mathbf{H} + \mathbf{J}$, preferred respectively by physicists and electrical engineers, are not exclusive. Both will be used. The magnetic constant μ_0 is $4\pi\ 10^{-7}$Tm/A. The polarization of a solid **J**, in teslas (T) is related to its magnetization **M** in $J/(\mathrm{Tm}^3)$ (equivalent units are A/m) by $\mathbf{J} = \mu_0\mathbf{M}$. Similarly, the free-space induction $\mathbf{B}_0$ in teslas (T) is related to the magnetic field strength **H** in A/m by $\mathbf{B}_0 = \mu_0\mathbf{H}$.

The tesla is a convenient magnitude in permanent magnetism, widely accepted both as a unit of polarization and flux density. It converts simply to the corresponding cgs units.* Magnetization is often multiplied by μ_0 and expressed in teslas, rather than using the slightly cumbersome units of kA/m or MA/m. Likewise for the magnetic field strength. It is unfortunate that J has to stand for so many quantities: polarization, total quantum number, angular momentum operator, exchange integral, current density, moment of inertia and the Joule, but all these can be distinguished by type styles, subscripts and common sense! In discussion of atomic-scale properties, use is made of the Bohr magneton ($\mu_B = 9.274\ 10^{-24}$ J/T) and the Bohr radius ($a_0 = 0.052\,92$ nm).

* The usual definition of induction in a magnet in the cgs system is $\mathbf{B} = \mu_0(\mathbf{H} + 4\pi\mathbf{M}) = \mu_0\mathbf{H} + 4\pi\mathbf{J}$, where $\mu_0 = 1$ G/Oe. The following four cgs → SI conversions are useful to bear in mind when reading literature written in cgs units; 1 kOe ≈ 80 kA/m 1 MG · Oe ≈ 8 kJ/m^3

1 kG = 0.1 T 1 emu/g = 1 J/(Tkg)

Quantities and units

Quantity	Symbol	SI unit	Equivalent cgs units
B-field; magnetic flux density	$\mathbf{B}$	tesla (T)	10^4 gauss (G)
H-field; magnetic field intensity	$\mathbf{H}$	A/m	$4\pi \cdot 10^{-3}$ oersted (Oe)
Energy density, energy product	$\mathbf{B} \cdot \mathbf{H}$	J/m^3	40π G · Oe
Magnetic moment	$\mathbf{m}$	J/T; Am^2	1000 emu (≡ erg/G)
Magnetization (magnetic moment/volume)	$\mathbf{M}$	A/m; $J/(T\ m^3)$	10^{-3} emu/cm^3 (≡ Oe)
Magnetization (magnetic moment/mass)	σ	Am^2/kg; J/(T kg)	emu/g
Demagnetizing factor	N	– $(0 \leq N \leq 1)$	– $(0 \leq N \leq 4\pi)$
Polarization	$\mathbf{J}$	T	$(10^4/4\pi)$ G

1
Introduction

J. M. D. Coey

A. BASIC CONCEPTS

The *magnetization* of a piece of ferromagnetic material has its origin in the spin and orbital magnetic moments of the atomic electrons. These moments arise from the angular momentum of the electrons. Nuclear magnetism is negligible by comparison. Summing the elementary contributions in a volume that is large on the atomic scale, one defines a local magnetization vector, **M**, that represents the magnetic moment per unit volume. In general, **M** is not uniform throughout a sample of macroscopic dimensions, but when the sum is carried out within a *domain*, the magnitude of **M**, known as the spontaneous magnetization M_s, is uniform; the direction of magnetization varies from domain to domain and typically lies along a crystallographic axis.

In equilibrium, the magnetization summed over the entire volume of the sample would normally be zero, but if the ferromagnetic sample can be prepared in a metastable state where it retains some net magnetization more or less independently of external magnetic fields, it is then known as a *permanent magnet*. Figure 1.1 illustrates the difference between the ferromagnet in its unmagnetized state and in the remanent state after saturation in a horizontal magnetic field which eliminates the domains in each crystallite.

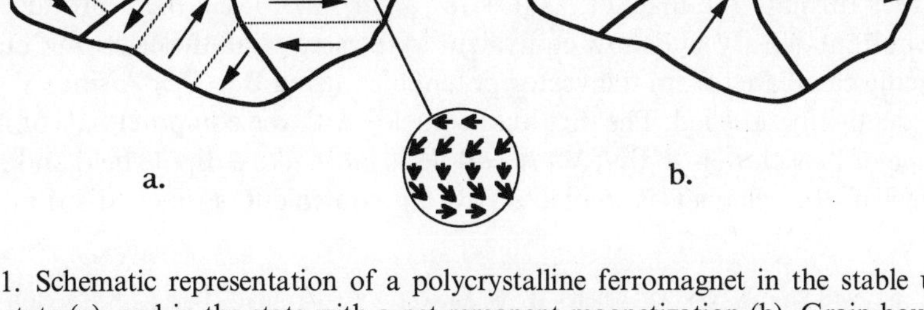

FIG. 1.1. Schematic representation of a polycrystalline ferromagnet in the stable unmagnetized state (a), and in the state with a net remanent magnetization (b). Grain boundaries and domain walls are represented by the solid and dotted lines respectively. Inserts show the alignment of the atomic moments within domains and in the domain walls.

1 B, H and M

In free space around the magnet, the B and H fields that appear in Maxwell's equations (measured in teslas and amperes per metre respectively) are parallel and proportional: $\mathbf{B}_0 = \mu_0\mathbf{H}$, where μ_0 is the magnetic constant defined to be exactly $4\pi \times 10^{-7}$ Tm/A in the Système International. Within the magnetic material, however, the relation is

$$\mathbf{B} = \mu_0(\mathbf{H} + \mathbf{M}) \tag{1.1}$$

so that **B** and **H** are not necessarily proportional or even parallel there. Units of M, like those of H, are A/m.

Useful information concerning **B** and **H** in the absence of external fields can be derived from an elementary argument. Consider the piece of magnetized material in empty space. From Maxwell's equations, the flux density **B**, also known as the induction, satisfies $\nabla \cdot \mathbf{B} = 0$, whereas the magnetic field strength **H** satisfies $\nabla \times \mathbf{H} = 0$ in the absence of conduction currents. It turns out[1] that $\int \mathbf{B} \cdot \mathbf{H}\, d\tau = 0$, when the integral is evaluated over all space. We now divide space into two sections, that occupied by the magnet and the remaining empty space; hence,

$$\int_{\text{empty space}} \mathbf{B} \cdot \mathbf{H} d\tau = -\int_{\text{magnet}} \mathbf{B} \cdot \mathbf{H} d\tau \tag{1.2}$$

The left-hand side of the equation, which is necessarily positive and equal to $\mu_0 \int H^2 d\tau$, is identified as twice the potential energy associated with the field set up by the magnet outside its own volume. The equilibrium domain configuration will tend to minimize this energy. An alternative view is to associate the energy with the magnet itself. Equation (1.2) then shows that the potential energy is $-1/2 \int_{\text{magnet}} \mathbf{B} \cdot \mathbf{H}\, d\tau$. Since the right-hand side of Eq. (1.2) must also be positive, it follows that **B** and **H** *tend to be antiparallel within the body of the magnet.* Figure 1.2 illustrates **B**, **H** and **M** for an isolated magnet in the absence of any external field. Boundary conditions are that the perpendicular component of **B** is continuous across an interface, whereas the parallel component of **H** is continuous there (in the absence of surface currents). The diagram also suggests methods of modelling the field produced by the magnet in the surrounding space. Since **B** is continuous through the magnet, as it is for a solenoid, it can be reproduced by a surface current density somehow equivalent to the sum of all the electronic currents. **B** is usually calculated from the vector potential **A** using $\mathbf{B} = \nabla \times \mathbf{A}$ since $\nabla \cdot \mathbf{B} = 0$ is then identically satisfied. The flux lines coincide with the equipotentials of **A** when the gauge of **A** is chosen so that $\nabla \cdot \mathbf{A} = 0$. **H** is more like a dipole field and may be estimated if the magnet is replaced by an equivalent surface distribution of

[1] **B** may be written as $\nabla \times \mathbf{A}$, where **A** is the vector potential. Using the vector identify $\nabla \cdot (\mathbf{A} \times \mathbf{H}) = (\nabla \times \mathbf{A}) \cdot \mathbf{H} - \mathbf{A} \cdot (\nabla \times \mathbf{H})$, it follows that $\int (\nabla \times \mathbf{A}) \cdot \mathbf{H} d\tau = \int \{(\nabla \times \mathbf{H}) \cdot \mathbf{A} d\tau + \int \nabla \cdot (\mathbf{A} \times \mathbf{H})\} dt$. The first integral is zero since $\nabla \times \mathbf{H} = 0$, the second is $\int\int (\mathbf{A} \times \mathbf{H}) dS$. Evaluated over a sphere of very large radius, this is also zero since $H \sim 1/r^3$, $A \sim 1/r^2$ and $S \sim r^2$.

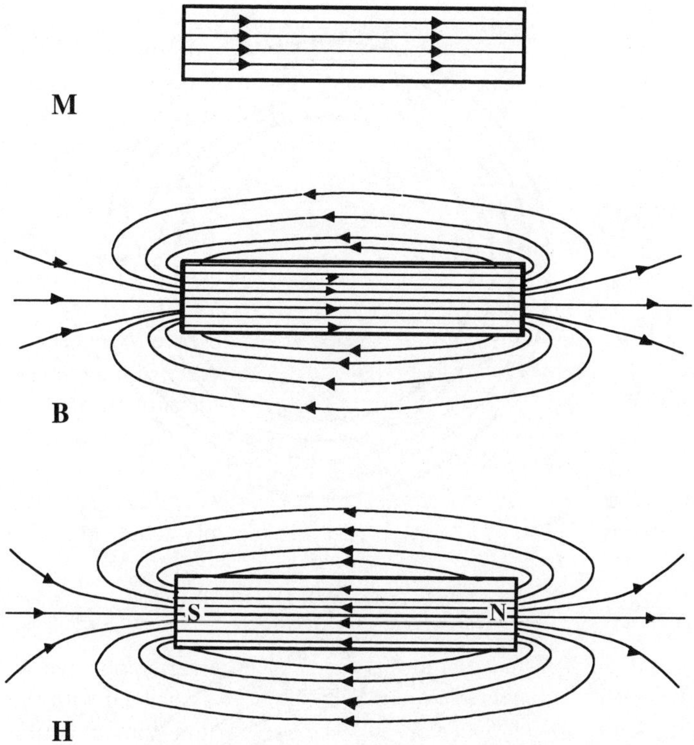

FIG. 1.2. Illustration of **M**, **B** and **H** for a permanent magnet.

magnetic 'charge' which acts as sources or sinks of **H**. In the absence of conduction currents, **H** may be calculated as the gradient of the scalar potential ψ associated with this surface charge since $\nabla \times (\nabla\psi) = 0$. The field lines are normal to the equipotentials of ψ. Magnetic poles in this sense are, of course, a fiction invented to sustain an analogy between magnetostatics and electrostatics. Another way to calculate the field produced by a magnet is to divide it into small elements $\delta\tau\ (r')$ and calculate the sum of dipole fields at a point $\mathbf{R} = \mathbf{r} + \mathbf{r}'$ from each element of magnetic moment $\mathbf{M}\ (\mathbf{r}')\ \delta\tau$:

$$\delta\mathbf{H}(\mathbf{R}) = (1/4\pi)\{3(\mathbf{M}\cdot\mathbf{r})\,\mathbf{r}/r^5 - \mathbf{M}/r^3\}\delta\tau \qquad (1.3)$$

The magnetization of an ideal permanent magnet should not be influenced by the magnetic field. In reality, **M** and **H** cannot be entirely independent; **M** depends on the thermomagnetic history of the sample. Outside the sample volume the dipole field is known as the *stray field*, but within the sample volume, **H** is known as the *demagnetizing field* since it tries to reduce **M**. However, it is now possible to prepare uniformly magnetized blocks of any desired shape with rare-earth iron or rare-earth cobalt alloys, something that was impossible with previous generations of metallic permanent magnets. H is uniform throughout the volume of any uniformly

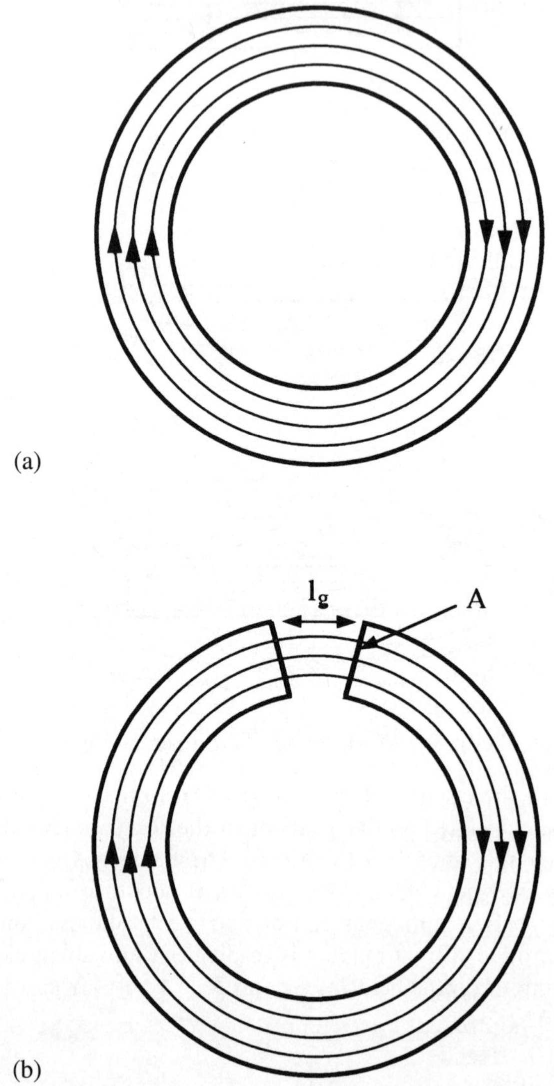

FIG. 1.3. Closed toroidal magnet (a), and toroid with an air gap (b). Lines of magnetic flux are shown.

magnetized ellipsoid[2] where **H** and **M** are related by the equation

$$\mathbf{H}_d = -N\mathbf{M} \tag{1.4}$$

[2]This statement is generally true for *soft* magnetic materials in an externally applied field H. It applies to permanent magnets in zero applied field when $M(r)$ is uniform.

when **M** lies along a principal axis of the magnet.

Here N is the *demagnetizing factor*, a pure number between 0 and 1 that depends solely on sample shape [Osb45]. For instance, ellipsoids of revolution with ratios of major to minor axes of 10, 1 and 0.1 have values of N equal to 0.02, 0.33 and 0.88 respectively. The sum of the demagnetizing factors in the three principal directions is unity. H_d becomes very small in long thin magnets, whereas in short, fat ones it follows from Eqs (1.1) and (1.4) that B is small. Magnetic thin films have $N = 1$ and $B = 0$ when magnetized perpendicular to the plane, whereas needles and toroids have $N = 0$ and $H = 0$ when magnetized along the axis. Practical magnets are frequently cylinders or blocks but it is possible to assimilate them to ellipsoids of comparable shape in order to estimate H_d. An upper limit to the demagnetizing field in a uniformly magnetized sample is M_s; in iron, for example ($\mu_0 M_s = 2.16$ T), H_d cannot exceed $2.16/\mu_0 = 1720$ kA/m.

The example of a toroidal permanent magnet is worth considering a little further. Unlike the magnet of Fig. 1.2, which is an open circuit producing a field throughout empty space, the toroid of Fig. 1.3a is a 'closed circuit'. Flux is all contained within the toroid and no field is created in the surrounding space. Applying Ampère's law $\int \mathbf{H} \cdot d\mathbf{l} = 0$ (the integral form of $\nabla \times \mathbf{H} = 0$) to a circular loop whose centre is on the toroid axis, it follows that $H = 0$ everywhere. The potential energy evaluated from the left- or right-hand side of Eq. (1.2) is zero. Now suppose that a small segment of the toroid is made of soft magnetic material of infinite permeability μ ($\mu = dB/dH = \infty$). Provided the spontaneous magnetization of the soft segment exceeds that of the permanent magnet, all the flux remains confined in the toroid. If we now want to remove the soft segment to infinity (Fig. 1.3b) we must do work against the force **F** exerted on it by the magnet. This work is equal to the energy associated with the field H_g that appears in the air gap, $(1/2)\mu_0 A l_g H_g^2$, where A is the cross-sectional area of the magnet or gap, and l_g is the gap length. The flux density B_g in a very narrow gap will be the same as it is in the magnet. An equivalent energy is associated with the magnet, because the field H_m in the magnet ($= H_d$) is no longer zero after the circuit is broken. According to Ampère's law, $H_g l_g = -H_m l_m$. The self-energy associated with the magnet is $-(1/2)\int \mathbf{B} \cdot \mathbf{H}\, d\tau = -(1/2) B_m H_m l_m A$ which is therefore equal to $(1/2)\mu_0 A l_g H_g^2$, since $B_m = B_g = \mu_0 H_g$. Hence the self-energy of the magnet is identical to the energy associated with the field in the air gap.

2 External fields

We consider next the response of a magnetic material to an externally applied field H_e (or B_0). We have seen that the lowest-energy state of a magnet in open circuit is normally one with no net magnetization. This state may be attained by thermal demagnetization. However, the existence of permanent magnets reflects the possibility of preparing material in a metastable state of higher energy, with M close to M_s.

(a)

(b)

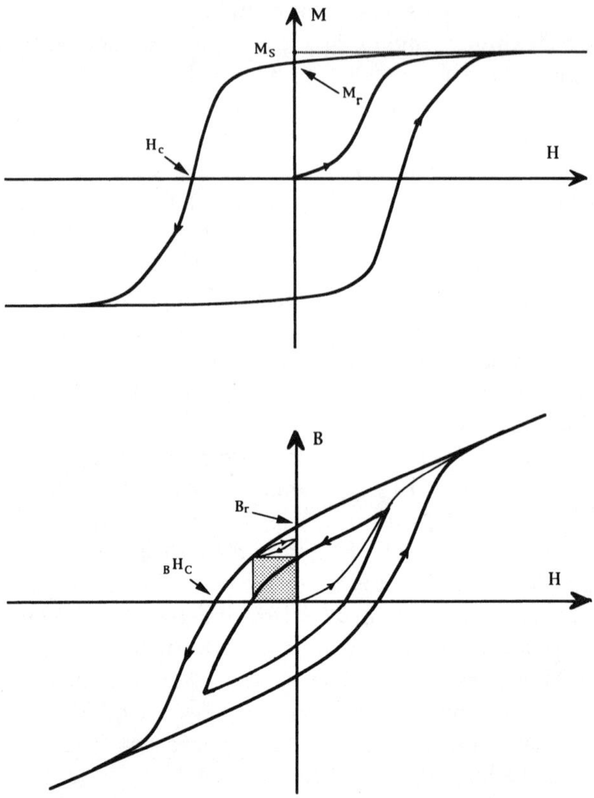

FIG. 1.4. Hysteresis loops for a permanent magnet; (a) M:H loop and (b) B:H loop. Virgin curves start at the origin. A minor loop and a recoil loop are shown in (b). The shaded rectangle of maximum area in the second quadrant represents the energy product $(BH)_{max}$.

The first consequence of setting a piece of uniformly magnetized material of volume V in an external field $\mathbf{B}_0$ is to change its energy by an amount

$$E = -V\mathbf{M} \cdot \mathbf{B}_0 \tag{1.5}$$

An alternative unit for $\mathbf{M}$, equivalent to A/m, is therefore J/(T m^3). Related to this energy are two effects that are exploited in certain permanent magnet applications:

1. When $\mathbf{B}_0$ and $\mathbf{M}$ are not parallel or antiparallel, a torque $\mathbf{T}$ is exerted on the magnet by the field:

$$\mathbf{T} = V\mathbf{M} \times \mathbf{B}_0 \tag{1.6}$$

2. When the field $\mathbf{B}_0$ is nonuniform, there is a net force $\mathbf{F}$ on the magnet directed along the field gradient:

$$\mathbf{F} = \nabla(V\mathbf{M} \times \mathbf{B}_0) \tag{1.7}$$

Since $\nabla \cdot \mathbf{B} = 0$ (i.e. $\partial B_x/\partial x + \partial B_y/\partial y + \partial B_z/\partial z = 0$) it is impossible to devise a static magnetic-field configuration when $V\mathbf{M}$ is constant in which the magnet can exist in a position of stable equilibrium at a given point. At best, two of the components of $\nabla \mathbf{B}_0$ in Eq. (1.7) ($\partial B_x/\partial x$, $\partial B_y/\partial y$, $\partial B_z/\partial z$) can be negative; the third must be positive. Therefore, on moving away from the equilibrium point where $\partial B_x/\partial x = \partial B_y/\partial y = \partial B_z/\partial z = 0$, there can be a restoring force in only two out of three perpendicular directions. This result, known as Earnshaw's theorem, implies the impossibility of the stable, passive magnetic levitation of a ferromagnet using permanent magnets without lateral constraint, and the impossibility of devising a static pattern of magnetic field that will direct magnetized particles to a desired spot in empty space (again without mechanical constraint). Stable levitation becomes possible if conduction currents are somehow involved, as in a superconductor with a magnet floating above it, or in an active electromagnetic circuit. Magnetic bearings and couplings are the subject of Chapter 9.

2.1 *Hysteresis*

An external field will usually change the state of magnetization of a ferromagnetic material in a way that is nonlinear and irreversible. A typical magnetization curve *M*(*H*) is illustrated in Fig. 1.4a.[3] The symmetric major hysteresis loop is traced reproducibly provided the applied field (corrected for the demagnetizing effect) is sufficient to achieve saturation in each direction, that is, $M = \pm M_s$. If the field is reversed short of saturation, then a minor loop wholly within the major loop is obtained, either centred at the origin for a virgin sample that has never been magnetized, or branching off at a point on the major loop. M_s is an intrinsic property of the ferromagnetic phase but the *remanence* M_r and *coercivity* H_c, defined as the points where the major loop cuts the axes, are not. They depend in a complex way on the particle size, shape and metallurgical microstructure of the magnet.

The *B*:*H* loop of Fig. 1.4b is related to the *M*:*H* loop by Eq. (1.1). The remanence B_r is just $\mu_0 M_r$, but the coercivity ${}_BH_c$ may now be smaller than before. H_c is sometimes referred to as intrinsic coercivity, an unfortunate usage because coercivity is not an intrinsic property of the ferromagnetic phase. *M*:*H* loops tend to be measured by physicists interested in magnetism,[4] whereas *B*:*H* loops are of more interest for engineering applications. Since H_d is less than M and opposite in direction, permanent magnets generally operate in the second quadrant of their *M*:*H* or *B*:*H* loops, hence this is the part featured on manufacturers' data sheets. The maximum potential energy (at least for an ellipsoidal sample) is obtained at the point on the loop where the product $-B \cdot H$ is maximized. The figure of merit,

[3]It is common engineering practice to quote the polarization *J*(*H*) instead of the magnetization *M*(*H*) of a magnet, where **J** in teslas is defined as $\mu_0\mathbf{M}$. Eq. (1.1) then reads $\mathbf{B} = \mu_0\mathbf{H} + \mathbf{J}$.

[4]Usually, the mass of a sample is better known than its volume, so that $\sigma : H$ loops are actually measured; $\sigma = M\rho$, where ρ is the density. Equivalent units for σ are $\mathrm{Am^2/kg}$ or J/(T kg). Ellipsoidally shaped samples with a known demagnetizing factor must be used when it is necessary to make a precise correction for the demagnetizing field.

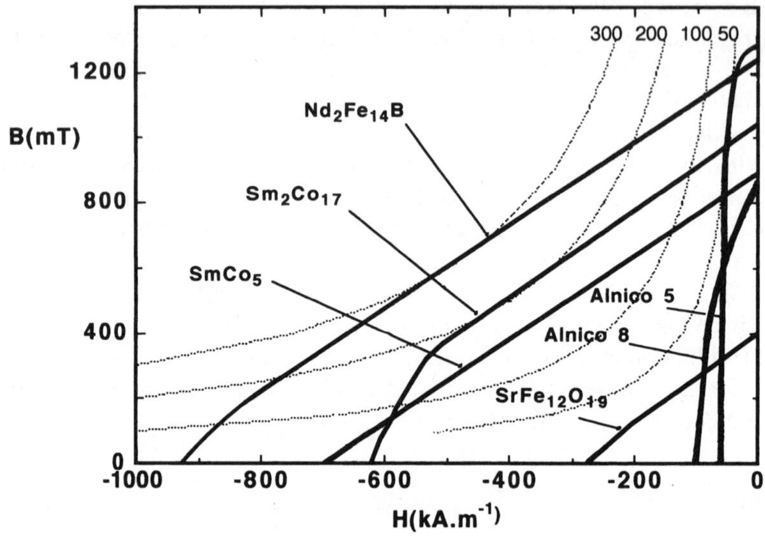

FIG. 1.5. Second quadrant B:H curves for typical Nd–Fe–B, Sm–Co Alnico and ferrite magnets. The dotted lines are contours of constant (BH).

$(BH)_{max}$, corresponding to the shaded area, is twice this maximum potential energy. Some typical B:H curves for different permanent magnets are presented in Fig. 1.5 and numerical data on M_r, H_c and $(BH)_{max}$ are given in Table 1.1. Note that the neodymium–iron–boron magnet has the greatest energy product. Note also that the second quadrant B:H curves for all the magnets except the Alnico are almost straight lines, with slope μ_0. Such is the behaviour of an *ideal permanent magnet*.

Ideal permanent magnets exhibit a magnetization curve which is a square loop like that shown in Fig. 1.6; the remanence has the largest possible value, $M_r = M_s$, and the magnetization remains independent of reverse field up to H_c. Some rare-earth magnets can have an intrinsic coercivity H_c that is much greater than M_s, but there is often nothing to be gained from this as we are usually interested only in the second quadrant of the B:H loop. Indeed, excessively high coercivity can make the materials difficult to magnetize.

The corresponding B:H loop, drawn on the same scale, is a line with slope μ_0 in the second quadrant. Any less-than-ideal material will exhibit loops lying within those shown in Fig. 1.6. It follows that $(BH)_{max} \leq B_r \cdot {}_BH_c/4$. Summarizing, we find the following intrinsic limitations on the performance of permanent magnets:

$$B_r \leq \mu_0 M_s \tag{1.8a}$$

$$_BH_c \leq M_s \tag{1.8b}$$

and

$$(BH)_{max} \leq \mu_0 {M_r}^2/4 \tag{1.8c}$$

Table 1.1 Typical properties of some oriented magnets

Type	Main phase	B_r (T)	$\mu_0 M_s$ (T)	H_c (kA/m)	${}_BH_c$ (kA/m)	$(BH)_{max}$ (kJ/m^3)
Ferrite	$SrFe_{12}O_{19}$	0.39	0.46	275	265	28
Alnico 5	$Fe_{48}Al_{16}Ni_{13}Co_{21}Cu_2^+$	1.28	1.40	54	52	43
Sm–Co	$SmCo_5$	0.88	0.95	1400	680	150
Sm–Co	Sm_2Co_{17}*	1.08	1.15	900	800	220
Nd–Fe–B	$Nd_2Fe_{14}B$	1.25	1.34	1000	920	300

+overall atomic composition.
* intergrown with 1:5 phase.

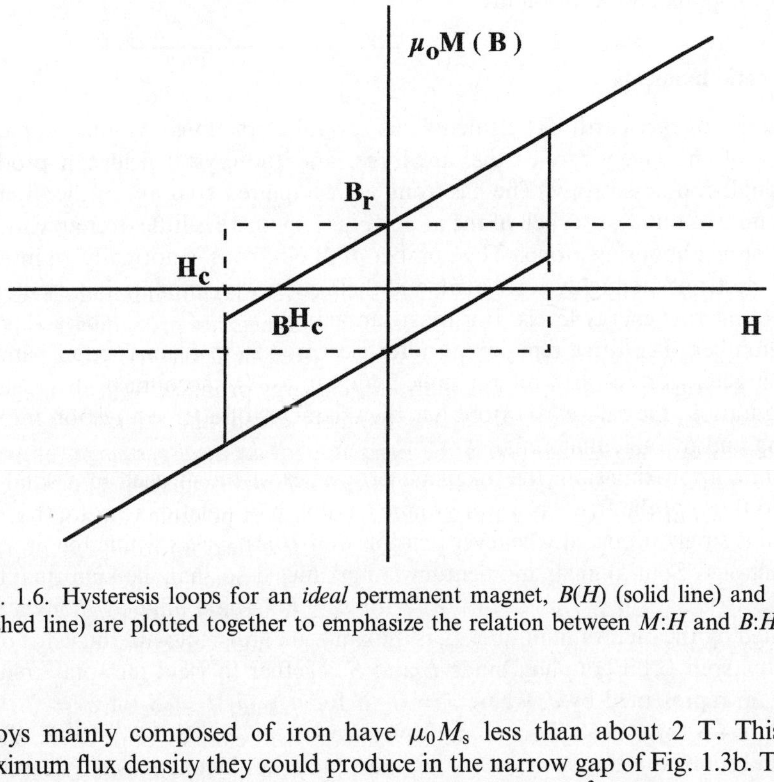

FIG. 1.6. Hysteresis loops for an *ideal* permanent magnet, $B(H)$ (solid line) and $\mu_0 M(H)$ (dashed line) are plotted together to emphasize the relation between M:H and B:H loops.

Alloys mainly composed of iron have $\mu_0 M_s$ less than about 2 T. This is the maximum flux density they could produce in the narrow gap of Fig. 1.3b. The limit on $(BH)_{max}$ corresponding to $\mu_0 M_s = 2$ T is ≈ 800 kJ/m^3, which sets a ceiling on the progress discussed in the Preface. Cobalt and nickel alloys have lower magnetizations, and correspondingly smaller upper limits to their energy products. Ferromagnetic alloys composed mainly of heavy rare-earths such as Dy or Ho which have the largest atomic moment ($10\mu_B$) of any element in the periodic table could in theory show energy products as high as 3000 kJ/m^3 and produce a flux density approaching 4 T. However, prospects are poor for realizing alloys of this type which would be ferromagnetic at ambient temperature and above. They are likely to be restricted to cryogenic applications, if they are ever used at all.

B. ATOMIC-SCALE MAGNETISM

The *intrinsic* magnetic properties of a material such as its spontaneous magnetization, Curie temperature and anisotropy field are determined by magnetic moments and interactions on the atomic scale. The *extrinsic* magnetic properties are those related to the hysteresis loop, which depend on the microstructure. Here we consider atomic magnetic moments, exchange interactions and anisotropy.

From Eqs (1.8a–c) it is obviously desirable to maximize the spontaneous magnetization, which means selecting a ferromagnetic compound with a Curie point well above ambient temperature, made up of atoms with a large magnetic moment in a small atomic volume.

1 Magnetic moments

The 4f series of rare-earth (R) elements has several attractions. Atomic moments, especially of the heavy rare-earths, are large, and the crystal field can produce strong single-ion anisotropy. The electrons with unpaired spin are in the 4f shell, which is not the outermost shell of the atom (Fig. 1.7); there is little overlap with the 4f shells of neighbouring atoms. The number of 4f electrons is normally an integer, n. These electrons are highly correlated, and their quantum states are represented as a series of narrow energy levels. For most rare-earths, $n = Z - 57$, where Z is the atomic number. The three remaining, outer electrons form a conduction band of 5d/6s character in rare-earth intermetallic compounds. On account of its localized 4f configuration, the rare-earth atom has the character of a tripositive ion in both insulating and metallic materials.

As a first approximation, the magnetic properties of the 4f shell in a solid are similar to those of the free ion. Intra-atomic Coulomb correlations ensure that any 4f orbital is singly occupied whenever possible and *Hund's rules* define the ground-state multiplet. Spin angular momentum is maximized so that the spin quantum number S is $n/2$ for $n \leq 7$ and $(14 - n)/2$ for $n \geq 7$. Orbital angular momentum, represented by the quantum number L, is maximized consistent with the value of S, and, lastly, spin–orbit coupling binds L and S together to yield the total angular momentum represented by J, where $J = L - S$ for $n < 7$, $J = S$ for $n = 7$ (Gd) and $J = L + S$ for $n > 7$. Excited multiplets with $J \neq L \pm S$ are 10^3–10^5 K above the ground state, and can be neglected for most purposes in magnetism. Exceptions are Sm^{3+} ($4f^5$) and Eu^{3+} ($4f^6$), where the first excited multiplets, $J = 7/2$ lying 1500 K above the $J = 5/2$ ground state for Sm^{3+} and $J = 1$ lying 530 K above the $J = 0$ ground state for Eu^{3+}, are sufficiently low-lying to be populated thermally or mixed into the ground state under the action of a magnetic field.

Magnetic moments are associated with both spin and orbital angular momenta. The spin magnetic moment of a single unpaired electron with no orbital angular momentum is $gS\mu_B$, or one Bohr magneton ($1\mu_B = 9.274 \times 10^{-24}$ J/T) since

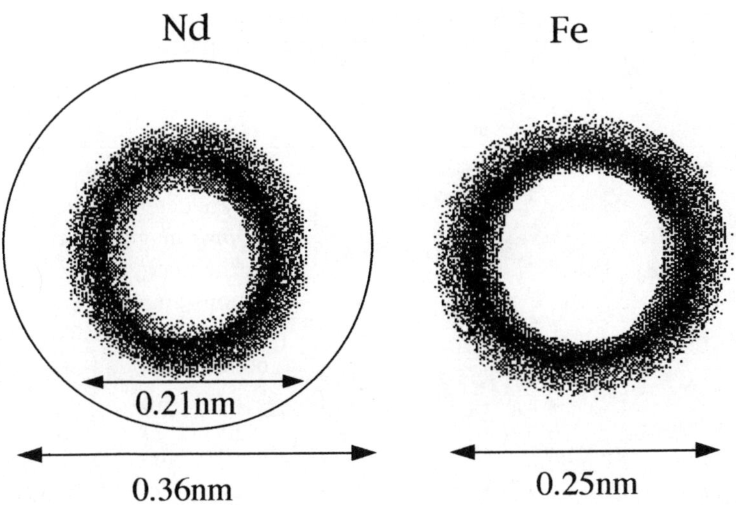

FIG. 1.7. Representation of a rare-earth atom (Nd) and a 3d transition atom (Fe), indicating the positions of the unpaired electron shells.

$S = 1/2$ and $g = 2$. The Landé g-factor for the rare-earth ion, defined as the ratio of the magnetic moment (in units of μ_B) to the angular momentum (in units of $\hbar$), is

$$g_R = 1 + \{J(J+1) + S(S+1) - L(L+1)\}/2J(J+1)$$

and the magnetic moment of the rare-earth ion is $g_R J \mu_B$. Table 1.2 lists n, S, L, J and g for the rare-earths in their most common electronic configurations. Also included are the quantities $\gamma_R = 2(g_R - 1)/g_R$, the ratio of spin moment to total moment of the ion, the total moment itself $m_R = g_R J \mu_B$, and the molar *Curie constant* $C_R = N_0 g_R^2 J(J+1)\mu_B^2/3k$. The largest moments appear for the heavy rare-earths, from Gd to Yb. The end-members La and Lu are nonmagnetic.

Unfortunately, the magnetic ground state of most rare-earth metals is not ferromagnetic, and only Gd has a Curie point (293 K) as high as ambient temperature. Hence the need to alloy rare-earths with the ferromagnetic transition metals to produce useful magnetic alloys. The saturation polarization $\mu_0 M_s$ of hexagonal close-packed Gd metal corresponding to its atomic moment of $7\mu_B$ is 2.5 T, little more than the 2.2 T found for iron, despite α–Fe having an atomic moment of only 2.2 μ_B and a bcc structure that is not dense-packed. The reason is that atomic densities for the 3d series are some three times greater than for the 4f series; metallic radii of the trivalent rare-earths decrease from 0.183 nm for La to 0.173 nm for Lu, whereas that of iron is 0.124 nm. Yttrium, with a metallic radius of 0.180 nm, serves as a nonmagnetic rare-earth analogue.

Magnetic moments of the ferromagnetic 3d transition (T) elements are unlike those of the 4f series in that the magnetic electrons are delocalized. The spin moment is very sensitive to the chemical environment, and the orbital contribution

Table 1.2 Magnetic properties of trivalent rare-earth ions

	n	S	L	J	g_R	γ_R	$g_R J$ (μ_B)	C_R (JK/T^{2}mole)
La	0	0	0	0	–	–	–	–
Ce	1	1/2	3	5/2	6/7	–1/3	2.14	8.0
Pr	2	1	5	4	4/5	–1/2	3.2	16.0
Nd	3	3/2	6	9/2	8/11	–3/4	3.27	16.4
Sm	5	5/2	5	5/2	2/7	–5	0.71	0.9
Eu	6	3	3	0	–	–	–	–
Gd	7	7/2	0	7/2	2	1	7	78.8
Tb	8	3	3	6	3/2	2/3	9	118.2
Dy	9	5/2	5	15/2	4/3	1/2	10	141.7
Ho	10	2	6	8	5/4	2/5	10	140.7
Er	11	3/2	6	15/2	6/5	1/3	9	114.8
Tm	12	1	5	6	7/6	2/7	7	71.5
Yb	13	1/2	3	7/2	8/7	1/4	4	25.7
Lu	14	0	0	0	–	–	–	–

is relatively insignificant. These differences arise because the 3d electrons are in an outer shell, and therefore participate in metallic bonding and electronic conduction by forming bands whose widths are of the order of 1 eV. Narrow d-bands can split and develop an unpaired spin moment if the gain in exchange energy outweights the loss in electronic energy due to the promotion of 3d$^{\downarrow}$ electrons into the 3d$^{\uparrow}$ subband (the Stoner criterion). Broader d-bands, like those of the early 3d transition elements Ti and V, do not split spontaneously; they are Pauli-paramagnetic. Futhermore, the 3d bands overlap with a partly-filled conduction band of mainly 4s character. The total number of 3d + 4s electrons is an integer, but roughly 0.6 electrons of the late transition elements are in the 4s band, so the number of 3d electrons is nonintegral.

Ferromagnets are classified as *strong* or *weak* depending on whether or not the majority-spin 3d$^{\uparrow}$ levels are completely occupied (Fig. 1.8). For cobalt and nickel, the 3d$^{\uparrow}$ bands are filled with five electrons so these elements are strong ferromagnets, whereas in iron there are some 3d$^{\uparrow}$ holes, so Fe is classified as a weak ferromagnet despite its relatively large moment. Weak ferromagnets tend to have a significant high-field slope to their magnetization curves in the saturated region, as the applied field changes the relative populations of ↑ and ↓ bands. The magnetic properties of the ferromagnetic transition elements are summarized in Table 1.3.

When alloyed with another element, the 4f atoms retain their magnetic moments, although there will be a modification of the crystal-field effects, as discussed in Section B3. However, the 3d moments can change substantially in metallic alloys. In an ionic compound with a strongly electronegative element such as O or F, there is transfer of the integral number of electrons needed to

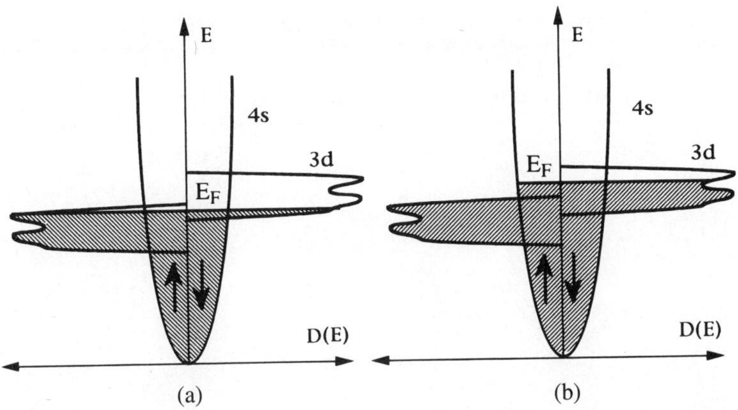

FIG. 1.8. Schematic band structures of (a) weak and (b) strong ferromagnets.

complete the ligand 2p shell, leading to compounds such as $Ni^{2+}F_2^-$, $Fe_2^{3+}O_3^{2-}$ or the hexagonal ferrite $Ba^{2+}Fe_{12}^{3+}O_{19}^{2-}$, which is widely used for inexpensive permanent magnets. In metallic alloys there is little charge transfer, but the outer electrons of the alloyed elements may form a common band. This was the idea behind the rigid band model of alloys of the 3d elements which led to the Slater–Pauling curve — a plot of ferromagnetic moments of the alloys as a function of the average number of valence electrons per atom. This curve has a slope -1 for strong ferromagnets.

These ideas were reformulated in terms of *magnetic valence* by Williams *et al.* [Wil83]. The bulk magnetization is considered in terms of an average magnetic moment per atom, derived simply from the total number of valence electrons in the alloy by assuming strong ferromagnetism. The chemical valence of an atom is $\mathscr{Z} = \mathscr{N}^\uparrow + \mathscr{N}^\downarrow$, where $\mathscr{N}^\uparrow$ and $\mathscr{N}^\downarrow$ are the numbers of valence electrons with spin up and spin down. The magnetic moment in units of μ_B is $m = \mathscr{N}^\uparrow - \mathscr{N}^\downarrow = 2\mathscr{N}^\uparrow - \mathscr{Z}$. Now $\mathscr{N}_d{}^\uparrow$ is precisely 5 for strong ferromagnets, so the magnetic valence $\mathscr{Z}_m$, defined by

$$\mathscr{Z}_m = 2\mathscr{N}_d{}^\uparrow - \mathscr{Z} \tag{1.9}$$

is an integer. The moment m is therefore

$$m = \mathscr{Z}_m + 2\mathscr{N}_s{}^\uparrow \tag{1.10}$$

where $\mathscr{N}^\uparrow = \mathscr{N}_d{}^\uparrow + \mathscr{N}_s{}^\uparrow$ is the number of electrons in the $\uparrow$ conduction band of mainly 4s character, which is practically unpolarized.

The average magnetic moment per atom in an alloy, $\langle m \rangle$, is deduced from Eq.(1.10) by replacing $\mathscr{Z}_m$ by its average value $\langle \mathscr{Z}_m \rangle$ over all the atoms present. $\mathscr{N}_d{}^\uparrow$ is 5 for atoms to the right of iron in the 3d series but it is zero for the early transition elements and the rare-earth series, because replacement of an atom of Fe,

Table 1.3 Magnetic properties of the ferromagnetic 3d elements

	n'	m^* (μ_B)	M_S † (kA/m)	$\mu_0 M_S$ † (T)	σ_S † (J/(T kg))	T_C (K)	Ferromagnetism
Fe	8	2.22	1720	2.16	218	1043	weak
Co	9	1.72	1370	1.76	162	1400	strong
Ni	10	0.61	485	0.61	56	635	strong

* at 0 K;
† at 20 °C.

Co or Ni by an atom of one of these elements removes five 3d states from below the Fermi level E_F and introduces five states above E_F. $\mathscr{Z}$ is 3 for B, Y and the lanthanides, 4 for C, Si, Ti,..., 8 for Fe, 9 for Co and 10 for Ni. $\mathscr{N}_s\!\uparrow$ is small, roughly 0.3. The model predicts that the average moment will be reduced with increasing rare-earth content in $R_{1-x}T_x$ binary alloys, falling to zero at a *critical concentration* x_c for the appearance of d-band magnetism which is 0.5 for iron, 0.6 for cobalt and 0.8 for nickel. The predicted average moment is moderately well followed by both crystalline and amorphous $R_{1-x}T_x$ alloys (Fig. 1.9) [Gav88, Coe85]. Consequently, if we wish to take advantage of the large magnetization of iron or cobalt in $R_{1-x}T_x$ alloys $\langle \mathscr{Z}_m \rangle$ should be greater than about 1, and x should be greater than 0.8, otherwise R will seriously reduce the magnetization.

The magnetic valence model is restricted to strong ferromagnets and it disregards the local chemical environment of the 3d element, which can have an important influence on the moment. Quite different atomic moments may be found in the same alloy on different crystallographic sites. A nice example is the hexagonal Laves phase $TiFe_2$, where iron on the 6c site has a moment of 0.6 μ_B, whereas iron on the 2a site is nonmagnetic [Wer70].

More information about 3d spin moments can be obtained from experiments such as Mössbauer spectroscopy, NMR, magnetic circular dichroism or neutron diffraction which probe the atomic sites. Self-consistent spin-polarized electronic structure calculations are becoming increasingly useful. Local spin-density calculations are available for the yttrium analogues of many of the compounds of interest here [Coe91], and they are often capable of reproducing the ground-state moments on different crystallographic sites to better than ±5%. Calculated densities of states for iron, cobalt and nickel are shown in Fig. 1.10. The Fermi level in iron and iron-rich intermetallics generally falls in a minimum in the ↓ density between bonding and antibonding states. Iron in the rare-earth intermetallics is approximately dense-packed, and its moment is then especially sensitive to interatomic spacing and lattice volume. Large magnetovolume anomalies are associated with the weak ferromagnetism of iron.

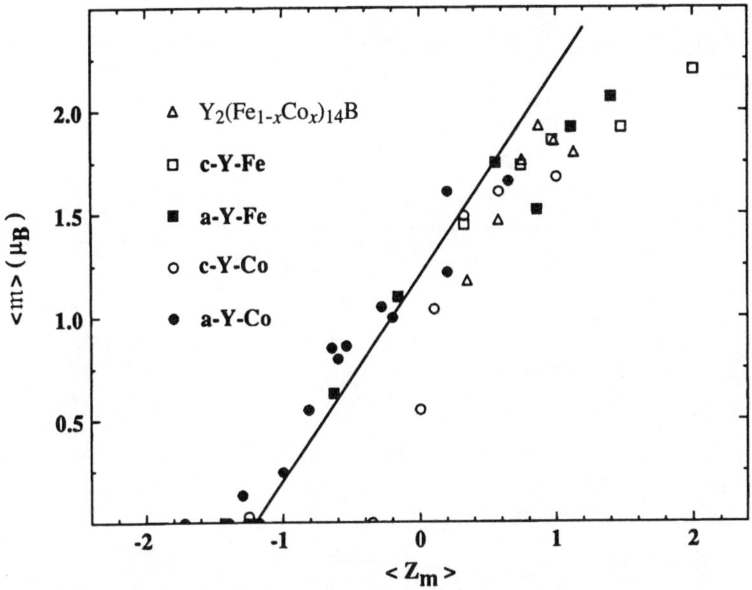

FIG. 1.9. Average magnetic moment per atom in crystalline and amorphous Y–Fe and Y–Co alloys, compared with the prediction of Eq. (1.10).

2 Exchange interactions

The magnetic coupling between the spins of two ions with spin quantum number S and no orbital moment may be represented by the Heisenberg Hamiltonian

$$\mathscr{H}_{ij} = -2J_{ij}\mathbf{S}_i \cdot \mathbf{S}_j \tag{1.11}$$

The exchange integral J_{ij} is positive for a ferromagnetic interaction and negative for an antiferromagnetic interaction between the two ions i and j, and it falls off rapidly with increasing distance between them. J/k is usually quoted in Kelvin, where k is Boltzmann's constant.

The principal magnetic interaction in an R–T alloy is that between transition-metal atoms, due to direct overlap of the 3d shells on neighbouring sites. That interaction is usually ferromagnetic, but depends sensitively on the interatomic spacing r_{ij}. There is evidence for iron that it changes sign from negative to positive when r_{ij} exceeds about 0.25 nm, which is a typical Fe–Fe distance in iron alloys. In any case, the ferromagnetic interaction is strongest for cobalt, and significantly weaker for iron or nickel.

Equation (1.11) cannot be applied directly to the T–T interaction because the 3d atoms are not ions with an integral number of electrons. An effective spin S_T can nevertheless be defined for the transition element from its atomic moment m by

$$m = gS_T\mu_B \tag{1.12}$$

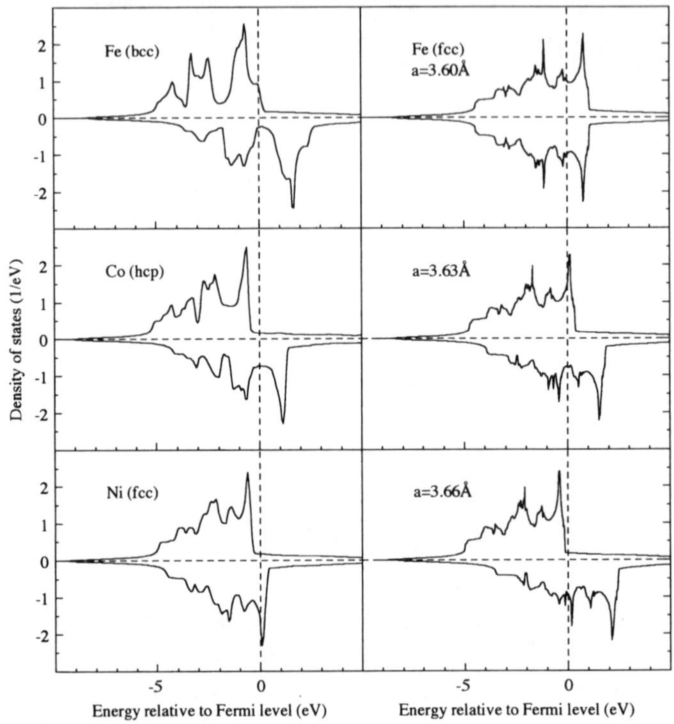

FIG. 1.10. Calculated density of states for α–Fe, Co and Ni. Iron is a weak ferromagnet; cobalt and nickel are strong. On the right, results are shown for γ–Fe with different lattice parameters, illustrating the sensitivity of the iron moment to unit cell volume in a dense-packed structure. The $\uparrow$ density is plotted positive and $\downarrow$ negative

with $g = 2$. The small orbital contribution to the 3d moment is neglected. To deal with the problem that S_T is not a proper quantum number, calculations involving S can be made from the Hamiltonian (1.11), with S_T substituted for S in the bottom line. Exchange coupling will be treated here in the molecular-field approximation [Sma66] introduced in Section B.4. Interactions of an atom with the rest of the crystal are represented by its interaction with an effective magnetic field. The Curie temperature, for instance, calculated from Eq. (1.11) with the assumption of only nearest-neighbour interactions is

$$T_C = 2ZJ\,S(S+1)/3k \qquad (1.13)$$

where Z is the number of nearest neighbours. This expression is adapted for 3d metals by replacing S by S_T.

The expression analogous to (1.11) for R–R interaction must be written

$$\mathcal{H}_{ij} = -2J_{ij}\,\mathbf{J}_i \cdot \mathbf{J}_i \qquad (1.14)$$

since J is the good quantum number, although this should not obscure the fact that the exchange interaction actually couples the electronic spins. Direct overlap between the 4f shells of adjacent rare-earth atoms is quite negligible, so the R–R exchange coupling is indirect. It can occur via spin polarization of the conduction band — the Ruderman–Kittel interaction. The corresponding Hamiltonian for the interaction of the two atoms with only spin moment (e.g. Gd^{3+}) via an s-band is

$$\mathscr{H}_{ij} = -(3\nu)^2(2\pi/E_F)^2 J_{sf}\, F(2k_F r_{ij})\, \mathbf{S}_i \cdot \mathbf{S}_i \tag{1.15}$$

where ν is the number of conduction electrons per atom and the function $F(x) = (x\cos x - \sin x)/x^4$. J_{ij} therefore falls off as $1/r_{ij}^3$, and oscillates in sign. The interaction J_{sf} between conduction and localized electrons is expected to be more or less constant in an isostructural series of compounds. To transform Eq. (1.15) to the form appropriate for the non-S state rare-earths requires that $\mathbf{S}$ be replaced by its projection along $\mathbf{J}$; $(g-1)J$. Hence the interaction energy should scale as $\{(g-1)^2 J(J+1)\}$, which is known as the *de Gennes factor*. This scaling applies to any interaction between localized 4f moments of the type $\mathbf{S}_i \cdot \mathbf{S}_j$, whether direct or indirect. R–R interactions in the rare-earth transition-metal intermetallic compounds of concern here are generally quite small by comparison with R–T and T–T interactions, and are often neglected.

The R–T interaction also proceeds via an indirect mechanism. Campbell [Cam72] proposed that 5d electrons are involved as the intermediary. There is intra-atomic exchange coupling between 4f and 5d electrons, and an interatomic interaction between 5d and 3d electrons. The 4f–5d interaction is ferromagnetic, but the 5d–3d interaction is antiferromagnetic when the 5d band is less than half full and the 3d band is more than half full, as they are for alloys of rare-earths with ferromagnetic transition metals. S_T therefore couples antiparallel to the rare-earth spin S_R, hence *the magnetizations of rare-earth and transition-metal sublattices couple parallel for the light rare-earth* ($J = L - S$) *and antiparallel for* Gd *and the other heavy rare-earths* ($J = L + S$). This rule has been confirmed by extensive measurements on R–T intermetallic compounds. It is the reason why this book focuses on alloys of iron and the light rare-earths, which have a ferromagnetic structure, rather than on alloys of iron and the heavy rare-earths, which are ferrimagnetic with a reduced total magnetization (Fig. 1.11).

3 Anisotropy

Easy-axis anisotropy is indispensable for maintaining the metastable domain configuration required of a permanent magnet. On an atomic scale, *single-ion anisotropy* arises through spin–orbit coupling of the orbital states that are stabilized by the electrostatic field, known as the *crystal electric field* (CEF) or *crystal field*, created at the atomic site by the rest of the crystal lattice. The atomic moment has corresponding preferred orientations or local *easy directions* that reflect the point symmetry of the site. The macroscopic *magnetocrystalline anisotropy* obtained by summing all the single-ion contributions reflects the overall crystal symmetry.

FIG. 1.11. Schematic magnetic structure of alloys of 3d transition metals with light rare-earths (left) and heavy rare-earths (right).

As with exchange, the microscopic origin of the phenomenon is best explained for the rare-earths which have a localized moment and a well-defined magnetic quantum number J. The Hamiltonian for a single ion is

$$\mathcal{H}_R = \mathcal{H}_c + \mathcal{H}_{so} + \mathcal{H}_{cf} + \mathcal{H}_{mag}$$

Here $\mathcal{H}_c$ includes the intra-atomic Coulomb interactions that result in states with well-defined L and S. $\mathcal{H}_{so}$ is the spin–orbit interaction, $-\lambda \mathbf{L} \cdot \mathbf{S}$, that couples $\mathbf{L}$ and $\mathbf{S}$ together to form $\mathbf{J}$. Since it is at least an order of magnitude larger than the crystal-field term $\mathcal{H}_{cf}$, the M_J states can be taken as the basis set for considering the crystal-field interaction. The final term, $\mathcal{H}_{mag} = -\mu_B(\mathbf{L} + 2\mathbf{S}) \cdot \mathbf{B}_0$, represents coupling of the rare-earth moment with an external or internal magnetic field, which may include the molecular field $\mathbf{B}_R^{mol}$ if only the one J multiplet is involved. But in general, and notably for Sm^{3+}, where J-mixing must be considered, the exchange is represented by a term $-2\mu_B \mathbf{S} \cdot \mathbf{B}_R^{ex}$, where the exchange field $\mathbf{B}_R^{ex} = \mathbf{B}_R^{mol}/\gamma$ acts only on the spin moment.

The crystal-field term $\mathcal{H}_{cf}$ is the Coulomb interaction of the charge distribution in the 4f shell with the potential created by the rest of the crystal. That potential $V(r\theta\phi)$ can be expressed as a sum of tesseral harmonics $Z_n^m(\theta, \phi)$

$$eV(r\theta\phi) = \sum_{n=0}^{6} \sum_{m=0}^{n} r^n K_n^m A_n^m Z_n^m(\theta\phi) \tag{1.16}$$

where all the information about the crystal is contained in the *crystal-field coefficients* A_n^m, which include the effect of screening of the crystal potential by the outer electrons of the rare-earth atom. The index m runs over integers $0 \leq m \leq n$, and covers terms in both sin $m\phi$ and cos $m\phi$ when $m \neq 0$ (an extra index s or c is sometimes included). n can take the values 2, 4 and 6 for the rare-earth series. The K_n^m are just numerical constants related to the normalization of the Z_n^m functions [Hut64]. A convenient way of evaluating the matrix elements of the potential for the wavefunctions corresponding to a particular value of J is the method of *operator equivalents*, introduced by Stevens, and described in some detail by Hutchings [Hut64]. The tesseral harmonics in the expansion of the potential are expressed in Cartesian coordinates, and then replaced by operators involving $\mathbf{J}_x$, $\mathbf{J}_y$, $\mathbf{J}_z$ and $\mathbf{J}^2$. For example, $K_2^0 r^2 Z_2^0$, which can be written as $3z^2 - r^2$, is replaced by $\boldsymbol{O}_2^0 = 3\mathbf{J}_z^2 - \mathbf{J}(\mathbf{J}+1)$. The crystal-field Hamiltonian then takes the form

$$\mathcal{H}_{cf} = \sum_{n=0}^{6} \sum_{m=0}^{n} B_n^m \boldsymbol{O}_n^m \tag{1.17}$$

The number of terms in the expansion is limited by the point symmetry of the rare-earth site, and some specific examples are given in Section C. The parameters B_n^m combine information on the potential created by the distribution of charge in the solid and the form of the charge distribution of the 4f shell.

Specifically

$$B_n^m = \theta_n < r^n > A_n^m \tag{1.18}$$

where θ_n is a reduced matrix element known as the Stevens coefficient for the rare-earth ion in the question (denoted α_J, β_J, γ_J for $n = 2, 4, 6$ respectively), $\langle r^n \rangle$ is the average over the 4f wavefunctions. The coefficients A_n^m describe the screened crystal field produced by the surroundings of the rare-earth ion, and should change little in an isostructural series of compounds with different rare-earths.

One way of evaluating A_n^m, which is of very limited use in a metal because of the presence of conduction electrons, is from the *point charge model*. Charges q_i are associated with the atomic sites $[R_i \Theta_i \Phi_i]$, and

$$A_n^m = \{4\pi e/(2n+1)\}\Sigma_i q_i K_n^m \, Z_n^m(\Theta_i, \Phi_i)/R_i^{n+1} \tag{1.19}$$

For example,

$$A_2^0 = (1/4)e\Sigma_i q_i (3\cos^2\Theta_i - 1)/R_i^3 \tag{1.20}$$

The point charge model can be improved by allowing for the screening effect of the outer electron shells, including the conduction electrons in the Thomas–Fermi approximation, for example, but the point charge model remains a doubtful approximation in metals where much of the crystal field may originate from the valence electrons of the rare-earth itself [Coe91]. At best it offers a way of estimating the ratios of terms of the same order n, when it is impossible to deduce them unambiguously from experiment.

The leading term $B_2^0 O_2^0$ in the expansion of the crystal-field interaction in uniaxial materials is particularly important, because it is the sign of this term, the product of the signs of, α_J and A_2^0, that determines whether the rare-earth has hard or easy-axis anisotropy. Easy-axis anisotropy is required for permanent magnets. The sign of A_2^0 is decided by the crystal structure, but that of α_J depends on whether the shape of the 4f charge distribution is prolate or oblate. The 4f charge distributions for the rare-earth ions are illustrated in Fig. 1.12. Values of α_J, β_J and γ_J for the rare-earth series are listed in Table 1.4. Note that the sign of α_J for Sm is opposite to that for Pr or Nd. A crystal with positive A_2^0 will exhibit easy-axis anisotropy when R = Pr or Nd but easy-plane anisotropy when R = Sm, and vice versa if A_2^0 is negative.

Another way of regarding the terms of different order in the crystal field Hamiltonian, Eq (1.17), is as the interaction between the multipole moments of the 4f distribution and the appropriate component of the potential created by the rest of the solid. Terms with $n = 2, 4, 6$ correspond to the quadrupole, hexadecapole and 64-pole interactions, respectively.

The relative strengths of the terms in the crystal field with $m = 0$ and $n = 2, 4$ and 6 for different rare-earths in a particular crystal structure are represented by the multipole moments $\theta_n \langle r^n \rangle \langle \boldsymbol{O}_n^0(J) \rangle$. These quantities, evaluated at $T = 0$ where,

$\langle J_z \rangle = J$, are displayed in Table 1.4. Also included there are the values of J_n, where $\langle O_2^0(J) \rangle_{T=0} = 2\,J_2$, $\langle O_4^0(J) \rangle_{T=0} = 8\,J_4$ and $\langle O_6^0(J) \rangle_{T=0} = 16\,J_6$. The sign of the fourth-order term, for example, will be relevant in determining, whether [100] or [110] is the easier direction in a tetragonal structure.

Fourth- and sixth-order terms are usually insignificant in comparison to the second-order terms at room temperature and above, because of the high powers of J_z that appear in the fourth- and sixth-order Stevens operators. $\langle J_z \rangle$ falls with increasing temperature, tending to zero at the Curie point. Figure 1.13a illustrates the point, comparing the temperature variation of the expectation values $\langle J_z \rangle$, $\langle O_2^0 \rangle$, $\langle O_4^0 \rangle$ and $\langle O_6^0 \rangle$ calculated for Er in $Er(Fe_{11}Ti)$ [Hu88]. However, the higher-order terms may be crucial for determining the ground-state magnetic structure and properties at low temperature.

Returning to the ferromagnetic transition metals, the crystal field acting on the bare 3d electrons is stronger than the spin–orbit interaction. The orbital moment is largely quenched, but single-ion anisotropy arises when some residual orbital contribution remains. The effect is typically an order of magnitude smaller than for the rare-earth series, but the contribution of the 3d element to the total magnetocrystalline anisotropy is significant because 3d atoms are the major constituents of the alloys of interest here. The 3d anisotropy can only be estimated in the framework of relativistic electronic structure calculations, so it is usually treated phenomenologically, as described in the next section. Temperature dependence of the 3d anisotropy is typically less pronounced than that of the second-order 4f term. It is often useful to distinguish three regions of temperature for rare-earth transition-metal intermetallic compounds where different terms in the anisotropy determine the magnetic structure, as shown in Fig. 1.13b.

At the lowest temperatures, the second-, fourth- and the sixth-order rare-earth anisotropy can all be important, and complex noncollinear magnetic structures may be the result. On raising the temperature, the influence of the fourth- and sixth-order terms will rapidly decline, leaving the second-order rare-earth terms as the principal source of magnetocrystalline anisotropy. This is often the situation at

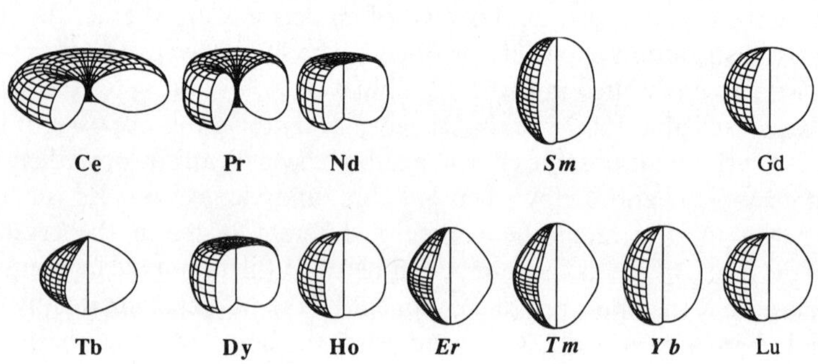

FIG. 1.12. 4f charge distributions for the tripositive rare-earth ions, distinguishing those ions with a positive quadrupole moment ($\alpha_J > 0$; italic type) from those with a negative quadrupole moment ($\alpha_J < 0$; roman type) [Coe91].

Table 1.4 Stevens coefficients, the constants J_n, the ratios B_2^m/A_2^m and relative strength of crystal field terms for the trivalent rare-earth series at zero temperature.

	J	α_J (10^{-2})	β_J (10^{-4})	γ_J (10^{-6})	J_2	J_4	J_6	$\alpha_J<r^2>O_2^0$ (a_0^2)*	$\beta_J<r^4>O_4$ (a_0^4)	$\gamma_J<r^6>O_6^0$ (a_0^6)	B_2^m/A_2^m ($10^{-2}a_0^2$)	B_2^m/A_2^m ($10^{-4}a_0^4$)	B_2^m/A_2^m ($10^{-6}a_0^6$)
Ce	5/2	–5.714	63.5	–	5	7.5	–	–0.748	1.51	–	–7.48	251.6	–
Pr	4	–2.101	–7.346	60.99	14	105	315	–0.713	–2.12	5.89	–2.55	–25.2	1168
Nd	9/2	–0.643	–2.911	–37.99	18	189	945	–0.258	–1.28	–8.63	–0.72	–8.47	–571
Sm	5/2	4.127	25.012	–	5	7.5	–	0.398	0.34	–	3.98	56.7	–
Tb	6	–1.010	1.224	–1.12	33	742.5	10395	–0.548	1.20	–1.28	–0.83	2.02	–7.69
Dy	15/2	–0.635	–0.592	1.04	52.5	2047.5	56306.25	–0.521	–1.46	5.64	–0.50	–0.89	6.26
Ho	8	–0.222	–0.333	–1.29	60	2730	90090	–0.199	–1.00	–10.03	–0.17	–0.46	–6.98
Er	15/2	0.254	0.444	2.07	52.5	2047.5	56306.25	0.190	0.92	8.98	0.18	0.56	9.97
Tm	6	1.010	1.633	–5.61	33	742.5	10395	0.454	1.14	–4.05	0.69	1.92	–24.4
Yb	7/2	3.175	–17.316	148.0	10.5	52.5	78.75	0.435	–0.79	0.73	2.07	–18.8	579

*$a_0 = 0.0529$ nm.

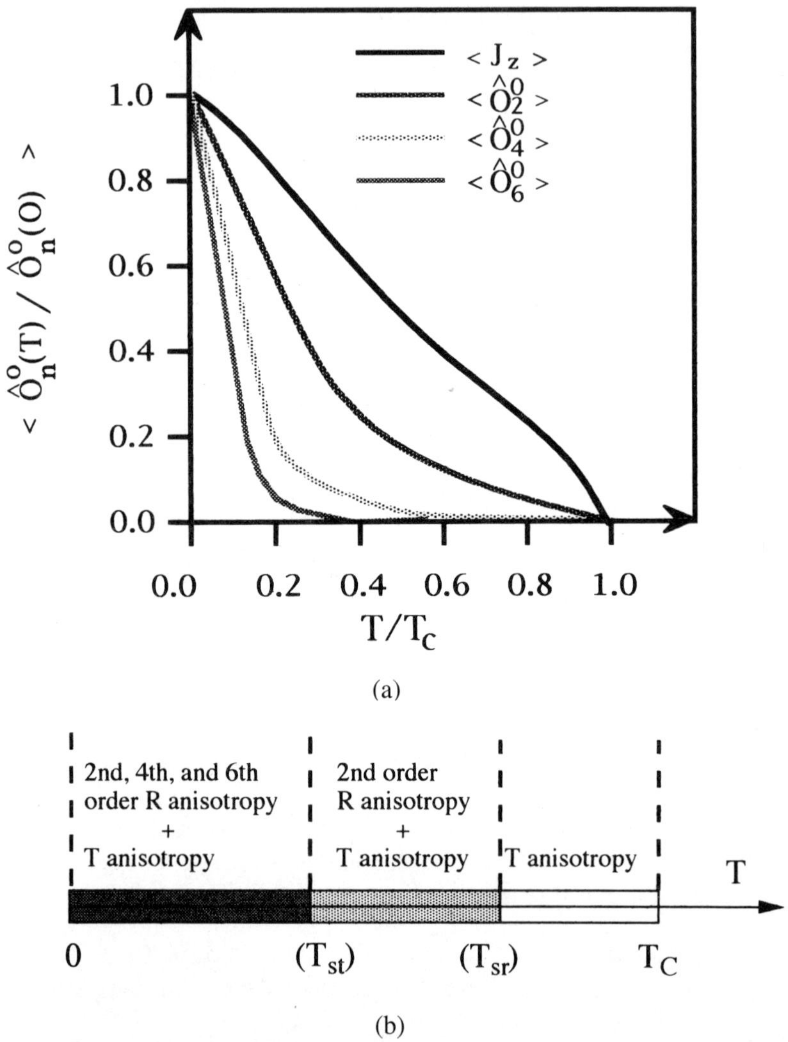

FIG. 1.13. (a) Temperature dependence of $<J_Z>$, $<O_2^0>$, $<O_4^0>$, $<O_6^0>$ for the rare-earth in a typical compound. (b) Regions of temperature where different terms dominate the anisotropy.

room temperature, and the second-order rare-earth anisotropy is the key to understanding the magnetism of most of the $R_2Fe_{14}B$ series. Finally, at higher temperatures the more rapid decline of the rare-earth magnetization compared with that of the transition metal can lead to a situation where the iron or cobalt anisotropy is the dominant term up to the Curie point, T_C.

There are two other potential sources of anisotropy. One arises because the magnetic dipole interaction is anisotropic. It can be seen from Eq. (1.3) that the mutual interaction of two dipoles depends on their orientation with respect to the vector joining them. If the dipoles lie parallel to one another, their interaction will be twice as great when $\mathbf{m} \parallel \mathbf{r}$ as when $\mathbf{m} \perp \mathbf{r}$. The sum of these two-ion contributions can give a contribution to the magnetocrystalline anisotropy in noncubic structures. Typical dipolar fields are < 1 T.

Finally, shape anisotropy should be mentioned, although it is not related to atomic-scale magnetism. An elongated particle will prefer, all else being equal, to have its magnetization lie along the long axis, thereby minimizing the self-energy in the demagnetizing field (Eq. (1.4)). The difference between the demagnetizing field in the easy and hard directions is less than $\mu_0 M_s (\sim 1\text{T})$. Alnicos owe their anisotropy to the shape effect associated with a special microstructure of elongated grains.

4 Modelling the interactions

In analysing the magnetic properties of R–T intermetallic compounds, we are obliged to resort to phenomenological models whose parameters are determined by experiment. One day it may be possible to predict the magnetic and structural properties of a crystal from a knowledge of its chemical composition by numerical solution of Schrödinger's equation. Then magnetism will have lost much of its appeal for physicists. In the meantime we endeavour to develop models with a small number of experimentally determined parameters which can describe phenomena such as Curie temperature, magnetic structure, temperature dependence of magnetization or anisotropy field for series of isostructural compounds.

Modelling of exchange interactions is done in the *molecular-field approximation* [Sma64]. This consists of replacing the interaction $-2J_{ij}\mathbf{S}_i \cdot \mathbf{S}_j$ of Eq. (1.11) summed over each pair of atoms by the interaction $-g\mu_B \mathbf{S}_i \cdot \mathbf{B}^{\text{mol}}$, of each moment with an effective molecular field $\mathbf{B}^{\text{mol}}$ acting at its site. The molecular field is determined self-consistently, and it is proportional to $\langle \mathbf{S}_z \rangle$.

The compounds that concern us here have two magnetic sublattices, composed of rare-earth and transition-metal atoms. This is already a simplification because there can be several sublattices of each kind if the structure includes inequivalent crystallographic sites. The effective fields acting on each sublattice are

$$B_T^{\text{mol}} = n_{TT} M_T + n_{RT} \gamma_R M_R \tag{1.21a}$$

and

$$B_R^{\text{mol}} = n_{RT} \gamma_R M_T + n_{RR} \gamma_R^2 M_R \tag{1.21b}$$

where the n_{TT}, n_{RT} and n_{RR} are molecular-field coefficients which reflect the strength of the interactions and $\gamma_R = 2(g_R - 1)/g_R$ is the ratio of the spin moment to the total moment of the rare-earth. The units of n_{AB} are the same as those of μ_0, and it is generally expressed as a multiple of μ_0.

The exchange energy of an atomic moment $g_R J \mu_B$ in the molecular field B_R^{mol} is $- g_R J \mu_B B_R^{\text{mol}}$. The temperature-dependence of the magnetization of the two sublattices is given by the Brillouin functions;

$$M_T(T) = M_T(0)\mathscr{B}_s(g_T S \mu_B B_T^{mol}/kT) \tag{1.22a}$$
$$M_R(T) = M_R(0)\mathscr{B}_J(g_R J \mu_B B_R^{mol}/kT) \tag{1.22b}$$

where

$$\mathscr{B}_J(x) = (1/2J)\{(2J+1)\coth[(2J+1)x/2J] - \coth[x/2J]\} \tag{1.23}$$

S in Eq. (1.22a) must be replaced by the effective spin S_T, defined by Eq. (1.12). Self consistent solutions of Eqs (1.21) and (1.22) are sought in terms of the molecular-field coefficients n_{TT}, n_{RT} and n_{RR}. Typical temperature dependence of the sublattice magnetizations for an R–T alloy with a light or a heavy rare-earth are shown in Fig. 1.14.

If the rare-earth is nonmagnetic, the Curie point is found by solving Eqs (1.21a) and (1.22a) in the limit $x << 1$. Then,

$$T_C = n_{TT}C_T, \text{ where } C_T = N_T g_T^2 S_T(S_T+1)\mu_B^2/3k \tag{1.24}$$

Here N_T is the number of atoms per cubic metre and C_T is the transition-metal Curie constant. Comparison with Eq. (1.13) shows that the relation between the Heisenberg and molecular field exchange parameters J and n_{TT} is $J = (N_T g^2 \mu_B^2/2Z)n_{TT}$. Also, $B^{mol} = 2ZSJ/g_T\mu_B$

When the rare-earth is magnetic, solution of Eqs (1.21) and (1.22) gives

$$T_C = (1/2)\{T_T + T_R + [(T_T - T_R)^2 + 4T_{RT}^2]^{1/2}\} \tag{1.25}$$

where $T_T = n_{TT}C_T$, $T_R = n_{RR}\gamma_R^2 C_R$ and $T_{RT} = n_{RT} \mid \gamma_R \mid (C_R C_T)^{1/2}$. C_R is the rare-earth Curie constant, defined as $N_R g_R^2 J(J+1)\mu_B^2/3k$, which was listed in Table 1.2 per mole of rare-earth.

The crystal field acting on the rare-earth will have some influence on the results of the molecular-field calculation. Instead of acting on a $(2J+1)$-fold degenerate energy level, the crystal-field ground state may be a doublet or even a singlet, but the effect on results such as those in Fig. 1.14 is usually slight. In any case, molecular-field theory takes no account of spin wave excitations, and provides an inaccurate, mean-field description of the behaviour in the critical region around T_C. Yet, despite its shortcomings, it offers the most useful basis for modelling magnetic properties of two-sublattice magnets.

The *anisotropy energy* of a crystal can be expressed as a series of terms involving the orientation of the total or sublattice magnetization with respect to the crystal axes. An expression that is commonly used when anisotropy is small compared to exchange in a tetragonal crystal is

$$E_a = K_1 \sin^2\theta + K_2 \sin^4\theta + K_3 \sin^6\theta + K_2' \sin^4\theta \cos 4\phi + K_3' \sin^6\theta \cos 4\phi \tag{1.26}$$

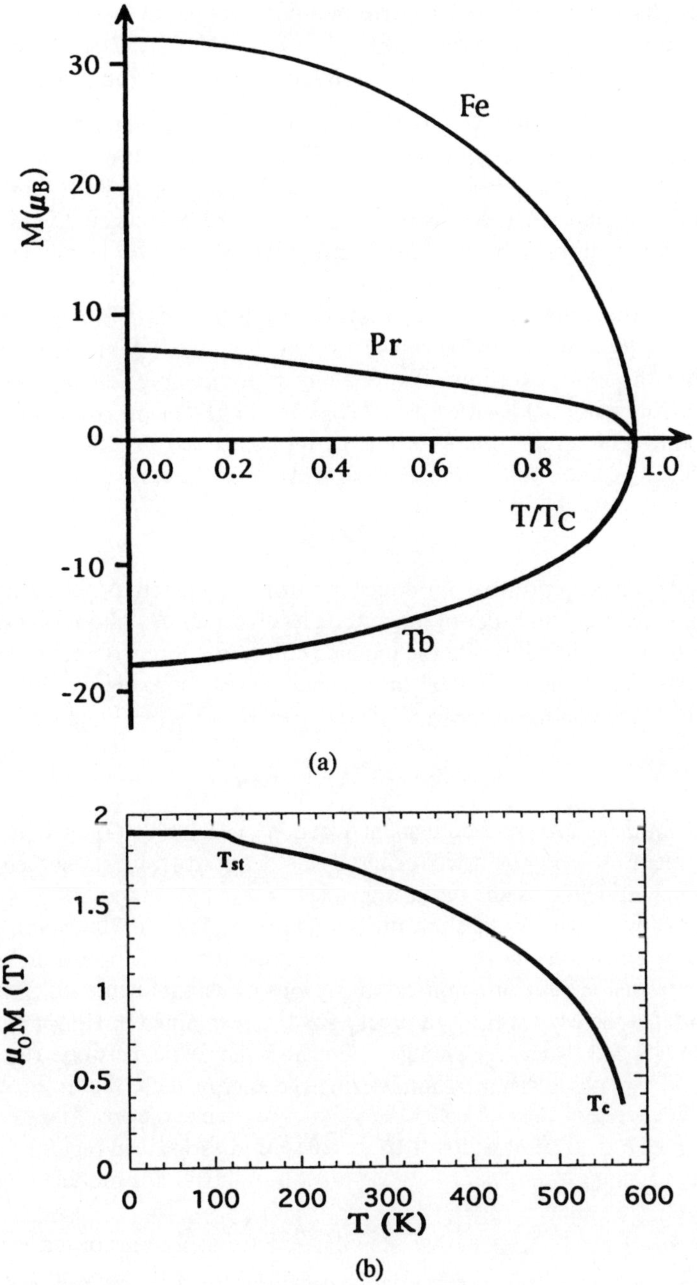

FIG. 1.14. (a) Temperature dependence of the rare-earth and iron sublattice moments in units of μ_B/ formula unit in $R_2Fe_{14}B$ compounds with a light rare-earth (Pr) and a heavy rare-earth (Tb). (b) Temperature dependence of the magnetization of $Nd_2Fe_{14}B$ in practical units. T_{st} and T_C mark the spin-tilt transition and Curie point respectively.

whereas in a hexagonal crystal the corresponding expression is

$$E_a = K_1 \sin^2\theta + K_2 \sin^4\theta + K_3 \sin^6\theta + K_3' \sin^6\theta \cos 6\phi \tag{1.27}$$

An alternative expression Eq. (2.29) gives the anisotropy energy as a sum of spherical harmonics. The relations between K_i and B_n^m are given by Eq. (2.38) and Eq. (2.45) or (2.46). The dominant term for the 3d anisotropy is $K_{1T}\sin^2\theta$, which describes an easy axis when K_{1T} is positive and a hard axis when K_{1T} is negative. Some of the other terms may be important for the rare-earths, particularly at low temperature. Units of E_a and K_i are J/m^3.

Related to the anisotropy energy is the concept of *anisotropy field*. The anisotropy energy is somehow equivalent to that produced by an effective field B_a acting along the easy direction. The analogy is useful, but inexact because the energy in a magnetic field varies as $-MB_a\cos\theta$. The anisotropy field is usually defined as that which reproduces the energy increase due to small deviations from the easy axis. The condition is $(dE_a/d\theta) = B_a M_s \sin\theta$, hence

$$B_a = 2K_1/M_s \tag{1.28}$$

The anisotropy field represents an upper limit on the coercivity $\mu_0 H_c$; achievable coercivities are usually an order of magnitude less than B_a. An alternative definition is the field B_K needed to saturate the magnetization in a direction perpendicular to the easy axis. The orientation of the magnetization is given by the condition $dE/d\theta = 0$, so the condition $\theta = \pi/2$ yields

$$B_K = 2(K_1 + 2K_2 + 3K_3)/M_s \tag{1.29}$$

The two definitions are only equivalent when $K_2 = K_3 = 0$. Otherwise, different results for the anisotropy field will generally be deduced from measuments in low and high applied fields, as shown in Fig. 1.15.

Magnetization curves in different directions on single-crystal samples, and spin reorientation transitions as a function of temperature or applied field can be analysed in terms of a set of empirical anisotropy constants from which it may be possible to infer the crystal-field parameters B_n^m, as explained in Chapter 2. A spin reorientation begins when K_1 changes sign, and the angle between the moment direction and the c-axis given by minimizing the energy E_a is

$$\sin\theta = \{(-K_2 \pm [K_2^2 - 3K_1K_3]^{1/2})/3K_3\}^{1/2} \tag{1.30a}$$

or, if K_3 is negligible, by

$$\sin\theta = (-K_1/2K_2)^{1/2} \tag{1.30b}$$

Since the number of exchange and crystal-field parameters needed to describe a single compound is quite large, it is very useful to be able to compare measurements on a series of isostructural compounds of different rare-earths and yttrium in order to develop an understanding of the interactions which are responsible for the intrinsic magnetic properties.

C. INTERMETALLIC COMPOUNDS

We now appreciate why $R_{1-x}T_x$ compounds having iron or cobalt as the major constituent and one of the magnetic light rare-earths Pr, Nd or Sm as a minor constituent are especially suited for use as high-performance permanent magnets. The best solution from the magnetic and economic points of view would be an alloy of Nd and Fe, but the Nd–Fe binary phase diagram, shown in Fig. 1.16a, is both simple and disappointing. There are two intermetallics, one of them, Nd_2Fe_{17}, an iron-rich compound with $x > 0.8$ crystalling in the rhombohedral Th_2Zn_{17} structure, but with a Curie point of only 326 K. What is worse, the sign of the anisotropy constant K_1 is negative, which means that the *c*-direction is hard, not easy.

If an intermetallic compound is to be considered as a candidate for making practical permanent magnets it needs to have:

1. a Curie point T_C above 500 K;
2. a low-symmetry crystal structure, preferably uniaxial (tetragonal, rhombohedral or hexagonal);
3. positive K_1, so that the *c*-axis is the easy direction;
4. a saturation polarization $J_s = \mu_0 M_s$ greater than 1.0 T.

Nd_2Fe_{17} satisfies only the second requirement. The sole Pr–Fe intermetallic, Pr_2Fe_{17}, is similar to Nd_2Fe_{17}. None of the three Sm–Fe binary intermetallics is suitable; $SmFe_2$ is cubic; Sm_2Fe_{17} has $T_C = 389$ K and K_1 is negative; $SmFe_3$ almost qualifies — it is rhombohedral with $T_C = 650$ K and K_1 is positive but the polarization is only 0.8 T on account of the high rare-earth content, $x = 0.75$.

The Sm–Co binary diagram, shown in Fig. 1.16b is much richer. There are seven intermetallic compounds; the first two of them, Sm_2Co_{17} and $SmCo_5$, are

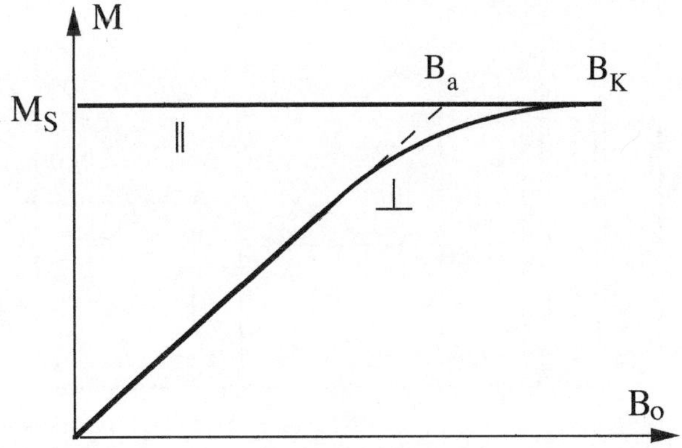

FIG. 1.15. Schematic magnetization curves for a compound with easy-axis anisotropy when the field is applied parallel or perpendicular to the easy axis, showing the anisotropy field B_a and the saturation field B_K.

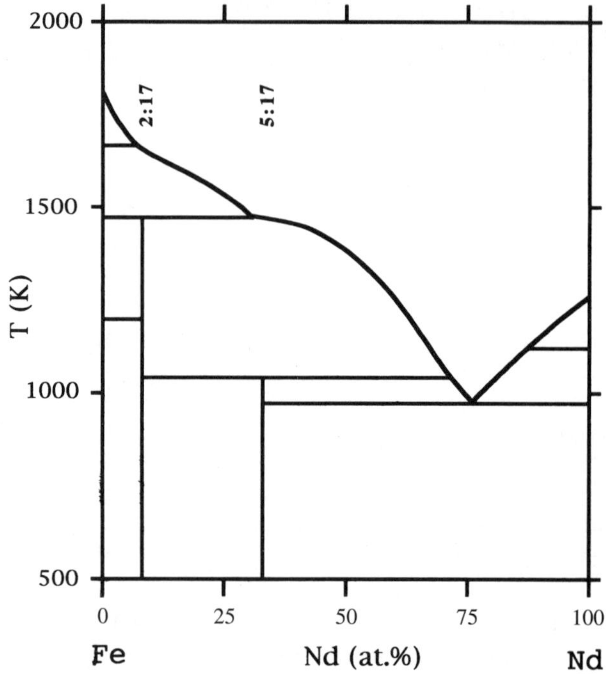

(a)

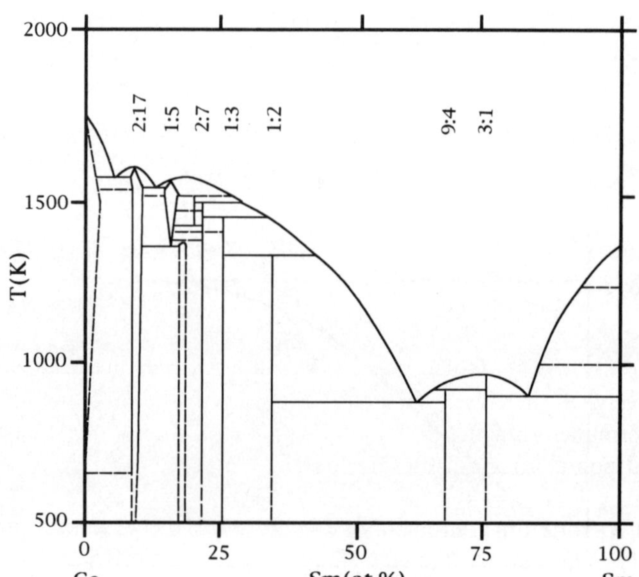

(b)

FIG. 1.16. Binary phase diagrams: (a) neodymium–iron; (b) samarium–cobalt.

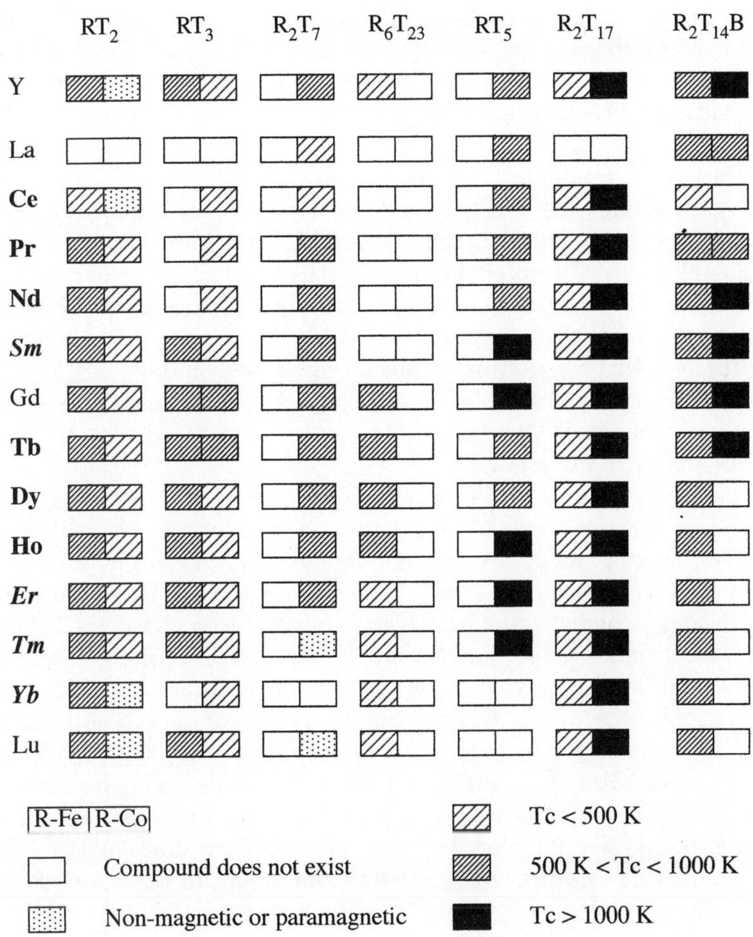

FIG. 1.17. Summary of rare-earth iron and rare-earth cobalt binary intermetallic phases.

hexagonal with positive K_1 and $\mu_0 M_s > 1$ T. Curie temperatures are 1100 K and 993 K respectively. Both have been used extensively in the manufacture of high-performance permanent magnets.

Some general points arising from this comparison are

1. The signs of K_1 for a Sm compound and its Pr and Nd counterparts are usually opposite. The reason is that the major contribution to K_1 at ambient temperature is from the rare-earth second-order crystal-field term $B_2^0 \boldsymbol{O}_2^0$; $K_1 \sim 3J^2 B_2^0$ (see Fig. 1.13b); the second-order Stevens coefficient α_j, and hence B_2^0, is opposite in sign for Sm and Pr or Nd (Table 1.4, Fig. 1.12).
2. The Curie temperatures of R–Co intermetallics tend to be much greater than

those of their iron counterparts. T_C of the iron end-member can often be improved by cobalt substitution.

3. More intermetallic phases form between Co (or Ni) and the light rare-earths then between Fe and light rare-earths.

Points (2) and (3) are illustrated further by Fig. 1.17.

If we want a rare-earth iron permanent magnet, we must therefore resort to a phase with *three* or more constituents. When the third component is nonmagnetic there should be as little of it in the alloy as possible, to avoid reducing the magnetization. One can seek a true ternary compound, such as $Nd_2Fe_{14}B$, an interstitially modified binary intermetallic such as $Sm_2Fe_{17}N_3$ or a pseudobinary of a structure type that has no iron end-member, such as $Sm(Fe_{12-x}Ti_x)$.

The remainder of this section is devoted to presenting the structures of intermetallic compounds that are of interest as permanent magnets. Data on these structures are listed in Table 1.5.

The tetragonal structure of $Nd_2Fe_{14}B$ [Her84, Giv84] is illustrated in Fig. 1.18; there are four formula units, hence 68 atoms in the unit cell. The main symmetry elements in the space group are the vertical 4_2 screw axes passing through 4c sites and two perpendicular mirror planes 110 and $\bar{1}10$ that are parallel to **c**. The two rare-earth sites, 4f and 4g, each have *mm* point symmetry, and Fig. 1.18b indicates how they are disposed in the planes at $z = 0$ and $z = 1/2$, which they share with the boron (4f) and one of the iron atoms (4c). Boron is coordinated by six irons that form a trigonal prism, a structural elements typical of transition-metal borides. Four of the remaining iron sites are sandwiched between these rare-earth boron planes. Iron atoms on 4e, $8j_1$, $16k_1$ and $16k_2$ sites form two kagomé networks of distorted hexagons and triangles, while the last iron site, $8j_2$, is sandwiched between them in a hexagonal antiprism. The rare-earth sites are surrounded by distorted hexagonal prisms of iron, such as the samarium in $SmCo_5$ ($CaCu_5$ structure).

Table 1.5 Some rare-earth intermetallic structure types

Type	Symmetry	Space group	Z	*a* (nm)	*c* (nm)	R-sites	T-sites
$Nd_2Fe_{14}B$	tetragonal	$P4_2/mnm$	4	0.88	1.22	4f,4g	4c,4e,$8j_1$, $8j_2$,$16k_1$,$16k_2$
$ThMn_{12}$	tetragonal	I4/mmm	2	0.87	0.50	2a	8f,8i,8j
$CaCu_5$	hexagonal	P6/mmm	1	0.51	0.41	1a	2c,3g
$CeCo_4B$	hexagonal	P6/mmm	2	0.50	0.69	1a,1b	2c,6i
Th_2Ni_{17}	hexagonal	$P6_3/mmc$	2	0.84	0.81	2b,2d	4f,6g,12j,12k
Th_2Zn_{17}	rhombo-hedral	$R\bar{3}m$	3*	0.83	1.21	$6c_2$	$6c_1$,9d,18f,18h

* hexagonal cell.

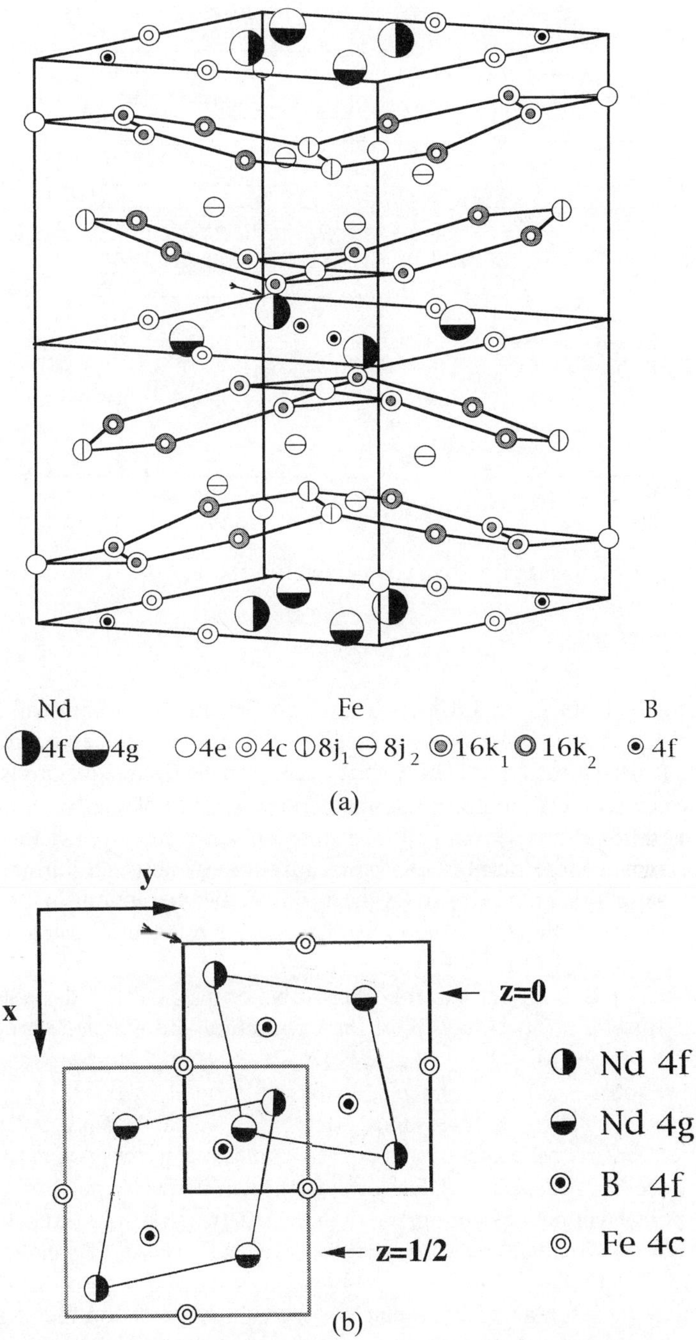

FIG. 1.18. (a) The tetragonal $Nd_2Fe_{14}B$ structure and (b) projections of the $z = 0$ and $z = 1/2$ layers onto the 001 plane. The origin is indicated by the arrows.

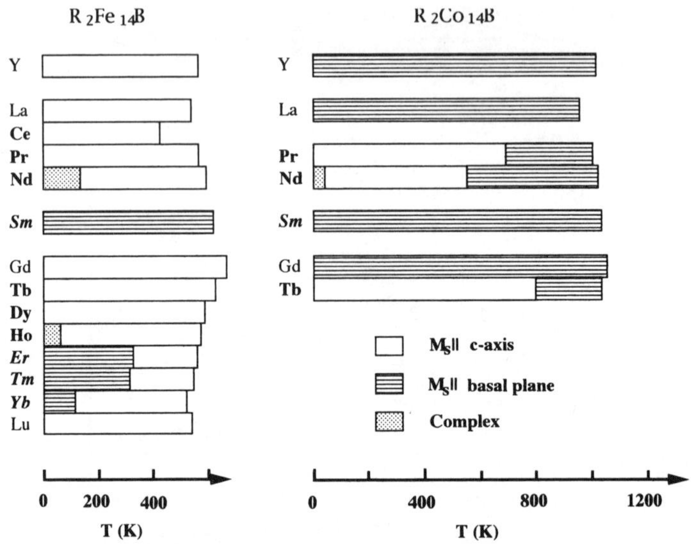

FIG. 1.19. Summary of magnetic transitions found in the $R_2T_{14}B$ series of compounds, T = Fe or Co.

All rare-earths from La to Lu, plus Y and Th (but not Eu or Sc) will form the $Nd_2Fe_{14}B$ phase, designated ϕ by metallurgists. Cobalt can replace iron completely for the lighter rare-earths, up to Tb. A wide range of partial substitutions is possible for all three constituents. Some of the magnetic properties of the $Nd_2Fe_{14}B$ series are summarized in Fig. 1.19, and they are presented in detail in Chapter 2. The existence of such a large number of isostructural compounds exhibiting different types of magnetic transition helps to fix the atomic-scale interactions in the complex structure. Indeed, $R_2Fe_{14}B$ is the only series of isostructural rare-earth iron compounds that exists from La to Lu.

Besides $Nd_2Fe_{14}B$, there are several other stable or metastable Nd–Fe–B ternary compounds [Bus86]. Information about their structural and magnetic properties is summarized in Table 1.6. The η phase, $Nd_{1+\varepsilon}Fe_4B_4$ [Giv85] appears as a minor constituent of some Nd–Fe–B magnets, as discussed in Chapter 3.

The hexagonal $CaCu_5$ structure is illustrated in Fig. 1.20a. This is the structure of the celebrated $SmCo_5$ permanent-magnet compound, but it does not exist for iron with any of the light rare-earths. The rare-earth is in a special position (1a) with hexagonal point symmetry (6/mmm) and there are two transition-metal sites (2c and 3g). Several related structure types, $CeCo_4B$, $TbCu_7$, Th_2Ni_{17}, Th_2Zn_{17} $ThMn_{12}$ and $Nd_3(Fe,Ti)_{29}$, can be thought of as modifications of $CaCu_5$. The $CeCo_4B$ structure is obtained by replacing one of the transition-metal atoms in alternate rare-earth layers by boron, as shown in Fig. 1.20b. The RFe_4B compound exists only for the heavy rare-earths, although it can be stabilized for R = Sm by the substitution of some Ni or Co for Fe, as in $Sm(Fe_{4-x}Co_x)B$.

Table 1.6 Nd–Fe–B intermetallic compounds

	Structure	Space group	a(nm)	c(nm)	T_C(K)	Remarks
$Nd_2Fe_{14}B$	Tetragonal	$P4_2/mnm$	0.879	1.218	585	ϕ phase
$Nd_{1+\varepsilon}Fe_4B_4$	Tetragonal	Pccn	0.712	3.501	13	η phase
$Nd_5Fe_2B_6$	Rhombohedral	R3m	0.55	2.43	64	ρ phase
$NdFe_{12}B_6$	Rhombohedral	$R\bar{3}m$	0.961	0.754	230	metastable
$Nd_2Fe_{23}B_3$	Cubic	$I\bar{4}3d$	1.419		655	metastable

The relation between the 1:5 and 2:17 structures is illustrated in Fig. 1.21. In either case, one-third of the rare-earth atoms is replaced by a closely spaced T–T dumbell; $3RT_5 - R + 2T = R_2T_{17}$. Replacing different fractions of the rare-earth atoms in RT_5 by T–T dumbells leads to the 1:7, 1:12 or 3:29 structures. In the rhombohedral version of the 2:17 structure (Th_2Zn_{17}) there are three formula units per unit cell and a single rare-earth site with point symmetry 3/m, whereas in the hexagonal version (Th_2Ni_{17}) there are two formula units per unit cell and two inequivalent rare-earth sites with point symmetry $\bar{6}$m2. Both polymorphs may be mixed on an atomic scale in certain 2:17 materials. Sm_2Co_{17} and its derivatives can be used to make excellent permanent magnets. The low Curie points of R_2Fe_{17} compounds can be remedied by introducing interstitial nitrogen or carbon into the R_2T_4 octahedral sites surrounding the rare-earth, indicated in Fig. 4.1, to yield $R_2Fe_{17}Z_3$ [Coe90]. $Sm_2Fe_{17}N_3$ has $T_C = 749$ K and a large positive value of K_1. Its intrinsic properties are favourable for making magnets. Interstitial intermetallics are the subject of Chapter 4.

Another potentially useful family of compounds has the tetragonal $ThMn_{12}$ structure, illustrated in Fig. 1.22 [Bus88, Li93]. This is a much simpler structure than $Nd_2Fe_{14}B$, with a single rare-earth site (2a) in a highly symmetrical special position (point symmetry 4/mmm), and three transition-metal sites (8f, 8i and 8j). The structure type does not exist for RFe_{12}, but it can be stabilized for $R(Fe_{12-x}M_x)$ where R is a light rare-earth and M is Ti, V, Cr, Mo, W or Al with x as little as 1–2. Curie temperatures are somewhat lower than those of $R_2Fe_{14}B$ (Fig. 2.50), but they can be improved by interstitial carbon or nitrogen [Hur92]. The magnetic properties of monoclinic compounds with crystal structures intermediate between 2:17 and 1:12, such as 3:29, can also be improved by interstitial modification [Cad94].

Finally, we consider the anisotropy of some of the main compounds of interest. The rare-earth anisotropy may be evaluated from the crystal-field Hamiltonian Eq. (1.17), which includes terms that reflect the point symmetry of the rare-earth site. The terms arising for the symmetries encountered above are

$$4/\text{mmm}: \mathcal{H}_{cf} = B_2^0 O_2^0 + B_4^0 O_4^0 + B_4^{4c} O_4^{4c} + B_6^0 O_6^0 + B_6^{4c} O_6^{4c} \tag{1.31a}$$

$$6/\text{mmm}: \mathcal{H}_{cf} = B_2^0 O_2^0 + B_4^0 O_4^0 + B_6^0 O_6^0 + B_6^6 O_6^6 \tag{1.31b}$$

or $\bar{6}$ m 2

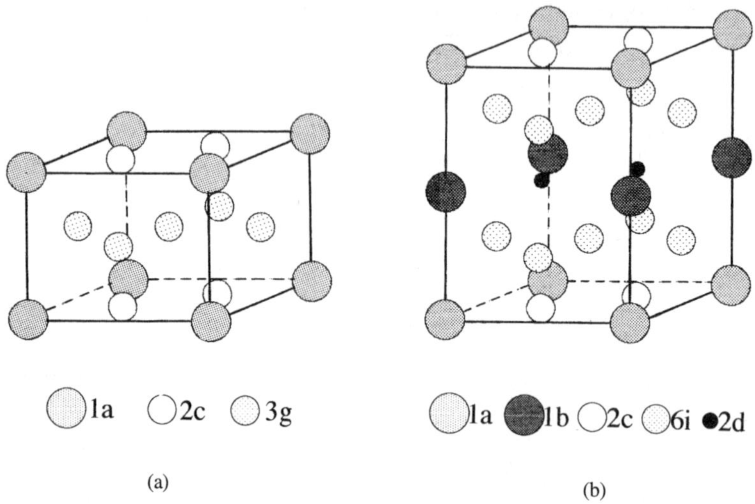

FIG. 1.20. (a) $CaCu_5$ structure. (b) $CsCo_4B$ structure.

$$3/\mathrm{m} : \mathscr{H}_{\mathrm{cf}} = B_2^0 O_2^0 + B_4^0 O_4^0 + B_4^3 O_4^3 + B_6^0 O_6^0 + B_6^3 O_6^3 + B_6^6 O_6^6 \tag{1.31c}$$

$$\begin{aligned}\mathrm{mm} : \mathscr{H}_{\mathrm{cf}} = {} & B_2^0 O_2^0 + B_2^{2s} O_2^{2s} + B_4^0 O_4^0 + B_4^{2s} O_4^{2s} + B_4^{4c} O_4^{4c} + B_6^0 O_6^0 \\ & + B_6^{2s} O_6^{2s} + B_6^{4c} O_6^{4c} + B_6^{6s} O_6^{6s}\end{aligned} \tag{1.31d}$$

Besides the contribution from the rare-earth, there is that from the transition-metal sublattices, for which the only important term is $K_{1T} \sin^2 \theta$. This is often deduced from measurements on the yttrium compound. The total anisotropy is then the sum of the contributions from the rare-earth and transition-metal sublattices. However, we have seen that the sign of the leading term in the total anisotropy at ambient temperature, $K_1 \sin^2 \theta$ will often reflect the sign of B_2^0. Depending on the sign of A_2^0, the electric-field gradient at the rare-earth site which is a function of the crystal structure, B_2^0 will be positive for rare-earths with positive α_J (Sm, Er, Tm) and negative for rare-earths with negative α_J (Pr, Nd, Tb ...) or *vice versa* (see Table 1.4). Table 1.7 lists the signs and magnitudes of A_2^0 and K_{1T} for compounds of interest. When A_2^0 is negative, it is the samarium compound that is suitable for magnets, whereas when A_2^0 is positive it is the praesoedymium or neodymium compound. Theoretical maximum energy products (Eq. (1.8c)) are also included in the table. The intrinsic magnetic properties of these compounds are discussed further in Chapter 2.

D. MAGNETS

The task of turning a material with good intrinsic properties into a permanent magnet with useful remanence and coercivity across a temperature range of interest

Table 1.7 Anisotropy and magnetization of permanent-magnet compounds

Compound	Structure type	A_2^0 (K/a$_0^2$)	K_{1T} (MJ/m^3)	K_1^{tot} (MJ/m^3)	B_a (T)	$\mu_0 M_s$ (T)	$\mu_0 M_s^2/4$ (kJ/m^3)
$Nd_2Fe_{14}B$	$Nd_2Fe_{14}B$	306	1.1	4.9	7.6	1.61	516
$Sm(Fe_{11}Ti)$	$ThMn_{12}$	–143	0.9	4.8	10.5	1.14	259
$Sm_2Fe_{17}N_3$	$Sm_2Fe_{17}N_3$	–242	–1.3	8.6	14	1.54	472
$SmCo_5$	$CaCu_5$	–200	6.5	17	40	1.05	219
Sm_2Co_{17}	Th_2Zn_{17}	–108	–0.4	3.3	6.5	1.30	336

for applications is a complex one. A microstructure is needed which incorporates energy barriers that can preserve the metastable, permanently magnetized state. There are specific solutions for each type of material, depending on the form of the phase diagram. Different aspects of the problem are discussed in Chapters 3, 5, 6 and 7, with special reference to Nd–Fe–B.

1 Processing

Two principal fabrication routes for Nd–Fe–B magnets are *powder metallurgy* [Sag84] and *melt spinning* [Cro84]. The powder route, which is also used for Sm–Co magnets, starts with ingot which is milled to a fine powder composed of monocrystalline particles a few microns in size, then aligned in a magnetic field, pressed, sintered at about 1100 °C and finally heat-treated to develop a microstructure which gives coercivity when the alloy is magnetized. A typical alloy composition is $Nd_{15}Fe_{77}B_8$, not the stoichiometric composition $Nd_{12}Fe_{82}B_6$. The final microstructure consists of ϕ grains partly enveloped by material derived from a Nd-rich liquid, with some inclusions (cf. Fig. 3.2a, Fig 6.19). Roughly 90% of the magnet volume is composed of the hard-magnetic ϕ phase.

Melt spinning involves a jet of molten alloy which is quenched on a rapidly rotating copper wheel to form ribbon composed of many randomly oriented nanocrystallites some 50 nm in size which is already coercive. The material may then be hot-worked to align the crystallites. Both routes are capable of yielding highly dense, oriented magnets with excellent energy products. Alternative approaches include mechanical alloying, hot extrusion and hydrogen processing. In any case, the aim is to create a metallurgical microstructure that can effectively resist the formation and propagation of domain walls in the magnetized state. The interstitially modified intermetallics such as $Sm_2Fe_{17}N_3$ may be produced by gas–solid reaction, then milled to a fine coercive powder. Gas–solid reactions are treated in Chapter 4 and magnet processing is the subject of Chapter 7.

2 Micromagnetics

Provided the magnetization vector $\mathbf{M}(\mathbf{r})$ changes slowly on the atomic scale, the micromagnetic problem involves finding the local minima of the free energy

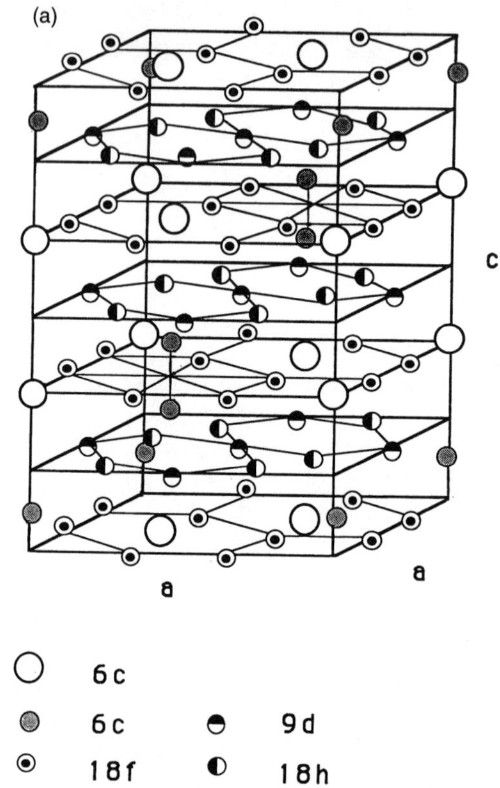
(a)
c
a
a
6c
6c
9d
18f
18h

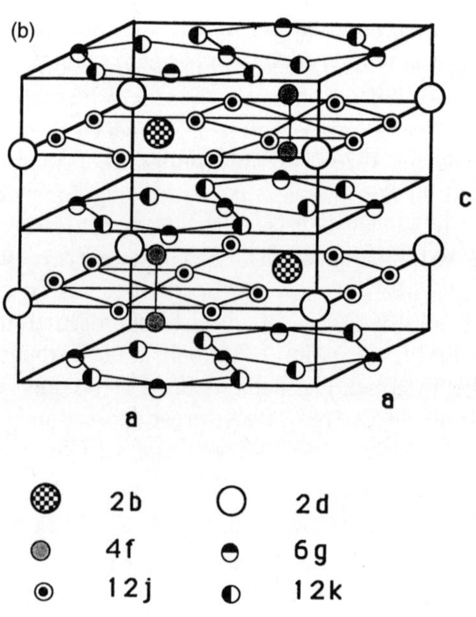
(b)
c
a
a
2b
2d
4f
6g
12j
12k

(c)

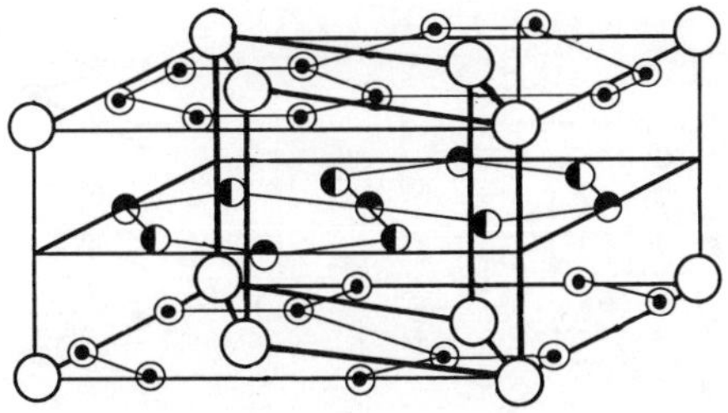

FIG. 1.21. Structure of rhombohedral Th_2Zn_{17} (a) and hexagonal Th_2Ni_{17} (b) structures, and the relation with the $CaCu_5$ (c) structure.

---

involving a sum of exchange, magnetostatic (Zeeman) and anisotropy terms

$$F = \int \{A_{ex}(\mathbf{r})[\nabla \mathbf{M}/M_s]^2 - \mu_0 \mathbf{M} \cdot \mathbf{H}_{loc} - K_1(\mathbf{r})[\mathbf{M} \cdot \mathbf{n}/M_s]^2 - \ldots\}\mathrm{d}\mathbf{r} \qquad (1.32)$$

for the given microstructure. Here M_s is the magnitude of the local magnetization vector $\mathbf{M}(\mathbf{r})$, A_{ex} is the temperature-dependent exchange stiffness measured in J/m. At zero temperature it is related to the exchange parameter J in Eq. (1.11) by the equation $A_{ex} = cJS^2/a$, where c is a constant of order unity depending on the structure (see Chapter 5, p. 247) and a is the interatomic spacing. $\mathbf{H}_{loc}$ is the local field, which is the sum of any applied field and half the dipole field produced by $\mathbf{M}(\mathbf{r})$. (The half appears because the energy in the dipole field is a self-energy.) The term in K_1 is the leading term in the magnetocrystalline anisotropy energy and $\mathbf{n}(\mathbf{r})$ is a unit vector specifying the local easy axis.

The general solution of Eq. (1.32) is a formidable problem. However, the solution when K_1 is sufficiently strong consists of uniformly magnetized *domains* where the magnetization lies along $\pm$ **n**, separated by narrow 180° *domain walls* (Fig. 1.23). The wall energy, γ_W, in J/m^2, is obtained by summing the exchange and anisotropy energies at every point across the wall, where the magnetization is inclined at an angle θ to the anisotropy axis:

$$\gamma_W = \int_{-\infty}^{\infty} \{A_{ex}(\mathrm{d}\theta/\mathrm{d}x)^2 + K_1 \sin^2\theta\}\mathrm{d}x. \qquad (1.33)$$

Minimizing the integral yields the Euler equation

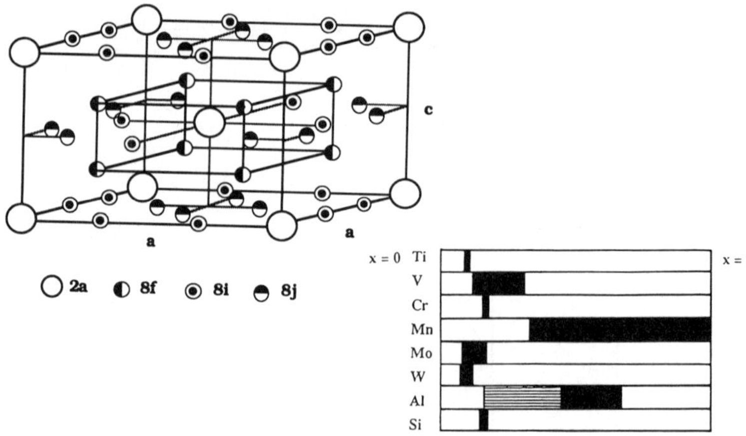

FIG. 1.22. The $ThMn_{12}$ structure. The stability ranges of the iron-rich phases are shown by the black bars.

$$2A_{ex}(d^2\theta/dx^2) = K_1 \sin 2\theta$$

with solution

$$\theta(x) = 2\tan^{-1}\exp\{x(K_1/A_{ex})^{1/2}\} \tag{1.34}$$

plotted in Fig. 1.23, where x is measured from the mid-point of the wall. The exchange and anisotropy terms are equal at every point and the wall is infinite in extent. About 160° of the rotation occurs within a distance δ_W, known as the domain wall width (see inset in Fig. 1.23):

$$\delta_w = \pi(A_{ex}/K_1)^{1/2} \tag{1.35}$$

The domain wall energy is evaluated from Eqs (1.33) and (1.34) as

$$\gamma_w = 4(A_{ex}K_1)^{1/2} \tag{1.36}$$

For many purposes it is convenient to deal with dimensionless scaling parameters. In Eq. (1.32) there are three material parameters, K_1, A_{ex}, and M_s. The interatomic spacing is negligible in a continuum theory, so the only dimensionless combination is $(K_1/\mu_0 M_s^2)^\lambda$. We set $\lambda = 1/2$ to define the dimensionless *hardness parameter*

$$\kappa = (K_1/\mu_0 M_s^2)^{1/2}$$

which must be greater than 1 for a permanent magnet. Typical values are given in Table 1.8 (p. 45).

Some characteristic micromagnetic phenomena may now be considered in terms of κ.

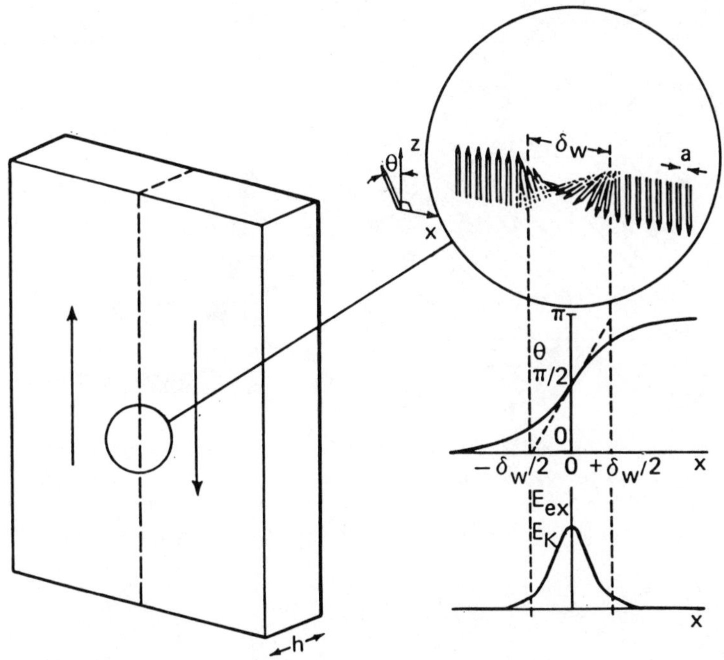

FIG. 1.23. Schematic representation of a 180° domain wall (Bloch wall) in a material with uniaxial anisotropy. The spin configuration in the wall is shown as well as the variation in orientation and exchange or anisotropy energy across the wall.

1. Imagine a soft magnet where a single spin is turned into an energetically unfavourable direction. The surrounding perturbation will decay over a characteristic scale known as the *exchange length*

$$l_{ex} = (A_{ex}/\mu_0 M_s^2)^{1/2}$$

which is about 3 nm for ferromagnetic iron- or cobalt-based alloys. The effect is sometimes referred to as the 'exchange spring' [Kne91]. Generally, l_{ex} is quite different from the domain wall width δ_W, which tends to infinity in an ideal soft material ($K_1 = 0$ in Eq. (1.35)). The domain wall width can be written in terms of l_{ex} as $\delta_W = \pi l_{ex}/\kappa$. Note that $\delta_W \approx l_{ex}$ in hard magnets, whereas $\delta_W >> l_{ex}$ in soft magnets.

2. Another important length is the critical single-domain particle size, d_c, the size of the largest crystallite for which the cost of creating a domain wall in equilibrium exceeds the gain in magnetostatic energy by so doing. For spherical particles ($N = 1/3$) with strong uniaxial anisotropy, a rough estimate of the critical diameter is obtained by equating the energy of a circular domain wall to half the magnetostatic energy of a uniform sphere, giving [Kit49]

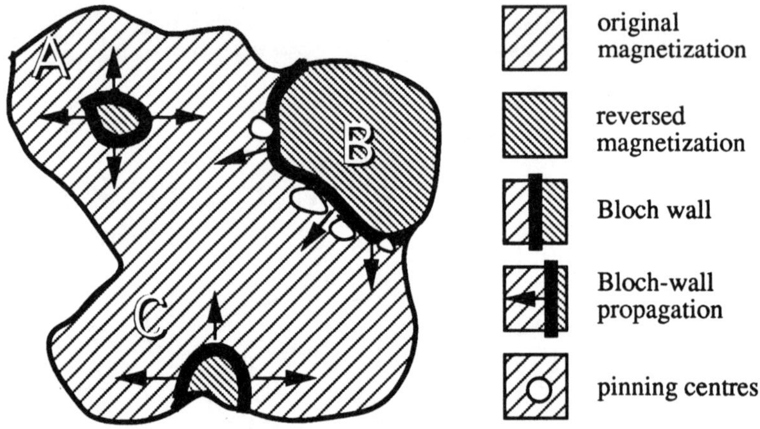

FIG. 1.24. Processes involved in magnetization reversal in the second quadrant of the hysteresis loop. A is a reverse domain which nucleates in the bulk at a defect or from a spontaneous thermal fluctuation. B is a reverse domain which has grown to a point where the wall is trapped by pinning centres. C is a reverse domain which nucleates at a rough surface.

$$d_c = 18\gamma_w/\mu_0 M_s^2 = 72\kappa l_{ex}$$

A typical value for rare-earth iron magnetic material is $\approx$ 200 nm. Since it is impracticable to mill particles to submicron dimensions, the grains in sintered magnets are not single-domain particles in the virgin state. Domain walls are easily swept across each grain in response to an applied field. Development of coercivity then hinges on preventing the nucleation and growth of reverse domains from the fully magnetized state. Single-domain grains are encountered in nanostructured material that can be produced by melt spinning or mechanical alloying.

3. In very small particles, magnetization reversal occurs 'coherently' in the sense that, during reversal, the magnetization remains homogeneous with the same magnitude and direction everywhere in the particle. But in larger particles, incoherent modes involving curling or fanning of the magnetization vector will dominate. The critical size above which incoherent reversal occurs scales as l_{ex}.

In a permanent magnet, it is essential to inhibit the spontaneous *nucleation* of reverse domains due to the action of the local field $\mathbf{H}_{loc}$, for example at irregularities on the surfaces of the magnetic grains. Nucleation occurs when the fully aligned solution of Eq. (1.32) first becomes unstable; it is necessary but not sufficient for magnetization reversal as the nucleus must then grow and propagate. If possible, it is helpful to *pin* whatever domain walls may be present to prevent the reverse domains from growing. Pinning centres can be planar defects or inclusions that are comparable in size to the domain walls. Processes involved in magnetization reversal are illustrated in Fig. 1.24. The two coercivity mechanisms, *nucleation* and *pinning* are readily distinguished from the shape of the initial magnetization curve

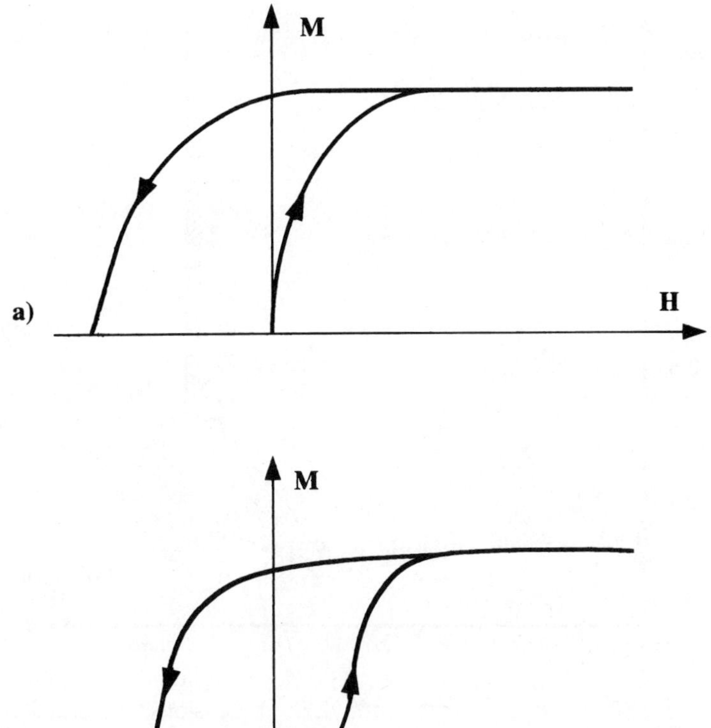

FIG. 1.25. Initial magnetization curves and the first two quadrants of the hysteresis loops of (a) a nucleation-type magnet and (b) a pinning-type magnet. The crystallites are supposed to be larger than the critical single-domain size d_c.

of grains which are multidomain in the thermally demagnetized, equilibrium state (Fig. 1.25).

A limit that the coercivity of any homogeneous ellipsoid must satisfy was derived by Brown [Bro57] from general energetic considerations independent of any model of magnetic reversal:

$$H_c \geq 2K_1/\mu_0 M_s - NM_s \tag{1.37}$$

The first term is the anisotropy field, and the second is small by comparison (Table 1.8) since $N \leq 1$ and $\mu_0 M_s \sim 1\text{T}$. Coercivity should therefore be comparable to or greater than the anisotropy field. In reality it is very much less, a discrepancy with theory known as *Brown's paradox*. The explanation is that real systems inevitably include inhomogeneities with spatial variations in anisotropy and exchange that give rise to the small coercivities observed in practice. Figure 1.26 shows the progress over the years in narrowing the gap between theory and practice.

The simplest model of magnetic reversal is that of Stoner and Wohlfarth, who considered coherent rotation of magnetization in an ellipsoidal single-domain

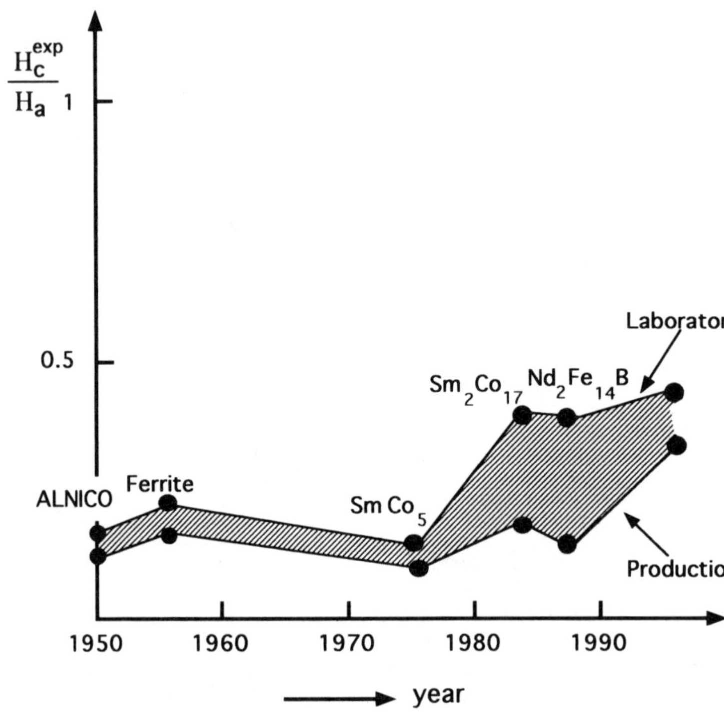

FIG. 1.26. Progress towards narrowing the gap between the anisotropy field H_a and the coercivity H_c. (After Kronmüller 1991.)

particle [Sto 48]. From Eq. (1.32) the angular-dependent part of the energy of the system is

$$E = K_1 \sin^2\theta - \mu_0 \mathbf{M} \cdot (\mathbf{H}_e + \tfrac{1}{2}\mathbf{H}_d)$$

where $\mathbf{H}_d$ is given by Eq. (1.4)

When the externally applied field $\mathbf{H}_e$ is along the major axis of the ellipsoid and this is also the anisotropy axis, there may be two energy minima, separated by a barrier, as shown in Fig. 1.27. A feature of the model is that the nucleation field H_n, where the magnetization first deviates from the axis, given by the condition $d^2E/d\theta^2 = 0$, is the same as the coercive field H_c, where the energy barrier disappears and magnetization reversal occurs:

$$H_c = 2K_1/\mu_0 M_s + [(1 - 3N)/2]M_s \tag{1.38}$$

The Stoner–Wohlfarth result is compatible with Brown's limit, since $0 \leq N \leq 1$. Note also that shape anisotropy is equivalent to magnetocrystalline anisotropy in the model with $K_1^{eff} = \mu_0 M_s^2(1 - 3N)/4$. The model is useful when considering coercivity in systems like Alnico with elongated single-domain particles where shape anisotropy is crucial. For example, a sufficiently small ellipsoid with

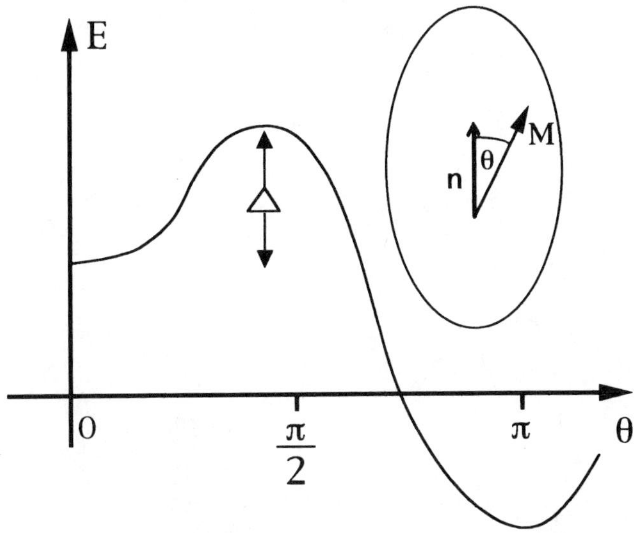

FIG. 1.27. Coherent reversal of the magnetization of an ellipsoidal crystallite. The anisotropy axis **n** (the *c*-axis of the crystallite) is taken to be the major axis. The energy is shown as a function of orientation of the anisotropy in a reverse field H_e. Δ is the energy barrier to reversal.

$N = 0.2$ ($c/a = 1.75$) made of a material with $K_1 \sim 0$ and $\mu_0 M_s \sim 2\text{T}$ is expected to have $\mu_0 H_c = 0.4\text{T}$, provided reversal does not occur via an incoherent mode such as fanning or curling.

The treatment of incoherent nucleation processes is difficult. Solutions are known in a small number of cases, such as uniformly magnetized ellipsoids [Aha63]. But the essential rôle of inhomogeneity in generating a coercivity lower than expected from Brown's limit by incoherent nucleation is illustrated in a simple model due to Skomski [Coe93]. The model, shown in Fig. 1.28, also demonstrates the qualitative differences that arise according to whether the scale of the inhomogeneity is greater or less than l_{ex}. There are two hemispheres α and β with the same M_s and A_{ex}, but different values of K_1. To check the stability of the system, it is sufficient to consider small deviations θ_α and θ_β of the magnetization vector in each hemisphere. The free energy of Eq. (1.32) becomes

$$F = [A_{ex}/R^2 - (1/24)\mu_0 M_s^2](\theta_\alpha - \theta_\beta)^2 + (1/4)\mu_0 M_s H(\theta_\alpha + \theta_\beta)^2 + (1/2)K_{1\alpha}\theta_\alpha^2 + (1/2)K_{1\beta}\theta_\beta^2 \quad (1.39)$$

The model has only two degrees of freedom, so the nucleation problem reduces to diagonalization of a 2×2 matrix. In the homogeneous case, $K_{1\alpha} = K_{1\beta} = K_1$, the two eigenvectors (nucleation modes) are (θ, θ) and $(\theta, -\theta)$, illustrated in Fig. 1.28b and c respectively. The first is coherent rotation, and the eigenvalue (nucleation field) at which F ceases to be positive is $2K_1/\mu_0 M_s$, as expected from Eq. (1.38) with $N = 1/3$. The second is an incoherent, curling-like mode with

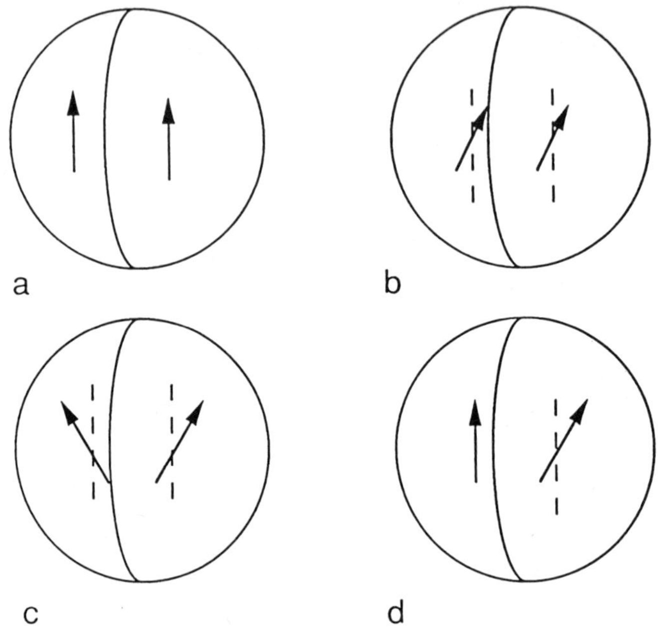

FIG. 1.28. Nucleation modes in a spherical sample (a) made up of two hemispheres α and ß which may have different anisotropy constants; (b) $K_{1\alpha} = K_{1\beta}$, coherent rotation; (c) $K_{1\alpha} = K_{1\beta}$, incoherent rotation (curling); (d) $K_{1\alpha} >> K_{1\beta}$, the soft hemisphere switches in a very low field unless its radius is $\lesssim l_{ex}$.

eigenvalue $2K_1/\mu_0 M_s + 8A_{ex}/\mu_0 M_s R^2 - M_s/3$. The nucleation field sets a lower limit to the coercivity. Note that Brown's inequality (Eq. (1.37)) is respected so long as the sphere is homogeneous. The first, coherent mode is lower in single-domain particles with very small R, but in particles with large R, the curling mode is lower. The changeover from coherent to incoherent reversal occurs when $R \gtrsim l_{ex}$, but the critical single-domain radius scales as κl_{ex}. This implies that it may be possible for the larger single-domain particles in hard magnets to reverse incoherently.

The inhomogeneous extreme is obtained when $K_{1\beta} << K_{1\alpha}$. Then the nucleation mode is localized (Fig. 1.28d) and the nucleation field is $K_{1\beta}/\mu_0 M_s + 4A_{ex}/\mu_0 M_s R^2 - M_s/6$. The soft hemisphere switches in a very low field unless its direction is stabilized by exchange coupling to the hard hemisphere in a tiny particle of radius less than about l_{ex}.

Calculations of the coercivity for real microstructures are impracticable, so one must resort to simplified models, computer simulation or a phenomenological approach. By analogy with Eq. (1.38), a plausible form for the coercivity is

$$H_c = \alpha 2K_1/\mu_0 M_s - N_{eff} M_s \tag{1.40}$$

where α is less than unity. The distribution of local fields can be deduced experimentally to yield the effective demagnetizing factor N_{eff}. Insight into the

Table 1.8 Micromagnetic properties of permanent magnet compounds

Compound	T_C (K)	A_{ex} (pJ/m)	M_s (MA/m)	l_{ex} (nm)	K_1 (MJ/m^3)	κ	δ_w (nm)	γ_w (mJ/m^2)	d_c (μm)
$BaFe_{12}O_{19}$	723	6	0.38	6.7	0.25	1.4	15.4	5	0.68
$SmCo_5$	993	24	0.84	5.3	17	4.4	3.7	81	1.68
Sm_2Co_{17}	1100	25	1.03	4.3	3.3	1.6	8.6	36	0.49
$Nd_2Fe_{14}B$	585	9	1.28	2.1	5	1.6	4.2	27	0.24
$Sm(Fe_{11}Ti)$	584	8	0.91	2.8	5	2.2	4.0	25	0.44
$Sm_2Fe_{17}N_3$	749	12	1.22	2.6	9	2.2	3.6	42	0.41

demagnetizing process can be achieved by comparing the temperature, time and angular dependence of H_c with model predictions, as discussed in Chapter 5. Real microstructures, observed by electron microscopy, are presented in Chapter 6.

The ideal microstructure for a uniaxial material is one where the *c*-axes of all the crystallites are aligned parallel. Remanence for this fully anisotropic texture can be as high as the spontaneous magnetization if the magnetic phase is fully dense. When there is any misorientation of the crystallites, the local magnetization will tend to revert to the nearest *c*-direction on decreasing field in the first quadrant. Provided there are no reverse domains,

$$M_r/M_s = \int P(\psi)\cos\psi \, d\psi \Big/ \int P(\psi)\, d\psi$$

where the function $P(\psi)$ represents the texture of the distribution of crystallite orientations; ψ is the angle between the *c*-axis and the magnetizing field. When the distribution is completely *isotropic*, $P(\psi) = \sin\psi$ and $M_r/M_s = 0.5$. A common type of magnet consists of randomly oriented crystallites bonded in a polymer matrix. While this may be cheap to produce, the discussion of Eq. (1.8) shows that $(BH)_{max} < \mu_0 M_r^2/4$ will be less than 0.25 of the value obtainable from a similar, fully oriented magnet.

The ratio M_r/M_s can be higher than 0.5 for random assemblies of cubic crystallites or uniaxial crystallites with a hard *c*-direction. Values are, respectively, 0.831, 0.866 and 0.912 for cubic materials with {100}, {111} and {110} easy axes, and 0.707 or 0.750 for a tetragonal or hexagonal material with {100} easy. Cubic or easy-plane compounds appear to present an advantage over the uniaxial ones, but this is misleading because it is very difficult to achieve usable coercivity in these materials at room temperature because of the low-energy barriers between the multiple easy directions.

Another way to increase the remanence ratio above 0.5 in isotropic uniaxial materials is to structure them on a nanometer scale so that there is exchange coupling across the grain boundaries. The magnetization will then deviate from **n**, the local easy axis, over a distance from the boundary of order of the domain wall

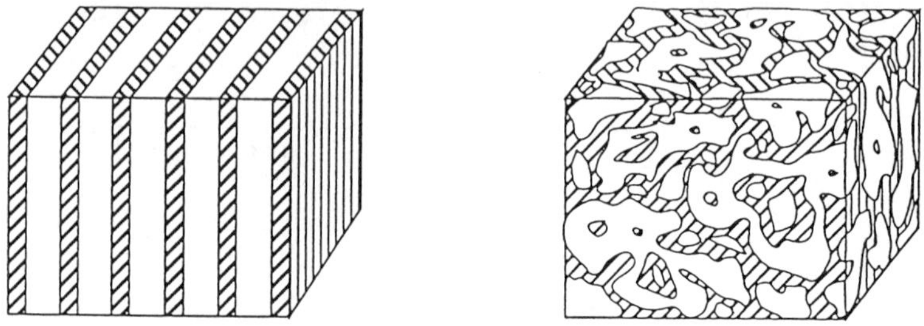

FIG. 1.29. Possible nanostructures for exchange-hardened two-phase systems [Sko93].

width. It is possible to achieve *remanence enhancement*, at the expense of a reduction in coercivity [Cle88].

Another way of achieving a hard magnet with improved remanence is by exchange coupling of hard and soft regions on a nanometer scale [Coe89, Kne91]. The prospects of finding a rare-earth intermetallic with a magnetization significantly larger than that of $Nd_2Fe_{14}B$ ($\mu_0 M_s = 1.6$ T) are poor, so the greatest achievable energy product may be about 500 kJ/m^3. However, some soft materials such as α–Fe and Fe–Co alloys have polarization in the range 2.1–2.3 T. Provided they can be incorporated in a two-phase nanostructure of hard and soft material in regions not much thicker than the exchange length or the domain wall width of the hard phase, as indicated by the discussion of Fig. 1.28d, it should be possible to stiffen them by exchange coupling to the hard phase. An as-yet hypothetical nanostructure of an oriented hard phase with very high anisotropy and a soft phase with high magnetization (e.g. $Sm_2Fe_{17}N_3/Fe_{65}Co_{35}$; Fig. 1.29) could prove to be the ultimate rare-earth iron permanent magnet, with an energy product in excess of 1 MJ/m^3 [Sko93]. To go beyond that, we would have to resort to cryogenic temperatures, using ferromagnetic alloys rich in heavy rare-earths or perhaps to superconductors with very high critical currents.

3 Magnetization and assembly

Magnets have to be shaped, magnetized and assembled before they can deliver the magnetic flux needed in an application. Sintered rare-earth magnets may be sliced and ground into simple shapes. Polymer bonding with compression or injection moulding allows more complex shapes to be produced to close tolerances, including components which serve both a magnetic and a mechanical function. The remanent magnetization of the composite is the product of the fill factor *f*, typically 0.8, and the remanence due to the distribution $P(\psi)$. Efficient use of rare-earth material demands the best possible degree of crystallite orientation, so there is a demand for monocrystalline or highly textured magnetic powder for bonding.

The ease of magnetization is related to the coercivity mechanism. Nucleation-type magnets where the domain walls are easily swept through the crystallites can be magnetized in a field of about 2 T, but when the walls are pinned at extended defects, including grain boundaries in magnets with a very small crystallite size $d < d_c$, then fields as high as 4 or 5 T may be required. These fields are usually produced by discharging a capacitor bank through a stout coil or magnetizing fixture designed to impart a special pole pattern to the magnet.

Forces between large magnetized blocks can pose problems during assembly, necessitating the use of special jigs. The forces between two contiguous 1 kg blocks of Nd–Fe–B can reach 1000 N!

At the other end of the scale, thin film magnets could find applications in miniature devices. Strictly speaking, uniformly magnetized thin films produce no usable external flux because the demagnetizing factor N is 1, so $B = 0$ for the component of magnetization perpendicular to the surface. If films are to be used as flux sources, the dimensions of the features must be within an order of magnitude of the film thickness.

E. APPLICATIONS

The applications of permanent magnets can be classified in several ways. One scheme, set out in Table 1.9, uses the physical effect that is being exploited — torque on a magnet in an external field, force on a magnet or on a piece of soft magnetic material in a nonuniform field, force on a current-carrying conductor or on charged particles moving in a magnetic field, induced emf, Zeeman splitting of energy levels and so on. A simpler scheme focuses attention on the characteristics of the magnet, distinguishing simply between *static applications*, where the magnet exists to produce a magnetic field in some region of space, where what happens has little effect on the flux density in the magnet itself, and *dynamic applications*, where the working point shifts along the $B{:}H$ curve during operation.

The unique attraction of permanent magnets (barring the eventuality of a room-temperature superconductor) is that they provide magnetic flux with no continuing expenditure of energy. Indeed, their ferromagnetism has its origin in resistanceless electronic currents circulating on the atomic scale. Furthermore, their ability to generate complex field patterns with intense spatial variations (large $\nabla\mathbf{B}$) is unrivalled by any electromagnetic device. The surface current that would be needed to generate a flux pattern similar to that of a long cylindrical magnet with $\mu_0 M = 1$ T is 796 kA/m. Solenoids, whether resistive or superconducting, able to produce such fields would have to be several centimetres in diameter to accomodate the necessary Ampère-turns.

Another useful feature of rare-earth permanent magnets is the large anisotropy field B_a, of order 10 T, which represents the strength of the interactions that fix the magnetization along the easy axis. It is therefore possible to arrange magnets with

Table 1.9 Applications of permanent magnets

Operation	Physical effect	Application
Static field		
	force on a current-carrying coil	actuators, loudspeakers, headphones, microphones, pick-ups
	forces on charged particles	multipoles for ion sources, wigglers for synchrotrons, free electron lasers, microwave sources
	Zeeman splitting (NMR)	magnetic resonance imagers
	Hall effect	sensors
	?	acupuncture magnets limescale control, oil dewaxing
	force in a field gradient	magnetic separation, levitated transportation, bearings, drives, couplings
Mechanical recoil		switchable holding and lifting magnets, variable flux sources
Active recoil	electromagnetic induction	generators, electrodynamic brakes
	force on a current-carrying conductor	PM brushed motors, synchronous motors, brushless motors, stepper-motors, actuators

their magnetization vectors aligned in practically any desired directions. The dimensionless transverse susceptibility is $\mu_0 M_s/B_a \sim 0.1$. Hence, for example, the direction of magnetization of two blocks in contact, with their easy directions perpendicular, will deviate by only a few degrees from the easy directions. As a consequence of the rigidity of the magnetization, the superposition of the flux density of rare-earth permanent magnets is linear and the magnetic material is effectively transparent, behaving like a vacuum, with permeability μ_0. This feature greatly simplifies the design of magnetic circuits. Electrical machines are usually modelled using the vector potential **A**, where as magnetostatic problems with no conduction currents can be modelled using the scalar potential ψ. Finite-difference and finite-element computer methods are now commonly used [How91, Cam94]. Design principles for static applications that exploit the high flux density and strong anisotropy of rare-earth iron magnets are discussed in Chapters 8 and 9.

Some ideas that are relevant for permanent-magnet applications can be introduced by considering the simple magnetic circuit of Fig. 1.30. The circuit directs flux produced by the permanent magnet of length l_m and cross-section A_m through the soft magnetic material to the air gap of length l_g and cross-section A_g. If there is no flux leakage:

$$B_m A_m = -B_g A_g \tag{1.41}$$

Furthermore, if the soft material is ideal insofar as it has infinite permeability, Ampère's law $\int \mathbf{H}\cdot d\mathbf{l} = 0$ gives

$$H_m l_m = H_g l_g \tag{1.42}$$

Multiplying the above two equations, and using the fact that $B_g = \mu_0 H_g$,

$$B_m H_m V_m = -B_g^2 V_g/\mu_0 \tag{1.43}$$

where V_m, V_g represent the volumes of the magnet and the gap. The flux in the air gap is maximized and optimum use is made of costly magnetic material when the product $B_m H_m$ is maximum (cf. Section A2). Dividing Eqs (1.41) and (1.42), we

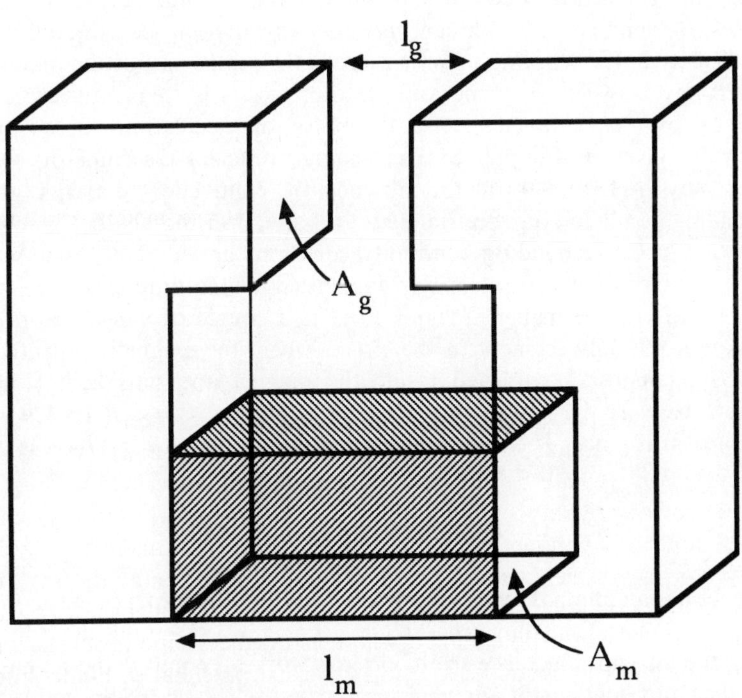

FIG. 1.30. A simple magnetic circuit consisting of a permanent magnet (shaded), soft magnetic material and an air gap.

obtain the equation for the *load line*:

$$B_m/H_m = -\mu_0 A_g l_m / A_m l_g \tag{1.44}$$

The slope of the load line ($-B_m/H_m$) is known as the *permeance coefficient*. The working point where the $B{:}H$ loop of a particular material intersects the load line is therefore determined by the geometrical dimensions of the magnet and air gap. As materials improve, the magnets grow shorter and fatter as they work near their $(BH)_{max}$ point. The traditional iron horseshoe magnet, so shaped to avoid demagnetization in its own demagnetizing field, is now a museum piece. Figure 1.31 vividly illustrates the influence of the permanent magnet material on device design. Not only can the device be made smaller, but the number of parts can be reduced, for example by moving the magnets of a brushless dc motor to the rotor and moulding the magnets together with the shaft and gear in one piece.

In practice, there will be some flux leakage in the circuit of Fig. 1.30, so a factor β is introduced on the left-hand side of Eq. (1.41). Furthermore, the soft material will not be ideal so another factor, α, is introduced on the left-hand side of Eq. (1.42). Typically, α is in the range 0.7–0.95, but β may be anywhere from 0.2 to 0.8. Rare-earth permanent magnets are particularly suited for use in ironless magnetic circuits where the flux is essentially confined to the working volume and to the magnets themselves. By suitable circuit design, *flux concentration* can be achieved whereby the flux density in the working volume exceeds the remanence of the magnet.

The analogy between magnetic and electric circuits is frequently exploited in calculations by magnetic engineers. Examples of calculations based on this approach are given in Chapter 8. The analogy rests on the similarity between Ampère's law $\int \mathbf{H} \cdot d\boldsymbol{l} = 0$ in the absence of conduction currents and the corresponding result for the electric field $\int \mathbf{E} \cdot d\boldsymbol{l} = 0$ in the absence of changing magnetic fields. Corresponding continuity equations are $\nabla \cdot \mathbf{B} = 0$ and $\nabla \cdot \mathbf{j} = 0$, where $\mathbf{j}$ is the electric current density. Magnetic potential difference is defined by $F_{ab} = \int_a^b \mathbf{H} \cdot d\boldsymbol{l}$; the permanent magnet itself is a source of *magnetomotive force* (mmf) as it is the only segment of the circuit where the magnetic potential rises, because H is oppositely directed inside the magnet and outside it (Fig. 1.2). Magnetic flux Φ is the equivalent of current, and *reluctance* R ($=F/\Phi$) is the equivalent of resistance. The reluctance of a short air gap, for instance, is

$$R_g = F_g/\Phi_g = l_g/\mu_0 A_g \tag{1.45}$$

The magnet has an internal reluctance determined by its working point on the $B{:}H$ loop; $R_m = F_m/\Phi_m$. The equivalent circuit of Fig. 1.32a illustrates the principle of matching the air gap reluctance to the desired working point of the magnet. Fig. 1.32b is the equivalent circuit allowing for nonzero reluctance of the soft segments and for flux losses. These losses are more severe in magnetic circuits than in their electrical counterparts because the relative permeability of iron is very much less

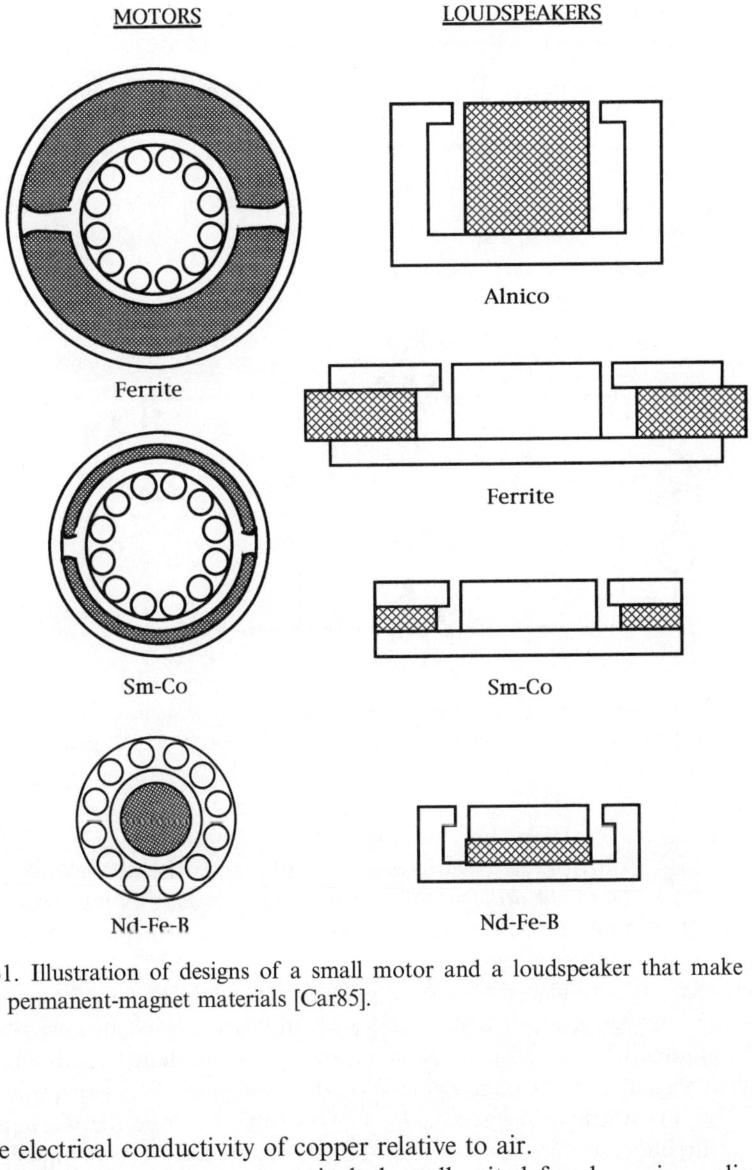

FIG. 1.31. Illustration of designs of a small motor and a loudspeaker that make use of different permanent-magnet materials [Car85].

than the electrical conductivity of copper relative to air.

Rare-earth iron magnets are particularly well suited for dynamic applications that involve changing flux density in the permanent magnet. For *mechanical recoil* (Fig. 1.33a), the air gap changes during operation from a narrow one with reluctance R_1 to a wider one with reluctance R_2. After several cycles, the working point follows a stable and almost linear trajectory, represented by the line P_2Q, whose slope is known as the *recoil permeability* μ_R. XY represents the useful flux change in the gap, and YZ the leakage flux that is wasted. The area $XX'Y'Y$ is a measure of $(BH)_u$, twice the useful recoil energy in the gap that is maximized when

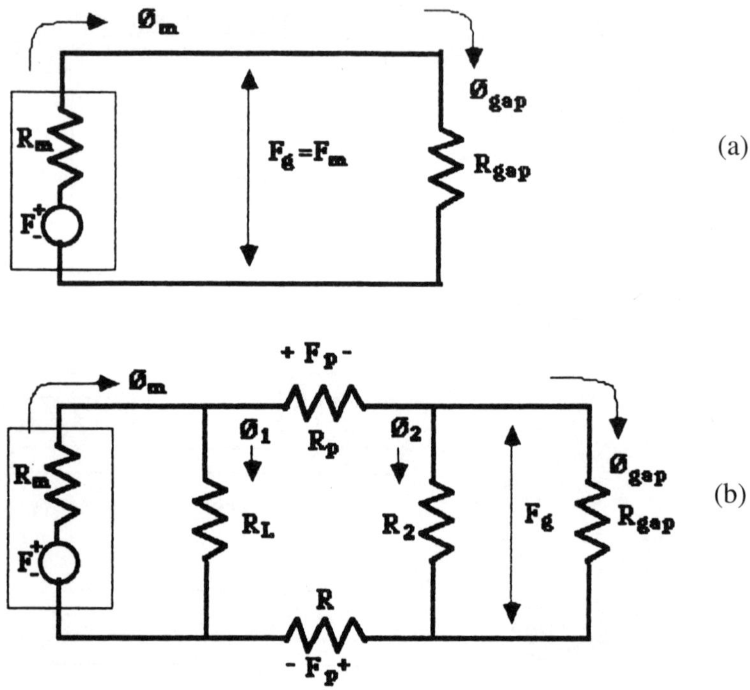

FIG. 1.32. Equivalent electrical circuits for the magnetic circuit of FIG. 30. (a) is the ideal circuit assuming no losses and infinite permeability of the soft segments, whereas (b) takes these effects into account.

X bisects P_2Q. The *recoil product* $(BH)_u$ will always be less than $(BH)_{max}$, but it approaches that limit in rare earth and ferrite materials with high coercivity where the recoil permeability approaches μ_0.

Active recoil occurs in permanent magnet motors and other devices where the magnets are subject to an H-field during operation as a result of currents in the copper windings. The field is greatest at startup, or in the stalled condition. Active recoil is represented by a displacement of the reluctance line along the H-axis (Fig. 1.33b) Provided $\mu_0 H_c$ exceeds B_r, it is possible to drive the working point momentarily into the third quadrant of the $B:H$ loop without demagnetizing the permanent magnet. The intrinsic coercivity may be as important a figure of merit as $(BH)_{max}$ in this type of application.

A number of examples of static and dynamic applications of high-performance permanent magnets were listed in Table 1.9. Holding devices, where a strong force is exerted on a piece of soft magnetic material in contact with the magnet, are examples of dynamic working with mechanical recoil, because the working point shifts from the open circuit point to the remanence point where $H = 0$ as the circuit is closed. The maximum force F that can be exerted at the face of a magnet may be

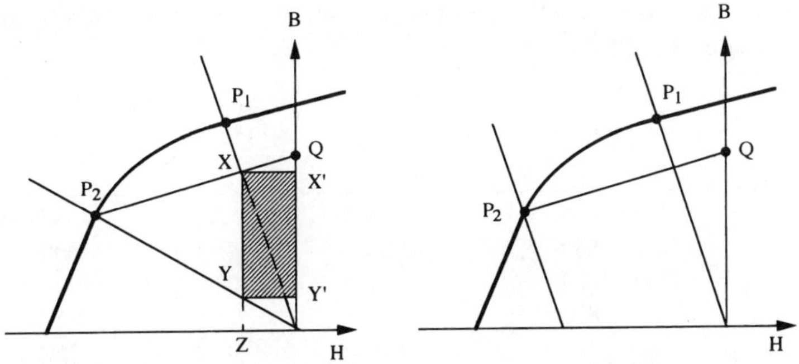

FIG. 1.33. Illustration of recoil in permanent-magnet circuits: (a) mechanical recoil between two different air gaps with working points P_1 and P_2. P_2Q is the recoil line, and the shaded area represents the useful energy product $(BH)_u$ and (b) active recoil, where the magnet is subjected to an intermittent H field. In an ideal permanent magnet, the demagnetization curve has slope μ_0 and Q coincides with the remanence point.

derived by considering the toroid of Fig. 1.3a. If it is cut into two C-shaped segments which are then separated by small distance δ, the energy appearing in the air gaps is $2 \times (1/2)\mu_0 H_g^2 A_g \delta = B_g^2 A_g \delta / \mu_0$. The work done separating the segments is $2F\delta$, hence the force per unit area is

$$F/A_g = B_g^2/2\mu_0 \tag{1.46}$$

Forces of up to 40 N cm^{-2} can be achieved for $B_g = 1$ T.

The most important commercial applications of high-performance permanent magnets are the motors, actuators and other electromagnetic machines discussed in Chapters 10 and 11. A vast range of motors can be designed with magnets, their power ranging from microwatts, for wristwatch motors, to hundreds of kilowatts, for industrial machines. The high energy product and high anisotropy of the rare-earth iron alloys makes it possible to realize compact, low-inertia, high-torque devices — stepper-motors, actuators, brushless DC motors and so on. Modern electronics is married to movement through such devices.

The magnets in electrical machines can be subject to temperatures of up to 100 °C or more. Magnetization and coercivity naturally decline as the Curie point is approached, and the temperature coefficients of these quantities around ambient temperature are listed in Table 1.10 for magnets made of different materials. Not all the loss is necessarily recoverable on returning to ambient temperature; there can be *irreversible losses* associated with thermal cycling. The maximum temperatures at which the materials can safely be used are also indicated in the table. The limit for $Nd_2Fe_{14}B$ is lower than for any of the others because of the relatively low Curie point, and the temperature coefficients are more severe. These difficulties must be overcome by astute engineering design, or by appropriate chemical substitution.

Table 1.10 Temperature coefficients and maximum operating temperatures of permanent magnets

	T_C (°C)	dM_S/dT (%/K)	dH_C/dT (%/K)	T_{max} (°C)
$SrFe_{12}O_{19}$	450	–0.20	0.45	250
Alnico 5	860	–0.02	0.03	500
$SmCo_5$	720	–0.04	–0.02	250
$Sm_2Co_{17}^*$	820	–0.03	–0.20	350
$Nd_2Fe_{14}B$	312	–0.13	–0.60	150

* intergrown with 1:5 phase.

Bibliography

Handbook of ferromagnetic materials, E. P. Wohlfarth and K. H. J. Buschow (editors), North Holland, Amsterdam, 1980 — a series of definitive reviews on all aspects of magnetic materials. Particularly relevant are the articles by Zijlstra on 'Permanent magnet theory' (vol. 3), Buschow on 'Permanent magnet materials based on 3d-rich ternary compounds' (vol. 4), Strnat on 'Rare-earth cobalt permanent magnets' (vol. 4), Li and Coey on 'Magnetic properties of ternary rare-earth transitional-metal compounds' (vol. 6) Franse and Radwański on 'Magnetic properties of binary rare-earth 3d-transition-metal intermetallic compounds' (vol. 7) and Fujii and Sun on 'Interstitially-modified intermetallics of rare-earth and 3d elements' (vol. 9).

Magnetic properties of metals; compounds between rare-earth elements and 3d, 4d or 5d elements, E. Burzo, A. Chelkowski and O. Kirchmayr in Landolt–Börnstein Tables 19:d2 (H. P. J. Wijn, editor) Springer, Berlin, 1990 — a definitive compilation of experimental data.

Physics of rare earth solids, K. N. R. Taylor and M. I. Darby, Chapman and Hall, London 1972 — a basic reference for the physics of rare-earth magnetism.

Rare earth intermetallics, W. E. Wallace, Academic Press, New York, 1973 — a classic early text, with emphasis on magnetic and thermal properties.

Magneti permanenti (2 vols), E. Burzo, Editura Academia *R. S.* Romania, Bucharest, 1986 — a comprehensive text in Romanian, with extensive references to the primary literature.

Permanent magnets, W. Wright and M. McCaig, Oxford University Press, London 1977 — a brief engineering design guide, with a summary of the main macroscopic concepts.

Permanent magnets in theory and practice (2nd edition), M. McCaig and A. G. Clegg, Petnech Press, London 1987 — updated edition of a summary of accumulated experience of the British Permanent Magnet Research Association's Central Research Laboratory.

Advances in permanent magnetism, R. J. Parker, Wiley, New York, 1990 — a comprehensive treatment of engineering applications of permanent magnets, with emphasis on rare-earth materials.

Permanent magnet materials and their applications, P. Cambell, Cambridge University Press, 1994 — an introduction for engineers with emphasis on magnet design, including modern finite-element calculation methods.

Nd–Fe permanent magnets, their present and future applications, I. V. Mitchell (editor), Elsevier, London, 1985 — Proceedings of a workshop held one year after the discovery of $Nd_2Fe_{14}B$. Covers an exceptionally wide variety of relevant topics from basic rare-earth magnetism to economic forecasts.

The concerted European action on magnets — final report, I. V. Mitchell, J. M. D. Coey, D. Givord, I. R. Harris and R. Hanitsch (editors), Elsevier, London, 1989 — Reports of the 58 academic and industrial laboratories forming the Concerted European Action on Magnets, with coordinators' summaries. More information on the general and topical meetings of CEAM will be found in the *CEAM Newsletter*, published quarterly.

Supermagnets — hard magnetic materials, G. J. Long and F. Grandjean (editors), Kluwer, 1991 — Proceedings of a 1990 NATO Summer School. Thirty chapters cover many topics, particularly the intrinsic magnetic properties of rare-earth iron compounds.

Review articles

$R_2Fe_{14}B$ materials: intrinsic properties and technological applications, J. F. Herbst, *Rev. Mod. Phys*, **63** (1991), 819–898.

New permanent magnet materials, K. H. J. Buschow, *Repts. Prog. Phys.*, **54** (1991), 1123–1214.

Modern permanent magnets for applications in electro-technology, K. J. Strnat, *Proc. IEEE*, **78** (1990), 923–946.

$RETM_5$ and RE_2TM_{17} permanent magnet development, K. Kumar, *J. Appl. Phys*, **63** (1988), R13–57.

Rare-earth transition metal permanent magnet materials, W. E. Wallace, *Prog. Solid State Chem.*, **16** (1985), 127–152.

Conference Proceedings

A series of specialized conferences dealing only with rare-earth magnets and their applications is held in alternate years. The Proceedings of the workshop on rare-earth magnets and associated Symposia on coercivity contain much primary material. Conferences in this series have been:

Year	Venue	Proceedings editor
1974	Dayton, Ohio	–
1976	Dayton, Ohio	K. J. Strnat
1978	San Diego, California	K. J. Strnat
1980	Hakone, Japan	H. Kaneko and T. Kurino
1981	Baden, Austria	J. Fidler
1983	Peking, China	X. Pan, W. Ho and C. Yu
1985	Dayton, Ohio	K. J. Strnat
1988	Bad Soden, Germany	C Herget and R. Poerschke
1989	Kyoto, Japan	T. Shinjo and T. Kurino
1990	Pittsburgh, Pennsylvania	S. G. Sankar
1992	Canberra, Australia	R. J. Street
1994	Birmingham, England	I. R. Harris
1996	Saõ Paulo, Brazil	F. Missel

Further invaluable primary sources are the Proceedings of the annual Magnetism and Magnetic Materials Conference (MMM) of the American Physical Society and the Applied Magnetics Conference (INTERMAG) of the IEEE. The annual meetings of the Magnetics Society of Japan, the biannial European Conference on Applied Magnetism (EMMA), and the triannial International Conferences on Magnetism (ICM) also contain much relevant material.

REFERENCES

[Aha 62] A. Aharoni, *Rev. Mod. Phys.*, **34** (1962), 227; *Phys. Rev.*, **131** (1963), 1478.

[Bro 57] W. F. Brown, *Phys. Rev.*, **105** (1957), 1479.

[Bus 86] D. B. de Mooij and K H J. Buschow, *Philips J, Res.*, **41** (1986), 400.

[Bus 88] K. H. J. Buschow, D. B. de Mooij, M. Brouha, H. H. A. Smit and R. C. Thiel, *IEEE Trans. Magn.*, **24** (1988).

[Cad94] J. M. Cadogan, H. S. Li, A. Margarian, J. B. Dunlop, D. H. Ryan, S. J. Collcott and R. L. Davies, *J. Appl. Phys.*, **76** (1994), 6138.

[Cam72] I. A. Campbell, *J. Phys.*, F **2** (1972), L47.

[Car 85] Cartocetti in *Nd–Fe permanent magnets, their present and future applications*, I. V. Mitchell, (ed.), Elsevier, London 1985.

[Cle88] G. B. Clemente, J. E. Keem and J. P. Bradley, *J. Appl. Phys.*, **64** (1988), 5299.

[Coe84] J. M. D. Coey and D. H. Ryan, *IEEE Trans. Magn.*, **20** (1984), 1278.

[Coe89] R. Coehoorn, D. B. de Mooij and J. de Waard, J. *Magn. Magn. Mater.*, **80** (1989), 101.

[Coe90] J. M. D. Coey and H. Sun, *J. Magn. Magn. Mater.*, **87** (1990), L251.

[Coe91] R. L. Coehoorn in *Supermagnets — hard magnetic materials*, Long, G. J. and Grandjean, F. (eds), Kluwer, 1991, Chapter 8.

[Coe93] J. M. D. Coey and R. Skomski, *Physica Scripta*, **T49** (1993), 315.

[Cro84] J. J. Croat, J. F. Herbst, R. W. Lee and F. E. Pinkerton, *J. Appl. Phys.*, **55** (1984), 2078.

[Gav88] J. P. Gavigan, D. Givord, H. S. Li and J. Voiron, *Physica*, B**149** (1988), 345.

[Giv84] D. Givord, H. S. Li and J. M. Moreau, *Solid State Commun.*, **50** (1984), 497.

[Giv85] D. Givord, J. M. Moreau and P. Tenaud, *Solid State Commun.*, **55** (1985), 303.

[Her84] J. F. Herbst, J. J. Croat, F. E. Pinkerton and W. B. Yelon, *Phys. Rev.*, B **29** (1984), 5176.

[How91] D. Howe, M. K. Jenkins and W. F. Low, in *Supermagnets — hard magnetic materials*, G. J. Long and F. Grandjean (eds), Kluwer, 1991, Chapter 25.

[Hu89] B. P. Hu, H. S. Li, J. P. Gavigan and J. M. D. Coey, *J. Phys. Condens. Matter*, **1** (1980), 755.

[Hur92] D. P. F. Hurley and J. M. D. Coey, *J. Phys. Condens. Matter*, **4** (1992), 5573.

[Hut66] M. T. Hutchings, *Solid State Physics*, **16** (1966), 227.

[Kit49] C. Kittel, *Rev. Mod. Phys.*, **21** (1949), 541.

[Kne91] E. F. Kneller and R. Hawig, *IEEE Trans. Magn.*, **27** (1991), 3588.

[Kro91] H. Kronmüller, in *Supermagnets — hard magnetic materials*, G. J. Long and F. Grandjean, (eds), Kluwer, 1991, Chapter 19.

[Osb45] J. A. Osborn., *Phys. Rev.*, **67** (1945), 351.

[Par62] R. J. Parker and R. J. Strudders, *Permanent magnets and their applications*, Wiley, New York, 1962.

[Rud85] C. Rudowicz, *J. Phys.*, C **18** (1985), 1415.

[Sag84] M. Sagawa, S. Fujimura, N. Togawa, H. Yamamoto and Y. Matsuura, *J. Appl. Phys.*, **55** (1984), 2083.

[Sko93] R. Skomski and J. M. D. Coey, *Phys. Rev.*, **B48** (1983), 15812.

[Sma66] J. S. Smart, *Effective field theories of magnetism*, Saunders, Philadelphia, 1966.
[Sto48] E. C. Stoner and E. P. Wohlfarth, *Phil. Trans. Roy. Soc.*, **240** (1948), 599.
[Wer70] G. K. Wertheim, D. N. E. Buchanan and J. H. Wernick, *Solid State Commun.*, **8** (1970), 2173.
[Wil83] A. R. Williams, V. L. Moruzzi, A. P. Malozemoff and K. Tekura, *IEEE Trans. Magn.*, **19** (1983), 1983.

2

Intrinsic magnetic properties

J. J. M. Franse and R. J. Radwański

This chapter focuses on the intrinsic magnetic properties of iron-rich rare-earth intermetallics. We are concerned with rare-earth metals R where the magnetic properties of the ion cores are well defined; the 4f level lies deep below the Fermi energy, and hybridization with the conduction-band electron states is negligible. This situation is realized for most iron and cobalt-based compounds with 4f elements, with the exception of some Ce compounds that exhibit anomalies which can be ascribed to a mixed-valence state of the cerium atom. The emphasis here is on $R_2Fe_{14}B$ compounds and on $Nd_2Fe_{14}B$ in particular, although other R_nT_m series are discussed as well. The $Nd_2Fe_{14}B$ family is rather extended. Isomorphous iron compounds exist for almost the whole rare-earth series, including Y and Th, while isomorphous cobalt compounds exist for most of the light rare-earth elements and for Gd and Tb. Likewise, there are extended isomorphous series for rare-earth iron compounds with 2 : 17 and 1 : 12 structures.

In these compounds there is an interplay between the localized magnetism of the rare-earth R atoms and the itinerant magnetism originating from the 3d electrons of the transition-metal T atoms. Moreover, both 3d and 4f moments interact with the spins of the 5d and 6s conduction electrons of the rare-earth atoms. As these problems have not been fully solved for elemental metals, a consistent theoretical description is still lacking. Despite these difficulties there is a rapidly growing body of experimental information on the magnetic and electronic properties of these compounds to guide our understanding.

The Curie temperature, T_C, of the $R_2Fe_{14}B$ series is presented in Fig. 2.1. It is approximately 550 K, remarkably higher than in R_2Fe_{17} due to the longer Fe–Fe distances. The spontaneous magnetization M_s at 4.2 K varies substantially over the $R_2Fe_{14}B$ series. Presented in terms of the magnetic moment per formula unit, however, M_s exhibits a regular variation with the R partner and reflects the ferromagnetic order for compounds with the light rare-earth elements (Ce, Pr, Nd, Sm) and the ferrimagnetic order for compounds with the heavy rare-earth elements, from Gd to Yb. To a first approximation, the spontaneous magnetization is a vector sum of the magnetization of the Fe sublattice, as found in the Y, La or Lu compounds, and that of the R sublattice with its typical variation over the rare-earth series, as shown in Fig. 2.2. The same systematics exist for other R_nT_m series where T = Fe, Co or Ni [Fra93].

One can get an impression about the wealth of magnetic phenomena in the $R_2Fe_{14}B$ compounds by comparing the magnetization curves of monocrystalline $Y_2Fe_{14}B$ and $Nd_2Fe_{14}B$, in Figs 2.3 and 2.4. In the former compound, the magnetic structure is collinear, with the magnetocrystalline anisotropy favouring the tetragonal *c*-axis at all temperatures below the magnetic ordering temperature.

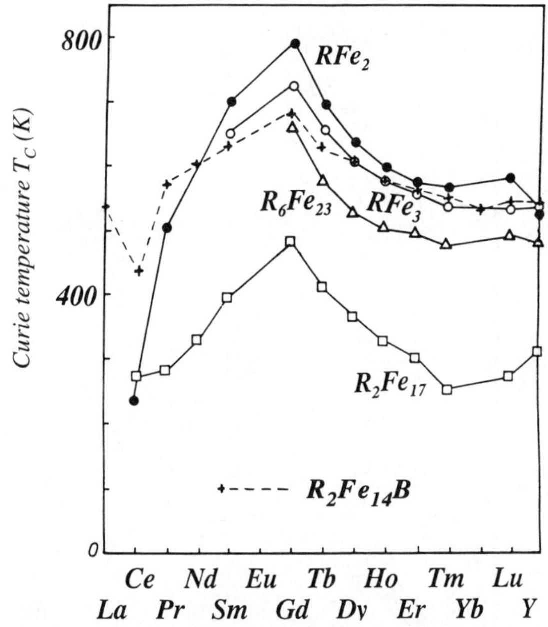

FIG. 2.1. Variation of the Curie temperature in the $R_2Fe_{14}B$ series with the R partner (dashed line); other R_nFe_m series are shown for comparison.

Along the easy axis, the magnetization saturates when the external field exceeds the demagnetizing field. The intricate magnetization curves of $Nd_2Fe_{14}B$ reflect a complex internal magnetic structure. At low temperatures $Nd_2Fe_{14}B$ exhibits a magnetic structure where the easy axis is tilted at an angle θ_s with respect to the c-axis. The tilt angle decreases in this case with increasing temperature, falling to zero at a temperature of 135 K, denoted by T_{st}. Similar behaviour occurs for $Ho_2Fe_{14}B$ and $Nd_2Co_{14}B$. At room temperature, the easy axis of magnetization for $Nd_2Fe_{14}B$ is directed along the tetragonal c-axis, resulting in the much simpler magnetization curves shown in Fig. 2.4b, which, in fact, are similar to those of $Y_2Fe_{14}B$ but with a larger anisotropy field.

The intrinsic magnetic properties of the rare-earth iron intermetallics can be understood in terms of exchange interactions and magnetocrystalline anisotropy. The *exchange interactions* involve all unpaired spins in the 3d–4f system. These interactions are generally considered to be of the Heisenberg type. Because there are three kinds of spin pairs in a 3d–4f intermetallic, the exchange interactions are formally described by three different exchange parameters: J_{TT}, J_{RT} and J_{RR}. The magnitude of these interactions decreases in the sequence mentioned for the iron- and cobalt-based compounds. J_{TT} is positive and leads to ferromagnetic coupling between the 3d spins. This interaction primarily governs the temperature dependence of the 3d moment and the value of the Curie temperature, T_C. The parameter J_{RT} is responsible for the internal magnetic structure. Its sign reflects

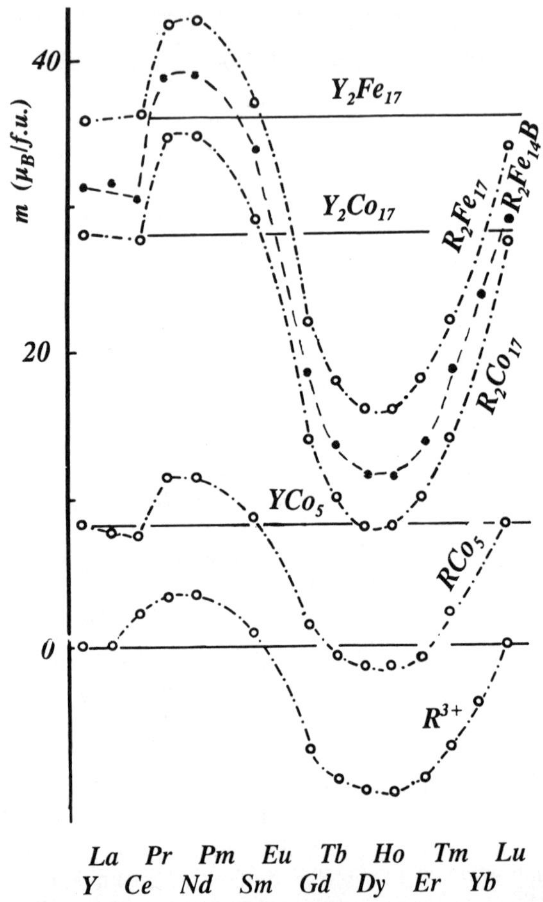

FIG. 2.2. Variation of the magnetic moment per formula unit at 4.2 K of the $R_2Fe_{14}B$, R_2Fe_{17}, R_2Co_{17} and RCo_5 series with the R partner. In the bottom part, the contribution of the R^{3+} moment to the magnetic moment is shown, taking into account the parallel or antiparallel moment configuration for the light or the heavy rare-earths relative to the 3d moment, respectively.

antiparallel coupling of the 4f and 3d spins which leads to parallel alignment of the 3d and 4f moments for light rare-earths and antiparallel alignment for heavy rare-earths. The 3d–4f interaction has some influence on the Curie temperature, especially for compounds rich in iron, and dominates the molecular field experienced by the rare-earth moments which, in turn, determines the temperature dependences of the magnetic moment and the magnetocrystalline anisotropy of the rare-earth ions. The R–R interaction between the 4f spins is generally neglected because values of J_{RR} are smaller than those of J_{RT} and, in addition, there are few R–R bonds in the iron-rich compounds.

The *anisotropy energy* determines the preference of the magnetization to be oriented along a specific crystallographic direction. The anisotropy field, B_a, can be

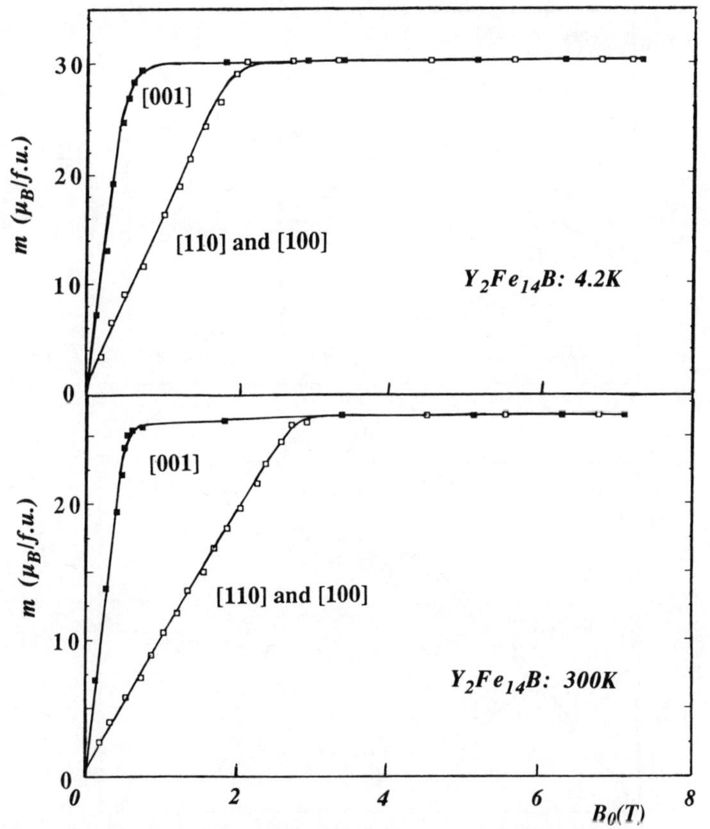

FIG. 2.3. Magnetization curves for monocrystalline $Y_2Fe_{14}B$ at 4.2 K and at 300 K. The larger anisotropy field at 300 K, compared to that at 4.2 K, is evident.

expressed in terms of the first anisotropy constant, K_1, and the spontaneous magnetization, M_s, as $2K_1/M_s$. In the present class of magnetic materials, the anisotropy energy originates from both the rare-earth and 3d sublattices. The 3d anisotropy can be deduced from compounds with non-magnetic rare-earth elements such as Y. The magnetization curves presented in Fig. 2.3 for external fields applied parallel and perpendicular to the *c*-axis of $Y_2Fe_{14}B$ show that at 4.2 K, B_a is 1.5 T and K_1 equals 0.8 MJ/m^3. In fact, the 3d anisotropy of all iron-based $R_2Fe_{14}B$ compounds increases between 4.2 K and 300 K and contributes substantially to the overall anisotropy at room temperature where $K_1^T \simeq 1.1$ MJ/m^3. The rare-earth contribution to the magnetocrystalline anisotropy is dominant at low temperatures whenever the 4f ions have an orbital moment. This contribution, however, decreases rapidly with increasing temperature and at room temperature it may already be comparable in magnitude to the 3d contribution for some rare-earths.

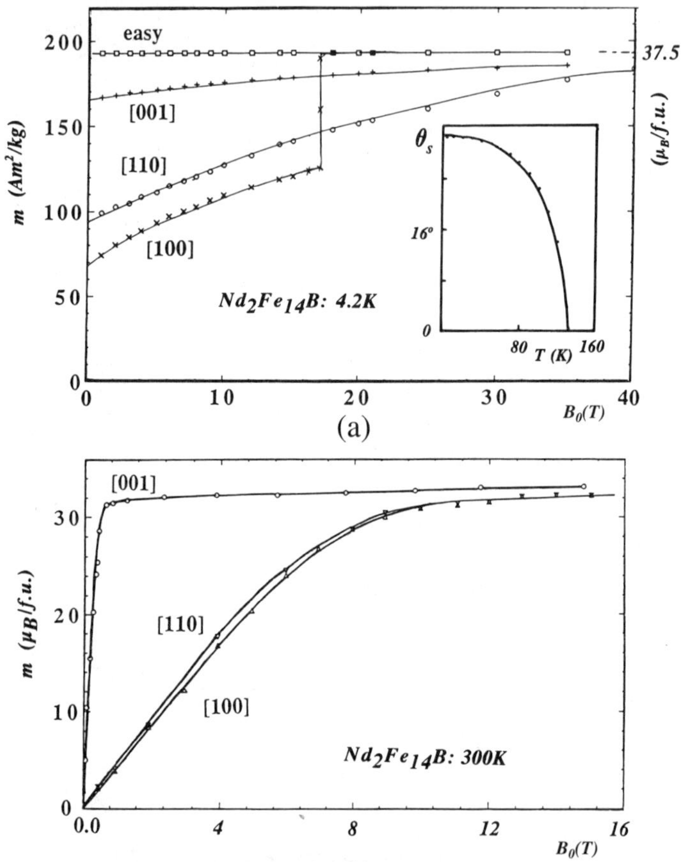

FIG. 2.4. Magnetization curves for monocrystalline $Nd_2Fe_{14}B$ at 4.2 K (a) and at room temperature (b); in the inset of (a) the temperature dependence of the tilt angle, θ_s is shown [Ver88].

For $Nd_2Fe_{14}B$ at room temperature, K_1 is 4.9 MJ/m^3 so that roughly three-quarters of the anisotropy is provided by the neodymium sublattice and one-quarter by the iron sublattice. In cases where the rare-earth and 3d anisotropies are of opposite sign, temperature-induced spin-reorientation phenomena can occur as a consequence of mutual cancellation of the anisotropy contributions from the two sublattices. At temperatures above the spin-reorientation temperature, T_{sr}, the anisotropy is dominated by the 3d contribution.

In the following sections, the magnetic properties of the 3d and rare-earth sublattices will be inspected more thoroughly. For a quantitative understanding of the magnetic properties of these materials the two key parameters are: the *R–T exchange interaction* and the *second-order crystal-field interaction* B_2^0 at the rare-earth sites. These parameters represent the interactions of a 4f ion with the

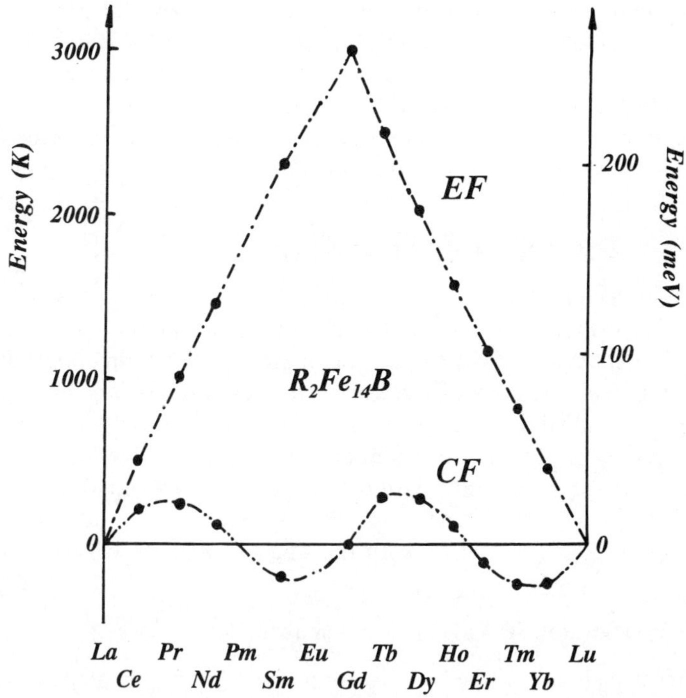

FIG. 2.5. Comparison of the crystal-field splitting, ΔE_{cf}, and exchange-field splitting, ΔE_{ef}, in the $R_2Fe_{14}B$ compounds [Rad87].

surrounding electronic spin moments and electronic charges, respectively. The former is described in terms of the intersublattice molecular-field coefficient n_{RT}, whereas the latter is expressed in terms of the crystalline electric field. A comparison of these two interactions across the $R_2Fe_{14}B$ series is given in Fig. 2.5. Most of the compounds belong, in first approximation, to the case where the magnetic interactions dominate the crystal-field splitting. Both interactions, however, are essential for a description of the rare-earth ions in 3d–4f intermetallics and of the magnetization curves in particular. The overall features are quite well understood for the whole series of $R_2T_{14}B$ compounds and for some of the other intermetallic series as well.

Units used in the magnetic literature are far from uniform. A comparison of the SI and cgs units was given in the Preface. Moreover, there is a difference between the macroscopic and atomic points of view. In the former, the parameters refer to the unit of volume, and magnetization is then expressed in A/m or (J/Tm^3). Alternatively, the magnetic polarization $\mu_0 M$ is expressed in teslas. In the atomic description, all units are related to a given number of atoms, ions or formula units. In that case one should speak of the magnetic moment expressed in μ_B/f.u. (J/T mol f.u.). The same holds for the anisotropy energy or the anisotropy constants where parameters in units of J/mol instead of J/m^3 better describe the intrinsic properties. There is a constant factor of 5.586 ($\approx N_A\mu_B$) between the magnetic moment in Am^2/mol f.u. and the magnetic moment in μ_B/f.u., where N_A is Avogadro's number. A factor of 8.31 ($\approx N_A k_B$) enters into the transformation of the anisotropy energy expressed in J/(mol f.u.) to K/f.u.. The density is always needed to convert from atomic-based units to the volume-based units used in technical calculations.

A. EXPERIMENTAL METHODS

Macroscopic techniques used for studying the magnetic parameters of R–T intermetallics include magnetization, susceptibility and thermodynamic studies. Neutron, Mössbauer and NMR experiments yield information about the properties on an atomic scale. Results of electronic structure calculations are included here since they are a valuable source of insight into the intrinsic magnetic properties. To illustrate the relevance of the different techniques, characteristic examples are presented of the type of information that can be obtained from each method. Monocrystalline samples provide a great advantage in studying the intrinsic magnetic properties, so we begin with the single-crystal growth of these materials.

1 Single-crystal growth of $R_2T_{14}B$ compounds

From extensive studies of the Nd–Fe–B phase diagram (Chapter 3C), it is known that $Nd_2Fe_{14}B$ does not melt congruently, unlike several of the R_2T_{17} compounds, but it is formed by a peritectic reaction. Thus, in order to grow monocrystalline material directly from the melt, an off-stoichiometric ratio of the starting materials has to be chosen. The most appropriate technique in that case is the travelling-solvent–floating-zone technique where the feed and seed materials have the proper stoichiometric composition while the solvent is off-stoichiometric. Successful attempts have also been reported with the Czochralski technique, starting with a large melt of off-stoichiometric material $Nd_{15}Fe_{77}B_8$ and applying a low growth rate of 3 mm/h using a 'necking' procedure to start the single-crystal growth [Men83]. Examples of monocrystals produced by the Czochralski technique are shown in Fig. 2.6. The Laue pattern of monocrystalline $Nd_2Fe_{14}B$ is presented in Fig. 2.7. Monocrystalline material has been produced for almost all existing $R_2(Fe, Co)_{14}B$ [Hir87, Kat88, Ver91] and $R_2(Fe, Co)_{17}$ [Sin88, Ver91] compounds. In Table 2.1, the lattice parameters, the calculated and the experimentally determined densities are presented for a selection of iron-based monocrystalline compounds which were grown by the Czochralski technique. Differences between the calculated and measured densities point to the presence of vacancies or to slight deviations from ideal stoichiometry.

2 Magnetization and magnetic susceptibility

The study of magnetization and susceptibility is the most direct way to investigate the magnetic parameters of the R–T intermetallics. The best information is obtained from monocrystalline samples in very large external fields. Magnetization curves like those in Fig. 2.4 reflect the interplay of crystalline–electric-field (CEF) and exchange interactions. Field-oriented frozen-powder samples provide a reasonable alternative to monocrystalline material for studying the magnetic anisotropy. Magnetization experiments on finely powdered samples that are free to rotate in the applied magnetic field are found to yield an accurate estimate for the intersublattice exchange interaction in the case of a ferrimagnetic structure, that is, for the heavy rare-earth compounds.

2.1 *High-field magnetization process*

Detailed high-field magnetization measurements can be performed in static or semicontinuous magnetic fields, but fields of 40 T, which are available in a few semicontinuous installations, are often insufficient to reach saturation along the hard magnetic axis. Experiments performed along the different crystallographic directions of a monocrystalline sample permit a full and quantitative analysis of the magnetization process. The simplest magnetization curves are obtained for compounds with nonmagnetic rare-earth elements. For the $Y_2Fe_{14}B$ sample of Fig. 2.3, the saturation magnetization at 4.2 K along the easy direction of magnetization is reached when the applied field exceeds the demagnetizing field, which is 0.5 T for a sphere of this compound (demagnetizing factor $N = 1/3$). For fields larger than the demagnetizing field, the magnetization along the easy direction is represented by

$$M(B_0, T) = M_s(0, T) + \chi_{hf}(T) \cdot B_0 \tag{2.1}$$

where M_s(0,T) is the spontaneous magnetization, χ_{hf}(T) is the high-field susceptibility, which is of the order of $10^{-3}\mu_B/T$ per 3d atom and B_0 is the applied field corrected for the demagnetizing field. Values for χ_{hf} for some selected 3d–4f compounds are given in Table 2.2. The high-field susceptibility is proportional to the density of states at the Fermi level. Values for easy-axis susceptibility similar to that of the yttrium compound are obtained at 4.2 K for many compounds with magnetic rare-earth elements. Apparently, the rare-earth moment is insensitive to external fields near zero temperature, indicating that the full trivalent moment of the rare-earth ion is realized. Enhanced values for the high-field susceptibility usually mark the development of field-dependent, noncollinear, internal magnetic structures. In the case of tilted magnetic structures, the susceptibility measured along any of the main crystallographic directions is substantial, but again it is small along the easy direction of magnetization, as for $Nd_2Fe_{14}B$ at 4.2 K.

The analysis of magnetization curves along the hard axis provides values for the anisotropy parameters appearing in the expansion of the magnetocrystalline anisotropy energy in terms of even powers of sin θ:

$$E_a = K_1 \sin^2\theta + K_2 \sin^4\theta + \dots \tag{2.2}$$

In order to derive values for the anisotropy constants K_n, B_0/M is usually plotted versus M^2, following Sucksmith and Thompson [Suc54]. The intercept of this plot

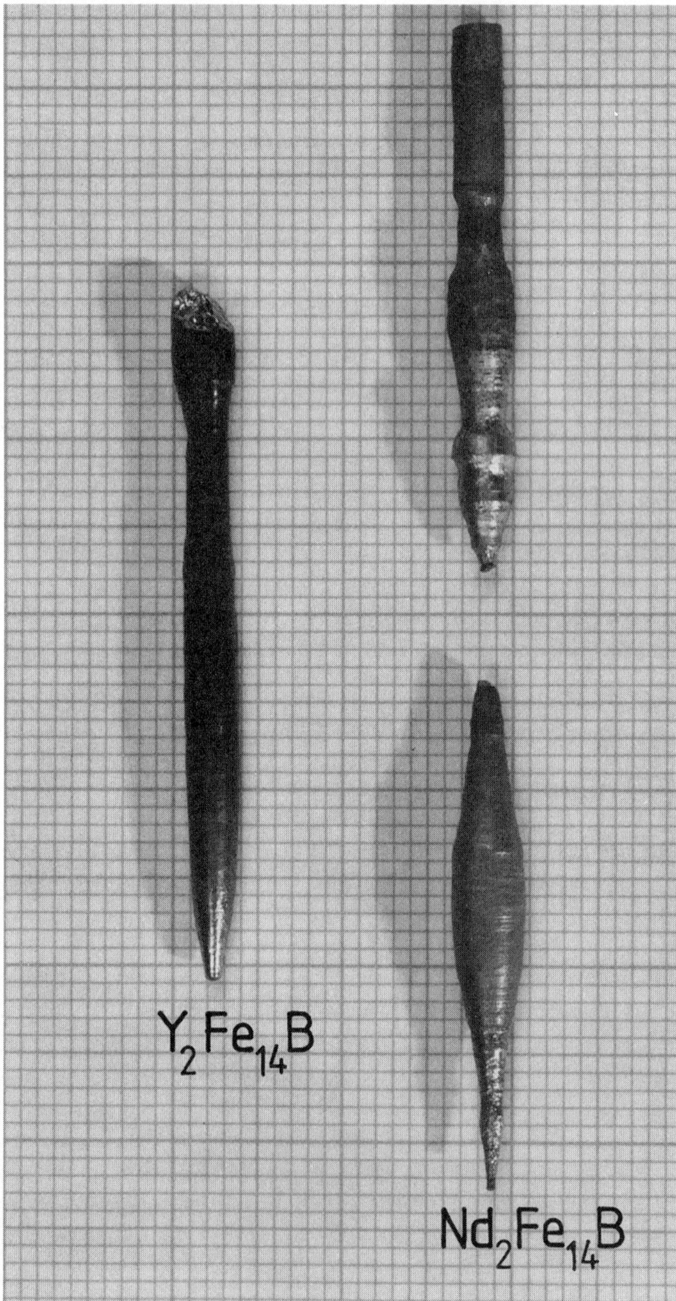

(a)

FIG. 2.6. (a) As-grown monocrystalline batches of $Y_2Fe_{14}B$ and $Nd_2Fe_{14}B$; (b) a crystal of $Dy_2Fe_{14}B$ cut by spark erosion in spheres and disks [Ver90]. Each square is 1 mm^2.

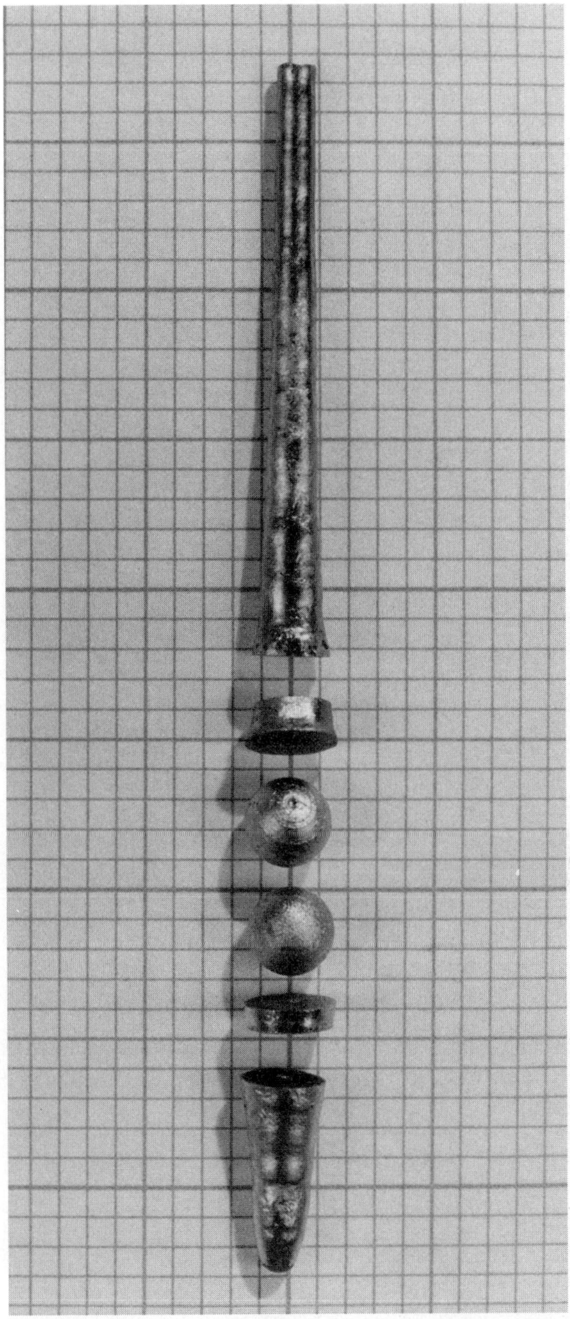

(b)

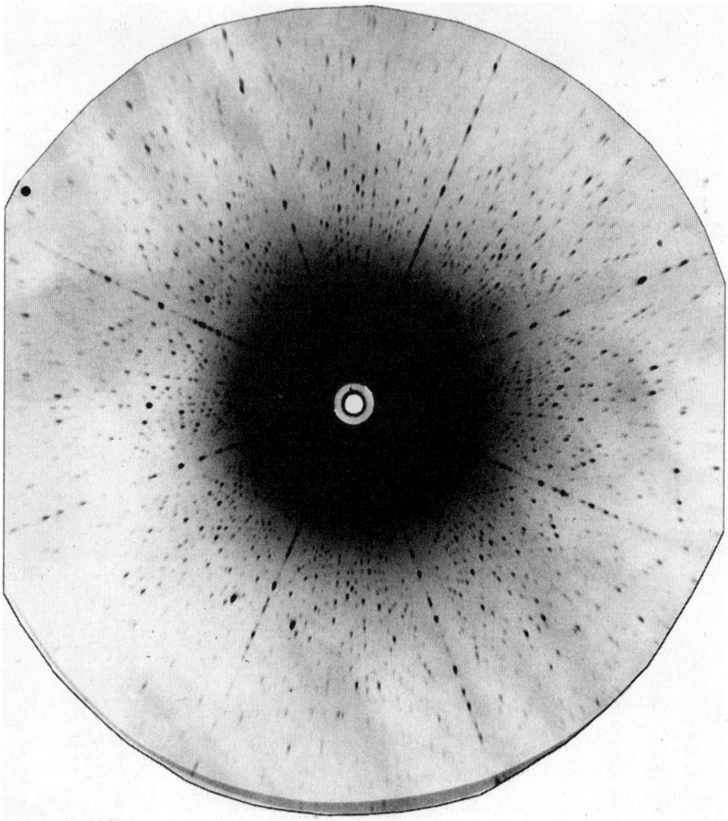

FIG. 2.7. Laue pattern of monocrystalline $Nd_2Fe_{14}B$ [Ver90].

Table 2.1 Values for the lattice parameters and calculated and experimental densities of the Czochralski-grown monocrystalline $R_2Fe_{14}B$ and R_2T_{17} compounds [Ver90]

Compound	Structure	Lattice parameters (nm)		Density (g/cm^3)	
		a	*c*	calc.	exp.
$Y_2Fe_{14}B$	tetr.	0.8732	1.1946	7.077	7.07(5)
$Nd_2Fe_{14}B$	tetr.	0.8766	1.2176	7.675	7.59(7)
$Gd_2Fe_{14}B$	tetr.	0.8758	1.2078	7.891	7.94(3)
Gd_2Fe_{17}	hex.	0.8498	0.8329	8.059	8.10(3)
Tb_2Fe_{17}	hex.	0.8488	0.8317	8.130	8.15(2)
Ho_2Co_{17}	hex.	0.8328	0.8140	9.088	–

Table 2.2 High-field susceptibility for some selected R-T intermetallics

Compound	High-field susceptibility χ_{hf} ($10^{-4}\mu_B$/T f.u.)	Ref.
$Y_2Fe_{14}B$	165	[Fra90]
$Gd_2Fe_{14}B$	100	[Rad89a]
$Lu_2Fe_{14}B$	280	[Rad89a]
Ho_2Fe_{17}	135	[Rad88b]
Fe	3.6	[Fra93]
YCo_5	27	[Ala81]
Ho_2Co_{17}	70	[Rad88b]

with the vertical axis provides a value for $2K_1/M_s^2$, whereas the slope yields $4K_2/M_s^4$. The Sucksmith–Thompson method is based on the minimization of the free energy that consists of the sum of magnetostatic and anisotropy energies, that is,

$$E = E_a - \mathbf{M}_s \cdot \mathbf{B}_0 \tag{2.3}$$

where M_s represents the value of the spontaneous magnetization observed along the easy axis. A deviation of the plot B/M vs M^2 from linearity points to the presence of higher-order anisotropy constants.

In this context, the concept of the *saturation field* B_K is introduced as the field at which the magnetization along the hard direction reaches the value measured along the easy direction. The saturation field along the hard axis of magnetization is related to the difference in anisotropy energy, ΔE_a, between the two magnetization directions:

$$\delta E_a = \int_0^{B_K} \mathbf{M} \cdot d\mathbf{B} \tag{2.4}$$

The *anisotropy field* B_a, by definition, is obtained as the crossing of the extrapolated low-field hard-axis magnetization curve with the easy-direction magnetization curve, as illustrated in Fig. 1.15. In the case of a uniaxial compound, characterized by a single second-order anisotropy constant K_1, B_K coincides with B_a. The anisotropy energy for most of the Y–3d compounds is well represented by the first term in the expression for E_a in Eq.(2.2). However, the Sucksmith–Thompson method is inapplicable for magnetic systems that are composed of different magnetic sublattices coupled by intersublattice exchange interactions because, apart from a rotation, the external field disturbs the internal magnetic structure. As a consequence, the value of M_s is not constant with increasing field. In that case, the K_i values derived are not simply related to the intrinsic magnetocrystalline anisotropy and should be considered as *effective anisotropy constants*. These effective parameters are still useful in a technical characterization of magnetic materials.

A phenomenological model has been developed by Asti [Ast89] to analyse anomalies found in the magnetization curves. *First-order magnetization process* (FOMP) transitions can be described by Eqs (2.2) and (2.3), if necessary by including terms in K_3, K_3', etc., to represent the higher-order anisotropy contributions. Noncollinear structures are not considered in the FOMP model as the exchange energy does not enter into the description. This neglect is equivalent to the assumption of an infinite value for the intersublattice coupling parameter J_{RT}. An example of a FOMP transition can be found in Fig. 2.4 near 16 T where the ⟨100⟩-axis magnetization curve jumps to its saturation value. It turns out that in the case of the light rare-earths with a ferromagnetic alignment of the two sublattice magnetizations, the decoupling of the Nd and Fe moments does not affect the magnetization curve substantially. The transition near 16 T is related to the higher-order CEF terms, as will be discussed later in this chapter.

Field-induced noncollinearities can be observed directly in the high-field magnetization curves of the Gd–Co compounds which have a small net moment due to the near-cancellation of contributions from the Gd and Co sublattices. The Gd ion itself is not expected to have a significant magnetocrystalline anisotropy because of the half-filled 4f shell and its ^{8}S ground state. The magnetization curves of $GdCo_5$ in Fig. 2.8 should be compared with those for YCo_5 in Fig. 2.9. The crossing of the easy-axis and hard-axis curves at low fields in Fig. 2.8 demonstrates the inapplicability of Eq. (2.3) to ferrimagnetic compounds because above this crossing field the hard-axis magnetization exceeds that along the easy direction. From these data it is found that the effective anisotropy constant K_1^{eff} of $GdCo_5$, derived by the Sucksmith–Thompson plot for fields below 5 T, is a factor of 3 smaller than that obtained for YCo_5. The magnetization curves calculated for $GdCo_3$, $GdCo_5$ and $Gd_2Co_{14}B$ in the full field range are shown in Figs 2.8 and 2.10. External fields approaching twice the intersublattice molecular field, $B_{\mathrm{R}}^{\mathrm{mol}}$ (220 T), are needed in order to attain the forced ferromagnetic state. The calculated curves are typical for the high-field magnetization curves of anisotropic ferrimagnetic compounds. Along the easy axis of $GdCo_5$, the ferrimagnetic structure is broken at the spin flop field of 67 T. Experiments on $Gd_2Co_{14}B$ show a spin flop field of 52 T [DeB89]. The corresponding fields for iron compounds are much greater, on account of their larger moments.

The calculation of the magnetization curves in Figs 2.8 and 2.10 has been carried out within a two-sublattice model where one considers the magnetic sublattices formed by the R and T moments. Sublattices are coupled antiferromagnetically for the heavy rare-earth elements due to the exchange interactions between 3d and 4f spins. As these interactions are of finite strength, an external field disturbs the collinear magnetic structure. The energy of the system in an external field, $\mathbf{B}_0$, takes the form [Rad86]:

$$E_{\mathrm{RT}} = E_{\mathrm{a}}^{\mathrm{T}} + E_{\mathrm{a}}^{\mathrm{R}} + n_{\mathrm{RT}}\gamma_{\mathrm{R}}\mathbf{M}_{\mathrm{T}} \cdot \mathbf{M}_{\mathrm{R}} - (\mathbf{M}_{\mathrm{T}} + \mathbf{M}_{\mathrm{R}}) \cdot \mathbf{B}_0 \tag{2.5}$$

Where $n_{\mathrm{RT}}\gamma_{\mathrm{R}}$ is a molecular-field coefficient, representing the exchange interactions between the R and T sublattices and:

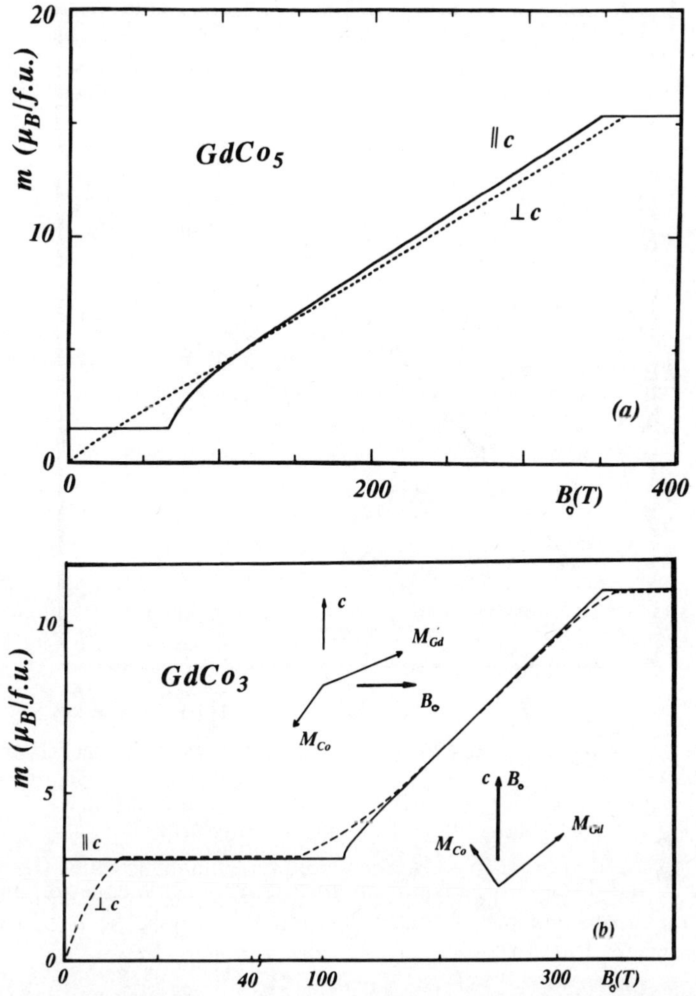

FIG. 2.8. Two examples of the calculated magnetization curves of ferrimagnetic compounds in the full field range with ($GdCo_3$) and without ($GdCo_5$) a saturated state in the antiparallel-moment configuration along the hard axis: (a) calculated with parameters relevant to $GdCo_5$ at 4.2 K [Rad92c]; and (b) with parameters relevant to $GdCo_3$ [Rad86]. For $GdCo_3$, the moment configurations for a field of 250 T applied parallel to the easy (hexagonal) axis and for a field of 100 T perpendicular to the easy axis are shown.

$$\gamma_R = 2(g_R - 1)/g_R \tag{2.6}$$

where g_R is the Landé g-factor for the rare-earth ion involved. Values for g_R and γ_R were given in Table 1.2. In Eq.(2.5), terms related to the intrasublattice interactions do not enter. This approximation is valid as long as the internal magnetic structure

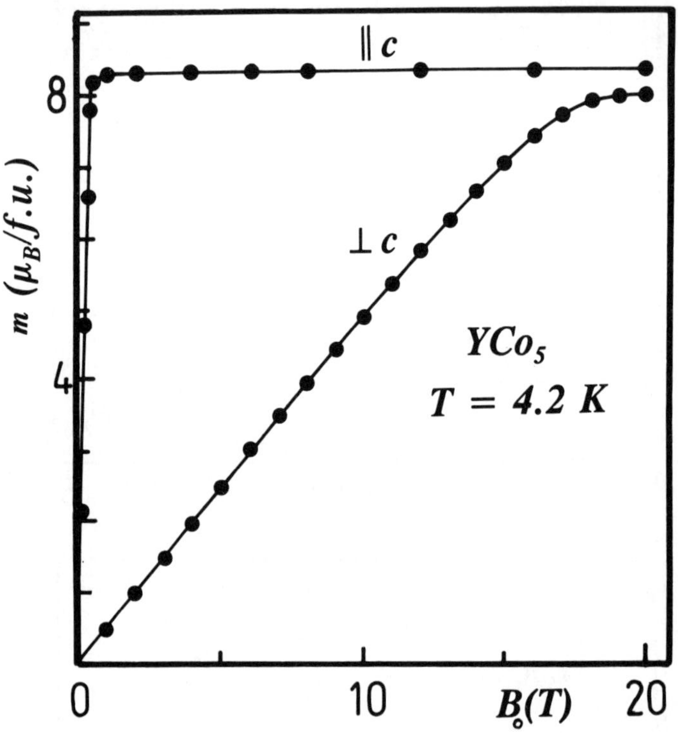

FIG. 2.9. Magnetization curves at 4.2 K of hexagonal YCo_5 along different crystallographic directions [Ala81].

within the R and T sublattices is not modified during the magnetization process. When no magnetocrystalline anisotropy is present, analysis shows that there is a field region where a noncollinear magnetic structure is realized:

$$n_{\mathrm{RT}}\gamma_{\mathrm{R}} \mid \mathbf{M}_{\mathrm{T}} - \mathbf{M}_{\mathrm{R}} \mid < B_0 < n_{RT}\gamma_{\mathrm{R}} \mid \mathbf{M}_{\mathrm{T}} + \mathbf{M}_{\mathrm{R}} \mid \tag{2.7}$$

Here, the magnetization curve is linear in field with a slope determined by the intersublattice interaction:

$$\mathrm{d}M/\mathrm{d}B = (\gamma_{\mathrm{R}} n_{\mathrm{RT}})^{-1} \tag{2.8}$$

The assumption of zero magnetocrystalline anisotropy is not realistic for R–T compounds but Verhoef *et al.* [Ver89] have shown that the magnetization curves of finely powdered polycrystalline samples of $Er_2(Fe, Mn)_{14}C$ with particles that are free to rotate in the applied magnetic field follow Eq.(2.8) quite well in sufficiently large fields. In fact, it turns out that the anisotropy does not enter into the free-energy expression because the fine particles can always orient their net magnetization direction along the direction of the external field. This can be realized in ferrimagnetic compounds where the anisotropy is associated with only

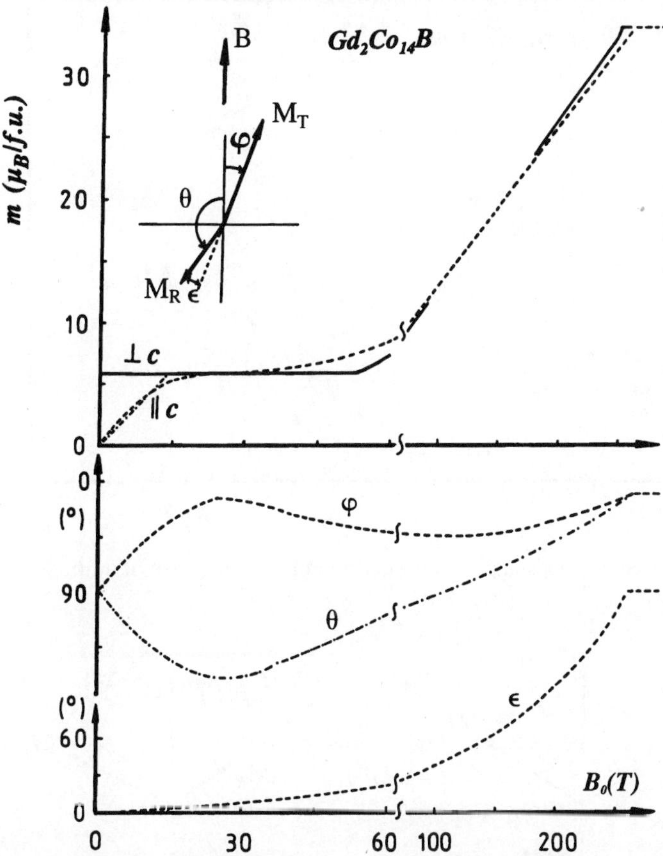

FIG. 2.10. Calculated magnetization curves of $Gd_2Co_{14}B$ at 4.2 K along different crystallographic directions in the full field range. The crossing of the two magnetization curves has been verified on aligned-powder samples and found to occur at 36 T [Rad90b].

one magnetic sublattice, including many of the 3d–4f compounds where the anisotropy of the R sublattice is an order of magnitude greater than the anisotropy of the 3d sublattice. Figure 2.11 shows the high-field magnetization curves of several $Er_2Fe_{14-x}Mn_xB$ compounds measured on fine particles obtained from polycrystalline material. These experiments reveal a wide field region with an almost linear magnetization curve. In this system, Mn substitution has been used in order to reduce the net magnetization and, consequently to shift the regime of linear magnetization to a lower field range.

A schematic picture for the fixed and the free sample methods is given in Fig. 2.12. In both experiments, the resultant moment of the ferrimagnetic system in the canted configuration reorients itself continuously during the magnetization process in order to follow the direction of the external field. In the case of free-

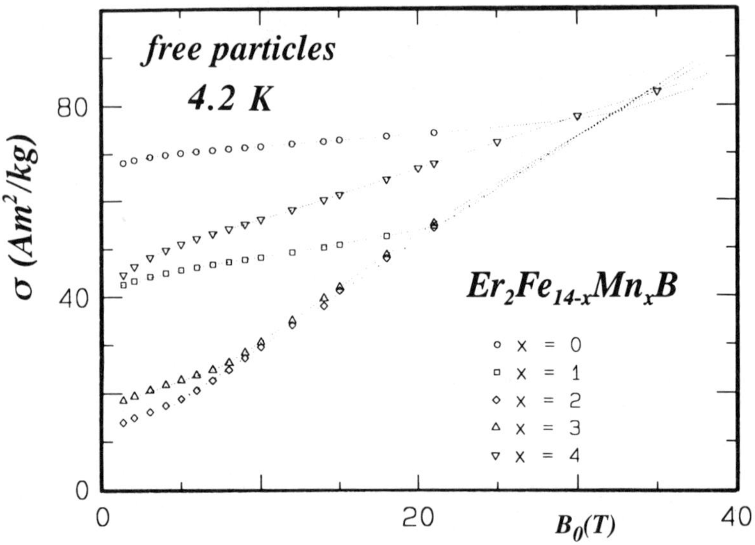

FIG. 2.11. 'Free-powder' magnetization curves of $Er_2Fe_{14-x}Mn_xB$ at 4.2 K[Ver90c].

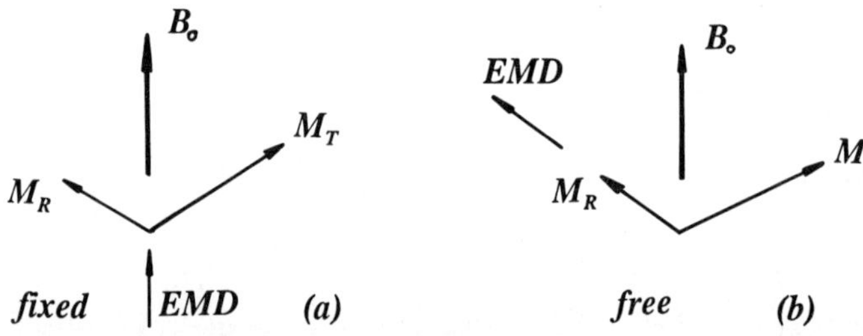

FIG. 2.12. Schematic picture of the fixed-powder (a) and free-powder (b) magnetization techniques in which the orientation of the external field, the sublattice moments and the easy direction of the R moment in the noncollinear region are shown. For fixed-sample experiments, the R sublattice magnetization is pushed out of the preferred direction, contrary to the free-sample experiments where the R sublattice magnetization always stays along the preferred direction. The anisotropy of the T sublattice is assumed to be negligible.

powder samples for which the anisotropy can be associated with the rare-earth sublattice only, the net magnetization direction coincides with the field direction. This is realized by an actual rotation of the powder particles which keeps M_R lying always along its easy direction. As a consequence, no information about the magnetic anisotropy of the R sublattice can be obtained from these experiments. In the case of fixed-sample experiments, both moments are pulled away from their easy direction through angles that depend on the strength of the applied field. Part of the angular dependence of the free energy is probed, from which the anisotropy parameters can be evaluated. The *free-powder technique* has been elaborated further by Zhang *et al.* [Zha94] and applied to R–T intermetallics in which both sublattice anisotropies are considered.

2.2 *Temperature dependence of magnetization*

The temperature dependence of the spontaneous magnetization of the R–T intermetallics provides another opportunity to evaluate the strength of the 3d–4f interaction. The temperature dependence of the magnetic moment of the 4f ion is described, in first approximation, by the Brillouin-function $\mathscr{B}_J$

$$M_R(T) = M_R(0)\mathscr{B}_J[x] \text{ with } x = g\mu_B J B_R^{mol}/kT \tag{2.9}$$

The molecular field experienced by the rare-earth ion contains contributions from the R–T and R–R interactions and can be written as

$$B_R^{mol} = B_{RT}^{mol} + B_{RR}^{mol} = n_{RT}\gamma_R M_T + n_{RR}\gamma_R^2 M_R \tag{2.10}$$

The net magnetization for the parallel (+) and antiparallel (-) configurations can be written as

$$M_{net} = | M_R \pm M_T |$$

The magnetization of the 3d sublattice is written as the sum of the magnetization of the yttrium counterpart and the contribution induced by the R–T interaction

$$M_T(T) = M_T^Y(T) + \chi_{hf} n_{RT}\gamma_R M_R$$

Values for the molecular-field coefficients can then be derived from

$$M_R(T) = g_R\mu_B J\mathscr{B}_J[g_R\mu_B J(n_{RT}\gamma_R M_T + n_{RR}\gamma_R^2 M_R)/k_B T] \tag{2.9$'$}$$

Gadolinium compounds are especially suited for an evaluation of the molecular field in this way due to the absence of CEF interactions. Still, the problem is to deduce the temperature dependence of the gadolinium sublattice from experimental magnetization data that always refer to the whole system. A rough approach is to subtract the transition-metal contribution determined from the yttrium counterpart, normalizing the temperature scale by T_C and neglecting the small contribution to M_T from the molecular field. The procedure is illustrated in Fig. 2.13 for $Gd_2Fe_{14}B$. The temperature variation of the Gd moment from 4.2 K to 300 K is well approximated by Eq.(2.9), yielding a value for B_{Gd}^{mol} of 275 T [Bog86].

In the case of compounds containing rare-earth ions with nonzero orbital moments, the molecular field B_R^{mol} derived from the temperature dependence of the magnetization is overestimated because the crystal-field interactions lead to a weaker temperature dependence of the rare-earth moment than is predicted by Eq.(2.9′).

In ferrimagnetic compounds, a *compensation temperature*, T_{comp}, may occur where the contributions from both sublattices to the magnetization cancel and result in a vanishing net magnetization. The compensation point is observed for ferrimagnetic R–T compounds where the rare-earth magnetization exceeds that of

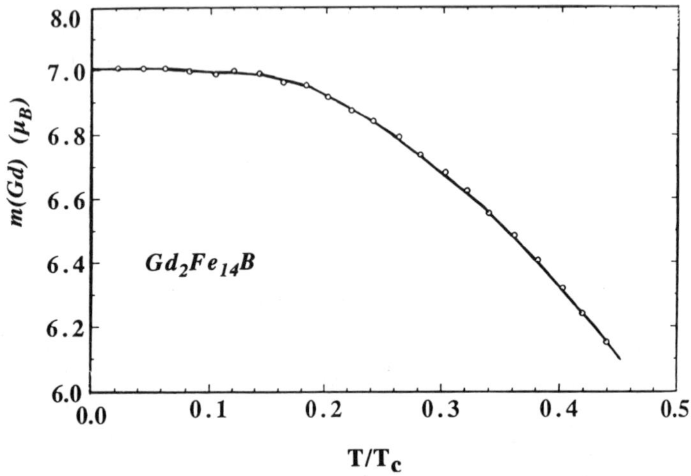

FIG. 2.13. Temperature dependence of the gadolinium magnetic moment ($T_C = 664$ K) in $Gd_2Fe_{14}B$; the solid line represents a fit of the data to the Brillouin function with $J = 7/2$.

the 3d sublattice at low temperatures. This results from the fact that the R-magnetization drops more rapidly with increasing temperature than that of the 3d sublattice. Thus, the temperature T_{comp} can be used to determine the temperature dependence of $M_R(T)$ and to estimate the molecular field experienced by the rare-earth ion. Results consistent with those from high-field magnetization have been obtained, in particular for compounds with low values for T_{comp}.

2.3 *Paramagnetic susceptibility*

Above T_C, the temperature dependence of the susceptibility can be very complex, in particular at lower temperatures. Only in some cases does the temperature dependence of the susceptibility follow a Curie–Weiss relation:

$$\chi = C/(T - \theta_p) \tag{2.11}$$

where C is the Curie constant. Values for C for the trivalent ions in the ground-state multiplet given by Hund's rules were presented in Table 1.2. More frequently, the magnetic susceptibility in the paramagnetic region can be represented by a modified Curie–Weiss (MC–W) law in the form

$$\chi = \chi_0 + C/(T - \theta_p) \tag{2.12}$$

where an extra, weakly temperature-dependent, term χ_0 appears associated with a paramagnetic background. An alternative expression is given by the more complex hyperbolic relation (HC–W) of the form

$$\chi^{-1} = \chi_0^{-1} + T/C - \sigma/(T - \theta_p) \tag{2.13}$$

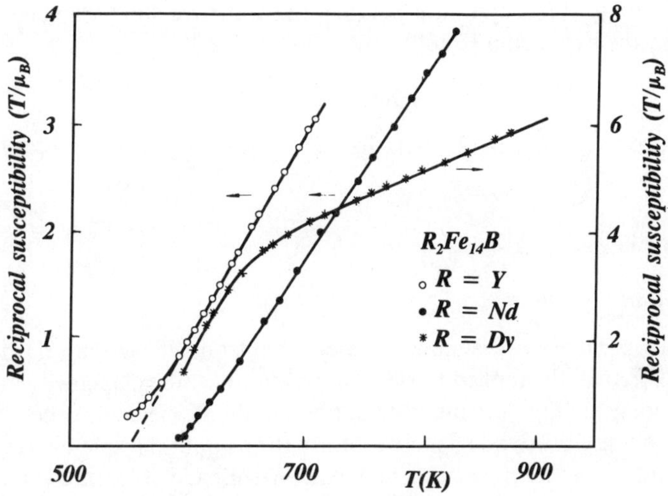

FIG. 2.14. Temperature dependence of the reciprocal susceptibility for ferromagnetic $Y_2Fe_{14}B$ and $Nd_2Fe_{14}B$ and for ferrimagnetic $Dy_2Fe_{14}B$ [Bur84].

The parameters χ_0, C, θ_p and σ occurring in Eq. (2.13) are often just fitting parameters. Generally, C–W behaviour is manifest in compounds such as YFe_2, YFe_3, Y_2Fe_{17} where only one type of magnetic moment is present. It is a general feature of itinerant ferromagnets that the atomic moment deduced from the Curie constant exceeds that measured in the ferromagnetic ground state [Rho63]. The MC–W behaviour is found for compounds where, besides the localized rare-earth moments, there is a paramagnetic medium. The hyperbolic expression is applicable for magnetic systems with two types of magnetic moment that continue to interact above the magnetic ordering temperature.

Experimental results for some $R_2Fe_{14}B$ compounds are presented in Fig. 2.14. The almost straight line for $Y_2Fe_{14}B$ fitted by Eq. (2.11) gives $C = 74\mu_B^2/\text{f.u.}$, which yields a value for the effective iron moment $g[S_T(S_T+1)]^{1/2}$ of 4.0 μ_B. Also for $Nd_2Fe_{14}B$, a straight line is obtained which is related to the ferromagnetic arrangement of the Fe and Nd moments. The slope of the curve gives a value for C of 85 $\mu_B^2/\text{f.u.}$, close to the value of 83 $\mu_B^2/\text{f.u.}$ obtained by assuming the same effective moment of Fe and a value for J of 9/2 for Nd. From Fig. 2.14 it can clearly be seen that the ferrimagnetic compounds $Dy_2Fe_{14}B$ and $Gd_2Fe_{14}B$ have to be fitted using Eq.(2.13). It turns out that the effective iron moment in these compounds calculated from the value of C in Eq. (2.13) is only slightly larger (4.1 μ_B in $Gd_2Fe_{14}B$) than for the yttrium counterpart.

None of the above expressions takes CEF interactions into account, which manifest themselves in the anisotropic susceptibility of monocrystals. The CEF interactions introduce an anisotropy of the paramagnetic Curie temperature. Assuming that the second-order CEF is dominant and that at sufficiently high temperature the $\chi^{-1}(T)$ curves are parallel, the value for the parameter B_2^0 can be derived from the expression [Rad95]:

$$B_2^0 = 2(a_{\parallel} - a_{\perp})/(3Ab)$$

where a, b are the fitting parameters in the expression $\chi^{-1}(T) = a + bT$ and A is $(2J-1)(2J+3)/5k_B$.

3 Neutron studies

3.1 *Neutron diffraction*

As the neutron carries a magnetic moment, neutron diffraction (elastic neutron scattering) is frequently applied to study magnetic structures, magnetic moments of the atoms and, generally, the spin distribution within the elementary cell. Studies on well-defined materials provide detailed information about the spin distribution. For example, studies on monocrystalline hexagonal Er_2Co_{17} led to values for the magnetic moments that turn out to be site-dependent [Moz92b]. The Er moments at the 2b site are only 8.3 μ_B compared with the full moment of 9.0 μ_B found at the 2d site. An even larger variation is found in the cobalt moments that range from 2.1 μ_B (4f) to 1.2 μ_B (12k). Polarized neutron experiments by Ritter [Rit89] on cubic YFe_2 indicate a moment at the Y site of -0.67 μ_B. By comparing spectra obtained for spin-up and spin-down diffraction at the [220] reflection which is exclusively due to Y, it is found that the Y moment is antiparallel to the Fe moment. Given the net moment of 2.9 μ_B/f.u. a value for the iron moment of 1.77 μ_B is deduced from magnetization studies. Givord *et al.* [Giv80] have found an orbital contribution to the iron moment of 0.07 μ_B in $LuFe_2$. These moments are well accounted for by band-structure calculations by Eriksson *et al.* [Eri90], who report an iron orbital moment of 0.07 μ_B and an induced yttrium moment of -0.47 μ_B.

Hexagonal YCo_5 is another compound that has been studied in order to determine the spin distribution over the elementary cell. The compound exhibits the strongest uniaxial anisotropy of all the 3d alloys, with $B_a = 17$ T (Fig. 2.9). The neutron-diffraction experiments reveal a substantial difference in the spatial spin distribution at the two Co sites [Sch80], which supports the earlier conclusion drawn by Streever [Str77] on the basis of nuclear-magnetic-resonance studies that the dominant contribution to the magnetocrystalline anisotropy arises from the cobalt 2c site, rather than from the 3g site.

3.2 *Inelastic neutron scattering*

Inelastic-magnetic-neutron scattering (INS) yields information on the dynamical magnetic response of both 3d and 4f systems. The technique can be used in principle to determine the CEF level scheme of the ground-state multiplet of a R^{3+} ion in a nonmagnetic compound as the available neutron energies (≈ 0.1 eV) are comparable to the crystal-field splitting. If the intermetallic compound is magnetically ordered, the molecular field, B_R^{mol}, splits the degenerate CEF levels. In the case of R = Gd, the molecular field, B_R^{mol}, is measured directly from the

Zeeman splitting. In general, the thermal-neutron magnetic-scattering cross-section $S(\mathbf{q}, \omega)$ for a system of N noninteracting ions in the dipole approximation for small values of the momentum transfer $\mathbf{q}$ and the energy transfer hω is given by

$$S(\mathbf{q}, \omega) \approx f^2(q)\Sigma e^{-\beta E_i} \cdot |< \mathrm{i} \mid J_\perp \mid \mathrm{f} >|^2 \, \delta(E_i - E_f - \hbar\omega) \qquad (2.14)$$

where $\hbar\omega$ is the neutron energy transfer, |i> and |f> are the initial and final 4f electron states with energy E_i and E_f, respectively, $J_\perp$ is the component of the total angular momentum operator perpendicular to the scattering vector $\mathbf{q}$.

Within the dipole approximation, the intensities for the transitions are determined by the corresponding dipole matrix elements, that is, transitions with changes of $J_\perp$ by 0 (the longitudinal mode) or $\pm$ 1 (the transverse mode). Full information about the energy and momentum transfer is obtained from studies on large crystals of a few cubic centimetres in a triple-axis experiment. Neutron absorption by the sample is a serious problem for some rare-earth compounds which can be solved, for instance, by using less absorbing isotopes of Gd, Dy or B. Useful information about dispersionless modes can be drawn from neutron-scattering spectra on polycrystalline samples. Peaks in the energy scan are assigned to localized excitations of the R ion involved.

A typical magnetic excitation scheme for monocrystalline, magnetically ordered rare-earth transition-metal intermetallics with a high-ordering temperature is shown in Fig. 2.15. Such spectra have been observed for the cubic Laves-phase compounds $HoFe_2$ and $ErFe_2$ [Ko81], as well as for Ho_2Co_{17} and Ho_2Fe_{17} [Cla82] and Dy_2Co_{17} [Col89], see Fig. 2.16. The flat mode, which manifests itself in the energy scale as a well-pronounced maximum, is related to single-ion excitations between the two lowest localized levels of the R ions. When the splitting is due solely to exchange interactions one obtains

$$\Delta E = g\mu_B B_R^{mol}.$$

This relation is correct for Gd compounds where CEF interactions are negligible. Strongly dispersive modes are spin-wave-like and give smooth contributions to the E–$\mathbf{q}$ spectra. The mode denoted T–T involves collective excitations of the T spins as observed in pure 3d elements. This spin-wave mode is usually difficult to observe, especially for $\mathbf{q}$ far from the centre of the Brillouin zone. The R–T mode involves excitations of the R and T spins.

The INS spectra of a material containing 4f ions depend very much on the relative strength of the CEF and exchange interactions. The spectra contain a few well-defined peaks related to excitations between the 4f levels. The number of transitions, as well as the transition probability, is determined by the matrix elements between the eigenstates which are governed by the local symmetry of the CEF interaction. These transitions are typically single-site excitations. With increasing strength of the exchange interactions, the INS spectra become simpler and for a dominant molecular field, the energy-level splitting becomes more and more Zeeman-like. In that case, the wave functions are essentially pure J_z states,

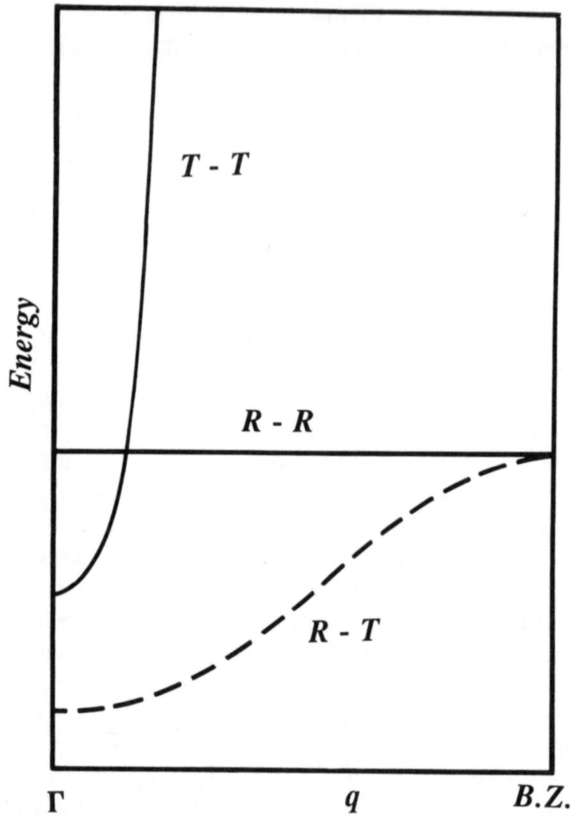

FIG. 2.15. Schematic representation of magnetic excitations in the ($\mathbf{q}$, E) space in ordered R–T intermetallics. These excitations are observable for monocrystalline samples in inelastic-neutron-scattering experiments, whereas for polycrystalline samples, only one peak associated with the flat R–R mode is observed. B. Z. denotes the Brillouin zone edge.

with the ground state having $J_z = J$. As a consequence, the probability for excitations between the adjacent levels in the ground-state J multiplet is much larger than those with $\Delta J_z > 1$. Neutron experiments at low temperatures give limited information as, often, only one transition is observed, associated with excitations from the ground state to the first excited level. Examples are given in Fig. 2.17 for $Dy_2Fe_{14}B$ and $Er_2Fe_{14}B$. Further excitations can be observed by performing experiments at elevated temperatures as excited states then become populated. Several transitions can be observed at low temperatures when the ground-state and excited-state wavefunctions are mixtures of several $| J_z >$ states, which arises when the rare-earth moment does not lie along a principal axis as in $Nd_2Fe_{14}B$ [Loe90, 91] and $DyFe_{11}Ti$ [Hur94], where tilted magnetic structures exist at low temperatures.

Spin-wave dispersion curves E($\mathbf{q}$) of $Nd_2Fe_{14}B$ for a wide $\mathbf{q}$-range measured by Mayer *et al.* [May92] have revealed (i) a strongly dispersive dependence with (ii) an

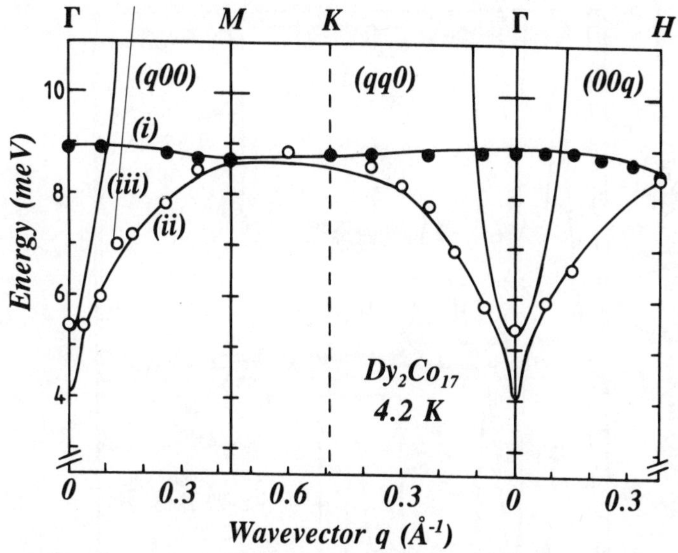

FIG. 2.16. Magnetic excitations at 4.2 K in Dy_2Co_{17} observed for a monocrystalline sample; the experimental point (○) and (●) have been observed around the (220) and (301) reciprocal lattice points respectively. The full lines represent the results of a linear spin-wave model. (i), (ii) and (iii) denote the excitation modes Dy–Dy, Dy–Co and Co–Co respectively [Col89a, b].

energy gap for $q = 0$ of 3.6 meV at 4.2 K that decreases to 1 meV at room temperature. The experimental results have been discussed in the frame of a classical spin-wave model.

4 Mössbauer and NMR spectroscopy

Mössbauer and NMR spectroscopy probes the magnetic and electronic properties on a microscopic scale via hyperfine interactions with the electric and magnetic moments of the nucleus. Information on at least two important parameters is provided: the magnetic hyperfine field, B_{hf}, and the quadrupolar interactions of the nucleus. The hyperfine field is related to the magnetic moment of the unfilled 3d or 4f shell, whereas the quadrupolar interactions are related to the electric-field gradient at the nucleus, V_{ij}, created by the unfilled 3d and 4f shell and the crystal-field potential at the site of the 3d or 4f shell due to all the other charges.

The hyperfine field is, in first approximation, assumed to be proportional to the magnetic moment according to the relation

$$B_{hf}^{T} = A_{hf} \cdot m_{T} \tag{2.18}$$

where the parameter A_{hf} is known as the hyperfine (hf) interaction constant. For iron, a value of $-15T/\mu_B$ for A_{hf} can be deduced from parameters relevant to α–Fe:

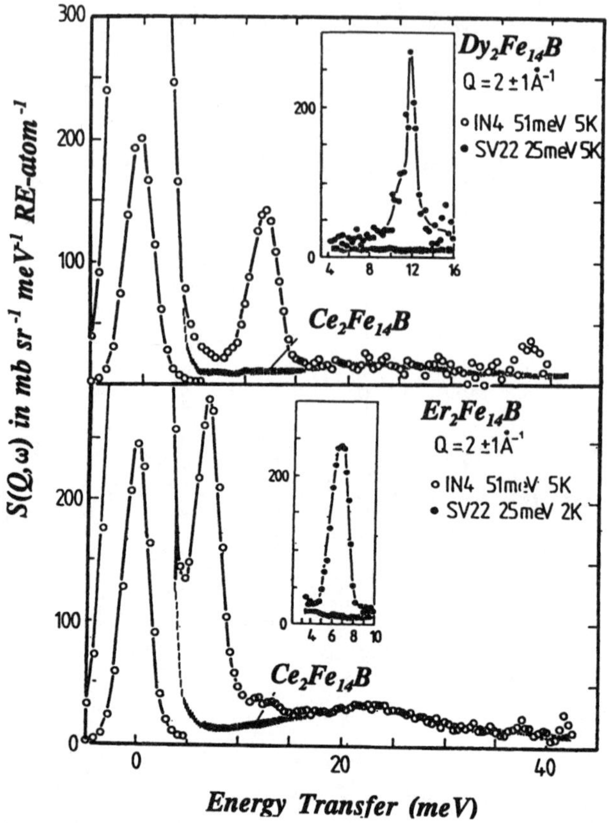

FIG. 2.17. Low-temperature INS spectra measured for polycrystalline $Dy_2Fe_{14}B$ and $Er_2Fe_{14}B$ revealing only one localized CEF excitation [Loe91].

$B_{\rm eff}^{\rm Fe} = -33$ T and $m = 2.22\ \mu_{\rm B}$. Mössbauer spectroscopy on ^{57}Fe nuclei in powder samples is a very convenient technique for estimating the local magnetic moment of the iron ions. The proportionality of Eq. (2.18) is found to work surprisingly well for the *average* moment in many intermetallic compounds. In Y_nFe_m, $A_{\rm hf}$ takes a value between -14.5 and $-15.0\ T/\mu_{\rm B}$. For cobalt, the hyperfine field, $B_{\rm hf}^{\rm Co}$, is detected by the NMR technique on ^{59}Co nuclei. In this case, the parameter $A_{\rm hf}$ takes a value of about $-12T/\mu_{\rm B}$ [Ber88].

For rare-earth nuclei in intermetallics with a large value of $T_{\rm C}$, the hyperfine field at the nucleus is close to the free-ion value that is known from atomic calculations. It indicates that the ground-state moment of the 4f shell has the maximal J_z value $(= J)$. An additional contribution to the rare-earth hyperfine field is transferred from the 3d sublattice. In an analysis of the different contributions to the hyperfine field of $SmCo_5$, the experimentally observed value of -342 T [Str75] is split into: -378 T, -12 T and $+48$ T as originating from the 4f shell, the polarization of Sm

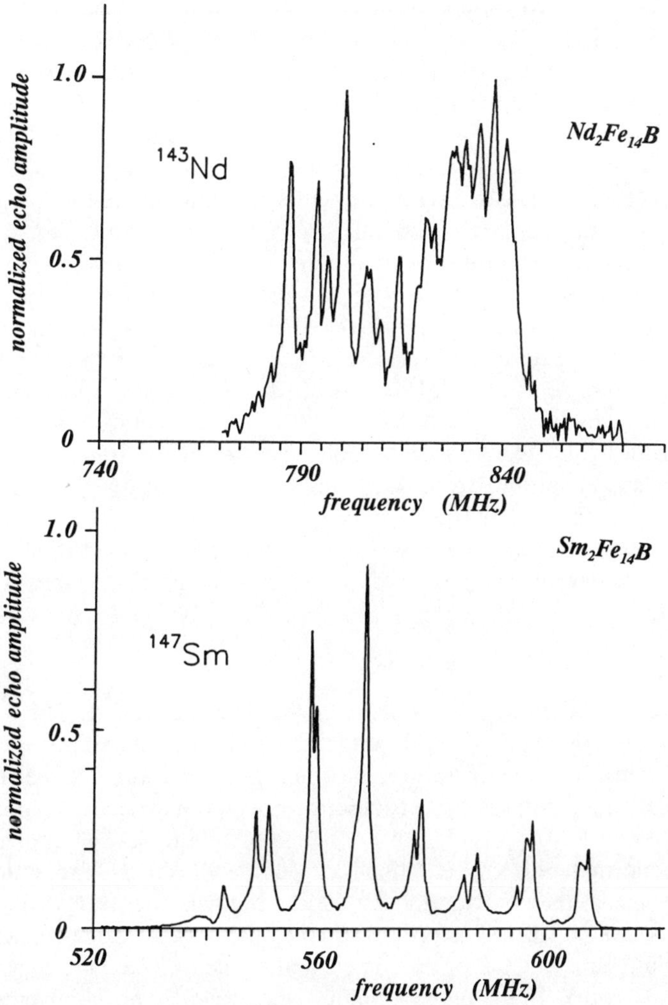

FIG. 2.18. ^{143}Nd NMR spectrum of $Nd_2Fe_{14}B$ at 4.2 K and ^{147}Sm NMR spectrum of $Sm_2Fe_{14}B$ at 4.2 K [Kap92].

core electrons and the conduction electrons respectively. A similar separation can be made for Sm_2Co_{17}. The 4f shell contribution is dominant. For the S-state Gd ion, the large orbital and dipolar terms from the 4f shell are absent, yielding rather small values for the observed hyperfine field. Examples of rare-earth NMR spectra for $Nd_2Fe_{14}B$ and $Sm_2Fe_{14}B$ are shown in Fig. 2.18.

The assumption in Eq.(2.18) that m_{T} is proportional to $B_{\mathrm{hf}}^{\mathrm{T}}$ is questionable. The effective hyperfine field contains several contributions and can be written as

$$B_{\mathrm{hf}} = B_{\mathrm{cp}} + B_{\mathrm{cond}} + B_{\mathrm{orb}} + B_{\mathrm{dip}} \tag{2.19}$$

where B_{cp} and B_{cond} represent the contribution from the core and conduction electrons respectively. The spin polarization of core B_{cp} and conduction B_{cond} electrons contribute differently to B_{hf}. The term B_{cp} is accurately proportional to the 3d spin moment ($-13.8\ \mathrm{T}/\mu_B$ for iron) [Coe89] but the term B_{cond} is mostly due to the small 4s polarization which is sensitive to charge transfer [Qi92c]. The last two terms are anisotropic and proportional to the orbital moment (B_{orb}) or to the dipolar field (B_{dip}). They can have a different sign compared to the dominant core-polarization contribution. For 3d metals, the last two terms in Eq. (2.19) are much smaller than those originating from the 3d and 4s spins leading to the rough proportionality between the ^{57}Fe hyperfine field and the iron moment found in practice. For the non-S-state rare-earths such as Sm and Nd, the anisotropic contributions are the dominant ones.

The temperature dependence of B_{hf} reflects the temperature dependence of the 4f or 3d magnetic moment. Hence, by analysing the temperature variation of the hyperfine field, the effective molecular fields can be derived using Eq.(2.9).

The quadrupolar splitting QS is due to interactions of the quadrupole moment of the nucleus with the electric field gradient (EFG), $V_{ij} = \partial^2 V/\partial x_i \partial x_j$. The electric field gradient at a rare-earth nucleus results partly from the quadrupole moment of the 4f shell and partly from that of the surrounding charges according to the formula

$$QS(T) = C < 3J_z^2 - J(J+1) > + QS^{\mathrm{latt}} \tag{2.20}$$

where C is a coefficient and $< \ >$ denotes the thermal average. The first term in Eq. (2.20) is strongly temperature-dependent, whereas the second one is expected to be largely insensitive to temperature, allowing for a separation between these two terms. In diagonal form, the EFG is described by two independent parameters V_{zz} and $\eta = | (V_{xx} - V_{yy})/V_{zz} |$. Using an estimate for the 4f-shell contribution (in hard magnetic materials the R^{3+} ions are in the $J_z = J$ ground state with the maximal moment) the contribution to the electric-field gradient arising from the lattice can be evaluated. V_{zz}^{latt} is proportional to the crystal-field coefficient A_2^0 with an appropriate screening factor and η is equal to the ratio A_2^2/A_2^0. This method of estimating A_2^{m} works best for gadolinium since there is no contribution to the electric-field gradient from the $4f^7$ half-filled shell. Higher-order CEF coefficients cannot be determined directly from hyperfine interactions, though knowledge of the energy level scheme of the 4f ion is hidden in the temperature dependence of the term $< 3J_z^2 - J(J+1) >$ [Rad94].

Band-structure calculations, by Coehoorn *et al.* [Coe90] reveal, however, a very limited validity of the proportionality between V_{zz} and A_2^0 in intermetallic compounds. The reason is that V_{zz} is determined mainly by the asphericity of the 6p shell, whose density near the nucleus is much larger than that of the 4f shell, whereas A_2^0 is determined by the asphericity of both 6p and 5d shells. In conclusion, the electric-field gradient given by Mössbauer and NMR spectroscopy, being evaluated at the position of the nuclei, is not simply related to the second-order CEF coefficients describing the electrical potential at the position of the 4f orbitals.

5 Magnetothermal effects

5.1 *Specific heat*

The specific heat of 3d–4f intermetallics contains electronic, C_{el}, lattice (phonon), C_{ph}, magnetic, C_m, and nuclear, C_n, contributions:

$$C = C_{el} + C_{ph} + C_n + C_m \tag{2.21}$$

The electronic part, γT, is linear in temperature, with the Sommerfeld coefficient γ yielding information on the integrated density of states at the Fermi level $N(E_F)$. As the high-field susceptibility χ_{hf} is also related to $N(E_F)$, the ratio of these two values plays an important role in understanding their microscopic origin. The phonon term is approximated by the Debye function. At low temperatures, this is represented by βT^3, where β is proportional to the inverse cube of the Debye temperature, θ_D. In compounds where magnetic- or crystal-field contributions are expected to be important, a knowledge of γ and θ_D is essential in order to separate the different contributions. Isostructural compounds with nonmagnetic rare-earth elements serve as references. The use of the Y compound, even with corrections for the atomic mass difference, does not always lead to a successful evaluation of the phonon contribution. There is a large scatter in values for θ_D as derived by different techniques or by the same method from different temperature regions. A single parameter is insufficient to account for the dynamics of the lattice over an extended temperature region. The experimental values for the Sommerfeld coefficient, γ, have been compared with values obtained from band-structure calculations [Fra93]. Enhancement of the experimental values for YCo_2 and $LuCo_2$ with respect to the calculated values is a manifestation of the moment instability in these compounds which are on the borderline for the appearance of magnetism (Fig. 1.9), and reflects the presence of spin-fluctuation contributions.

Hyperfine interactions are manifest in specific-heat experiments by an upturn in the heat capacity at low temperatures. This upturn of C_n is extremely pronounced in the case of the magnetically ordered Ho compounds at temperatures below 5 K. For Ho_2Co_{17}, for instance, the heat capacity at 1.3 K is as much as 2.3 J/(K mol Ho) [Rad88d].

The magnetic contribution, C_m, in compounds with a magnetic rare-earth element, is associated with the increasing population of excited localized states. These localized states are due to CEF and molecular-field interactions of the 4f ion which lift the $(2J + 1)$-fold degeneracy of the ground-state multiplet. Having the energy-level scheme available, it is straightforward to evaluate this contribution. The inverse procedure is troublesome as other contributions have to be evaluated and subtracted accurately. Moreover, the CEF contributions at higher temperatures tend to be hidden by the phonon contributions, making the evaluation of C_m problematic. Specific-heat experiments are most informative in the case of a small energy separation between the ground state and the first excited state. Then, a Schottky-like contribution to the specific heat is detectable at low temperatures and the multiplicity of the levels and their energy separation can be derived. This energy separation is 144 K for $ErFe_2$ and 82 K for Ho_2Co_{17}, for example. The different

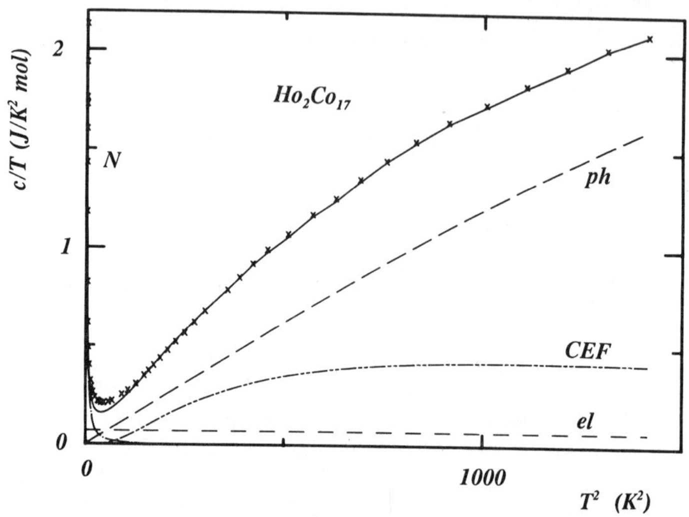

FIG. 2.19. Low-temperature specific heat for Ho_2Co_{17} separated into the electronic, phonon, nuclear and f-subsystem contributions [Rad88d].

contributions are illustrated in Fig. 2.19 for Ho_2Co_{17}. Experiments on monocrystalline samples in external magnetic fields can give further information as the applied field additionally shifts the energy levels depending on its direction with respect to the crystal axes.

5.2 *Magnetovolume effects and magnetostriction*

Magnetostriction effects are linear or volume strains in a crystal which are associated with the strength of the magnetic interactions, the amplitude of the magnetic moment or the direction of the magnetization vector. Volume magnetostriction appears when the exchange energy or the amplitude of the magnetic moments is sensitive to interatomic distances. It may exceed 1% in iron-based intermetallics, and can reach 5% in RMn_2 but the effect is generally much smaller for cobalt-rich compounds.

The spontaneous volume magnetostriction, ω_s, is estimated by comparing the temperature dependence of the volume of a magnetic intermetallic with that of a nonmagnetic analogue. In weak ferromagnets, an applied magnetic field can induce an extra forced volume magnetostriction associated with the magnetic susceptibility χ_{hf} of Eq.(2.1). The volume magnetostriction in RCo_2 compounds has been studied extensively. It turns out from the experiments that the volume follows the square of the cobalt moment induced by the rare-earth partner. This interpretation is in contrast with a discussion of magnetovolume effects in terms of a strain dependence of the exchange interactions which is frequently invoked for iron-based compounds. A strong dependence of the iron–iron exchange integral on the interatomic distance, which may eventually lead to a change in sign, is thought to be the origin of the strong pressure dependence of the Curie temperature Y_2Fe_{17} and $Y_2Fe_{14}B$ that is shown in Fig. 2.23 in the next section. The effects of interstitials on the value of T_C of the R_2T_{17} and $R(Fe, T)_{12}$ series is also usually discussed in terms of the lattice expansion.

The anisotropic magnetostriction is associated with the direction of the magnetization and is several orders of magnitude smaller than the spontaneous magnetostriction. Special attention has been devoted to the giant anisotropic magnetostriction in the Laves phase RFe_2 compounds reviewed by Clark [Cla79, 80]. The anisotropic magnetostriction leads to uniaxial distortion of the elementary cell of the crystal that can be studied by X-ray diffraction even on polycrystalline samples.

A single-ion model, related to rare-earth ions, has been employed to explain the sign and the magnitude of the anisotropic magnetostriction in rare-earth intermetallics. According to this model the anisotropic magnetostriction constants are represented by

$$\lambda = \Lambda \alpha_J < r_{4f}^2 > J_2 \tag{2.25}$$

Where Λ is a numerical factor that is expected to be much the same for all R ions within an isostructural series and α_J is the second-order Stevens factor. The last three factors are related to single-ion characteristics of the R ion involved. The anisotropic magnetostriction constant, therefore, should be proportional to $\alpha_J < r_{4f}^2 > J_2$, values of which have been tabulated in Table 1.4. Note that a similar dependence is expected for the first anisotropy coefficient of the rare-earth sublattice. The large value for the parameter λ_{111} of $TbFe_2$ ($4.4.10^{-3}$) follows these trends as the Tb^{3+} ion exhibits the most pronounced anisotropic single-ion properties of the heavy rare-earths. In the isostructural compound $GdFe_2$, λ_{100} is less than $5 \cdot 10^{-5}$. Most of the results on the magnetostriction of the 3d–4f intermetallics have been analysed within the single-ion model: for the RCo_5 series by Andreev *et al.* [And83] and for the $R_2Fe_{14}B$ series by Ibarra *et al.* [Iba86, Iba92] and del Moral [delM90].

B. 3D MAGNETISM IN $R_2T_{14}B$ AND RELATED COMPOUNDS

In general, alloying of 3d elements with rare-earth metals weakens the 3d magnetism. The magnetic ordering temperatures decrease significantly for Co and Ni compounds on increasing the R concentration; see Fig. 2.20. The anomalously low values of T_C for the iron-rich compounds are reflected in the strong volume dependence of the Fe moment. The 3d moment decreases gradually with increasing yttrium concentration, as is shown in Fig. 2.21, but the presence of a magnetic rare-earth partner reinforces the 3d magnetism. T_C and the 3d moment are always higher in the Gd compound as compared with the Y-compound with the same rare-earth concentration. The presence of Gd transforms the Pauli-like paramagnet YCo_2 into a ferromagnet. Band-structure calculations performed for Y–Fe, Y–Co

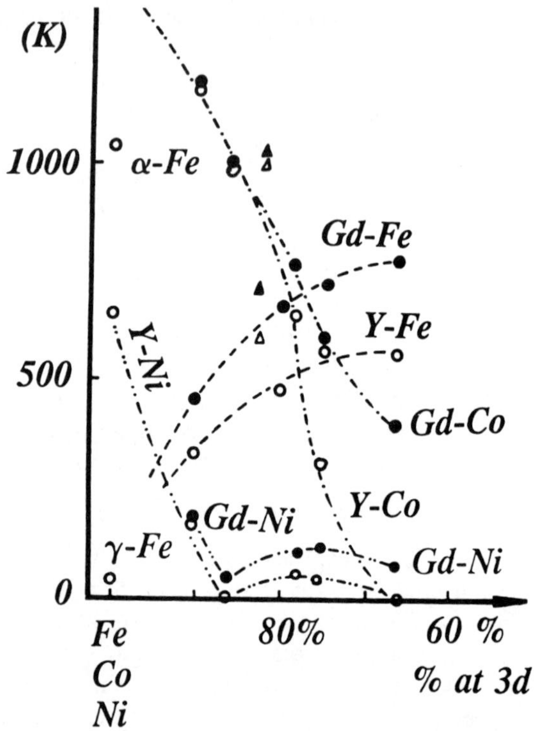

FIG. 2.20. Curie temperature, T_C, of $Y_2T_{14}B$ (△, T = Fe, Co) and of $Gd_2T_{14}B$ (▲, T = Fe, Co), together with the magnetic ordering temperature of the Y–Fe series (Y_2Fe_{17}, Y_6Fe_{23}, YFe_3, YFe_2), the Gd–Fe series, the Y–Co series (Y_2Co_{17}, YCo_5, Y_2Co_7, YCo_3, YCo_2) and the Gd–Co series as well as the Y–Ni series (Y_2Ni_{17}, YNi_5, YNi_3, YNi_2) and the Gd–Ni series as a function of the 3d transition-metal atomic concentration. In addition, the magnetic ordering temperatures of α–Fe, γ–Fe, hcp–Co and fcc–Ni are shown.

and Y–Ni compounds [Coe89, Daa91, Fäh93] reproduce the basic experimental features, for example the magnetic moment (dominated by the spin part) and its concentration dependence, the general shape of the density-of-states curves, and so on. There is significant hybridization of the 4d and 5s conduction electrons of Y with the 3d electrons of the T elements. This hybridization becomes more effective for smaller concentrations of the 3d metals. Band-structure calculations reveal that the small moment on the Y site, coupled antiparallel to the overall 3d moment, increases in value with increasing Y concentration from $-0.20\ \mu_B$ in Y_2Fe_{17} to $-0.47\ \mu_B$ in YFe_2.

In performing band-structure calculations, the mutual interatomic 3d interactions and intra-atomic correlations have to be considered. When these calculations are performed within a molecular-field approximation, a very large n_{TT} coefficient

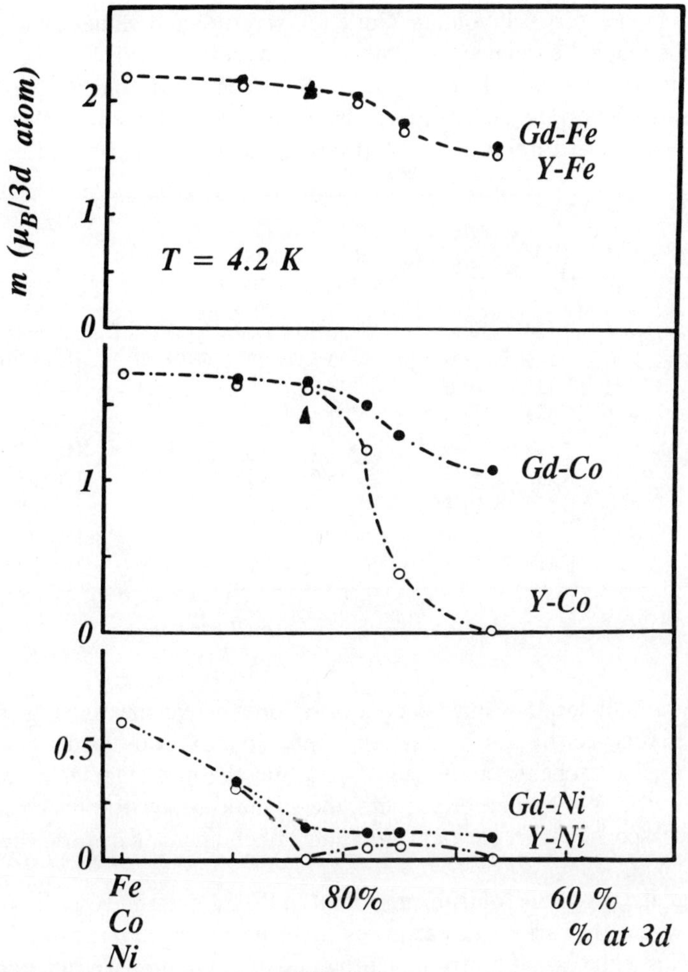

FIG. 2.21. The 3d magnetic moment as a function of the atomic concentration of the 3d element in alloys with Y or Gd: ▲ marks $Y_2T_{14}B$.

is obtained. The value of n_{TT} for Y_2Fe_{17} is 5940 T/μ_B. The 3d-band splitting I of 0.925 eV due to Fe–Fe interactions is equivalent to $n_{TT} = 8000\ T/\mu_B$ [Rad93]. Such values for n_{TT}, however, lead to T_C-values that greatly exceed the experimentally observed ones. This discrepancy reveals the shortcomings of the mean-field approach to 3d magnetism. On the other hand, starting from the experimental values for T_C and accepting the mean-field relations to be applicable to 3d magnetism (with $g = 2$ and $J = S$), values for n_{TT} and for the molecular field B^{mol} can be derived. In the case of the Y–Co compounds values for B^{mol} up to 1450 T (in Y_2Co_{17}) and n_{TT} of 900 T/μ_B are found. In compounds with cobalt, B^{mol}

Table 2.3 Elementary cell volume, Curie temperature and values at 4.2 K for the spontaneous magnetic moment, m_s, mean ^{57}Fe hyperfine field, B_{hf}, the anisotropy field, B_a, and the first anisotropy constant, K_1, of $R_2Fe_{14}B$ and $R_2Co_{14}B$ compounds with non-magnetic rare-earths; values for metallic iron (bcc structure) and cobalt (hcp structure) refer to 14 atoms to facilitate the comparison

Compound	V (Å^3)	T_C (K)	m_s (μ_B/f.u.)	B_{hf} (T)	B_a (T)	K_1 (MJ/m^3)	Ref.
$Y_2Fe_{14}B$	923	566	30.6	–33.9	1.5	0.8	[Yam86, Bol87b]
$La_2Fe_{14}B$	966	543	31.0	–33.5	1.3	0.7	[Bol87b]
$Ce_2Fe_{14}B$	928	427	30.0	–31.7	3.0	1.8	[Bol87b]
$Lu_2Fe_{14}B$	901	534	28.5	–31.0	2.4	1.3	[Bus85, Bol87b]
$Th_2Fe_{14}B$	926	479	29.8	–31.7	2.3	1.6	[Bol87b]
Fe	–	1043	31.0	–33.0	0.05	0.04	[Woh80]
$Y_2Co_{14}B$	866	1015	19.4	–	–	–1.4	[Bus85, Bol87a]
$La_2Co_{14}B$	903	955	20.4	–	–	–1.4	[Bus85, Yam86,87]
Co	–	1413	23.9	–	1.1	0.8	[Woh80]

decreases gradually on lowering Co concentration whereas the exchange parameter J_{Co-Co} remains constant. A different behaviour is observed for the iron compounds. The exchange parameter J_{Fe-Fe} and the molecular field are smallest for Y_2Fe_{17}, the Y–Fe compound with the highest concentration of iron. This anomaly is associated with the magnetovolume phenomena in iron-rich compounds [Rad86].

The magnetic properties of iron and cobalt in R–T compounds are best examined for alloys where the rare-earth partner is trivalent and does not carry a magnetic moment. This is the case for yttrium, lanthanum and lutetium. Cerium and thorium atoms have no magnetic moments in their tetravalent state, but the magnetic properties of Ce–T or Th–T compounds are not representative for the remaining R–T series because of the extra electron per rare-earth atom that is introduced into the d-band. Some basic magnetic properties of iron and cobalt compounds with a nonmagnetic rare-earth partner are collected in Table 2.3. Yttrium is considered to be the most representative nonmagnetic rare-earth on account of its size, which is similar to those of elements in the middle of the 4f series.

1 Magnetic moment of the 3d sublattice with non-magnetic rare-earth elements

The magnetic moment of the iron sublattice at 4.2 K is close to 30 μ_B/f.u. in all $R_2T_{14}B$ compounds. It results in a polarization for $Y_2Fe_{14}B$ of 1.44 T or a magnetization of 1.15 MA/m. The iron moment in $Y_2Fe_{14}B$ of 2.2 μ_B is in agreement with the systematics observed for Y–Fe intermetallics of different

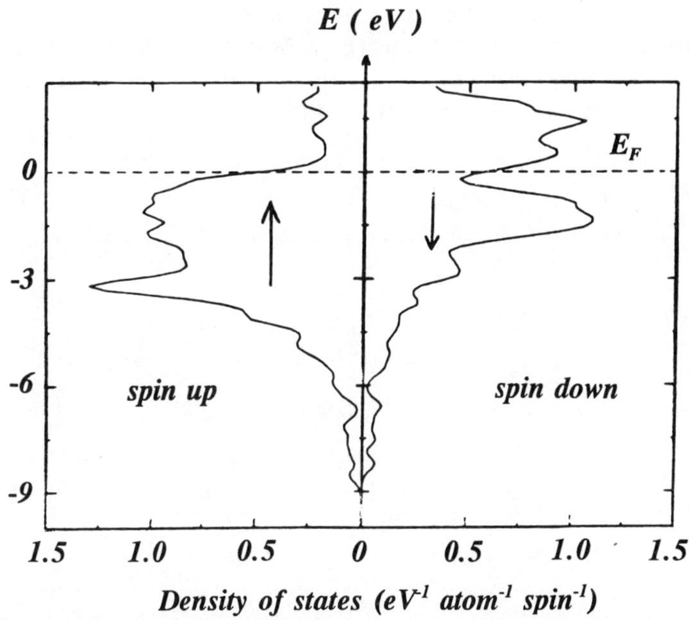

FIG. 2.22. Calculated spin-polarized density of states in $Y_2Fe_{14}B$, showing an exchange splitting of spin-up and spin-down 3d sub-bands of about 2eV [Sel88].

compositions; see Fig. 2.21. As the yttrium concentration increases, the average magnetic valence falls, resulting in lower values for the moment per atom The average iron moment of 2.2 μ_B in $Y_2Fe_{14}B$ is very close to that of metallic iron. This value must be regarded as an average over the six ion sites but one that is typical for the $R_2Fe_{14}B$ series.

Self-consistent spin-polarized band-structure calculations [Sel88, Coe89, Nor91, Fäh93] confirm that the iron-rich intermetallics, like iron itself, are ferromagnets with an almost full $3d^{\uparrow}$ subband. The total density of states for $Y_2Fe_{14}B$ is illustrated in Fig. 2.22. The exchange splitting between the spin-up and spin-down 3d subbands amounts to about 2 eV ($\sim$ 23 000 K). There is much evidence from band calculations, Mössbauer spectroscopy and neutron-diffraction measurements that the value of the iron moment varies significantly from one crystallographic site to another, depending on the local atomic environment. For example, the hyperfine fields experienced by the ^{57}Fe nuclei in $Y_2Fe_{14}B$ at 4.2 K have an average value which is close to 33.5 T observed for bcc iron but vary from one site to another between 30.3 T and 36.9 T [Fru87, Lon91]. In Table 2.4, a comparison is made between values for the iron magnetic moments deduced from Mössbauer measurements, assuming proportionality between the values of the hyperfine field and the magnetic moment, polarized-neutron experiments and self-consistent electronic structure calculations. All the data confirm a considerable variation of the moment from site to site. According to the neutron experiments, the largest iron

Table 2.4 Iron magnetic moment (μ_B) in $Y_2Fe_{14}B$ at 4.2 K at the different crystallographic sites, as deduced from ^{57}Fe Mössbauer spectroscopy, neutron scattering and electronic-structure calculations and the calculated induced moment on Y and B. The crystal structure is shown in Fig. 1.18.

Site	Mössbauer [Fru87]	Neutron [Giv85]	Band-structure calculations [Sel88]	[Coe89]	[Nor91]
$16k_1$	2.10	2.25	2.14	2.11	2.25
$16k_2$	2.23	2.25	2.24	2.31	2.35
$8j_1$	2.31	2.40	2.10	2.22	2.10
$8j_2$	2.40	2.80	2.83	2.51	2.86
4c	1.90	1.95	2.55	2.40	2.45
4e	2.30	2.15	2.20	2.01	2.15
B	–	–	–	–0.15	–
Y	–	–	–	–0.40	–

moment in $Y_2Fe_{14}B$, 2.80 μ_B, resides on the $8j_2$ site (Fig 1.18). This site has the largest number of iron neighbours (12) and there are no yttrium neighbours. The smallest iron moment, 1.95 μ_B, is found on the 4c site which has only eight iron neighbours and four yttrium neighbours. A further example is provided by $Y_2Fe_{17}N_3$. Here, the values for the moments range from 2.6 μ_B on 4f and 6g sites to 2.1 μ_B on 12j sites. These values depend sensitively on the iron and rare-earth coordination numbers and the bond lengths.

The influence of the rare-earth neighbours on the 3d magnetic moment is through hybridization of 3d and 5d/6s states, which leads to a broadening of the 3d band and a little charge transfer. The splitting of the 3d subbands also diminishes in the case of a reduced number of interacting magnetic partners. This destructive role of the rare-earth neighbours on the 3d moments is common to all R–T compounds, both crystalline and amorphous. There is also a small and negative induced moment of 4d character at the yttrium site ($\approx -0.4\ \mu_B$) in the 2 : 14 : 1 compound [Coe89]. A similar effect of 5d character for the 4f elements is responsible for the R–T exchange coupling. Studies of the $R_2Fe_{14}B$ and R_2Fe_{17} compounds with nonmagnetic rare-earth elements reveal that there is a tendency for the iron sublattice magnetization and for the average hyperfine field at 4.2 K to decrease going from La to Lu. Superimposed on this trend, there will be a small effect of the exchange field from the magnetic rare-earth partner on the iron moment and hyperfine field which is greatest for R = Gd [Fru87].

As a consequence of the relatively low Curie temperatures of the iron-rich intermetallics, the iron moment at room temperature is noticeably smaller than at 4.2 K, and there is a significant temperature variation of the magnetization. The iron moment of $Y_2Fe_{14}B$ decreases by 15% between 4.2 K and room temperature and the temperature dependence of the iron-sublattice magnetization up to room temperature is well approximated by

$$M_{Fe}(T) = M_{Fe}(0)\{1 - a(T/T_C)^{3/2}\} \quad (2.25)$$

where $a = 0.43$ [Gav88].

The average cobalt moment in $Y_2Co_{14}B$ derived from magnetization measurements is 1.40 μ_B at helium temperature. It is considerably lower than the value of 1.71 μ_B observed in metallic cobalt. The dilution of Co atoms in 3d–4f systems leads to a more rapid decrease of the cobalt magnetic moment than is the case for iron. Nuclear magnetic resonance studies on ^{59}Co nuclei, as well as first-principles calculations [Gu87], again indicate large variations in the values of the cobalt moment over the different sites in the compound.

2 Curie temperature

The value of the Curie temperature, T_C, for $Y_2Fe_{14}B$ (566 K) is only half that of metallic iron (1043 K). A lower value (358 K) is found for Y_2Fe_{17}, the Y–Fe compound with the highest concentration of iron. In the molecular-field approximation, the Curie temperature is related to J_{TT}, the exchange–interaction parameter between iron spins, by Eq.(1.13). Despite the difficulty of applying a localized moment description to metallic systems, the values of J_{TT} describe the strength of the 3d–3d interactions when an effective spin S_T is associated with the 3d atom. Values of J_{TT}/k_B in Y–Fe alloys decrease sharply with increasing iron content from 103 K for YFe_2 to 23 K for Y_2Fe_{17}, and these extrapolate to a low and negative value for γ-Fe [Rad85]. $Y_2Fe_{14}B$ has an intermediate iron content and the value for J_{TT}/k_B is 39 K. The corresponding values for the molecular field, B_{Fe}^{mol}, at zero temperature are 691 T, 396 T and 591 T respectively. In compounds with cobalt, T_C decreases gradually with decreasing cobalt concentration. The calculated molecular field, B_{Co}^{mol}, decreases smoothly from 1700 T for metallic cobalt to 565 T for YCo_3. For $Y_2Co_{14}B$ it amounts to 1330 T. Hence, the variation in T_C and B_{Co}^{mol} is entirely due to the variation in cobalt moment. The exchange parameter J_{TT} remains virtually unchanged at 115 K for all the cobalt compounds [Rad86].

The variations of T_C as a function of iron or cobalt content are, therefore, quite different and the difference stems from the strong dependence of J_{Fe-Fe} on the local environment and the lattice volume, which dependence is absent for J_{Co-Co}. Most striking is the comparison of antiferromagnetism of face-centred γ–Fe with $Z = 12$ and ferromagnetism of body-centred α–Fe with $Z = 8$. With iron we have to look for the best compromise between a large moment and a high Curie temperature, which is achieved with about 80% iron in a dense-packed alloy, whereas, for cobalt, the moment and Curie temperature increase hand in hand.

The great sensitivity of the Curie temperature to the lattice volume in iron-rich intermetallics is well known from studies of Y_2Fe_{17}. Interstitial hydrogen, carbon or nitrogen can be used to expand the unit cell by up to 7%. $Y_2Fe_{17}N_3$, for example, has a value for T_C of about 695 K, twice as large as its parent Y_2Fe_{17}. A

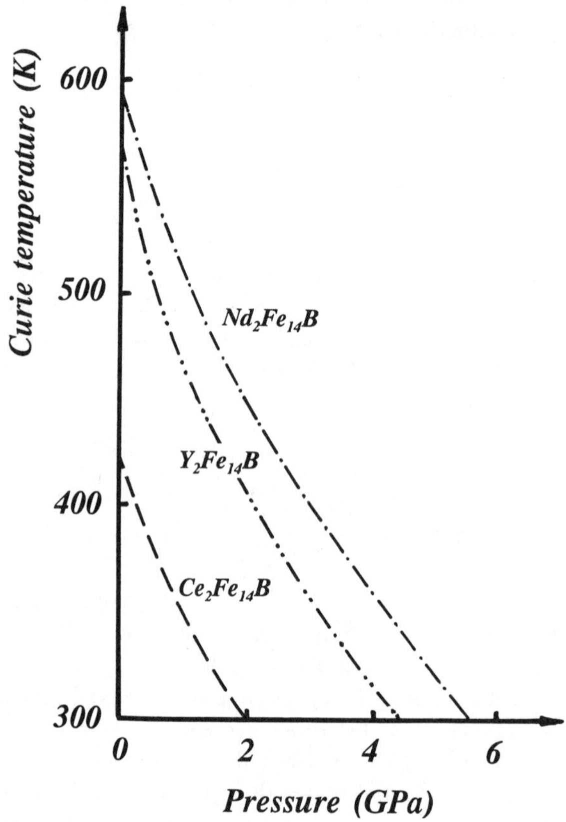

FIG. 2.23. Pressure dependence of the Curie temperature of $R_2Fe_{14}B$ with R = Y, Nd and Ce [Nag87].

strong volume dependence of T_C is also observed for $Y_2Fe_{14}B$. Applying an external pressure of 4.2 GPa reduces T_C of $Y_2Fe_{14}B$ from 566 K to 316 K, that is, by nearly 50% [Nag87]. It results in a value of $-0.13\ \mathrm{GPa}^{-1}$ for $\partial \ln T_C/\partial p$. With a value of $0.8\ \mathrm{Mbar}^{-1}$ for the compressibility, one finds a value of -16 for $\partial \ln T_C/\partial \ln V$. The pressure results for $Y_2Fe_{14}B$ are presented in Fig. 2.23, together with data for $Nd_2Fe_{14}B$ and $Ce_2Fe_{14}B$. Hydrogenation of $Y_2Fe_{14}B$ causes an increase of T_C. The additional volume per hydrogen atom in $Y_2Fe_{14}BH_x$ is about $1.8{\cdot}10^{-3}\ \mathrm{nm}^3$. For $Y_2Fe_{14}BH_{3.4}$, the Curie temperature is 60 K greater than that for $Y_2Fe_{14}B$. The variation of Curie temperature with hydrogen expansion of the $Y_2Fe_{14}B$ lattice is again consistent with the contraction under pressure. The changes in the Curie temperature for the hydrogenation process lead to a value for $\partial \ln T_C/\partial \ln V$ of -14, close to that derived from the pressure dependence of T_C. The variations of T_C for the $R_2Fe_{14}B$ compounds have been attributed to changes in the average iron moment, rather than to a change in the exchange interaction J_{TT} on the basis of the pressure dependence of the magnetic moment of $Y_2Fe_{14}B$ [Gav88,

Coe86]. The iron-sublattice moment is indeed largely reduced with pressure. For $Y_2Fe_{14}B$ $\partial \ln m_{Fe}/\partial p$ amounts to $-16 \cdot 10^{-3}$ GPa^{-1} [Gav88].

The abnormal sensitivity of the exchange energy to the local environment is manifest in the magnetovolume effect that is a common feature of compounds with a high iron concentration. Within a fixed localized-moment picture, the reduction of the Curie temperatures in iron-rich compounds and increase with lattice expansion have been attributed to a strong dependence of the parameter J_{TT} on interatomic distances. In dense-packed structures, a distance shorter than about 250 pm may give rise to antiferromagnetic interactions and thereby weaken the ferromagnetic coupling between the iron spins. The exchange energy actually contributes to stabilize the bcc structure of α–Fe. Within the band model this reduction is thought to be due to a strong volume dependence of the density of states. In spin-fluctuation theory, the Curie temperature is related to the spin-up and spin-down density of states at the Fermi level. This theory, however, is unable to reproduce the values of T_C quantitatively in ferromagnets with large moments [Qi94].

3 Magnetovolume effect

The magnetovolume effect is apparent in the thermal expansion — the temperature dependence of the lattice parameters shows a significant upturn below T_C [Gav84, And85, Bus86, Fuj87]. By extrapolating the thermal expansion data of the paramagnetic region into the magnetically ordered state, the spontaneous volume magnetostriction, ω_s, can be evaluated. It amounts to 2.4% at 0 K and varies with temperature approximately as the square of the magnetization [Fuj87]. Effects of similar magnitude are observed for other $R_2Fe_{14}B$ and R_2Fe_{17} compounds and even for amorphous iron-rich alloys, confirming that the magnetovolume effect is a characteristic feature of the iron sublattice. By following ω_s throughout the $R_2Fe_{14}B$ series the contributions originating from the iron sublattice and the rare-earth sublattice have been separated. The latter effect, being roughly proportional to the rare-earth moment, does not exceed one-quarter of that originating from the iron sublattice. It is interesting to note that the values for the spontaneous volume magnetostriction in $R_2Fe_{14}B$ are among the highest observed for any ferromagnetic material. The giant volume anomaly, unlike the invar effect in cubic alloys, is anisotropic and is mainly due to the variation in the a-parameter, along which direction the distance between neighbouring iron atoms is shortest.

4 Magnetocrystalline anisotropy of the 3d sublattice

Analysis of the magnetization curves of a ferromagnetic monocrystalline sample is generally carried out in terms of a phenomenological model where the total energy per unit volume of the system with magnetization M_s in an external field is written as the sum of the anisotropy energy, E_a, and the magnetostatic energy $\mathbf{M}_s \cdot \mathbf{B}_0$, where $\mathbf{B}_0$ is the applied field corrected for the demagnetizing effect (Eq. (2.3)). The anisotropy energy, E_a, is expressed in terms of the anisotropy constants K_i. In general, $E_a = K_1 \sin^2\theta + K_2 \sin^4\theta$ suffices to give a proper description of the 3d anisotropy. The magnetization curves of $Y_2Fe_{14}B$ (Fig. 2.3) are typical for a ferromagnet. For fields directed along the *c*-axis, which is the easy axis of magnetization, the curve is flat. For fields applied along a perpendicular hard direction, the magnetization curve is nearly linear up to the anisotropy field, indicating that the first anisotropy constant is dominant. This corresponds to a positive value of K_1 with a value of 0.8 MJ/m^3 at 4.2 K. A field of 3 T is sufficient to rotate the magnetization completely from its preferred direction into the hard direction. Above this field both curves almost coincide; a small anisotropy of the magnetic moment of the order of 2% is indicated. Values for the anisotropy constants as well as for the anisotropy field $B_a(= 2K_1/M_s)$ are included in Table 2.3 for the iron and cobalt compounds with nonmagnetic rare-earth elements at 4.2 K. The anisotropy energy in $Y_2Fe_{14}B$ is more than an order of magnitude larger than that of metallic iron which is cubic, whereas that of $Y_2Co_{14}B$ is comparable in magnitude and opposite in sign to that of metallic cobalt, which is hexagonal. Substantial 3d anisotropies also occur for YCo_5, Y_2Fe_{17} and Y_2Co_{17} compounds. The large 3d anisotropy in uniaxial intermetallics must be due to the orbital contribution to the iron moment, of order 0.1 μ_B, at the low-symmetry transition-metal sites. Spin–orbit coupling with this partially unquenched orbital moment is mainly responsible for the magnetocrystalline anisotropy of the iron and cobalt sublattices. For $Y_2Co_{14}B$, the value of K_1 is negative and the anisotropy has easy-plane character, contrary to the easy-axis character of $Y_2Fe_{14}B$. An examination of the anisotropy of Y_nCo_m and Y_nFe_m intermetallics shows that the leading term in the anisotropy energy is in most cases opposite in sign for isostructural iron and cobalt compounds. Thuy *et al.* [Thu88] have related this systematic behaviour to opposite signs of the second-order Stevens factor for iron and cobalt ions.

The temperature dependence of the iron-sublattice anisotropy in $Y_2Fe_{14}B$ is unusual. It increases with increasing temperature, reaching a maximum in the vicinity of room temperature. This phenomenon may be associated with the anomalous temperature variation of the ratio of the lattice parameters c/a. An alternative interpretation is that the contributions arising from the six different iron sites could have different temperature dependencies, the bulk anisotropy being the resultant of contributions of opposite sign that largely cancel.

Magnetism of the 3d atoms is best discussed in terms of band magnetism, but band-structure calculations for the magnetocrystalline anisotropy of the 3d atoms are difficult to perform. The magnetocrystalline anisotropy is a property of fundamental importance for applications but from the energy-calculation point of view it is a tiny effect, of order 1 K/atom. Results for YCo_5 obtained by means of relativistic linear muffin-tin band-structure calculations [Coe89] show that the MCA energy depends on the band structure in a very subtle way. For a reasonable band-filling value of 48 electrons, three of Y and nine of Co, the correct easy direction of the Co moment, as well as the proper MCA energy value (within 20%), is obtained for YCo_5. These calculations predict a MCA energy of the 3d sublattice that is 2.5 times larger than that observed in YCo_5, a compound with $CaCu_5$ structure but with a lower, by 3%, electron concentration. This lower electron

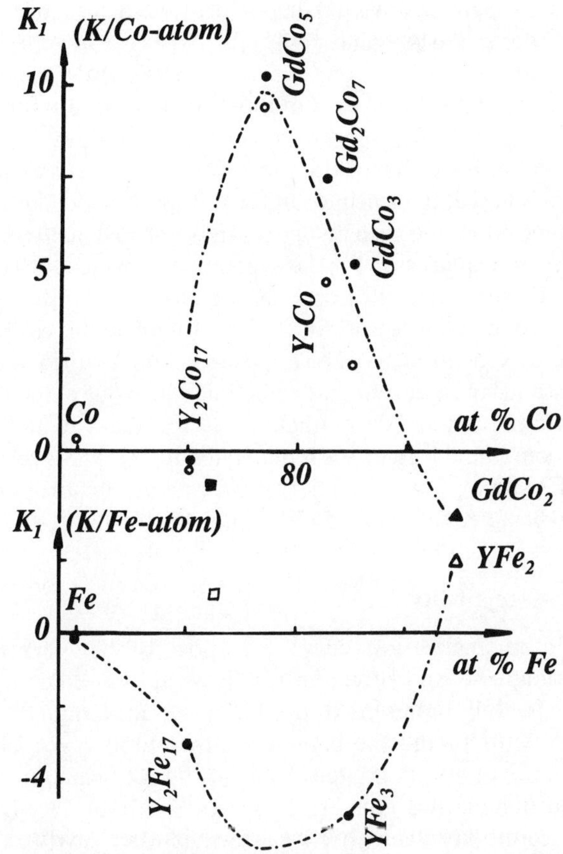

FIG. 2.24. Magnetocrystalline anisotropy constant K_1 of tetragonal $Y_2Fe_{14}B$ (□) and $Y_2Co_{14}B$ (■) compounds, together with the variation of K_1 for the 3d sublattice in the hexagonal Y–Co (○) Gd–Co (●) and Y–Fe (●) compounds. Positive values point to a uniaxial anisotropy [Rad93].

concentration can be realized by 25–30% Fe substitutions, for instance. Here, it is worth remembering that YFe_5 does not form, but partial Fe substitution in YCo_5 is possible.

From an experimental point of view, the MCA energy of the 3d sublattice is deduced from compounds with R ions having $L = 0$, that is, Y, La, Lu, Gd. For compounds with Gd, the intersublattice coupling has to be taken into account in order to evaluate the Co/Fe sublattice anisotropy. In most compounds, the saturation along the hard axis is reached at rather low field values. The outstanding value of 17 T that is observed for YCo_5 at 4.2 K for the saturation field corresponds to a value for the MCA energy of 7.3 MJm^{-3} (47 K/f.u.). A phenomenological explanation of the anisotropy energy in YCo_5 has been given in terms of the ratio of the lattice parameters c/a using the single-ion model, where the value for the second-order CEF parameter for the hcp structure is determined by the deviation of the c/a ratio from the ideal value.

The individual-site contributions have been evaluated from the compositional dependence of the MCA energy in a number of isostructural series $Y_n(Co_xFe_{1-x})_m$ [Thu86]. Contributions to the anisotropy energy of opposite sign are obtained for Fe and Co ions in all different 3d sites. The individual-site-anisotropy [Thu86, Thu88] picture is based, in principle, on the localized model developed for the rare-earth anisotropy. Although such an approach is not well justified for 3d magnetism, it offers an elegant explanation for the experimental evidence that the MCA energy of the Co and Fe sublattices is of opposite sign for most of the series, see Fig. 2.24. (The unexpected planar anisotropy in Y_2Co_{17} is weak and is easily transformed into a uniaxial one by substitutions.) The opposite values for the Co and Fe anisotropy can be understood by accepting that some ingredients of a $3d^7$ ionic configuration for the Fe atoms and a $3d^8$ configuration for the Co atoms are present in intermetallic compounds. In these ionic states the relevant second-order Stevens factor of the Fe and Co ions, characterizing the charge quadrupolar moment, are of opposite sign, being equal, $-2/105$ ($3d^7$) and $+2/105$ ($3d^8$), respectively.

5 High-field susceptibility

The high-field magnetization curve for most of the ferromagnetic $R_2T_{14}B$ compounds along the easy direction for fields above the demagnetizing field is quite flat. It is well represented by Eq. (2.1) in terms of the spontaneous magnetization $M_s(0,T)$ and the high-field susceptibility χ_{hf}. The value of χ_{hf} in $Y_2Fe_{14}B$ amounts to approximately $10^{-2}\mu_B/(T\ f.u.)$, which means that in a field of 30 T the iron sublattice moment increases by about 1%. Detailed studies of high-field susceptibility have been performed for $Gd_2Fe_{14}C$ and $Lu_2Fe_{14}C$ [Rad89a].

6 3d magnetic moment in intermetallics with magnetic rare-earth elements

The 3d sublattice magnetization in compounds with magnetic rare-earth elements experiences an extra molecular field arising from the surrounding 4f spins. The presence of gadolinium moments, for instance, increases the iron-sublattice magnetic moment in $Gd_2Fe_{14}B$ to 32.0 μ_B/f.u. compared with 30.6 μ_B in $Y_2Fe_{14}B$. This effect can be analysed within the free-energy expression of the 3d sublattice written in a phenomenological way as

$$E = E_a - \mathbf{M}_T \cdot (n_{RT}\gamma_R\mathbf{M}_R + \mathbf{B}_0) \qquad (2.26)$$

where an extra molecular field $\mathbf{B}_T^{RT}(= n_{RT}\gamma_R\mathbf{M}_R)$ is incorporated in the Zeeman term in Eq.(2.2). In the case of $Gd_2Fe_{14}B$, the intersublattice exchange is equivalent to an extra field B^{RT} of 140 T acting on the Fe moment. Applying Eq. (2.1) with a value for χ_{hf} of $9 \cdot 10^{-3}$ μ_B/f.u., an additional moment of 1.25 μ_B/f.u. is calculated that largely explains the difference between the Gd and Y compounds. The remainder could be due to a polarization of the 5d electrons of Gd. The increase in the 3d magnetization is expected to be most pronounced in gadolinium compounds as the gadolinium ion carries the largest spin. Anyhow, the presence of magnetic rare-earth elements does not influence the magnetization of the 3d sublattice in the R–T intermetallics by more than a few percent.

From band-structure calculations, substituting Gd for Y, Coehoorn [Coe92] concluded that the Fe local moments depend only slightly on the R atoms. He suggested that large differences between moments on corresponding Fe sites in the Y and Nd compounds, revealed in neutron-diffraction experiments, indicate a systematic error in the analysis. Also, the comparison of the calculated Fe magnetic moments in $Y_2Fe_{14}B$ [Sel88] and in $Nd_2Fe_{14}B$ [Nor91] does not allow conclusions to be drawn, as the effect discussed is much smaller than the accuracy of the calculations. In summary, the average effect of a magnetic rare-earth on the Fe moment is less than 0.1 μ_B, and it is comparable in magnitude to the effect of a volume change of the Fe site on substitution.

C. $R_2Fe_{14}B$ COMPOUNDS WITH MAGNETIC RARE-EARTH ELEMENTS

1 $R_2Fe_{14}B$ compounds

Values for the magnetic parameters of the $R_2Fe_{14}B$ and $R_2Co_{14}B$ series are given in Table 2.5. The spontaneous magnetic moments point to ferromagnetic ($R < Gd$) or ferrimagnetic ($R \geq Gd$) order of the iron and rare-earth moments, with values for the R moments close to the free-ion values $g_R J \mu_B$. The temperature-dependence of the spontaneous magnetization of crystals of $R_2Fe_{14}B$ for light and heavy rare-earths is shown in Fig. 2.25 [Hir86].

We begin the analysis of compounds with a magnetic rare-earth element with gadolinium where the rare-earth anisotropy is expected to be negligible due to the S-state of the Gd^{3+} ion. The spontaneous magnetic moment of $Gd_2Fe_{14}B$ is much lower than that of the yttrium counterpart, due to the antiparallel alignment of the gadolinium and the transition-metal moments. Taking the full trivalent moment ($g_R J \mu_B = 7 \mu_B$) for gadolinium, one obtains a value for m_T of 32 μ_B in $Gd_2Fe_{14}B$ that is quite close to that of $Y_2Fe_{14}B$ ($m_T = 30.6\ \mu_B$). The magnetization curves of $Gd_2Fe_{14}B$ are simple, see Fig. 2.25. The hard-axis curve starts linearly in the field, reaching the value of spontaneous magnetization at 1.7 T, leading to a value of 0.58 MJ/m^3 (10.1 K/f.u.) for the anisotropy constant K_1. The higher-order anisotropy constants have been found by magnetic torque measurements to be two orders of magnitude smaller. The reduction in the anisotropy energy of $Gd_2Fe_{14}B$ compared to that observed for $Y_2Fe_{14}B$ (0.8 MJ/m^3 or 13.9 K/f.u.) may be attributed to dipole–dipole interactions or to band effects connected with the 5d electrons.

The effects of the strong magnetic anisotropy of the rare-earth sublattice are illustrated by the magnetization curves of $Nd_2Fe_{14}B$ (Fig. 2.4). The low and high-field behaviour, as measured along the three main crystallographic directions in fields up to 35 T, is quite complex. It is clear that a field as large as 35 T is not sufficient to saturate the magnetization along the < 110 > and < 001 > directions.

Surprisingly, saturation is reached along the < 100 > direction, which is the hard axis at low fields, after an abrupt transition at 16 T. At low fields, the easy direction of magnetization does not lie along any of the main crystallographic directions as nonzero values of the spontaneous magnetization are observed for all three main crystallographic directions. Free-sample measurements reveal the full value of the spontaneous magnetization. It is inferred from the magnetization values at zero field that the magnetization of $Nd_2Fe_{14}B$ is titled from the tetragonal c-axis by 30°. The tilt angle, θ_s, reduces with increasing temperature, falling to zero at the spin-tilt temperature $T_{st} = 135$ K.

As an example of a ferrimagnetic compound with strong anisotropy, the magnetization curves of $Dy_2Fe_{14}B$ are presented in Fig. 2.26. The spontaneous magnetic moment along the easy axis (c-axis) results in a value of 11.40 μ_B/f.u. (57.0 Am^2/kg), with a negligibly small value for the high-field differential susceptibility. For the field applied along the hard axis the magnetic isotherm is rather linear in field up to 10 T, resulting in an effective anisotropy field, B_a, of 23 T. The slope provides a value for the effective anisotropy constant K_1^{eff} of 5.3 MJ/m^3. A significant feature of ferrimagnetic system at relatively high external fields is the crossing of easy- and hard-axis magnetization curves. This crossing is often misleadingly identified as the anisotropy field. In fact, this crossing, which determines the effective anisotropy field and the effective anisotropy constants is not governed so much by the intrinsic anisotropy [Rad87] as by the canting of the ferrimagnetic moment configuration which is controlled by the intersublattice coupling parameter $n_{RT}\gamma_R$. (See Section A.2.1)

1.1 *Phenomenological approach: one-sublattice model*

As a first approach to the analysis of the magnetization curves of the $R_2T_{14}B$ compounds, a phenomenological model is used for which the total energy is given by Eq.(2.3), with the anisotropy energy E_a written in the form appropriate for the tetragonal symmetry:

$$E_a = K_1 \sin^2\theta + (K_2 + K_2' \cos 4\varphi)\sin^4\theta + (K_3 + K_3' \cos 4\varphi)\sin^6\theta \tag{2.27}$$

In this approach, the value of M_s is considered to be constant and independent of the value of the applied magnetic field. The anisotropy constants for $Nd_2Fe_{14}B$ and other compounds at 4.2 K, deduced by fitting the magnetization curves within this procedure, are listed in Table 2.5. More detailed studies reveal, however, that there is a strong dependence of the value for the anisotropy constants on the field interval in which the magnetization data are analysed (Fig. 2.27). This effect is the reason for a large scatter in the experimentally evaluated anisotropy constants. The phenomenological anisotropy analysis does not take into account the field-induced changes in the complex magnetic structures of the 3d–4f hard magnetic materials. The temperature variation of the effective anisotropy constants, K_i, for $Nd_2Fe_{14}B$, obtained from an analysis of magnetization data in fields up to 18 T, is presented in Fig. 2.27. K_2 and K_3 decrease monotonically with increasing temperature, while K_1

Table 2.5 Values for the Curie temperature, T_C (K), the magnetic moment, m_s (μ_B), the magnetic polarization, $\mu_0 M_s$(T), the anisotropy field, B_a (T), the anisotropy constants, K_i (MJ/m^3), and the energy difference between easy and hard directions, ΔE_a (MJ/m^3) for $R_2Fe_{14}B$ and $R_2Co_{14}B$ compounds at 4.2 K; the easy magnetization direction (EMD, # indicates a tilted easy direction) is indicated, as well as the sign of the Stevens factor α_J

Compound	T_C	m_s	$\mu_0 M_s$	B_a	K_1	K_2^*	K_3^*	ΔE_a	EMD	α_J	Ref
$Y_2Fe_{14}B$	566	30.6	1.55	1.5	+0.8	–	–	+0.8	$\parallel c$	0	a,b
$La_2Fe_{14}B$	543	31.0	1.50	1.3	+0.7	–	–	+0.7	$\parallel c$	0	a
$Ce_2Fe_{14}B$	427	30.0	1.52	3.0	+1.8	–	–	+1.8	$\parallel c$	0	a
$Pr_2Fe_{14}B$	565	37.6	1.76	29	+24	–	–	+24.0	$\parallel c$	–	a
					+55	–87	+62	+30.0			c
$Nd_2Fe_{14}B$	585	37.7	1.86	35#	–16.0	+28.0	+0.5	+12.5	$\parallel c^{\#}$	–	a
					–16.6	+39.3	–13.0	+9.7			b
					–14.7	+32.9	–6.5	+11.7			c
$Sm_2Fe_{14}B$	621	33.3	1.57	40	–26.0	–	–	–26.0	$\perp c$	+	d
$Gd_2Fe_{14}B$	664	18.0	0.90	1.7	+0.5	–	–	+0.5	$\parallel c$	0	e
$Tb_2Fe_{14}B$	620	13.2	0.65	40	+6.9	–	–	+6.9	$\parallel c$	–	a
$Dy_2Fe_{14}B$	598	11.4	0.58	23	+3.8	–	–	+3.8	$\parallel c$	–	a
					+4.8	+1.1	–0.2	+5.7			c
$Ho_2Fe_{14}B$	569	11.2	0.57	18#	–1.1	+4.4	+0.1	+3.4	$\parallel c^{\#}$	–	a
					–1.3	+4.6	+0.1	+3.4			f
					–1.6	+5.4	–1.8	+2.0			c
$Er_2Fe_{14}B$	554	13.1	0.66	16	–1.4	–	–0.3	–1.7	$\perp c$	+	a
$Tm_2Fe_{14}B$	545	18.1	0.93	17	–3.6	–	+0.7	–2.9	$\perp c$	+	a
$Yb_2Fe_{14}B$	527	~ 23	–	–	–	–	–	–	$\perp c$	+	
$Lu_2Fe_{14}B$	535	28.2	–	2.4	–	–	–	–		0	
$Y_2Co_{14}B$	1015	19.2	1.0	3.0	–1.3	–	–	–	$\perp c$	0	g
$La_2Co_{14}B$	955	20.4	1.1	3.4	–1.4	–	–	–	$\perp c$	0	g
$Pr_2Co_{14}B$	995	24.8	1.3	75	–	–	–	–	$\parallel c$	–	g
$Nd_2Co_{14}B$	1007	25.7	1.3	30#	–1.4	+19.0	–	+17.6	$\parallel c^{\#}$	–	h
$Sm_2Co_{14}B$	1029	18.1	1.0	–	–	–	–	–	$\perp c$	+	
$Gd_2Co_{14}B$	1050	5.4	0.3	14	–1.4	–	–	–1.4	$\perp c$	0	i
$Tb_2Co_{14}B$	1035	(3)	0.1	–	–	–	–	–	$\parallel c$	–	g

a. [Sin84] b. [Giv84] c. [Ver88] d. [Kid87] e. [Bol87b]
f. [Fuj87] g. [Bus85] h. [Yam87] i. [Hir87]

changes sign near 140 K, at the spin-tilt temperature, T_{st}. The in-plane anisotropy constants, K_2' and K_3', decrease monotonically with increasing temperature. At low temperatures, the two in-plane anisotropy constants turn out to be of opposite sign and nearly equal in magnitude, leading to a relatively small resultant in-plane anisotropy. Their values are comparable with that of K_1. A description along these lines of the magnetization process in $Nd_2Fe_{14}B$, as well as the effect of temperature

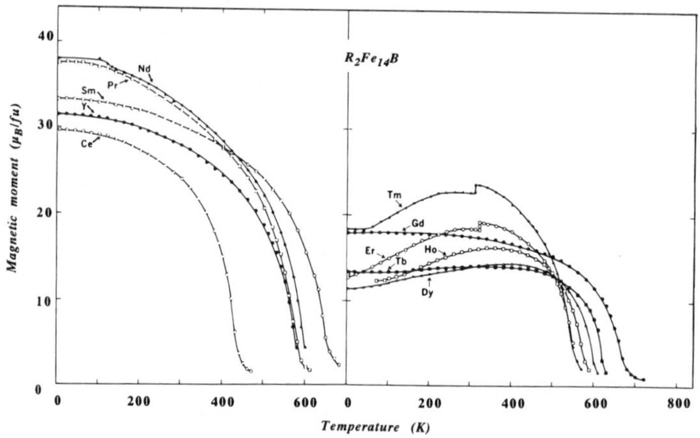

FIG. 2.25. Temperature dependence of the magnetic moment of $R_2Fe_{14}B$ crystals for the light and heavy rare-earths [Hir86a].

on the FOMP transition in this compound, has been given by Asti [Ast90]. The energy difference between the states where the magnetization lies along the c-axis or in the tetragonal place, ΔE_a, is given by

$$\Delta E_a = K_1 + K_2^* + K_3^* \tag{2.28}$$

where the asterisks indicate that the appropriate in-plane anisotropy constants are included in the second and third terms, that is, $K_2^* = K_2 - K_2'$ and $K_3^* = K_3 - K_3'$. There is a one-to-one correspondence between the signs of ΔE_a at 4.2 K and the second-order Stevens factor of the rare-earth ion α_J (see Table 2.5, last two columns), proving the dominant role of the rare-earth ion in the anisotropy at 4.2 K; but no such correlation exists between the signs of K_1 and α_J. It will be shown later that this correlation holds if one uses the anisotropy coefficients κ_i. Also, the values for ΔE_a do not follow the single-ion origin of the anisotropy. In the single-ion approach one would expect a very strong anisotropy for the Tb and Dy compounds, see Table 1.4, whereas the values for ΔE_a, given in Table 2.5, are quite modest for these compounds. Induced noncollinearity (canting) of the transition-metal and rare-earth sublattice magnetizations in applied fields is the main reason for the low values of the effective anisotropy constants compared with those of the intrinsic parameters. This effect will be discussed in more detail in Section C. 1.7. Values for the magnetic polarization and effective values for the anisotropy field and anisotropy constant K_1 at room temperature are given in Table 2.6.

The parametrization of magnetization curves in terms of the anisotropy constants K_i is useful in a phenomenological description of magnetic materials, but does not provide much insight into the origin of the anisotropy energy. Nor can the analysis reproduce magnetization values larger than the spontaneous magnetization along the easy axis, as observed, for instance, in $Dy_2Fe_{14}B$ when the field is applied along the hard axis. The crossing of the easy- and hard-axes

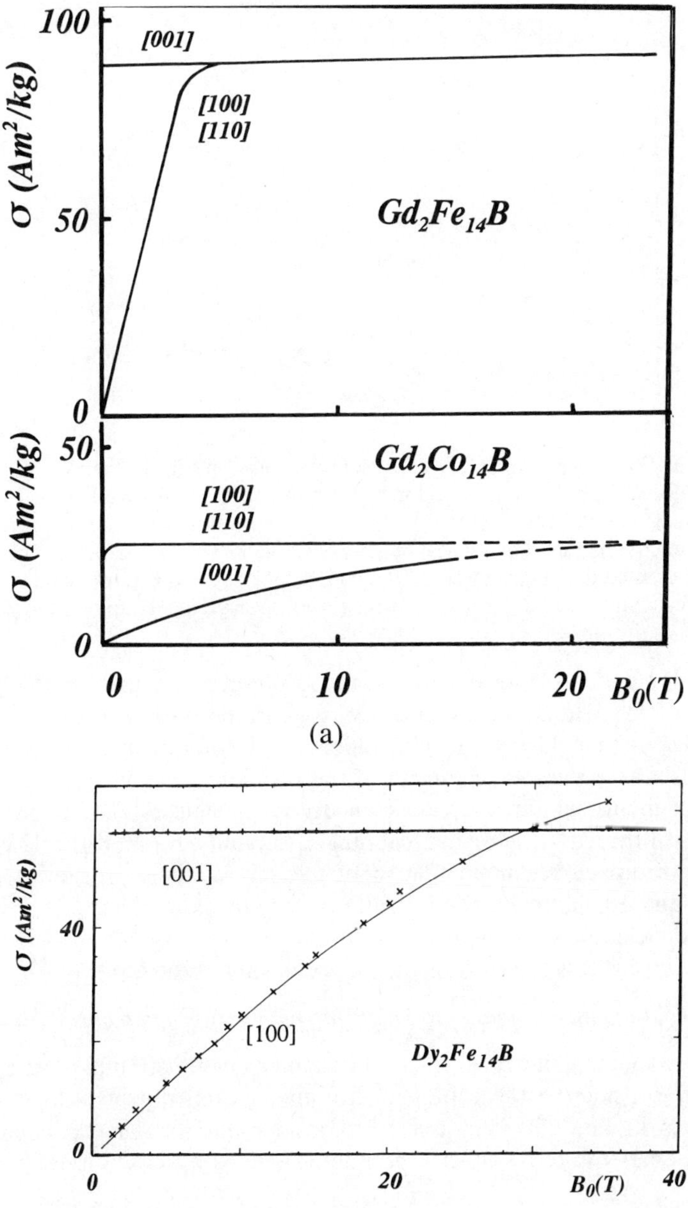

FIG. 2.26. The magnetization curves at 4.2 K of (a) monocrystalline $Gd_2Fe_{14}B$ and $Gd_2Co_{14}B$; and (b) monocrystalline $Dy_2Fe_{14}B$ in fields up to 35 T. The crossing of the two magnetization curves is the consequence of a noncollinear magnetic structure of Dy and Fe moments in fields perpendicular to the easy direction [Ver88].

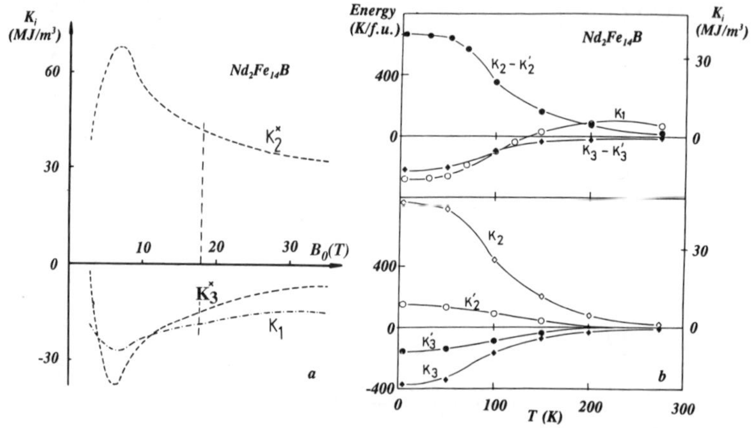

FIG. 2.27. (a) Field dependence of the anisotropy constants K_i, defined by Eq. (2.27), for $Nd_2Fe_{14}B$ as evaluated from high-field magnetization curves at 4.2 K. The field value along the horizontal axis indicates the field up to which the magnetization curves have been considered in the analysis. The vertical line denotes the field of 18 T at which the anisotropy constants presented in (b) have been evaluated [Rad88]. (b) Temperature dependence of the anisotropy constants for $Nd_2Fe_{14}B$ as derived from measurements up to 18 T [Cad88].

magnetization curves demonstrates that the resultant magnetization changes under the action of the external field due to a change in the internal magnetic structure.

The effect of a field-induced noncollinearity is of much less importance for compounds with a parallel alignment of the two sublattice moments and it is the reason that a one-sublattice model provides reasonable estimates for the MCA parameters of the light rare-earth compounds. In reality, for a proper description of the magnetization curves, in high magnetic fields in particular, the directions of the rare-earth and 3d sublattice magnetizations have to be calculated at any value of the applied field by a simultaneous solution of the equations that describe the magnetization process for each of the magnetic sublattices.

1.2 *High-field magnetization of $R_2Fe_{14}B$ ferrimagnetic compounds*

Field-induced noncollinearities in the internal magnetic structure are taken into account by considering the expression for the free energy of a 3d–4f magnetic system given in Eq.(2.5). It is useful to consider the angular dependence of the anisotropy energy in terms of the Legendre functions $P_n^m(\cos\theta)$ in the form

$$E_a(\theta, \varphi) = \Sigma\Sigma\kappa_n^m P_n^m(\cos\theta)\cos(m\varphi) \tag{2.29}$$

The equilibrium state of the magnetic system is described by the polar (θ) and the azimuthal (φ) angles for *each* of the sublattice magnetizations that are coupled by exchange interactions. These angles provide the components of the sublattice magnetization along the applied field, usually directed along one of the main crystallographic directions. In a two-sublattice model it is assumed that all T moments, due to strong 3d–3d interactions, form one T sublattice with resultant

Table 2.6 Magnetic polarization, $\mu_0 M_s$, anisotropy field, B_a, and effective anisotropy constant K_1 for $R_2Fe_{14}B$ and $R_2Co_{14}B$ compounds at room temperature

R	$\mu_0 M_s$ (T)	B_a (T)	K_1 (MJ/m^3)	Ref.
$Y_2Fe_{14}B$	1.44	2.0	+ 1.1	[Bol87a]
La	1.39	2.0	+ 1.1	[Yam86]
Ce	1.17	3.5	+ 1.8	[Yam86, Sag85]
Pr	1.55	8.0	+ 5.0	[Bol87a, Sag85]
Nd	1.61	7.6	+ 4.9	[Giv85, Aba86]
Sm	1.51	12.5	– 12.0	[Sag85]
Gd	0.83	3.1	+ 1.0	[Aba86]
Tb	0.92	13.9	+ 3.4	[Aba86]
Dy	0.72	15	+ 4.2	[Hir86]
Ho	0.80	7.1	+ 2.0	[Aba86]
Er	0.90	5.5	+ 1.3	[Aba86]
Tm	1.15	–	–	–
Lu	1.13	2.6	+ 1.5	[Bol87a]
Th	1.25	2.0	+ 1.3	[Bol87a]
$Y_2Co_{14}B$	1.00	3.5	– 1.4	[Bol87a]
La	1.00	3.5	– 1.4	[Yam86]
Pr	1.09	13.0	+ 6.0	[Bol87a]
Nd	1.06	5.2	+ 2.2	[Yam87]
Gd	–	–	– 1.3	[Bol87a]

values for M_T and E_a^T. The R sublattice is characterized by M_R and E_a^R and is composed of R moments that are expected to rotate coherently in applied external fields. In a two-sublattice model the equilibrium state is found by minimizing the energy of Eq.(2.5) with respect to the angles θ and φ. By comparing the calculated and experimental magnetization curves, the set of anisotropy coefficients of the rare-earth sublattices and the intersublattice R–T molecular-field coefficient are simultaneously determined. The anisotropy energy of the iron sublattice E_a^T, which is an order of magnitude smaller than that of the rare-earth sublattice, is taken to be the same as in $Y_2Fe_{14}B$. Note that Eq. (2.29) $E_a(\kappa_n^M)$ is an alternative to Eq. (1.26–7) or (2.27), $E_a(K_i)$.

Owing to the two different crystallographic R sites and the low local symmetry, a five- (one iron- plus four rare-earth-) sublattice model is needed for the analysis of the magnetic behaviour of the $R_2Fe_{14}B$ compounds (see Fig. 2.28). A two-sublattice model, however, already gives good insight into the magnetization process. This view is further supported by the fact that, despite a large difference in the values for the A_2^2 coefficients of the two different crystallographic sites, the leading CEF parameters A_2^0 are of almost equal magnitude for the two rare-earth sites in the $Nd_2Fe_{14}B$ structure, see Table 2.7. The reduction in the number of fitting

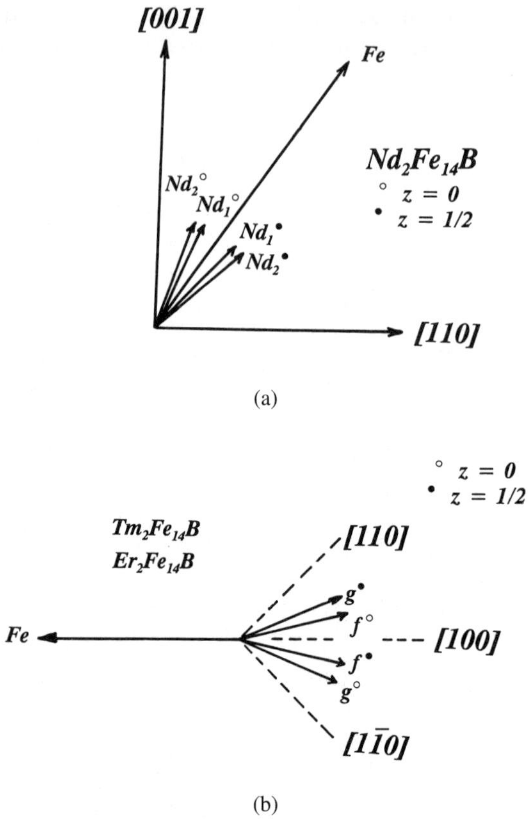

FIG. 2.28. (a) Magnetic structure of $Nd_2Fe_{14}B$ at low temperature in the absence of external fields [Cad88]; the ferromagnetic fan structure is composed of one iron and four neodymium moments. (b) Magnetic structure in the basal plane of planar $Tm_2Fe_{14}B$ and $Er_2Fe_{14}B$ at low temperatures.

parameters is essential if meaningful values are to be obtained. The set of anisotropy coefficients for the neodymium sublattice, derived from the high-field magnetization measurements on $Nd_2Fe_{14}B$ in fields up to 35 T, presented in Fig. 2.4, is given in Table 2.8. The angular dependence of the anisotropy energy is shown in Fig. 2.29. There is a minimum in the anisotropy energy for an angle of 30°, reproducing the experimental value for the tilt angle very well. This minimum is formed by significant contributions from the anisotropy coefficients κ_4^0 and κ_6^0, clearly indicating the important role played by higher-order CEF interactions. The coefficient κ_6^4 plays a major role in the formation of the energy minimum at 90° in the $<100>$ direction. The existence of this minimum causes the abrupt field-induced moment-reorientation transition at 16.3 T for the field applied along this direction. Indeed, detailed measurements by Verhoef *et al.* [Ver88] show that the transition at 16.3 T in $Nd_2Fe_{14}B$ is of first order, with a jump in the magnetization, whereas the increase in the magnetization in case of $Pr_2Fe_{14}B$ is continuous; see

Table 2.7 Crystal-field coefficients for $R_2Fe_{14}B$ [Cad88]; the parameters A_n^m are expressed in units Ka_0^{-n}; the first set of values is for the 4f site, the second set is for the 4g site

R	A_2^0	A_2^2	A_4^0	A_4^4	A_6^0	A_6^4
			4f site			
Pr	176	–116	3	–9	–7	–42
Nd	304	–200	–15	43	–2	–33
Tb	304	–200	–15	43	–2	–33
Dy	292	–192	–14	42	–1	–20
Ho	298	–196	–9	26	–1	–19
Er	292	–222	–16	37	–1	–6
Tm	258	–205	–12	35	–2	–27
Yb	151	–100	–7	22	–1	–17
			4g site			
Pr	179	351	3	8	–7	–4
Nd	308	605	–13	–41	–2	–13
Tb	308	605	–13	–41	–2	–13
Dy	296	581	–12	–39	–1	–8
Ho	302	593	–8	–24	–1	–7
Er	296	689	–14	–39	–1	–3
Tm	262	617	–11	–33	–2	–10
Yb	154	303	–6	–20	–1	–6

Fig. 2.30. At a field of 16 T, applied along the <100> direction, $Pr_2Fe_{14}B$ undergoes a metamagnetic-like transition. The angular dependence of the magnetocrystalline anisotropy energy of the praseodymium sublattice is shown in Fig. 2.31. The higher-order anisotropy coefficients form a valley in the angular dependence that, in the presence of the external field, becomes a broad plateau from 16° to 50°, allowing the magnetization to pass smoothly at 16 T to a new configuration, tilted by about 50° from the *c*-axis. A peculiar property of the $Pr_2Fe_{14}B$ curve is that the magnetization after the metamagnetic-like transition does not lie along a main crystallographic direction.

Taking into account the crystal-field and exchange-field interactions for the whole series of $R_2Fe_{14}B$ compounds, it is expected that a competition between the second-order and the fourth-order anisotropy terms should also occur for the Pr and Dy compounds just as for $Nd_2Fe_{14}B$ and $Ho_2Fe_{14}B$. However, the relatively large value of the second-order crystal-field term for $Pr_2Fe_{14}B$ and $Dy_2Fe_{14}B$ overrules the higher-order terms and both compounds retain uniaxial anisotropy at

Table 2.8 Magnetocrystalline anisotropy coefficients, κ_n^m, of the Nd sublattice in $Nd_2Fe_{14}B$ as derived from high-field magnetization measurements and values for the crystal-field parameters as derived from these anisotropy coefficients; the corresponding values for the crystal-field coefficients A_n^m are given in the last three columns for Nd, Ho and Pr, respectively; [Rad90c]

Order	κ_n^m		B_n^m	A_n^m(Nd)	A_n^m(Ho)	A_n^m(Pr)
n, m	(K/f.u.)	(MJ/m^3)	(K/ion)	(Ka$_0^{-n}$)	(Ka$_0^{-n}$)	(Ka$_0^{-n}$)
2,0	−166	−9.7	−2.36	+330	+334	+220
4,0	+38	+2.2	$+1.35 \times 10^{-2}$	−16	−15	−15
6,0	+36	+2.1	$+1.16 \times 10^{-4}$	−2.9	−1.1	−6.6
4,4	−1	−0.06	-33×10^{-4}	+3.9	+3.9	+3.9
6,4	+28	+1.6	69×10^{-4}	−12.0	−12.0	−12.0
2,2(f)	±25	±1.5	±1.46	±210	±210	±140
2,2(g)	∓77	∓4.5	∓4.53	∓660	∓660	+440

low temperatures. An analysis of magnetization curves, similar to that performed for $Nd_2Fe_{14}B$, has been carried out for $Ho_2Fe_{14}B$ [Rad88a]. The magnetization of $Ho_2Fe_{14}B$ is tilted below 57 K. The CEF calculations, with values for the CEF parameters that are consistent with those obtained for $Nd_2Fe_{14}B$, have revealed a minimum in the angular dependence of the holmium anisotropy energy in the vicinity of 24°. It means that this direction is preferred by the holmium magnetic moment. Owing to the large values of the crystal-field parameters A_n^2 and the finite value of the intersublattice interaction, the magnetic structure of $Ho_2Fe_{14}B$ has been inferred to be a noncoplanar antiferromagnetic fan structure. The four holmium sublattice moments are spread around a direction antiparallel to the iron magnetization that is tilted by 24.3° from the *c*-axis. Indeed, this complex structure has been verified by polarized-neutron experiments. The spread of holmium moments is no more than a few degrees [Fru88].

Tracing the rôle of the low-symmetry parameter A_2^2 one finds that it is most effective in the formation of fan structures for easy-plane systems in the case of relatively weak intersublattice interactions. Indeed, in the erbium and thulium compounds, planar fan structures, as presented in Fig. 2.28, have been found by means of neutron diffraction [Yam85]. The behaviour of this fan structure under the action of large external fields has been studied. Calculations predict an abrupt formation of a field-induced, noncollinear magnetic structure for fields above 42 T applied along the *a*-direction. This transition has been observed on field-oriented, fixed-powder specimens at the proper field value [Rad92a].

1.3 *Temperature-induced reorientation of the easy direction*

In the $R_2Fe_{14}B$ compounds, there are two different types of reorientation of the easy direction as a function of temperature. The transitions observed at the spin-tilt

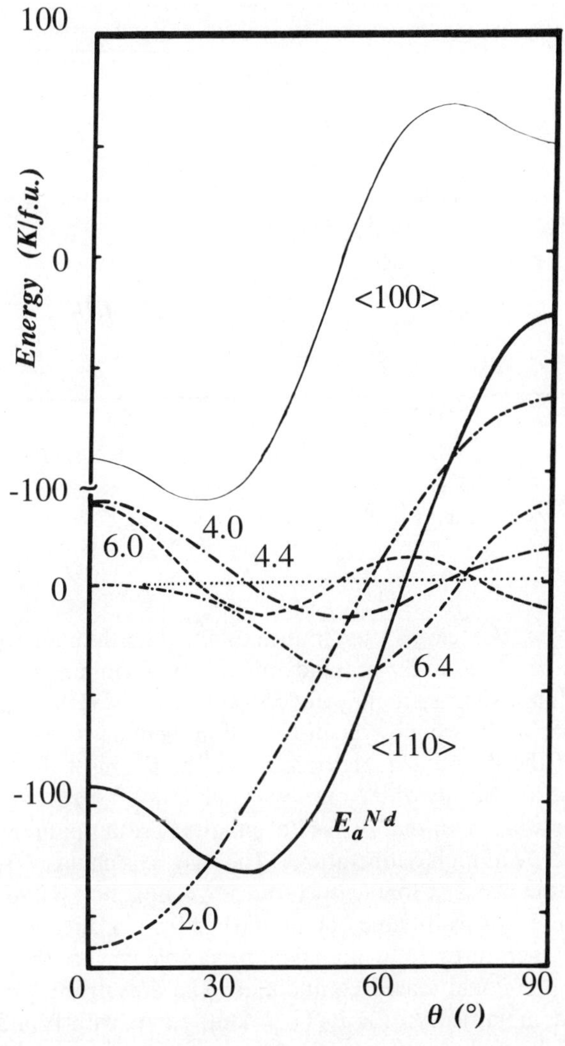

FIG. 2.29. Angular dependence of the anisotropy energy of the neodymium sublattice in the planes (110) (a) and (100) (b), as derived from the analysis of high-field magnetization curves measured up to 35 T at 4.2 K; the different multipole contributions κ_n^m to the anisotropy energy are shown [Rad89c].

temperature, T_{st}, for $Nd_2Fe_{14}B$, $Nd_2Co_{14}B$ and $Ho_2Fe_{14}B$ (Fig. 2.32) turn out to be due to a peculiarity of the CEF interactions of the rare-earth ion involved and are associated with a competition between the second- and higher-order CEF terms. Spin reorientations of a second type are due to the competition between the rare-earth and 3d anisotropies. This type of transition is observed at the temperature T_{sr},

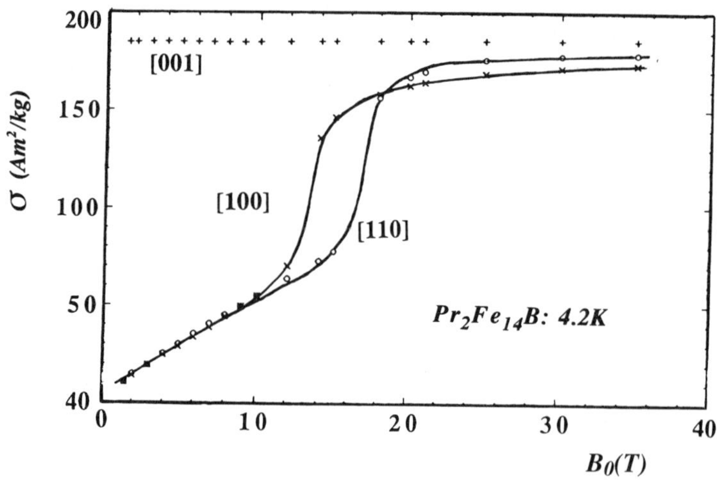

FIG. 2.30. Magnetization curves of $Pr_2Fe_{14}B$ at 4.2 K along three main crystallographic directions [Ver88].

which is just above room temperature in the erbium– and thulium–iron compounds and at higher temperatures in the praseodymium– , neodymium– or terbium–cobalt compounds. Both transitions are fully understood in terms of the CEF interactions. For the occurrence of the second type of transition there has to be a competition in the anisotropy of the R and T sublattices where the R anisotropy prevails at low temperatures. The anisotropy of the rare-earth sublattice decreases in general much faster with temperature than that of the 3d sublattice, resulting in a dominant role for the 3d anisotropy at high temperatures. The easy axis changes from a direction perpendicular to the *c*-axis at low temperatures to a direction parallel to the *c*-axis at high temperatures for iron-based compounds or from a direction parallel to the *c*-axis at low temperatures to a direction perpendicular to the *c*-axis at high temperatures for the cobalt-based compounds. The anisotropy preference of the rare-earth and 3d sublattices in the $R_2T_{14}B$ compounds with iron and cobalt are summarized in Table 2.9, together with values for the temperatures T_{st} and T_{sr}. These reorientation phenomena can clearly be seen in Fig. 2.33, where the temperature dependence of the anisotropy energy of some rare-earth sublattices in $R_2Fe_{14}B$ is shown, together with that of the iron-sublattice.

1.4 *The 3d–4f intersublattice exchange interactions*

The molecular field is a way of representing exchange interactions between unpaired spins. The molecular field experienced by a rare-earth *moment* has been written in Eq. (2.10) as the sum of two contributions due to the 3d and 4f surroundings. The first term greatly exceeds the second one in T-rich intermetallics. The product $n_{RT}\gamma_R$ in Eq. (2.10) has the meaning of an intersublattice molecular-

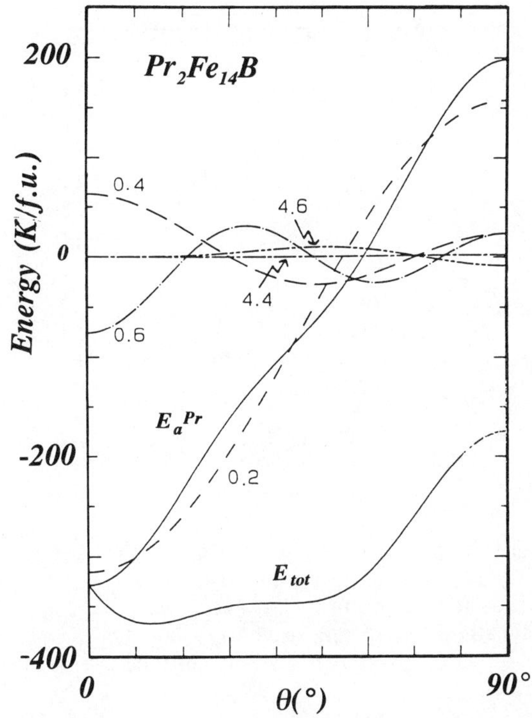

FIG. 2.31. The angular dependence of the magnetocrystalline anisotropy energy of the praseodymium sublattice in $Pr_2Fe_{14}B$ in the (110) plane at 4.2 K, as derived from an analysis of high-field magnetization curves; the different multipole contributions to the anisotropy energy are shown. The total free energy, E_{tot}, of $Pr_2Fe_{14}B$, shown at the field of 15 T just below the transition, illustrates the formation of a broad plateau allowing for smooth rotation of the magnetization at 16.3 T [Rad90c].

field coefficient. The factor γ_R is needed to take into account that the interactions between the 3d and 4f sublattices are, in fact, interactions between the spins. Hence, the exchange field, experienced by the rare-earth *spin* moment is a more fundamental parameter than the molecular field. The exchange field arising from the 3d-spin matrix:

$$\mathbf{B}_R^{ex} = n_{RT}\mathbf{M}_T \tag{2.30}$$

and the molecular field, $\mathbf{B}_R^{mol}$, that represents the intersublattice coupling between the rare-earth and 3d magnetic moments are related for a given rare-earth ion by

$$\mathbf{B}_R^{mol} = \gamma_R \mathbf{B}_R^{ex} \tag{2.31}$$

The strength of the molecular field B_R^{mol} or the intersublattice molecular-field coefficient, $n_{RT}\gamma_R$, can be estimated in various ways. The most direct way is to

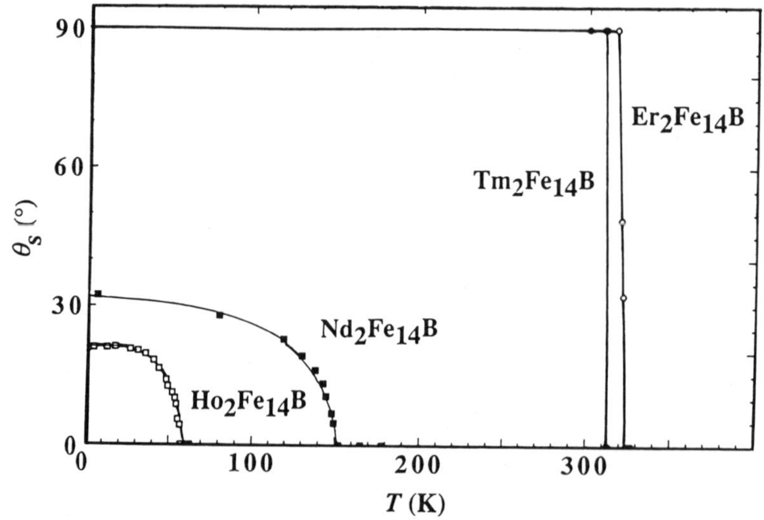

FIG. 2.32. Temperature dependence of the tilt angle θ_s of the magnetization with respect to the c-axis in $Nd_2Fe_{14}B$ and $Ho_2Fe_{14}B$, two compounds showing a tilted magnetic structure at low temperatures due to a competition between lower- and higher-order anisotropy constants, and in $Er_2Fe_{14}B$ and $Tm_2Fe_{14}B$, two compounds showing a temperature-induced moment reorientation at high temperatures due to competition between the rare-earth and iron anisotropies [Hir86].

measure the $|\,7/2> \rightarrow |\,5/2>$ localized excitations by inelastic neutron scattering on the gadolinium compound. This experiment is difficult to perform on account of the enormous neutron absorption cross-section of the Gd nuclei. Nevertheless, results have been reported for $Gd_2Fe_{14}B$ by Loewenhaupt *et al.* [Loe91] which reveal a sharp peak at 37.5 meV, yielding a value for the molecular field of 324 ± 30 T ($2\mu_B B_{ex}^R/k_B = 435$ K). The analysis of the temperature dependence of the rare-earth magnetization in the 3d–4f intermetallics is an easier, but less straightforward, way to derive a value for the 3d–4f molecular field, provided the temperature variation of the 3d sublattice is known from the Y compound. This procedure, discussed in Section A.2.2, yields for $Gd_2Fe_{14}B$ a value for B_{Gd}^{mol} of 275 T [Bog86]. Values for the parameter n_{RT} and the molecular field, derived from high-field magnetization studies on polycrystalline samples in the noncollinear range, are given in Table 2.10 for ferrimagnetic $R_2(Fe, A)_{14}B$ systems, where A is a non-magnetic element that reduces the spontaneous magnetization and consequently brings the noncollinear structure into the accessible field range. Analysis of the values for the Curie temperatures through the 3d–4f series provides another estimate for n_{RT}, although the results are less accurate.

Results for the exchange fields for the $R_2Fe_{14}B$ compounds are summarized in Fig. 2.34. These data have been derived in a molecular-field approximation from the Curie temperature, from the temperature dependence of the magnetization and from high-field magnetization studies. Except for the latter, these values contain

Table 2.9 The expected preference of the rare-earth and the 3d sublattice magnetization for the crystallographic direction in the $R_2Fe_{14}B$ and $R_2Co_{14}B$ compounds compared with the experimental information on the easy direction at 4.2 K. The preference of the rare-earth sublattice follows the sign of the second-order Stevens factor, α_J. Spin-reorientation transitions at T_{st} result from a competition of second- and higher-order crystal-field interactions of the rare-earth ions, whereas the transitions at T_{sr} result from the competing rare-earth and 3d anisotropies; # denotes a tilted direction

Compound	α_J	Expected preference R-subl.	T-subl.	Experimental data Tilt angle (°)	T_{st} (K)	T_{sr} (K)
$Y_2Fe_{14}B$	0	–	$\parallel c$	0	–	–
La	0	–	$\parallel c$	0	–	–
Ce	0	–	$\parallel c$	0	–	–
Pr	–	$\parallel c^{\#}$	$\parallel c$	0	–	–
Nd	–	$\parallel c^{\#}$	$\parallel c$	30	135	–
Sm	+	$\perp c$	$\parallel c$	90	–	$\sim T_c$
Gd	0	–	$\parallel c$	0	–	–
Tb	–	$\parallel c$	$\parallel c$	0	–	–
Dy	–	$\parallel c^{\#}$	$\parallel c$	0	–	–
Ho	–	$\parallel c^{\#}$	$\parallel c$	24	57	–
Er	+	$\perp c$	$\parallel c$	90	–	328
Tm	+	$\perp c$	$\parallel c$	90	–	314
Yb	+	$\perp c$	$\parallel c$	90	–	120
Lu	0	–	$\parallel c$	0	–	–
$Y_2Co_{14}B$	0	–	$\perp c$	90	–	–
La	0	–	$\perp c$	90	–	–
Ce	0	–	$\perp c$	90	–	–
Pr	–	$\parallel c^{\#}$	$\perp c$	0	–	660
Nd	–	$\parallel c^{\#}$	$\perp c$	12	37	545
Sm	+	$\perp c$	$\perp c$	90	–	–
Gd	0	–	$\perp c$	90	–	–
Tb	–	$\parallel c$	$\perp c$	0	–	795

contributions originating from the R–R interactions. By inspecting Fig. 2.34, one notices that there is a tendency for the exchange field B_R^{ex} or n_{RT} to decrease from left to right across the the lanthanide series [Cad87]. The interaction is actually more than twice as large for a light rare-earth element such as Nd than for the heavy rare-earths, a tendency which is common to other R_nT_m series as well [Bel87, Fig. 2.34] and which has been related to the decrease in the radius of the 4f shell going from La to Lu. Due to the localization of the 4f electrons, the interaction between the 3d and 4f spins cannot be a direct one. According to Campbell

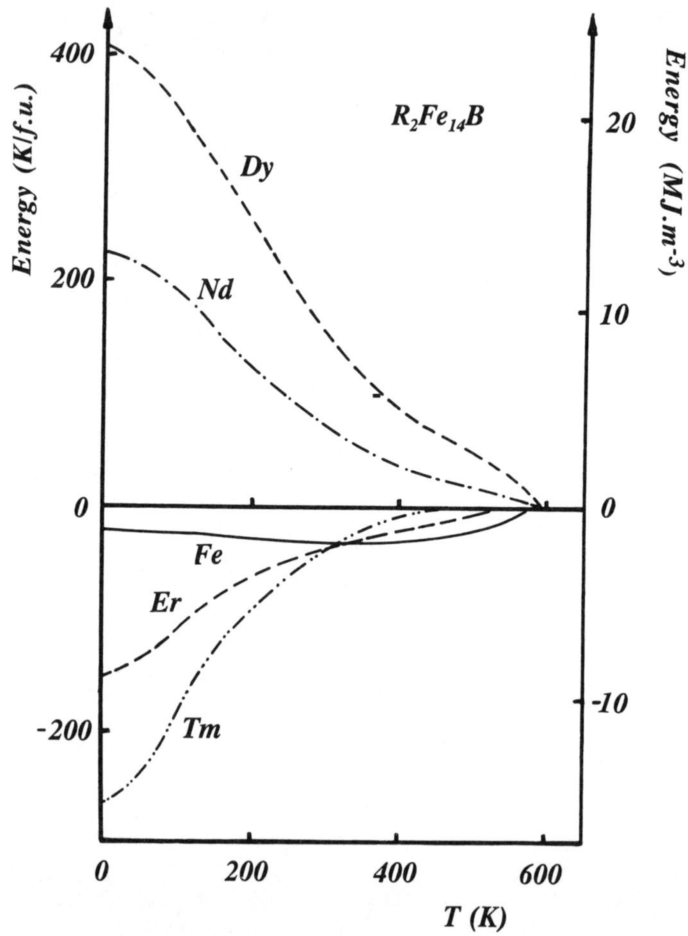

FIG. 2.33. Temperature dependence of the rare-earth contribution to the MCA energy of some $R_2Fe_{14}B$ compounds; the uniaxial anisotropy of the iron sublattice, drawn here with the opposite sign, dominates the planar anisotropy of rare-earth ions Er and Tm at temperatures above the spin-reorientation temperature, T_{sr}.

[Cam72], the R–T interaction proceeds via the 5d electrons. The 3d–4f interaction parameter in this view represents the combined effect of interatomic 3d–5d and intra-atomic 5d–4f interactions. The 3d–4f exchange interactions decrease across the 4f series because of a decreasing overlap between the 4f and 5d shell following the lanthanide contraction.

The exchange field enables us to compare, in a quantitative way, the strength of the exchange interactions in different 3d–4f compounds. Attributing the exchange field in first approximation to the 3d nearest neighbours, one can correlate the macroscopic exchange field with the microscopic exchange coupling parameter,

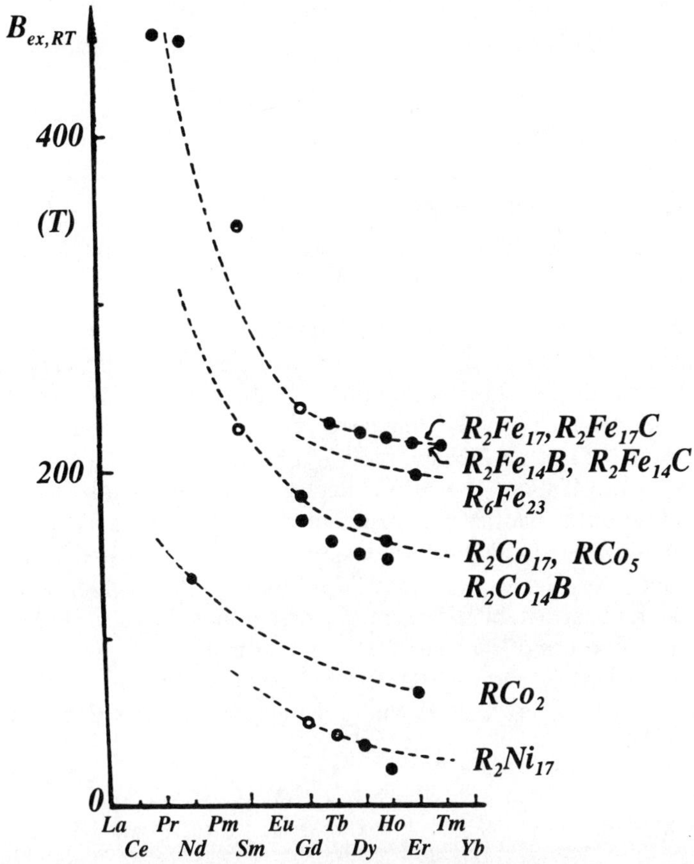

FIG. 2.34. Variation of the exchange field, B^{R}_{ex}, experienced by the 4f spin moment for the $R_2Fe_{14}B$ series. Data are derived from various kinds of experiment, mostly from studies in high magnetic fields and from inelastic-neutron-scattering experiments [Rad90a].

Table 2.10 Experimental values for the coefficients n_{RT} ($T\,kg/Am^2$) and values for J_{RT} (10^{-22}J) derived from these results, together with values for the molecular field B^{mol} (T) and the exchange field B^{ex} (T); after Verhoef *et al.* [Ver90a]

R	Gd	Tb	Dy	Ho	Er	Tm
n_{RT}	1.72	1.19	0.82	0.61	0.45	0.41
J_{RT}	1.25	1.29	1.17	1.18	0.95	1.03
B^{mol}	274	186	130	95	70	63
B^{ex}	274	279	260	238	210	220

J_{RT}, between the 3d and 4f spins in the Heisenberg-type of Hamiltonian in the following way [Rad85, Rad86]:

$$B_R^{ex} = -2Z_{RT}S_T J_{RT}/g\mu_B \tag{2.32}$$

where Z_{RT} represents the number of 3d atoms surrounding the rare-earth (18 for the $Nd_2Fe_{14}B$ structure) and S_T is the effective spin of the 3d ion deduced from its ordered magnetic moment. Values for J_{RT} are included in Table 2.10.

1.5 *Crystal-field effects on the 4f ions*

In order to arrive an appropriate description of magnetization curves of 3d–4f intermetallics, we have to consider the internal magnetic structure as formed by the different (rare-earth and 3d-transition-metal) sublattices. This internal magnetic structure is determined by the magnetic moment and the magnetocrystalline anisotropy of the 4f and the 3d sublattices and by the mutual exchange interactions between the 3d and 4f spins. The 3d–3d interactions are sufficiently strong to couple all 3d spins into one sublattice with resultant values for the magnetic moment and the MCA energy. The mutual interactions between the 4f spins are considered to be negligibly small. As a consequence, the 3d–4f exchange interactions and the MCA energy of the rare-earth sublattice are the essential ingredients for a description of the magnetic properties of rare-earth ions in a 3d matrix.

The Hamiltonian for a rare-earth ion includes the intra-atomic Coulomb and spin–orbit interactions, as well as the exchange, crystal-field and magnetostatic interactions:

$$\mathcal{H}_R = \mathcal{H}_{so} + \mathcal{H}_{cf} + \mathcal{H}_{ex} + \mathcal{H}_z \tag{2.33}$$

where the terms account for the spin–orbit, crystal-field and exchange interactions. $\mathcal{H}_z$ stands for the Zeeman magnetostatic energy term in an external field. These terms are written in sequence of their importance. This Hamiltonian takes the explicit form

$$\mathcal{H}_R = -\lambda \mathbf{L} \cdot \mathbf{S} + \mathcal{H}_{cf} - g_s\mu_B \mathbf{S} \cdot \mathbf{B}_R^{ex} - \mu_B(\mathbf{L} + 2\mathbf{S}) \cdot \mathbf{B}_0 \tag{2.34}$$

where λ denotes the spin–orbit coupling parameter. As the intra-atomic Coulomb and spin–orbit interactions are much stronger than the others, it is usually sufficient to consider the ground-state multiplet only. In that case, the spin–orbit interaction determines the total angular momentum **J** of the ground state according to Hund's third rule. Subsequently, the molecular field, mainly due to 3d–4f exchange interactions, stabilizes its maximal component $J_z = J$. A comparison of the crystal-field energy and the molecular-field interaction for the different rare-earth ions has been shown in Fig. 2.5. These energies have to be compared with the spin–orbit splitting Δ_{so} that is typically of the order of 3000 K. The low-lying excited multiplets, however, have to be taken into account in the case of Eu and Sm ions where there is significant J-mixing as an excited multiplet lies closer, at 420 K and 1440 K respectively. The J-mixing is sometimes suggested to play a rôle for Ce ($\Delta_{so} = 3170$ K), Pr (3020 K),

Tm (11 940 K) and Yb (14 820 K) compounds but, due to the large intermultiplet distance, mixed-valence-type processes are more probable.

Because of the small size of the 4f shell, the 4f electrons are rather isolated from their surroundings. The mean value of the 4f radius amounts to 40 pm in the case of Gd, less than 15% of the distance to the nearest neighbours in $Gd_2Fe_{14}B$. To describe the electrostatic interactions for the 4f electrons, the crystal-field potential is expanded in a series of spherical harmonics associated with the different multipoles of the electrostatic interactions. The 4f charge cloud is aspherical for most rare-earth ions, and the multipole moments of the 4f-shell charge distribution are characterized by the product of Stevens factors of the second- (α_J), fourth- (β_J) and sixth- (γ_J) order and the mean value of the 4f-shell radius, $< r^n_{4f} >$. These latter values have been computed by Freeman and Desclaux [Fre79]. The multipole parameters for all trivalent 4f ions were given in Table 1.4 (p. 21). For ions with a positive value of α_J (Sm, Er, Tm and Yb) the electron-density distribution has a prolate shape, whereas for ions with a negative value of α_J (Pr, Nd, Tb, Dy, Ho) it is oblate (Fig. 1.12). The crystal-field parameters, B^m_n, that enter into the Hamiltonian $\mathscr{H}_R$ carry information about the multipole moments of the 4f ion under investigation as well as about the crystal-field potential produced by the charge distribution within the lattice. They can be written as

$$B^m_n = \theta_n < r^n_{4f} > A^m_n \tag{2.35}$$

where $\theta_2 = \alpha_J$, $\theta_4 = \beta_J$ and $\theta_6 = \gamma_J$. The products of the first two terms represent the multipole moments of the 4f shell, whereas the crystal-field coefficients, A^m_n, are characteristic of the crystallographic structure under consideration.

A restriction to the lowest multiplet simplifies the computations. Then, the rare-earth Hamiltonian can be written as the sum of the exchange and crystal-field interactions only:

$$\mathscr{H}_R = \mathscr{H}_{cf} + g_J\mu_B \mathbf{J}\cdot(\mathbf{B}^{mol}_R + \mathbf{B}_0) \tag{2.36}$$

The exchange interactions of the 4f spin are approximated by the molecular field experienced by the rare-earth moment, B^{mol}_R. The crystal-field Hamiltonian is conventionally expressed in terms of Stevens operators, $\mathscr{H}_{cf} = \Sigma B^m_n \boldsymbol{O}^m_n$.

In case of the $Nd_2Fe_{14}B$ structure (Fig. 1.18), the second-order coefficients, A^0_2, are positive at the two different crystallographic sites, which causes the c-axis to be the easy direction of magnetization for the oblate ions. The 4f-electron density is restricted in that case to layers containing the rare-earth atoms, and much of the crystal field acting on the 4f electrons arises from the rare-earth atoms in the same atomic layer. Interatomic distances between the rare-earth ions in the tetragonal planes are substantially shorter than the spacing between R ions along the c-axis (360 pm and 600 pm respectively).

A quite different atomic arrangement is realized in the $CaCu_5$-type of structure (Fig. 1.20). The compound $SmCo_5$, another outstanding magnetic material, has this structure but here the nearest R neighbours of the rare-earth ion are found along the c-axis and the crystal-field coefficient A^0_2 is negative.

For ions with a prolate 4f electron density (R = Sm, Er, Tm and Yb), the easy direction in the $R_2Fe_{14}B$ structure lies in the tetragonal plane. The magnetic structure can be quite complex due to the in-plane anisotropy. Fan structures of one iron and four differently oriented rare-earth moments, as presented in Fig. 2.28b, have been observed in neutron-diffraction experiments performed on single-crystalline $Tm_2Fe_{14}B$ [Yam85]. The fan structures result from an interplay of exchange interactions between 3d and 4f spins and the basal-plane anisotropy contributions connected with the off-diagonal terms in the crystal-field Hamiltonian, especially A_2^2. In $Nd_2Fe_{14}B$ at low temperature, the four neodymium moments form a ferromagnetic fan structure together with the iron moment, which is tilted at an angle of about 30° with respect to the c-axis (Fig. 2.28a). The spread in orientations in the fan structure of the neodymium moments with respect to the iron moments does not exceed a few degrees due to strong ferromagnetic Nd–Fe interactions.

In order to give a description of the crystalline electric field at the rare-earth sites in the tetragonal $Nd_2Fe_{14}B$ structure up to order 6, 36 parameters are needed. They are associated with four inequivalent rare-earth sites with local orthorhombic symmetry. The $Nd_2Fe_{14}B$ structure belongs to the tetragonal space group $P4_2/mnm$, where the point symmetry of the 4f and 4g sites occupied by the R ions is only mm. These sites are positioned in the elementary cell in planes at $z = 0$ and $z = 1/2$ (Fig. 1.18). There are two f and two g rare-earth sites denoted by f_1, f_2 and g_1, g_2, respectively, which are related by the 4_2 tetragonal-screw axis directed along the c-axis; the surroundings of the ions located in the plane $z = 1/2$ are rotated by 90° around the c-axis compared with the surroundings of the ions situated in the plane $z = 0$. The crystal-field Hamiltonian at each of the four rare-earth sites involves nine terms (Eq. (1.31d)):

$$\mathcal{H}_{\mathrm{cf}} = B_2^0 O_2^0 + B_2^{2,\mathrm{s}} O_2^{2,\mathrm{s}} + B_4^0 O_4^0 + B_4^{2,\mathrm{s}} O_4^{2,\mathrm{s}} + B_4^{4,\mathrm{c}} O_4^{4,\mathrm{c}} + B_6^0 O_6^0 + B_6^{2,\mathrm{s}} O_6^{2,\mathrm{s}} + B_6^{4,\mathrm{c}} O_6^{4,\mathrm{c}} + B_6^{6,\mathrm{s}} O_6^{6,\mathrm{s}} \tag{2.37}$$

The z-axis of this Hamiltonian is parallel to the $<001>$ direction and the x-axis is along the $<100>$ direction. Symmetry considerations lead to the following constraints for the CEF parameters:

$$\begin{aligned} B_n^0(z=0) &= B_n^0(z=1/2) \\ B_n^{4,\mathrm{c}}(z=0) &= B_n^{4,\mathrm{c}}(z=1/2) \end{aligned} \tag{2.37a}$$

and

$$B_n^{2,\mathrm{s}}(z=0) = -B_n^{2,\mathrm{s}}(z=1/2) \tag{2.37b}$$

Given this large number of crystal-field variables, it is helpful if some relations between them can be established unambiguously by experiment. Estimates of the second-order crystal-field parameters have been made from measurements of the electric-field gradient at the rare-earth nucleus [Li91] using Mössbauer spectroscopy or NMR. This method provides information about the electric field gradient at the rare-earth nucleus, whereas for understanding the magnetic anisotropy we are interested in the electric field gradient experienced by the electrons of the 4f shell. The two quantities differ by some orders of magnitude because of the shielding effects of the atomic electrons, and it is questionable to what extent they should be proportional.

In the $R_2Fe_{14}B$ family of compounds, Mössbauer measurements have also been reported by ^{161}Dy, ^{166}Er, ^{169}Tm and ^{174}Yb [Bog86, Ber86, Fru87, Mey88]. Values for the hyperfine fields are very close to the free-ion values calculated for the full trivalent rare-earth moment, which implies that the ground state corresponds to a state in which $<J_z>$ is equal to J. This is in agreement with the magnetization data and is strong evidence that the exchange field in Eq. (2.34) dominates the crystal-field interactions. However, Mössbauer experiments by Bogé *et al.* [Bog86] on monocrystalline $Gd_2Fe_{14}B$ revealed that the EFG experienced by the Gd nucleus is highly asymmetric and substantially different at the two gadolinium sites. The large asymmetry of the electric-field gradient $V_{ij} = \partial^2/\partial x_i \partial x_j$ is caused by the low local symmetry. Then, the EFG is described by the component V_{zz} and the asymmetry parameter $\eta = |(V_{xx} - V_{yy})|/V_{zz}$. It has been found that the z-axis of the EFG, along which it has its largest absolute value, is the crystallographic c-axis for the 4f site but the 4g site it lies in the plane perpendicular to the c-axis, at 45° to the a-axis. The EFG parameters at the two sites, V_{zz} and η, can be transformed into the crystal-field coefficients A_2^0 and A_2^2. It turns out that the second-order terms A_2^0 are positive for both sites and almost equal, although the off-diagonal term A_2^2 is three times larger for the 4g site than for the 4f site. Thus for uniaxial systems, the two rare-earth sites may be considered to contribute in the same sense to the rare-earth sublattice anisotropy.

1.6 *Macro- and microscopic anisotropy parameters*

The anisotropy coefficients, κ_n^m, of the rare-earth sublattice directly mirror the crystal field parameters, B_n^m, of the rare-earth ion involved. In the large-exchange limit, where the crystal-field splitting, ΔE_{cf}, is much smaller than the exchange-field splitting, ΔE_{ef}, the Legendre functions are just the classical equivalents of the Stevens operators, $\boldsymbol{O}_n^m(J_z)$. The strong molecular field leads to a $|J, J_z = -J>$ ground state and the relations between the coefficients, κ_n^m, and the crystal-field parameters, B_n^m, can be written as follows:

$$\kappa_2^0 = B_2^0\langle \boldsymbol{O}_2^0(J)\rangle, \kappa_4^0 = B_4^0\langle \boldsymbol{O}_4^0(J)\rangle, \kappa_6^0 = B_6^0\langle \boldsymbol{O}_6^0(J)\rangle \tag{2.38a}$$

$$\kappa_4^4 = 1/8\, B_4^4\langle \boldsymbol{O}_4^4(J)\rangle, \kappa_6^4 = 1/16\, B_6^4\langle \boldsymbol{O}_6^4(J)\rangle \tag{2.38b}$$

where the values of $\langle \boldsymbol{O}_n^0(J)\rangle$ are the expectation values of the Stevens operators $\boldsymbol{O}_n^m(J_z)$ for $J_z = J$

$$J_2 = \langle \boldsymbol{O}_2^0(J_z = J)\rangle = < 3J^2 - J(J+1) > \tag{2.39}$$

and so on. Values of $\langle \boldsymbol{O}_n^0(J)\rangle$ are collected for all rare-earth trivalent ions in Table 1.4. The large-exchange limit is quite well fulfilled for many 3d-based hard-magnetic intermetallic compounds characterized by a high value of T_C (Fig. 2.5). For $Nd_2Fe_{14}B$, for instance, ΔE_{ex} is 1500 K, whereas ΔE_{cf} is only 150 K. Within the single-ion picture, the anisotropy coefficients of the rare-earth sublattice should follow the values of the product $\theta_n < r_{4f}^n > J_n$ throughout the isostructural series as long as the crystal-field coefficients A_n^m may be regarded as constant. This view is partly supported by the experimental results shown in Table 2.8, where it appears that the CEF coefficients for Nd and Ho ions are quite similar. It is the great advantage of using the coefficients κ_n^m rather than the anisotropy constants K_i that one can easily deduce them all from one given R compound within an isostructural series.

In reality, the crystal-field interactions are small but not negligible compared to the molecular-field interactions, and the relations in Eq. (2.38) must then be regarded as quite good but still approximate relations. It means that for each case the crystal-field parameters must be refined in order to reproduce the described phenomena exactly. A set of crystal-field coefficients, A_n^m, of the trivalent neodymium ion in $Nd_2Fe_{14}B$, calculated from the anisotropy coefficients κ_n^m, is included in Table 2.8. Although this set of CEF coefficients is not complete it is quite adequate as it turns out that several of the higher-order parameters are less relevant. The published set of parameters [Cad88, Yam88, Rad89b] are in fair agreement. The last set reproduces the energy separations in the energy-level scheme determined by inelastic neutron scattering. The crystal-field coefficients in the $R_2Fe_{14}B$ compounds are very much the same, giving strong evidence for the single-ion origin of the magnetocrystalline anisotropy. There is, however, a significant reduction of the leading term A_2^0 for the praseodymium and ytterbium compounds. This reduction has been attributed to an incipient mixed valence of the rare-earth ions [Gav88]. Although there is undoubtedly some ambiguity in the crystal-field parameter sets, the analysis of a complete set of isostructural compounds with different rare-earths which respond quite differently to the second-, fourth- and sixth-order terms in the crystal field, have allowed a substantial resolution of what seemed at first sight to be a problem of intractable complexity. Using the parameters of Tables 2.7 and 2.8, the exchange field and the temperature variation of the magnetic properties of the iron sublattice, it is now possible to describe or predict the magnetic properties of the $R_2Fe_{14}B$ compounds in rather fine detail.

1.7 *Electronic structure of rare-earth ions*

The energy-level scheme of the rare-earth ions is a part of the electronic structure of the intermetallic. In order to account for properties of intermetallic compounds in the electronic band-structure calculations two approaches are made. In the first, the f states are considered as part of the band structure [Bro91] whereas in the second approach the 4f states are treated as part of the core which is compatible with their localized status. The second approach is more natural since we know from photoemission experiments that the f level lies 2–8 eV below the Fermi level E_F. The Fermi level lies in the conduction-electron band formed in the 3d–4f intermetallics by the 6s/5d rare-earth and 3d/4s transition-metal electrons. The f level consists, in fact, of a large number of localized states known as the multiplet structure. For rare-earth ions the multiplet structure is well accounted for by the Russel–Saunders L–S scheme and Hund's rules. The Russel–Saunders scheme couples orbital and spin momenta of n individual f electrons into quantum states **L** and **S** referring to the whole f^n system and to the total angular momentum **J** varying from $|L-S|$ to $L+S$. Hund's third rule yields the ground multiplet with $J=|L-S|$ for light rare-earths and $J=L+S$ for heavy rare-earths. The multiplet structure ranges from 0.2 eV to 5 eV and can be measured by photoemission, inelastic-neutron-scattering and light-scattering experiments. The position of the first-excited multiplet depends on the strength of the spin–orbit interactions λ and lies at 3020 K and 2740 K and the Pr^{3+} and Nd^{3+} ions. The $(2J+1)$ degeneracy of the ground multiplet is removed by CEF and exchange interactions. The evaluation of the energy-level scheme, that is, the splitting of the ground-state multiplet, on the basis of experimental observations is the goal of theoretical studies of 4f intermetallics, and these considerations are predominant in this chapter.

The restriction to the ground-state multiplet facilitates computations enormously. In that case, the Stevens equivalent operators can be used for the CEF Hamiltonian. Within the ground-state multiplet the Landé factor g_R, the angular momentum operators J_z, J_+ and J_- are defined and the Hamiltonian of Eq. (2.34) can be rewritten as

$$H_R = \sum_{n=0}^{6}\sum_{m=0}^{n} B_n^m O_n^m(J_+, J_-, J_z) + g_R\mu_B \mathbf{J}_R \cdot (\mathbf{B}_R^{mol} + \mathbf{B}_0) \tag{2.40}$$

Eigenvalues and eigenfunctions result in a straightforward way from the diagonalization of this Hamiltonian, provided that the relevant CEF and exchange field parameters are known. The eigenfunctions Γ_i can be expressed as a linear combination of the eigenfunctions of the free ion associated with the different z components of **J**, $|LSJM>$:

$$|\Gamma_i> = \sum_{M=-J}^{J} a_{iM} \,|M> \tag{2.41}$$

From the eigenfunctions, different characteristics of the state i can be calculated. The expectation value of the magnetic moment μ_i associated with each eigenstate can be expressed as

$$\mu_i = -g_R\mu_B < \Gamma_i \,|\, \mathbf{J} \,|\, \Gamma_i > \tag{2.42}$$

Having the eigenfunctions and eigenenergies, the magnetic and electronic properties of the R ion, can then be calculated. First of all, the free energy is given by

$$F_R(T, \mathbf{n}) = -kT \ln Z_R(\mathbf{n}) \tag{2.43}$$

where Z_R is the partition function over the available energy levels:

$$Z_R(\mathbf{n}) = \mathrm{Tr}\, \exp(-H_R(\mathbf{n})/kT) = \sum_{i=0}^{2J+1} \exp(-E_i^R(\mathbf{n})/kT) \tag{2.44}$$

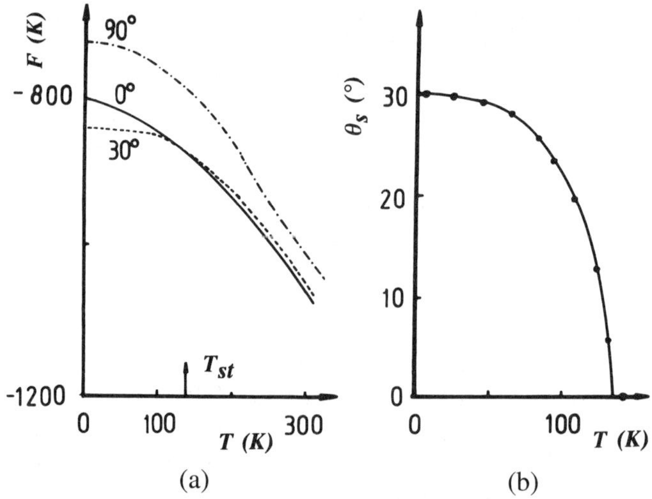

FIG. 2.35. (a) Temperature dependence of the free energy of the Nd^{3+} ion in $Nd_2Fe_{14}B$ for different directions of the molecular field $B_{\mathrm{Nd}}^{\mathrm{mol}}$; the angle is given with respect to the *c*-axis; at low temperatures the configuration of the neodymium moment tilted by 30° has the lowest energy; above T_{st} the neodymium moment favours the *c*-direction as do the Fe moments. (b) Calculated temperature dependence of the tilt angle θ_s (full curve) and experimental data (●).

E_i^{R} denotes the energy of the localized *i* levels with respect to the ground level and **n** denotes the direction of the molecular field. The free energy is a function of temperature and the direction **n**, and it is different for the different crystallographic directions. The temperature dependence is illustrated in Fig. 2.35, whereas the energy-level scheme for the Nd^{3+} ion in $Nd_2Fe_{14}B$ is presented in Fig. 2.36.

The Hamiltonian in Eq. (2.34) enables us to evaluate the ground state and the energy-level scheme of a 4f ion under the combined action of the crystal-field and molecular-field interactions. It provides values for the magnetic moment and the local magnetocrystalline anistropy energy at any temperature and allows us to trace the effects of the external magnetic field. The energy-level scheme in Fig. 2.36 is calculated for different directions **n** of the internal molecular field. It turns out that the sequence of levels is fully governed by the molecular field. The crystal field, however, modifies the energy separation between the levels, as well as their angular dependences, substantially. The angular dependence of the ground-state level is of most importance as it directly reflects the local anisotropy energy of the rare-earth ion. It exhibits an energy minimum at 30°, indicating that this direction is favoured by the Nd-sublattice leading to the tilted magnetic structure (Fig. 2.35). Inelastic-neutron-scattering results of Loewenhaupt *et al.* [Loe88], presented in Fig. 2.37, reveal at 4.2 K two pronounced excitations with energies of 22 meV (= 250 K) and 36 meV (= 418 K). These excitations are attributed to the transitions from the ground state to the first and the second excited-state levels, in good agreement with the energy-level scheme presented in Fig. 2.36. Moreover, with increase in temperature, the position of the first peak shifts down in energy whereas the second peak remains

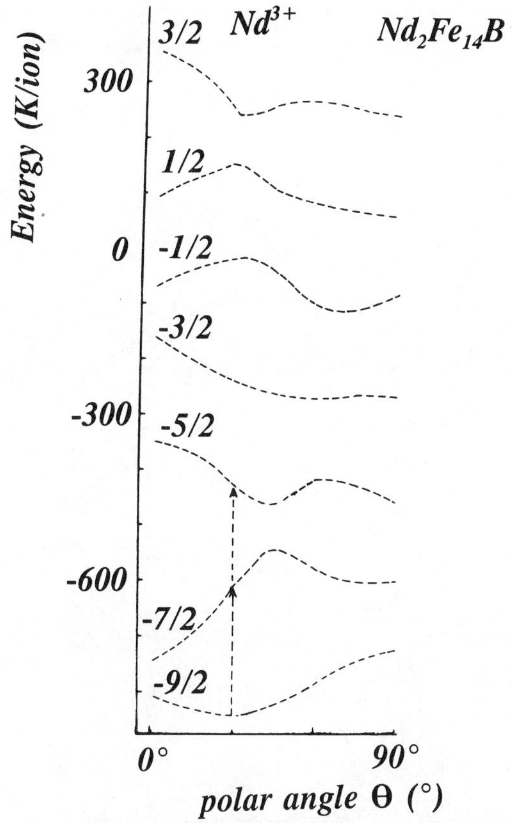

FIG. 2.36. The energy-level scheme of the Nd^{3+} ion in $Nd_2Fe_{14}B$ at 4.2 K under the combined action of the crystalline electric and molecular fields as a function of θ, the angle of the internal molecular field with respect to the tetragonal c-axis. The levels are labelled with the main $| J_z >$ contribution to the eigenfunction. The minimum at 30° indicates that this direction is energetically favoured and explains the occurrence of a tilted magnetic structure of $Nd_2Fe_{14}B$ at low temperatures. The arrows indicate excitations observed in inelastic-neutron-scattering experiments [Rad89c].

practically unchanged. This can be understood from Fig. 2.36 since with increase of temperature the tilt angle θ_s decreases and the distance to the first excited state is reduced, whereas the distance to the second level remains practically unchanged. The same experiments performed for $Dy_2Fe_{14}B$ and $E_2Fe_{14}B$ (Fig. 2.17) reveal in both cases only one excitation at energies equal to 12 meV and 6.8 meV respectively. These are interpreted as excitations to the first excited level at 140 K and 80 K, respectively.

2 $R_2(Fe,T)_{14}B$ solid solutions

A large variety of solid solutions based on the $Nd_2Fe_{14}B$ type of structure has been investigated. All components of $Nd_2Fe_{14}B$ can be partly substituted by metals or

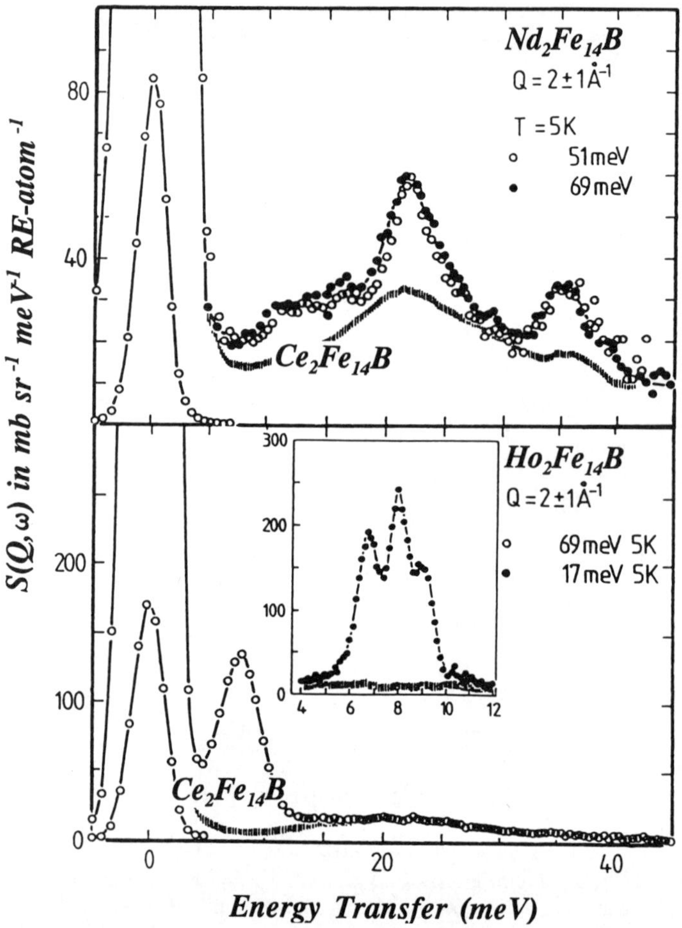

FIG. 2.37. Inelastic-neutron-scattering spectra for $Nd_2Fe_{14}B$ showing two well-defined peaks at 22 (235 K) and 36 (420 K) meV, as well as for $Ho_2Fe_{14}B$ showing a three-peak structure [Loe88].

metalloids of a similar nature. Substitutions are applied either to improve the microstructure and performance of magnets (Chapter 6) or to obtain a better insight into the microscopic interactions of the isomorphous compounds. Since $Nd_2Fe_{14}B$ exhibits a rather large spontaneous magnetization, substitutions to improve intrinsic properties are aimed at increasing the anisotropy field and improving the thermal stability of the magnetic properties above room temperature.

2.1 *Substitutions for the rare-earths*

Values for the Curie temperature of all compounds of the $R_2Fe_{14}B$ type lie within a rather narrow temperature range, so a substantial increase of T_C on substituting Nd

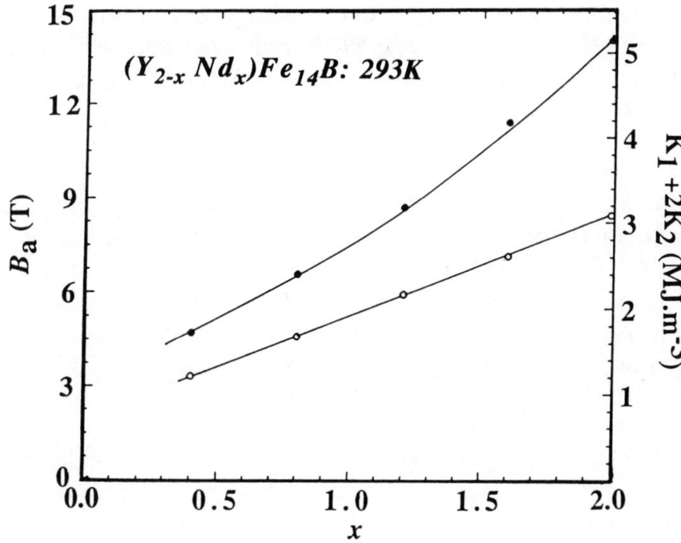

(a)

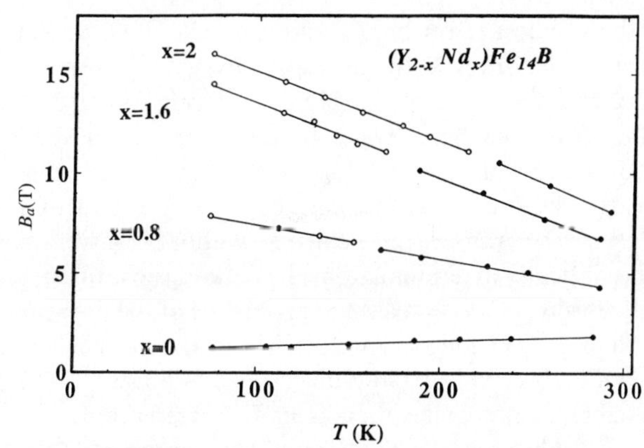

(b)

FIG. 2.38. (a) Composition dependence of the anisotropy field values B_a (●) and the computed anisotropy constants $K_1 + 2K_2$ (o) at 293 K for $Y_{2-x}Nd_xFe_{14}B$. (b) Temperature dependence of the anisotropy field, B_a (●), and the critical field, B_{cr} (○), for various compositions in $Y_{2-x}Nd_xFe_{14}B$ [Bol87c].

by other rare-earth elements is not expected. The slight increase in T_C on substitution of Nd by rare-earth elements from the middle of the 4f series (Gd, Tb, Dy) is accompanied by a considerable reduction of the spontaneous magnetization due to the antiparallel coupling of the magnetic moments of these elements to the iron moment. For (NdDy)$Fe_{14}B$, T_C is 598 K compared with 585 K in $Nd_2Fe_{14}B$, but the value for M_s at room temperature amounts to 0.85 MA/m, a much smaller value than for the pure compound (1.28 MA/m) [Yel86]. Nevertheless, Tb and Dy substitutions offer remarkable increases in the anisotropy field as the dysprosium and terbium compounds have the largest uniaxial anisotropy energy due to their strong second-order crystal-field interactions, see Table 1.4. In practical applications, dysprosium is the preferred additive because it is cheaper. The FOMP transition at 16.5 T in $Nd_2Fe_{14}B$ is shifted towards higher field values by a partial substitution of Nd by Dy as in $Nd_{1.8}Dy_{0.2}Fe_{14}B$ [Ver90]. With increasing yttrium substitution in $Nd_{2-x}Y_xFe_{14}B$, the anisotropy field and the anisotropy energy, as well as the critical field and the temperature region in which the FOMP transition in $Nd_2Fe_{14}B$ occurs, decrease in a predictable way, see Fig. 2.38.

Substitutions by other rare-earths have been used in the study of the magnetic phase diagram. Yttrium substitutions indicate that the tilted structure in $Nd_2Fe_{14}B$ persists across the $Nd_{2-y}Y_yFe_{14}B$ series at 4.2 K. The temperature T_{st} decreases slowly with substitutions up to $y = 1.8$ and then falls rapidly, vanishing just before $Y_2Fe_{14}B$. All ions with $\alpha_J < 0$, which have a c-axis preference of the moment, diminish the tilt angle and the spin-tilt temperature. Terbium and dysprosium, with their strong c-axis preference, reduce T_{st} most strongly [Yel86]. Holmium substitutions have the smallest effect on T_{st} and θ_s as the tilted structure is also a characteristic feature of $Ho_2Fe_{14}B$. In general, ions with positive values of α_J increase the tilt angle and the value of T_{st}. Here, samarium is most effective.

Spin reorientations from the basal plane to the c-axis occur for Er and Tm compounds with iron (see Table 2.9 and Fig. 2.32). Again, these phenomena are modified by rare-earth substitutions and can be understood within the model developed above. For the rare-earth elements without anisotropy, the spin reorientation vanishes if nearly all erbium atoms have been substituted. Substitutions by ions with a negative value of α_J decrease the preference of the rare-earth sublattice for the tetragonal plane and, in consequence, lead to a much faster decrease of T_{sr}. The compound (ErPr)$Fe_{14}B$ already exhibits an easy c-axis down to 4.2 K, indicating that the Pr c-axis preference dominates the planar Er preference.

Neutron-diffraction studies reveal a preferential occupation of Dy atoms for the 4f sites [Yel86]. This tendency has been associated with the somewhat larger space available for the bigger neodymium ions at the 4g sites, and it results in a nonlinear concentration dependence of the anisotropy field and the anisotropy energy.

2.2 *Substitutions for iron by cobalt*

The formation of pure $R_2Co_{14}B$ compounds is restricted to the light rare-earth elements and to Gd and Tb. The spontaneous magnetization is much lower than in the Fe counterparts, due to the lower value of the cobalt moment. The magnetic moment of the cobalt sublattice is approximately 20 μ_B/f.u. On the other hand, the compounds with cobalt exhibit much higher Curie temperatures compared with the iron compounds (Table 2.5). The anisotropy energy originating from the 3d

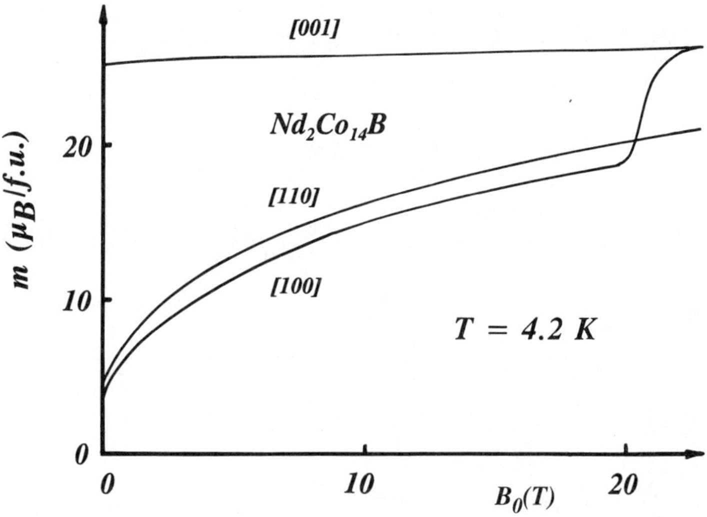

FIG. 2.39. Magnetization curves of single-crystalline $Nd_2Co_{14}B$ at 4.2 K [Hir87].

sublattice in the cobalt and iron compounds is of the same order of magnitude but K_1 is of opposite sign. The cobalt magnetization favours the tetragonal plane with $K_1 = -17$ K/f.u.

The phenomena occurring in the $R_2Co_{14}B$ compounds are very similar to those in the iron counterparts. For $Nd_2Co_{14}B$ at 4.2 K, there is a tilted structure with a tilt angle of about 12° with respect to the *c*-axis towards the < 110 > direction [Hir87a, Hir87b]. The tilted structure vanishes at about 37 K. $Nd_2Co_{14}B$, like $Nd_2Fe_{14}B$, exhibits a FOMP transition when a magnetic field is applied along the hard axis in the plane, the < 100 > direction; see Fig. 2.39. At 4.2 K, the critical field of 20.4 T for $Nd_2Co_{14}B$ is higher than that of 16.3 T for $Nd_2Fe_{14}B$. Unlike $Nd_2Fe_{14}B$, however, $Nd_2Co_{14}B$ exhibits a further spin-reorientation transition from the *c*-axis to the tetragonal plane at $T_{sr} = 545$ K, due to a competition between the axial neodymium preference and the planar preference of the cobalt magnetization. Exchange and crystal-field interactions in $Nd_2Co_{14}B$, as studied by Li *et al.* [Li88], show that the exchange field, B_R^{ex}, in the $Nd_2Co_{14}B$ is much smaller than in $Nd_2Fe_{14}B$, but the difference is accounted for by the smaller cobalt spin, $S_T^{Co} = 0.7$, compared with $S_T^{Fe} = 1.1$. A consequence of the smaller exchange field is the more rapid fall of anisotropy field with temperature in $Nd_2Co_{14}B$, as compared with $Nd_2Fe_{14}B$ (Fig. 2.40). Since the tilt angle at helium temperature is determined by the relative magnitudes of the crystal-field parameters (or anisotropy coefficients), the much smaller tilt angle indicates reduced values for higher-order terms. This demonstrates the influence of 3d atoms on the crystal field at the rare-earth site. Partial substitution of iron by cobalt in $Nd_2Fe_{14}B$ introduces, apart from a FOMP transition along the [100] direction, an additional FOMP transition for the magnetization curve along the [110] direction.

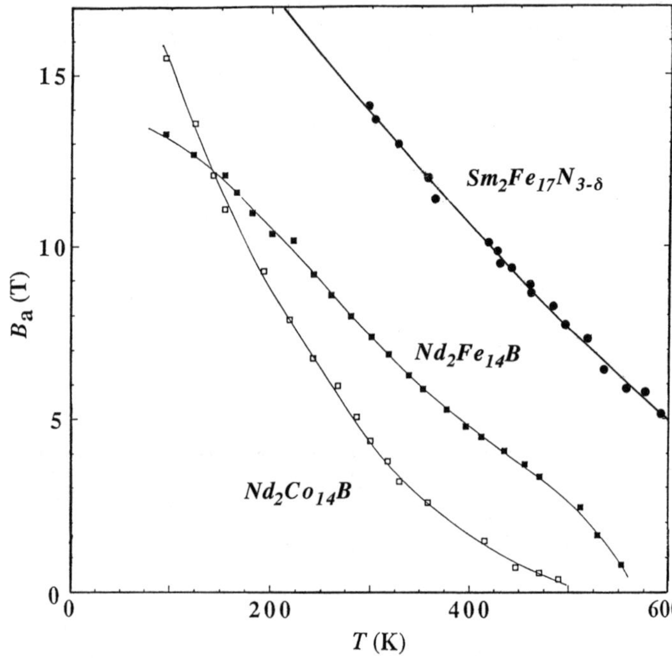

FIG. 2.40. The temperature dependence of the anisotropy field B_a in $Nd_2Co_{14}B$ (□), $Nd_2Fe_{14}B$ (■) [Grö86] and $Sm_2Fe_{17}N_{3-\delta}$ [Kat91].

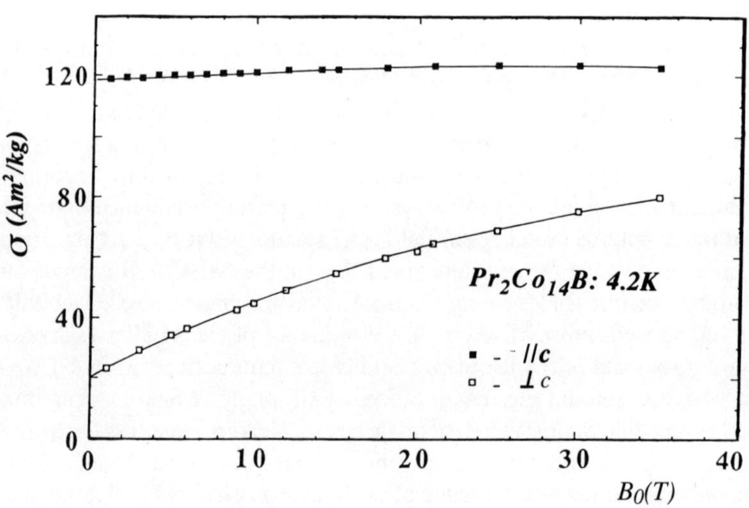

FIG. 2.41. Magnetization curves of a magnetically aligned polycrystalline sample of $Pr_2Co_{14}B$ measured parallel and perpendicular to the alignment direction [Bus85].

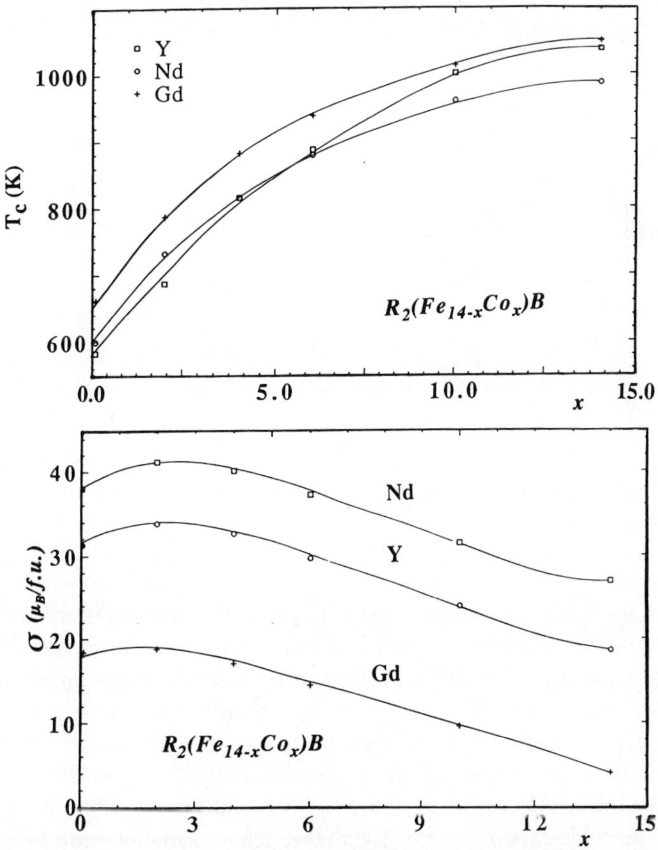

FIG. 2.42. Dependence of the Curie temperature and the saturation moment at 4.2 K in $R_2Fe_{14-x}Co_xB$ (R – Y, Nd and Gd) on the cobalt concentration [Hu86].

An extraordinarily large 'anisotropy field' of 14 T [DeB88] is observed for $Gd_2Co_{14}B$ (Fig. 2.10). An even larger value of 36 T is measured for the saturation field B_K. However, this anisotropy field is an effective anisotropy field as the magnetization does not saturate at this field where the hard-axis magnetization crosses the easy-axis magnetization. The anisotropy energies of $Gd_2Co_{14}B$ and $Gd_2Fe_{14}B$ are actually the same magnitude in spite of the factor of 10 differences in the effective anisotropy field between the $Gd_2Co_{14}B$ and $Gd_2Fe_{14}B$ compounds, which reflects the low resultant magnetization of $Gd_2Co_{14}B$, ten times smaller compared to the iron counterpart.

$Pr_2Co_{14}B$ is a compound showing enormous anisotropy; see Fig. 2.41. Magnetic measurements performed on magnetically aligned polycrystalline samples show that only 60% of the saturation magnetization is reached in a field of 35 T [Bus85]. On this basis a value of 75 T has been derived for the anisotropy field at 4.2 K. The anisotropy energy rapidly decreases with temperature. Nevertheless, the anisotropy

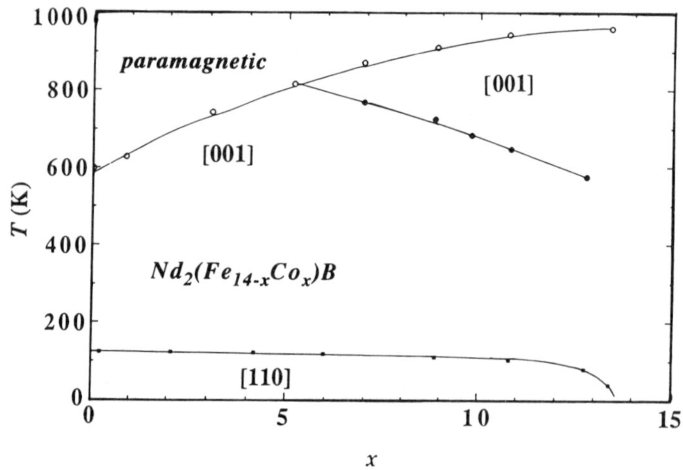

FIG. 2.43. Magnetic phase diagram of the $Nd_2Fe_{14-x}Co_xB$ system; the different easy directions of magnetization are indicated.

energy of $Pr_2Co_{14}B$ at room temperature is still slightly larger than that of $Pr_2Fe_{14}B$ at the same temperature. Temperature-induced spin-reorientation transitions from the *c*-axis to the tetragonal plane are observed for cobalt-based compounds with a negative value of α_J (Pr, Nd and Tb). The values for the spin-reorientation temperature, T_{sr}, are quite high and well above room temperature; see Table 2.9.

The temperature dependence of the rare-earth magnetic moment and the local magnetocrystalline anisotropy in $R_2Co_{14}B$ falls much faster with temperature, due to the molecular fields experienced by the rare-earth ions in compounds, with cobalt being only about 60% of those in compounds with iron [Li88, Rad90b]; see Fig. 2.34.

The magnetization of $Nd_2(Fe, Co)_{14}B$ passes through a slight maximum at $x = 0.2$, (Fig. 2.42) and follows the Slater–Pauling curve [Grö86]. In $Nd_2(Fe, Co)_{14}B$, the values for T_{st} decrease monotonically with cobalt concentration from 135 K to 37 K. The corresponding phase diagram for $Nd_2(Fe, Co)_{14}B$ is shown in Fig. 2.43. At the Co-rich side, a spin-reorientation from the c-axis to the basal plane occurs at higher temperatures. At 4.2 K, the anisotropy energy of $Nd_2Co_{14}B$ is larger than that of the iron compound, but at room temperature the situation is reversed due to the faster decrease of the anisotropy energy with temperature of $Nd_2Co_{14}B$, shown in Fig. 2.40, reflecting the weaker R–Co interactions. At room temperature, the anisotropy field of the $Pr_2(Fe, Co)_{14}B$ series is comparable to that observed in $Nd_2(Fe, Co)_{14}B$, making these compounds equally attractive for applications. ^{57}Fe Mössbauer studies of $Nd_2(Fe, Co)_{14}B$ reveal a preferential occupation of the $8j_2$ sites by iron atoms. Investigations of pseudobinary $Y_2Fe_{14-x}Co_xB$ compounds are usually undertaken in order to understand the 3d anisotropy in these materials [Bol87b]. The dependence of the anisotropy energy on the cobalt concentration is not linear. At 4.2 K, it passes through a maximum of $x = 6$ and changes sign at $x = 10$. The quite complex composition, as well as temperature dependencies are thought to be due to competing effects arising from the different 3d crystallographic sites [Hon88]. Results show that the composition dependence can be quantitatively understood by taking into account the preferential occupation of the different crystallographic sites by iron and cobalt atoms and the different signs of the second-order Stevens factor attributed to the iron and cobalt ions. Similar behaviour of the anisotropy energy is observed for the $La_2Fe_{14-x}Co_xB$ series [Grö88].

Summarizing the influence of the cobalt substitutions on the magnetic properties of $Nd_2Fe_{14}B$, we note that, although there is a decrease in the anisotropy energy, a small replacement of Fe by Co improves the performance of these materials by a rapid increase in the Curie temperature and by a slight initial increase in the saturation magnetization.

2.3 *Other substitutions for iron*

A large number of other substitutions for iron have been studied with the aim of improving the properties of the 3d sublattice. The effects on T_C and M_s in the case of $Nd_2Fe_{14}B$ are summarized in Table 2.11. A substantial preferential site occupation of manganese and chromium on the six nonequivalent transition-metal sites has been revealed by neutron-diffraction studies [Moz88]. In particular, the $8j_2$ site is found to be favoured by Mn and Cr, whereas iron atoms stay preferentially at the 4e and 4c sites. This situation differs from that observed for cobalt substitutions,

Table 2.11 Effects of substitutions for iron on the Curie temperature, T_C, on the anisotropy field, B_a and on the iron sublattice magnetization, M_{Fe}, at helium temperatures for the $Nd_2(Fe_{14-x}T_x)B$ compounds [Coe87, Ku87]; * denotes substitution for *B*, # interstitial atom

T	d ln T_C/dx (%)	d ln B_a/dx (%)	d ln M_{Fe}/dx (%)
Al	–6	–4	–27
Si	+3	–16	–17
Sc	+3	–24	–17
Ti	–20	–37	–26
V	–19	–50	–28
Cr	–8	–8	–17
Mn	–14	–6	–16
Co	+12	–1	+1
Ni	+4	–1	–4
Cu	+1	–	–10
Ru	–19	–39	–27
C*	–8	+27	–7
H#	+5	–15	+3

where a preferential occupation of iron atoms on the $8j_2$ site is observed. Substitution of iron by cobalt, nickel, manganese and chromium affects the volume magnetostriction markedly. This can subsequently affect the magnetoelastic coupling, giving rise to magnetoelastic contributions to the anisotropy energy which eventually exceed the direct contributions to the anisotropy energy from the various individual 3d ions. By inspection of Table 2.11 one can conclude that no substantial improvement of the basic characteristics of the $Nd_2Fe_{14}B$ compound has been found with these substitutions, although unexpected effects, such as an increase in the Curie temperature with replacement of iron atoms by nonmagnetic silicon, have been discovered.

2.4 *Substitutions for boron*

Investigations of the isostructural $R_2Fe_{14}C$ compounds started immediately after the discovery of $Nd_2Fe_{14}B$ [Bol85, Aba85, Liu86]. The magnetic properties of the $R_2Fe_{14}C$ and $R_2Fe_{14}B$ series are very similar. The value for the magnetic moment of $Gd_2Fe_{14}C$ is found to be 18.1 μ_B/f.u. [DeB88b], almost the same as for $Gd_2Fe_{14}B$ [Bog85]. The value of the magnetic moment is lower for $Lu_2Fe_{14}C$ as compared with that for $La_2Fe_{14}C$, in analogy with the $R_2Fe_{14}B$ series. The volume of the elementary cell of the carbon series is lower by about 1.5%, mainly due to the lower value of the *c*-parameter. The Curie temperatures are approximately 40 K lower than those of the corresponding $R_2Fe_{14}B$ compounds. This is most likely due to the

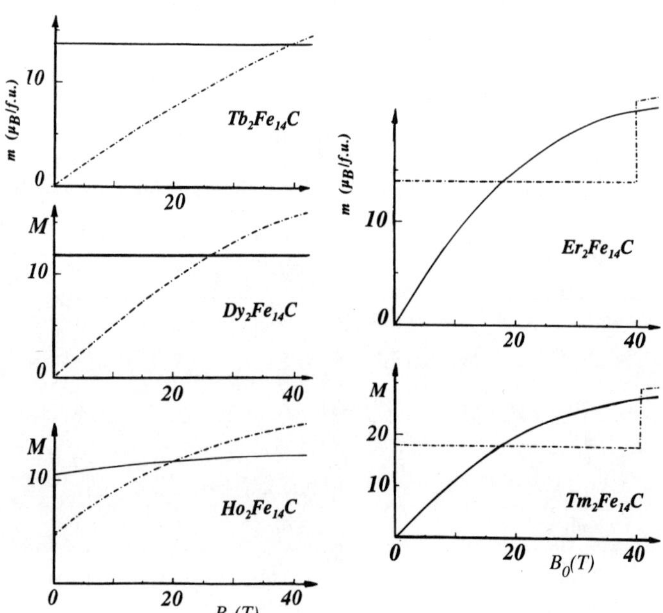

FIG. 2.44. Calculated high-field magnetization curves for $R_2Fe_{14}C$ compounds with the heavy rare-earths along different crystallographic directions [Rad89a].

magnetovolume effect. Values of T_C throughout the series for the heavy rare-earth elements follow the systematics of the de Gennes factor quite well, leading to a value for the exchange-coupling parameter J_{RT} and the exchange fields that are very close to those in the $R_2Fe_{14}B$ series [Rad90b, Liu93].

$Nd_2Fe_{14}C$ also exhibits a tilted magnetic structure at low temperatures; the tilt angle is estimated to be 26° [DeB88b] and T_{st} is found to be 120 K. Measurements on magnetically-aligned powder samples are quite similar to those observed for the corresponding boron compounds. So, carbon substitutions will not introduce changes in the anisotropy fields and the anisotropy energy. It means that the exchange and crystal-field interactions are not substantially affected by the replacement of boron by carbon.

High-field magnetization studies for field-oriented frozen-powder samples of ferrimagnetic compounds have been performed by Radwanski *et al.* [Rad89a]. These studies resulted in the evaluation of the intersublattice molecular-field coefficients and of the molecular field B_{mol}^{R} for heavy rare-earths. The value of 73 T for B_{mol}^{Er} corresponds to 219 T for the exchange field. The resulting 3d–4f exchange-coupling parameter J_{RT} amounts to -7.4 K. The parameter n_{RT} governs the behaviour in the high-field region. The magnetization curves, calculated within the two-sublattice model for ferrimagnetic $R_2Fe_{14}C$ compounds are presented in Fig. 2.44. The crossing of the curves is related to the field-induced noncollinearity of the R and Fe sublattice magnetizations. For the planar Er and Tm compounds the first-order moment reorientation is expected to occur within the tetragonal plane. These transitions have indeed been verified experimentally [Rad92a].

2.5 *Hydrogenation*

By heating in hydrogen or by applying hydrogen pressure, hydrides of the $R_2Fe_{14}B$ compounds can be obtained [Her84]. The hydrides are thermally unstable and decompose on heating [Cad86]. Hydrogenation of the $R_2Fe_{14}B$ compounds is of technological interest in magnet processing, as discussed in Chapter 7. $Nd_2Fe_{14}B$ readily takes up hydrogen at room temperature under 50 bar H_2 atmosphere to a composition $Nd_2Fe_{14}BH_{2.7}$. The maximum uptake of hydrogen across the series ranges from two to five atoms per formula unit. Hydrogenation improves two parameters: the Curie temperature T_C and the spontaneous magnetization M_s at room temperature in particular. The Curie temperature increases in $Y_2Fe_{14}BH_{3.4}$, for instance by about 60 K, although the value of T_C cannot be determined precisely as the compounds lose hydrogen above about 450 K before the paramagnetic state is reached [Coe86]. At room temperature the value of M_s is about 10% larger. The effect of hydrogenation on the magnetocrystalline anisotropy and spin-reorientation phenomena is quite interesting but not promising for applications. Hydrogenation of $Nd_2Fe_{14}B$ suppresses the anisotropy. At room temperature, B_a is reduced to a value of only 2 T. A hydrogen-induced tilted structure has been reported for $Dy_2Fe_{14}BH_{4.7}$ [Coe87]. The anisotropy constants in this compound are drastically changed as compared with those of the pure compound. K_1 changes sign with increase in temperature to a positive value in the vicinity of $T_{st} = 75$ K.

Table 2.12 Magnetic parameters at 4.2 K for monocrystalline R_2Fe_{17} and R_2Co_{17} compounds; [Sin88; Ver90]

Compound	Magnetization σ (Am^2/kg)	m (μ_B/f. u.)	T_C (K)	B_R^{ex} (T)
Y_2Fe_{17}	174.2	35.2	327	–
Ce_2Fe_{17}	136.2	30.0	225	–
Nd_2Fe_{17}	171.4	38.0	326	–
Sm_2Fe_{17}	151.9	34.0	389	–
Gd_2Fe_{17}	101.4	22.9	478	290
Tb_2Fe_{17}	81.9	18.6	410	–
Dy_2Fe_{17}	74.0	16.9	370	214
Ho_2Fe_{17}	81.5	18.7	335	219
Er_2Fe_{17}	75.9	17.5	310	190
Tm_2Fe_{17}	99.7	23.0	306	–
Y_2Co_{17}	132.5	28.0	1080	–
Pr_2Co_{17}	145.6	33.5	1170	–
Nd_2Co_{17}	146.8	33.9	1155	–
Sm_2Co_{17}	101.2	23.6	1190	340
Gd_2Co_{17}	58.1	13.7	1210	–
Tb_2Co_{17}	42.5	10.0	1185	189
Dy_2Co_{17}	35.5	8.4	1165	166
Ho_2Co_{17}	33.6	8.0	1180	159
Er_2Co_{17}	48.3	11.6	1180	150
Tm_2Co_{17}	53.2	13.6	1185	–

In $Er_2Fe_{14}B$, the addition of 2.5 hydrogen atoms increases the spin-reorientation temperature from 325 K to 365 K [Reg88]. Hydrogen atoms influence the crystalline-electric field at the rare-earth sites due to their occupation of adjacent tetrahedral interstitial sites but this effect has not been quantified because of the lack of monocrystalline samples.

3 R_2Fe_{17} and related interstitial compounds

3.1 *R_2Fe_{17} and R_2Co_{17}*

The 2 : 17 structure is formed for rare-earths with the 3d metals iron, cobalt and nickel. The structure is closely related to the $CaCu_5$-type structure and is obtained by replacing one-third of the rare-earth atoms by pairs of 3d atoms. The symmetry is either hexagonal or rhombohedral (Fig. 1.21). A collection of the magnetic parameters of the R_2Fe_{17} and R_2Co_{17} compounds can be found in Table 2.12. All

Table 2.13 Correlation between the experimentally observed easy direction of magnetization at low temperatures for the R_2Fe_{17} series and the Stevens factors α_J and γ_J

Compound	Easy-direction experiment	α_J	EMD in plane experiment	γ_J
Y	plane	0		0
Pr	plane (t)	–	complex	+
Nd	plane (t)	–	b	–
Sm	plane	+		0
Gd	plane	0		0
Tb	plane	–	b	–
Dy	plane	–	a	+
Ho	plane (t)	–	b	–
Er	plane	+	a	+
Tm	*c*-axis ($<$ 72 K)	+		–

(t) Easy direction tilted out of the plane.

the iron compounds have a low Curie temperature. The magnetic anisotropy of Y_2Fe_{17} is rather large and easy-plane, as is the case for Y_2Co_{17}. Insight into 3d anisotropy has been gained by studying the nonlinear variation for the pseudo-binary $Y_2(Fe, Co)_{17}$ series [Thu88]. On most 3d sites the iron and cobalt contributions to the overall anisotropy have opposite signs, resulting for most Y–T intermetallics in an opposite sign of K_1 for the iron and cobalt compounds. Apparently, Y_2T_{17} is an exception since both Y_2Fe_{17} and Y_2Co_{17} are easy-plane systems. The reason is that the overall 3d anisotropy is the resultant of contributions from the four different crystallographic sites that largely cancel but leave a significant energy difference between the hard- and easy-magnetization directions. Electronic structure calculations of Y_2Fe_{17} [Coe89; Beu91] show a strong volume-dependence of the 3d magnetism and give a partial density of states for each of the four different iron sites, yielding iron moments of 2.41 μ_B (4f), 2.35 μ_B (12j), 2.12 μ_B (12k) and 1.95 μ_B (6g).

The easy-magnetization direction at 4.2 K roughly follows the sign of α_J for much of the R_2Fe_{17} series, as shown in Table 2.13. A negative value of α_J in these compounds corresponds to an easy-plane ground state, with the exception of Sm_2Fe_{17} and Er_2Fe_{17}. Easy axis anisotropy and a marked increase in T_C can, however, be achieved by susbstituting some Al or Ga for Fe in Sm_2Fe_{17}. For easy-plane systems, the basal-plane preference of the magnetization follows the sixth-order Stevens factor γ_J. As a consequence of the large easy-plane anisotropy of the iron sublattice, the expected temperature-induced reorientation of the magnetization of Sm_2Fe_{17} and Er_2Fe_{17} does not occur, but for Tm_2Fe_{17}, the reorientation takes place around 72 K. Single-crystal magnetization curves have been measured for R_2Fe_{17} compounds by Andreev *et al.* [And85] for Y, Gd, Tb, Dy, Er and Tm in fields up to 6 T, by Sinnema *et al.* [Sin88] for compounds with Y, Dy, Ho and Er up to 35 T and by Verhoef *et al.* [Ver90] on compounds with Gd, Tb and Er in fields up to 50 T. Data on Tb_2Fe_{17} are shown in Fig. 2.45. Subsequent studies by Franse *et al.* [Fra94] on Er_2Fe_{17} revealed moment reorientations in the magnetization measurements at 4.2 K that were due to a break in the antiparallel configuration of the erbium and iron moments at 36 T along the easy direction and at 44 T along the hard direction in the easy plane. For the remaining ferrimagnetic R_2Fe_{17} compounds, these transitions have been calculated to occur in fields above 40 T [Fra 88]. High-field magnetization

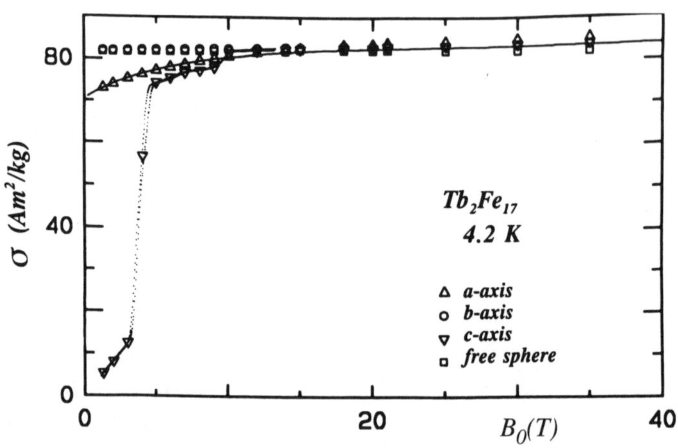

FIG. 2.45. Magnetization curves at 4.2 K for a Tb_2Fe_{17} monocrystalline sphere [Ver92b].

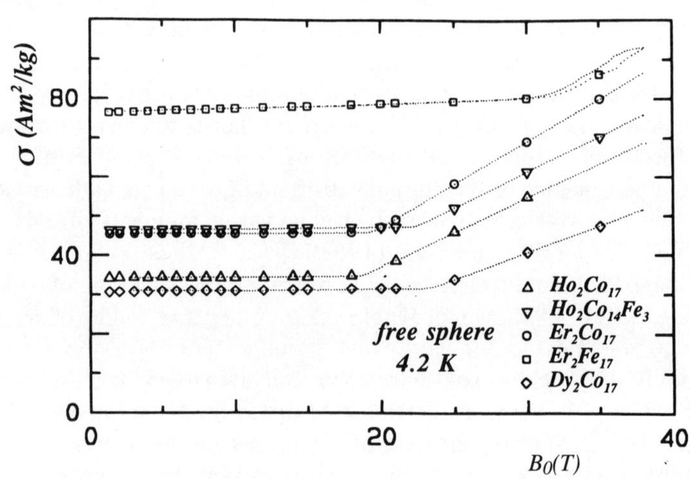

FIG. 2.46. High-field magnetization curves at 4.2 K of several R_2T_{17} monocrystalline spheres that are free to orient themselves in the applied magnetic field [Ver92a].

Table 2.14 Values of the exchange parameters for some R_2Co_{17} compounds, determined by fitting magnetization of monocrystalline spheres that are free to orient themselves in the applied magnetic field

Compound	n_{RT} (kgT/Am^2)	J_{RT} (10^{-22} J)	B_R^{mol} (T)	B_R^{exc} (T)
Dy_2Co_{17}	0.730	1.020	87	174
Ho_2Co_{17}	0.562	0.977	64	160
Er_2Co_{17}	0.456	0.949	50	150

Table 2.15 Values for the crystal-field parameters, B_n^m, and crystal-field coefficients, A_n^m, of some R_2Co_{17} compounds

Compound	Site	B_2^0 (K)	A_2^0 (Ka$_0^{-2}$)	B_4^0 (mK)	A_4^0 (Ka$_0^{-4}$)	B_6^0 (μK)	A_6^0 (Ka$_0^{-6}$)	B_6^6 (mK)	A_6^6 (Ka$_0^{-6}$)	Ref.
Nd_2Co_{17}*		+1.21	−169	+12	−14	−20	+0.04	+8	−14	[Rad88c]
Ho_2Co_{17}	2b	+0.58	−342	+0.60	−13	−13.5	+2.0	+0.14	−20	[Rad89b]
Ho_2Co_{17}	2d	+0.20	−119	+0.82	+18	−18.1	+2.6	+0.14	−20	[Rad89b]
Dy_2Co_{17}		+0.35	−70	+0.28	−3	0	0	−0.28	−44	[Col89a,b]

*Rhombohedral structure with $B_4^3 = -4$ mK and $B_6^3 = -3$ mK.

studies on monocrystalline samples have resulted in an accurate evaluation of the 3d–4f interactions, in particular for ferrimagnetic compounds where the high-field magnetization curves measured on a free-to-rotate monocrystalline sample, is a perfectly straight line passing through zero (Fig. 2.46). The inverse of this slope is a direct measure of the parameter n_{RT}. Values for the parameters n_{RT} and B_R^{mol} thus obtained for some R_2Co_{17} compounds are given in Table 2.14.

The unambiguous evaluation of CEF parameters is hampered by the presence of two inequivalent rare-earth sites in the Th_2Ni_{17} structure which is formed for the heavy rare-earth compounds. A full set of CEF parameters with rhombohedral symmetry has been evaluated for Nd_2Co_{17} where only one rare-earth site exists. Published data for the crystal-field parameters B_n^m are given in Table 2.15.

The existence of localized states in the R_2T_{17} compounds has been clearly evidenced by inelastic-neutron-scattering measurements, which demonstrate the consistency of macro- and microscopic experimental results. INS measurements have been performed on the compounds, Ho_2Co_{17} and Ho_2Fe_{17} [Cla83], Dy_2Co_{17} [Col89] and Nd_2Co_{17} [Col94]. Results were shown for Dy_2Co_{17} in Fig. 2.16, where the data points have been fitted to three different dispersion modes: (i) a weakly dispersive mode due to the R–R interactions; (ii) a dispersive mode due to the R–T interactions; and (iii) a highly dispersive mode related to the T–T interactions. The following values for the exchange parameters have been deduced: $J_{TT} = (17 \pm 5)$ meV, $J_{RT} = (0.21 \pm 0.05)$ meV, $J_{RR} = (0.002 \pm 0.003)$ meV. High-field magnetization data at 4.2 K on monocrystalline Dy_2Co_{17} give a value for J_{RT} of 0.20 meV, in close agreement with the neutron data. On the basis of this successful analysis of micro- and macroscopic experimental results one can conclude that the magnetocrystalline anisotropy in 2 : 17 compounds with Fe and Co is largely produced by the rare-earth sublattice that is of single-ion origin. The value of A_2^0 is $\sim -110\ \mathrm{K}a_0^{-2}$ in the rhombohedral 2 : 17 cobalt compounds such as Sm_2Co_{17}, but its magnitude is significantly less in the corresponding iron compounds.

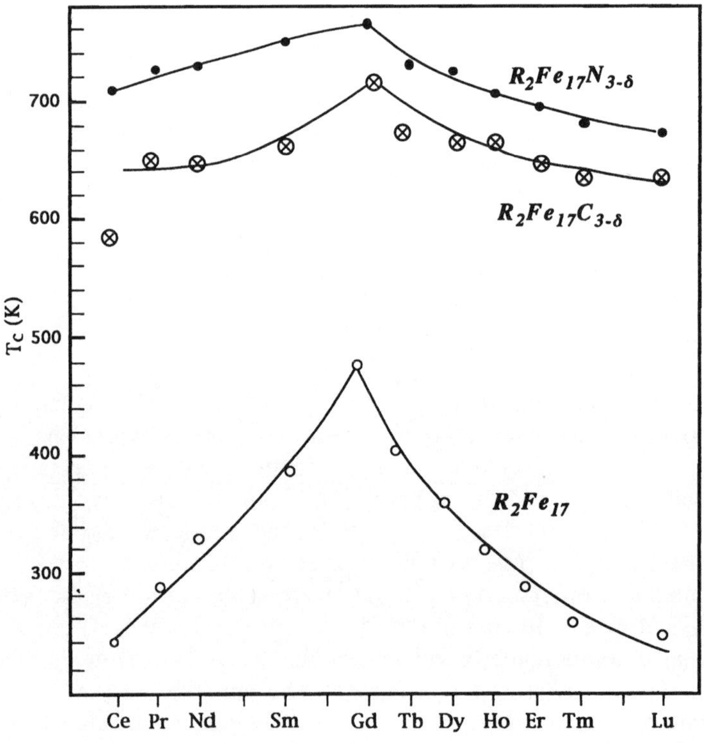

FIG. 2.47. Variation of the Curie temperature across the series $R_2Fe_{17}C_x$, $R_2Fe_{17}N_x (x \sim 3)$ and R_2Fe_{17} [Coe91a].

Table 2.16 Structural and magnetic properties of $R_2Fe_{17}C_{3-\delta}$ compounds ($\delta \approx 0.5$). Lattice parameters are evaluated at room temperature. Magnetic-moment values are measured at 4.2 K; h and r refer to the hexagonal Th_2Ni_{17} and the rhombohedral Th_2Zn_{17} structure respectively

R	Structure	*a* (nm)	*c* (nm)	T_C (K)	Anisotropy	M(μ_B/f.u.)
Y	h	0.886	0.840	668	plane	35.8
Ce	r	0.873	1.256	589	plane	33.8
Pr	r	0.880	1.259	653	plane	34.5
Nd	r	0.879	1.260	659	plane	35.1
Sm	r	0.875	1.257	668	*c*-axis	34.5
Gd	r	0.870	1.261	711	plane	20.1
Tb	r	0.867	1.264	680	plane	21.3
Dy	h	0.865	0.840	674	plane	17.1
Ho	h	0.861	0.843	672	plane	17.5
Er	h	0.860	0.841	663	*c*-axis < 124K plane > 124K	17.9
Tm	h	0.860	0.843	656	*c*-axis < 226K plane > 226K	21.2
Lu	h	0.857	0.842	657	plane	36.4

Table 2.17 Comparison between the structure and the room-temperature magnetic properties of Sm_2Fe_{17} and $Sm_2Fe_{17}Z_{3-\delta}$ [Hur92]

Compound	*a* (nm)	*c* (nm)	$\Delta V/V$ (%)	T_C (K)	M_s (T)	Easy direction	B_a (T)
Sm_2Fe_{17}	0.854	1.234	–	389	1.17	$\perp c$	–
$Sm_2Fe_{17}N_{2.5}$	0.873	1.264	6.3	749	1.54	$\parallel c$	14
$Sm_2Fe_{17}C_{2.2}$	0.875	1.257	6.2	668	1.43	$\parallel c$	13

3.2 $R_2Fe_{17}C_x$

Direct preparation of interstitial $R_2Fe_{17}C_x$ carbides with $x < 1.5$ from the melt was reported by de Mooij and Buschow [DeM88]. The carbon content may be increased by substituting some Al or Ga for Fe [She93]. Subsequently Coey *et al.* showed that, by exposing a finely ground powder of R_2Fe_{17} to 1 bar of a hydrocarbon gas such as butane at a temperature of 750–820 K for one to two hours, interstitial $R_2Fe_{17}C_x$ compounds with $x > 2$ are formed [Coe91]. This procedure allows more carbon atoms to be incorporated into the 2:17 lattice, for reasons discussed in Chapter 4. In any case, the carbon atoms enter to form $R_2Fe_{17}C_x$ with $x < 3$. By carbonation the Curie temperature of R_2Fe_{17} compounds can increase by about 300 K (Fig. 2.47). As T_C is raised far above room temperature, carbonated R_2Fe_{17} compounds have attracted interest as promising iron-based permanent-magnet materials. Good properties are obtained for the interstitial carbides $R_2Fe_{17}C_{3-\delta}$ with $\delta \approx 0.7$, for which composition T_C increases by an average of 330 K. These carbides are formed for the rare-earths from Ce to Lu. The average iron moment at helium temperature increases by roughly 10% [Qi92], so the increase of T_C ($\sim$ 100%) is mainly due to an enhancement of the Fe–Fe exchange associated with the lattice expansion produced by the interstitials. The variation of the plane-axis anisotropy across the $R_2Fe_{17}C_x$ series ($x \simeq 2.5$) is shown in Table 2.16. Corresponding data for the $R_2Fe_{17}N_x$ series are shown in Table 4.3. One can see that easy-axis anisotropy appears at room temperature only in the Sm compound and it is just this compound that can be considered for use in permanent magnets. The magnetic parameters for Sm_2Fe_{17} compounds with N and C interstitials, including the volume expansion $\Delta V/V$, are collected in Table 2.17.

3.3 $R_2Fe_{17}N_x$

Nitrogen atoms can be introduced into the R_2Fe_{17} intermetallic compounds by exposing a powder to a nitrogen-containing gas at approximately 700 K (see Chapter 4). The nitrogen causes even more drastic changes in the magnetic properties than does carbon [Coe90a, Coe90b]. It was found that on nitrogenation the Curie temperature of the R_2Fe_{17} series increases on average by almost 400 K, Fig. 2.47. As nitrogen atoms enter the octahedral interstitial sites of the R_2Fe_{17} lattice (Fig. 4.1), there is a uniform expansion up to $x = 3$ (Fig. 4.4). Preparation, structural and magnetic properties and thermal stability of Sm_2Fe_{17}-carbides, nitrides and hydrides are discussed in Chapter 4. These issues have been reviewed by Skomski [Sko94a, b] and there are some shorter surveys [Coe91, Coe92, Chr93].

Neutron diffraction [Isn92] has established that nitrogen, like carbon, occupies the large (R_2T_4) octahedral interstitial sites called 9e sites in the Th_2Zn_{17} structure and 6h sites in the Th_2Ni_{17} structure. This position, adjacent to the rare-earth atoms, greatly influences the crystal-field potential experienced by the 4f orbitals.

Crystal-field parameters of the parent compounds are radically changed by the occupation of the three interstitial sites around the rare-earth. An estimate of the leading second-order CEF term in $Sm_2Fe_{17}N_3$ from the B_a value by Li and Coey [Li91] results in a value of A_2^0 of -242 Ka_0^{-2}, which is close to that found in $SmCo_5$, -200 Ka_0^{-2}. An analysis of high-field magnetization curves gives $A_2^0 = -333$ Ka_0^{-2} [Kuz94]. ^{155}Gd Mössbauer spectroscopy gives an electric-field gradient V_{zz} of $+12 \cdot 10^{21}$ V/m^2, from which an even larger value has been inferred. The substantial increase of the anisotropy field of Sm_2Fe_{17} after loading with interstitials has been explained satisfactorily within the single-iron approach [Li92, Sko94a]. The value of B_a at 4.2 K is more than 20 T [Kat93]. It is substantially greater than that of $Nd_2Fe_{14}B$ throughout the range of temperature of interest for practical magnet applications (Fig. 2.40).

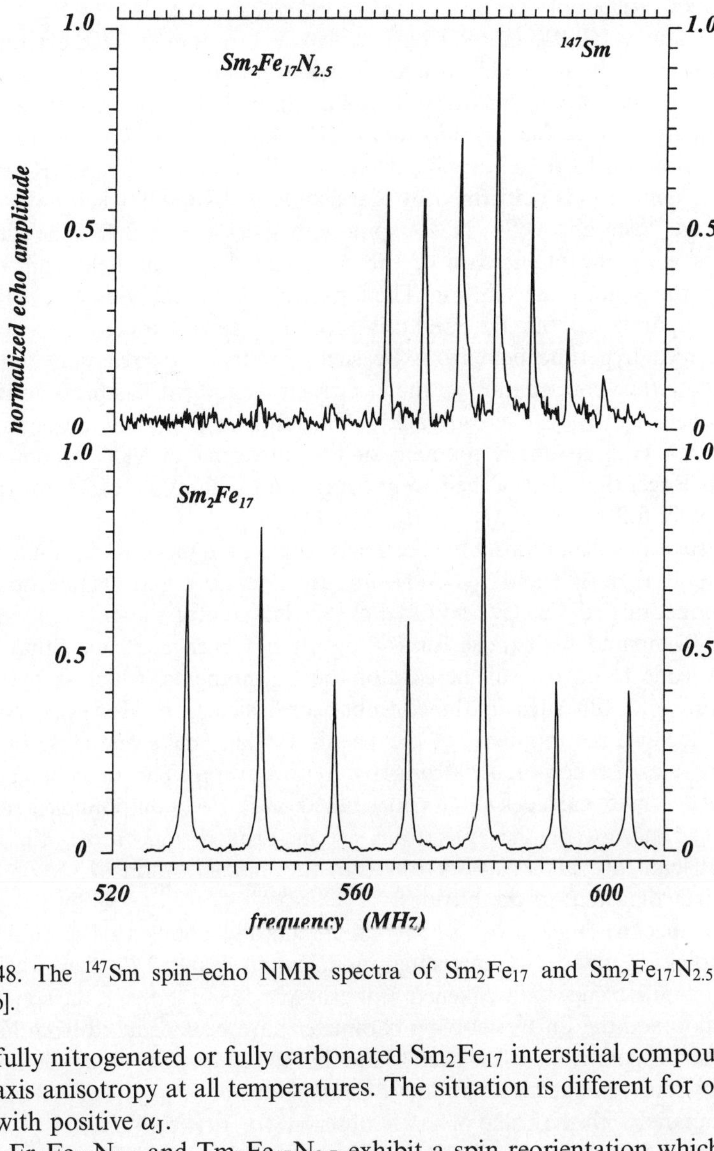

FIG. 2.48. The ^{147}Sm spin–echo NMR spectra of Sm_2Fe_{17} and $Sm_2Fe_{17}N_{2.5}$ at 4.2 K [Kap92b].

The fully nitrogenated or fully carbonated Sm_2Fe_{17} interstitial compounds show easy *c*-axis anisotropy at all temperatures. The situation is different for other rare-earths with positive α_J.

Both $Er_2Fe_{17}N_{2.7}$ and $Tm_2Fe_{17}N_{2.7}$ exhibit a spin reorientation which emerges from a competition of the rare-earth and iron anisotropies at temperatures of 120 K and 220 K, respectively. The rare-earth anisotropy in these two compounds favours the *c*-direction, whereas the iron sublattice prefers the easy direction to be in the basal plane. The fact that the spin reorientation in $Er_2Fe_{17}N_{2.7}$ and $Tm_2Fe_{17}N_{2.7}$ occurs at a much higher temperature than in Er_2Fe_{17} (where it is absent) and Tm_2Fe_{17} (where T_{sr} is 72 K), indicates that the Er and Tm *c*-axis anisotropies are strongly enhanced in the presence of interstitial nitrogen atoms. Similar spin reorientations occur in the carbon compounds [Qi92d] (Table 2.16).

Mössbauer studies of $R_2Fe_{17}C_x$ and $R_2Fe_{17}N_x$ by means of ^{57}Fe, ^{166}Er and ^{169}Tm resonances [Hu91, Gub92] have revealed an anomaly in the temperature variation of the hyperfine field around the spin-reorientation temperatures. From the rare-earth quadrupolar splitting it was confirmed that the interstitial C or N atoms strongly increase the second-order CEF coefficient. The large increase in the lattice contribution to the electric-field gradient is confirmed by NMR studies of $Sm_2Fe_{17}N_x$ compounds performed by Kapusta *et al.* [Kap92] which have revealed well-resolved ^{147}Sm and ^{149}Sm NMR spin–echo spectra, Fig. 2.48. The nitrogenation causes a significant increase of the samarium hyperfine field and a drastic decrease of the quadrupole splitting. The hyperfine field in $Sm_2Fe_{17}N_{2.5}$ is larger by 7.9 T as compared with Sm_2Fe_{17} and this effect has been attributed to a decrease of the transferred hyperfine field from Fe sites. From the decrease in quadrupole splitting the lattice contribution to the electric-field gradient has been derived, but the CEF coefficient A_2^0 is overestimated, as usual from hyperfine measurements.

The spontaneous magnetic moment of the nitrogenated Y_2Fe_{17} compound is about 15% larger than that of the parent compound. The values of M_s for the series approximately follow $M_s = M_{Fe} \pm M_R$ [Sun90].

Band-structure calculations have been performed by Bauerle and Fähnle [Bau92] for hexagonal $Y_2Fe_{17}C_3$ and $Y_2Fe_{17}N_3$ and by Coehoorn and Daalderop [Coe92] for rhombohedral $Gd_2Fe_{17}C_3$ and $Gd_2Fe_{17}N_3$. Results for the magnetic moments have been compared with those for R_2Fe_{17}. It has been confirmed that volume expansion leads to an overall increase of the Fe moments, which is particularly large for 9d and 18h sites in the rhombohedral structure. However, N and C interstitials reduce the moments of the 18f sites, which have one close interstitial neighbour at a distance of 0.190 nm, by hybridization. The magnetocrystalline anisotropy energy of carbides and nitrides based on R_2Fe_{17} compounds is related to the chemical nature of the interstitial atoms around the rare-earth. Carbon interstitials lead to a higher anisotropy, but the magnetization of the carbides is slightly lower than that of the nitrides.

The exchange parameters n_{TT} or J_{TT} are practically doubled as a result of the volume expansion induced by interstitial modification, Table 2.17. Nevertheless, the R–T interaction is somewhat weakened. For example, for $Er_2Fe_{17}C_x$, the spontaneous magnetization and the Er–Fe coupling parameter have been deduced from high-field magnetization experiments. It has been found that the coupling between the Er and Fe moments is somewhat reduced by the interstitial carbon: J_{Er-Fe} falls to 5.7 K for $x = 3$, compared with the value of 6.4 K observed for Er_2Fe_{17} [Liu94].

The effect of nitrogenation on the magnetostriction of Y_2Fe_{17}, Pr_2Fe_{17} and Nd_2Fe_{17} has been studied by Clark *et al.* [Cla92]. It turns out that in contrast to the Curie temperature, the magnetization and the magnetic anisotropy which are all substantially increased after nitrogenation, the magnetostriction remains unchanged or even slightly reduced. In $Y_2Fe_{17}N_x$ and $Nd_2Fe_{17}N_x$, the magnetostriction is positive at high temperatures and decreases with decreasing temperatures. For the nitrogenated Pr_2Fe_{17} the volume magnetostriction has been evaluated as $+0.37 \cdot 10^{-2}$, slightly larger than the value of $+0.31 \cdot 10^{-2}$ found in Pr_2Fe_{17} and consistent with the increased value of the iron moment.

3.4 $R_2Fe_{17}C_xN_y$

The compound $Sm_2Fe_{17}N_3$ is the only one in the $R_2Fe_{17}N_3$ series with uniaxial anisotropy at all temperatures below T_C. It is a metastable compound and the decomposition of the ternary nitride into α-Fe and Sm limits the chemical stability of the phase. From this point of view, samples of $Sm_2Fe_{17}C_xN_y$ have been prepared. Homogenized particles of melt-processed $Sm_2Fe_{17}C_{x(x\sim1)}$ can be charged with nitrogen, or the carbonitrides can be made by a two-step gas–solid reaction. When the intrinsic magnetic properties of nitrides and carbides are compared from a permanent-magnet point of view, $Sm_2Fe_{17}C_{3-\delta}$ is slightly less favourable than the nitride compound owing to the larger value of the anisotropy field, $B_a = 14$ T for the nitride. The saturation field is higher, $B_K = 22$ T, due to a significant K_2 contribution to B_K [Kat91]. However, the carbides have better resistance to higher temperatures. An optimum composition may include both C and N.

For more details concerning intrinsic properties of interstitial intermetallics, see the review of Fujii and Sun [Fuj95].

4 R($Fe_{12-x}M_x$) and related interstitial compounds

4.1 $R(Fe_{12-x}M_x)$

The RT_{12} compounds crystallize in the $ThMn_{12}$ type of structure (Fig. 1.21). The elementary cell of this structure is tetragonal and contains two formulae units. It belongs to the space group I4/mmm. The R atoms occupy the 2a Th site, with the tetragonal point symmetry 4/mmm, whereas the 3d atoms are distributed over the three different manganese sites 8i, 8j and 8f. The $ThMn_{12}$ type of structure is related to the $CaCu_5$ structure by replacing half the R atoms by pairs of transition-metal atoms, according to the formula [Yan88]

$$2RT_5 - R + 2T \rightarrow RT_{12}$$

The hexagonal symmetry of the $CaCu_5$ elementary cell is not preserved. The *c*-axis of the hexagonal $CaCu_5$ cell becomes the *a*-axis of the tetragonal $ThMn_{12}$ cell, with the result that the *c*-axis of the $ThMn_{12}$ structure is perpendicular to that of the $CaCu_5$ structure.

Binary iron compounds of the RFe_{12} type are not formed. The RFe_{12} phase, however, is stabilized by substituting small amounts of metals such as Si, V, Cr, Mo and W for Fe, according to the formula $R(Fe_{12-x}M_x)$. When M = Si, V, Ti and Cr, the minimum x required is close to 1, whereas for M = W and Mo, it is closer to 2.

The $R(Fe_{12-x}M_x)$ compounds are ferromagnetic. Curie temperatures for the $RFe_{11}Ti$ compounds are in the range of 488 K (Lu) to 607 K (Gd). These values for T_C are approximately 200 K larger than for the R_2Fe_{17} compounds. The Curie temperatures can be increased further by substitution of cobalt for iron. The iron sublattice magnetization in $R(Fe_{12-x}M_x)$ exhibits a preference for the tetragonal

c-axis, as can be learned from magnetic studies on aligned powder of $Y(Fe_{10}V_2)$ [DeB88a]. For this latter compound the saturation field along the hard axis at 4.2 K has been estimated to be 4 T. In $R(Fe_{10}Si_2)$ compounds T_C increases linearly with the de Gennes factor of the heavy R-ion involved, see Fig. 2.49, indicating that the R–R interaction proceeds as usual via the R spins rather than the R moments.

The saturation moment of 17 μ_B/f.u., observed for $Y(Fe_{10}V_2)$ at 4.2 K, indicates an iron moment of 1.7 μ_B, which is substantially smaller than that observed for other iron-rich compounds such as Y_2Fe_{17} or $Y_2Fe_{14}B$. The smaller magnetization data are consistent with the Mössbauer result on ^{57}Fe nuclei that provide a value of 23 T for the average hyperfine field [Gub88]. Values of 2.04, 1.77 and 1.56 μ_B have been reported for the Fe moment in $Y(Fe_{11}Ti)$ at 15 K by Qi *et al.* [Qi92b] for the 8i, 8j and 8f sites, respectively. A similar variation of the iron moment at different sites has also been revealed by neutron scattering [Yan88b] and in electronic structure calculations of the magnetic moments [Coe90b; Qi22b]. Mössbauer studies on the rare-earth nuclei (^{155}Gd, ^{161}Dy, ^{166}Er and ^{169}Tm) by Gubbens and van der Kraan [Gub88] in $R(Fe_{10}V_2)$ and by Dirken *et al.* [Dir89] in compounds with Ti and W, have revealed values for the hyperfine fields that are quite close to the free-ion value, pointing to a dominant role of the exchange interactions. Indeed, the full R^{3+}-ion moment is found in magnetic measurements.

Magnetic measurements on monocrystalline $Dy(Fe_{11}Ti)$ by Hu *et al.* [Hu90] have resulted in the determination of the following set of CEF coefficients: $A_2^0 = -32.3$ Ka_0^{-2}, $A_4^0 = -12.4$ Ka_0^{-4}, $A_4^4 = -118$ Ka_0^{-4}, $A_6^0 = +2.56$ Ka_0^{-6}, $A_6^6 = 0.64$ Ka_0^{-6}. The easy direction for the ground state of $Dy(Fe_{11}Ti)$ lies within the tetragonal plane. $Dy(Fe_{11}Ti)$ exhibits two temperature-induced moment reorientations. Below 58 K the spontaneous magnetization is parallel to the <110> direction, whereas

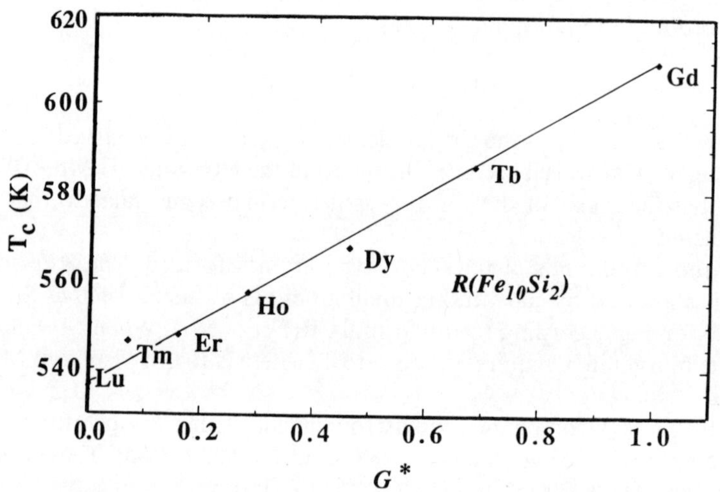

FIG. 2.49. Curie temperature of $RFe_{10}Si_2$ versus the normalized de Gennes factor $G^* = G_R/G_{Gd}$ [Bus88].

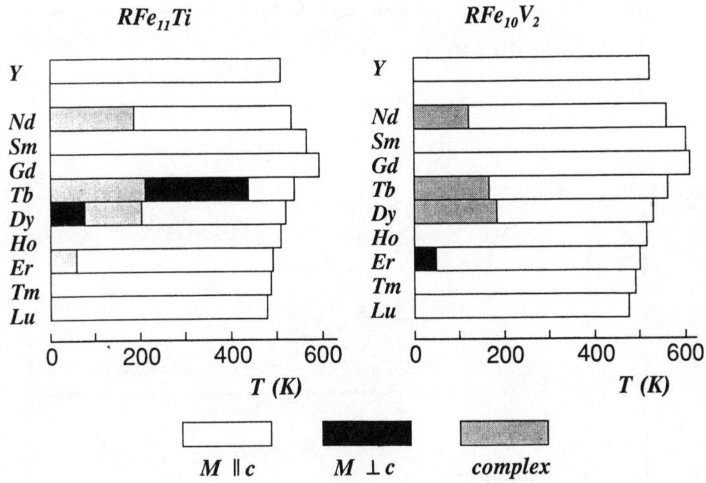

FIG. 2.50. Temperature variation of the easy magnetic direction in the $R(Fe_{11}Ti)$ and $R(Fe_{10}V_2)$ series.

above 200 K it lies along the tetragonal axis. Between these temperatures a complex canted magnetic structure occurs [Kou93]. The magnetic phase diagrams for the $R(Fe_{11}Ti)$ and $R(Fe_{10}V_2)$ series shown in Fig. 2.50 indicate the significance of higher-order CEF interactions since A_2^0 in these compounds is rather small.

Inelastic-neutron-scattering studies have been performed on several 1:12 compounds. The $R(Ni_{10}Si_2)$ materials are more suitable for an evaluation of CEF parameters than the $R(Fe_{10}Si_2)$ compounds. These materials order at very low temperatures due to the absence of a significant magnetic moment at the Ni sites and, hence, the magnetic behaviour at low temperatures is dominated by the CEF interactions. At 20 K, CEF excitations have been observed for $Er(Ni_{10}Si_2)$ with energies of 30, 45, 60 and 66 K. These transitions are expected to be transitions from the Kramers-doublet ground state to excited doublets. The CEF parameters, however, could not be evaluated unambiguously. Nevertheless these studies indicate that the CEF interactions are relatively weak, with an overall splitting of the order of 100 K. This value is typical for erbium compounds. In $ErNi_5$ the overall splitting amounts to 155 K [Rad92b].

The R–T exchange has been measured directly by inelastic neutron scattering from $Gd(Fe_{11}Ti)$ [Hur94]. The exchange field $B_R^{ex} = 255$ T. A somewhat smaller (220 T) value is found in the interstitial nitride $Gd(Fe_{11}Ti)N$ (Fig. 2.51).

The magnetic isotherms for free particles in the $R(Fe_{12-x}V_x)$ (R = Ho and Er) compounds at 4.2 K in fields up 38 T have been measured by Zhong *et al.* [Zho90b]. The magnetization curves for compensated ferrimagnetic compounds, that is, with x close to 3.5, are presented in Fig. 2.52. The parameter n_{RT} as derived from the slope in the high-field region has been found to be 5.8 and 4.2 T f.u./μ_B for the Ho and Er compounds respectively. The same authors have derived values of + 52 K and –8.5 K for the exchange parameters J_{Fe-Fe} and J_{R-Fe} respectively. Both values are close

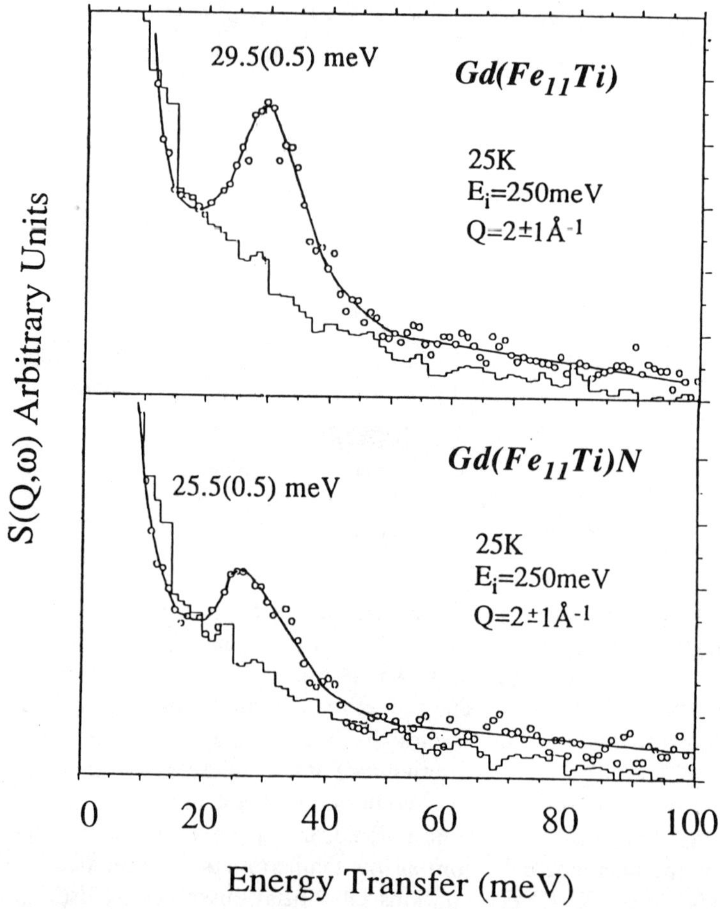

FIG. 2.51. Inelastic neutron spectra showing the $|7/2\rangle \rightarrow |5/2\rangle$ transition for Gd^{3+} in $Gd(Fe_{11}Ti)$ and $Gd(Fe_{11}Ti)\,N_1$ from which B_R^{ex} is obtained directly [Hur94].

to those observed in other binary compounds. For more details regarding the magnetic properties of ternary 1:12 phases, the reader is referred to the review of Li and Coey [Li91].

4.2 $R(Fe_{12-x}M_x)Z, Z = N$ or C

Similarly to the 2:17 compounds, the 1:12 compounds can absorb nitrogen, hydrogen or carbon. The gas-phase interstitial modification (GIM) process was extended to $RFe_{11}Ti$ compounds by Yang *et al.* [Yan91a, b]. Again, it turns out that when these small metalloid atoms enter the lattice interstitially, the tetragonal structure is retained but the lattice parameters increase slightly. Neutron-diffraction studies have revealed that the nitrogen atoms occupy the interstitial 2b sites [Yan91b]; see Fig. 4.1b.

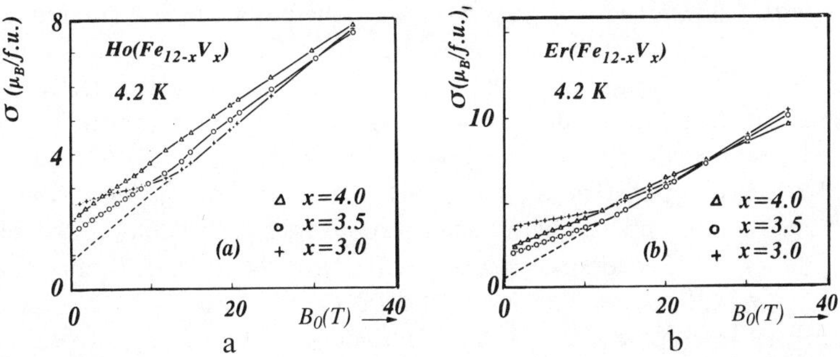

FIG. 2.52. Field dependence of the magnetization at 4.2 K in the nearly compensated ferrimagnetic $Ho(Fe_{12-x}V_x)$ (a) and $Er(Fe_{12-x}V_x)$ (b) compounds obtained on free-powder particles [Zho90b].

The effects of interstitial atoms on the magnetic properties of the $RFe_{11}Ti$ compounds are very similar to those found in the R_2Fe_{17} compounds. Firstly, the uptake of N, C or H expands the lattice which causes the Fe–Fe ferromagnetic interactions to increase by about 50%, thereby augmenting T_C. Secondly, these atoms modify the CEF experienced by 4f electrons substantially, by taking positions close to the rare-earth atoms. A_2^0 is approximately 300 $\mathrm{K}a_0^{-2}$ when $Z = N$ [Hur93], opposite in sign to that for the interstitial 2:17 compounds on account of the arrangement of the interstitials in a dumbell. There is also a large enhancement of the sixth-order crystal field coefficient A_6^0 [Hur95]. As a consequence, there are substantial modifications of the anisotropic behaviour of the rare-earth sublattice. For instance, $HoFe_{11}Ti$ shows c-axis preference in the whole temperature range, whereas the hydrogenated sample $HoFe_{11}TiH_{1.2}$ shows an anomaly in the ac-susceptibility curve at 150 K which is characteristic for a spin-reorientation transition [Tom93]. On the other hand, the spin reorientation in $NdFe_{10.2}Mo_{1.8}$ at 175 K disappears for the nitrogenated sample $Nd(Fe_{10.2}Mo_{1.8})N_{0.8}$.

The effect of carbon on $R(Fe_{11}Ti)$ compounds has been thoroughly investigated by Hurley and Coey [Hur92]. The carbides have been prepared for rare-earths from Nd to Lu and for Y by a reaction of intermetallic powders with butane (C_4H_{10}) at temperatures close to 500 °C. The content of carbon expressed in values for x is estimated between 0.8 and 0.9. The unit volume increases on average by 1.8%, although the main expansion goes on within the basal plane, with $\Delta a/a \approx 0.8\%$. $\Delta c/c$ is found to be much smaller, $\approx 0.2\%$ only. Values for T_C are about 160 K higher than those for the parent compounds, being at the level of 670–734 K. Correlating these values one obtains a value for the magnetic Grüneissen parameter Γ_m of -14.8, in general agreement with values found in Fe-rich intermetallics [Rad88b]. Slightly larger values of Γ_m found in the $RFe_{11}Ti$-nitrides as values for T_C are approximately 50 K higher. The expansion of the lattice enhances the Fe–Fe interactions which is evident when comparing $Y(Fe_{11}Ti)$ and the carbonated or nitrogenated compounds [Qi92]. In analysing the values for T_C across the heavy rare-earth $R(Fe_{11}Ti)$ compounds with $Z = C$, Hurley and Coey [Hur92] estimated the molecular-field coefficients n_{Fe-Fe} and n_{R-Fe} as 471 μ_0 and 144 μ_0, respectively, compared with values of 334 μ_0 and 200 μ_0 for the parent series. The reduction in the value of n_{R-Fe} in the 1:12 interstitial compounds (Fig. 2.51) is analogous to that found in the 2:17 materials. It has been attributed to the valence electrons of the rare-earth atoms forming covalent bonds with the nitrogen or carbon 2p electrons, by which the rare-earth (5d)–iron (3d) wave function overlaps is reduced.

^{57}Fe Mössbauer spectra of nitrides reveal that the average hyperfine field is increased by 3 T as compared with $R(Fe_{11}Ti)$ but there is no comparable effect in carbides [Li92, Qi92b] accompanying the increase of the average Fe moment.

From an application point of view, $Nd(Fe_{11}Ti)C_{1-\delta}$ seems to be a plausible candidate for development as a hard magnetic material due to favourable combination of high Curie temperature, large anisotropy, and high saturation magnetization. The room-temperature anisotropy field is promising for a high coercivity, although the intrinsic magnetic properties of $Nd(Fe_{11}Ti)Z_{1-\delta}$ $(Z = C, N)$ compounds are inferior to those of $Sm_2Fe_{17}Z_{3-\delta}$ $(Z = C, N)$ compounds.

Conclusions

The intrinsic magnetic properties of the intermetallic compounds $R_2Fe_{14}B$, and $Nd_2Fe_{14}B$ in particular, are now quite well understood. High-magnetic-field, Mössbauer, inelastic-neutron-scattering and other experiments provide a consistent picture for the magnetic interactions. Surprisingly, all features can be understood on the basis of isotropic exchange interactions between the 3d and 4f spins, 3d anisotropy and crystal-field interactions of the rare-earth ions. The tilted magnetic structures observed in the neodymium and holmium compounds at low temperatures reveal the importance of the higher-order CEF interactions.

Due to the large number of rare-earth transition-metal intermetallics (more than 200 compounds have been reported with different compositions involving various rare-earth elements), modifications in the magnetic properties of the 3d transition-metal and rare-earth ions can be investigated systematically. The observed systematics are helpful in developing a unified and consistent picture of the 3d and 4f magnetism. The progress that has been made in the understanding of the electronic and magnetic properties was made possible by the improved sample quality, more powerful experimental techniques and extended computation facilities, all developed in the 1980s. Significant correlations have been obtained between the results of macroscopic and microscopic measurements such as inelastic-neutron-scattering experiments and high-field magnetization studies.

It is worth noting that the range of angles over which the R moment in most hard magnetic materials rotates under the action of the applied field is quite limited. In the case of $Ho_2Fe_{14}B$, the holmium moment rotates over an angle of only 8° in fields up to 40 T applied along the <110> direction. This restricts the precision with which the crystal-field parameters for 3d–4f compounds can be evaluated from high-field measurements. On the one hand, there is a need for higher magnetic fields but the combination of different techniques can lead to a consistent picture for the crystal-field parameters of an isostructural series.

Although the main magnetic properties of 3d–4f compounds, the $R_2T_{14}B$, R_2T_{17} and $R(Fe_{12-x}M_x)$ series in particular, are quite well known, the role played by the conduction electrons in the exchange and crystal-field interactions is still poorly understood. An understanding of this role could stimulate the search for new magnetic materials. Interstitial modification with N or C has extended the range of available intermetallics and enhanced our ability to tailor their magnetic properties.

In discussing the electronic and magnetic properties of the R_nT_m systems, two interacting electronic subsystems have been taken into account. The physical properties of the rare-earth subsystem result from the interplay of exchange and CEF interactions, and the compounds cover the whole spectrum from exchange-dominated to CEF-dominated systems. The exchange interactions between the 3d and 4f spins evaluated for $R_2Fe_{14}B$ are similar to those for several other series of R_nT_m intermetallics. The crystalline–electric-field interactions have been investigated for all rare-earth ions with nonzero orbital moments and systematic trends are observed in several isostructural series of compounds. Thanks to the efforts aimed at understanding the properties of a technically useful material, $Nd_2Fe_{14}B$ is certainly one of the intermetallic compounds with the best-known CEF and exchange parameters.

APPENDIX A

Legendre functions

The Legendre functions $P_n^m(\cos\theta)$ are used in this chapter for the description of the magnetocrystalline anisotropy energy as given in Eq. (2.29). These functions are after Hutchings (1964). The normalization factors in the functions with nonzero values of m are put equal to 1 for convenience:

$$P_2^0(\cos\theta) = \frac{1}{2}(3\cos^2\theta - 1),$$

$$P_2^2(\cos\theta) = (1 - \cos^2\theta),$$

$$P_4^0(\cos\theta) = \frac{1}{8}(35\cos^4\theta - 30\cos^2\theta + 3),$$

$$P_4^2(\cos\theta) = (1 - \cos^2\theta)(7cos^2\theta - 1),$$

$$P_4^3(\cos\theta) = (1 - \cos^2\theta)^{\frac{3}{2}}\cos\theta,$$

$$P_4^4(\cos\theta) = (1 - \cos^2\theta)^2,$$

$$P_6^0(\cos\theta) = \frac{1}{16}(231cos^6\theta - 315\cos^4\theta + 105\cos^2\theta - 5),$$

$$P_6^2(\cos\theta) = (1 - \cos^2\theta)(33\cos^4\theta - 18\cos^2\theta + 1),$$

$$P_6^3(\cos\theta) = (1 - \cos^2\theta)^{\frac{3}{2}}(11\cos^3\theta - 3\cos\theta),$$

$$P_6^4(\cos\theta) = (1 - \cos^2\theta)^2(11\cos^2\theta - 1),$$

$$P_6^6(\cos\theta) = (1 - \cos^2\theta)^3.$$

APPENDIX B

Anisotropy coefficients and anisotropy constants

By making use of the definition given by Eqs (2.27) and (2.29), relations between the anisotropy coefficients κ_n^m and the anisotropy constants can be found. In the case of tetragonal symmetry the anisotropy energy is

$$E_a(\theta, \phi) = K_1 \sin^2\theta + K_2 \sin^4\theta + K_3 \sin^6\theta + (K_2' \sin^4\theta + K_3' \sin^6\theta)\cos 4\phi \qquad (1.26)$$

and the relations between the K_i and κ_n^m are given by

$$K_1 = -\frac{1}{2}(3\kappa_2^0 + 10\kappa_4^0 + 21\kappa_6^0) \qquad (2.45)$$

$$K_2 = \frac{1}{8}(35\kappa_4^0 + 189\kappa_6^0)$$

$$K_3 = -(231/16)\kappa_6^0$$

$$K_2' = \kappa_4^4 + 10\kappa_6^4$$

$$K_3' = -11\kappa_6^4$$

In the case of hexagonal symmetry the anisotropy energy is

$$E_a(\theta, \phi) = K_1 \sin^2\theta + K_2 \sin^4\theta + K_3 \sin^6\theta + K_3'' \sin^6\theta \cos 6\phi \qquad (1.27)$$

and the relations between the K_i and κ_n^m are given by

$$K_1 = -\frac{1}{2}(3\kappa_2^0 + 10\kappa_4^0 + 21\kappa_6^0) \qquad (2.46)$$

$$K_2 = \frac{1}{8}(35\kappa_4^0 + 189\kappa_6^0)$$

$$K_3 = -\left(\frac{231}{6}\right)\kappa_6^0$$

$$K_3'' = \kappa_6^6$$

Bibliography

[Bus88] 'Permanent magnet materials based on 3d-rich ternary compounds', K. H. J. Buschow in: *Ferromagnetic materials*, vol. 4, E. P. Wohlfarth and K. H. J. Buschow (eds), North-Holland, Amsterdam (1988), pp. 1–130.

[Bur90] 'Magnetic properties of metals', E. Burzo, A. Chelkowski and H. R. Kirchmayr in: *Landolt–Börnstein new series* III 19d2, H. P. J. Wijn (ed.), Springer, Berlin (1990), pp. 1–469.

[Li91] 'Magnetic properties of ternary rare-earth transition-metal compounds', H. S. Li and J. M. D. Coey in: *Handbook of magnetic materials*, vol. 6, K. H. J. Buschow (ed.), North-Holland, Amsterdam (1991), pp. 1–84.

[Her91] '$R_2Fe_{14}B$ materials: intrinsic properties and technological aspects', J. F. Herbst, *Reviews of Modern Physics*, **63** (1991), 819–898.

[Fra93] 'Magnetic properties of binary rare-earth 3d-transition-metal intermetallic compounds', J. J. M. Franse and R. J. Radwański in: *Handbook of magnetic materials*, vol. 7, K. H. J. Buschow (ed.), North-Holland, Amsterdam (1993), pp. 307–502.

[Fuj95] Interstitially modified intermetallics of rare earth and 3d elements, H. Fujii and H. Sun, in: *Handbook of magnetic materials*. vol. 9, K. H. J. Buschow (ed.) Elsevier, Amsterdam (1995), pp. 303–404.

REFERENCES

[Aba85] C. Abache and H. Oesterreicher, *J. Appl. Phys.* **57** (1985), 4112.

[Aba86] C. Abache and H. Oesterreicher, *J. Appl. Phys.* **60** (1986), 3675.

[And85] A. V. Andreev, A. V. Deryagin, S. M. Zadvorkin and S. V. Terentyev, *Sov. Phys. Solid State*, **27** (1985), 987.

[And88] A. V. Andreev, V. Sechovsky, N. V. Kudrevatykh, S. S. Sigaev, E. N. Taronov, *J. Less-Common Metals* **144** (1988), L21.

[Ast93] G. Asti, in: *Handbook of magnetic materials*, vol. 5, K. H. J. Buschow and E. P. Wohlfarth (eds), North-Holland, Amsterdam, 1993, p. 398.

[Bel87] E. Belorizky, M. A. Frémy, J. P. Gavigan, D. Givord and H. S. Li, *J. Appl. Phys.*, **61** (1987), 3971.

[Ber86] Y. Berthier, M. Bogé, G. Czjek, D. Givord, C. Jeandey, H. S. Li and J. L. Oddou, *J. Magn. Magn. Mater.*, **54–57** (1986), 589.

[Ber88] Y. Berthier, N. Nassar and T. Vadieu, *J. Phys.* (Paris), **49** (1988), C8–585.

[Beu91] T. Beuerle, P. Braun and M. Fähnle, *J. Magn. Magn. Mater*, **94** (1991), L11.

[Bog86] M. Bogé, G. Czjzek, D. Givord, C. Jeandey, H. S. Li and J. L. Oddou, *J. Phys. F: Met. Phys.*, **16** (1986), L67.

[Bol85] F. Bolzoni, F. Leccaube, L. Pareti and J. L. Sanchez, *J. Phys.*, **46** (1985), C6–305.

[Bol86] F. Bolzoni, A. Deriu, F. Leccaube, L. Pareti and J. L. Sanchez, *J. Magn. Magn. Mater.*, **54–57** (1986), 595.

[Bol87] F. Bolzoni, F. Leccaube, O. Moze, L. Pareti and M. Solzi, *J. Magn. Magn. Mater.*, **67** (1987), 373.

[Bol87a] F. Bolzoni, J. M. D. Coey, J. Gavigan, D. Givord, O. Moze, L. Pareti and T. Viadieu, *J. Magn. Magn. Mater.*, **65** (1987), 123.

[Bol87b] F. Bolzoni, J. P. Gavigan, D. Givord, H. S. Li, O. Moze and L. Pareti, *J. Magn. Magn. Mater.*, **66** (1987), 158.

[Bol87c] F. Bolzoni, O. Moze and L. Pareti, *J. Appl. Phys.*, **62** (1987), 615.

[Bro93] M. S. S. Brooks and B. Johansson, in: *Handbook of magnetic materials*, vol. 7, K. H. J. Buschow (ed.), North-Holland, Amsterdam 1993, p. 139.

[Bur84] E. Burzo, in: *Physics of magnetic materials*, J. Rauluszkiewicz, H. Szymczak and H. K. Lachowicz (eds), World Scientific, Singapore 1984, p. 304.

[Bur85] E. Burzo, L. Stanciu and W. E. Wallace, *J. Less-Common Metals*, **111** (1985), 83.
[Bur86a] P. Burlet, J. M. D. Coey, J. P. Gavigan, D. Givord and C. Meyer, *Solid State Commun.*, **60** (1986), 723.
[Bur86b] E. Burzo, N. Plugaru, V. Pop, L. Stanciu and W. E. Wallace, *Solid State Commun.*, **58** (1986), 803.
[Bur87] E. Burzo, N. Plugaru and V. Pop, *J. Magn. Magn. Mater.*, **70** (1987), 343.
[Bus85] K. H. J. Buschow, D. B. de Mooij, S. Sinnema, R. J. Radwański and J. J. M. Franse, *J. Magn. Magn. Mater.*, **51** (1985), 211.
[Bus86] K. H. J. Buschow, *J. Less-Common Metals*, **118** (1986), 349.
[Bus86b] K. H. J. Buschow, D. B. de Mooij and H. M. van Noort, *J. Less-Common Metals*, **125** (1986), 135.
[Bus88a] K. H. J. Buschow, B. B. de Mooij, M. Brouha, H. H. A. Smit and R. C. Thiel, *IEEE Trans. Mag.*, **MAG-24** (1988), 1611.
[Bus88b] K. H. J. Buschow, *J. Appl. Phys.* **63** (1988), 3130.
[Bus90] K. H. J. Buschow, R. Coehoorn, D. B. de Mooij, K. de Waard and T. H. Jacobs, *J. Magn. Magn. Mater.*, **92** (1990), L35.
[Bus91] K. H. J. Buschow, *Rep. Prog. Phys.*, **54** (1991), 1123.
[Cad84] J. M. Cadogan and J. M. D. Coey, *Phys. Rev.*, **30** (1984), 7326.
[Cad86] J. M. Cadogan and J. M. D. Coey, *Appl. Phys. Lett.*, **48** (1986), 442.
[Cad87a] J. M. Cadogan, *J. Less-Common Metals*, **135** (1987), 269.
[Cad87b] J. M. Cadogan, J. M. D. Coey, J. P. Gavigan, D. Givord and H. S. Li, *J. Appl. Phys.*, **61** (1987), 3975.
[Cad88] J. M. Cadogan, J. P. Gavigan, D. Givord and H. S. Li, *J. Phys. F: Metal Phys.*, **18** (1988), 779.
[Cal66] H. B. Callen and E. R. Callen, *J. Phys. Chem. Solids*, **27** (1966), 1271.
[Cam72] I. A. Campbell, *J. Phys: Metal Phys.*, **2** (1972), L47.
[Chr93] C. N. Christodoulou and T. Takeshita, *J. Alloys and Compounds*, **198** (1993), 1.
[Coe86a] J. M. D. Coey, *J. Less-Common Metals*, **126** (1986a), 21.
[Coe86b] J. M. D. Coey, A. Yaouanc and D. Fruchart, *Solid State Commun.*, **58** (1986), 857.
[Coe87a] J. M. D. Coey, A. Yaouanc, D. Fruchart, and P. l'Héritier, *J. Less-Common Metals*, **131** (1987a), 4.19.
[Coe87b] J. M. D. Coey, *Physica Scripta*, T**19** (1987), 426.
[Coe89] R. Coehoorn, *Phys. Rev.* B **39** (1989), 13072.
[Coe90] J. M. D. Coey and Hong Sun, *J. Magn. Magn. Mater.*, **87** (1990), L251.
[Coe90b] R. Coehoorn, *Phys. Rev.* B **41** (1990), 11790.
[Coe91] J. M. D. Coey, Hong Sun, Y. Otani and D. P. F. Hurley, *J. Magn. Magn. Mater.*, **98** (1991), 76.
[Coe91a] J. M. D. Coey, *Physica Scripta*, T **39** (1991), 21.
[Coe92a] R. Coehoorn and G. H. O. Daalderop, *J. Magn. Magn. Mater.*, **104–107** (1992), 1081.
[Coe92b] J. M. D. Coey and D. P. F. Hurley, *J. Magn. Magn. Mater.*, **104–107** (1992), 1098.
[Col89a] J. H. P. Colpa, S. Sinnema, J. J. M. Franse, R. J. Radwański and P. H. Frings, *Physica*, B **156–157** (1989), 731.
[Col89b] J. H. P. Colpa, S. Sinnema, P. H. Frings, J. J. M. Franse and R. J. Radwański, *J. Phys.: Condens. Matter*, **1** (1989b), 2047.
[Col94] J. H. P. Colpa, P. H. Frings, R. J. Radwański, J. J. M. Franse and M. Alba, *J. Magn. Magn. Mater.*, **131** (1994), 167.
[Daa91] G. H. O. Daalderop, P. J. Kelly and M. F. H. Schuurmans, *Phys. Rev.*, B **44** (1991), 12054.
[Dar74] M. I. Darby and E. D. Issac, *IEEE Trans. Mag.*, **MAG-10** (1974), 259.

[DeB87] F. R. de Boer, Huang Ying-kai, D. B. de Mooij and K. H. J. Buschow, *J. Less-Common Metals*, **135** (1987), 199.

[DeB88a] F. R. de Boer, Huang Ying-kai, Zhang Zhi-dong, D. B. de Mooij and K. H. J. Buschow, *J. Magn. Magn. Mater.*, **72** (1988), 167.

[DeB88b] F. R. de Boer, R. Verhoef, Zhang Zhi-dong, D. B. de Mooij and K. H. J. Buschow, *J. Magn. Magn. Mater.*, **73** (1988), 263.

[DeB91] A. Del Moral, M. R. Ibarra, P. A. Algarabel and J. I. Arnaudas, 1991, in: *Physics of magnetic materials*, W. Gorzkowski, M. Gutowski, H. K. Lachowicz and H. Szymczak (eds), World Scientific, Singapore, p. 90.

[DeM87] D. B. De Mooij, J. L. C. Daams and K. H. J. Buschow, *Phil. J. Research*, **42** (1987), 339.

[DeM88a] D. B. De Mooij and K. H. J. Buschow, *J. Less-Common Metals*, **136** (1988), 207.

[DeM88b] D. B. De Mooij and K. H. J. Buschow, *J. Less-Common Metals*, **142** (1988), 349.

[Dep87] P. Deppe, M. Rosenberg, S. Hirosawa and M. Sagawa, *J. Appl. Phys.*, **61** (1987), 4337.

[Erd87] K. Erdmann, P. Deppe, M. Rosenberg and K. H. J. Buschow, *J. Appl. Phys.*, **61** (1987), 4340.

[Eri90] O. Eriksson, B. Johansson, R. C. Albers, A. M. Boring and M. S. S. Brooks, *Phys. Rev.*, B **42** (1990), 2707.

[Fäh93] M. Fähnle, K. Hummler, M. Liebs and T. Beuerle, *Appl. Phys.*, A**57** (1993), 67.

[Fra85] J. J. M. Franse, F. R. de Boer, P. H. Frings, R. Gersdorf, A. Menovsky, F. A. Muller, R. J. Radwański and S. Sinnema, *Phys. Rev.*, **B31** (1985), 4347.

[Fra88] J. J. M. Franse, R. J. Radwański and S. Sinnema, *J. Phys.*, **49** (1988), C8-505.

[Fra90] J. J. M. Franse, R. J. Radwański and R. Verhoef, *J. Magn. Magn. Mater.*, **84**, (1990), 299.

[Fru85] D. Fruchart, P. Wolfers, P. Vulliet, A. Yaouanc, R. Fruchart and P. l'Heritier in *Nd–Fe permanent magnets: Their present and future applications*, I. V. Mitchell (ed.), Elsevier, London, p. 173.

[Fru87] R. Fruchart, P. L'Heritier, P. Dalmas de Reotier, D. Fruchart, P. Wolfers, J. M. D. Coey, L. P. Ferreira, R. Guillen, P. Vulliet and A. Yaouanc, *J. Phys. F: Met. Phys.*, **17** (1987), 483.

[Fue87] C. D. Fuerst, G. P. Meisner, F. E. Pinkerton and W. B. Yelon, *J. Appl. Phys.*, **61** (1987), 2314.

[Fuj87a] H. Fujii, H. Nagata, Y. Uwatoko, T. Okamoto, H. Yamamoto and M. Sagawa, *J. Magn. Magn. Mater.*, **70** (1987a), 331.

[Fuj87b] A. Fujita, H. Onodera, H. Yamauchi, M. Yamada, S. Hirosawa and M. Sagawa, *J. Magn. Magn. Mater.*, **69** (1987b), 267.

[Ful87] P. Fulde and M. Loewenhaupt, *Adv. Phys.*, **34** (1987), 589.

[Gav88] J. P. Gavigan, D. Givord, H. S. Li and J. Voiron, *Physica*, B **149** (1988), 345.

[Gav88a] J. P. Gavigan, D. Givord, H. S. Li, O. Yamada, H. Maruyama, M. Sagawa and S. Hirosawa, *J. Magn. Magn. Mater.* (1988).

[Giv74] D. Givord and R. Lemaire, *IEEE Trans. Mag.*, **MAG-10** (1974), 109.

[Giv84] D. Givord, H. S. Li and R. Perrier de la Bathie, *Solid State Commun.*, **51** (1984), 857.

[Giv85] D. Givord, H. S. Li and F. Tasset, *J. Appl. Phys.*, **57** (1985), 4100.

[Giv86] D. Givord, H. S. Li, J. M. Moreau and P. Tenaud, *J. Magn. Magn. Mater.*, **54–57** (1986), 445.

[Giv88] D. Givord, H. S. Li, J. M. Cadogan, J. M. D. Coey, J. P. Gavigan, O. Yamada, H. Maruyama, M. Sagawa and S. Hirosawa, *J. Appl. Phys.*, **63** (1988), 3713.

[Grö86] R. Grössinger, R. Krewenka, X. K. Sun, R. Eibler, H. R. Kirchmayr and K. H. J. Buschow, *J. Less-Common Metals*, **124** (1986), 165.

[Grö87] R. Grössinger, R. Krewenka, R. Eibler, S. Sinnema, Yang Fu-Ming, Huang Ying-Kai, F. R. de Boer and K. H. J. Buschow, *J. Less-Common Metals*, **132** (1987), 265.

[Grö88] R. Grössinger, H. Kirchmayr and K. H. J. Buschow, *J. Less-Common Metals*, **136** (1988), 367.

[Gu87] Z. Q. Gu and W. Y. Ching, *Phys. Rev.*, **B36** (1987), 8530.

[Gue87] M. Gueramian, A. Bezinge, K. Yvon and J. Muller, *Solid State Commun.*, **64** (1987), 639.

[Gub92] P. C. M. Gubbens, A. A. Moolenaar, T. H. Jacobs and K. H. J. Buschow, *J. Magn. Magn. Mater,*, **104–107** (1992), 1113.

[Har87] I. R. Harris, P. J. McGuiness, D. G. R. Jones and J. S. Abell, *Physica Scripta*, T**19** (1987), 435.

[Her85] J. F. Herbst, J. J. Croat and W. B. Yelon, *J. Appl. Phys.*, **57** (1985), 4086.

[Her86] J. F. Herbst and W. B. Yelon, *J. Appl. Phys.*, **57** (1985), 2343; also: *J. Magn. Magn. Mater.*, **54–57** (1986), 570.

[Hir85] S. Hirosawa and M. Sagawa, *Solid State Commun.*, **54** (1985), 335.

[Hir86a] S. Hirosawa, Y. Matsuura, H. Yamamoto, S. Fujimura, M. Sagawa and H. Yamauchi, *J. Appl. Phys.*, **59** (1986), 873.

[Hir86b] S. Hirosawa, E. Oswald and W. E. Wallace, *J. Magn. Magn. Mater.*, **59** (1986), 185.

[Hir87a] S. Hirosawa, K. Tokuhara, H. Yamamoto, S. Fujimura, M. Sagawa and H. Yamauchi, *J. Appl. Phys.*, **61** (1987), 3571.

[Hir87b] H. Hiroyoshi, M. Yamada, N. Saito, H. Kato, Y. Nakagawa, S. Hirosawa and M. Sagawa, *J. Magn. Magn. Mater.*, **70** (1987), 337.

[Hon90] Hong Sun, J. M. D. Coey, Y. Otani and D. P. F. Hurley, *J. Phys.: Condens. Matter*, **2** (1990), 6465.

[Hon92] Hong Sun, Y. Otani and J. M. D. Coey, *J. Magn. Magn. Mater.*, **104–107** (1992), 1439.

[Hu90] B. P. Hu, H. S. Li, J. M. D. Coey and J. P. Gavigan *Phys. Rev.*, B**41** (1990), 2221.

[Hu91] B. P. Hu, H. S. Li, H. Sun and J. M. D. Coey, *J. Phys: Condensed Matter*, **3** (1991), 3983.

[Hua86] M. Q. Huang, E. B. Boltich, W. E. Wallace and E. Oswald, *J. Magn. Magn. Mater.*, **60** (1986), 270; also: *J. Less-Common Metals*, **124** (1986), 55.

[Hum92] K. Hummler and M. Fähnle, *Phys. Rev.*, B **45** (1992), 3161.

[Hur93] D. P. F. Hurley, Thesis, University of Dublin, 1993.

[Hur92] D. P. F. Hurley and J. M. D. Coey, *J. Phys.: Condens. Matter*, **4** (1992), 5573.

[Hur94] D. P. F. Hurley, J. M. Coey, T. Osakabe, Y. Kuroiwat and M. Rohgi, *J. Phys. Soc. Japan*, **63** (1994), 3048.

[Hur95] D. P. F. Hurley, M. Kuz'min, J. M. D. Coey and M. Kohgi, *J. Magn. Magn. Mater.*, **140–144** (1995), 1027.

[Iba87] M. R. Ibarra, P. A. Algarabel, A. Alberdi, J.Bartolome and A. del Moral, *J. Appl. Phys.*, **61** (1987), 3451.

[Iba92] M. R. Ibarra, Z. Arnold, P. A. Algarabel, L. Morellon and J. Kamarád, *J. Phys.: Condens. Matter*, **4** (1992), 9721.

[Ino85] J. Inoue and M. Shimizu, *J. Phys. F: Met. Phys.*, **15** (1985), 1511.

[Ins92a] O. Isnard, S. Miraglia, D. Fruchart and J. Deportes, *J. Magn. Magn. Mater.*, **103** (1992), 157.

[Isn92b] O. Isnard, S. Miraglia, J. L. Soubeyroux, J. Pannetier and D. Fruchart, *Phys. Rev.*, B**45** (1992), 2920.

[Jur86] M. Jurczyk and W. E. Wallace, *J. Magn. Magn. Mater.* **59** (1986), L182.

[Kam87] J. Kamarad, Z. Arnold and J. Schneider, *J. Magn. Magn. Mater.*, **67** (1987), 29.

[Kap92a] Cz. Kapusta, M. Rosenberg, P. C. Riedi, J. Zukrowski, G. Stoch, H. Figiel, T. H. Jacobs and K. H. J. Buschow, *J. Alloys and Compounds*, **182** (1992), 331.

[Kap92b] Cz. Kapusta, M. Rosenberg, R. G. Graham, P. C. Riedi, T. H. Jacobs and K. H. J. Buschow, *J. Magn. Magn. Mater.*, **104–107** (1992), 1333.
[Kap93] Cz. Kapusta, H. Figiel, G. Stoch, J. S. Lord and P. C. Riedi, *IEEE Trans.Magn.*, **29**, (1993), 2893.
[Kat91] M. Katter, J. Wecker, L. Schultz and R. Grössinger, *J. Magn. Magn. Mater.*, **92** (1991), L14.
[Kat93] H. Kato, M. Yamada, G. Kido, Y. Nakagawa, T. Iriyama and K. Kobayashi, *J. Appl. Phys.*, **73** (1993), 6931.
[Kay91] F. E. Kayzel, R. Verhoef, S. Q. Li, J. J. M. Franse, R. J. Radwański, S. Hoch and H. Kronmüller, *J. Magn. Magn. Mater.*, **101** (1991), 424.
[Kid87] G. Kido, H. Kato, M. Yamada, Y. Nakagawa, S. Hirosawa and M. Sagawa, *J. Phys. Soc. Japan*, **56** (1987), 4635.
[Koh91] T. Kohashi, M. Ono, M. Date, A. Yamagushi, X. P. Zhong, Q. Wang, F. M. Yang, R. J. Radwański and F. R. de Boer, *J. Appl. Phys.*, **69** (1991), 5542.
[Kou93] X. C. Kou, T. S. Zhao, R. Grössinger, H. R. Kirchmayr, X. Li and F. R. de Boer, *Phys. Rev.*, B **47** (1993), 3231.
[Ku87] H. C. Ku and L. S. Yen, *J. Less-Common Metals*, **127** (1987), 43.
[Kuz94] M. D. Kuz'min and J. M. D. Coey, *Phys. Rev.*, B**50** (1994), 12533.
[LHe84] P. L'Héritier, P. Chaudouet, R. Madar, A. Rouault, J. P. Senateur and R. Fruchart, *C. R. Acad. Sci.*, Ser. II, **299** (1984), 849.
[Li88] H. S. Li, J. P. Gavigan, J. M. Cadogan, D. Givord and J. M. D. Coey, *J. Magn. Magn. Mater.*, **72** (1988), L241.
[Li92a] H. S. Li, Y. P. Li and J. M. D. Coey, *J. Magn. Magn. Mater.*, **104–107** (1992), 1444.
[Li92b] Z. W. Li, X. Z. Zhou and A. H. Morrish, *J. Phys.: Condens. Matter*, **4** (1992), 10409.
[Li92c] H. S. Li and J. M.Cadogan in: *Proc. 7th Int. Symp. on Magnetic Anisotropy and Coercivity in RE-TM Alloys*, Perth, 1992, p. 185.
[Liu86] N. C. Liu and Stadelmaier, *Mat. Lett.*, **4** (1986), 377.
[Liu94] J. P. Liu, N.Tang, D. C. Zeng, X. P. Zhong, F. R. de Boer and K. H. J. Buschow, *IEEE Trans Magn.*, **30** (1994), 849.
[Loe88] M. Loewenhaupt, M. Pranger, A. P. Murani and H. E. Hoening, *J. Magn. Magn. Mater.*, **76–77** (1988), 408.
[Loe90] M. Loewenhaupt, I. Sosnowska and B. Frück, *Phys. Rev.*, B **42** (1990), 3866.
[Loe91] M. Loewenhaupt, I. Sosnowska, A. Taylor and R. Osborn, *J. Appl. Phys.*, **69** (1991), 5593.
[Mat85] Y. Matsuura, S. Hirosawa, H. Yamamoto, S. Fujimura and M. Sagawa, *Appl. Phys. Lett.*, **46** (1985), 308.
[May92] H. M. Mayer, M. Steiner, N. Stüber, H. Weinfurter, B. Dorner, P. A. Lindgård, K. N. Clausen, S. Hock and R. Verhoef, *J. Magn. Magn. Mater.*, **104–107** (1992), 1295.
[McE88] K. A. McEwen, *J. Magn. Magn. Mater.*, **76–77** (1988), 391.
[Men83] A. Menovsky and J. J. M. Franse, *J. Cryst. Growth*, **65** (1983), 286.
[Moh87] P. Mohn and E. P. Wohlfarth, *J. Phys. F: Met. Phys.*, **17** (1987), 2421.
[Moz88] O. Moze, L. Pareti, M. Solzi, F. Bolzoni, W. I. F. David, W. T. A. Harrison and A. W. Hewat, *J. Less-Common Metals*, **136** (1988), 375.
[Moz92a] O. Moze, R. Caciuffo, K. H. J. Buschow and G. Amoretti, *J. Magn. Magn. Mater.*, **104–107** (1992), 1391.
[Moz92b] O. Moze, R. Caciuffo, B. Gillon and F. E. Kayzel, *J. Magn. Magn. Mater.*, **104–107** (1992), 1394.
[Nag87] H. Nagata, S. Hirosawa, M. Sagawa, A. Ishikawa and S. Endo, *J. Magn. Magn. Mater.*, **70** (1987), 334.

[Nor91] L. Nordström, B. Johansson and M. S. S. Brooks, *J. Appl. Phys.*, **69** (1991), 5708.
[Par87] L. Pareti, M. Solzi, F. Bolzoni, O. Moze and R. Panizzieri, *Solid State Commun.*, **61** (1987), 761.
[Ped86] A. T. Pedziwiatr and W. E. Wallace, *J. Less-Common Metals*, **126** (1986), 41.
[Ped87] A. T. Pedziwiatr. and W. E. Wallace, *Solid State Commun.*, **64** (1987) 1017; also: *J. Magn. Magn. Mater.*, **65** (1987), 139.
[Piq94] C. Piqué and R. Burriel, *IEEE Trans. Magn.*, **30** (1994), 604.
[Qi92a] Q. N. Qi, H. Sun, R. Skomski and J. M. D. Coey, *Phys.*, B**45** (1992), 12278.
[Qi92b] Q. N. Qi, Y. P. Li and J. M. D. Coey, *J. Phys.: Condens. Matter*, **4** (1992), 8209.
[Qi92c] Q. N. Qi, H. Sun, R. Skomski and J. M. D. Coey, *Phys. Rev.*, B**45** (1992), 12278.
[Qi92d] Q. N. Qi, . D. Kuzmin, H. Sun and J. M. D. Coey, *J. Alloys Compounds*, **182** (1982), 313.
[Qi94] Q. N. Qi, R. Skomski and J. M. D. Coey, *J. Phys.: Condens. Matter*, **6** (1994), 3245.
[Rad85] R. J. Radwański, J. J. M. Franse and S. Sinnema, *J. Magn. Magn. Mater.*, **51** (1985), 175.
[Rad86] R. J. Radwański, *Physica*, B **142** (1986), 57.
[Rad87] R. J. Radwański, J. J. M. Franse, *Phys. Rev.*, B**36** (1987), 8616.
[Rad88a] R. J. Radwański, J. J. M. Franse, *J. Magn. Magn. Mater.*, **74** (1988), 43.
[Rad88b] R. J. Radwański, J. J. M. Franse and K. Krop, *Physica*, B **149** (1988), 306.
[Rad88c] R. J. Radwański, J. J. M. Franse and S. Sinnema, H. J. M. Heerooms and J. H. P. Colpa, *J. Magn. Magn. Mater.*, **76–77** (1988), 182.
[Rad89a] R. J. Radwański, F. R. de Boer, J. J. M. Franse and K. H. J. Buschow, *Physica*, B **159** (1989), 311.
[Rad89b] R. J. Radwański and J. J. M. Franse *Physica*, B **154** (1989), 181.
[Rad89c] R. J. Radwański and J. J. M. Franse *J. Magn. Magn. Mater.*, **80** (1989), 14.
[Rad89d] R. J. Radwański and J. J. M. Franse, J. C. P. Klaasse and S. Sinnema, *J. Phys.*, **49** (1988), C8–531.
[Rad90a] R. J. Radwański, J. J. M. Franse and R. Verhoef, *J. Magn. Magn. Mater.*, **83** (1990), 127.
[Rad90b] R. J. Radwański, X. P. Zhong, F. R. de Boer, J. J. M. Franse and K. H. J. Buschow, *Physica*, B **164** (1990), 131.
[Rad90c] R. J. Radwański, J. J. M. Franse and R. Verhoef, *J. Magn. Magn.* Mater., **83** (1990), 141.
[Rad92a] R. J. Radwański, X. P. Zhong, F. R. de Boer, F. M. Yang, J. Y. Li, T. Kohashi, M. Ono, M. Date and A. Yamaguishi, *J. Magn. Magn. Mater.*, **104–107** (1992), 1139.
[Rad92b] R. J. Radwański, N. H. Kim-Ngan, F. E. Kayzel, J. J. M. Franse, D. Gignoux, D. Schmitt and F. Y. Zhang, *J. Phys.: Condens. Matter*, **4** (1992), 8859.
[Rad92c] R. J. Radwański, J. J. M. Franse, P. H. Quang and F. E. Kayzel, *J. Magn. Magn. Mater.*, **104–107** (1992), 1321.
[Rad92d] R. J. Radwański, *Physica*, B **182** (1992), 209.
[Rad93] R. J. Radwański and J. J. M. Franse, in: *Physics of trans. metals*, P. M. Oppeneer and J. Kübler (eds), World Scientific, 1993, vol. II, p. 782.
[Rad94] R. J. Radwański, *J. Phys.: Condens. Matter* **6** (1994), L15.
[Rad95] R. J. Radwański and N. H. Kim-Ngan, *J. Magn. Magn. Mater.*, **140–144** (1995), 1373.
[Reg87] J. R. Regnard, A. Yaouanc, D. Fruchart, P. l'Héritier, J. M. D. Coey and J. P. Gavigan, *J. Appl. Phys.*, **61** (1987), 3565.
[Rit89] C. Ritter, *J. Phys.: Condens. Matter*, **1** (1989), 2765.
[Sag85] M. Sagawa, S. Fujimura, H. Yamamoto, Y. Matsuura, and S. Hirosawa, *J. Appl. Phys.*, **57** (1985), 4094; also: M. Sagawa, *J. Magn. Soc. Japan*, **9** (1985), 25.
[Sel88] D. J. Sellmyer, M. A. Engelhardt, S. S. Jaswal and A. J. Arko, *Phys. Rev. Lett.*, **60** (1988), 2077.

[She93] B. G. Shen, F. W. Wang, L. S. Kong, L. Cao and H. Q. Guo, *J. Magn. Magn. Mater.*, **127** (1993), L267.

[Shi69] M. Shiga and Y. Nakamura, *J. Phys. Soc. Japan*, **26** (1969), 24.

[Shi93] K. Shimizu, *J. Magn. Magn. Mat.*, **123** (1993), L9; also: *errata J. Magn. Magn. Mater.*, **125** (1993), 372.

[Sin84] S. Sinnema, R. J. Radwański, J. J. M. Franse, D. B. de Mooij and K. H. J. Buschow, *J. Magn. Magn. Mater.*, **44** (1984), 333.

[Sin87] S. Sinnema, J. J. M. Franse, R. J. Radwański, A. Menovsky and F. R. de Boer, *J. Phys. F: Met. Phys.*, **17** (1987), 233.

[Sin89] S. Sinnema, Thesis, University of Amsterdam (1988).

[Sko94a] R. Skomski in: *Interstitial alloys for reduced energy consumption and pollution*, G. J. Long *et al.* (eds), Kluwer, Dordrecht, 1994.

[Sko94b] R. Skomski, S. Brennar and S. Wisth in: *Interstitial alloys for reduced energy consumption and pollution*, G. J. Long *et al.* (eds), Kluwer, Dordrecht, 1994a.

[Str75] R. L. Streever, *Phys. Rev.*, B **12** (1975), 4653.

[Suc54] A. Sucksmith and J. E. Thompson, *Proc. Roy. Soc.*, **225** (1954), 362.

[Sun90] H. Sun, J. M. D. Coey, Y. Otani and D. E. F. Hurley, *J. Phys., Condens. Matter* (1990), 6465.

[Szp85] B. Szpunar and J. A. Szpunar, *J. Appl. Phys.*, **57** (1985), 4130.

[Thu86] N. P. Thuy and J. J. M. Franse, *J. Magn. Magn. Mater.*, **54–57** (1986), 915.

[Thu88] N. P. Thuy, J. J. M. Franse, N. M. Hong and T. D. Hien, *J. Phys. (Paris)*, **49** (1988), C8-499.

[Tom93] E. Tomey, O. Isnard, A. Fagan, C. Desmoulins, S. Miraglia, J. L. Soubeyroux and D. Fruchart, *J. Alloys and Compounds*, **191** (1993), 233.

[Try92] J. Trygg, B. Johansson, M. S. S. Brooks, *J. Magn. Magn. Mater.*, **104–107** (1992), 1447.

[VNo85] H. M. van Noort and K. H. J. Buschow, *J. Less-Common Metals*, **113** (1985), L9.

[Vas85] A. Vasquez, J. M. Friedt, J. P. Sanchez, P. L'Heritier and R. Fruchart, *Solid State Commun.*, **55** (1985), 783.

[Vas87] A. Vasquez, J. P. Sanchez, *J. Less-Common Metals*, **127** (1987), 71.

[Ver88] R. Verhoef, J. J. M. Franse, A. A. Menovsky, R. J. Radwański, S. Q. Ji, F. M. Yang, H. S. Li and J. P. Gavigan, *J. Phys. (Paris)*, **49** (1988), C8-565.

[Ver90a] R. Verhoef, P. H. Quang, J. J. M. Franse and R. J. Radwański, *J. Magn. Magn. Mater.*, **83** (1990), 139.

[Ver90b] R. Verhoef, R. J. Radwański and J. J. M. Franse, *J. Magn. Magn. Mater.*, **89** (1990), 176.

[Ver90c] R. Verhoef, P. H. Quang, J. J. M. Franse and R. J. Radwański *J. Appl. Phys.*, **67** (1990), 4771.

[Ver90d] R. Verhoef, Thesis, University of Amsterdam (1990).

[Ver92a] R. Verhoef, S. Sinnema, P. H. Quang and J. J. M. Franse, *J. Magn. Magn. Mater.*, **104–107** (1992), 1325.

[Ver92b] R. Verhoef, P. H. Quang, R. J. Radwański, C. Marquina and J. J. M. Franse, *J. Magn. Magn. Mater.*, **104–107** (1992), 1473.

[Wan92] Y. Z. Wang, G. C. Hadjipanayis, A. Kim, D. J. Sellmyer and W. B. Yelon, *J. Magn. Magn. Mater.*, **104–107** (1992), 1132.

[Yel86] W. Y. Yelon, B. Foley, C. Abache and H. Oesterreicher, *J. Appl. Phys.*, **60** (1986), 2982.

[Yam85] M. Yamada, Y. Yamaguchi, H. Kato, H. Yamamoto, Y. Nakagawa, S. Hirosawa and M. Sagawa, *Solid State Commun.*, **56** (1985), 663.

[Yam86] H. Yamauchi, M. Yamada, Y. Yamaguchi, H. Yamamoto, S. Hirosawa and M. Sagawa, *J. Magn. Magn. Mater.*, **54–57** (1986), 575.

[Yam87a] M. Yamada, H. Kato, H. Hiroyoshi, H. Yamamoto and Y. Nakagawa, *J. Magn. Magn. Mater.*, **70** (1987), 328.

[Yam87b] H. Yamauchi, H. Yamamoto, S. Hirosawa and M. Sagawa, *J. Magn. Magn. Mater.*, **70** (1987), 340.

[Yam88] M. Yamada, H. Kato, H. Yamamoto and Y. Nakagawa, *Phys. Rev.*, B **38** (1988), 620.

[Yan88a] Y. C. Yang, L. S. Wang, S. H. Sun and D. M. Gu, *J. Appl. Phys.*, **63** (1988), 3702.

[Yan88b] Y. C. Yang, H. Sun, L. S. Kong, J. L. Yang, Y. F. Ding, B. S. Zhang, C. T. Ye, L. Jin and H. M. Zhou, *J. Appl. Phys.*, **61** (1988), 5968.

[Yan91a] Y. C. Yang, X. D. Zhang, L. S. Kong, Q. Pan and S. L. Ge, *Appl. Phys. Lett.*, **58** (1991), 2042.

[Yan91b] Y. C. Yang, X. D. Zhang, L. S. Kong, Qi Pan, S. L. Ge, J. L. Yang, Y. F. Ding, B. S. Zhang, C. T. Ye and Lan Jin, *Solid State Commun.*, **78** (1991), 313.

[Zha91] Z. D. Zhang, X. K. Sun, Y. C. Chuang, F. R. de Boer and R. J. Radwański, *Phys. Rev.*, B **43** (1991), 555.

[Zho90a] X. P. Zhong, R. J. Radwański, F. R. de Boer, R. Verhoef, T. H. Jacobs and K. H. J. Buschow, *J. Magn. Magn. Mater.*, **83** (1990), 143.

[Zho90b] X. P. Zhong, R. J. Radwański, F. R. de Boer, T. H. Jacobs and K. H. J. Buschow, *J. Magn. Magn. Mater.*, **86** (1990), 333.

3

Phase relations

K. G. Knoch

Metallurgical phase diagrams illustrate which phases coexist in equilibrium at a particular temperature and elemental composition. Knowledge of phase equilibria is important for an understanding of the processes taking place during casting, sintering and annealing of alloys for permanent magnets, and the resulting microstructures. Systems of most interest here are the binaries R–Fe, R = Pr, Nd or Sm, ternaries, especially Nd–Fe–B, and quaternary and higher systems where additional elements are incorporated in an effort to improve magnetic properties, corrosion resistance and processability.

Diagrams of binary systems A_xB_{100-x} are represented as two-dimensional composition: temperature x:T plots, where the compositional range is divided into a number of single-phase fields and two-phase regions. The uppermost line at any temperature is the liquidus, which shows where the liquid (l) first begins to crystallize into solid (s). According to the Gibbs phase rule[1] [Pri66], no more than three phases can coexist in equilibrium at any point on the diagram. Crystallization $l \rightarrow s$ normally proceeds via a $l + s$ two-phase region where both the liquid and the solid phases change their compositions with cooling. Many intermetallics have a well-defined stoichiometry and are represented by vertical lines, but a compound can sometimes exist over a limited range of composition, when it appears as a narrow single-phase field on the diagram.

The three-phase solidification reactions are the *eutectic* reaction, e, where a liquid freezes to give a mixture of two solids with different compositions $l \rightarrow s_1 + s_2$, and the *peritectic* reaction p, where the liquid and solid react to yield a solid of different composition $l + s_1 \rightarrow s_2$. These reactions appear as horizontal line segments on the phase diagram, whose positions are determined by differential thermal analysis (DTA). Examples of both reactions can be seen in the Nd–Fe phase diagram shown in Fig. 3.1 [Lan90]. The phases present at any point are usually determined by metallography or scanning electron microscopy and X-ray diffraction, after quenching to room temperature. Any horizontal line spanning a two-phase region on the diagram is known as a *tie line*. The relative proportion of the two phases coexisting at a point on the tie line whose compositions are given by its extremities is given by the lever rule [Pri66].

Ternary diagrams are three-dimensional x:y:T plots, where x and y are the two compositional variables necessary to specify the ternary composition $A_xB_yC_z$. The third variable z is $100-x-y$. All three can be represented in triangular coordinates on a horizontal plane, so the phase diagram is a triangular prism with T as the

[1] The phase rule for a system at constant pressure is $F = C + 1 - P$; it relates the number of degrees of freedom F to the number of components C ($C = 2$ for binary systems) and the number of coexisting phases P.

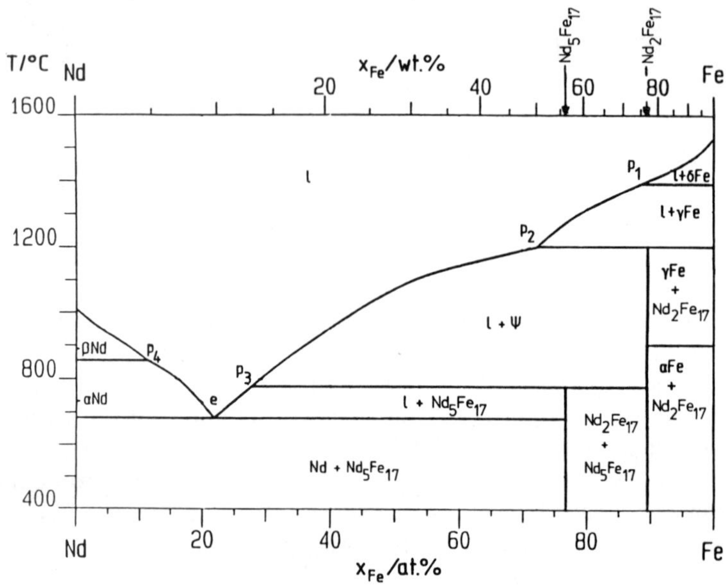

FIG. 3.1. Nd–Fe phase diagram [Lan90].

vertical axis. In order to present the diagram on paper, one takes isotherms (horizontal sections of constant T) or isopleths (vertical sections of constant x, y, z, x/y, y/z or z/x. Another useful representation is the vertical projection of the liquidus surface, showing the composition of the liquid which is in equilibrium with two solid phases as lines of double saturation and the solidification reactions as points where three lines meet. Arrows on the lines of double saturation show the direction in which the composition of the liquid changes with decreasing temperature.

According to the phase rule, no more than four phases can now coexist in equilibrium, so there are three possible solidification reactions, the *eutectic* reaction $E: L \rightarrow S_1 + S_2 + S_3$, the *transition* reaction $U: L + S_1 \rightarrow S_2 + S_3$ and the *peritectic* reaction $P: L + S_1 + S_2 \rightarrow S_3$. Reactions are now represented by segments of horizontal planes.

Quaternary and higher-order systems are more complicated thermodynamically, and require more than three dimensions to represent their phase diagrams. Only very special sections can be illustrated. When the amount of a fourth element is small, the region of interest of a quaternary system can lie close to one of the boundary ternary systems, and it may be possible to examine the phase diagram in this region.

Reaction kinetics and diffusion depend exponentially on temperature, so at temperatures below about 500 °C it may be practically impossible to access the

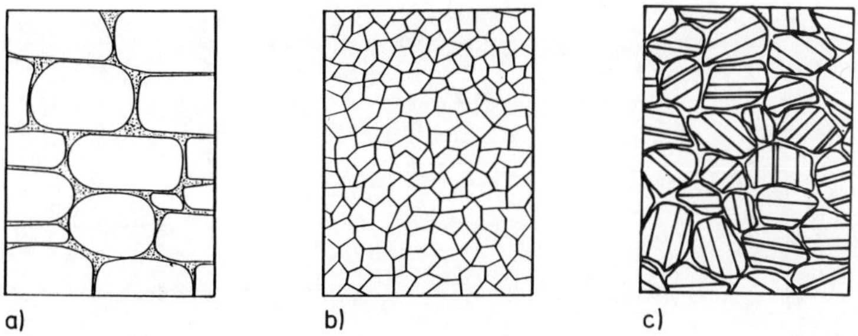

FIG. 3.2. Some ideal microstructures for permanent magnets.

equilibrium corresponding to global rather than local minima of the free energy. The microstructure of an alloy therefore reflects the phase equilibria, diffusion kinetics and thermal history of the sample. In a permanent magnet, a special phase microstructure is sought which inhibits the nucleation of reverse domains or pins the domain walls. Some examples of ideal microstructures are shown in Fig. 3.2:

1. This is a dispersion of cystallites of the magnetically hard phase in an enveloping, nonferromagnetic, intergranular phase, which can be achieved by sintering at a temperature where a liquid is present to wet the solid grains [Sch90]. Liquid-phase sintering yields high-density material, and also tends to remove sharp edges from the grains if this will reduce the interface energy.
2. This is a nanocrystalline structure with little or no grain boundary phase. Each grain is smaller than the single-domain limit d_c, and the domain walls are pinned at the boundaries. Such structures may be obtained directly by rapid quenching, or by annealing amorphous material [Iwa89, Cla92].
3. This is an intimate mixture of two phases interspersed by a platelet phase [Fid87]. Of the former two phases, one is the magnetically hard phase, and the other is a minority phase with different magnetic properties which is dispersed as a very fine precipitate with at least one dimension comparable to the domain wall thickness δ_w. Sm_2Co_{17}-type magnets are precipitation-hardened in this way.

1 Binary systems

All rare-earth iron phase diagrams R–Fe [Kub82, Mas90] exhibit a deep eutectic at the rare-earth side, near $x \approx 75$ at.% and 600 °C $\leq T \leq$ 900 °C. A R_2Fe_{17} compound (ψ) exists for the series of rare-earths from Ce to Lu, but very few other intermetallics form between iron and the light rare-earths. None whatever form between Fe and La.

Praseodymium. The rhombohedral Pr_2Fe_{17} compound is the only stable intermetallic [Nei90]. It forms peritectically at 1100 °C, which is 300 °C below the liquidus temperature [Mas90]. As with other light rare-earths, Pr_2Fe_{17} has a

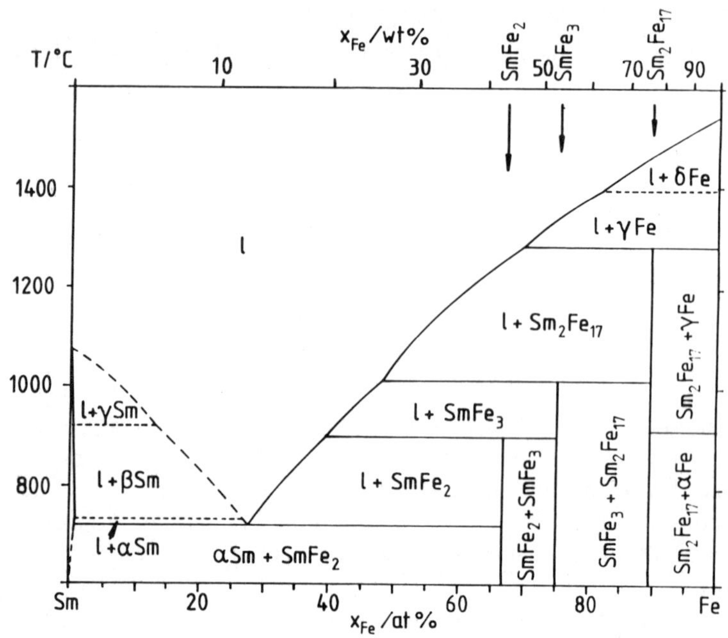

FIG. 3.3. Sm–Fe phase diagram [Kub82].

limited homogeneity range on the rare-earth side, which corresponds to replacing an Fe_2 dumbell in the structure by a rare-earth atom. Three metastable phases, A_1, A_1' and A_3 were found [Nei90] but compositions are unknown.

Neodymium. The Nd–Fe phase diagram [Lan90,Hen93] was shown in Fig. 3.1. It was long thought that Nd_2Fe_{17} was the only stable phase but it is now known that another phase, Nd_5Fe_{17}, forms peritectically and is stable below about 780 °C. $NdFe_2$ can be formed under high pressure only [Can72]. A metastable phase A_1 of an amorphous or nanocrystalline structure [Giv92] was found [Sch88, Lan90] which is likely to play a key role in the ternary Nd–Fe–B system where it transforms into $Nd_2Fe_{14}B$ and may explain the significant improvement of the coercivity on annealing.

Samarium. In the Sm–Fe diagram shown in Fig. 3.3 [Kub82], there are three intermetallic phases: rhombohedral Sm_2Fe_{17}, rhombohedral $SmFe_3$, and the cubic Laves phase $SmFe_2$. All form peritectically, although the 2:17 phase lies closer to the liquidus than for Pr or Nd. In fact, this compound melts congruently for Dy and Ho.

The crystallographic and magnetic properties of the binary compounds are summarized in Table 3.1a and b respectively. Some of the metastable phases have also been observed in amorphous material formed by quenching the eutectic melt [Iwa89, Bra90, Yan91]; they are generally poorly characterized and may be stabilized by oxygen.

Table 3.1 Some data for pure metals and binary alloys

(a) Structural data (T_C values marked by * denote Néel temperatures, ** denotes no magnetic ordering)

Phase	Symmetry	Space group	Structure type	Lattice parameters (nm)	T_C(K)
α–Fe	bcc	$Im\bar{3}m$	W	$a = 0.28665$	1043
Pr	hex.	$P6_3/mmc$	α–La	$a = 0.36714, c = 1.18310$	**
Pr_2Fe_{17} (ψ)	rhomb.	$R\bar{3}m$	Th_2Zn_{17}	$a = 0.8585,\ c = 1.2464$	283
Nd	hex.	$P6_3/mmc$	α–La	$a = 0.36566, c = 1.17983$	*7.5
Nd_2Fe_{17} (ψ)	rhomb.	$R\bar{3}m$	Th_2Zn_{17}	$a = 0.8571,\ c = 1.2445$	327
Nd_5Fe_{17}	hex.	$P6_3/mcm$	Nd_5Fe_{17}	$a = 0.8574,\ c = 1.235$	503
Sm	rhomb.	$R\bar{3}m$	α–Sm	$a = 0.86281, c = 2.6233$	*14.8
Sm_2Fe_{17} (ψ)	rhomb.	$R\bar{3}m$	Th_2Zn_{17}	$a = 0.8570,\ c = 1.2440$	385
$SmFe_3$	rhomb.	$R\bar{3}m$	Be_3Nb	$a = 0.5787,\ c = 2.491$	650
$SmFe_2$	fcc	$Fd\bar{3}m$	$MgCu_2$	$a = 0.7417$	675
Metastable					
A_1 (Nd)	amorph.				518
A_1 (Pr)					498
A_1' (Pr)					≈460
A_3 (Pr)					537
$NdFe_7$	rhomb.			$a = 0.738, c = 1.072$	543
Nd_7Fe_3	tetrag.			$a = 0.512, c = 0.801$	

(b) Magnetic data (J_s at 1.2 K)

Phase	J_s(T)
Nd_2Fe_{17}	1.66
Nd_5Fe_{17}	1.50
Fe	2.22
Pr_2Fe_{17}	1.69
Sm_2Fe_{17}	1.51
$SmFe_3$	0.80
$SmFe_2$	0.61

2 The ternary Nd–Fe–B System

The first information on the ternary Nd–Fe–B phase diagram was provided by Chaban *et al.* in 1979 [Cha79]. Several years elapsed before it was discovered that the $Nd_2Fe_{14}B$ phase (ϕ) can be used to produce permanent magnets. Chaban *et al.* showed an isothermal section at 600 °C with all three stable ternary phases in this system, namely, $Nd_2Fe_{14}B$ (ϕ), $Nd_{1.1}Fe_4B_4$ (η) and $Nd_5Fe_2B_6$ (ρ), even though the

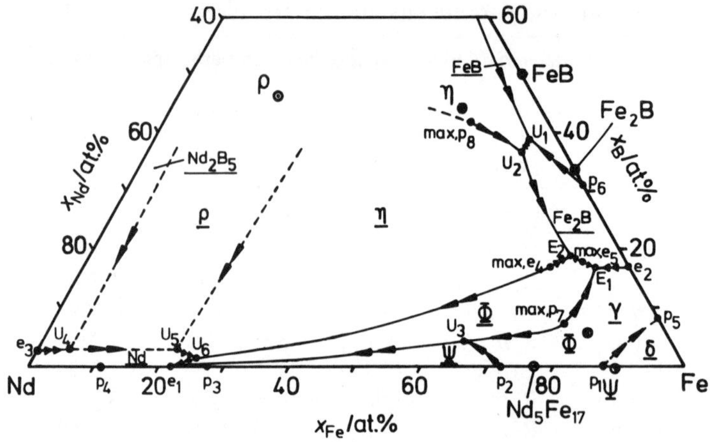

FIG. 3.4. Liquidus projection of the ternary phase diagram Nd–Fe–B.

Table 3.2 Structural and magnetic data for ternary phases in the Nd–Fe–B system (J_s values at 4.2 K)

Phase	Symmetry	Space group	Structure type	Lattice parameters (nm)	T_C(K)	J_s(T)
$Nd_2Fe_{14}B$ (ϕ)	tetrag.	$P4_2/mnm$	$Nd_2Fe_{14}B$	$a = 0.8792$, $c = 1.2177$	578	1.86
$Nd_{1.1}Fe_4B_4$ (η)	orthorh.	Pccn		$a = b = 0.7117$, $c = 3.501$	13	0.13
$Nd_5Fe_2B_6$ (ρ)	rhomb.	R3m		$a = 0.55$, $c = 2.43$	64	0.15
Metastable						
$Nd_2Fe_{23}B_3$	bcc	$I\bar{4}3d$		$a = 1.419$	655	2.57
$NdFe_{12}B_6$	rhomb.	R3m	$SrNi_{12}B_6$	$a = 0.9605$, $c = 0.7541$	230	1.01
$Nd_2Fe_{17}B$						
ε					523	
A_1						

compositions were not quite correct. Some data for these phases are included in Table 3.2, where some metastable phases are listed as well [Bus86, Iwa89, Lin92]. The liquidus projection shown in Fig. 3.4 is based on the work of Matsuura *et al.* [Mat85], Schneider *et al.* [Sch86], Landgraf *et al.* [Lan91] and Knoch *et al.* [Kno94]. There is still some uncertainty in this diagram, particularly in the neighbourhood of p_3 and the related Nd_5Fe_{17} which are both marked in the boundary system but not taken into account in the ternary system; Nd_5Fe_{17} has not been observed in ternary

Table 3.3 Reactions in the ternary system Nd–Fe–B

Symbol	Reaction	T (°C)
p_1	$l + \delta\text{Fe} \rightarrow \gamma\text{Fe}$	1392
p_2	$l + \gamma\text{Fe} \rightarrow \psi$	1185
p_3	$l + \psi \rightarrow \text{Nd}_5\text{Fe}_{17}$	780
p_4	$l + \beta\text{Nd} \rightarrow \alpha\text{Nd}$	856
p_5	$l + \delta\text{Fe} \rightarrow \gamma\text{Fe}$	1381
p_6	$l + \text{FeB} \rightarrow \text{Fe}_2\text{B}$	1389
p_7	$l + \gamma\text{Fe} \rightarrow \phi$	1180
p_8	$l + \text{NdB}_4 \rightarrow \eta$	1345
e_1	$l \rightarrow \text{Nd}_5\text{Fe}_{17} + \text{Nd}$	685
e_2	$l \rightarrow \gamma\text{Fe} + \text{Fe}_2\text{B}$	1177
e_3	$l \rightarrow \text{Nd} + \text{Nd}_2\text{B}_5$	1000
e_4	$l \rightarrow \eta + \phi$	1115
e_5	$l \rightarrow \text{Fe}_2\text{B} + \phi$	1110
U_1	$L + \text{FeB} \rightarrow \text{Fe}_2\text{B} + \text{NdB}_4$	?
U_2	$L + \text{NdB}_4 \rightarrow \eta + \text{Fe}_2\text{B}$	?
U_3	$L + \gamma\text{Fe} \rightarrow \phi + \psi$	1130
U_4	$L + \text{Nd}_2\text{B}_5 \rightarrow \text{Nd} + \rho$	?
U_5	$L + \rho \rightarrow \text{Nd} + \eta$	(745)
U_6	$L + \eta \rightarrow \text{Nd} + \phi$	710
E_1	$L \rightarrow \phi + \gamma\text{Fe} + \text{Fe}_2\text{B}$	1105
E_2	$L \rightarrow \eta + \phi + \text{Fe}_2\text{B}$	1095

alloys [Sch89]. Hallemans *et al.* [Hal95] calculated the phase diagram and found excellent agreement with the experimental data as presented here for Nd contents below 60 at.%. Discrepancies for higher Nd contents are attributable to uncertainties for both the theoretical and the experimental data. It is known that Nd_2Fe_{17} forms from the metastable A_1 phase which exists both in the binary and in the ternary systems. However, when a ternary alloy containing A_1 is annealed, A_1 transforms into ϕ and not into Nd_5Fe_{17} as in the binary case. Investigations by Tsai *et al.* [Tsa87] supported by the results of Knoch *et al.* [Kno94] necessitate changes in the lines of double saturation max, $p_7 \rightarrow U_3 \rightarrow e_1$ and $p_2 \rightarrow U_3$ compared to [Mat85] and [Sch86]. Table 3.3 lists the reactions as indicated in the phase diagram with the corresponding temperatures. Figure 3.5 shows the important isothermal section at 1050 °C [Kno94]. This is approximately the temperature used for liquid-phase sintering of Nd–Fe–B magnets. There is a two-phase region where ϕ is in equilibrium with liquid, and a three-phase region where ϕ is in equilibrium with liquid, and η. The commonly used magnet composition Nd15–Fe77–B8 falls in the three-phase region. The presence of the η phase is associated with sharp edges on the ϕ grains which contribute to the dipolar stray fields, thereby tending to reduce the coercivity. η does not form on sintering the composition Nd18.5–Fe73–B6.5 which falls in the two-phase region. In the other three-phase regions ϕ coexists with at least one ferromagnetically soft phase, Nd_2Fe_{17}, Fe or Fe_2B which tend to

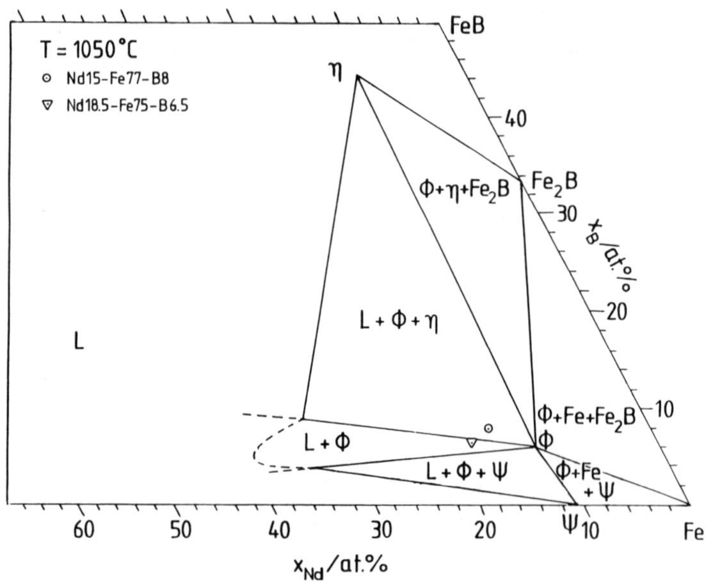

FIG. 3.5. Isothermal section of the Nd–Fe–B phase diagram at 1050 °C.

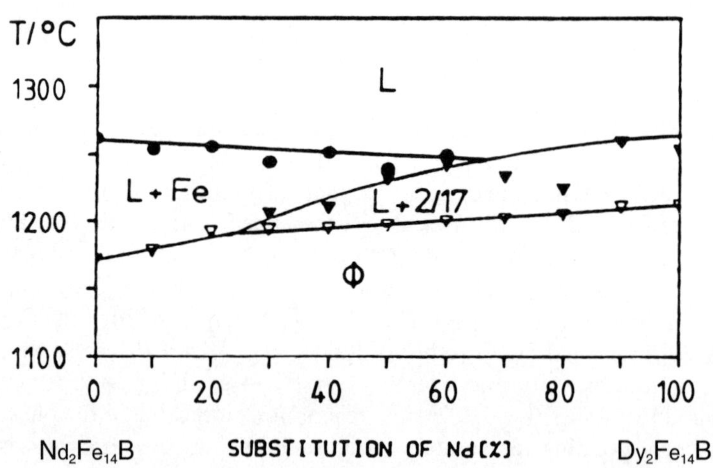

FIG. 3.6. Primary and secondary phases with the substitution of Nd by Dy. (From [Gri89].)

destroy the coercivity. Compositions lying in these regions are unsuitable for magnet production, unless they are nanocrystalline.

The Nd-rich phase may be hexagonal (a = 0.39, c = 0.69 nm) when the iron content is less than 3%, but an fcc form with a = 0.52 nm contains more iron, up to 9% [Fid89]. These results originate from analytical transmission electron microscopic studies; however, from analytical scanning electron microscopic investigations no solubility of Fe in Nd is reported.

3 Quaternary systems: additions to Nd–Fe–B

Many metallic elements can be added to the ternary Nd–Fe–B system. They may be grouped according to whether they dissolve in ϕ or in the Nd-rich melt or form new tenary compounds.

3.1 *Rare-earth substitution*

The ϕ phase exists for rare-earths from Ce to Lu, so all these elements may be substituted for Nd. Small Dy additions are particularly useful for improving coercivity, and the Nd/Dy section is shown in Fig. 3.6 [Gri89]. The primary crystallization product is γ–Fe for a Dy content up to 65%.

The similarity of Pr and Nd ensures that the magnetic properties are basically unchanged by Pr-substitution [Zha85].

3.2 *Transition-metal substitution*

The ϕ phase exists for cobalt for rare-earths from La to Tb. There is a continuous solid solution between $Nd_2Fe_{14}B$ and $Nd_2Co_{14}B$ with monotonically increasing Curie temperature [Vil88]. However, cobalt additions can stabilize the $Nd(Fe, Co)_2$ pseudobinary Laves phase, which is disadvantageous for coercivity. Further, A1 addition renders this phase nonmagnetic at room temperature [Miz87, Tsu88, All89].

Manganese and nickel have limited solubility in $Nd_2Fe_{14}B$; for Mn Zhang *et al.* [Zha85] report up to about 3 at.% whereas a substitution of Ni for Fe of up to 12% is reported [Aba86, Bol87]. Magnetic properties are generally impaired.

3.3 *Metalloid substitution*

The boron in $Nd_2Fe_{14}B$ can be completely substituted by carbon, and the ternary carbides exist for rare-earths from La to Lu. The complete structure was first determined for $La_2Fe_{14}C$ [Mar85]. The $R_2Fe_{14}C$ compounds form from Fe, χ (R_4FeC_6) and ψ ($R_2Fe_{17}C_x$) by an annealing treatment. Very long anneals lead to an isotropic cellular microstructure that may involve $R_2Fe_{14}C$ and $R_2Fe_{17}C_x$. The transformation temperature rises from 880 °C for Nd to 1100 °C for the heavy rare-earth metals [Moo88], but it may be increased by further additions such as B, Cu or Mn [Eis91, Gri91]. This is advantageous because the time required for the transformation decreases significantly at higher temperatures, particularly as soon as ϕ is in equilibrium with liquid. By careful control of the microstructure it is possible to obtain coercive material directly by heat treating Nd–Fe–C ingot. A useful section of the Nd–Fe–C diagram between 600 °C and 700 °C is shown in Fig. 3.7 [Gri90, Bus91]. The ψ phase can dissolve interstitial carbon, but the equilibrium homogeneity range of $Nd_2Fe_{17}C_x$ extends only to $x \leq 1.5$ at these temperatures.

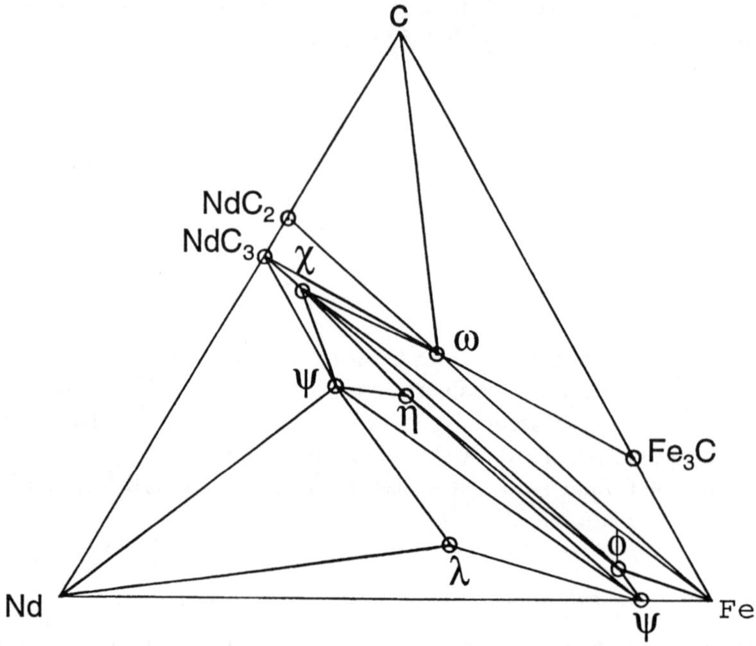

FIG. 3.7. Isothermal section for the system Nd–Fe–C between 600 °C and 700 °C showing the phase equilibria [Gri90, Bus91].

3.4 *Noble metal additions*

Copper and silver additions to Nd–Fe–B do not affect the ϕ phase, but interact with Nd to reduce the temperature of the Nd-rich eutectic [Cha90]. It is normally at 685 °C, but can be lowered to 485 °C by adding copper [Shi88]. This greatly facilitates hot rolling of Nd–Fe–B and Pr–Fe–B magnets [Shi90]. Copper additions permit a significant reduction of the boron content to at least 5 at.% without forming the Nd_2Fe_{17} phase [Kia91]; even lower B contents proved successful in the case of magnets containing a Nd–Pr mixture [Kre91]. A $Nd_6Fe_{13}Cu$ phase (δ) having a tetragonal structure [Kno93a] related to that of $La_6Co_{11}Ga_3$ exists in the ternary Nd–Fe–Cu system [Mül92], as shown in Fig. 3.8a. The compound orders magnetically at 190 °C with a small net polarization $J_s = 10$ mT, but it possesses exceptionally strong magnetocrystalline anisotropy.

3.5 *Al, Ga additions*

These elements substitute both for iron in the ϕ phase and in the Nd-rich liquid, where they reduce the eutectic temperature [Tok89]. Aluminium and gallium also reduce the surface tension of the liquid and thereby improve the microstructure

[Kno90]. Variation of wetting angle with Al and Ga content is illustrated in Fig. 3.9. Both form a δ phase ($Nd_6Fe_{11}Al_3$, $Nd_6Fe_{11}Ga_3$) with an extended homogeneity range, as shown in Fig. 3.8b and c; Al also forms a Nd (Fe, Al)$_2$ (μ) pseudobinary [Gri90a].

3.6 *Nb, Mo, Ti, V and Zr*

These elements are mostly used with other additions to Nd–Fe–B and they form ternary borides such as Mo_2FeB_2 [Hu88, Hir90], $Nb_3Fe_3B_4$ [All89], V_2FeB_2 [Sag90], VFeB [Yin92], and ZrB_2 [All89].

For Nb, which degrades the magnetic properties slightly, there also appears $NbFe_2$ and a phase with approximate composition $NdNb_7Fe_7$. The former creates paramagnetic precipitates inside ϕ which may be small enough to act as pinning centres for magnetic domains [Hu88].

Besides improving coercivity, Mo and V also have a useful effect on the corrosion properties of the Nd-rich phase [Hir90a, Ten90]. On the introduction of extra boron, the formation of other borides inhibits grain growth of ϕ.

Apparently, Ti does not improve any magnetic properties when it is the only additive in sintered magnets [Zha85] but significant improvements of H_c were obtained for melt-spun material [Kiy89] Improvement of $(BH)_{max}$ in sintered magnets was achieved by using Ti together with several other additives, where it also contributes to the corrosion resistance [Shi90]. Ti is mainly found as TiB_2, which inhibits grain growth [Kit91].

Zr is found to degrade the magnetic properties of sintered magnets [Leo89], although the reason for this is not fully understood. It forms coherent precipitates in ϕ, but is mainly dissolved in η and an additional Zr-rich intergranular phase of lamellar structure [Pol88]. However, in HDDR processed material, very small amounts of Zr, typically 0.1 at.%, produce a spectacular improvement in magnetic properties of Nd Fe B [Tak90, McG92]. The HDDR process is described in Chapter 7. Just how the Zr acts in this material is not fully understood but this example illustrates that phase relations and microstructure in multicomponent systems are pre-eminently a subject for painstaking experimentation, the results of which cannot always be foreseen.

3.7 *Oxygen*

Most rare-earth alloys contain oxygen which is introduced unintentionally when they are processed. Surfaces are usually coated by an oxide layer that forms under conditions of ambient humidity, with some hydrogen entering the intermetallic lattice simultaneously. Small quantities of oxygen are not easy to detect, so its significance is imperfectly understood. Several ternary Nd–Fe–O compounds with an O-content ≤ 5 at.% have been reported [Sta85, Sch89a].

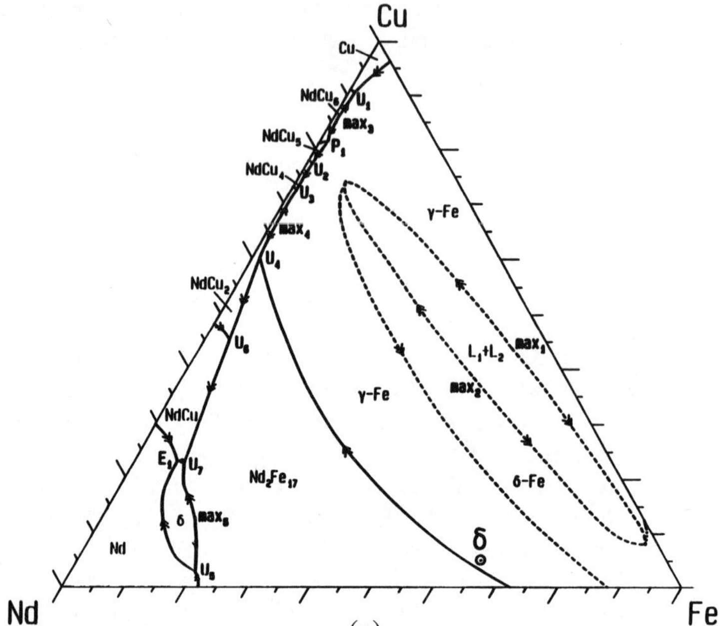

(a)

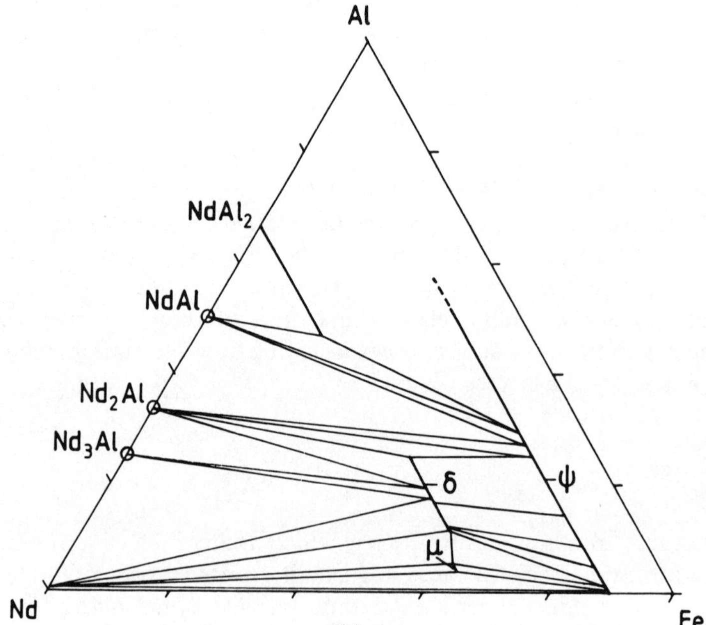

(b)

FIG. 3.8. (a) Projection of the liquidus surface for the Nd–Fe–Cu system [Mül92]. (b) Isothermal section at 600 °C of the Nd–Fe–Al system [Gri90a]. (c) Isothermal section at 600 °C of the Nd–Fe–Ga system [Gri90a].

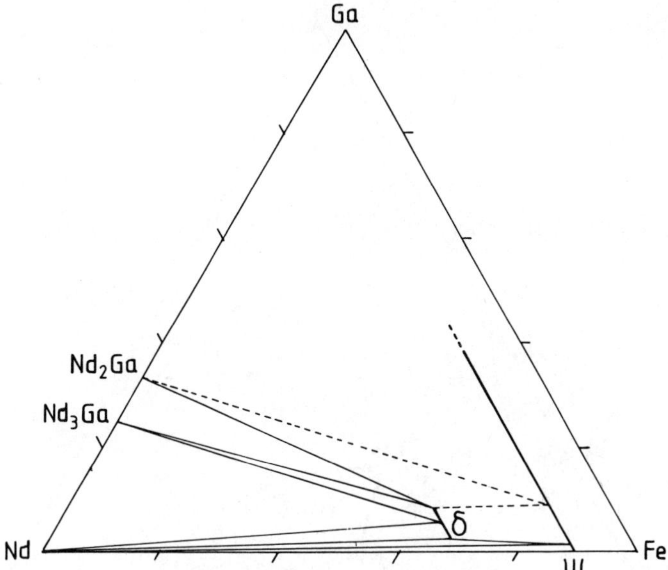

(c)

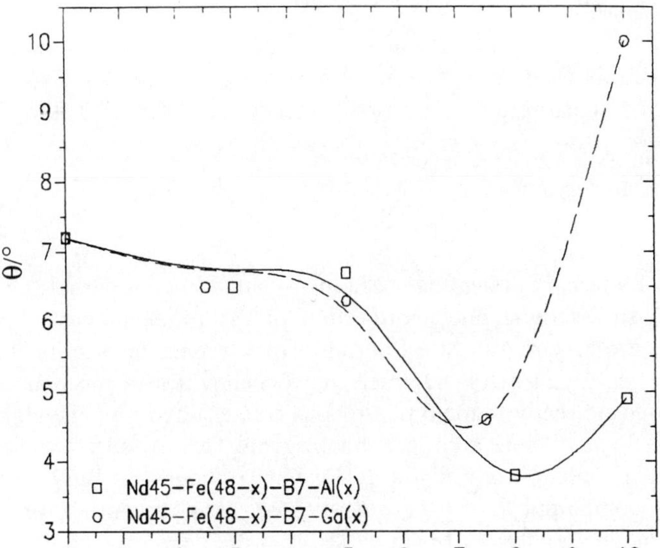

FIG. 3.9. Dependence of the wetting angle θ for sample compositions $Nd_{45}Fe_{(48-x)}B_7Al_x$ and $Nd_{45}Fe_{(48-x)}B_7Ga_x$ on the content of Al x_{Al} and Ga x_{Ga} respectively [Kno90].

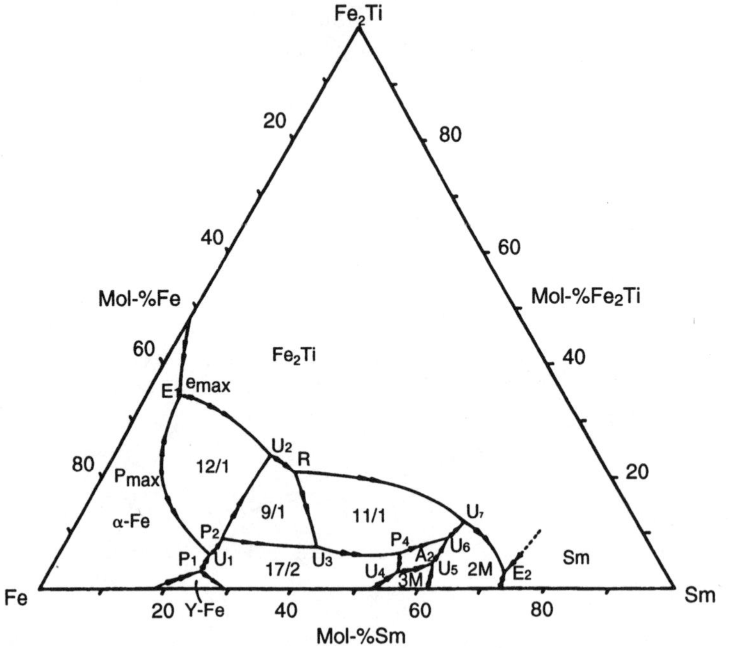

FIG. 3.10. Projection of the liquidus surface for the system Sm–Ti–Fe, Sm-rich side not included [Rei92]. The scale is in mole% for clarity.

Oxides such as Dy_2O_3, Al_2O_3 and others are sometimes used to inhibit grain growth and to ameliorate corrosion resistance [Gha87, Gri88, Tak89, Bes92].

4 Additions to Sm–Fe

4.1 *Metals*

The Sm–Ti–Fe ternary system has been investigated in considerable detail [Rei92]. Figure 3.10 shows the liquidus projection of the phase diagram, omitting the regions that are very rich in Sm. Interest in this system stems from the report by Schnitzke *et al.* [Sch90a] of extremely high coercivities in mechanically alloyed magnets based on the hexagonal phase $Sm_5(Ti, Fe)_{17}$ and also the ability of small quantities of Ti to stabilize an iron-rich tetragonal $ThMn_{12}$-structure pseudobinary phase which is absent from all the R–Fe binary diagrams [Bus77, Oha87]. The $Sm(TiFe_{11})$ compound has been investigated as a potential rare-earth–iron-permanent magnet material. It is discussed in more detail in Chapter 6.

A problem arising from the peritectic formation of Sm_2Fe_{17} is the appearance of α–Fe in cast ingots, and the need to anneal them below 1280 °C to achieve pure ψ phase for subsequent nitrogenation (see below).

Very little is known about the additive Nb. The addition of 4 at.% to 5 at.% Nb leads to basically Fe-free Sm_2Fe_{17} as-cast material by the formation of

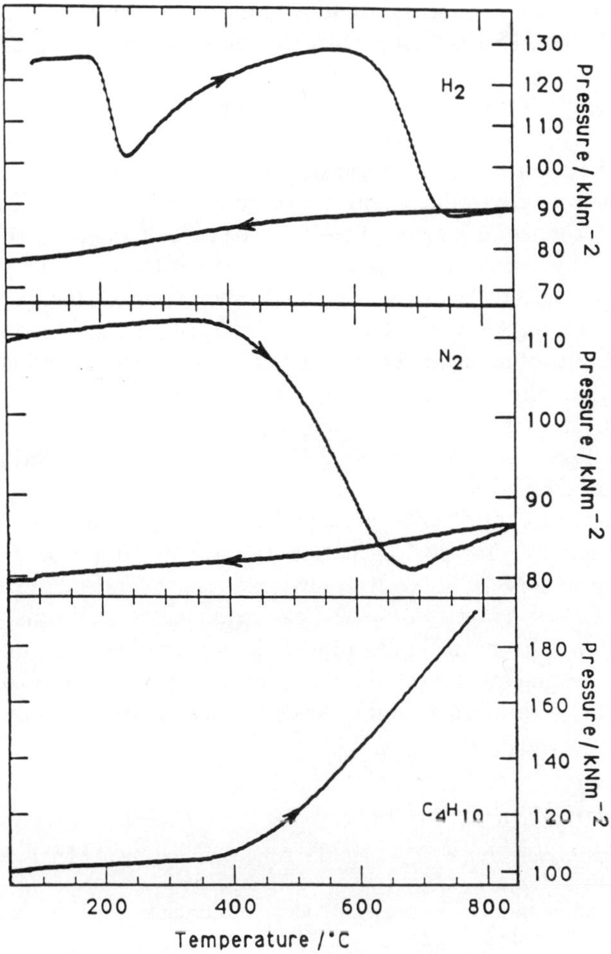

FIG. 3.11. Gas absorption curves for Y_2Fe_{17} for different gases: H_2, N_2 and C_4H_{10} [Sun92].

paramagnetic $NbFe_2$. Normally, the harmful occurrence of free Fe originates from the peritectic formation of Sm_2Fe_{17} [Pla92]. The addition of V to the rhombohedral compound Sm_2Fe_{17} yields an orthorhombic distortion of the hexagonal $CaCu_5$ structure, but does not make the material suitable for permanent magnets. Prospects arise in the case of nitrogenation [Shc92]. Co addition generally yields higher Curie temperatures in 1:12 and 2:17 compounds.

4.2 *Gas-phase interstitial modification*

The magnetic properties of iron-rich phases can be significantly modified by the change of lattice parameters which occurs with the incorporation of small atoms from the gas phase. Unlike the high-temperature liquid–solid reactions of traditional metallurgy, these gas–solid reactions take place at only a few hundred degrees centigrade. The interstitial elements which can be introduced in this way are principally hydrogen, nitrogen and carbon. They can give rise to solid solutions or ternary compounds which can be stable or metastable, with the diffusion of the metallic species being very limited at the temperature used for the gas-phase reactions. These issues are examined in the following chapter.

Hydrogen can be introduced into most binary rare-earth iron intermetallics, as well as into ternaries such as $Nd_2Fe_{14}B$. In the case of Sm_2Fe_{17}, solutions with up to about five hydrogen atoms per formula exist with somewhat enhanced Curie temperatures. The limit with nitrogen and carbon is three atoms per formula unit, but the influence on the magnetic properties of Sm_2Fe_{17} is much more pronounced, with Curie temperature increases of over 300 K. Nitrogen gas and NH_3 are possible carriers for the nitrogenation reaction [Coe90, Ota91, Yan91a], whereas many hydrocarbon gases and vapours such as methane or benzene are suitable for carbonation [Coe91, Chr92]. Table 2.17 listed some data showing the effect of nitrogen and carbon.

Typical gas absorption curves measured with a thermopiezic analyser (TPA) are shown in Fig. 3.11 for hydrogenation (a), carbonation with C_4H_{10} (b) and nitrogenation with N_2 (c) [Sun92]. Hydrogenation evidently shows a reversible gas absorption at low temperatures, whereas in all cases reversible gas absorption finally ends in an irreversible disproportionation into SmH_x, SmN or SmC and α–Fe. The thermodynamics of the gas–solid reaction is discussed in Chapter 4. Nitrogenation of $SmFe_2$ and $SmFe_3$ leads initially to an amorphous product.

Bibliography

Binary alloy phase diagrams, T. B. Massalski, American Society of Metals, Metals Park Ohio 1990 — a complete collection of binary diagrams, with commentary.

Alloy phase equilibria, A. Prince, Elsevier, Amsterdam, 1966 — a presentation of thermodynamic principles and results.

REFERENCES

[Aba86] C. Abache and H. Oesterreicher, *J. Appl. Phys.*, **60** (1986), 1114.

[All89] C. H. Allibert, *Concerted European action on magnets*, I. V. Mitchell, J. M. D. Coey, D. Givord, I. R. Harris and R. Hanitsch (eds), Elsevier Applied Science, 1989, p. 358.

[Bes92] S. Beseničar, B. Salje, G. Dražič and J. Holc, *J. Magn. Magn. Mat.,* **104–107** (1992), 1175.

[Bol87] F. Bolzoni, F. Leccabue, O. Moze, L. Pareti and M. Solzi, *J. Magn. Magn. Mater.*, **67** (1987), 373.

[Bra90] J. R. Bradley, R. K. Mishra, E. Pinkerton and T. Schroeder, *J. Magn. Magn. Mater.*, **86** (1990), 44.

[Bus77] K. H. J. Buschow, *Rep. Prog. Phys.* **40** (1977), 1179.

[Bus86] K. H. J. Buschow, D. B. de Mooij and H. M. van Noort, *J. Less-Common Met.*, **125** (1986), 135.

[Bus91] K. H. J. Buschow, *Rep. Prog. Phys.*, **54** (1991), 1123.

[Can72] J. F. Cannon, D. L. Robertson and H. T. Hall, *Mat. Res. Bull.* **7** (1972), 5.

[Cha79] N. F. Chaban, Y. B. Kuz'ma, N. S. Bilanizhko, O. O. Kachmar and N. V. Petriv, *Dopov. Akad. Nauk USSR Ser. A: Fiz. Mat. Tekh. Nauki*, **10** (1979), 875.

[Cha90] W. C. Chang, C. K. Paik, H. Nakamira, N. Takahasi, S. Sugimoto, M. Okada, M. Homma, *IEEE Trans. Mag.*, **MAG-26** (1990), 2684.

[Chr92] C. N. Christodoulou and T. T. Takeshita, *JALCOM*, **190** (1992), 91.

[Cla92] M. T. Clavaguera-Mora, J. A. Diego, M. D. Baró, S. Suriñach, N. Clavaguera and F. Cebollada, *JALCOM*, **182** (1992), 211.

[Coe90] J. M. D. Coey, H. Sun, *J. Magn. Magn. Mat.*, **89** (1990), L251.

[Coe91] J. M. D. Coey and D. P. F. Hurley, *J. Magn. Magn. Mater.*, **104–107** (1991), 1098.

[Eis91] J. Eisses, D. B. de Mooij, K. H. J. Buschow and G. Martinek, *J. Less-Common Met.*, **171** (1991), 17.

[Fid87] J. Fidler, J. Bernardi and P. Skalicky, *Mat. Res. Soc. Proc.*, **96** (1987), 181.

[Fid89] J. Fidler, K. G. Knoch, H. Kronmüller and E. -Th. Henig, *J. Mat. Res.*, **4** (1989), 806.

[Gha87] M. Ghandehari and J. Fidler, *Mat. Lett.*, **5** (1987), 285.

[Giv92] D. Givord, J. P. Nozières, M. F. Rossignol, D. W. Taylor, I. R. Harris, D. Fruchart and S. Miraglia, *JALCOM*, **176** (1991), L5.

[Gri88] B. Grieb, K. G. Knoch, E.-Th. Henig and G. Petzow, *J. Magn. Magn. Mater.*, **80** (1989), 75.

[Gri89] B. Grieb, E.-Th. Henig, G. Schneider and G. Petzow, *Z. Metallkd.*, **80** (1989), 95.

[Gri90] B. Grieb, E.-Th. Henig, G. Schneider, G. Knoch, G. Petzow and D. de Mooij, *Proc. User Aspects of Phase Diagrams*, Petten, F. H. Hayes (eds), The Institute of Metals, London, 1991, p. 169.

[Gri90a] B. Grieb, E.-Th. Henig, G. Martinek, H. H. Stadelmaier and G. Petzow, *IEEE Trans. Magn*, **MAG-26** (1990), 1367.

[Gri91] B. Grieb, K. Fritz and E.-Th. Henig, *J. Appl. Phys.*, **70** (1991), 6447.

[Hal95] B. Hallemans, P. Wollants, and J.R. Roos, *J. Phase Equilibria*, **16** (1995), 137.

[Hen93] K. Hennemann, H. L. Lukas, and H.-J. Schaller, Z. *Metallkde.*, **84** (1993) 668.

[Hir90] S. Hirosawa, H. Tomizawa, S. Mino and A. Hamamura, *IEEE Trans.Magn.*, **MAG-26** (1990), 1960.

[Hir90a] S. Hirosawa, S. Mino and H. Tomizawa, *J. Appl. Phys.*, **69** (1990), 5844.

[Hu88] J. Hu, Y. Wang, X. Li, L. Yin, M. Feng, D. Dai, T. Wang, J. G. Zhao and Z. Wang, *J. de Phys. Coll.*, **C8** (1988), 601.

[Iwa89] E. Iwamura, K. Nagayama, T. Suzuki and T. Umeda, *10th Int. Workshop on Rare Earth Magnets and their Applications*, Kyoto (Japan), 1989, p. 293.

[Kia91] A. Kianvash and I. R. Harris, *J. Appl. Phys.*, **70** (1991), 6453.

[Kit91] Y. Kitano, J. Shimomura and M. Shimotomai, *J. Appl. Phys.*, **69** (1991), 6055.

[Kiy89] T. Kiyomiya, H. Kohmura and K. Matsui, *10th Int. Workshop on Rare Earth Magnets and their Applications*, Kyoto (Japan), 1989, p. 551.

[Kno90] K. G. Knoch, B. Grieb, E.-Th. Henig, H. Kronmüller and G. Petzow, *IEEE Trans. Magn.*, **MAG-26** (1990), 1951.

[Kno93a] K. G. Knoch, A. Le Calvez, Q. Qi, A. Leithe-Jasper and J. M. D. Coey, *J. Appl. Phys.*, **73**, (1993), 1.

[Kno94] K. G. Knoch, B. Reinsch and G. Petzow, *Z. Metallkd.*, **85** (1994), 350.

[Kre91] J. E. Krentz, R. W. Lee and R. A. Waldo, *Proc. 5th Joint Magnetism and Magnetic Materials Conf.*, Pittsburgh, PA (USA), 1991.

[Kub82] O. Kubaschewski, *Iron — binary phase diagrams*, Springer Verlag, Verlag Stahleisen mbH, Düsseldorf, 1982.

[Lan90] F. J. G. Landgraf, G. S. Schneider, V. Villas-Boas and F. P. Missell, *J. Less-Common Met.*, **163** (1990), 209.

[Lan91] F. J. G. Landgraf, F. P. Missell, K. G. Knoch, B. Grieb and E.-Th. Henig, *J. Appl. Phys.*, **70** (1991), 6107.

[Leo89] M. Leonowicz, *J. Magn. Magn. Mater.*, **83** (1990), 211.

[Lin92] Ya. L. Linetsky, V. M. Raigorodsky and V. Yu. Tsvetknov, *JALCOM*, **184** (1992), 35.

[Mar85] E. B. Marasin, O. I. Boduk, A. O. Tsokol' and V. S. Fundamenskii, *Sov. Phys. Crystall.*, **30** (1985), 338.

[Mas90] T. B. Massalski, *Binary alloy phase diagrams*, ASM International, 1990.

[Mat85] Y. Matsuura, S. Hirosawa, H. Yamamoto, S. Fujimura, M. Sagawa and K. Osamura, *Jap. J. Appl. Phys.*, **24** (1985), L635.

[McG92] P. J. McGuiness, C. Short, A. F. Wilson and I. R. Harris, *JALCOM*, **184** (1992), 243.

[Miz87] T. Mizoguchi, I. Sakai, N. Niu and K. Inomata, *IEEE Trans. Mag.*, **MAG-23** (1987), 2281.

[Moo88] D. B. de Mooij and K. H. J. Buschow, *J. Less-Common Met.*, **142** (1988), 349.

[Mül92] C. Müller, B. Reinsch and G. Petzow, *Z. Metallkd.*, **83** (1992), 845.

[Nei90] A. C. Neiva, T. Yonamine, F. J. G. Landgraf and F. P. Missell, *6th Int. Symp. on Magn. Anisotropy and Coercivity in RE–TM Alloys*, Pittsburgh, 1990, p. 236.

[Oha87] K. Ohashi, T. Yokohama, R. Osugi and Y. Tawara, *IEEE Trans.Mag.*, **MAG-23** (1987), 3101.

[Ota91] Y. Otani, D. P. F. Hurley, H. Sun and J. M. D. Coey, *J. Appl. Phys.*, **69** (1991), 5584.

[Pla92] A. E. Platts, I. R. Harris and J. M. D. Coey, *JALCOM*, **185** (1992), 251.

[Pol88] R. J. Pollard, P. J. Grundy, S. F. H. Parker and D. G. Lord, *IEEE Trans. Magn.*, **MAG-24** (1988), 1626.

[Pri66] A. Prince, *Alloy phase equilibria*, Elsevier Publ. Comp., Amsterdam-London-New York, 1966.

[Rei92] B. Reinsch, B. Grieb, E.-Th. Henig and G. Petzow, *IEEE Trans. Magn.*, **MAG-28** (1992), 2832.

[Sag90] M. Sagawa, P. Tenaud, F. Vial and K. Hiraga, *IEEE Trans. Magn.*, **MAG-26** (1990), 1957.

[Sch86] G. Schneider, E.-Th. Henig, G. Petzow and H. H. Stadelmaier, *Z. Metallkde.*, **77** (1986), 755.

[Sch88] G. Schneider, G. Martinek, H. H. Stadelmaier and G. Petzow, *Mater. Lett.*, **7** (1988), 215.

[Sch89] G. Schneider, F. J. G. Landgraf and F. P. Missell, *J. Less-Common Met.*, **153** (1989), 169.

[Sch89a] G. Schneider, E.-Th. Henig, B. Grieb and G. Knoch, *Concerted European action on magnets*, I. V. Mitchell, J. M. D. Coey, D. Givord, I. R. Harris and R. Hanitsch (eds), Elsevier Applied Science, 1989, p. 335.

[Sch90] G. Schneider, E.-Th. Henig, F. P. Missell and G. Petzow, *Z. Metallkd.*, **81** (1990), 322.

[Shc92] Ye. V. Shcherbakova, G. V. Ivanova, A. S. Yermolenko, Ye. V. Belozerov and V. S. Gaviko, *JALCOM*, **182** (1992), 199.

[Shi88] T. Shimoda, K. Akioka, O. Kobayashi and T. Yamagami, *J. Appl. Phys.*, **84** (1988), 5290.

[Shi90] T. Shimoda, T. Akioka, O. Kobayashi, T. Yamagami and A. Arai, *11th Int. Workshop on Rare Earth Magnets and their Applications*, Pittsburgh (USA), 1990, p. 17.

[Sta85] H. H. Stadelmaier and A. El-Masry, *Proc. 4th Int. Symp. on Magnetic Anisotropy and Coercivity in Rare Earth — Transition Metal Alloys*, Dayton (USA), 1985, p. 613.

[Sun92] H. Sun, Thesis, Dept. of Pure and Applied Physics, Trinity College Dublin, 1992.

[Tak89] T. Takeshita and A. Watanabe, *10th Int. Workshop on Rare Earth Magnets and their Applications*, Kyoto (Japan), 1989, p. 477.

[Tak90] T. Takeshita and R. Nakayama, *11th Int. Workshop on Rare Earth Magnets and their Applications*, Pittsburgh (USA), 1990, p. 49.

[Ten90] P. Tenaud, F. Vial and M. Sagawa, *IEEE Trans. Magn.*, **MAG-26** (1990), No. 5, 1930.

[Tok89] M. Tokunaga, Y. Nozawa, K. Iwasaki, M. Endoh, S. Tanigawa and H. Harada, *IEEE Trans. Magn.*, **MAG-25** (1989), 3561.

[Tsa87] D. S. Tsai, T. S. Chin, S. E. Hsu and M. P. Hung, *IEEE Trans. Magn.*, **MAG-23** (1987), 3607.

[Tsu88] Y. Tsubokawa, R. Shimizu, S. Hirosawa and M. Sagawa, *J. Appl. Phys.*, **63** (1988), 3319.

[Vil88] V. Villas-Boas, F. P. Missell and S. F. da Cunha, *J. Appl. Phys.*, **64** (1988), 5549.

[Yan91] C. J. Yang, W. Y. Lee, S. D. Choi and Y. H. Jeong, *Mat. Lett.*, **12** (1991), 233.

[Yan91a] Y.-C. Yang, X.-D. Zhang, L.-S. Kong, Q. Pan, Y.-T. Hou, S. Huang and L. Yang, *J. Less-Common Met.*, **170** (1991), 37.

[Yin92] X. J. Yin, I. P. Jones and I. R. Harris, *J. Magn. Magn. Mat.*, **116** (1992), L325.

[Zha85] M. Zhang, D. Ma, X. Jiang and S. Liu, *8th Int. Workshop on Rare Earth-Magnets and their Applications*, Dayton (USA), 1985, p. 541.

4

Interstitial modification

R. Skomski

A. INTRODUCTION

In 1990, Coey and Sun [Coe90] discovered that the magnetic properties of Sm_2Fe_{17} are dramatically altered by nitrogen introduced from a gas in a low-temperature reaction. In the simplest case, gas-phase nitrogenation is conducted on finely ground Sm_2Fe_{17} powder in a pressure of about 1 bar nitrogen gas; at typical nitrogenation temperatures (400–500 °C) the nitrogenation reaction takes a few hours, depending on the particle size.

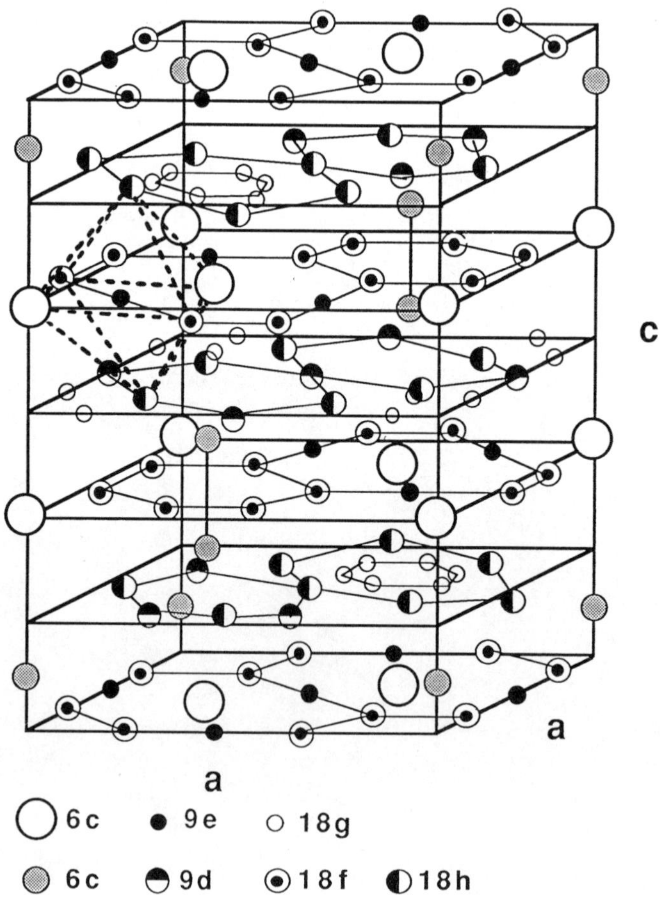

(a)

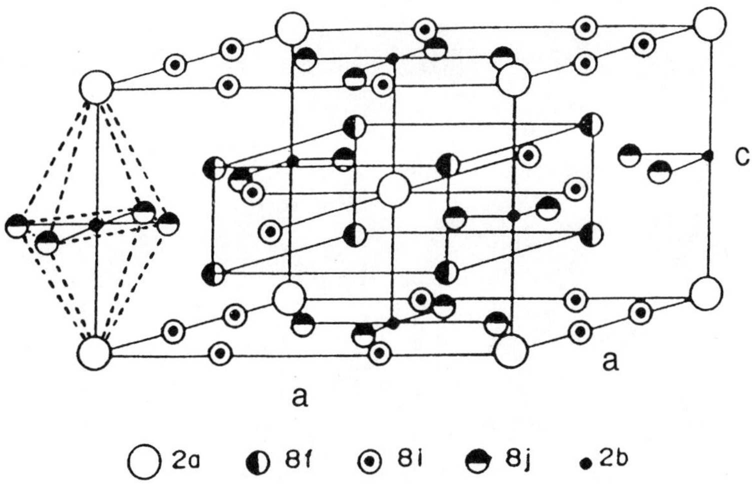

(b)

FIG. 4.1. Crystal structures of interstitial intermetallics (a) R_2Fe_{17} (rhombohedral). The rare-earth occupies 6c sites, nitrogen 9e sites, and hydrogen 9e and some 18g sites. All other sites are occupied by iron. (b) $R(Fe_{12-x}M_x)$ (tetragonal). The rare-earth occupies 2a sites. Iron and the stabilizing element M occupy 8f, 8i and 8j sites and nitrogen occupies 2b sites. The R_2T_4 octahedral sites occupied by nitrogen are shown by the dashed lines.

Upon nitrogenation, the lattice expands by about 6 vol.% and the Curie temperature increases from 116 °C to 476 °C, which is related to rather general trends observed in dense-packed iron-rich intermetallics. Furthermore, due to the crystal-electric field created by the nitrogen atoms, the magnetocrystalline anisotropy changes from easy-plane to strong easy-axis [Coe90, Sun90]. This means that the interstitial nitride $Sm_2Fe_{17}N_x$ ($x \approx 3$) is a promising permanent magnetic material, in striking contrast to Sm_2Fe_{17}, which is not suitable for permanent magnetic applications.

Since 1990, gas-phase interstitial modification has not only been extended to other rhombohedral R_2Fe_{17} (R = light rare-earth) and hexagonal R_2Fe_{17} (R = heavy rare-earth and yttrium) compounds, but also to several other intermetallic structure types including 1:12 compounds [Yan91]. A common feature of all those nitrides is that nitrogen occupies large *interstitial* sites, as opposed to substitutional modification [Men86], which is less effective in raising the Curie temperature and improving the magnetocrystalline anisotropy. The interstitially modified 2:17 and 1:12 structures are illustrated in Fig. 4.1. The nitrogen source may be molecular N_2, or another nitrogen-containing gas such as ammonia [Coe90].

Interstitial modification is not restricted to nitrogen. Carbon can equally well be introduced by gas-phase reaction using hydrocarbon gases [Coe91a, 92a, Hur92] or from the solid state [Sko93a, b] to yield carbides $Sm_2Fe_{17}C_x$ ($x \approx 2.2$). Note that

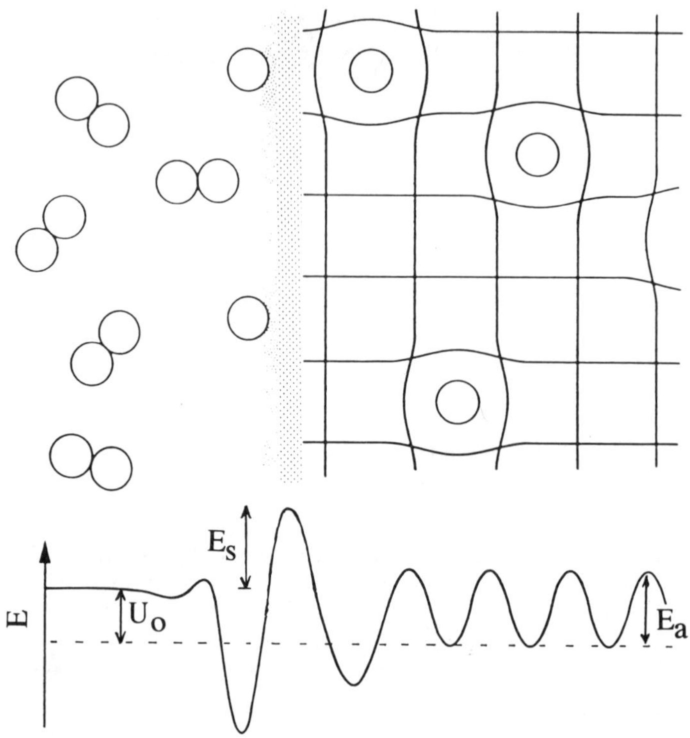

FIG. 4.2. Illustration of the gas-phase interstitial modification process. Molecular nitrogen split at the surface overcomes the surface energy barrier E_s and diffuses into the bulk where the activation energy is E_a. Interactions between interstitial atoms on different sites in the intermetallic lattice are mainly elastic in nature.

the intrinsic magnetic properties of the carbides (Curie temperature, magnetocrystalline anisotropy) are nearly as good as those of the nitrides. Addition of carbon into the melt can also be used to produce interstitial carbides, but here the carbon content does not exceed the value $x = 1.5$ [Zho90]. Finally, hydrogen has long been known as an interstitial in rare-earth intermetallics. At temperatures of about 250 °C hydrogen gas easily reacts with Sm_2Fe_{17}, but the anisotropy remains easy-plane, with much less increase in Curie temperature [Wan88]. Typical reactions of gases with Y_2Fe_{17} were illustrated in Fig. 3.11.

The gas-phase interstitial modification process is illustrated in Fig. 4.2. The gas molecules dissociate at the particle surface, and subsequent diffusion throughout the intermetallic lattice yields the interstitially modified intermetallic. Like hydrogenation, nitrogenation of intermetallics is typically a quasiequilibrium process whose product is thermodynamically unstable with respect to further reaction products such SmN, Fe_4N and α–iron in the case of $Sm_2Fe_{17}N_x$ [Coe94]. This means that the process is only feasible if there is a temperature range where the

Table 4.1 Atomic radii and electronegativities [Ems89]

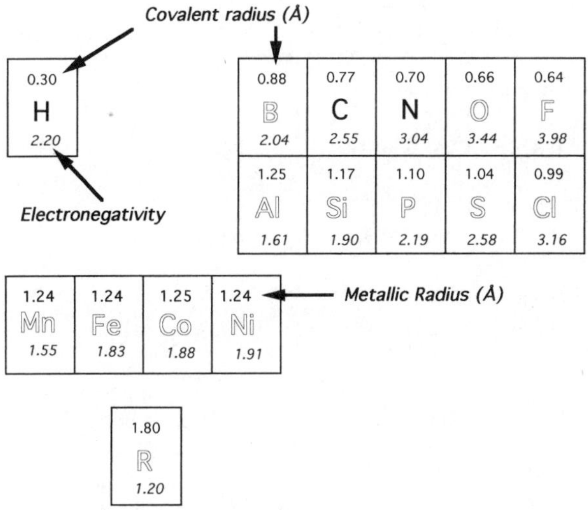

Table 4.2 Structural properties of interstitially modified samarium and yttrium intermetallics [Sun90, Coe92c, Qi92a] [Qi92b, Bre94]; ΔV is the macroscopic volume expansion per interstitial atom

	a (nm)	c (nm)	$\Delta V/V$ (%)	Δv (nm^3)
Sm_2Fe_{17}	0.854	1.243	–	–
$Sm_2Fe_{17}N_{2.5}$	0.873	1.264	6.3	0.0066
$Sm_2Fe_{17}N_{3.7}$	0.875	1.283	8.3	0.0060
$Sm_2Fe_{17}C_{2.2}$	0.875	1.257	6.2	0.0073
$Sm_2Fe_{17}H_x$				
Y_2Fe_{17}	0.848	0.826	–	–
$Y_2Fe_{17}N_{2.6}$	0.865	0.844	6.3	0.0062
$Y_2Fe_{17}C_{2.2}$	0.866	0.840	6.1	0.0071
$Y_2Fe_{17}H_{2.7}$	0.854	0.829	1.8	0.0017
$Y_2(Fe_{11}Ti)$	0.850	0.480	–	–
$Y_2(Fe_{11}Ti)N_{0.8}$	0.858	0.480	1.8	0.0039
$Y_2(Fe_{11}Ti)C_{0.9}$	0.858	0.480	1.7	0.0033
$Y_2(Fe_{11}Ti)H_{1.1}$	0.851	0.479	0.2	0.0003

interstitials can diffuse throughout the intermetallic lattice, but where atomic-scale diffusion of the metal atoms is blocked.

Small, light atoms are candidates for use in interstitial modification (Table 4.1). In view of the size of the interstices in rare-earth transition-metal intermetallics, the atomic radius should be less than about 1 Å. Hydrogen and the 2p elements fit the

Table 4.3 Structural and magnetic properties of $R_2Fe_{17}N_{3-\delta}$ compounds [Sun90]. Magnetic moments are measured at 4.2 K

	a (nm)	c (nm)	$\Delta V/V$ (%)	Δv (nm^3)	Easy direction	T_c(K)	m(μ_B)
Rhombohedral Th_2Zn_{17} structure							
Ce_2Fe_{17}	0.847	1.232			plane	241	35.2
$Ce_2Fe_{17}N_{2.8}$	0.873	1.265	9.1	0.0083	plane	713	38.1
Pr_2Fe_{17}	0.857	1.242			plane	290	30.0
$Pr_2Fe_{17}N_{2.5}$	0.877	1.264	6.6	0.0069	plane	728	38.3
Nd_2Fe_{17}	0.856	1.244			plane	326	37.0
$Nd_2Fe_{17}N_{2.3}$	0.876	1.263	6.3	0.0073	plane	732	41.3
Sm_2Fe_{17}	0.954	1.243			plane	389	38.0
$Sm_2Fe_{17}N_{2.3}$	0.873	1.264	6.3	0.0071	c-axis	749	43.9
Gd_2Fe_{17}	0.851	1.243			plane	477	34.0
$Gd_2Fe_{17}N_{2.4}$	0.869	1.266	6.2	0.0067	plane	758	38.2
Tb_2Fe_{17}	0.845	1.241			plane	410	22.9
$Tb_2Fe_{17}N_{2.3}$	0.866	1.266	7.2	0.0080	plane	733	27.3
Hexagonal Th_2Ni_{17} structure							
Dy_2Fe_{17}	0.845	0.830			plane	367	18.6
$Dy_2Fe_{17}N_{2.8}$	0.864	0.845	6.4	0.0059	plane	725	25.0
Ho_2Fe_{17}	0.844	0.828			plane	335	18.2
$Ho_2Fe_{17}N_{3.0}$	0.862	0.845	6.5	0.0055	plane	709	24.5
Er_2Fe_{17}	0.842	0.827			plane	310	17.5
$Er_2Fe_{17}N_{2.7}$	0.861	0.846	7.0	0.0066	c-axis $T < 120$ K	697	20.5
Tm_2Fe_{17}	0.840	0.828			c-axis $T < 72$ K	306	23.0
$Tm_2Fe_{17}N_{2.7}$	0.858	0.847	6.7	0.0063	c-axis $T < 220$ K	690	30.5
Lu_2Fe_{17}	0.839	0.826			plane	290	34.7
$Lu_2Fe_{17}N_{2.7}$	0.857	0.848	7.1	0.0066	plane	678	38.5

bill. Boron and carbon can be introduced from the melt to form ternary compounds such as $Nd_2Fe_{14}B$, but boron is not easy to introduce into pre-existing structures by low-temperature reaction. Oxygen and fluorine are extremely electronegative and tend to provoke immediate disproportionation into oxides and fluorides. We are left with hydrogen, nitrogen and carbon. Tables 4.2 and 4.3 illustrate the structural changes that are brought by interstitial modification of some iron-rich rare-earth intermetallics.

If Sm_2Fe_{17} is heated in nitrogen, the nominal composition of the quasi-equilibrium compound is $Sm_2Fe_{17}N_x$ ($x < 3$), and nitrogen occupies almost exclusively the 9e R_2T_4 octahedral sites in the rhombohedral Th_2Zn_{17} structure (Fig. 4.1a) [Coe91b, Isn92a]. The reaction can be written as

$$Sm_2Fe_{17} + \frac{x}{2}N_2 \Leftrightarrow Sm_2Fe_{17}N_x \tag{4.1}$$

with the nitrogen content $x = 3 - \delta$. The nonintegral nitrogen content indicates an incomplete filling of the octahedral sites so the nitrides are actually nonstoichiometric compounds (solid solutions). Typically, $\delta \approx 0.3$, which corresponds to a quasiequilibrium site occupancy $c = x/3$ of 0.9. The incomplete filling of interstitial sites is due to thermal activation: full interstitial occupation represents an ordered state which is thermodynamically unfavourable at high temperatures.

Diffusion kinetics are sluggish at typical nitrogenation temperatures, but if the temperature is increased the competing disproportionation reaction of the $Sm_2Fe_{17}N_x$ lattice into SmN and α–Fe intervenes [Coe90, 94, Sun90]. Even small amounts of α–Fe are able to destroy coercivity, so partly nitrided material cannot be used as a permanent magnet. If the nitrogenation proceeds, equilibrium is reached, and the total reaction reads

$$Sm_2Fe_{17} + \frac{25}{8}N_2 \rightarrow 2SmN + \frac{17}{4}Fe_4N$$

$Sm_2Fe_{17}N_x$ is merely a metastable intermediate product in this equation.

Disproportionation kinetics show a pronounced temperature dependence which is reasonably well described by [Sko93e]

$$t = t_0 e^{E_0/kT}$$

where t is the time after which the material can no longer be used as a permanent magnet. This equation originates from the Arrhenius-type temperature dependence $D = D_0 \exp(-E_0/kT)$ of atomic diffusivity in solids.

In interstitial nitrides, iron diffusion is responsible for the disproportionation kinetics [Kat92], and the activation energy E_0 refers to the diffusivity D_{Fe} of iron [Sko93e]. Indeed, assuming a critical length scale $L \approx 3$ nm of the α–Fe precipitates we obtain the right order of magnitude for t. Precipitates larger than 3 nm are (i) likely to act as nucleation centres and (ii) can be clearly identified by X-ray diffraction. Based on an estimated activation energy for iron diffusion of $E_0 = 220$ kJ/mole* [Sko93e] we obtain with $D_0 = 1$ mm^2/s and $L = \sqrt{(2D_{Fe}t)}$ the disproportionation times $t(600\ °C) \approx 1$ min, $t(500\ °C) \approx 1$ h, and $t(400\ °C) \approx 6$ days. Within this estimation, a time of 100 years is reached at 280 °C. Taking account of the uncertainty in the estimate of E_0, these values indicate that $Sm_2Fe_{17}N_3$ can be regarded as stable below about 250–300 °C.

* 1 kJ/mole is equal to a thermal energy $E/k = 120.3$ K per atom.

1 Statistical mechanics of gases in metals

From the point of view of statistical thermodynamics, gas–metal systems are very close to the Ising model and to gas–liquid systems. As pointed out by Lee and Yang [Lee52], the two states of a gas–liquid lattice gas (occupied and vacant) correspond to the two states ↑ (spin up) and ↓ (spin down) of the $S = 1/2$ Ising model. Later on, Alefeld [Ale69] generalized this model to gas–solid systems where the total number of hydrogen atoms in the solid is parametrized by a chemical potential.

To discuss the behaviour of interstitial nitrides it is useful to draw a comparison with interstitial hydrides. Metallic hydrides, such as the classical gas–solid system $PdH_{1-\delta}$ and the intermetallic system $LaNi_5H_x$, have attracted much attention as potential hydrogen storage media and are comparatively well understood [Ale78, Mot36, Wic78]. An essential feature is that the hydrides above a critical temperature (e.g. $T_c = 292\ °C$ for palladium hydride) form a gas–solid solution with randomly occupied interstitial sites. In this case the gas uptake is determined by the net reaction energy (absorption energy) U_0. Metals with $U_0 < 0$ (exothermic reaction) such as Pd or V show high hydrogen solubilities even at elevated temperatures, whereas the hydrogen solubility in metals with $U_0 > 0$ (endothermic reaction e.g. Fe or Co) never exceeds the high-temperature value $c(T \to \infty) \approx 0.2\%$ [Fas76].

Below T_c phase segregation into a hydrogen-poor phase (e.g. α palladium hydride) and a hydrogen-rich phase (e.g. β palladium hydride) occurs. The question arises whether nitrogen in Sm_2Fe_{17} yields a simple gas–solid solution or a mixture of nitrogen-poor $\alpha{-}Sm_2Fe_{17}N_x$ and nitrogen-rich $\beta{-}Sm_2Fe_{17}N_x$ phases. The weight of experimental and theoretical evidence favours a *gas–solid solution* at typical nitrogenation temperatures where quasiequilibrium can be established [Coe92b, Sko93a, b]. Strictly speaking, $Sm_2Fe_{17}N_x$ should therefore be termed a 'nitrogen-containing compound' but for convenience we will call it a 'nitride'.

At typical nitrogenation temperatures, the gas–solid reaction proceeds by thermally activated bulk diffusion. Assuming that $Sm_2Fe_{17}N_x$ is a simple gas–solid solution, the temperature dependence of the diffusion constant and nitrogenation rate is given by the Arrhenius law:

$$D = D_0 \exp(-E_a/kT) \tag{4.2}$$

High temperatures enhance the reactivity with nitrogen, but also permit the onset of iron diffusion above about 600 °C, although $E_a < E_0$. The instability of the nitride becomes apparent and it disproportionates into SmN and bcc iron [Coe90, 91b, Sko93e]

$$Sm_2Fe_{17}N_3 \to 2SmN + Fe_4N + 13Fe$$

Hydrogen-loaded intermetallics are also metastable and disproportionate in a similar way. This is the basis of the HDDR process described in Chapter 7.

1.1 *Master equation*

To illustrate some principles of statistical mechanics and to introduce the behaviour of gases in metals, we start with the one-dimensional gas–solid model. As we will see later on, this model reproduces many features of real gas–solid systems. The phase space of the lattice-gas model consists of N discrete positions of a single nitrogen or hydrogen atom. There are N_g states in the gas phase and N_s interstitial sites, each separated by activated 'saddle-point' states, with an activated surface state between the solid and the gas (Fig 4.2). Typically, the number of gas states is much larger than the number of interstitial sites: 1 cm^3 of Sm_2Fe_{17} at 500 °C can absorb as much as 510 cm^3 of molecular nitrogen at 1 bar.

To calculate thermodynamical mean values:

$$\langle A(t)\rangle = \sum_{i=1}^{N} P_i(t) A_i$$

we have to find the probabilities $P_i(t)$ of finding the gas atom at the position x_i. The complete many-body problem, which includes the degrees of freedom of the lattice, is very difficult to solve so we use a simplified equation of motion (master equation). For this purpose we assume that the knowledge of the probabilities $P_i(t-\tau)$ is sufficient to determine the probability $P_i(t)$, where τ is a small time interval (Markov condition). This assumption works fairly well for dissipative phenomena such as internal friction in liquids but cannot be used, for instance, to describe periodic phenomena such as sound propagation in solids. For instance, the solution of the equation of motion $\partial^2 x/\partial t^2 + kx = 0$ requires the knowledge of two initial values or, equivalently, of $P_i(t-\tau)$ and $P_i(t-2\tau)$.

Using the series expansion $P_i(t+\tau) = P_i(t) + \tau\, \partial P_i/\partial t$ and bearing in mind that $\Sigma P_i(t) = 1$, the Markov condition can be written as [Ebe86, Hak90]

$$\frac{\partial P_i}{\partial t} = \sum_{j=1}^{N} [W_{ij}P_j - W_{ji}P_i] \tag{4.3}$$

This master equation is widely used in statistical physics; an example is the electrical conductivity in metals [Mot36], where the transition rates W_{ij} represent quantum-mechanical scattering probabilities.

The next problem is to fix the transition probabilities $W_{ij} = W(x_j \to x_i)$. The simplest way is to choose Boltzmann factors

$$W_{ij} = \Gamma_0 e^{-(U_{ij} - U_j)/kT} \tag{4.4}$$

Here Γ_0 is an (optical phonon) attempt frequency of order 10^{14} Hz. The $U_{ij} = U_{ji}$ are the energies of the intermediate states between x_i and x_j, that is, U_a for the saddle-point states and U_B for the surface barrier state. Only transitions between neighbouring states are considered.

1.2 *Solubility*

Putting $\partial P_i/\partial t = 0$ in Eq. (4.3) yields the equilibrium thermodynamics of the system. In reality, this condition is realized by vanishing net transition rates (detailed balance principle):

$$W_{ij}P_j^0 - W_{ji}P_i^0 = 0 \tag{4.5}$$

Putting Eq. (4.4) into Eq. (4.5) we find that the U_{ij} terms cancel each other and obtain the equilibrium probabilities

$$P_i^0 = \frac{1}{Z}\mathrm{e}^{-U_i/kT} \tag{4.6}$$

The normalization factor Z (partition function) is given by the condition $\Sigma P_i^0 = 1$ and reads

$$Z = \sum_{i=1}^{N} \mathrm{e}^{-U_i/kT}$$

In the model of Fig. 4.1, there are N_g gas states and N_s interstitial states so

$$Z = N_g \mathrm{e}^{-U_g/kT} + N_s \mathrm{e}^{-U_s/kT} \tag{4.7}$$

Equations (4.6) and (4.7) yield the equilibrium probability $\langle c_s \rangle_0$ of finding the gas atom in the solid:

$$\langle c_s \rangle_0 = \frac{N_s \exp(-U_s/kT)}{N_g \exp(-U_g/kT) + N_s \exp(-U_s/kT)} \tag{4.8}$$

All equilibrium properties of the system can be calculated directly from Z. For instance, introducing the net reaction energy (absorption energy) $U_0 = U_s - U_g$, the result, Eq. (4.8), is reproduced by

$$\langle c_s \rangle_0 = -kT \frac{\partial \ln Z}{\partial U_0} \tag{4.9}$$

Note that the equilibrium concentration, Eq. (4.8), can be written in the more convenient form

$$\langle c_s \rangle_0 = \frac{1}{1 + (N_g/N_s)\,\mathrm{e}^{U_0/kT}} \tag{4.10}$$

It is worth noting that the number N_g of gas sites increases with the gas volume $V \alpha 1/p$.

Equation (4.10) allows us to draw the following conclusions:

1. The driving force of the gas–solid reaction is the net reaction energy (absorption energy) U_0. Since $N_s \ll N_g$, the solubility $\langle c_s \rangle_0$ is very small for endothermic reactions ($U_0 > 0$).

2. Solubility decreases with increasing number of gas sites, that is, with decreasing pressure. At very low pressure, where the gas entropy is largest, the entropy contribution in $F = U - TS$ dominates and the atom is almost certainly outside the solid. Note that the master equation includes the entropic contributions.
3. In the high-temperature limit $T \to \infty$, the entropy dominates as well but here the solubility approaches the finite value $\langle c_s \rangle_0 = N_s/(N_g + N_s)$.
4. In the low-temperature limit $T \to 0$, we have $\langle c_s \rangle_0 \approx 1$ for exothermic reactions ($U_0 < 0$) and $\langle c_s \rangle_0 \approx 0$ for endothermic reactions ($U_0 > 0$).

1.3 *Diffusion*

To study the nonequilibrium thermodynamics of the system we assume that the gas atom is initially on an interstitial site at position x_{i0}. Driven by thermal collisions it will start a random walk in the lattice (diffusion), until the homogeneous distribution $\langle c_i \rangle = \langle c \rangle_0$ is reached. For discrete jumps with equal probabilities in both directions, the probabilities $P_i(t)$ are given by Pascal's triangle (Fig. 4.3), which becomes a Gaussian function for very large N_s (continuum approximation). In reality, a large variety of probability profiles is realized, but all these profiles can be interpreted as overlays of Gaussian profiles.

To describe the diffusion process in the lattice, we start again with the master equation, Eq. (4.3), and identify the probability $P_i(t)$ with the concentration $\langle c_i(t) \rangle$. There are direct jumps between neighbouring interstitial sites only, so Eq. (4.3) reads

$$\frac{\partial \langle c_i \rangle}{\partial t} = W_{i,i+1}\langle c_{i+1} \rangle + W_{i,i-1}\langle c_{i-1} \rangle - W_{i+1,i}\langle c_i \rangle - W_{i-1,i}\langle c_i \rangle \tag{4.11}$$

$$\frac{1}{1}$$

$$\frac{1}{2} \qquad \frac{1}{2}$$

$$\frac{1}{4} \qquad \frac{2}{4} \qquad \frac{1}{4}$$

$$\frac{1}{8} \qquad \frac{3}{8} \qquad \frac{3}{8} \qquad \frac{1}{8}$$

$$\frac{1}{16} \qquad \frac{4}{16} \qquad \frac{6}{16} \qquad \frac{4}{16} \qquad \frac{1}{16}$$

FIG. 4.3. Pascal's triangle, normalized by the factor $1/2^N$.

With $U_{ij} - U_i = E_a$ we obtain from Eq. (4.4) the constant expression

$$W_{i,i\pm 1} = \Gamma_0 e^{-E_a/kT} \tag{4.12}$$

so the discrete diffusion equation reads

$$\frac{\partial \langle c_i \rangle}{\partial t} = \Gamma_0 e^{-E_a/kT} [\langle c_{i+1} \rangle + \langle c_{i-1} \rangle - 2\langle c_i \rangle] \tag{4.13}$$

This equation shows that the effect of the activation energy reduces to a renormalization of the diffusion time. In particular, an attractive site-bonding interaction between diffusing atoms and interstitial sites will modify the activation energy E_a, but it does *not* change the concentration profiles $\langle c_i \rangle$.

Series expansion yields the concentration in the continuum approximation

$$\langle c_{i\pm 1} \rangle = \langle c_i \rangle \pm \frac{\partial \langle c_i \rangle}{\partial x} a + \frac{1}{2} \frac{\partial^2 \langle c_i \rangle}{\partial x^2} a^2 \pm \ldots \tag{4.14}$$

where a is the lattice constant of the one-dimensional lattice: $\langle c_{i\pm 1} \rangle = \langle c(x_i \pm a) \rangle$. Putting Eq. (4.14) into Eq. (4.13) we finally obtain the diffusion equation

$$\frac{\partial \langle c_i \rangle}{\partial t} = D \frac{\partial^2 \langle c_i \rangle}{\partial x^2} \tag{4.15}$$

with

$$D = \Gamma_0 a^2 e^{-E_a/kT} \tag{4.16}$$

It is worth while making the following comments:

1. Comparison of Eq. (4.16) with Eq. (4.2) yields $D_0 = \Gamma_0 a^2 \sim 10^{-6}$ m^2/s*. This approximate relation, which remains valid for higher-dimensional lattices [Fas76, Boc78], can be used to check the reliability of diffusion measurements. Large deviations from this value often indicate that the temperature range of the measurements is too small to obtain reliable Arrhenius fits. An exact determination of D_0 for the diffusion paths in real intermetallic lattices is quite a complicated problem [Chr94].
2. As far as the temperature dependence of D is concerned, the Boltzmann factor is the leading term. As a crude rule, E_a increases quadratically with the size of the diffusing atoms [Boc78, Fas76]. This explains the low diffusivity of nitrogen and carbon at typical modification temperatures ($D \approx 10^{-15}$ m^2/s at 500 °C) compared to hydrogen ($D \approx 10^{-9}$ m^2/s at 500 °C). In other words, the diffusion length $L \approx \sqrt{(Dt)}$ of hydrogen in a given time is a thousand times larger than that of nitrogen, and the hydrogen equilibrium concentration, which is very small at typical nitrogenation temperatures, is reached almost instantaneously in micron-size particles.

* D_0 includes a numerical factor to take account of lattice geometry.

B. NITRIDES, CARBIDES, AND HYDRIDES

1 Gas–solid solutions

1.1 *Phase structure*

It can be shown from Eq. (4.16), with a typical nitrogen activation energy $E_a \sim$ 130 kJ/mole ($\sim$ 1.3 eV), that the room-temperature nitrogen diffusivity is far too low to yield any changes of the nitrogen distribution, so the quenched magnet keeps the interstitial structure acquired at the nitrogenation temperature.

Most of the evidence available for $Sm_2Fe_{17}N_{3-\delta}$ favours a gas–solid solution at typical nitrogenation temperatures. The evidence can be summarized as follows:

1. It is possible to observe intermediate lattice parameters (Fig. 4.4), which are typical of gas–solid solutions, by X-ray diffraction analysis [Coe92b, Bre95].
2. Samples with intermediate nitrogen contents exhibit intermediate Curie temperatures [Kat92].
3. Homogenization of partly nitrided grains yields intermediate nitrogen concentrations [Muk92].
4. Domain-size observations [Mül92] can be explained by intermediate anisotropy constants only.

Similiar arguments apply to R_2Fe_{17} carbides and to $R(Fe_{12-x}M_x)$ nitrides and carbides. It is therefore reasonable to treat the interstitially modified intermetallics in question as gas–solid solutions. Note that NMR measurements indicate variable rare-earth environments in 2:17 compounds [Kap92], and hence support the solid–solution theory, although it is difficult to draw quantitative conclusions concerning intermetallics such as Y_2Fe_{17} which incorporate a degree of structural disorder [Moz94].

Table 4.1. shows that nitrogen and carbon are large atoms, compared to hydrogen, so the existence of large interstices is a precondition for thorough interstitial modification. It turns out that rare-earth intermetallics with Th_2Zn_{17}, Th_2Ni_{17}, $ThMn_{12}$ $Nd_3(FeTi)_{29}$, $NaZn_{13}$, and $BaCd_{11}$ structures have large R_2T_4 octahedra, which accounts for the large solubility of nitrogen and carbon in iron-rich intermetallics with these structures.

However, the energy which is necessary to distort the intermetallic lattice around the interstial atoms is not the only contribution to the net reaction energy U_0. During the interstitial modification process, the nitrogen triple bond must be broken (Fig. 4.2), but energy is regained by forming bonds between the interstitial atoms and the intermetallic lattice. In particular, there is a considerable electronegativity difference between iron and rare-earths on the one hand, and nitrogen or carbon on the other hand, which contributes to the energy balance. The resulting net reaction energy represents a sum of contributions which largely cancel each other, and it is difficult to predict U_0 for a given gas–solid system. Here we treat U_0 as an empirical parameter.

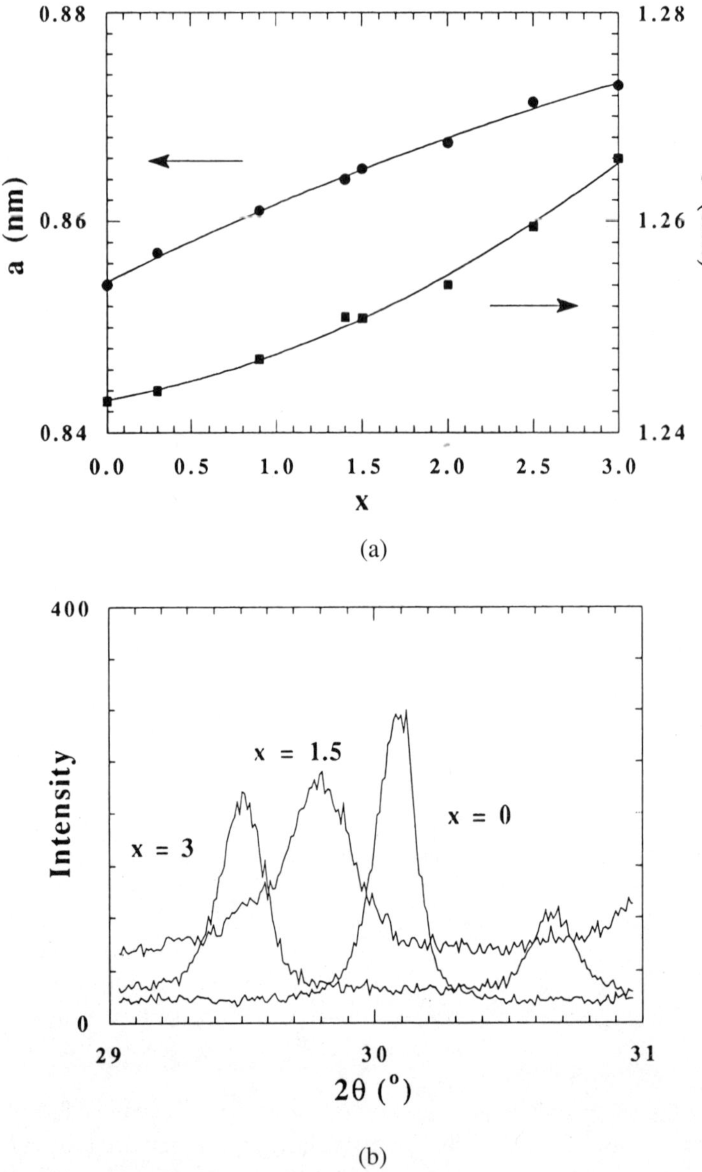

FIG. 4.4. (a) Lattice parameters for $Sm_2Fe_{17}N_x$ as a function of x. (b) X-ray diffraction from the (113) planes is shown for three values of x [Bre95].

To obtain a formula similiar to Eq. (4.10) we use the lattice-gas model Fig. 4.5. The gas-phase interstitial modification is conducted in a sufficiently large reaction volume which contains all necessary reactants. The gas volume is divided into V/V_0 lattice cells, where $V_0 \approx (2R_N)^3$ is the cell volume.

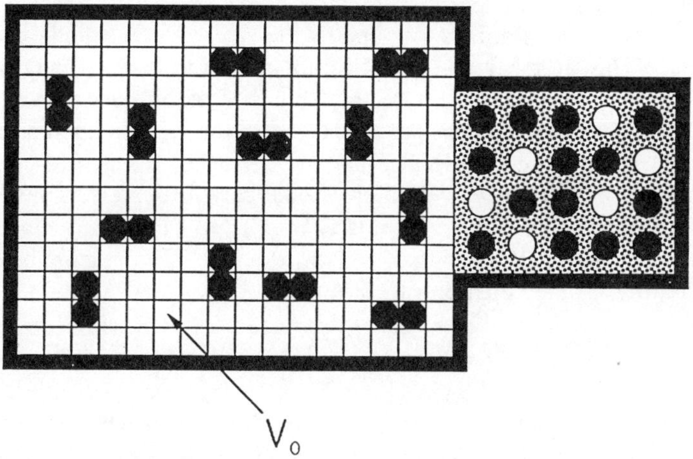

FIG. 4.5. Schematic illustration of a lattice-gas model. Nitrogen (●) is found in the gas-phase (left) and in the intermetallic lattice (right).

The calculation of the partition function [Sko93a] is similiar to Eq. (4.7), but requires careful counting of all different lattice–gas configurations with the same energy. For ideal gases the calculation can be done using elementary combinations and permutations. In the case of two states {empty, occupied} per interstitial site the main step is the application of the binomial formula, in the case of three states, for example, {empty, occupied by hydrogen, occupied by nitrogen}, the trinomial formula must be used, and so on.

Motional degrees of freedom [Ale69] are neglected. A good example of this purely configurational approach is the ideal atomic gas where the energy of the gas is neglected ($U = 0$). Then the partition function simply reads $Z = \Omega$, where $\Omega = (V/V_0)^n$ is the number of different ways of distributing n atoms over V/V_0 lattice sites. The relations $F = -kT\ln Z$ and $p = -\partial F/\partial V$ immediately lead to the equation of state $pV = nkT$: neither potential nor kinetic energy are necessary to explain the pressure exerted by an ideal gas. Another example is the entropic elasticity of rubber: the energy of a piece of rubber is practically independent of its deformation state [Tre43].

1.2 *Molecular nitrogen*

When any R_2Fe_{17} compound (R = rare-earth or yttrium) is heated in molecular nitrogen, the nonstoichiometric nitride $R_2Fe_{17}N_{3-\delta}$ is formed where nitrogen occupies the large octahedral 9e sites (Fig. 4.2) practically exclusively [Isn92a]. From the theoretical point of view, the very small Boltzmann factors for other interstices may permit a small occupancy of these sites, but there is no clear experimental evidence on this point.

In order to determine the partition function of the lattice-gas model, Fig. 4.5, we have to calculate the numbers N_g and N_s of the gas- and solid-phase states respectively. The number of identical-looking ideal-gas configurations is $N_g = (3V/V_0)^{(N-\nu_i)/2}$. Here N is the total number of nitrogen atoms, ν_i is the number of nitrogen atoms in the solid, and the factor 3 is the number of possible orientations per nitrogen molecule. The ν_i nitrogen atoms in the solid are distributed over $n_s!/[(n_s - \nu_i)!\, \nu_i!]$ different configurations, so that the number Ω_i of states with the energy $\mathbb{H} = \nu_i U_0$ reads $\Omega_i = n_s!/[(n_s - \nu_i)!\nu_i!]\,(3V/V_0)^{(N-\nu_i)/2}$.

As shown in [Sko93a], the sum $Z = \Sigma_i\, \Omega_i \exp(-\mathbb{H}/kT)$ can now be calculated analytically to yield the partition function

$$Z = \left(1 + \sqrt{\left(\frac{p}{p_0}\right)} e^{-U_0/kT}\right)^{N_s} \tag{4.17}$$

with the pressure parameter $p_0 = 3kT/V_0 \approx 100$ kbar. N_s is the total number of 9e sites in the lattice and p the nitrogen pressure in the closed reaction volume. Using $N_s\langle c\rangle_0 = -kT\, \partial \ln Z/\partial U_0$ (cf. Eq. (4.9)) we obtain the equilibrium interstitial concentration [Sko93a]:

$$\langle c\rangle_0 = \frac{1}{1 + \sqrt{(p_0/p)}\, e^{U_0/kT}} \tag{4.18}$$

For *small* concentrations this equation of state yields the simple square-root dependence on p known as *Sieverts' law* [Ale78].

The constant p_0 is about the same for all metals [Fas76] so the net reaction energy (absorption energy) U_0 is obtained from the nitrogen solubility in well-defined

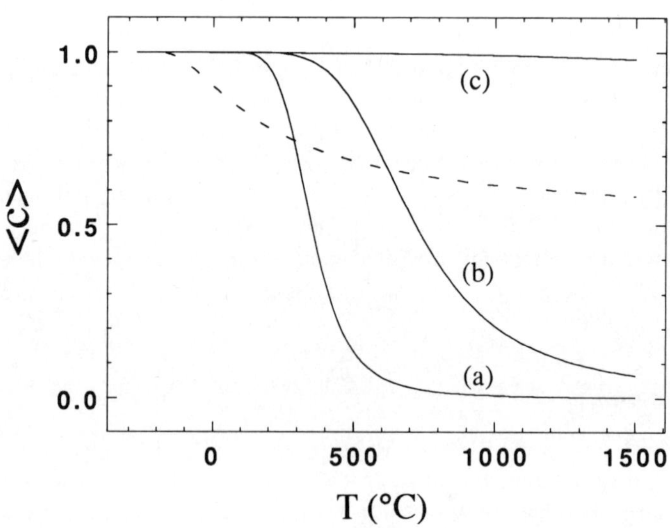

FIG. 4.6. Temperature dependence of the interstitial modification: (a) atomic gas; (b) molecular gas; and (c) solid with $U_0 = -50$ kJ/mole. The dashed line is for a solid with $U_0 = -5$ kJ/mole, which is an estimate of the net reaction energy for carbon in Sm_2Fe_{17}.

conditions. Long-term isothermal absorption experiments [Coe92b, Sko93a] yield the moderately strong exothermic net reaction energy $U_0 = -57 \pm 10$ kJ/mole (~ 0.6 eV) for $Sm_2Fe_{17}N_{3-\delta}$ (Fig. 4.6b); the solubility of nitrogen in other R_2Fe_{17} compounds is similiar to that in Sm_2Fe_{17}. Note that the accessible temperature window in which it is possible to establish quasiequilibrium is restricted: above about 550 °C the nitride disproportionates, whereas the equilibration times become very long below about 350 °C. Substituting other 3d elements for iron leads to a greater (less exothermic) reaction energy for heavier elements such as cobalt, but a lower (more exothermic) U_0 for lighter elements such as manganese.

The reaction energy for nitrogen in unalloyed iron is endothermic, $U_0 \sim 10$ kJ/mole. Consequently, the equilibrium nitrogen content in iron is very small, but the activation energy E_a for diffusion (Table 4.6) is about 50–60 kJ/mole less than for R_2Fe_{17} [Fas76].

1.3 *Mixed nitrogen–hydrogen atmosphere*

The interstitial modification of rare-earth intermetallics may be carried out using a mixture of nitrogen and hydrogen gases. Under typical nitrogenation conditions, the partial pressure of ammonia (NH_3) in *equilibrium* with N_2 and H_2 is negligibly small [Mac83, Spe83], so ammonia need not be considered separately. Note, however, that the catalytic decomposition of ammonia at the particle surface is a nonequilibrium process which strongly affects the nitrogenation behaviour (see Section B.1.5.).

If Sm_2Fe_{17} is heated in a mixture of molecular nitrogen and hydrogen, the high diffusivity of hydrogen leads to a rapid hydrogen uptake which is followed by a much slower nitrogenation reaction. However, due to the low hydrogen solubility at typical nitrogenation temperatures of order 700 K, the interstitially modified intermetallic has the nominal quasiequilibrium composition $Sm_2Fe_{17}H_xN_y$ with $x \ll y$. In this compound, nitrogen occupies the 9e sites nearly exclusively (cf. Fig. 4.2), while hydrogen is found on both octahedral 9e sites and tetrahedral 18g sites [Isn92a, b]. The hydrogen content can be written as $x = x_{oct} + x_{tetr}$ with $x_{oct} < 3$ and $x_{tetr} < 2$, so $x < 5$. This means that only two out of six tetrahedral sites per formula unit are actually occupied by hydrogen, which indicates strong interatomic repulsion on the 18g octahedron (see Section B.3.3).

To simplify the calculation of x and y we make the assumption that we have only two available tetrahedral sites per formula unit. With the hydrogen net reaction energies U_{oct} and U_{tet} we obtain the partition function

$$Z = \left(1 + \sqrt{\left(\frac{p_N}{p_0}\right)} e^{-U_0/kT} + \sqrt{\left(\frac{p_H}{p_0}\right)} e^{-U_{oct}/kT}\right)^{N_0} \left(1 + \sqrt{\left(\frac{p_H}{p_0}\right)} e^{-U_{tetr}/kT}\right)^{2N_0/3}$$

from which the mean occupancies are obtained analogously to Eq. (4.9). In particular, for the nitrogen concentration we find

$$y = \frac{3}{1 + \sqrt{(p_0/p_N)}\, e^{U_0/kT}(1 + \sqrt{(p_H/p_0)}\, e^{-U_{oct}/kT})} \tag{4.19}$$

Comparing Eqs. (4.18) and (4.19) we see that hydrogen displaces nitrogen from the 9e sites. To determine the order of magnitude of this displacement we take into account that the one-bar outgassing temperature of hydrogen, at which about half of all octahedra are still occupied, is of the order of 500 K [Isn92b]. Application of Eq. (4.18) to hydrogen then yields the estimate $U_{oct} = -25$ kJ/mole. This means that hydrogen reduces the nitrogen solubility by only about 1%. On the other hand, there is a hydrogen-induced pre-expansion of the lattice which facilitates the nitrogen uptake (cf. Sections B.3 and C.2). Also, the generation of microcracks in the pre-expanded lattice accelerates the nitrogenation reaction.

Octahedral and tetrahedral occupancies in R_2Fe_{17} intermetallics have been monitored by neutron diffraction during deuterium desorption experiments [Isn92b]; from these data an energy U_{tetr} of order -15 kJ/mole has been deduced [Sko94a].

The lattice-gas model, Fig. 4.5, can be generalized to interstitial modification involving atomic gases or solids. In the hypothetical case of an atomic gas, we obtain the equilibrium solubility (Fig. 4.6a):

$$\langle c \rangle_0 = \frac{1}{1 + (p_0/p)\, e^{U_0/kT}} \tag{4.20}$$

For noble gases, U_0 is always positive so their equilibrium solubility in metals is negligible (cf. Section 3.5). Such atoms may be introduced by ion implantation, which is a nonequilibrium method of introducing interstitial atoms. The concentration of implanted ions does not only depend on the crystal temperature, but also on the velocity and direction of the arriving ions. Ion implantation tends to destroy the host.

1.4 *Solid–solid reaction*

As discussed by Skomski *et al.* [Sko93b], the solubility becomes pressure independent in the case of a solid–solid reaction:

$$\langle c \rangle_0 = \frac{1}{1 + e^{U_0/kT}} \tag{4.21}$$

On reacting Sm_2Fe_{17} with graphite at 500 °C, lattice expansions up to 5% have been achieved [Sko93b]. In fact, the standard carbonation reactions using hydrocarbon gases are probably solid–solid reactions since the reagents decompose at the surface to leave a thin film of solid carbon which then diffuses into the intermetallic.

Figure 4.6 compares the temperature dependence of the solubilities Eqns (4.18), (4.20) and (4.21). At a given value U_0, the lower entropy of a solid compared to a gas causes the solubility to be highest for solid–solid reactions. Analogously, the configurational entropy of a molecular gas is lower than that of an atomic one, which explains the higher solubility of molecular gases at fixed U_0.

Note that Eq. (4.21) predicts half-occupation at very high temperatures, which is in agreement with the experimental fact that the interstitial concentration of carbon introduced from the melt does not exceed about $x = 1.5$ [Bus91]. The $R_2Fe_{17}C_x$ compounds are actually stable at high temperature when $x \leq 1.5$ (Chapter 2C), whereas the nitrogen-containing compounds appear to be metastable for all x. Attempts to introduce other solid elements, such as boron or silicon, interstitially into Sm_2Fe_{17} by low-temperature reaction have been largely unsuccessful [Sko93b].

1.5 *Nonequilibrium interstitial modification*

The formation of ammonia from molecular nitrogen and hydrogen is slightly exothermic ($\Delta H = -46$ kJ/mole) but the large entropy difference makes the formation of ammonia thermodynamically unfavourable under typical nitrogenation conditions [Mac83, Spe83]. In other words, the reaction $N_2 + 3H_2 \rightarrow 2NH_3$ reduces the number of gas atoms and with it the entropy of the system, so nitrogen in equilibrium with hydrogen at about 500 °C and 1 bar pressure can be regarded as largely ammonia-free (less than 1%). A reasonable turnover in industrial ammonia generators requires high pressures and low temperatures.

However, if a high ammonia concentration is maintained artificially, the ammonia decomposes catalytically at the metallic surface, which strongly influences the nitrogenation behaviour of the metallic lattice. In particular, ammonia decomposition on transition-metal-rich intermetallics includes the formation of intermediate nitrides, which is equivalent for the metal to a huge nitrogen pressure. Once inside the metal, the nitrogen atoms are subject to bulk diffusion and occupy the sites which are energetically most favourable. An example of this process is the production of Fe_4N by heating iron in ammonia.

When Sm_2Fe_{17} is heated in ammonia, lattice expansions of up to 8.3 vol.% have been observed [Iri92, Bre94]. Here it is very likely that in addition to occupying three 9e sites per formula unit some of the nitrogen enters the 18g hexagon of tetrahedral sites:

1. Compared to the nitride with $x = 3$, the additional expansion in the crystallographic c-direction is much larger than the in-plane contribution.
2. Assuming a constant volume expansion of roughly 2% per nitrogen atom (Fig. 4.4), the value 8.3 vol.% corresponds to about $x = 4$ nitrogen atoms per formula unit. This is compatible with nitrogen occupancy of the 3b interstitial site at the centre of the 18g hexagon, or of one out of six of the 18g tetrahedra, in addition to the 9e octahedra.

A simple view of nitrogenation using ammonia is based on the (equilibrium) reaction constant $K = [N_2]^{0.5}[H_2]^{1.5}/[NH_3]$, where [ ] represents the partial pressure. This illustrates that a moderate increase in ammonia concentration is equivalent to a large increase in nitrogen concentration (nitrogen pressure). However, the reaction constant explanation obscures the nonequilibrium character of the reaction, which is thermodynamically closer to ion implantation than to conventional gas-phase interstitial modification.

1.6 *Microscopic approaches*

The calculation of U_0 from first principles is difficult, even for hydrogen–metal systems. Many models are purely empirical, but here are the basic ideas of some typical approaches.

1. *Explanations based on band-structure considerations.* The classical example of a gas–solid system is hydrogen in palladium. Hydrogen occupies the octahedral interstices in fcc palladium so the ideal composition of the hydride is PdH. In fact, a composition PdH_x with $x < 0.6$ is easily achieved, but hydrides with $x > 0.6$ require disproportionately large pressures. This behaviour is explained by the fact that the palladium 4f band remains incompletely filled for up to 0.6 hydrogen atoms per formula unit, whereas higher hydrogen contents require the energetically much less favourable filling of the 5s band [Wic78].

   In [Gri84] the energy difference between Fermi energy and the lowest-lying s-like conduction band is made responsible for the observed net reaction energies. The idea behind this semiempirical approach is that a small difference between that s-band and the Fermi level indicates a large energy gain due to hybridization.
2. *Jellium-like models.* The reaction is assumed to be exothermic if the embedding energy of the interstitial atoms in an electron gas of appropriate density is larger than the corresponding dissociation energy (218 kJ/mole · H for H_2 and 374 kJ/mole · N for N_2). It turns out ([Nør82] and references therein) that the reaction between an electron gas and noble gases is always endothermic, whereas the net reaction for other interstitials can have either sign. However, the exactness of these methods is limited: errors up to about ±70 kJ/mole occur.
3. *The Miedema model.* Here the electron density n_{WS} at the Wigner–Seitz cell boundary and the electronegativity Φ^* are used to predict binding energies [Mie80], while the elastic energy necessary to expand the intermetallic lattice around the interstitial atoms is neglected. The basic parameters are P and Q which give the relative contributions of ionic (Φ^*) and covalent (n_{WS}) binding effects respectively. Unfortunately, P and Q are not constant but depend to some extent on the elements in question, so it is difficult to derive reliable quantitative predictions. A crude estimation, based on data given in [Mie80], yields a largely ionic character of the bonds Z–T and Z–R(Z = N, C; T = Fe, Co; R = Y, La), but this is inconsistent with the idea that covalent effects dominate the crystal-field contribution of carbon in 2–17 intermetallics ([Bus91] and references therein).

2 Interstitial diffusion

2.1 *Diffusion in spherical grains*

At typical nitrogenation temperatures (400–500 °C), the nitrogenation reaction proceeds by thermally activated bulk diffusion. (Quantum tunnelling may become significant at low temperature [Keh78, Sko95a].) To describe the diffusion we first neglect short-circuit diffusion along grain boundaries or microcracks and assume that our metal powder represents an ensemble of isotropic spherical grains. The relevant parameters are thus temperature T, time t and particle radius R.

The temperature dependence of the diffusion time is given by the Arrhenius law, Eqs (4.2) and (4.12). Beside the prefactor D_0, which is quite universal, the diffusivity is determined by the activation energy E_a, which is required to move the metal atoms apart to generate an outlet for the diffusing atom. Unlike the net reaction energy U_0, which can be either positive or negative, the activation energy E_a is always positive and its magnitude is determined by the elastic properties of the host metal and the size and nature of the diffusing atoms as well as the site-binding energy E_a itself [Boc78, Fas76]. Due to its elastic origin the activation energy scales roughly as the square of the radius of the diffusing atom. Typical values are $E_a = 250$ kJ/mole for the self-diffusion of metals, $E_a = 40$ kJ/mole for the diffusion of hydrogen in metals, and $E_a = 120$ kJ/mole for the diffusion of carbon, oxygen, or nitrogen in metals [Boc78, Fas76]. For nitrogen in Sm_2Fe_{17}, values $D_0 = 1.0\ \mathrm{mm}^2/\mathrm{s}$ and $E_a = 133 \pm 15$ kJ/mol are found [Coe92b, Sko93a]. Other authors report E_a of 100–163 kJ/mole [Uch93].

To calculate the diffusion times we have to solve the three-dimensional equation

$$\frac{\partial c}{\partial t} = D\nabla^2 c \tag{4.22}$$

subject to appropriate boundary conditions. Usually, we can assume that the gas supply through the particle surface is sufficiently high to ensure $c(R, t) = \langle c \rangle_0$. An exception is the surface-controlled reaction, where the energy barrier at the particle surface dominates the reaction kinetics.

In the bulk-controlled limit the solution of Eq. (4.22) reads [Cra75, Boc78]

$$c(r,t) = \langle c \rangle_0 \left(1 + \frac{2R}{\pi r} \sum_{m=1}^{\infty} \frac{(-1)^m}{m} \sin\frac{m\pi r}{R} \exp\left(-\frac{m^2\pi^2 Dt}{R^2}\right)\right) \tag{4.23}$$

Typical nitrogenation profiles are shown in Fig. 4.7. Note that the profile depends on the mean concentration only, but not explicitly on t or R.

Integration of Eq. (4.23) yields the nitrogen content x on the 9e sites:

$$x = 3\langle c \rangle_0 \left(1 - \frac{6}{\pi^2} \sum_{m=1}^{\infty} \frac{1}{m^2} \exp\left(-\frac{m^2\pi^2 Dt}{R^2}\right)\right) \tag{4.24}$$

In the long-time limit Eq. (4.24) reduces to the simple exponential law

$$x = 3\langle c \rangle_0 \left(1 - \frac{6}{\pi^2} \exp\left(-\frac{\pi^2 Dt}{R^2}\right)\right) \tag{4.25}$$

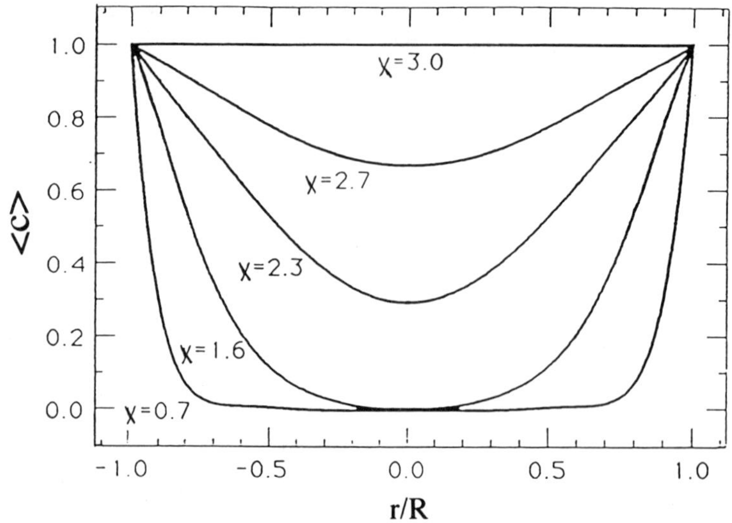

FIG. 4.7. Nitrogen concentration profiles for spherical particles (Eq. 4.23). The equilibrium nitrogen content is taken as 3.0.

2.2 *Real grains*

In practice there are significant deviations from the idealized model described in the preceding section:

1. Particles are usually non-spherical and it is difficult to solve Eq. (4.22) for real-grain geometries. One exception is for large flat plates [Boc78] where the concentration profile becomes

$$c(z,t) = \langle c \rangle_0 \left(1 - \frac{4}{\pi} \sum_{m=0}^{\infty} \frac{1}{2m+1} \sin \frac{(2m+1)\pi z}{L} \exp\left(- \frac{(2m+1)^2 \pi^2 Dt}{L^2} \right) \right)$$

where L is the thickness of the plate.

However, it is possible to evaluate the initial stage of diffusion for particles of arbitrary shape, where the mean concentration is given by $x = 3\ \langle c \rangle_0\ A\ L_0(t)/V$. Here A and V are the total powder surface area and the total powder volume, respectively, and L_0 is the effective diffusion length

$$L_0(t) = \sqrt{\left(\frac{4Dt}{\pi} \right)}$$

The square-root dependence $x \propto \sqrt{t}$ in the initial stage of nitrogenation has been used to determine the diffusion constant of nitrogen in iron-rich rare-earth intermetallics [Coe92b, Sko93a], whereas U_0 is determined from the quasiequilibrium nitrogen content using Eq. (4.18). Results are shown for $Sm_2Fe_{17}N_3$

Table 4.4 Diffusion parameters and net reaction energies for nitrogen in Sm_2Fe_{17} and $Nd(Fe_{11}Ti)$ [Coe92b, Sko93a]

	$Sm_2Fe_{17}N_3$	$Nd(Fe_{11}Ti)N$
D_0 (mm^2/s)	1.0	6.5
E_a (kJ/mole)	133 ± 15	158 ± 20
U_0 (kJ/mole)	57 ± 5	51 ± 10

Table 4.5 Nitrogenation nomograph. Any straight line connects possible nitrogenation temperatures, times and radii of Sm_2Fe_{17} particles for 90% nitrogenation. To avoid disproportionation the line must lie beneath the black dot.

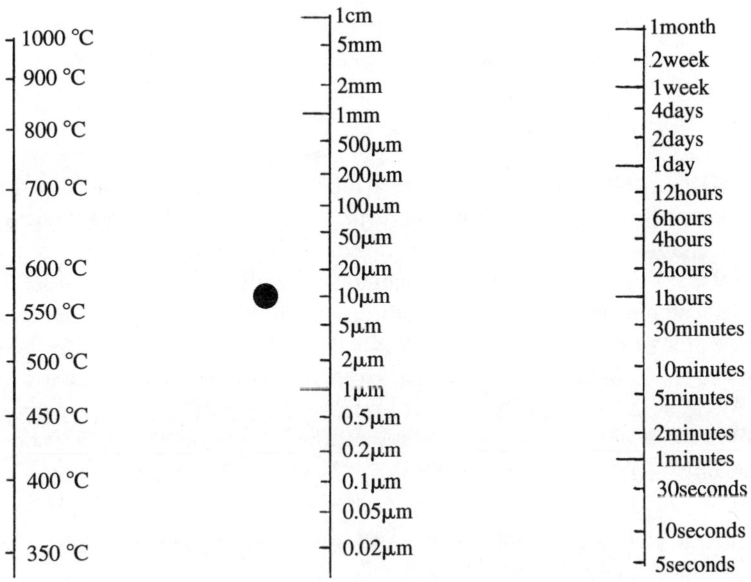

Table 4.6 Diffusivity of 2p atoms in bcc metals [Fas76]

System	E_a (kJ/mole)	D_0 (mm^2/s)
C in Fe	80.3	0.39
N in Fe	76.8	0.49
O in Fe	97.5	3.7
C in Nb	138.1	0.40
N in Nb	146.0	0.86
O in Nb	112.6	2.1

and $Nd(Fe_{11}Ti)$ in Table 4.4. Table 4.5 is a nomograph which relates the size, temperature and time needed for 90% complete nitrogenation of spherical Sm_2Fe_{17} grains. Table 4.6 contains some comparative data for interstitial diffusion in metals.

2. Diffusion in noncubic grains is, at least slightly, anisotropic, so Eq. (4.22) must be replaced by

$$\frac{\partial c}{\partial t} = D_\parallel \frac{\partial^2 c}{\partial z^2} + D_\perp \left(\frac{\partial^2 c}{\partial x^2} + \frac{\partial^2 c}{\partial y^2} \right)$$

In the case of diffusion perpendicular to a flat surface, the diffusion coefficient reads $D(\theta) = D_\parallel \cos^2\theta + D_\perp \sin^2\theta$, where θ is the angle between the surface normal and crystallographic c-axis. Averaging this expression over a random distribution of planes we obtain the effective diffusion constant $D_{\text{eff}} = 2D_\perp/3 + D_\perp/3$. Anisotropic effects can be taken into account approximately by replacing D by D_{eff} in Eqs (4.23–25).

3. At low temperature, the nitrogen uptake is surface-dominated if the energy barrier E_s is large (see Fig. 4.2). In this regime the nitrogen flux through the grain surface is limited so the nitrogenation times are much larger than estimated from Eq. (4.24). Taking this constant-flux assumption as the boundary condition for Eq. (4.22), we obtain an average concentration which increases linearly with time and a parabolic concentration profile; the nitrogenation process is similiar to the filling of a liquid in an empty jar [Sko94a]. Values of E_s are quite sample-dependent, but of the order of 80 kJ/mole. The physical reason for a high-energy barrier E_s may be a layer of surface oxide or adsorbed water which inhibits the catalytic dissociation of the nitrogen molecules. E_s determines the cross-over temperature below which the surface energy barrier controls the nitrogen uptake. A typical temperature above which the nitrogen uptake in Sm_2Fe_{17} changes from $x \propto t$ to $x \propto \sqrt{t}$ is about 350 °C.

4. The nitrogenation behaviour strongly depends on the size distribution of the grains. The diffusion time increases quadratically with the particle size so that smaller grains may be fully nitrided, or already disproportionating, while bigger ones still exhibit an unitrided core (see Section C1.2).

5. Finally, perhaps the most important feature is the inhomogeneity of real particles, which leads to phenomena such as short-circuit diffusion along grain boundaries or microcracks (see Section C.2.2). Similarly, there is the possibility of immobilization diffusion [Cra75] where non-9e sites act as easy pathways in the bulk without changing the phase structure of the nitride. The 9e nitrogen is immobile. This mechanism leads to deviations from Eq. (4.23) [Zha 95].

3 Interaction effects

3.1 *The Ising-model analogy*

In the preceding sections, the only interactions considered were those within gas molecules, the site-binding interaction between interstitial atoms and the host lattice, and the hard-core repulsion which forbids multiple occupation of interstitial sites. Lattice-gas systems where interatomic interaction between atoms on *different* interstitial sites is neglected are conventionally termed 'interactions-free'. The reason for this somewhat misleading name is that the statistical mechanics of these systems is equivalent to that of interaction-free Ising models — at least so far as the interstitial occupation is concerned. For instance, the model with the equation of state Eq. (4.10) can be mapped onto the interaction-free spin-1/2 Ising model if we identify the set $c_i = \{0, 1\}$ with $s_i = \{-1, 1\}$. Correspondingly, the interaction-free spin-1 Ising model [Yeo92] can be used to describe the site-blocking between hydrogen and nitrogen atoms if we identify the possible occupancies $\{0, 1_{\mathrm{H}}, 1_{\mathrm{N}}\}$ with $s_i = \{-1, 0, 1\}$.

However, calculations such as those presented in the preceding sections represent only one application of lattice-gas statistics. Since 1952, when Lee and Yang [Lee52] used the spin-1/2 Ising model to discuss the gas–liquid phase transition, lattice-gas calculations have become a standard method for studying interaction effects and phase transitions. An example is the use of the Ising-model analogy to discuss the phase structure of hydrogen in metals [Ale69]. Typically, the number of hydrogen atoms is kept constant by a chemical potential, which acts as a parameter, and the question is asked how these atoms distribute in the lattice. In the following section we use a slightly different but physically equivalent [Sko95a] approach to discuss interaction effects in rare-earth intermetallics.

3.2 *Phase transitions*

In Section B.1.1 we have treated the nitrides and carbides in question as interstitial gas–solid solutions. It is important to realize that systems where the surfaces are nitrogen-rich, or grains with many defects and microcracks or polydisperse powders where the small grains are more thoroughly nitrided than the large ones may appear to present a discontinuous nitrogen content [Col92] without representing a two-phase system [Sko95a].

Nevertheless, the question remains under what conditions the formation of a two-phase nitride is possible. A typical gas–solid phase diagram is shown in Fig. 4.8: below a critical temperature T_{c} the solid-solution nitride is expected to segregate into a two-phase mixture of nitrogen-poor α and nitrogen-rich β phases. In fact, the very low diffusivity of nitrogen below 400 °C makes it practically impossible to obtain information about phase equilibrium there, and it has not been possible to determine the critical temperature T_{c}. Furthermore, low-temperature phase diagrams tend to be rather complicated. For example, niobium forms as many as 14 different phases with hydrogen [Fut86]. In other words, phase segregation is an essential feature of gas–metal systems and it is worth discussing its physical implications.

Let us first recall the molecular-field approximation of the spin-1/2 Ising model [Mac63, Sko94a, Yeo92]. Starting with the Hamiltonian

$$\mathcal{H} = -H \sum_i s_i - \sum_{i>k} J_{ik} s_i s_k$$

and using $\Sigma_k J_{ik} = J_0$ we introduce the molecular field $H_{\text{eff}} = H + J_0\langle s\rangle$, which replaces the external field in the paramagnetic equation of state. It is well known that this approximation yields the mean-field equation of state

$$\langle s\rangle = \tanh\left(\frac{H + J_0\langle s\rangle}{kT}\right) \tag{4.26}$$

with a magnetic phase transition at $H = 0$ and $T = J_0/k$.

This procedure can be generalized to gas–metal systems. The Hamiltonian reads

$$\mathscr{H} = b\sum_i c_i - \sum_{i>k} A_{ik}c_ic_k$$

where the coefficients A_{ik} describe the interaction between different sites $i \neq j$. With $A_0 = \Sigma_k A_{ik}$ the molecular-field interaction reads $b_{\text{eff}} = b_0 + A_0\langle c\rangle_0$ and Eq. (4.18) becomes

$$\langle c\rangle_0 = \frac{1}{1 + \mathrm{e}^{[(b_0 - A_0\langle c\rangle_0/kT]}} \tag{4.27}$$

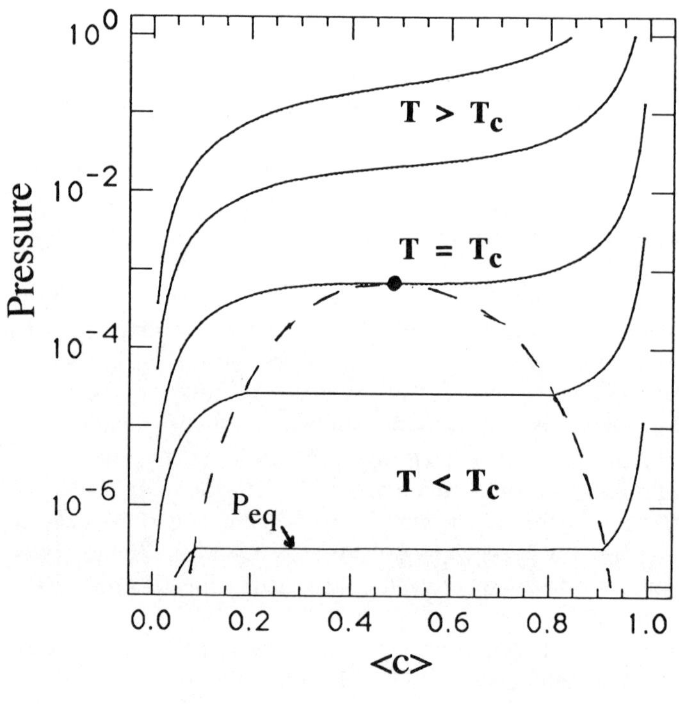

FIG. 4.8. Schematic phase diagram of a gas–solid system. The dot marks the critical point, the solid lines are isotherms and the dashed line marks the limit of the two-phase region where gas-rich and gas-poor phases coexist. P_{eq} is a plateau pressure.

In the interaction-free limit $A_0 = 0$ this equation has to reproduce Eq. (4.18), hence $b_0 = U_0 - 0.5kT\ln(p/p_0)$.

To discuss the phase diagram generated by Eq. (4.27), we put the identity $c_i = (1 + s_i)/2$ into Eq. (4.27). This reproduces Eq. (4.26) with $J_0 = A_0/4$ and $H = -U_0/2 - A_0/4 - kT\ln(p/p_0)/4$. It is known how the behaviour of the Ising model depends on J_0 and H, so we can conclude:

1. Phase segregation is found below the critical temperature $T_c = A_0/4$, so A_0 is positive in systems where phase segregation occurs (attractive interaction) [Fla88].
2. Neglecting hysteresis effects, the phase transition occurs at $H = 0$, that is, at $p = p_0 \exp(-4)\exp((2U_0 + A_0)/kT)$. This temperature dependence of the so-called plateau pressure (equilibrium pressure) $p = p_{eq}$ is commonly referred to as the *van't Hoff relation* [Sch88]. For a discussion of hysteresis effects see [Sko95a, Wag78].

Equation (4.27) gives a simple and illustrative idea of the α–β (α– α') transition in gas–metals systems. We should, however, keep in mind that the analogy between Ising magnets and gas–lattice systems is not complete. For instance, there is no magnetic analogy for the so-called onion-skin effect which leads to the relief of coherent stresses [Wag74, 78].

Note that these considerations are purely phenomenological; we made no assumption concerning the microscopic origin of A_0. In the original Ising model, interaction means quantum-mechanical exchange coupling between neighbouring spins. In the case of gases in metals the situation is more complex: interstitial atoms are subject to both electronic and elastic interaction, and both mechanisms must be considered.

3.3 *Electronic interaction*

Gas atoms in metals not only interact with atoms at neighbouring interstices but are also subject to electronic interaction with the host lattice. Strictly speaking, the site-blocking repulsion on a given interstial site can be regarded as electronic as well, but this interaction is 'diagonal' in A_{ik} and *does not influence the tendency towards phase segregation.* An attractive interatomic interaction is needed for a phase transition. Interatomic electronic interaction is characterized by two properties [Wag74, 78]. (i) Charge-density inhomogenities in metals are strongly screened by conduction electrons so the electronic interaction can be regarded as short-ranged. (ii) The electronic interaction, which involves the orthogonalization of electronic orbitals, is mainly repulsive ($A < 0$) and therefore cannot be responsible for an α–β phase transition, which requires $A > 0$. There is an empirical rule for hydrogen in metals (*Westlake's rule*) which requires a minimum separation of about 2.1 Å between two hydrogen atoms [Yvo88].

A simple example of electronic interaction effects is $Sm_2Fe_{17}H_x$ where hydrogen occupies the octahedral 9e sites and some of the tetrahedral 18g sites (Fig. 4.1a). At

FIG. 4.9. The 18g hexagon (cf. FIG. 4.1a): (a) empty hexagon; (b) single occupation; and (c) double occupation.

low temperatures the nominal composition approaches $Sm_2Fe_{17}H_5$ [Isn92b], so only two out of six 18g sites per formula unit can actually be occupied. The partition function depends on the detailed distribution of the hydrogen atoms on the 18g hexagon (Fig. 4.9). It is likely that the hydride contains empty hexagons (Fig. 4.9a), one-hydrogen states (Fig. 4.9b) and hydrogen dumbells (Fig. 4.9c). Under these conditions the partition function for the 18g subsystem reads

$$Z = \left(1 + 6\sqrt{\left(\frac{p_0}{p}\right)}\mathrm{e}^{-U_t/kT} + 3\frac{p}{p_0}\mathrm{e}^{-(2U_t + \Delta U)/kT}\right)^{N'}$$

where U_{tet} is the net reaction energy for the tetrahedrals, ΔU is an interaction energy, and N' is the number of 2:17 formula units. The numbers 6 and 3 are the number of different single and dumbell configurations respectively. Note that it is not possible to represent this partition function in the interaction-free form of Eq. (4.17). The maximum hydrogen content $x = 5$ indicates that ΔU is only weakly positive — otherwise the maximum hydrogen content $x = 4$ would be observed.

3.4 *Elastic interaction*

The main contribution to the interatomic interaction is elastic: the occupation of an interstitial site leads to a small volume expansion and to a corresponding long-range lattice distortion around the interstitial site which interacts with other interstitial atoms [Wag74, 78]. Illustratively speaking, the gas atoms enlarge the interstitial holes in their neighbourhood which can now be occupied more easily. This is felt by the individual interstitial atoms as attractive interaction. It is worth while mentioning that the sign of A_0 depends on the boundary condition at the crystal surface: the interaction is attractive in the case of a free surface, but repulsive in the case of a clamped surface [Wag74, 78].

Above T_c the random thermal motion of the gas atoms dominates the interaction, but below T_c the formation of clusters and phase segregation is observed (see Fig. 4.8). The elastic contribution to A_0 scales as $E(\Delta v)^2/\Omega$, where E is Young's modulus, Ω the lattice volume per interstitial site and Δv the volume expansion per interstitial atom [Pei78, Wag74, 78]. Typically, $\Delta v \approx 3$ Å^3 for hydrogen and $\Delta v \approx 6$ Å^3 for nitrogen and carbon (Tables 4.2 and 4.3). For nitrogen in Sm_2Fe_{17}, this formula yields a critical temperature which is a factor of 1.5–2 smaller than that of palladium hydride ($T_c = 292$ °C for PdH_x, [Wic78]). Alternatively, there are about six times fewer gas atoms per metal atom in Sm_2Fe_{17} nitride than in PdH_x, which yields a large value of Ω for the nitride and explains the gas–solid solution character of $Sm_2Fe_{17}N_x$.

In crystals with inhomogeneous nitrogen concentration the critical temperature depends on the macroscopic shape of the crystal [Wag74, 78]. This remarkable property is a direct consequence of the long-range character of the elastic interaction and has a certain analogy in the macroscopic demagnetization factor in permanent magnets — in both cases the local energy balance is affected by the sample geometry.

It is worth while mentioning that in most cases the molecular-field approximation breaks down as the critical point is approached. Near the critical point the thermodynamical fluctuations are very strong so that the introduction of averaged quantities such as $\langle s \rangle$ is inaccurate. The deviations from the mean-field behaviour depend on whether the interaction involves a finite number of neighbours (short-range interaction), or an infinitely large number of neighbours (long-range interaction). As argued by Wagner [Wag74, 78] gas–metal systems are almost unique due to their molecular-field behaviour as a consequence of long-range correlation. Even superconductors exhibit a large but *finite* range of interaction above 0 K.

Note, finally, that the lattice expansion in $Sm_2Fe_{17}N_{3-\delta}$ is partly due to spontaneous magnetostriction (about 2 vol.%). Above the Curie temperature this contribution disappears, so the critical temperature depends indirectly on the nitrogenation temperature.

3.5 *Diffusion*

Comparing real gas–metal systems with the model system of Section A1.3, where there is only one gas atom in the lattice, we can imagine that interstitial diffusion is affected by the presence of other atoms. Surprisingly, it turns out that site-blocking interaction, that is, the single-occupancy condition, does not reduce the bulk diffusion constant. Only the tracer diffusion constant, which describes the mobility of individually marked atoms, is reduced by site-blocking [Gra90, Kut82].

To explain this paradox, we write the transition rate Eq. (4.12), which is independent of the occupancy of the final state (index i), in a modified form as $W_{i,i\pm1} = \Gamma_0 \exp(-E_a/kT)(1 - \langle c_i \rangle)$. It is obvious that this modified equation includes site-blocking interaction: if the final state is occupied ($c_i = 1$), then the transition rate $W_{i,i\pm1} = W(i \pm 1 \to i)$ vanishes. Putting the corresponding transition rates into Eq. (4.13) we find that all quadratic terms cancel each other so the diffusion equations Eqs (4.13) and (4.15) remain unchanged. In particular, the diffusion constant D is *independent* of the interstitial concentration. This behaviour is similiar to the scattering of quasifree conduction electrons in a dense metallic electron gas [Mot36].

However, if there is any attractive interatomic interaction, the situation becomes complicated. Attractive interaction means a tendency towards clustering and formation of sharp interfaces, whereas diffusion always tends to destroy clusters and sharp interfaces. Examples are the smooth, bath-tub-like diffusion profiles, Fig. 4.7, as opposed to a sharp transition between the two phases with $\langle c(\mathbf{r})\rangle \approx 0$ and $\langle c(\mathbf{r})\rangle \approx 1$. A molecular-field result for the diffusion constant is [Kut82]

$$D = D_f\left(1 - 4c(1-c)\frac{T_c}{T}\right)$$

where D_f is the interaction-free diffusion constant. This implies $D = 0$ at the critical point wherever $c = 0.5$. Below T_c the diffusion constant can be negative, which means that a given concentration inhomogeneity is not smoothed out but enhanced.

Alternatively, the motion of interacting atoms is not purely random but includes an additional deterministic motion towards regions with high interstitial concentration. This 'pre-expansion' effect yields a small contribution to the nitrogenation kinetics in 2:17 rare-earth intermetallics.

C. PROPERTIES OF INTERSTITIAL PERMANENT MAGNETS

1 Magnetic properties

The improvement of permanent magnetic properties due to interstitial modification is one of the most captivating trends in modern materials science. This refers in particular to nitrogen in Sm_2Fe_{17}: upon nitrogenation, we have seen that the Curie temperature increases from 116 °C to 476 °C (Fig. 2.47 and Table 4.3), the magnetization is enhanced by about 50%, and the magnetocrystalline anisotropy changes from easy-plane to easy-axis.

Sometimes, the intrinsic magnetic properties are changed in an unfavourable sense upon nitrogenation. A good example is nitrogen in $SmFe_{11}Ti$, where Curie temperature and magnetization improve, but the anisotropy changes from easy-axis to easy-plane [Coe91c]. Generally, the permanent magnetic properties of the carbides tend to be slightly worse than those of the isostructural nitrides, and for hydrogen, which has long been known as an interstitial in rare-earth transition-metal intermetallics, changes of the magnetic properties are much less pronounced.

The main figure of merit of a permanent magnet is the energy product $(BH)_{max}$, which never exceeds the value $\mu_0 M_s^2/4$ corresponding to an ideal rectangular hysteresis loop. This means that $Sm_2Fe_{17}N_3$ with $\mu_0 M_s \approx 1.54$ T is nearly as good a potential high-performance permanent magnet as the present room-temperature record holder $Nd_2Fe_{14}B$ with $\mu_0 M_s \approx 1.61$ T. The low Curie temperature $T_C = 312$ °C of $Nd_2Fe_{14}B$ compared with that of $Sm_2Fe_{17}N_3$, $T_C = 476$ °C means that the latter is potentially the better permanent magnetic material in a range $T > 150$ °C. Moreover, the magnetocrystalline anisotropy of $Sm_2Fe_{17}N_3$ exceeds that of $Nd_2Fe_{14}B$ at all temperatures (Fig. 2.40). It has been possible to realize coercivities as high as 4.36 T [Kuh92] in nanocrystalline $Sm_2Fe_{17}N_3$ magnets made by mechanical alloying.

Table 4.7 Curie temperature and magnetization of selected 2:17 compounds

	T_C (°C)	$\mu_0 M_s$ (RT) (T)	Ref.
Sm_2Fe_{17}	116	1.00	[Sun90]
$Sm_2Fe_{17}N_{2.5}$	476	1.54	[Coe91c]
$Sm_2Fe_{17}C_{1.1}$	279	1.24	[Coe91d]
$Sm_2Fe_{17}C_{2.2}$	395	1.43	[Coe92c]
Ce_2Fe_{17}	–32	0.00	[Sun90]
$Ce_2Fe_{17}N_{2.8}$	440	1.52	[Sun90]
Nd_2Fe_{17}	57	0.76	[Sun90]
$Nd_2Fe_{17}N_{2.3}$	459	1.69	[Sun90]
Gd_2Fe_{17}	204	0.47	[Sun90]
$Gd_2Fe_{17}N_{2.4}$	485	1.33	[Sun90]
Dy_2Fe_{17}	94	0.52	[Sun90]
$Dy_2Fe_{17}N_{2.8}$	452	1.15	[Sun90]
Er_2Fe_{17}	37	0.34	[Sun90]
$Er_2Fe_{17}N_{2.7}$	424	1.36	[Sun90]

On the other hand, due to the metastability of the nitride it is not possible to use conventional methods such as high-temperature liquid-phase sintering to process $Sm_2Fe_{17}N_3$ magnets, and it is difficult in any case to compete with $Nd_2Fe_{14}B$, which combines excellent performance with moderate processing and raw material costs.

1.1 *Curie temperature*

Compared with the R_2Fe_{17} parent compounds, the room-temperature magnetization of the respective nitrides is enhanced by at least 50% (Table 4.7). This cannot be ascribed to the 15% increase in magnetic moment, particularly since the 6% volume increase reduces M, the magnetic moment per unit volume. In fact, the large increase in room-temperature magnetization is mainly due to the enhanced Curie temperature of the nitride.

As opposed to cobalt-rich rare-earth intermetallics (see Chapter 2), whose Curie temperatures approach that of pure cobalt, the Curie temperatures of iron-rich rare-earth intermetallics are much lower than that of pure iron. For instance, $Nd_2Fe_{14}B$ consists largely of iron, which is very favourable from the points of view of raw material cost and magnetic moment, but its Curie temperature is lower than might be expected from the volume fraction of iron. On the other hand, the low Curie temperatures of iron-rich rare-earth intermetallics are accompanied by a strongly negative pressure dependence dT_C/dp [Bro73] (Fig. 2.23), which leads us to the conclusion that the Curie-temperature increase upon interstitial modification originates from the volume expansion [Coe91c].

This behaviour can be rationalized phenomenologically in terms of the Bethe–Slater curve [Qi92a]. A more microscopic approach is the Mohn–Wohlfarth theory [Moh87], which has been used to explain the Curie-temperature increase upon nitrogenation [Jas91]. Actually, the Mohn–Wohlfarth approach is a renormalized Stoner theory, applicable to very weak itinerant ferromagnets but not to transition-metal-rich rare-earth intermetallics. [Qi94, Sko94a]. A more likely explanation of the Curie-temperature enhancement is that the parent compounds are fairly close to the onset of ferromagnetism described by the Stoner criterion (quasi-weak ferromagnetism, [Sko95a, Rao94]), whereas the nitrides and carbides are nearly-strong ferromagnets.

In inhomogeneously nitrided grains, the local volume expansion does not only depend on the local nitrogen concentration but also on the average nitrogen concentration in the lattice. This means that the Curie temperature becomes a 'global' quantity which turns out to be more homogeneous than expected from the local nitrogen concentration [Coe92b, Sko93a]. In particular, the expanded outer regions cause the centre of the particle to expand epitaxially and to exibit an increase in Curie temperature. Even in the initial stage of nitrogenation there is a finite Curie temperature increase at the particle centre, though there is nearly no nitrogen there.

1.2 *Anisotropy and coercivity*

In good permanent magnets, the first-order anisotropy constant K_1 is large and positive, which fixes the magnetization direction parallel to the crystallographic *c* axis (easy-axis anisotropy, Table 4.8). All R_2Fe_{17} parent compounds are easy-plane,

Table 4.8 Crystal-field related properties of permanent magnets (for detailed references see [Sko95a])

	B_a(RT) T	K_1(RT) MJ/m^3	K_2(RT) MJ/m^3	A_2^0 K/a_0^2
$Sm_2Fe_{17}N_{3-\delta}$	14	8.6	1.5	–240
$Sm_2Fe_{17}C_{1.0}$	5			–130
$Sm_2Fe_{17}C_{2.2}$	13	7.4	0.74	–200
$Sm(Fe_{11}Ti)$	10			–140
$SmCo_5$	40	17		–200
Sm_2Co_{17}	6.5	3.3		–140
$Nd_2Fe_{14}B$	7.6	4.9	0.66	300

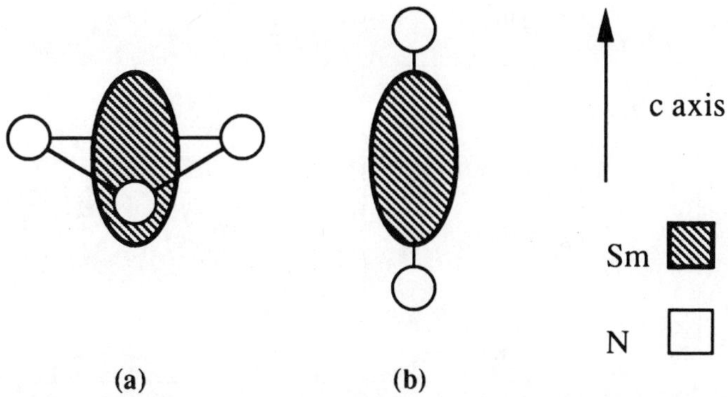

FIG. 4.10. Interstitial environment of the rare-earth in the 2:17 and 1:12 compounds, illustrating the opposite sign of the second-order crystal-field interaction in these structures.

while $Sm_2Fe_{17}N_x$ and $Sm_2Fe_{17}C_x$ exhibit the easy-axis anisotropy at room temperature. Samarium 1:12 compounds such as $Sm(Fe_{11}Ti)$ are reasonably interesting permanent magnetic materials, but their nitrides and carbides are easy-plane. On the other hand, the negative K_1 of easy-plane compounds such as $Nd(Fe_{11}Ti)$ or $Pr(Fe_{10}Mo_2)$ becomes positive upon nitrogenation [Coe91c, Yan93].

The magnetocrystalline anisotropy of rare-earth permanent magnets and their interstitially modified derivates is mainly provided by the rare-earth sublattice. Interstitial atoms modify the elecrostatic crystal field around the rare-earth atoms and cause the aspherical 4f charge clouds to change their orientation. Depending on whether the 4f orbitals are prolate or oblate (Fig. 1.12), the change in magnetocrystalline anisotropy produced by the interstitials can have either sign. In the case of $Sm_2Fe_{17}N_3$, the large increase in magnetocrystalline anisotropy is a combined effect of the prolate shape of the Sm 4f shell, the negative crystal-field charge of nitrogen, and the in-plane coordination of the interstial 9e site [Sko93c]. The interstitial environments of the rare-earth sites in the 2:17 and 1:12 compounds are sketched in Fig. 4.10

The simplest approach to calculating the single-ion rare-earth anisotropy contribution is the point-charge model, which assumes that the crystal-field generating charges $\rho(\mathbf{r})$ are sufficiently far away from the rare-earth atom so that $\rho(\mathbf{r})$ can be replaced by $\rho(\mathbf{r}) = \Sigma_i Q_i \delta(\mathbf{r} - \mathbf{R})$ [Hut64]. Since the point-charge model neglects the screening of localized charges by conduction electrons, the crystal-field interaction is greatly overestimated if chemically reasonable point charges are used [Sko93c]. Thomas–Fermi screening by the conduction electrons causes the crystal-field parameters A_n^m to decrease as $\exp(-qR)/R$, where R is the distance between rare-earth atom and crystal-field creating charge [Sko94]. The respective distance dependence of the point-charge model is $1/R^{n+1}$.

The coercivity of $Sm_2Fe_{17}N_3$ magnets is nucleation-controlled, that is, it depends on the absence of soft regions such as micron-scale α–Fe inhomogenities (see Chapter 5). This means that the formation of α–Fe due to the excess strain and the

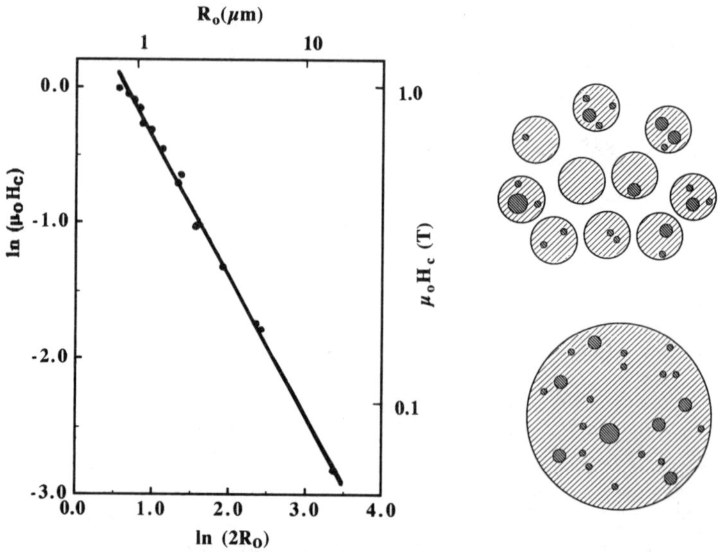

FIG. 4.11. Particle-size dependence of the coercivity in $Sm_2Fe_{17}N_3$ milled for different times after nitrogenation [Kob95]. The explanation illustrated is that the larger nucleation centres are isolated in a few fragments after milling.

plastic (incoherent) deformation of the lattice near the particles' surfaces is very harmful to coercivity. In this case coercivity can be recovered by zinc bonding [Ota91], which removes the free iron from the surface.

Figure 4.7 shows that the particle core is nearly nitrogen-free for nominal compositions $x < 2$. In the case of an ideal linear concentration dependence of the anisotropy field, regions with $x(\mathbf{r})$ less than about 0.5 show easy-plane anisotropy. This 'soft core' acts as the nucleus for magnetic reversal, which easily propagates into the hard regions of the grain and destroys coercivity [Sko93d]. The existence of a soft centre is avoided if the average nitrogen content exceeds $x = 2.1$. Another way is to remill the powder after nitrogenation [Fur91, Wen94], which isolates the soft core of large particles from their well-nitrided outer shell.

Data on coercivity as a function of particle size R_0 shown in Fig. 4.11 indicate an inverse dependence $H_c = 2K_1L/\mu_0 M_s R_0$ [Kob95]. To explain this result we suppose that the particles have a well-defined volume $4\pi R_0^3/3$ but contain randomly distributed soft-magnetic nuclei. Ultimately, the size distribution of these nuclei determines the hysteresis loop, since large nuclei are more harmful to coercivity. A power-law size distribution which satisfies the physically reasonable condition that the average size of the nuclei is much larger than the Bloch-wall width of the hard matrix phase but much smaller than R_0 may be assumed. As discussed in [Kob95], these assumptions lead to the observed $1/R_0$ dependence of the coercivity, where the quantity L turns out to be the average distance between neighbouring soft inclusions. This statistical coercivity model underpins the idea that milling after nitrogenation enhances the coercivity by isolating harmful nuclei.

2 Processing-related properties

2.1 *Stress and strain*

To discuss how stress and strain in inhomogeneously nitrided grains influence the magnetic properties of the powder, we assume that the equilibrium lattice expansion ε_{ik} depends linearly on the interstitial concentration $\langle c(\mathbf{r})\rangle$

$$\varepsilon_{ik} = \langle c\rangle \varepsilon_0 \delta_{ik}$$

In R_2Fe_{17} nitrides the volume expansion $\Delta V/V = \varepsilon_v \approx 6\%$, hence $\varepsilon_0 = \varepsilon_v/3 = 0.02$. To calculate the stress $\sigma_{ik}(\mathbf{r})$ which is caused by an inhomogeneous nitrogenation profile we use Hooke's approximation:

$$\sigma_{ik} = \frac{E}{1+\nu}(\varepsilon_{ik} - \varepsilon_{ik}^0) + \frac{\nu E}{(1+\nu)(1-2\nu)}\delta_{ik}\mathrm{Tr}(\varepsilon_{ik} - \varepsilon_{ik}^0)$$

Here E is Young's modulus and ν is Poisson's ratio; the equilibrium expansion is given by $\varepsilon_{ik}^0 = \langle c\rangle \varepsilon_0 \delta_{ik}$. The grains are treated as isotropic spheres. The local expansion ε_{ik} is then obtained by minimizing the expression

$$U_E = \frac{1}{2}\sum_{ik} \sigma_{ik}(\varepsilon_{ik} - \varepsilon_{ik}^0)$$

Typical stress and strain profiles are given in [Coe92b, Sko93a,b]. Main results are:

1. In the initial stage of nitrogenation, the *radial* strain at the particle surface $\varepsilon_r(R) = \varepsilon_0(1+\nu)/(1-\nu)$. This means that $\varepsilon_r(R)$ exceeds the equilibrium expansion by about 1%, which actually throws serious doubt on the validity of Hooke's law. This excess strain $\varepsilon_r(R) = \varepsilon_0$ is due to the fact that the outer nitrided regions cannot expand laterally because they are connected to the unexpanded inner core. To realize this constraint a large lateral compression is necessary which yields, via Poisson's ratio, an excess radial expansion.
2. At the particle centre, there is a volume expansion $\Delta\varepsilon_v = 2(1-2\nu)\langle c\rangle/(1-\nu)$, even when there is no nitrogen. This contribution is caused by the expanded outer regions of the grain and vanishes only for $\nu = \frac{1}{2}$ (incompressibility).
3. Smooth diffusion profiles such as these shown in Fig. 4.7 always cause smooth, stress and strain profiles, while discontinuous concentration profiles $\langle c(\mathbf{r})\rangle$ cause discontinuous stress and strain profiles.

For flat particles (thin plates) similiar results are obtained.

The energy U_E associated with the inhomogeneous stress and strain profiles is, in both cases,

$$U_E = \frac{\nu E}{1-\nu}\varepsilon_0^2(\langle c^2\rangle - \langle c\rangle^2)$$

Assuming $E = 150$ GPa and $\nu = \frac{1}{3}$ we obtain an energy of less than 3 kJ/mole nitrogen atoms, which is small in comparison with the net reaction energy $|U_0| = 57$ kJ/mole. This pre-expansion energy is regained during nitrogenation and corresponds to a small attractive interaction.

2.2 *Nitrogenation conditions*

The gas-phase interstitial modification using N_2 gas, which yields nitrides $Sm_2Fe_{17}N_x$ with $x \approx 3$, can be regarded as the archetypical nitrogenation process. Nitrogenation using ammonia (Section B1.5) is physically more complicated and may yield nitrides with the nominal composition $Sm_2Fe_{17}N_x$, with $x > 3$. The question arises whether there is any practical advantage in producing nitrides with nitrogen contents much larger than $x = 3$.

It seems that it is neither possible nor desirable to increase the nitrogen content of the 2:17 structure beyond $x \sim 4$. Greater nitrogen absorption (up to $x = 6$ [Iri92]) is associated with a large amorphous X-ray diffraction background which indicates that the material has disproportionated. When $x > 4$, there is no concomitant volume expansion. Note that one hydrogen atom in metals very often yields a lattice expansion of about 2.9 Å^3, whereas the corresponding value for nitrogen is of the order of 6 Å^3 [Coe90, Gri84, Pei78, Sun90].

Overloaded nitrides, with $x \approx 4$, exhibit intrinsic magnetic properties which are poorer than those of $Sm_2Fe_{17}N_3$. In particular, there is experimental evidence that the magnetocrystalline anisotropy deteriorates upon further nitrogenation [Iri92]. This can be explained by applying the superposition principle ([Sko93c] and references therein) to $Sm_2Fe_{17}N_4$: provided the additional nitrogen atoms occupy the 3b site at the centre of the 18g hexagon (Fig. 4.2), the axial coordination of the excess nitrogen must be expected to reduce the anisotropy compared to $Sm_2Fe_{17}N_3$.

A major practical problem is to find optimum nitrogenation conditions. Due to the low nitrogen diffusivity of order 10^{-15} m^2/s at typical nitrogenation conditions,

Table 4.9 Diffusion constants deduced from $D_0 = 1.02 \times 10^{-6}$ mm^2/s and $E_a = 133$ kJ/mole [Coe92b, Sko93a]

Temperature °C	D (μm^2/h)	D (m^2/s)
400	0.17	4.38×10^{-17}
425	0.41	1.13×10^{-16}
450	0.90	2.50×10^{-16}
475	1.88	5.24×10^{-16}
500	3.76	1.05×10^{-15}
525	7.20	2.00×10^{-15}
550	13.24	3.68×10^{-15}
575	23.48	6.52×10^{-15}
600	40.31	1.12×10^{-14}

the initial stage of diffusion can be studied by monitoring the nitrogen pressure in a closed volume (thermopiezic analysis, TPA) [Coe92, Sko93a]. The main result of these measurements is an activation energy E_a of approximately 130 kJ/mole, which falls in the range of values typical of nitrogen (and carbon) in metals (see Table 4.4). Table 4.9 shows diffusion constants at some temperatures of interest.

Let us use Eqs (4.23) and (4.24) to predict the nitrogenation behaviour of Sm_2Fe_{17}. The most convenient way is to start from a reference time $t_0 = R^2/D$, where D is taken from Table 4.9. This reference time fixes the scale of the following easy-to-rationalize nitrogenation scenario:

1. initial stage, defined by $\langle c \rangle = 10\%$, at $t = 0.001\ t_0$,
2. half-nitrided particle, defined by $\langle c \rangle = 50\%$, at $t = 0.03\ t_0$,
3. some nitrogen at the particle centre, that is, $c(0) = 10\%$, at $t = 0.07\ t_0$,
4. half-empty particle, defined by $c(0) = 50\%$, at $t = 0.14\ t_0$,
5. nearly complete nitrogenation, defined by $\langle c \rangle = 90\%$, at $t = 0.18\ t_0$,
6. practically homogeneous nitrogenation, that is, $c(0) = 90\%$, at $t = 0.30\ t_0$.

As in Section B.2, $\langle c \rangle$ and $c(0)$ denote the average interstitial concentration and the local interstitial concentration at the particle centre, respectively. Experiment shows that the behaviour of carbon in 2:17 intermetallics is not dissimilar to that of nitrogen, and Table 4.9 may be used to estimate the diffusivity of carbon as well.

The nitrogenation time increases quadratically with the particle size so that the nitrogenation behaviour strongly depends on the size distribution of the particles. Small particles easily absorb nitrogen but, because of their large specific surface, they are sensitive to surface oxidation, whereas large particles exhibit unitrided cores and require long nitrogenation times, when the nitride tends to disproportionate. In more detail, a broad particle-size distribution leads to an extended power-law dependence of the long-time nitrogenation time, as opposed to the exponential dependence Eq. (4.25) [Sko93a]. One way to improve the particle-size distribution is by jet milling [Wal94]. Broad particle-size distributions not only degrade the processing properties of the interstitial nitrides, but also affect the investigation of the material by methods such as X-ray diffraction analysis and NMR spectroscopy: the mixture of nitrided and unnitrided particles then mimics a two-phase system.

In reality, the observed nitrogenation behaviour is affected by short-circuit diffusion along microcracks and lattice defects [Col92, Fuk92, Mül92], which reduces the nitrogenation time. Easy pathways may be, for instance, grain boundaries between microcrystallites, so that bulk Sm_2Fe_{17} behaves as a powder consisting of many microcrystallites rather than a piece of single-crystalline material. In nanocrystalline two-phase mixtures of $Sm_2Fe_{17}N_3$ and α–Fe, the iron provides the rapid diffusion pathways. Similiar effects may be caused by the inhomogeneous lattice pre-expansion upon hydrogen or ammonia treatment. However, ultimately, the diffusion has to proceed through the bulk of each microcrystallite, and we can replace the particle size R in Section B.2 or Table 4.4 by an effective particle size, that is, the size of the relevant microcrystalline units.

Due to the metastability of $Sm_2Fe_{17}N_3$, conventional high-temperature sintering techniques cannot be used to process these materials. Furthermore, the low diffusivity of nitrogen in Sm_2Fe_{17} excludes interstitial modification of powder particles larger than a few micrometres. This causes a number of technological problems: Sm_2Fe_{17} pieces have to be milled and sieved in an efficient manner, nitrogenation and bonding have to be conducted in a way which assures a sufficiently high coercivity, and proper alignment and compaction are necessary to utilize the appreciable saturation magnetization.

The most promising processing routes are polymer and zinc bonding of oriented $Sm_2Fe_{17}N_3$ powders, but in practice it is difficult to achieve good orientation of very fine particles, and the packing fractions are poor. In future it may be possible to avoid the need for very fine powder by precipitation hardening [Suz94]. In the case of nanostructured magnets the isotropic character of the material yields a rather low energy product, but the excellent coercivity may find niche applications.

Conclusions

Interstitially modified intermetallics such as $Sm_2Fe_{17}N_3$ can be regarded as archetypical lattice–gas systems. At typical nitrogenation temperatures $T = 400-500$ °C they are solid solutions, but due to the low diffusivity of nitrogen below 400 °C it is difficult to measure a critical temperature. In reality, quenching fixes the solid-solution structure at room temperature. This behaviour is independent of whether the interstitial atoms are introduced using molecular nitrogen, ammonia, hydrocarbon gases or solid carbon.

Apart from secondary factors such as surface activity or easy-pathway transport along grain boundaries and microcracks, the interstitial modification reaction proceeds by thermally activated bulk diffusion. The ultimate concentration profiles are smooth, unlike two-phase systems, where the concentration profiles are expected to be discontinuous. Using the experimentally determined diffusion parameters $E_a = 133$ kJ/mole and $D_0 = 1.0$ mm^2/s it is possible to predict the nitrogenation behaviour of the Sm_2Fe_{17} grains.

$Sm_2Fe_{17}N_3$ is a metastable compound which disproportionates at $T \geq 600$ °C, but below 250 °C the nitride $Sm_2Fe_{17}N_3$ can be regarded as stable.

The magnetic and electronic properties of interstitially modified transition-metal-rich rare-earth intermetallics are a scientifically fascinating and technologically important subject. The increase in room-temperature magnetization mainly originates from the Curie temperature enhancement; changes in the ground-state magnetic moment are less pronounced. A likely explanation of the enhanced Curie temperature is the tendency towards strong ferromagnetism in the expanded nitrides and carbides. Interstitial atoms modify the magnetocrystalline anisotropy directly by changing the electrostatic crystal field acting on the rare-earth 4f electrons. Theories as simple as the point-charge and screened-charge models provide qualitative and semiquantitative explanations, respectively, but a detailed quantitative understanding has not yet been achieved.

The intrinsic magnetic properties of $Sm_2Fe_{17}N_3$ stand comparison with any other permanent magnetic material and demonstrate how interstitial modification can be used to improve material properties in a systematic manner. The room-temperature energy product of $Sm_2Fe_{17}N_3$ is inferior to that of the present record holder $Nd_2Fe_{14}B$, but in the technically important region $150 < T < 200$ °C $Sm_2Fe_{17}N_3$ is potentially the better permanent magnetic material. It is difficult to achieve dense packing fractions combined with a high degree of orientation of the powder, which are necessary to exploit the high saturation magnetization of the material. The most promising processing routes are polymer and zinc bonding of oriented $Sm_2Fe_{17}N_3$ powders, but the enormous coercivity of isotropic nanostructured magnets may find specialized applications.

Bibliography

Gases in metals, J. D. Fast, Macmillan, London 1976 — an excellent introduction to the behaviour of gases in metals. Emphasis is on the behaviour of hydrogen in metallic elements such as nickel and palladium.

REFERENCES

[Ale69] G. Alefeld, *Phys. Stat. Sol.*, **32** (1969), 67.

[Ale78] G. Alefeld and J. Völkl, in: *Hydrogen in metals* I, Springer, Berlin 1978, p. 1.

[Boc78] B. S. Bochstein, *Diffusion in metals* (in Russian), Metallurgiya, Moscow, 1978.

[Bre94] S. Brennan, R. Skomski and J. M. D. Coey, *IEEE Trans. Magn.*, **30** (1994), 571; R. Skomski, S. Brennan and J. M. D. Coey, *Phys. Stat. Sol.* (a), **139** (1993), K11.

[Bre95] S. Brennan, R. Skomski, Q. -N. Qi and J. M. D. Coey, *J. Magn. Magn. Mater.* **140–144** (1995), 999.

[Bro73] M. Brouha and K. H. J. Buschow, *J. Appl. Phys.*, **44** (1973), 1813.

[Bus91] K. H. J. Buschow, *Rep. Prog. Phys.*, **54** (1991), 1123.

[Chr94] C. N. Christodoulou and N. Komada, *J. Alloys Comp.*, **206** (1994), 1.

[Coe90] J. M. D. Coey and H. Sun, *J. Magn. Magn. Mater.*, **87** (1990), L251.

[Coe91a] J. M. D. Coey, H. Sun, Y. Otani and D. P. F. Hurley, *J. Magn. Magn. Mater.*, **98** (1991), 76.

[Coe91b] J. M. D. Coey, J. F. Lawler, H. Sun and J. E. M. Allen, *J. Appl. Phys.*, **69** (1991), 3007.

[Coe91c] J. M. D. Coey, *Physica Scripta*, **T39** (1991), 21.

[Coe91d] J. M. D. Coey and Y. Otani, *J. Mag. Soc. Japan*, **15** (1991), 677.

[Coe91e] J. M. D. Coey, H. Sun and D. P. F. Hurley, *J. Magn. Mater.*, **101** (1991), 310.

[Coe92a] J. M. D. Coey, H. Sun, Y. Otani and D. P. F. Hurley, *J. Magn. Magn. Mater.*, **104–107** (1992), 1439.

[Coe92b] J. M. D. Coey, R. Skomski, and S. Wirth, *IEEE Trans. Magn.* **28** (1992), 2332.

[Coe92c] J. M. D. Coey and D. P. F. Hurley, *J. Magn. Mater.*, **104** (1992), 1098.

[Coe93] J. M. D. Coey and R. Skomski, *Physica Scripta*, **T49** (1993), 315.

[Coe94] J. M. D. Coey, *Festschrift for K. H. J. Buschow* (1994), in press.

[Col92] C. C. Colucci, S. Gama and F. A. O. Cabral, *IEEE Trans. Magn.*, **28** (1992), 2578.

[Cra75] J. Crank, *The mathematics of diffusion*, OUP, Oxford 1975.

[Ebe86] W. Ebeling and R. Feistel, *Physik der Selbstorganisation und Evolution*, Akademie-Verlag, Berlin, 1986.

[Ems89] J. Emsley, *The elements*, Clarendon Press, Oxford, 1989.
[Fas76] J. D. Fast, *Gases in metals*, Macmillan, London, 1976.
[Fla88] T. B. Flanagan and W. A. Oates, *Hydrogen in intermetallic compounds*, I, L. Schlapbach (ed.), Springer, Berlin, 1988, p. 49.
[Fuk91] A. Fukuno, C. Ishizawa and T. Yoneyama, *J. Appl. Phys.*, **70** (1991), 6021.
[Fuk92] A. Fukuno, C. Ishizawa, H. Chihara and T. Yoneyama, *IEEE Trans. Magn.*, **28** (1992), 2575.
[Fut86] M. Futran and C. K. Hall, *J. Chem. Phys.*, **85** (1986), 1539.
[Gra90] R. Granek and A. Nitzan, *J. Chem. Phys.*, **92** (1990), 1329.
[Gri84] R. Griessen and A. Driessen, *Phys. Rev.*, B **30** (1984), 4372.
[Hak90] H. Haken, *Synergetics*, Springer, Berlin 1990.
[Hur92] D. P. F. Hurley and J. M. D. Coey, *J. Phys.*, CM **4** (1992), 5573.
[Hut64] M. T. Hutchings, *Solid State Phys.*, **16** (1964), 227.
[Iri92] T. Iriyama, K. Kobayashi, N. Imaoka, T. Fukuda, H. Kato and Y. Nakagawa, *IEEE Trans. Magn.*, **28** (1992), 2326.
[Isn92a] O. Isnard, S. Miraglia, J. L. Soubeyroux, D. Fruchart and J. Pannetier, *Phys. Rev.*, B **45** (1992), 2920.
[Isn92b] O. Isnard, J. L. Soubeyroux, D. Fruchart, L. M. Gacia and J. Bartolomé, *Physica* B **180**, **181** (1992), 629.
[Jas91] S. S. Jaswal, W. B. Yelon, G. C. Hadjipanayis, Y. Z. Wang and D. J. Sellmyer, *Phys. Rev. Lett.*, **67** (1991), 644.
[Kap92] C. Kapusta, R. J. Zhou, M. Rosenberg, P. C. Riedi and K. H. J. Buschow, *J. Alloys Compounds*, **178** (1992), 139.
[Kat92] M. Katter, J. Wecker, C. Kuhrt, L. Schultz and R. Grössinger, *J. Magn. Magn. Mater.*, **117** (1992), 419.
[Keh78] K. W. Kehr, in: *Hydrogen in metals* I, G. Alefeld and J. Völkl (eds.), Springer, Berlin, 1978, p. 197.
[Kob95] K. Kobayashi, R. Skomski and J. M. D. Coey, *J. Alloys and Compounds*, **222** (1995), 1.
[Kuh92] C. Kuhrt, K. O'Donnell, M. Katter, J. Wecker, K. Schnitzke and L. Schultz, *Appl. Phys. Lett.*, **60** (1992), 3316.
[Kut82] R. Kutner, K. Binder and K. W. Kehr, *Phys. Rev.*, B **26** (1982), 2967.
[Lee52] T. D. Lee and C. N. Yang, *Phys. Rev.*, **87** (1952), 410.
[Mac63] W. Macke, *Thermodynamik und Statistik*, Geest and Poortig, Leipzig, 1963.
[Men86] R. van Mens, *J. Magn. Magn. Mater.*, **61** (1986), 24.
[Mie80] A. R. Miedema, P. F. de Châtel and F. R. de Boer, *Physica*, B **100** (1980), 1.
[Moh87] P. Mohn and E. P. Wohlfarth, *J. Phys.*, F **17** (1987), 2421.
[Mot36] N. F. Mott and H. Jones, *The theory of metals and alloys*, Oxford University Press, Oxford, 1936.
[Moz94] O. Moze, R. Caciuffo, B. Gillon, G. Calestani, F. E. Kayzel and J. J. M. Franse, *Phys. Rev.*, B **50** (1994), 9293.
[Muk92] T. Mukai and T. Fujimoto, *J. Magn. Magn. Mater.*, **103** (1992), 165.
[Mül92] K.-H. Müller, P. A. P. Wendhausen, D. Eckert and A. Handstein, *Pro. 7th Int. Symp. Magn. Anisotropy and Coercivity in RE–TM Alloys*, Canberra, 1992, p. 34.
[Nør82] J. Nørskov, *Phys. Rev.*, B **26** (1982), 2875.
[Ota91] Y. Otani, A. Moukarika, H. Sun, J. M. D. Coey, E. Derlin and I. R. Harris, *J. Appl. Phys.*, **69** (1991), 6735.
[ODo94] K. O'Donnell, C. Kuhrt and J. M. D. Coey, *J. Appl Phys* (1994).
[Pei78] H. Peisl, in: *Hydrogen in metals* I, G. Alefeld and J. Völkl (eds), Springer, Berlin, 1978, p. 53.
[Qi92a] Qi-nian Qi, H. Sun, R. Skomski and J. M. D. Coey, *Phys. Rev.*, B **45** (1992), 12278.

[Qi92b] Q.-N, Qi, Y.-P. Li and J. M. D. Coey, *J. Phys.: Condens. Matter*, **4** (1992), 8209.

[Qi94] Q.-N. Qi, R. Skomski and J. M. D. Coey, *J. Phys. Condens. Matter*, **6** (1994), 3245.

[Rao94] X.-L. Rao, R. Skomski, Q.-N. Qi and J. M. D. Coey, *Phys. Stat. Sol.* (b), **186** (1994), K23.

[Sch88] L. Schlapbach, in: *Hydrogen in intermetallic compounds* I, L. Schlapbach (ed.), Springer, Berlin, 1988, p. 1.

[Sko93a] R. Skomski and J. M. D. Coey, *J. Appl. Phys.*, **73** (1993), 7602.

[Sko93b] R. Skomski, C. Murray, S. Brennan and J. M. D. Coey, *J. Appl. Phys.*, **73** (1993), 6940.

[Sko93c] R. Skomski, M. D. Kuz'min and J. M. D. Coey, *J. Appl. Phys.*, **73** (1993), 6934.

[Sko93d] R. Skomski, K.-H. Müller, P. A. P. Wendhausen and J. M. D. Coey, *J. Appl. Phys.*, **73** (1993), 6047.

[Sko93e] R. Skomski and J. M. D. Coey, *J. Mag. Eng. and Perf.*, **2** (1993), 241.

[Sko94] R. Skomski, *Phil. Mag.*, B **70** (1994), 175; R. Skomski, in: *Proc. 8th Int. Symp. Magn. Anisotropy and Coercivity in RE–TM Alloys*, C. A. F. Manwaring *et al.* (eds), University of Birmingham, 1994, p. 25.

[Sko95a] R. Skomski, S. Wirth and S. Brennan, in: *Institial alloys for reduced energy consumption and pollution*, G. J. Long, F. Grandjean and K. H. J. Buschow (eds), Kluwer, Dordrecht, 1995, R. Skomski (*ibid*); R. Skomski and N. M. Dempsey (*ibid*).

[Sko95b] R. Skomski, K. Kobayashi, S. Brennan and J. M. D. Coey, *J. Magn. Magn. Mater.*, **140–144**, (1995), 1079.

[Spe83] N. D. Spencer and G. A. Somorjai, *Rep. Prog. Phys.* **46** (1983), 1.

[Sun90] H. Sun, J. M. D. Coey, Y. Otani and D. P. F. Hurley, *J. Phys.*, CM **2** (1990), 6465.

[Suz94] S. Suzuki, S. Suzuki and M. Kawasaki, *J. Appl. Phys.*, **76** (1994), 6708.

[Tre43] L. R. G. Treloar, *Rep. Prog. Phys.*, **9** (1943), 113.

[Uch93] H.-H. Uchida, H. Uchida, T. Yanagisawa, S. Kise, T. Suzuki, Y. Matsumura, U. Koike, K. Kamada, T. Kurino and H. Kaneko, *J. Alloys and Compounds*, **196** (1993), 71.

[Wag74] H. Wagner and H. Horner, *Adv. Phys.*, **23** (1974), 587.

[Wag78] H. Wagner, in: *Hydrogen in metals* I, G. Alefeld and J Völkl (eds), Springer, Berlin, 1978, p. 5.

[Wal94] B. Wall, M. Katter, W. Rodewald and M. Velicescu, *IEEE Trans. Magn.*, **30** (1994), 675.

[Wan88] X.-Z. Wang, K. Donnelly, J. M. D. Coey, B. Chevalier, J. Etourneau and T. Berlureau, *J. Mater. Sci.*, **28** (1988), 329.

[Wen94] P. A. P. Wendhausen, B. Gebel, D. Eckert and V. H. Müller, *J. Appl. Phys.*, **75** (1994), 6018.

[Wic78] E. Wicke and H. Brodowsky, in: *Hydrogen in Metals* II, G. Alefeld and J. Völkl (eds), Springer, Berlin, 1978, p. 73.

[Yan91] Y.-C. Yang, X.-D. Zhang, S.-L. Ge, Q. Pan, Gong, H.-L. Li, J.-L. Yang, B.-S. Zhang, C.-T. Ye and L. Jin, *J. Appl. Phys.*, **70** (1991), 6001.

[Yan93] Y.-C. Yang, Q. Pan, X.-D. Zhang, M.-H. Zhang, C.-L. Yang, Y. Li, S.-L. Ge and B.-F. Zhang, *J. Appl. Phys.*, **74** (1993) 4066.

[Yeo92] J. M. Yeomans, *Statistical mechanics of phase transitions*, Clarendon Press, Oxford 1992.

[Yvo88] K. Yvon and P. Fischer, in: *Hydrogen in intermetallic Compounds* I, L. Schlapbach (eds), Springer, Berlin, 1988, p. 87.

[Zha95] Y. D. Zhang, J. I. Budnick, and W. A. Hines, *Appl. Phys. Lett.* **67** (1995), 208.

[Zho90] X. P. Zhong, R. Radwanski, F. R. de Boer, T. H. Jakobs and K. H. J. Buschow, *J. Magn. Magn. Mater.* **86** (1990), 333.

5
Coercivity

D. Givord and M. F. Rossignol

Coercivity is the characteristic of hard magnetic materials where magnetization reversal does not take place immediately the applied magnetic field is reversed. The applied field needed to reverse the magnetization is the *coercive field* H_c.

Hence, the state of a magnet placed in a reverse field, less than H_c, corresponds to a metastable energy minimum. The other state, where the magnetization is parallel to the direction of the applied field, is more stable. An energy barrier exists between these two energy minima — as sketched in Fig. 5.1. This barrier is the origin of coercivity.

The existence of the energy barrier implies an anisotropy of the magnetic energy which can have two origins: either atomic-scale anisotropy, acting on the individual magnetic moments, or macroscopic-scale anisotropy, acting on the resultant magnetization. The former is due to local dipolar or magnetocrystalline interactions; the latter originates from the anisotropy of the demagnetizing field, related to the shape of the grains.

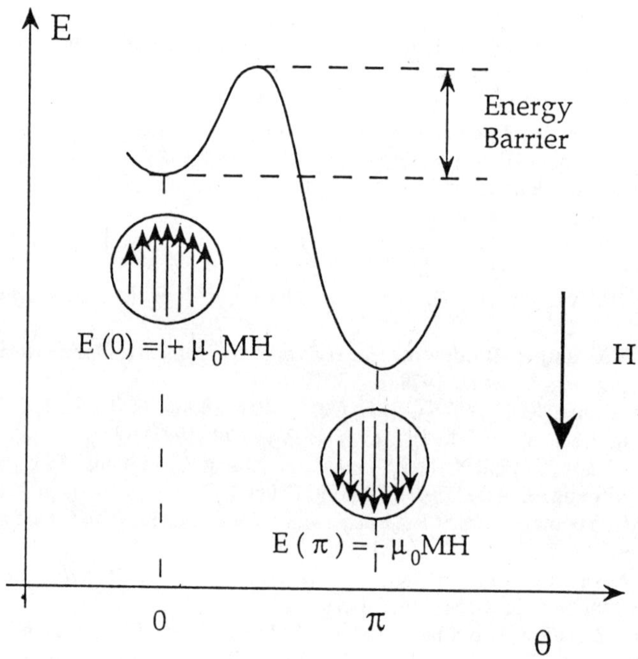

FIG. 5.1. The metastable coercive state (for $\theta = 0$) and the energy barrier for a magnet.

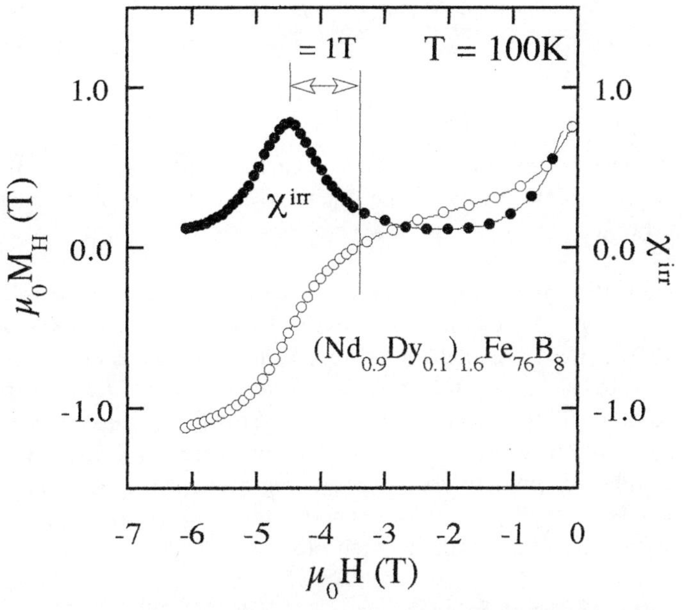

FIG. 5.2. Coercive field in an isotropic magnet: comparison of the two definitions. *Open circles*: magnetization. *Black circles*: irreversible susceptibility.

The aim of this chapter is to provide insight into why the magnetization reverses at a given coercive field. To answer this, one needs to evaluate H_c theoretically and relate it to experimentally measured values. In Section A, the coercive behaviour of a single grain will be considered as a model system. The subsequent sections B and C will be devoted to modelling and experimental analysis of real systems.

From an experimental point of view, H_c is deduced from the hysteresis cycle $M_H(H)$, where M_H is the projection of the resultant magnetization in the direction of the applied field. We define H_c as the field where the largest number of grains reverse their magnetization, that is, where $dM_H^{irr}/dH (= \chi^{irr})$ is maximum, rather than the field where the magnetization becomes zero (usual definition). The two definitions often give the same result, but they can differ strongly when reversible rotation (in isotropic magnets, for example) or free-domain wall motion lead the magnetization to become negative before the energy barrier is overcome (Fig. 5.2).

A. BASIC IDEAS: THE SINGLE GRAIN AS A MODEL SYSTEM

1 Ideal homogeneous material

In this first part we consider the magnetization reversal of a single grain of an ideal homogeneous material at zero Kelvin, simply under the action of the reverse applied field. We will distinguish between systems with strong magnetocrystalline anisotropy, and systems with no magnetocrystalline anisotropy, where only shape anisotropy exists.

1.1 *Systems with strong magnetocrystalline anisotropy*

Since magnetocrystalline anisotropy acts on each atomic moment, the movement of any of the moments away from their equilibrium position costs energy. Consequently:

1. If magnetization reversal is accomplished by *uniform coherent rotation* of the moments — where they all remain parallel and rotate in unison — they must pass by the hard magnetization direction and the coercive field is then equivalent to the magnetocrystalline anisotropy field H_a. This process was analysed in detail by Stoner and Wohlfarth [Sto48] and constitutes the reference for magnetization reversal. Their model is presented in Section B. 2.1.1.
2. Any nonuniform process costs more energy than coherent rotation, because additional exchange and dipolar energy terms come into play as soon as moments cease to be parallel to each other.

These remarks illustrate Brown's inequality [Bro63] (Eq. 1.37 with no shape anisotropy):

$$| H_c | \geq 2K_1/\mu_0 M_s$$

valid for any reversal process, in a uniform ideal system with uniaxial anisotropy. Here $2K_1/\mu_0 M_s$, where K_1 and M_s are, respectively, the magnetocrystalline anisotropy energy and the spontaneous magnetization, is the simplest expression for the anisotropy field H_a in uniaxial systems.

Note that the term 'magnetocrystalline' used above to qualify H_a and K_1 is not actually entirely appropriate since these terms contain other anisotropy contributions arising from local dipolar or exchange interactions (see, for instance, [Née54])

1.2 *Systems with no magnetocrystalline anisotropy*

Let us consider an ellipsoidal monocrystalline grain of ideal homogeneous material, initially saturated. Shape anisotropy is due to the fact that the energy of the sample in its demagnetizing field is greatest when **M** is perpendicular to the major axis, the c-axis, of the ellipsoid and least when it is parallel to it. Consequently, in zero applied field the magnetization lies along the major axis. When this ellipsoid is placed in an increasing reverse field, the coercive field, *for uniform coherent rotation*, is equal to the shape anisotropy field $H_a^{sh} = (N_\perp - N_\parallel)M_S$, where $N_\parallel$ and $N_\perp$ are de-magnetizing factors parallel and perpendicular to the *c*-axis.

Shape anisotropy depends on the resultant magnetization, a macroscopic physical property, and it acts only when the magnetization departs from the direction of the major axis of ellipsoid. However, we can imagine magnetization

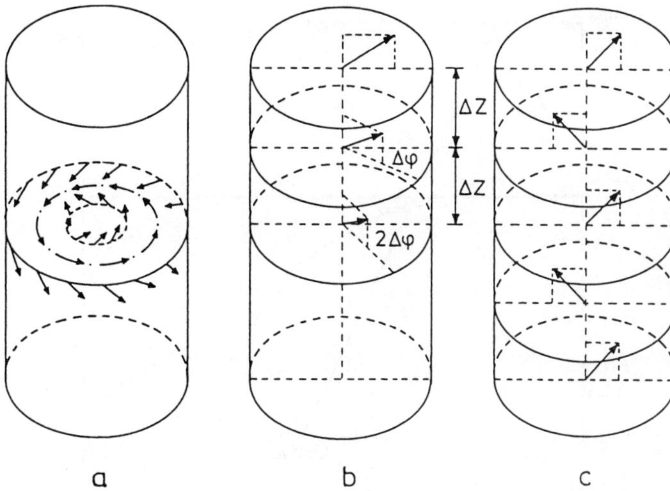

FIG. 5.3. Demagnetization modes of the infinite cylinder: (a) curling; (b) twisting; (c) buckling. (After [Zij82].)

reversal occurring through alternative nonparallel configurations of magnetic moments in which the resultant magnetization orientation remains along its preferred axis. 'Curling', 'buckling', 'twisting', etc. (Fig. 5.3) are possible processes (they are *nonuniform collective processes*). Due to the nonuniform magnetization, these configurations will give rise to a loss of exchange energy. A barrier characteristic of each of these configurations must replace the barrier associated with uniform rotation. The effective process brought into play will be that for which the corresponding energy barrier is the lowest. This depends on the diameter of the elongated grains concerned. In the case where one or other of these nonuniform processes may be more economical than uniform rotation, the coercive field is less than the shape anisotropy field H_a^{sh}.

For example, Frei *et al.* [Fre57] and Aharoni *et al.* [Aha58, Aha65] compare coherent rotation and nonuniform processes (curling, buckling, and twisting) in the case of a prolate ellipsoid. For nonuniform processes, the critical field depends on the dimensions of the ellipsoid [Aha66]. A certain critical minor radius, R_c, exists, below which magnetization reversal occurs by uniform rotation, whereas for $R > R_c$ the curling mode is generally favoured (for an infinite cylinder of iron $R_c \simeq 9$nm [Zij82]). For a minor radius of the order of 10 R_c, H_{crit} is typically 1/10 of the critical field obtained for uniform rotation (i.e. H_a^{sh}). Figure 5.4 compares the coercive fields associated with the uniform and nonuniform processes as a function of the diameter of the elongated grain [Aha58].

For other grain shapes, other processes may be involved in magnetization reversal [Zij82, Aha86], but, in every case, uniform rotation tends to occur for small grain volumes, and nonuniform processes for large volumes.

When a prolate grain ($R \simeq 10R_c$) is placed in a reverse external field applied at an angle θ_{CH} from the c-axis of the ellipsoid (Fig. 5.5 inset), irreversible magnetization

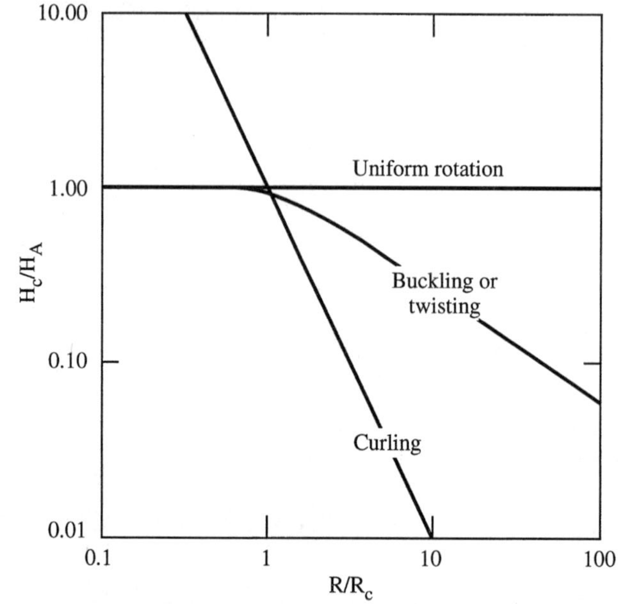

FIG. 5.4. Comparison between the critical fields associated with the uniform, 'curling' and 'buckling' processes, as a function of the radius of a prolate grain. (After [Zij82].)

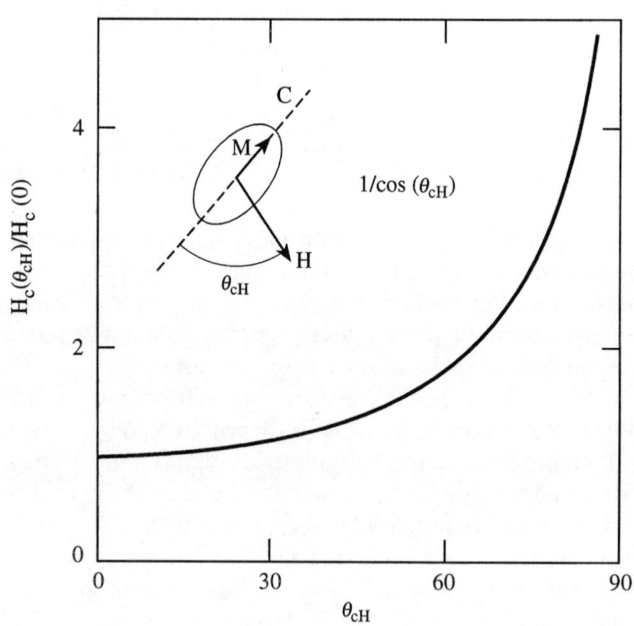

FIG. 5.5. Calculated angular dependence of the coercive field without any rotation of the moments ($1/\cos\theta_{cH}$ law).

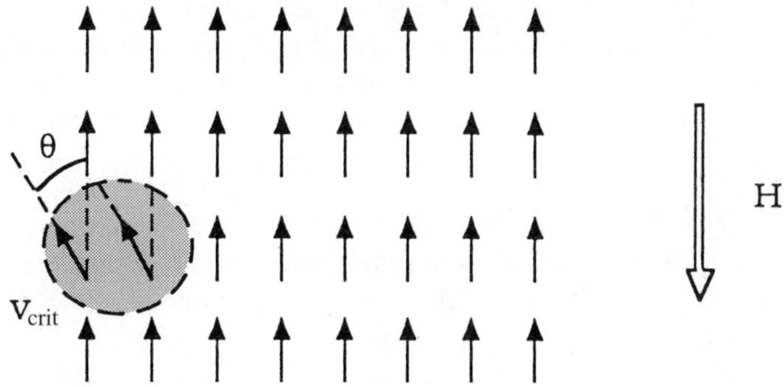

FIG. 5.6. A local deviation from the saturation state in the volume v_{at}.

reversal occurs for field values which are a function of the angle θ_{cH}; uniform and nonuniform processes lead to very different angular dependences.

The effect of the field component perpendicular to the c-axis is always a reversible rotation of the moments. However, the reversible rotation of the moments will be very small, up to the magnetization reversal, if H_c is much smaller than the anisotropy field. This is generally the case when the reversal mode is nonuniform. Therefore, the field component parallel to the c-axis, $H \cos\theta_{cH}$, remains almost antiparallel to the magnetic moments and produces their reversal when $H \cos\theta_{cH} \approx H_c$.

The angular dependence of the coercive field, for a nonuniform curling mode of magnetization reversal [Fre 57], must then resemble a $1/\cos\theta_{cH}$ law (Fig. 5.5). This behaviour, often observed experimentally, actually characterizes the fact that the coercive field is much weaker than the anisotropy field. It is not observed for uniform coherent rotation: cf. the Stoner–Wohlfarth model presented in Section B, p. 229.

1.3 *Conclusion*

In fact, magnetocrystalline and shape anisotropies coexist in the majority of magnets. In a homogeneous material to which energy is provided uniformly by an applied field, magnetization reversal, through the uniform coherent rotation process, must occur in a field equal to the total anisotropy field $H_a^{tot} = H_a + H_a^{sh}$.

Alternative nonuniform processes may be found (depending on the sample dimensions) allowing the shape anisotropy barrier to be by-passed less expensively; then, an alternative term replaces H_a^{sh}. However, in all cases, it is theoretically impossible, in a homogeneous sample, to reverse magnetization in a field smaller than the magnetocrystalline anisotropy field H_a.

Nevertheless, in all magnets possessing a very large intrinsic magnetocrystalline anisotropy (rare-earth magnets, for example), the measured coercive fields are always found to be much lower than the H_a. This discrepancy (Fig. 1.26) is usually referred to as 'Brown's paradox' [Bro45].

The explanation for Brown's paradox is that coercivity does not reflect the intrinsic properties of homogeneous material. Real magnet systems never are homogeneous. Structural heterogeneities of the material, with spatial variations of anisotropy and exchange, lead to the existence of small critical volumes where magnetization reversal is nucleated (V_{crit} in Fig. 5.6), before spreading, more or less easily, across the whole system; *a noncollective process* is then brought into play.

2 Real heterogeneous material

From now on, we consider only systems with strong magnetocrystalline anisotropy.

Structural heterogeneities may influence magnetization reversal in each grain of a real material via:

1. Spatial distribution of unequal energy barriers associated with different types of structural defects.
2. Inhomogeneous lowering of the barriers due to the local effect of dipolar interactions.

In the second case the internal field actually felt by the magnetic moments differs from the applied field mainly because of the local magnetic dipolar interactions. In the following, the internal field which causes the reversal of **M** is called *the critical field* and is denoted $\mathbf{H}_{crit}$. Neglecting the macrosopic demagnetizing field, the relation between the (applied) coercive field $\mathbf{H}_c$ and the internal critical field is the following:

$$\mathbf{H}_{crit} = \mathbf{H}_c + \mathbf{H}_{dip.loc}$$

2.1 *Reversal process by stages*

In saturated systems, under the influence of a reverse applied field, magnetization reversal is an event concerning, simultaneously, both the principal anisotropic phase and the defects present in the grain. Starting with the nucleation of a domain of reverse magnetization on a certain defect, this event is completed when the domain wall created by this nucleation disappears, having traversed the whole grain. The reversal process may be schematically divided into four steps corresponding to characteristic mechanisms. A critical field is associated with each of them, which allows one to proceed to the next stage. These mechanisms are the following:

1. *Nucleation*: formation of an inverse domain and the emergence of a domain wall, at the defect where the anisotropy barrier is the lowest; characteristic field denoted H_N.
2. *Passage*: passage of the domain wall from the defect to the principal phase: characteristic field denoted H_p.
3. *Expansion*: expansion of the domain wall in the principal phase; characteristic field denoted H_{exp}.
4. *Pinning*: possible pinning, and subsequent depinning, of the domain wall on defects—pinning sites (magnetic heterogeneities); characteristic propagation field denoted H_{pr}.

The largest of these critical fields is the one we call H_{crit}, determining magnetization reversal at H_c.

2.2 *Temperature and thermal activation effects*

Firstly, temperature changes the intrinsic magnetic properties (K_1, M_s) defining the height of the energy barriers. Secondly, there is the phenomenon of thermal activation which provides a mechanism to overcome these energy barriers. It is the latter that we will now consider in more detail.

Contrary to the external field, which only supplies an energy density, the temperature provides a fixed energy (proportional to $k_B T$) which can be concentrated into volumes of different sizes, so allowing some energy barriers to be overcome locally; the smaller the volume concerned, the larger the effects of thermal activation.

At finite temperature, the thermal activation effects have to be added to those of the heterogeneities already mentioned inducing noncollective magnetization reversal. Evidence for thermal activation effects is given by the magnetic after-effect occurring for fields near the coercive field; in this field region small energy barriers exist which can be crossed by thermal activation after a sufficiently long time.

2.3 *Characteristic dimensions*

All the mechanisms taking place successively in noncollective magnetization reversal process come into play on some characteristic length scale.

The first nucleus formed with reversed magnetization constitutes a local magnetic deviation from saturation and, consequently, the characteristic length for such a deviation is the domain wall width δ_w, which, in hard magnetic materials, is approximately equal to the exchange length l_{ex} (see Table 1.8).

The other successive mechanisms necessary for achieving magnetization reversal bring into play the displacement of a domain wall through an energy barrier: barriers between defects and the principal phase (including pinning/depinning mechanism), or drastic increases of the domain wall area and therefore of the domain wall energy (expansion mechanism). In all these cases the critical dimension of the displacement is again the domain wall width itself.

In summary δ_w appears to be the typical characteristic dimension for all the mechanisms involved in magnetization reversal. In $Nd_2Fe_{14}B$ at 300 K, for instance, the domain wall width is 4 nm; for comparison, the critical single-domain particle size d_c is around 200 nm.

B. THEORETICAL DESCRIPTION

Permanent magnets are complex systems whose magnetic properties are governed by numerous intrinsic as well as extrinsic parameters. The understanding of coercivity rests on the modelling of the various processes involved in magnetization reversal, under the influence of temperature and reverse applied field.

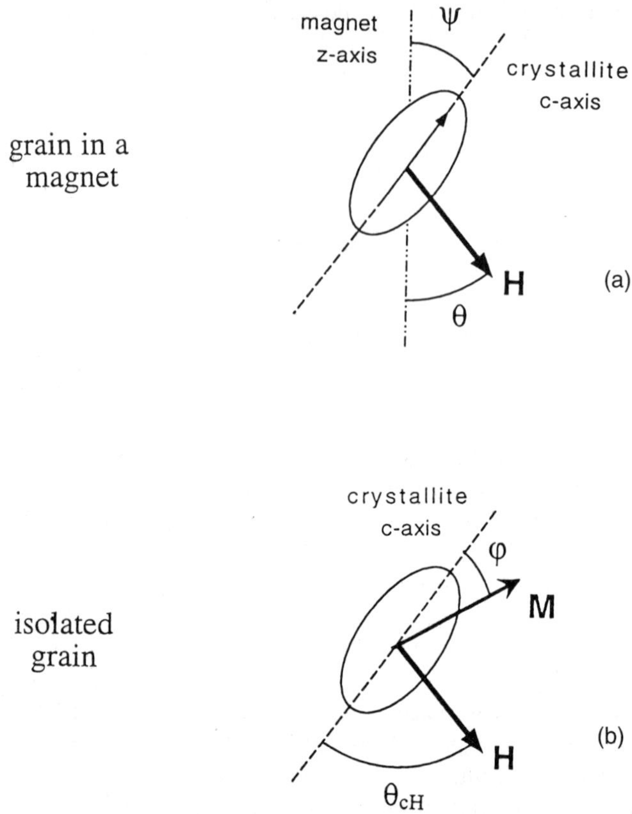

FIG. 5.7. Definition of the angles involved in the analysis. (Note that ϕ here is the angle θ of Chapters 1 and 2.)

1 Magnetic parameters

Magnets are composed of individual single crystalline grains of the hard magnetic phase, of variable size, shape and orientation. The grains, being more or less dispersed in a nonmagnetic phase, may or may not be coupled.

Describing such a system requires several sets of parameters:

1. *Intrinsic parameters of the hard magnetic phase*: the basic ones are the spontaneous magnetization M_s, the anisotropy constants K_1, K_2 ... and the Curie temperature T_C. From these, others can be deduced such as the exchange coefficient A_{ex}, the domain wall energy γ_w, the domain wall width δ_w, and the anisotropy field H_a (for simple uniaxial systems: $H_a = 2K_1/\mu_0 M_s$).
2. *Microstructural parameters* which differ depending on the preparation route for each type of magnet. These are essentially the volume fraction, v, of the hard magnetic phase and the angular distribution, $f(\psi)$, of the easy directions of the

grains with respect to the alignment direction (see the definition of the angles in Fig. 5.7). The nature and amount of complementary phases are other important parameters.

3. *Quality parameters of the magnet*: the remanent magnetization M_r obtained in zero internal field, and the coercive field H_c are the main parameters. The maximum energy product $(BH)_{max}$, and the limit H_K, beyond which irreversible losses in magnetization are unacceptable, are very useful parameters for magnet users.
4. *Magnetization reversal parameters*, among which are the critical volume, v, of the nucleus determining magnetization reversal and the effective demagnetizing factor, N_{eff}, due both to the self-demagnetizing field of the individual grains and to the dipolar interactions between grains. Other parameters are required for describing the complex link between microstructure and coercivity, but unfortunately they are unavailable to experiment as they are associated with mechanisms happening on a nanometer scale; later on we will meet the phenomenological parameters α or α_K with which one tries to characterize the magnetic effect of the defect determining magnetization reversal.

2 Models

Two main approaches have been proposed to describe coercivity in rare-earth permanent magnets. One is a micromagnetic approach which has been developed particularly by Aharoni [Aha60, Abr60] and, more recently, by Kronmüller *et al.* [Kro87a, 87b, 88]; the other is a phenomenological 'global' approach developed by Givord *et al.* [Giv88a]. Although both approaches have been applied successfully for R–Fe–B magnets, their respective philosophies are different.

The *micromagnetic model* aims at determining the external conditions of field and temperature required to produce magnetization reversal in a critical region of a grain for a given mechanism. This static model uses an energy barrier proportional to the magnetocrystalline anisotropy energy in the hard magnetic phase. Nucleation is considered to occur at the first local departure of some moments from saturation. After that, Kronmüller and coworkers assume that uniform rotation takes place within the region of the defect; consequently, the Stoner–Wohlfarth model is used as a constant reference.

The critical phenomenon in magnetization reversal in the *'global' model*, whatever the actual mechanism involved, is the formation of a magnetization heterogeneity. The typical heterogeneity in magnetic materials is the domain wall. Therefore, the energy brought into play is the energy of a domain wall in the region of the grain concerned. This energy is taken as proportional to that of the domain wall inside the hard magnetic phase. Similarly, the size of the reversed nucleus must be related to the characteristic domain wall width within the same region. One will see later on that its size may be evaluated from experiment (cf. Section C. 2.2.4).

Both the above models have a common weakness: they use the energy defined inside the hard magnetic phase as the energy reference. This is because the physical characteristics of the defects are unknown. The differences are (i) the micromagnetic approach claims to differentiate between the reversal mechanisms, whereas the global approach cannot, (ii) thermal activation effects are intrinsically considered in the 'global' model, while they are neglected in the micromagnetic approach.

2.1 *The micromagnetic approach*

Provided the magnetization vector $\mathbf{M}(\mathbf{r})$ changes slowly on the atomic scale, a theoretical treatment of magnetization processes in each small grain starts from the Gibbs free energy Φ for a uniaxial crystal:

$$\Phi = \int \{A_{\text{ex}}(\nabla\varphi(\mathbf{r}))^2 + K_1 \sin^2\varphi(\mathbf{r}) + K_2 \sin^4\varphi(\mathbf{r}) - \frac{1}{2}\mu_0\mathbf{H}_{\text{s}}(\mathbf{r})\cdot\mathbf{M}_{\text{s}}(\mathbf{r}) - \mu_0\mathbf{H}_{\text{ext}}\cdot\mathbf{M}_{\text{s}}(\mathbf{r})\}\text{d}^3r \tag{5.1}$$

where $\varphi(\mathbf{r})$ is the angle between $\mathbf{M}_{\text{s}}(\mathbf{r})$ and the c-axis of the grain, and $\mathbf{H}_{\text{s}}(\mathbf{r})$ is the magnetic dipolar field derived from a scalar potential U by $\mathbf{H}_{\text{s}}(\mathbf{r}) = -\nabla U(\mathbf{r})$, where $U(\mathbf{r})$ obeys Poisson's equation $\nabla^2 U(\mathbf{r}) = \mu_0 \operatorname{div} \mathbf{M}_{\text{s}}(\mathbf{r})$.

Micromagnetic equilibrium equations, with respect to the angle φ, are obtained from the variational calculation $\delta_\varphi\Phi = 0$ leading to

$$2A_{\text{ex}}\nabla^2\varphi(\mathbf{r})^2 - 2K_1 \sin\varphi(\mathbf{r})\cos\varphi(\mathbf{r}) - 4K_2\sin^3\varphi(\mathbf{r})\cos\varphi(\mathbf{r}) + \mu_0 H_{\text{ext}}M_{\text{s}}\sin[\theta_{\text{cH}} - \varphi(\mathbf{r})] + \mu_0 H_{\text{s}}M_{\text{s}}\sin[\theta_{\text{cH}} - \varphi(\mathbf{r})] = 0 \tag{5.2}$$

Equation (5.2) gives the internal field for which magnetization reversal occurs. However, as a rigorous treatment of equilibrium equations would be extremely difficult, several assumptions are made to simplify the problem:

1. limitation to a one-dimensional process, which may be described by the rotation angle $\varphi(z)$ if the rotation takes place in the (x, z)-plane or by the rotation angle $\varphi(r)$ in the case of a problem with radial symmetry.
2. nucleation assumed to occur at the first local deviation from saturation (in a field defined as the nucleation field H_{N}), enabling linearization of the equations,
3. simple analytical profiles of the anisotropy defect described by $K_i(z)$, in a given region of space.

2.1.1 *Application to homogeneous systems: The Stoner–Wohlfarth model* [Sto48]

For an ellipsoidal single-domain particle of homogeneous material with demagnetizing factors $N_\parallel$ and $N_\perp$ parallel and perpendicular to the c-axis, the equilibrium equations can be easily solved. The coercive field, when $K_2 = 0$ and the field is applied along the c-axis, is given by

$$H_{\text{c}} = \frac{2K_1}{\mu_0 M_{\text{s}}} - N_{\text{eff}}M_{\text{s}} \quad \text{with} \quad N_{\text{eff}} = N_\parallel - N_\perp$$

This expression is equivalent to Eq. (1.38) with $N = N_\parallel$ since $N_\parallel + 2N_\perp = 1$. Here $2K_1/\mu_0 M_{\text{s}}$ is the magnetocrystalline anisotropy field, and $N_{\text{eff}}M_{\text{s}}$ is the shape anistropy field of the grain. The above equation is an explicit expression of the

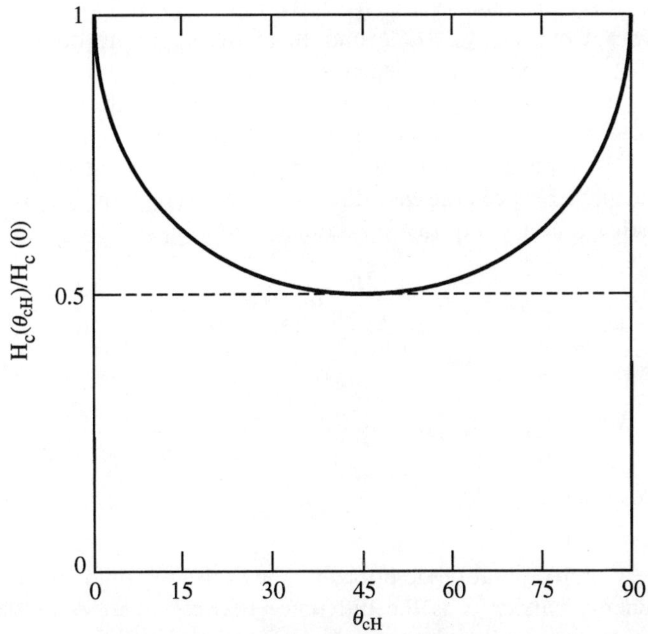

FIG. 5.8. Calculated angular dependence of the coercive field allowing for uniform rotation of the moments (Stoner–Wohlfarth model).

results given in Section A, where the coercive field for coherent rotation was stated to be equal to the total anistropy field H_a^{tot} (sum of both magnetocrystalline anisotropy field and shape anisotropy field).

Angular dependence of coercivity: When the field is applied at an angle θ_{cH} from the easy axis of the grain (Fig. 5.7), $H_c(\theta_{cH})$ is written as follows:

$$H_c(\theta_{cH}) = \left(\frac{2K_1}{\mu_0 M_s} - (N_\parallel - N_\perp)M_s\right)\frac{1}{(\cos^{2/3}\theta_{cH} + \sin^{2/3}\theta_{cH})^{3/2}} \tag{5.3}$$

or

$$H_c(\theta_{cH}) = \frac{2K_1}{\mu_0 M_s}\alpha_{\theta_{cH}} - N'_{eff}M_s \tag{5.4a}$$

with

$$N'_{eff} = (N_\parallel - N_\perp)\alpha_{\theta_{cH}} \quad \text{and} \quad \alpha_{\theta_{cH}} = \frac{1}{(\cos^{2/3}\theta_{cH} + \sin^{2/3}\theta_{cH})^{3/2}} \tag{5.5a}$$

In this case, on the one hand, the coupling energy of the magnetization with the field is reduced by a factor $\cos\theta_{cH}$; on the other hand, due to the noncollinearity between **M** and **H**, a torque is exerted which may induce a certain magnetization rotation towards the field direction, that is, the physical system is modified and this may further affect the value of H_c. As θ_{cH} varies between 0 and $\pi/2$, the coercive field decreases from H_a to $H_a/2$ and increases again up to its initial value (Fig. 5.8). Note that this angular dependence is quite different from that in Fig. 5.5.

Note also that a real magnet is composed of an assembly of grains not fully oriented. The coercive field is then affected by grain orientation, even when the external field is applied opposite to the crystallite alignment direction Oz. If ψ measures the angle between the easy direction of a given grain and Oz (see Fig. 5.7) then, for each grain at ψ Eq. (5.4a) becomes

$$H_c = \frac{2K_1}{\mu_0 M_s}\alpha_\Psi - N'_{\text{eff}} M_s \tag{5.4b}$$

where:

$$N'_{\text{eff}} = (N_\parallel - N_\perp)\alpha_\Psi \quad \text{and} \quad \alpha_\Psi = \frac{1}{(\cos^{2/3}\Psi + \sin^{2/3}\Psi)^{3/2}} \tag{5.5b}$$

Two further remarks:

1. In the very simple situation described by the Stoner–Wohlfarth model, the value of the energy barrier Δ is the difference between energies of the maximum energy state and the initial metastable state. For the single domain particle of total uniaxial anisotropy energy, $K = K_1 + \frac{1}{2}\mu_0 M_s^2(N_\perp - N_\parallel)$, with the field applied along the easy axis of magnetization and antiparallel to the magnetization, one obtains

$$\Delta = KV(1 - H/H_a^{\text{tot}})^2 \tag{5.6}$$

   where V is the particle volume. The coercive field is the field where $\Delta = 0$, hence $H_c = H_a^{\text{tot}}$. From $H_a^{\text{tot}} = 2K/\mu_0 M_s$ and from Eq. (5.6) one derives

$$(\partial\Delta/\partial H)_T = -\mu_0 M_s V(1 - H/H_a^{\text{tot}}) \tag{5.7}$$

   $(\partial\Delta/\partial H)_T$ is involved in the expression for the activation volume (see p. 270 Section C.2.2.4).

   Note also that the energy maximum corresponds to the magnetization being at an angle $\cos^{-1}(-H/H_a)$: this angle varies from $\pi/2$ up to 0 as H varies from 0 to H_a.

2. When the second anisotropy constant K_2 is included in the Stoner–Wohlfarth model, the α_ψ - parameter is given by

$$\alpha_\Psi = \frac{1}{(\cos^{2/3}\Psi + \sin^{2/3}\Psi)^{3/2}}\left(1 + \frac{2K_2}{K_1 + K_d}\frac{\tan^{2/3}\Psi}{1 + \tan^{2/3}\Psi}\right) \tag{5.8}$$

   where

$$K_d = \frac{1}{2}(N_\perp - N_\parallel)\mu_0 M_s^2 \text{ (shape anisotropy field coefficient)}$$

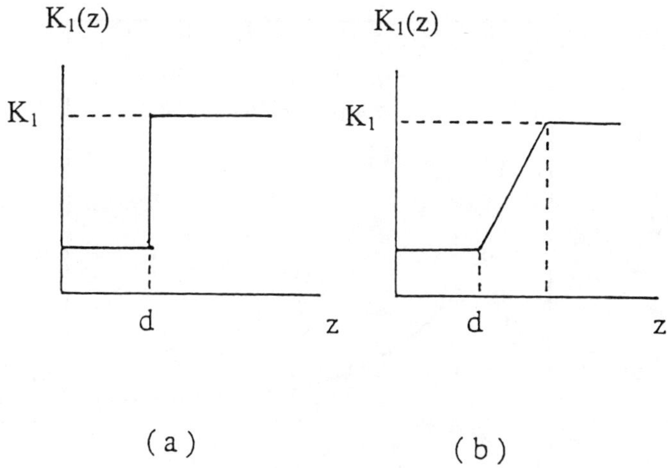

FIG. 5.9. Schematic representation of different profiles of anisotropy defects: (a) Step-like discontinuity [Aha60]; (b) Constant gradient variation [Abr60].

Comparison with Eq. (5.5b) shows that, for $K_2 > 0$, the second term in Eq. (5.8) flattens the angular dependence of H_c. In particular, H_c obtained for $\Psi = 45°$ is given by [Kro87b, 91]

$$H_c(45°) = \frac{K_1 + K_2 + K_d}{\mu_0 M_s} \tag{5.9}$$

It is approximately the minimum coercive field.

2.1.2 *The effect of nanostructural inhomogeneities: nucleation and passage into the hard magnetic phase* The above results refer to ideal homogeneous materials. In real materials anisotropy and/or exchange inhomogeneities are present. The differential equilibrium equations (equivalent to Eq. (5.2)) are then characterized by nonconstant coefficients. Solutions of such equations are difficult. In order to simplify the problem, only anisotropy defects are considered in most models. Simple analytical profiles of the anisotropy defects are usually assumed, such as step like profile, linear variation or quasiharmonic anisotropy perturbation.

**Step-like anisotropy defect.* Aharoni [Aha60] considers a simple step-like anisotropy defect (Fig. 5.9a), with $K_1(z) = 0$ for $z < d$ and $K_1(z) = K_1$ for $z > d$, where d is the width of the defect and K_1 is the anisotropy energy in the unperturbed region. Such a profile, although overly simplistic, enables the essence of the mechanism by which defects influence coercivity to be understood. Minimizing the free energy (Eq. (5.1)), $\mathrm{d}\Phi/\mathrm{d}\varphi = 0$ with such an anisotropy profile yields

$$2A_{ex}\mathrm{d}^2\varphi/\mathrm{d}z^2 + K_1 \sin 2\varphi - \mu_0 M_s H \sin\varphi = 0 \quad \text{for } z > d$$

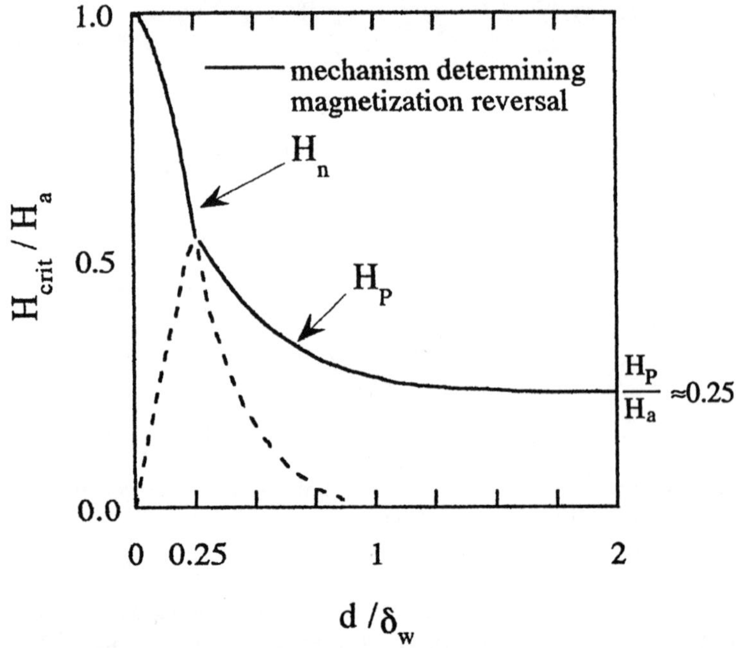

FIG. 5.10. Critical field as a function of the defect width for a step-like anisotropy discontinuity (nucleation and passage mechanisms). (After [Aha60].)

and

$$2A_{ex}d^2\varphi/dz^2 - \mu_0 M_s H \sin\varphi = 0 \qquad \text{for } z < d$$

To obtain the nucleation field H_N, for the first local deviation from saturation, we assume $\varphi \approx 0$ so that the equations can be linearized. When d is much smaller than the domain wall width δ_w, H_N is large ($\approx H_a$) and it decreases as the width of the soft region increases. For $d = \delta_w/4$, H_N reaches 0.5 H_a and it vanishes for $d \approx \delta_w$ (Fig. 5.10). The calculation of the nucleation field has been extended to a three-dimensional step-like defect [Sko94].

Let us consider now the field H_p, which allows the passage of the nucleated domain wall from the soft phase into the hard one. For $d \approx \delta_w/4$, H_p is of the order of 0.5 H_a (then $H_p \approx H_N$). For d less than $\delta_w/4$, H_p is less than H_N. This implies that nucleation leads to complete reversal of the magnetization. For d larger than $\delta_w/4$, H_p decreases from 0.5 H_a up to 0.25 H_a; it is then larger than H_N. Both H_N and H_p are compared in Fig. 5.10.

* *Linear variation of the anisotropy between soft and hard magnetic regions.* A slightly more realistic profile of the anisotropy defect, considered by Abraham and Aharoni [Abr60], assumes the anisotropy to vary linearly between a region of

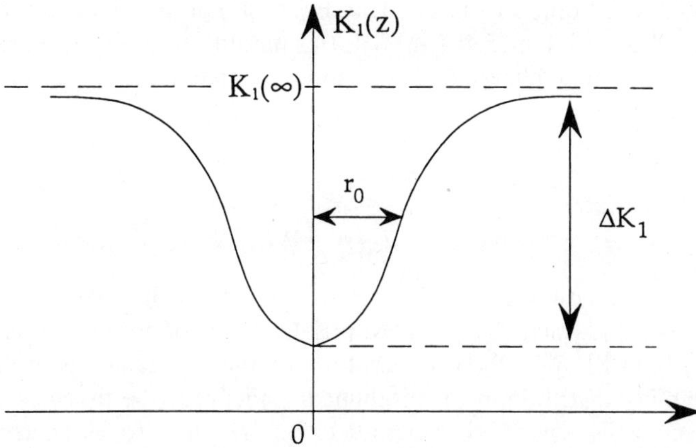

FIG. 5.11. Schematic representation of a quasi-harmonic anisotropy defect. (After [Kro87].)

perfect material and a region with no anisotropy (Fig. 5.9b). The nucleation process is similar to the one described above, but the passage field H_p is smaller and can be one order of magnitude less than H_a.

* *Quasiharmonic inhomogeneity.* In a similar manner, Kronmüller [Kro87a] has considered a planar region of halfwidth r_0, where the anisotropy follows a spatial variation of the form (Fig. 5.11)

$$K_1(z) = K_1(\infty) - \frac{\Delta K_1}{\cosh^2(z/r_0)} \tag{5.10}$$

where $K_1(\infty)$ denotes the anisotropy constant in the unperturbed matrix and ΔK_1 the reduction of K_1 at the centre of the inhomogeneous strip. Such an anisotropy profile is appropriate to describe, for example, a planar grain surface.

When the field is applied in the opposite direction, the solution of Eq. (5.2) for such a profile and for $K_2 = 0$ is

$$H_c = \frac{2K_1}{\mu_0 M_s}\alpha_K - N_{eff} M_s = H_N - N_{eff} M_s$$

which must include a misalignment parameter α_Ψ, averaged or not, when an assembly of grains is considered (see Section C, p. 253 for application to sintered magnets):

$$H_c = \frac{2K_1}{\mu_0 M_s}\alpha_\Psi \alpha_K - N_{eff} M_s \tag{5.11}$$

Here α_K, the microstructural parameter, is given by

$$\alpha_K(r_0) = 1 - \frac{1}{4\pi^2}\frac{\delta'^2_B}{r_0^2}\left[1 - \sqrt{\left(1 + \frac{4\pi^2 r_0^2}{\delta'^2_B}\right)}\right]^2 \tag{5.12}$$

with $\delta'_B = \pi\sqrt{(A_{ex}/\Delta K_1)}$.

It may be noted that Eq. (5.11) differs from Eq. (5.4a) or (5.4b) only in the value of α_K. Whereas in Eq. (5.4a) or (5.4b) $\alpha_K = 1$, this parameter, in Eqs (5.11) and (5.12), depends on r_0/δ'_B in a characteristic manner. Three limiting regions may be considered:

$$\begin{array}{lll} 1. & r_0 << \delta'_B, & \alpha_K \approx 1 - \pi^2 (r_0/\delta'_B)^2 \\ 2. & 2\pi r_0 \geq \delta'_B, & \alpha_K \approx \delta'_B/\pi r_0 \\ 3. & 2\pi r_0 >> \delta'_B, & \alpha_K \approx 1 - \Delta K_1/K_1 \end{array} \tag{5.13}$$

In region 3, the nucleation field is the anisotropy field at the centre of the defect.

The parameter α_Ψ which describes the misalignment of grains is assumed to be given by Eq. (5.8) with $K_d = 0$. However, for a quantitative analysis of the coercive field, a probability distribution of misaligned grains must be taken into account, and the effective α_Ψ value representative for H_c has to be determined by appropriate averaging. This will depend on whether or not the grains are magnetically coupled.

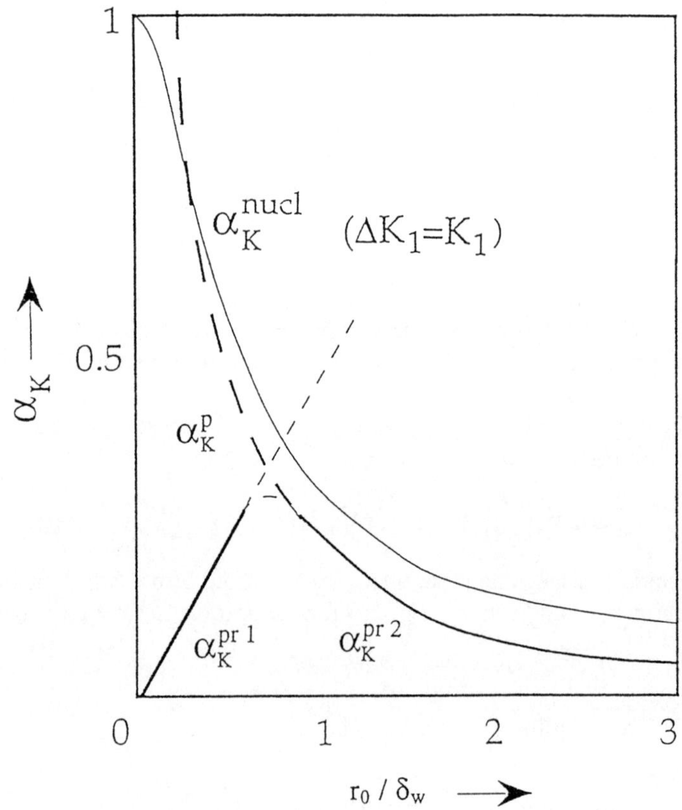

FIG. 5.12. Expected variations for α_K depending on the coercivity mechanism and on the dimension of the defect. (After [Kro87].)

With this type of anisotropy profile, the field H_p, which allows passage of the domain wall into the hard magnetic phase, is found to be

$$H_p = H_a \frac{2\delta_w}{3\pi r_0} \quad (\text{i.e. } \alpha_K^p = \frac{2\delta_w}{3\pi r_0})$$

when (5.14)

$$\Delta K_1 \approx K_1 \text{ and } r_0 > \delta_w \quad (\text{in this case } \delta'_B = \delta_w)$$

According to Eq. (5.14), H_p decreases as the defect width increases; H_p is actually proportional to the gradient $\delta K_1/\delta z$. But for a narrow defect ($r_0 << \delta'_B$), a thin planar barrier model, which will be presented further, leads to a linear variation of H_p versus r_0/δ_w (Fig. 5.12).

2.1.3 *Pinning at magnetic heterogeneities in the hard magnetic phase* The usual way of describing interactions of domain walls with pinning sites is to treat the wall as a deformable membrane pinned at points on its surface (Fig. 5.13). The pinning sites are characterized by the maximum restoring force, $f = (\mathrm{d}e/\mathrm{d}z)_{max}$, which a pin, of interaction energy $e(z)$, can exert on a wall forced to move in the z-direction by a given field. Let S be the domain wall area which is under the influence of the force exerted by a given pin site. For a given value of the restoring force f, the effectiveness of the pinning increases as S decreases. The critical propagation field required to unpin a domain wall from a unique pin site is [Ker43]

$$H_{pr} = f/2\mu_0 M_s S$$

S decreases as the pin site density, ρ, increases. For large coercivities it is therefore essential to have a large number of traps or, eventually, a planar defect.

On applying a field H parallel to one of the magnetization directions, the wall bows out, resulting in an increase in the interaction area S with each pin site, which

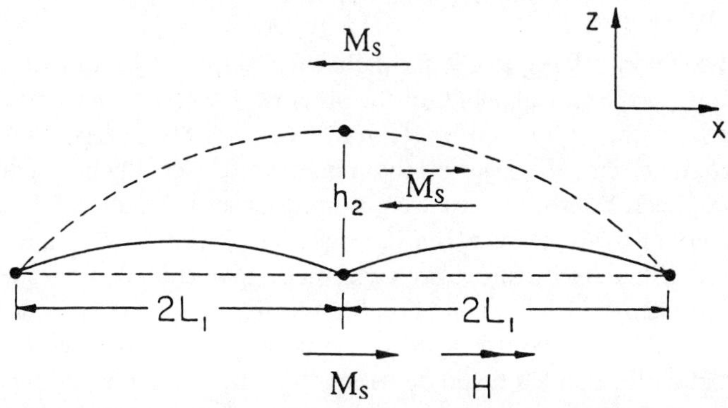

FIG. 5.13. Cross-section of bowed wall in an applied field H. The solid curves indicate the initial wall position, the broken curve shows the wall after breakaway from the central pin. (After [Gau83].)

reduces the local value of the propagation field. When H increases sufficiently, the wall breaks away from the pin centre and sweeps out a certain volume until it encounters another pin (Fig. 5.13). If the value of S is then reduced the wall will be pinned, and the field must be further increased to obtain further propagation of the wall.

The relevant parameter in the model is the area S, which depends on the mechanism by which a given distribution of pinning sites interacts with the walls. Gaunt [Gau83] has adopted a model developed by Friedel to describe the displacement of dislocations [Fri63]. S is obtained as

$$S = \sqrt{(\pi\gamma_w/3\mu_0 M_s H\rho)}$$

and hence the propagation field is

$$H_{pr} = \frac{3\rho f^2}{4\pi\gamma_w\mu_0 M_s} = 0.24\frac{\rho f^2}{\gamma_w\mu_0 M_s} \tag{5.15}$$

The model developed here describes the case of *strong pinning*. It is valid when a parameter β defined as $\beta = 3f/8\pi\gamma_w b$ is larger than 1 (4b is the interaction range of the pin, close to the domain wall width δ_w). β less than 1 describes the case of *weak pinning* and the above model breaks down. Calculation of the propagation field for weak pinning gives a value which is, however, similar to that calculated by Eq. (5.15).

The energy barrier Δ associated with pinning mechanisms depends on whether the pinning is strong or weak. In an applied field $H < H_{pr}$,

$$\Delta = \frac{4}{3}fb\left[1 - \sqrt{\left(\frac{H}{H_{pr}}\right)}\right]^{\frac{3}{2}} \tag{5.16}$$

for strong pins, and

$$\Delta = 31\ \gamma_w b^2\left[1 - \frac{H}{H_{pr}}\right] \tag{5.17}$$

for weak pins. These two formulae give different forms for the variation of H_{pr} with temperature.

Numerical solutions of Eq. (5.15) show that the pin density required to obtain a significant value for the critical field is of the order of $\rho \approx 4 \cdot 10^{20}$ per cm^3 ($4 \cdot 10^8$ per μm^3), which is of the order of 1% of atomic density. Such large values are not realistic for point defects. This suggests that pinning must actually be due to extended heterogeneities. Such heterogeneities are observed in $Sm(Co\ Cu\ Fe\ Zr)_{7-8}$ magnets, where a cellular microstructure exists (see Section C.4.3.2 p. 279 and Chapter 6 pp. 297–8). A cellular structure also guarantees that the pinning sites are not uniformly distributed, which is a condition for effective pinning (otherwise the energy of the system would be independent of the position of the domain walls [Kro78]).

For a system with extended heterogeneities, H_{pr} may be calculated by expressing the energy of a 180° domain wall in a one-dimensional model within the framework of the micromagnetic theory. Consider a defect region of width d, with exchange $A \pm \Delta A$ and anisotropy $K_1 \pm \Delta K_1$ (Fig. 5.14). If the anisotropy in the defect region

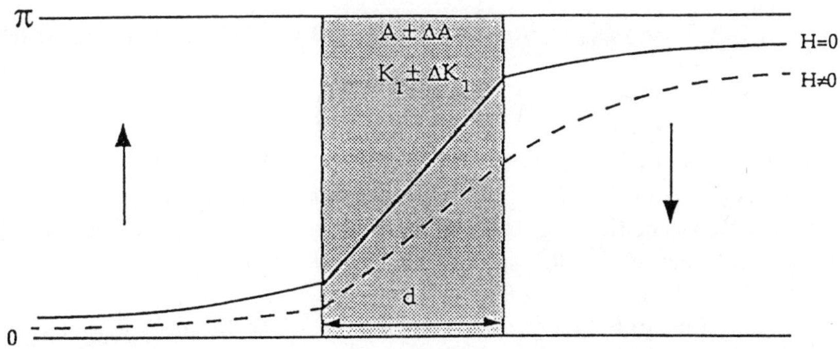

FIG. 5.14. Schematic representation for pinning of a domain wall by an extended planar defect. (After [Zij82].)

is given by $K_1 - \Delta K_1$, the minimum energy state, in zero field, corresponds to the wall being centred on the defect. When a weak magnetic field is applied, the wall becomes asymmetrical in order to reduce the magnetostatic energy of the atom chain. The field leading to domain wall propagation in the case of thin planar heterogeneities ($d < \delta_w$) is [Hil75, 77, Kro78, Zij82]

$$H_{pr} \approx H_a \frac{d}{\delta_w}\left[\frac{\Delta A}{A} + \frac{\Delta K_1}{K_1}\right]$$

In the case where $d >> \delta_w$:

$$H_{pr} \approx H_a \frac{1}{4}\left[\frac{\Delta A}{A} + \frac{\Delta K_1}{K_1}\right]$$

showing that H_{pr} no longer depends on d.

One may note that:

1. In the former case (thin heterogeneities), the factor d/δ_w indicates that the pinning efficiency depends on the size of the defect with respect to the wall thickness (see Fig. 5.12). The propagation field due to planar narrow heterogeneities [Hil75] of width $2r_0 = d < \delta_w$ is found to be

$$H_{pr} = H_a \frac{\pi}{3\sqrt{3}} \frac{r_0}{\delta_w} \quad (\text{i.e. } \alpha_K^{pr\,1} = \frac{\pi}{3\sqrt{3}} \frac{r_0}{\delta_w}) \tag{5.18}$$

2. In the latter case (extended planar heterogeneities), the wall is trapped at the interface between the defect and the perfect material. The propagation field then corresponds to that calculated for a step-like discontinuity: $H_{pr} = H_p \approx H_a/4$ when $\Delta A = 0$ and $\Delta K_1 = K_1$.

   An anisotropy perturbation such as that given by Eq. (5.10), with $\Delta K_1 = K_1$, is a more realistic assumption compared with the step-like discontinuity assumed above. The propagation field becomes (case $r_0 > \delta_W$, with $r_0 \approx d/2$)

$$H_{\mathrm{pr}} = H_{\mathrm{a}} \frac{2\delta_{\mathrm{w}}}{3\pi r_0}, \quad (\text{i.e. } \alpha_{\mathrm{K}}^{\mathrm{pr}\,2} = \frac{2\delta_{\mathrm{w}}}{3\pi r_0}) \tag{5.19}$$

and it is equal to H_{p} given by Eq. (5.14). H_{pr} ($= H_{\mathrm{p}}$) decreases as the extension of the defect increases (Fig. 5.12). As explained above, this comes from the fact that the gradient of the anisotropy tends towards zero.

3. Relations (5.18) and (5.19) indicate that the pinning efficiency varies as r_0/δ_{w} for narrow heterogeneities and as δ_{w}/r_0 for wide heterogeneities. When $2r_0 \approx \delta_{\mathrm{w}}$ a maximum value of the propagation field is obtained, which is of the order of 0.3 H_{a} (Fig. 5.12).

   As the temperature is increased δ_{w} will vary, whilst the radius r_0 of each pin remains constant. Therefore, the pin sites of maximum effectiveness (i.e. corresponding to $2r_0 = \delta_{\mathrm{w}}$) will change with temperature. This behaviour is unlike the one expected for nucleation. There, coercivity is determined by the sites where H_{N} is minimum and the same sites may be involved at different temperatures.
4. Comparison of Eq. (5.19) and Eq. (5.11) gives $\alpha_{\Psi}^{\mathrm{pin}} = 1$ and $\alpha_{\mathrm{K}}^{\mathrm{pin}} = 2\delta_{\mathrm{w}}/3\pi r_0$ [Kro88]. From the above discussion of pinning, it appears that the critical field is always much weaker than the anisotropy field. The angular dependence of the critical field must then resemble a $1/\cos\theta_{\mathrm{cH}}$ law (see Section A.1.2 p. 223), and $\alpha_{\Psi}^{\mathrm{pin}}$ must be equal to $1/\cos\Psi$. As for the case of nucleation, the effective value of $\alpha_{\Psi}^{\mathrm{pin}}$ involved depends on whether or not the grains are coupled. The values of $\alpha_{\Psi}^{\mathrm{pin}}$ then obtained are, respectively, the minimum of $1/\cos\Psi$ (coupled) or its mean value (uncoupled). However, in both cases, due to the grain orientation distribution in real magnets, $\alpha_{\Psi}^{\mathrm{pin}} \approx 1$ (see Section C, p. 253, for application to sintered magnets).

2.1.4 *The limitation of the micromagnetic approach* The theoretical treatment of the coercivity problem is complicated because the corresponding micromagnetic equations are nonlinear. The linearized micromagnetic equations have been used above to calculate critical fields for one- or two-dimensional model microstructures. The restriction to one- or two-dimensional heterogeneities is a severe approximation. The question arises whether, in three dimensions, the nucleus, after having penetrated into the hard magnetic phase, can propagate and lead to the reversal of the magnetization in the whole grain. In fact, the increase in wall area associated with expansion in three dimensions tends indeed to impede wall propagation, which could provide another coercivity mechanism for which the critical field has to be evaluated.

2.2 *Three-dimensional approach for expansion mechanism*

Assuming that the material is homogeneous, the pinning of domain walls by heterogeneities can now be neglected and the energy of the system (reversed nucleus + domain wall in a reverse magnetic field *H*) may be written as

$$E = -2\mu_0 M_{\mathrm{s}} H v + \gamma_{\mathrm{w}} s$$

where v and s are the volume and the surface of the nucleus respectively.

The basis of the model is to assume that the domain wall comes into the homogeneous hard magnetic phase through an area of given size, for instance via a disc of radius r_0, from which the nucleus was formed. For simplicity, one may consider a nucleus of conical shape and then the energy is given by

$$E = -\frac{2}{3}\pi r^2 z\mu_0 M_s H + \pi r\gamma_w\sqrt{(z^2 + r^2)}$$

where r is the radius of the disc at the base of the cone and z is its height.

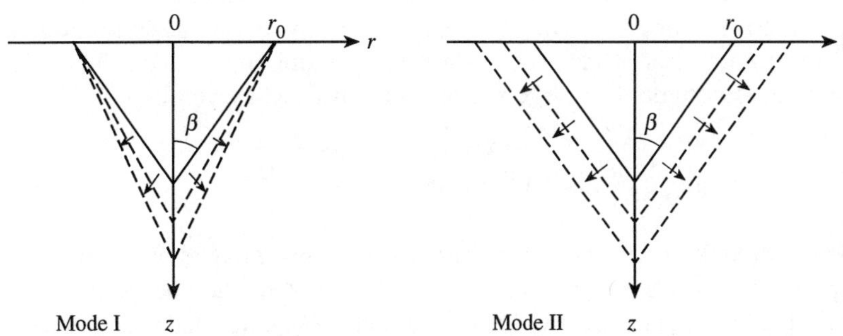

FIG. 5.15. Growth modes for a conical nucleus. Mode *I*:z increases at constant r_0; Mode II: z and r increase together at constant z/r ratio ($\beta = 35°$).

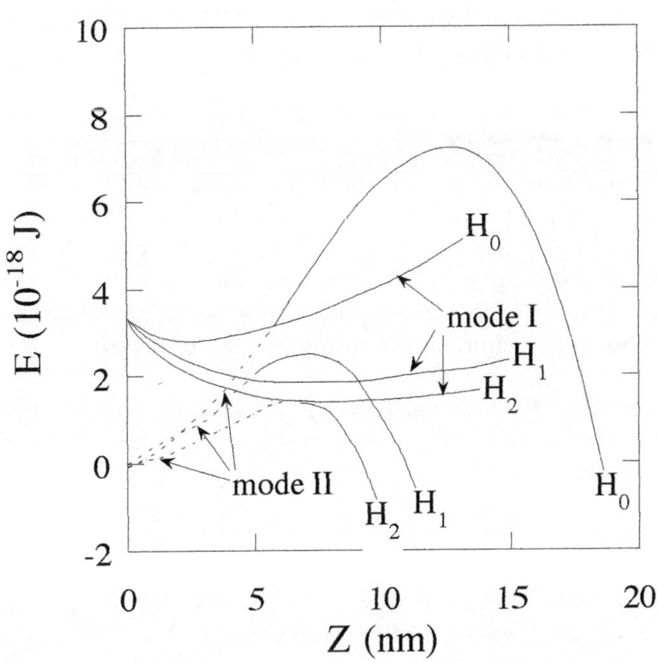

FIG. 5.16. Energy variation as a function of z for the two growth modes of a conical nucleus.

Two modes of nucleus growth may be considered:

Mode I: the nucleus grows at constant $r(= r_0)$ (Fig. 5.15a), that is, the base remains the same.
Mode II: z and r increase together at a constant z/r ratio (Fig. 5.15b). This mode corresponds to the classical idea of nucleation.

The energy variation as a function of z is presented in Fig. 5.16, for different values of the applied field for both modes. Mode I leads to an energy minimum; the nucleus penetrates progressively into the bulk with increasing field. In mode II, an energy maximum exists, since the surface term is dominant at small z while the volume term is dominant at large z. For an applied field given by

$$H = \sqrt{\left(\frac{3}{2}\right)} \frac{\gamma_W}{\mu_0 M_s r_0} = \frac{\sqrt{6}}{\pi} \frac{\delta_W}{r_0} H_a$$

the curves describing the energy variations are tangential at a point P, which corresponds to the energy minimum in Mode I and to the energy maximum in mode II (Fig. 5.16). This defines the critical field, H_{exp}, at which the nucleus may expand to infinity. At the point P, the half-angle of the cone (β) is equal to 35.26° ($\cos\beta = \sqrt{(2/3)}$), which corresponds to a cone with maximum volume for a given surface. The ratio between energy gained in the field and domain wall energy lost is thus optimized.

For $r_0 \approx \delta_W$, H_{exp} amounts to around $0.4H_a$, which shows that a significant critical field may indeed be obtained within this model.

2.3 *Phenomenological approaches*

Because exact description of the magnetic behaviour of hard magnetic systems through modelling the link between magnetic properties and microstructure is not possible, phenomenological approaches have been developed which do not pretend to provide a detailed description of the phenomenon involved but only offer a dimensional description of these effects. Two approaches have been proposed, which differ mainly in the nature of the energy barrier considered.

2.3.1 *Barrier related to the anisotropy energy* The simplest of the phenomenological approaches of coercivity is to relate its temperature dependence to that of H_a and M_s as follows:

$$H_c(T) = \alpha(T)H_a(T) - N_{eff}M_s(T) \tag{5.20}$$

Experimental data can be analysed by means of Eq. (5.20) and taking α and N_{eff} to be fitting parameters. Such an analysis has been performed for Nd–Fe–B magnets by Sagawa *et al.* [Sag87, 88] and Hirosawa *et al.* [Hir89], where a good fit was obtained by taking α to be temperature independent.

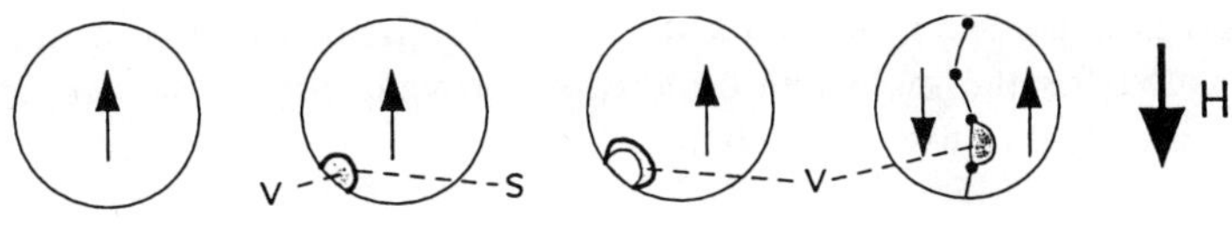

FIG. 5.17. Stages in magnetization reversal bringing into play a critical volume v.

2.3.2 *Barrier related to the domain wall energy: 'global' approach* This approach has been developed in parallel with the micromagnetic one [Ten88, Giv88a]. It is based on the fact that the same energy terms are essentially involved in all magnetization reversal processes. Whatever the mechanism involved (Fig. 5.17) (a defect acting as a nucleation site, or the expansion up to its critical volume of an already reversed nucleus in the main phase, or the break-away of domain walls from their pinning sites after swelling), a nucleus of critical volume v is formed with reversed magnetization which induces complete magnetization reversal.

The value of the energy barrier Δ_0 must be proportional to the increase in the associated domain wall energy. It may be written as: $\Delta_0 \propto \gamma'_{\mathrm{w}} s$, where s is the surface area of the nucleus.

From dimensional considerations, the domain wall energy may be written as $\gamma'_{\mathrm{w}} s = \alpha \gamma_{\mathrm{w}} v^{2/3}$, where α is a geometrical parameter which relates s and $v^{2/3}$ and quantifies the relationship between the domain wall energy in the principal phase γ_{w} and within the nucleus γ'_{w}.

In an applied field (equal to the coercive field H_{c}), the energy barrier Δ_0 can be written as

$$\Delta_0 \approx \mu_0 v \mathbf{M}_{\mathrm{S}} \cdot \mathbf{H}_{\mathrm{c}} + \mu_0 v N_{\mathrm{eff}} M_{\mathrm{S}}^2 + \Delta_{\mathrm{act}}$$

The first two terms on the right-hand side of this relation express the interaction energy with the applied field and the dipolar interaction energy respectively; the third term, Δ_{act}, is the thermal activation energy.

Thermal activation is discussed in further detail in Section C.2.2.4, pp. 267–72, where it is shown that $\Delta_{\mathrm{act}} = kT \ln(t/\tau)$.

For typical values of t, between 1 s and 1000 s, the logarithmic factor is approximately constant and equal to 25. It follows that $\Delta_{\mathrm{act}} \approx 25$ kT. At a given temperature T, the expression for the coercive field within this model is thus

$$H_{\mathrm{c}} = \alpha \frac{\gamma_{\mathrm{w}}}{\mu_0 M_{\mathrm{s}} v^{1/3}} - N_{\mathrm{eff}} M_{\mathrm{s}} - 25 S_{\mathrm{v}} \tag{5.21}$$

where $S_{\mathrm{v}} = kT/\mu_0 M_{\mathrm{s}} v$ is the magnetic viscosity deduced from after-effect measurements.

An important feature of this model is that the critical nucleus v in Eq. (5.21) is assumed to be the same as the activation volume (v_{a}) which can be determined experimentally (see Section C.2.2.4). The formation of the reversed nucleus corresponds to the creation of a nonuniform state of magnetization. The characteristic dimension for such a state is δ_w^3. Comparison of v_a with the value of δ_w^3 calculated for the main phase indicates to what degree the properties in the nucleus differ from those of the main phase.

2.4 *Comparison between models*

Between both phenomenological approaches, the comparison of the first terms $\alpha_{(1)}H_a$ — according to Eq. (5.20) — and $\alpha_{(2)}(\gamma_w/\mu_0 M_s v^{1/3})$ — according to Eq. (5.21) — can be made.

If one considers that the anisotropy energy is sufficiently well described by the one parameter K_1, then: $H_a = 2K_1/\mu_0 M_s$ and $\gamma_w = 4\sqrt{(A_{ex}K_1)}$, so the relation obtained between $\alpha_{(1)}$ and $\alpha_{(2)}$ is

$$\alpha_{(1)} = \alpha_{(2)} \frac{1}{\pi} \frac{\delta_w}{v^{1/3}} \tag{5.22}$$

Between the micromagnetic and phenomenological approaches. On the one hand, obviously $\alpha_{(1)} = \alpha_K \alpha_\Psi$ (comparison between relations (5.11) and (5.20)).

On the other hand, substituting the α_K value given by relation (5.13) (when $2\pi r_0 \geq \delta_w$) in Eq. (5.11) gives, for $\Delta K_1 = K_1$ and $\gamma_w = 4\sqrt{(A_{ex}K_1)}$

$$H_c = \frac{\alpha_\Psi}{2} \frac{\gamma_w}{\mu_0 M_s r_0} - N_{eff} M_s \tag{5.23}$$

This relation can be compared to Eq. (5.21). They are equivalent at $T = 0$ if

$$\alpha_{(2)} = \alpha_\psi (v^{1/3}/2r_0)$$

3 Numerical micromagnetic investigations

With the increase in the calculation power of computers, numerical approaches to magnetization reversal are developing very rapidly. These calculations start from the total magnetic Gibbs free energy, Φ, as expressed in Eq. (5.1).

The main problem is to evaluate the dipolar fields. Different numerical techniques are used. The finite difference method provides the easier approach, but it is limited to simple descriptions of the physical system. The finite element techniques are more difficult to handle but allow more complex shapes to be described.

The other problem is the minimization of the energy Φ. The conjugate gradient method is often used. It is however not well suited to describing unsymetrical processes which occur in symetrical systems during magnetization reversal. For that purpose the use of a damping factor in the Gilbert–Landau equation is more appropriate.

Some numerical simulations of microstructural effects in appropriate model systems are now presented.

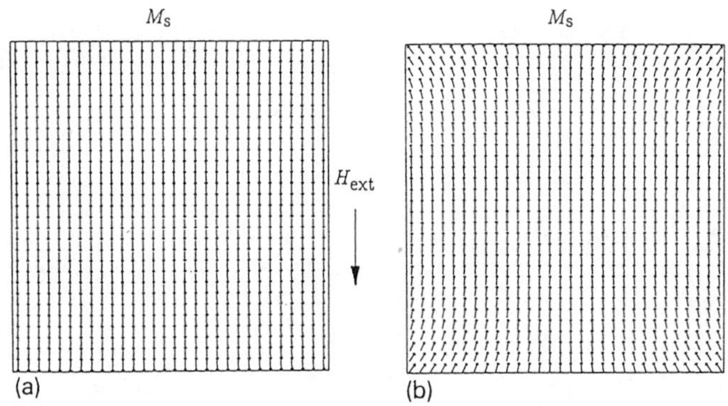

FIG. 5.18. Spin configuration in a prism of $Nd_2Fe_{14}B$ of square cross section (edge length 60 nm): (a) ground state for $H_{ext} = 0$; (b) spin configuration at the nucleation field. (After [Sch91a].)

3.1 *The rôle of demagnetizing fields in polyhedral particles*

The demagnetizing field of nonellipsoidal particles with sharp edges is inhomogeneous. The edges lead to the existence of volume 'charges' which have been studied for a parallelepiped with a square cross-section (edge length equals d) [Sch91a, Sch93a]. The simulated magnetization for different applied fields is shown in Fig. 5.18. At a field much smaller than the ideal nucleation field, $2K_1/\mu_0 M_s$, the inhomogeneous critical magnetic state becomes unstable and the magnetization reverses its orientation in the entire sample. The critical field is found to decrease logarithmically as the grain size, d, increases: $H_{crit} = C - \ln d$, and its angular dependence is flattened as compared to that expected for a Stoner–Wohlfarth particle. For a set of such grains with $d > 10\ \mu m$, strong demagnetizing fields are found to reduce the critical field to about 50% of the ideal nucleation field.

3.2 *Intergranular coupling at grain boundaries*

The effect of exchange and/or dipolar interactions between two neighbouring grains of different orientations has been investigated.

The grain boundary may be assumed to be formed by an infinite plane [Her92b; Gon93a]. The behaviour of the system is then represented by a chain of magnetic moments perpendicular to the plane. Atomic pairs are assumed to be exchange-coupled across the interface with an exchange coefficient γA_{ex} (γ being a phenomenological coefficient accounting for the reduction in the exchange interactions across the interface). Figure 5.19 presents the variation of the reduced critical field ($H_{crit}/H_{crit}^{theor}$) as a function of the material parameters. For a given anisotropy value, the critical field decreases with increasing exchange interaction, both inside the grains (parameter A_{ex}) and at the interface (parameter γA_{ex}). The reduction in the critical field depends on the nature of the intergranular phase. For a paramagnetic (or nonmagnetic) phase of thickness d, the critical field increased

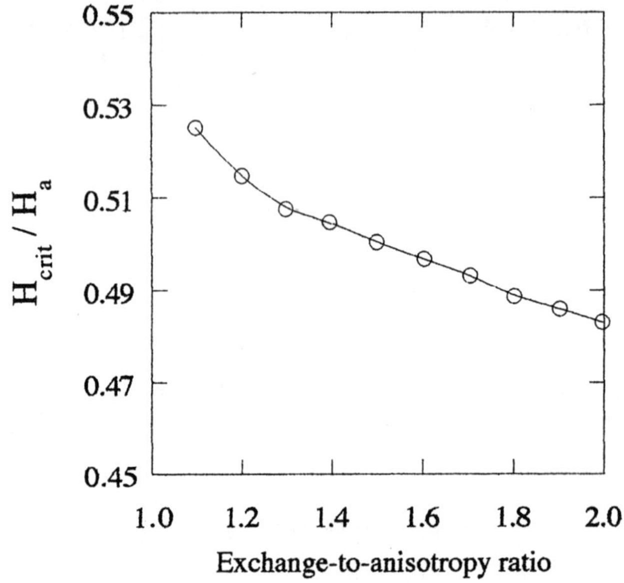
0.55
0.53
0.51
0.49
0.47
0.45
H_{crit} / H_a
1.0
1.2
1.4
1.6
1.8
2.0
Exchange-to-anisotropy ratio

(a)

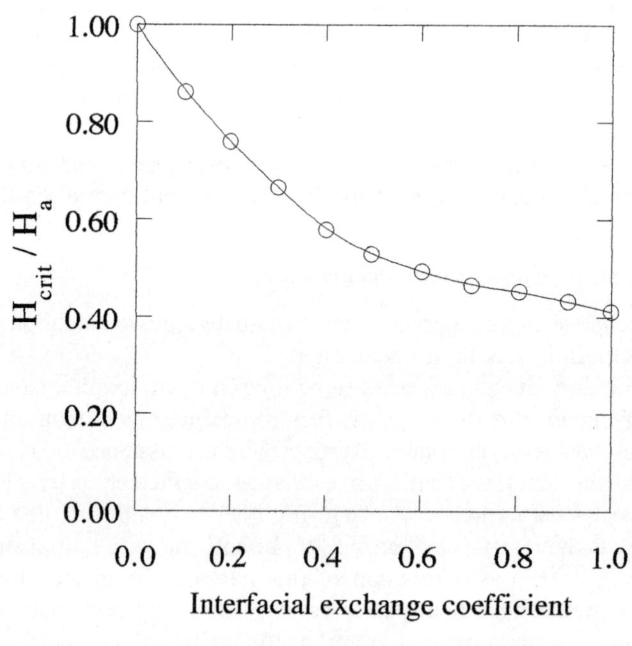
1.00
0.80
0.60
0.40
0.20
0.00
H_{crit} / H_a
0.0
0.2
0.4
0.6
0.8
1.0
Interfacial exchange coefficient

(b)

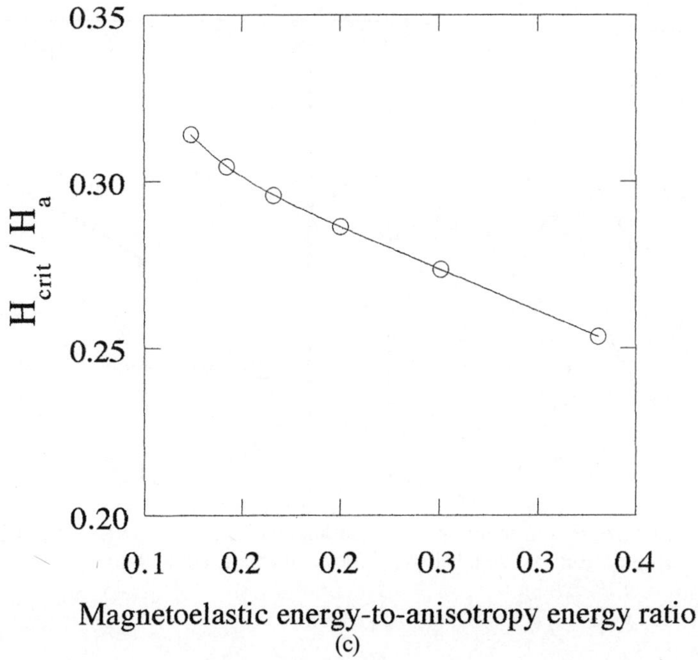

FIG. 5.19. Dependence of the critical field on the (a) exchange-to-anisotropy ratio ($\gamma = 0.6$), (b) magnitude of the interfacial exchange coupling (exchange-to-anisotropy ratio = 1.5), (c) magnetostatic energy-to-anisotropy ratio (exchange-to-anisotropy ratio = 1.8, $\gamma = 0.8$). (After [Gon94b])

with d until d is of the order of twice δ_w ($\approx$ the penetration length of exchange interactions); for $d > \delta_w$, H_{crit} is equal to H_a. On the contrary, for a soft ferromagnetic phase, the critical field decreases as d increases. For $d \approx \delta_w$, the critical field is still of the order of the coercive field of the hard magnetic phase.

The rôles of exchange and dipolar interactions are considered in another approach [Kro94] by modelling the behaviour of two coupled hexagonal grains, assuming that one grain is misaligned by an angle of 65° with respect to the applied field. The calculation has been performed using the finite-element method. The simulated magnetization as a function of the reduced external field $H_{ext}/(2K_1/\mu_0 M_s)$ is shown in Fig. 5.20 for various particular assumptions: isolated grain (A), magnetostatically coupled grains separated by a nonmagnetic boundary phase of thickness 5 nm (B) and grains coupled by exchange and magnetostatic interactions (C). In all cases, a large reduction of the ideal coercive field is obtained. In case (B), magnetization reversal occurs in two steps, the first one corresponding to reversal of the misoriented grains and the second one corresponding to the reversal of the ideally oriented grain. The coercive field of grains coupled by dipolar and exchange interactions decreases for large misalignment (Fig. 5.21), in contrast with the Stoner–Wohlfarth model.

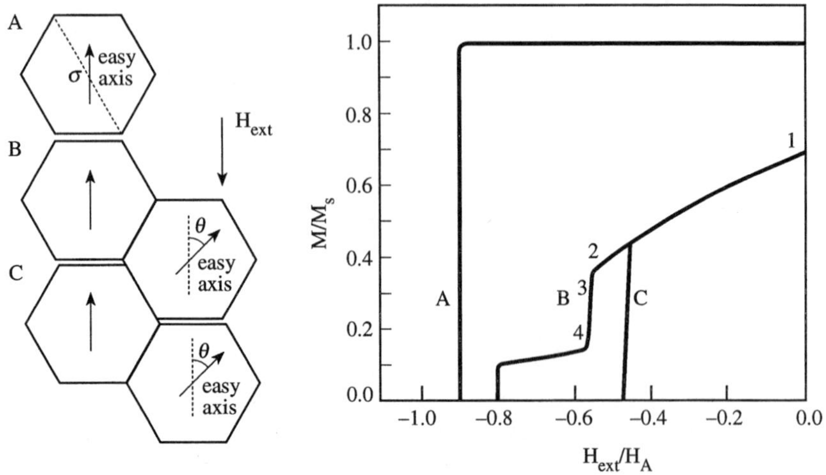

FIG. 5.20. The demagnetization curves for a misaligned isolated grain (A), dipolar (B) and exchange (C) coupled grains. The labels refer to the magnetic states shown on left-hand side of the figure. (After [Kro94].)

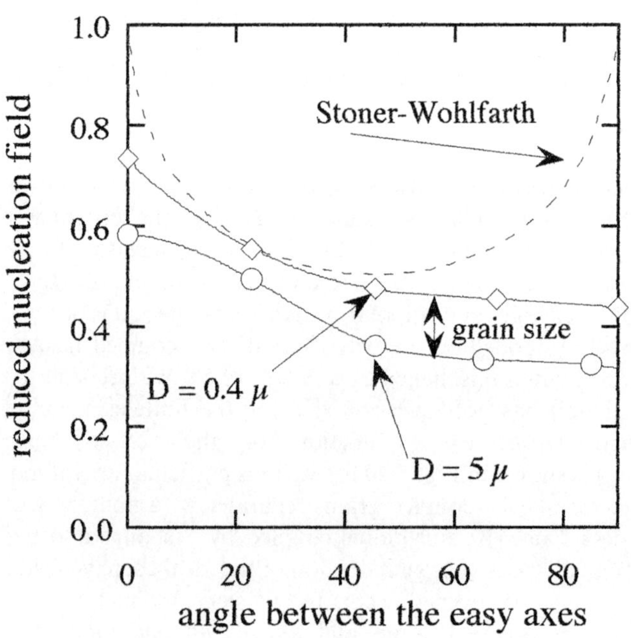

FIG. 5.21. Effect of misalignment on the nucleation field of exchange coupled grains for two grain sizes $D = 0.4\ \mu$m and $D = 5\ \mu$m. The reduced nucleation field, $H_N/(2K_1/\mu_0 M_s)$, is plotted as a function of the angle between the easy axes of the grains. (After [Kro94].)

3.3 *Cascade-type demagnetization in grain ensembles*

Grains of hard magnetic phase are always coupled by long-range magnetostatic interactions even if they are completely separated by nonmagnetic phases. From numerical calculations it is found [Sch93b] that a reversed grain significantly increases the demagnetizing field in some grains of its immediate neighbourhood. Consequently, reversal by cascade follows initial reversal in a grain with lower coercivity. Such processes are favoured at large grain size (see Fig. 5.21 for two grain sizes $D = 0.4\mu$ and $D = 5\mu$). For grain diameter less than 10 nm, cascade-type demagnetizing processes are suppressed [Sch91a].

C. EXPERIMENTAL ANALYSIS OF MAGNETIZATION REVERSAL IN R–Fe–B MAGNETS

The analysis of coercivity mechanisms rests on a knowledge of the various physical parameters listed in Section B.1. Some parameters (magnetization, anisotropy, exchange constant, domain wall energy and domain wall width) characterize the main hard phase $R_2Fe_{14}B$, at the origin of coercivity for all magnets in this family, whereas others depend on the specific microstructure characterizing each type of magnet.

In this section, these various parameters are determined, experimental methods are described, and the main coercivity processes in different types of R–Fe–B magnets are discussed in the light of quantitative results.

1 The $R_2Fe_{14}B$ hard magnetic phase

Classical methods to obtain the relevant intrinsic parameters are given in Chapter 2, as well as experimental values measured over a large range of temperatures. In the case of $Nd_2Fe_{14}B$ at 300 K, $\mu_0 M_s = 1.65$ T and $K_1 = 5.06 \times 10^6$ J/m^3, from which a reference anisotropy field $H_a = 6.13$ MA/m ($\mu_0 H_a - 7.7$ T) is deduced.

The Curie temperature of $Nd_2Fe_{14}B$ is 592 K. From this, one can deduce the zero temperature value of the exchange stiffness A_{ex}. This appears, for example, in the exchange energy density $A_{ex}\Sigma_i(\alpha_i)^2$ where α_i are the direction cosines of the local magnetization vector. $A_{ex}(0)$ may be expressed as

$$A_{ex}(0) = (z'/2)(JS^2 d^2/v_{at})$$

where d is the interplanar spacing, z', the number of nearest-neighbour atoms in the adjacent planes of a given central atom and v_{at}, the atomic volume. Considering the dense character of the $R_2Fe_{14}B$ structure, relations which apply for close-packed fcc or hcp structures will be approximately valid, that is, $z' = 12, d = \sqrt{(2/3)}\xi$ and $v_{at} = \xi^2/\sqrt{2}$ ($\xi \approx 0.25$ nm is the interatomic distance between iron nearest-neighbours). In the classical limit, $J = 3k_B T_C/2zS^2$, where z is the total number of nearest-neighbours; hence

$$A_{ex}(0) = k_B T_c/2\sqrt{2}\xi$$

The value of $A_{ex}(0)$ obtained is 11.3×10^{-12} J/m. At finite temperature, $A_{ex}(T)$ may be assumed to vary as $[M(T)/M_s(0)]^2$. The thermal variation of the spontaneous magnetization [Hir86] was shown in Fig. 2.25. It follows that at room temperature $A_{ex}(300\ \mathrm{K}) = 8.9 \times 10^{-12}$ J/m. A similar value is expected for other $R_2Fe_{14}B$ compounds. These values of $A_{ex}(T)$ are used in the following sections to evaluate the domain wall energy γ_w and width δ_w.

Experimental determination of γ_w *and* δ_w *in R–T alloys.* In a system formed of only one type of magnetic moment, the most general expressions for γ_w and δ_w, within the continuous medium approximation, are, respectively,

$$\gamma_w = 2\surd(A_{ex}) \int_{\varphi_i}^{\varphi_f} \surd(E(\varphi))\mathrm{d}\varphi \tag{5.24}$$

$$\delta_w = \surd(A_{ex}) \int_{\varphi_i}^{\varphi_f} \frac{1}{\surd(E(\varphi))}\mathrm{d}\varphi \tag{5.25}$$

where $E(\varphi)$ is the magnetocrystalline anisotropy energy for a moment at an angle φ from the easy axis. φ_i and φ_f are, respectively, the initial and final angles of the moments in the wall with respect to the easy direction.

The width δ_w obtained from Eq. (5.25) is infinite. It is usual to approximate δ_w by assuming that the profile of the wall is the same everywhere as at the wall centre (Fig. 1.23), that is,

$$\delta_w = (\varphi_f - \varphi_i)\surd\left(\frac{A_{ex}}{E(\pi/2)}\right).$$

For uniaxial systems, $E(\varphi)$ may be simply expressed as $K_1 \sin^2 \varphi$, and therefore $E(\pi/2) = K_1$. It follows that, for a 180° domain wall, $\delta_w = \pi\surd(A_{ex}/K_1)$. 92% of the total wall energy is contained within δ_w.

An original approach to the evaluation of γ_w and δ_w, particularly suited for R–T alloys where several magnetic sublattices are involved with different local magnetizations and anisotropies (and, in a more general way, suited for high-anisotropy two-sublattices systems), has been proposed by Cadogan *et al.* [Cad92]. When the resultant magnetization is not aligned along a symmetry direction, particularly within the wall, the T and R moments are not parallel to each other. The direction of the resultant magnetization is not the relevant parameter to describe exchange and anisotropy within the wall. In these systems, it may, however, be assumed that the exchange constant A_{ex} is determined essentially by T–T interactions which dominate over R–T and R–R interactions. The balance between exchange and anisotropy, at any position in the wall, may thus be considered by taking the T moments as a reference (angle φ_T from the easy axis). This means that the anisotropy energy, expressed as $E(\varphi_T)$, must include in addition to the T anisotropy, the R anisotropy and exchange interactions due to canting between R and T moments. So, for R–T systems, provided φ is replaced by φ_T, γ_w and δ_w can be directly derived from the general expressions (5.24) and (5.25) [Cad92], provided $E(\varphi_T)$ is known. $E(\varphi_T)$ is obtained from CEF analysis of magnetization measurements on single crystals [Li87, Cad88] as explained in Chapter 2. Fig. 5.22 shows data on $E(\varphi_T)$ for $Nd_2Fe_{14}B$.

The domain wall energy and width obtained in $Nd_2Fe_{14}B$, $Pr_2Fe_{14}B$ and $Dy_2Fe_{14}B$ are shown in Fig. 5.23 [Cad92]. Values obtained in $(Nd–Dy)_2Fe_{14}B$ solid solutions interpolate between the end members [Tay92].

2 Sintered R–Fe–B magnets

2.1 *General description*

Magnets prepared by powder sintering constitute the largest class among the various R–Fe–B magnets. They are formed of $Nd_2Fe_{14}B$ crystallites with a typical size of 5–15 μm, which are highly oriented (M_r/M_s varies between 85% and 96%) and surrounded by various intergranular phases. The magnetic parameters of these magnets differ slightly according to the nature and amount of additives: at 300 K, typically $\mu_o M_r \approx 1.2$ T (i.e. it is about 30% smaller than the spontaneous magnetization of $Nd_2Fe_{14}B$), whereas $\mu_0 H_c \approx 1.4$ T (about 20% of $\mu_0 H_a$ the anisotropy field of the $Nd_2Fe_{14}B$ phase). A typical hysteresis curve is shown in Fig. 5.24.

Distribution of grain orientation. Incomplete grain alignment leads to broadening of the coercive field distribution through the angular dependence of coercivity at the scale of each crystallite. Let Ψ be the angle between the c-axis in a given grain and z,

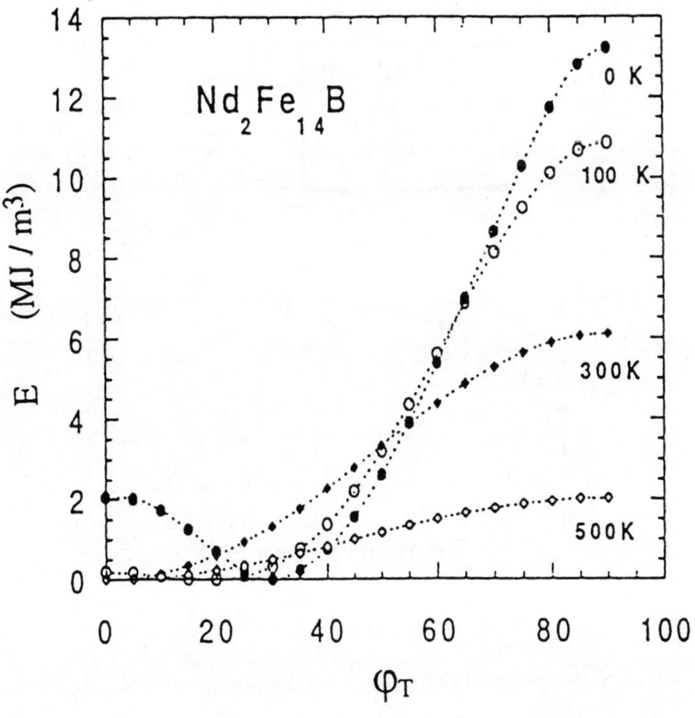

FIG. 5.22. Anisotropy energy $E(\varphi_T)$ in $Nd_2Fe_{14}B$ at different temperatures.

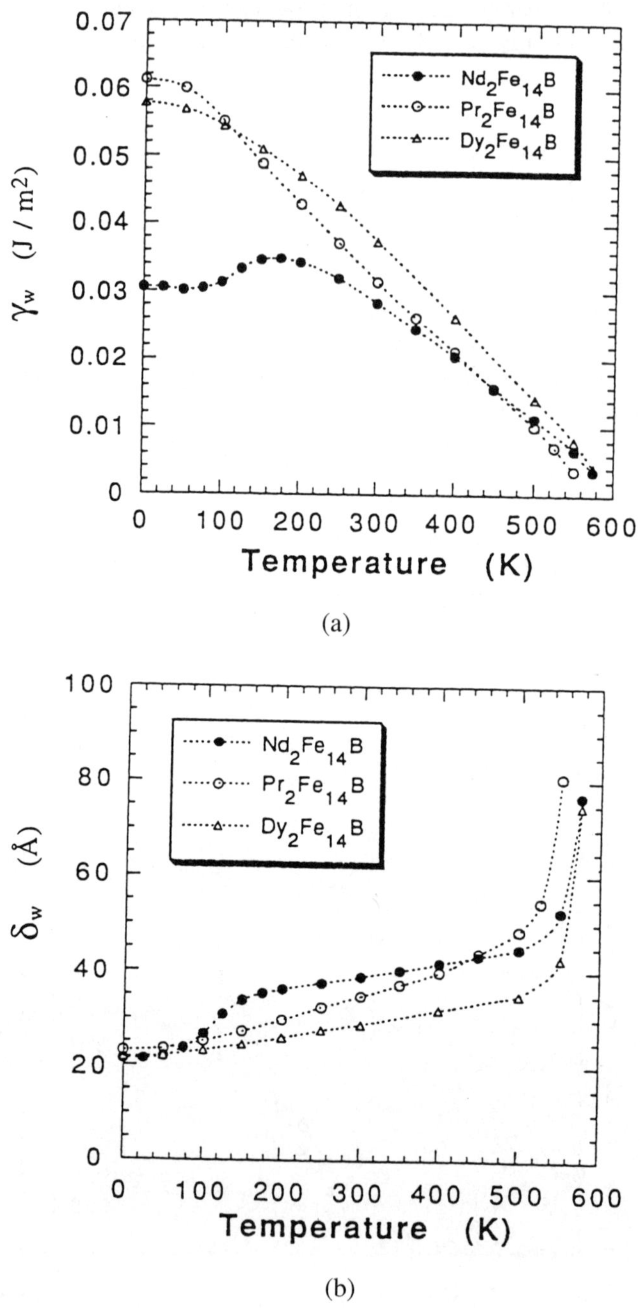

FIG. 5.23. Calculated domain wall energy γ_w (a) and domain wall width δ_w (b) in $Nd_2Fe_{14}B$, $Pr_2Fe_{14}B$ and $Dy_2Fe_{14}B$.

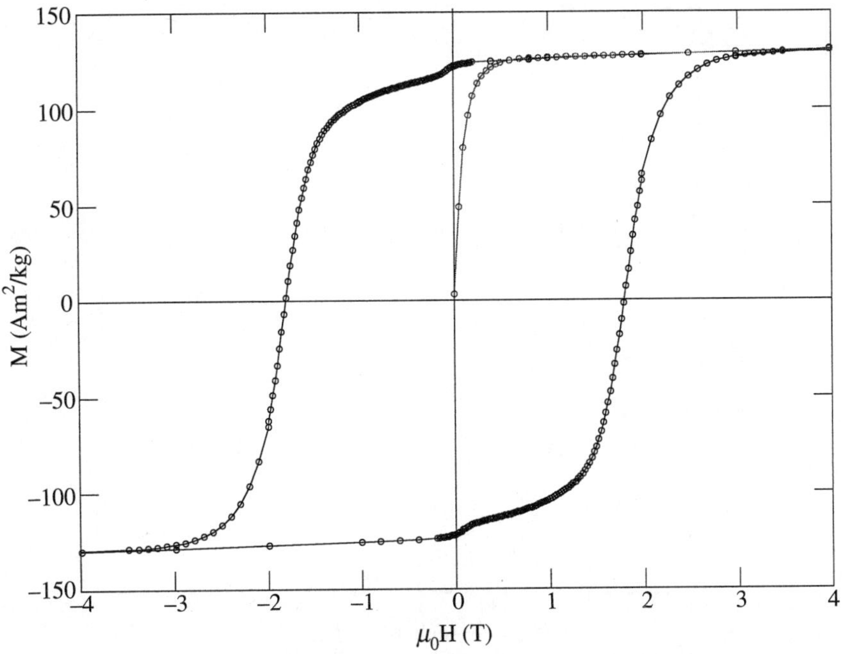

FIG. 5.24. Hysteresis loop of sintered $Fe_{17}Nd_{19}B_7GaNb_2$, showing the initial magnetization curve.

the mean direction of grain orientation (see Fig. 5.7). The angular distribution function $f(\Psi)$ may be assumed to be independent of the azimuthal angle and the probability for c-axes to lie in the range Ψ to $\Psi + \mathrm{d}\Psi$ is $p(\Psi)\mathrm{d}\Psi = f(\Psi) \sin \Psi \, d\Psi$. Müller *et al.* [Mül94] have shown that experimentally obtained distribution functions are best accounted for by Gaussian functions. It is, however, almost equivalent and simpler from the point of view of mathematical treatment, to express the misalignment by an expansion of the form

$$f(\Psi) = \Sigma_n A_n \cos^n(\Psi) \tag{5.26}$$

where A_n are normalization factors such that $\Sigma_n A_n/(n+1) = 1$.

Several experimental methods have been used to determine $f(\Psi)$ in Nd–Fe–B sintered magnets, including remanence measurements [Kro87b], fits of demagnetization curves [Eck92] and X-ray texture analysis [Ten88]. Asti *et al.* [Ast93a] have developed an original approach using the single-point detection (SPD) technique. A pulsed magnetic field H is applied at successive various angles with respect to the z-axis of the magnet. For each measurement, the peak height in the experimentally measured second derivative of the magnetization gives the proportion of crystallites oriented perpendicular to H.

Figure 5.25 compares the texture distribution function for two Nd–Fe–B sintered magnets with different degrees of crystallite orientation (two methods of texture determination are also compared). In the highly oriented M1 magnet, the texture

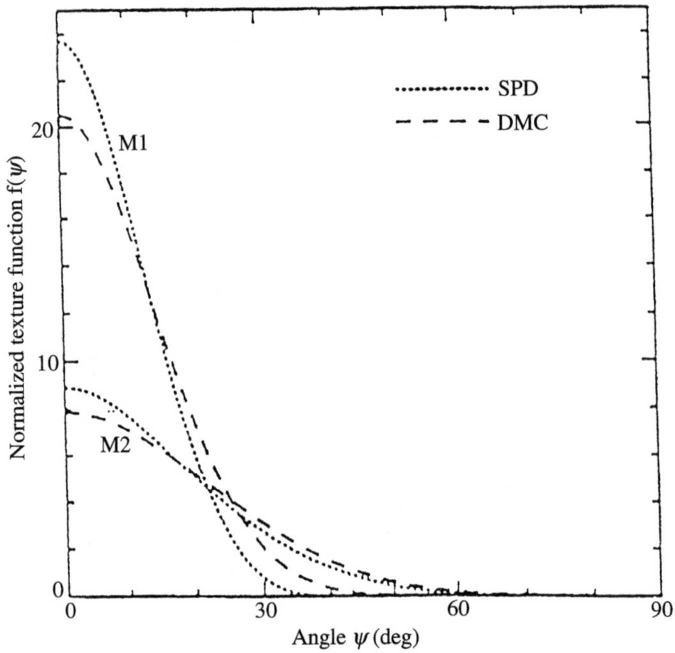

FIG. 5.25. Texture functions for the sintered $Nd_{16}Fe_{78}B_8$ samples M1 and M2 obtained by fitting of demagnetization curves (DMC) compared with functions calculated from SPD data [Ast93b].

function is $-3.2\cos^{16}\Psi + 5.1\cos^{18}\Psi$ and the maximum misalignment is 30°. In the M2 magnet, where the degree of crystallite orientation is reduced, the function is $0.47\cos^{6}\Psi + 0.24\cos^{18}\Psi$ and the population of crystallites at Ψ angles larger than 30° is approximately 20%.

2.2 *Elements of the analysis*

The analysis of coercivity mechanisms involves a synthesis of various experimental observations. For some of them the analysis necessitates a comparison with model predictions. Essential experimental observations are the thermal variation and the angular dependence of the coercive field H_c, as well as the thermal variation of the magnetic viscosity coefficient S_v.

2.2.1 *Initial magnetization curves* The initial magnetization curve of a thermally demagnetized Nd–Fe–B sintered magnet is shown in Fig. 5.24. An initial large susceptibility characterizing domain wall motion is observed in low fields; above $\mu_0 H = 0.6$ T, the magnetization is nearly saturated and the susceptibility is weak. This curve confirms that the size of the grains constituting the magnet is larger than the single-domain size and it indicates that free-domain wall motion occurs within the grains in the virgin state.

2.2.2 *Temperature dependence of H_c: comparison with theoretical models* The temperature dependence of H_c in some R–Fe–B sintered magnets is shown in Fig. 5.26. A continuous decrease is observed with increasing temperature and H_c vanishes about 100 K below the Curie point of $Nd_2Fe_{14}B$.

It was stressed in the preceding sections that whatever the exact coercivity mechanism may be (i) coercivity is intrinsically linked to anisotropy and (ii) large dipolar interactions exist in permanent magnets due to structural heterogeneities. Based on these ideas, Hirosawa *et al.* [Hir89] have analysed the temperature dependence of coercivity on the basis of the phenomenological expression (5.20).

An approximately linear relationship exists between H_c/M_s and H_a/M_s (Fig. 5.27). The slope α of the straight line is around 0.35–0.4. N_{eff}, the negative intercept on the vertical axis is around 1 (i.e. $\mu_0 H_{dip} \approx 1.6$ T), in good agreement with values derived from other evaluations in this section.

Within the *micromagnetic approach*, the analysis of coercivity is made from relation (5.11) according to the following procedure. First of all, a value for α_Ψ must be obtained, based on initial hypotheses concerning magnetic coupling between grains and magnetization reversal mechanisms under consideration, respectively.

1. *Magnetic coupling*. The alternatives are the following: either reversal occurs independently in the different crystallites, then α_Ψ^{eff} is the average of α_Ψ over the different crystallite orientations, or avalanche processes occur, reversal begins in a region with minimum coercivity from where it propagates to the whole ensemble of grains, and α_Ψ^{eff} is the minimum of α_Ψ.
2. *Magnetization reversal mechanism*. The two magnetization reversal mechanisms considered are nucleation and pinning. In the case of nucleation, coherent rotation is assumed in the activation volume. α_Ψ is given by the Stoner–Wohlfarth relation (5.5b). Its value averaged over different crystallite orientations is 0.65 and its minimum value is 0.5. In the case of pinning, α_Ψ is $1/\cos\Psi$, and both its averaged and minimum values are approximately equal to 1. The compatibility of the temperature dependence of $H_c(T)$ with the mechanism considered must then be analysed.

The method may be illustrated with reference to the example of Pr–Fe–B magnets [Kou94]. Assuming that pinning is involved, α_Ψ^{eff} is taken equal to 1. In Fig. 5.28a and b respectively, H_c/M_s is plotted versus $(\pi H_a/3\sqrt{3}\,\delta_w M_s)$ and versus $(2\,\delta_w H_a/3\pi M_s)$, which corresponds to tests for pinning by thin and extended heterogeneities respectively (see Eq. (5.18) and (5.19)). In neither case is a good linearity obtained, except in a very limited temperature range. Moreover, the value deduced for α_K is larger than 0.3, which is the maximum possible value for a pinning mechanism (see Section B.2.1.3, p. 238).

Assuming that nucleation is involved, two values are possible: $\alpha_\Psi^{eff} \approx 0.5$ (minimum value for a cascade process) or $\alpha_\Psi^{eff} \approx 0.65$ (averaged value for individual processes). Note that α_Ψ^{eff} is actually slightly temperature dependent as it depends on the ratio between anisotropy constants of different orders. H_c/M_s is plotted versus

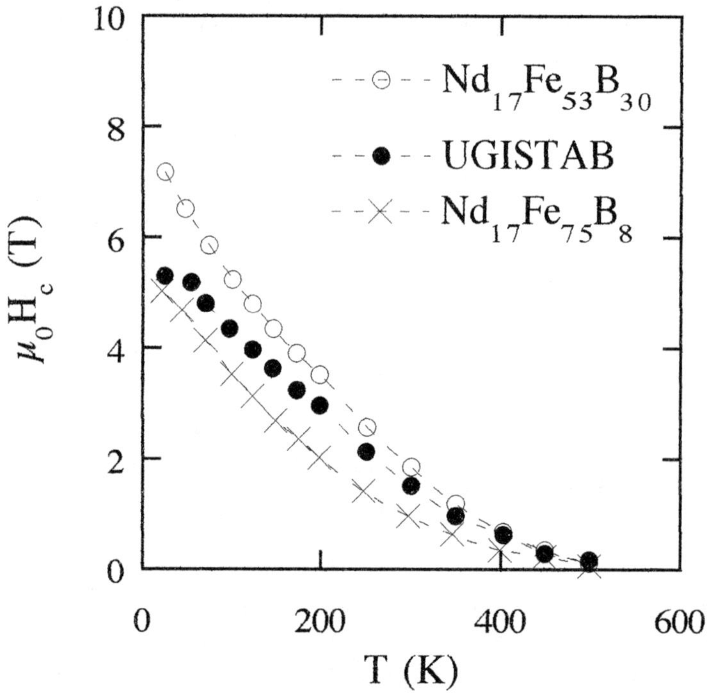

(a)

$\alpha_{\Psi}^{\rm eff} H_{\rm a}$ in Fig. 5.29a (avalanche process: minimum value of $\alpha_{\Psi}^{\rm eff}$ used) and 5.29b (averaged value of $\alpha_{\Psi}^{\rm eff}$: independent grains). A better fit is obtained for the avalanche process. The value of $\alpha_{\rm K}$ deduced is 0.86, that of $N_{\rm eff}$ is almost 2. From these results, the micromagnetic approach suggests that nucleation mechanisms determine the coercivity of R–Fe–B magnets. Note, however, that in this analysis, $\alpha_{\rm K}$ was taken to be temperature independent. According to relation (5.12), $\alpha_{\rm K}$ should vary with T through the temperature dependence of $\delta_{\rm B}'$. However, any T-dependence in $\alpha_{\rm K}$ led in the present system to a worse fit to the data. To account for $\alpha_{\rm K}$ being T-independent, it has been suggested [Kou94] that r_0 (the half-width of the defect determining coercivity) could vary as $\delta_{\rm B}'$.

The values of $\alpha_{\rm K}$ and $N_{\rm eff}$ obtained in a large variety of sintered Nd–Fe–B magnets are shown in Fig. 5.30. The values of $\alpha_{\rm K}$ range from 0.65 to 0.95, those of $N_{\rm eff}$ from 1 to 1.3. $\alpha_{\rm K}$ is related to the size of some microstructural defects, through relation (5.12). From this, values of r_0 of the order of 1 nm are deduced. It may be noted that this value is significantly less than $\delta_{\rm w}(\approx 4$ nm).

Consider now an analysis of coercivity within the *global approach* [Giv88a, Tay92, 94]. The activation volume, $v_{\rm a} = v$, must be determined prior to the analysis of

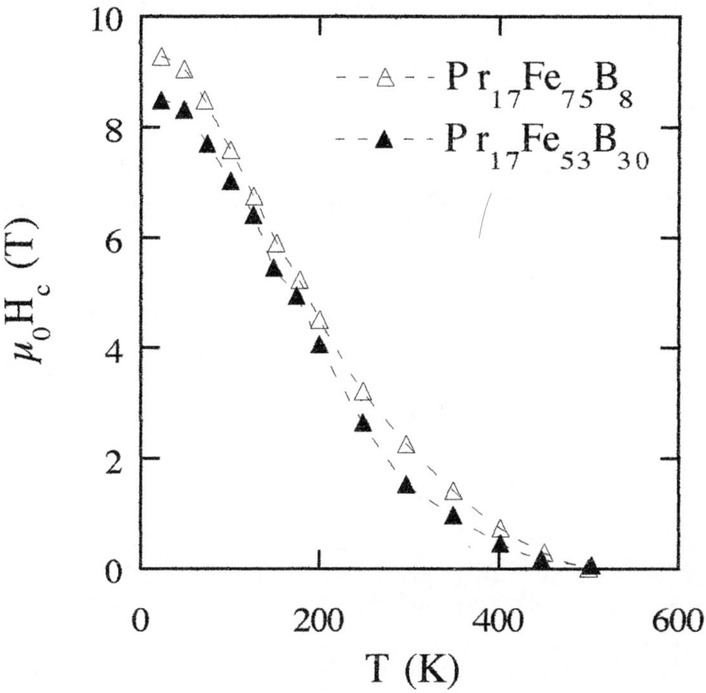

(b)

FIG. 5.26. Temperature dependence of H_c in some R–Fe–B sintered magnets. UGISTAB = $Nd_{15}Fe_{67.5}Co_5V_4B_8Al_{0.5}$.

$H_c(T)$; the determination method will be discussed below. The intrinsic coercive field is obtained by subtracting the contribution of thermal activation from the experimental coercive field: $H_0 = H_c - 25k_BT/vM_s$. The plot of H_0/M_s as a function of $(\gamma_w/\mu_0 M_s^2 v^{\frac{1}{3}})$ for different R–Fe–B magnets is shown in Fig. 5.31. A linear relationship is obtained. This is true even for the Nd–Fe–B magnets down to very low temperatures, despite the magnetization reorientation occurring at 135 K, which results from a competition between local anisotropy terms of different orders. In fact, the influence of higher-order terms is implicitly taken into account in the model which allows γ_w to be evaluated. No adjustable parameter is needed to obtain this linearity. All the physical parameters (H_c, M_s, γ_w and v) are deduced directly (at all temperatures) from measurements. Note also that this linearity can no longer be obtained when one substitutes, for the activation volume v, a term proportional to the cube of the domain wall width, δ_w^3, in the $R_2Fe_{14}B$ phase (Fig. 5.31 *c*). The identification of the activation volume v_a (see Section C.2.2.4, p. 269) with the critical volume v involved in magnetization reversal therefore appears to be justified.

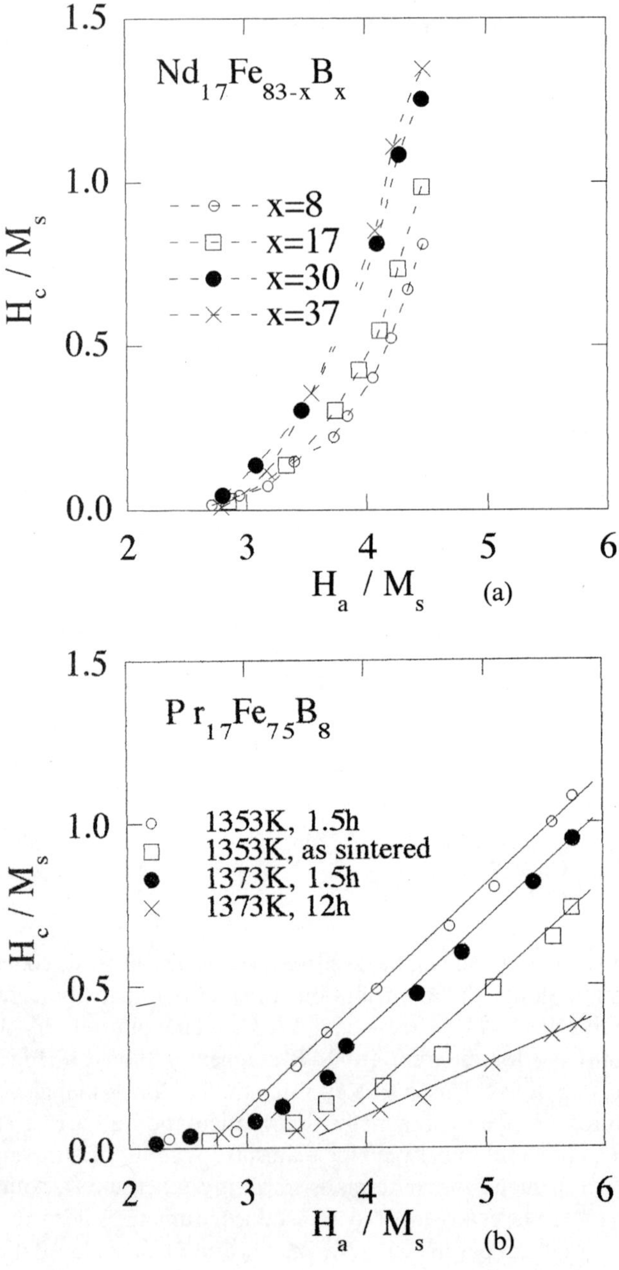

FIG. 5.27. Approach of Hirosawa and Sagawa. The H_c/M_s versus H_A/M_s plots for the sintered $Nd_{17}Fe_{83-x}B_x$ and $Pr_{17}Fe_{75}B_8$ magnets. The Pr magnets are sintered under different conditions. [Hir89]

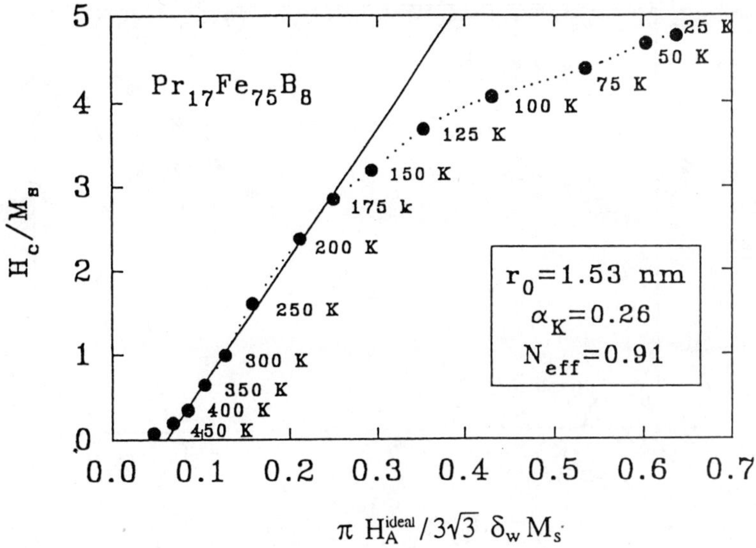

(a)

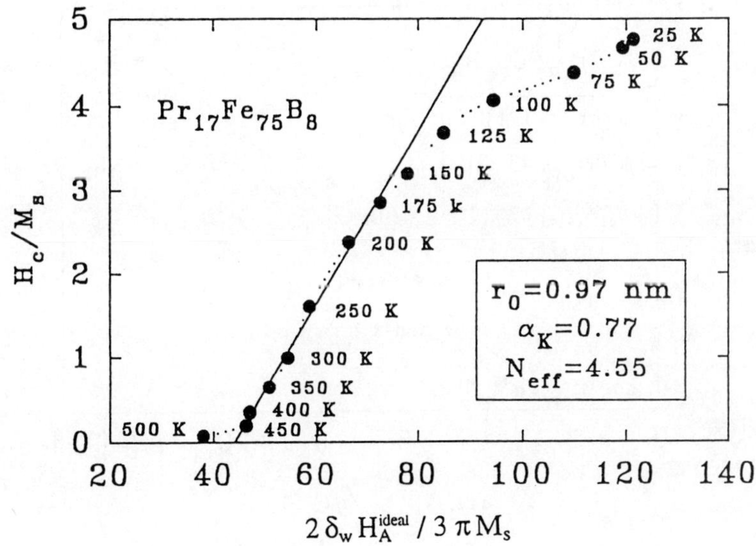

(b)

FIG. 5.28. Micromagnetic approach. The plots of H_c/M_s, versus $\pi H_a^{\text{ideal}}/3\sqrt{3}\delta_w M_s$ (a), and versus $2\delta_w H_a^{\text{ideal}}/3\pi M_s$ (b), for $Pr_{17}Fe_{75}B_8$($H_a^{\text{ideal}} = 2K_1/\mu_0 M_s$).

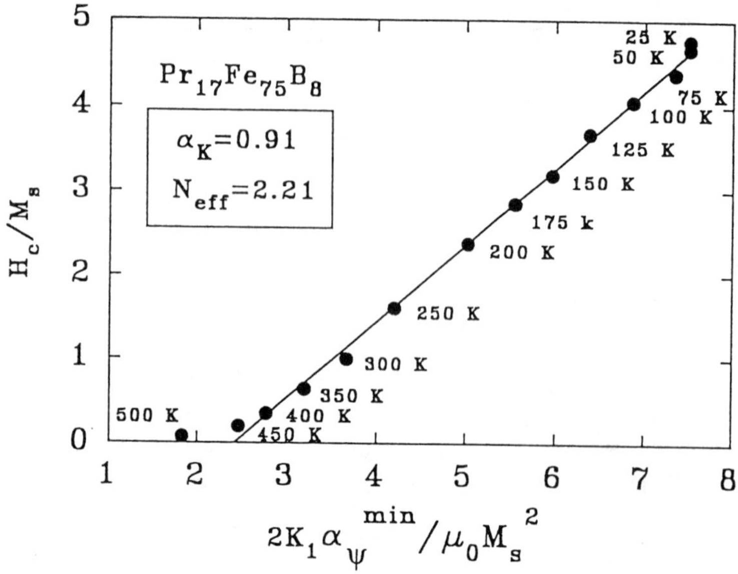

(a)

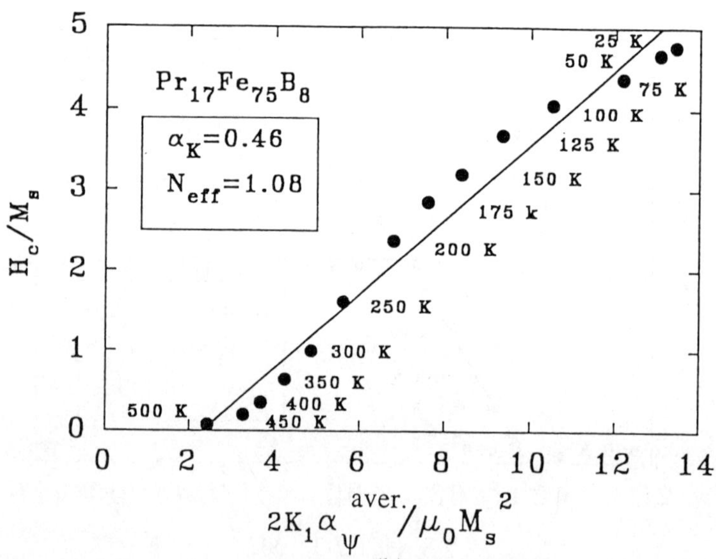

(b)

FIG. 5.29. Micromagnetic approach. The plots of H_c/M_s versus $\alpha_\Psi^{\mathrm{eff}} H_a^{\mathrm{ideal}}/M_s$ for $Pr_{17}Fe_{75}B_8$: (a) $\alpha_\Psi^{\mathrm{eff}} = \alpha_\Psi^{\mathrm{min}}$; (b) $\alpha_\Psi^{\mathrm{eff}} = \alpha_\Psi^{\mathrm{aver}}$; ($H_a^{\mathrm{ideal}} = 2K_1/\mu_0 M_s$).

The coefficient α is of the order of 1 and the values of N_{eff} tend to be smaller than in the micromagnetic approach (compare Fig. 5.32 and Fig. 5.30). Furthermore, the

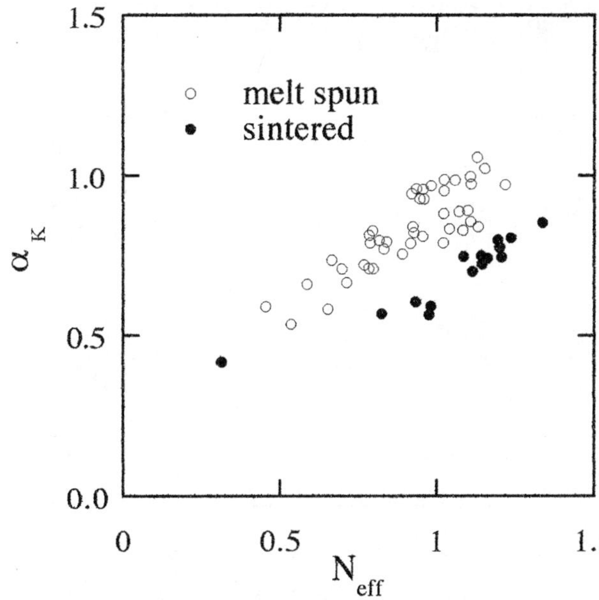

FIG. 5.30. Micromagnetic approach. The microstructural parameters α_K versus N_{eff} in a series of R–Fe–B magnets.

values of effective dipolar fields obtained are confirmed by a direct evaluation (see Section C.2.3 pp. 272–5).

2.2.3 *Angular dependence of H_c* The study of the angular dependence of coercivity $H_c(\theta)$ provides another approach to the experimental analysis of magnetization reversal mechanisms.

To measure the angular dependence of the coercive field, the sample is cut in the shape of a thin cylinder whose axis is perpendicular to the z-axis of the magnet. The magnetization is saturated in a positive field applied along z. A negative field is then applied along a direction in the plane of the cylinder, at an angle θ from z (see Fig. 5.7). The observed angular dependence of $H_c(\theta)$ is different from that of an individual crystallite $h_c(\theta_{cH})$ due to the distribution of crystallite orientation.

In order to model the angular dependence $H_c(\theta)$, the above alternative assumptions (see p. 253) are considered [Elb91].

1. According to the first assumption, magnetization reversal occurs independently in each crystallite. The contribution to the irreversible susceptibility of crystallites at a given θ_{cH} is

$$\chi^{irr} = 2M_s \cos\theta_{cH} N_\theta(\theta_{cH})/(dh_c/d\theta_{cH}), \tag{5.27}$$

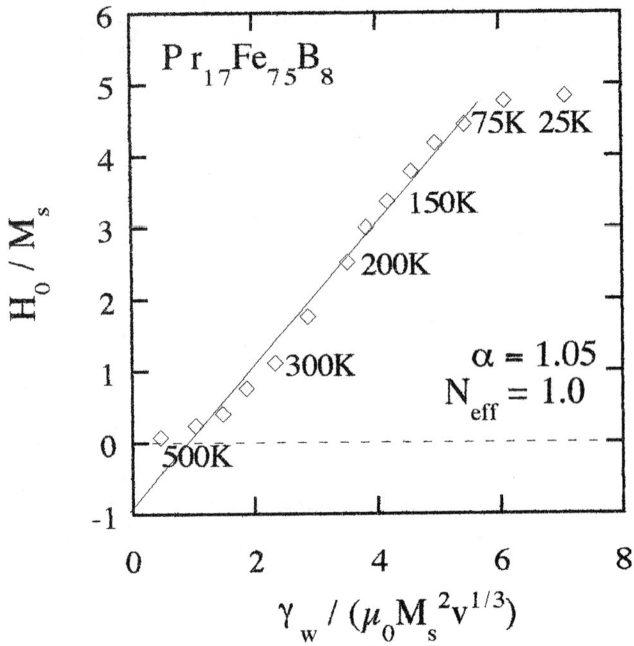
$Pr_{17}Fe_{75}B_8$
75K
25K
150K
200K
300K
500K
α = 1.05
N_{eff} = 1.0
H_0 / M_s
$\gamma_w / (\mu_0 M_s^2 v^{1/3})$

(a)

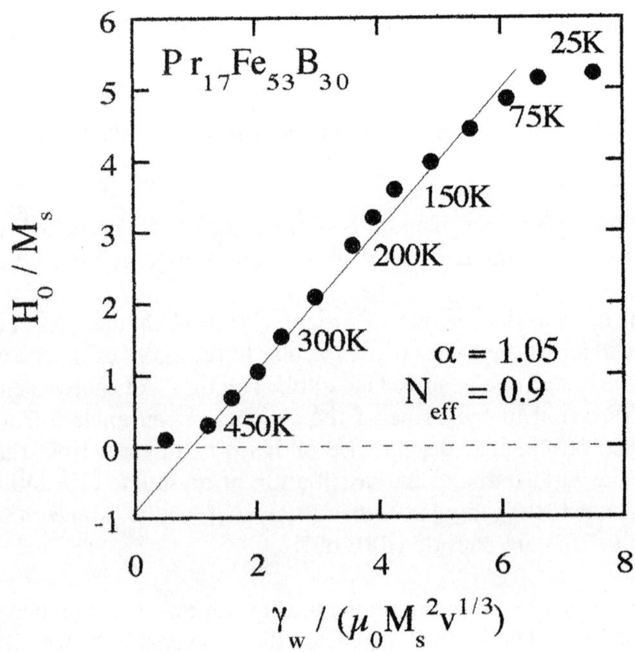
$Pr_{17}Fe_{53}B_{30}$
25K
75K
150K
200K
300K
450K
α = 1.05
N_{eff} = 0.9
H_0 / M_s
$\gamma_w / (\mu_0 M_s^2 v^{1/3})$

(b)

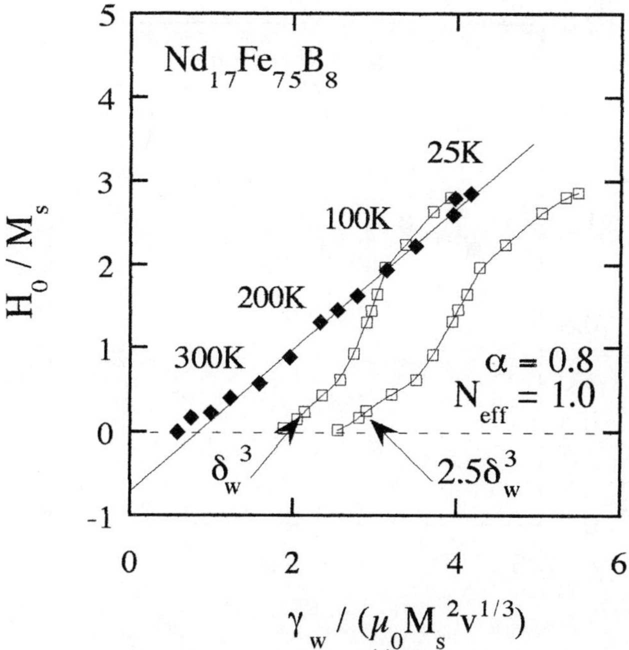

(c)

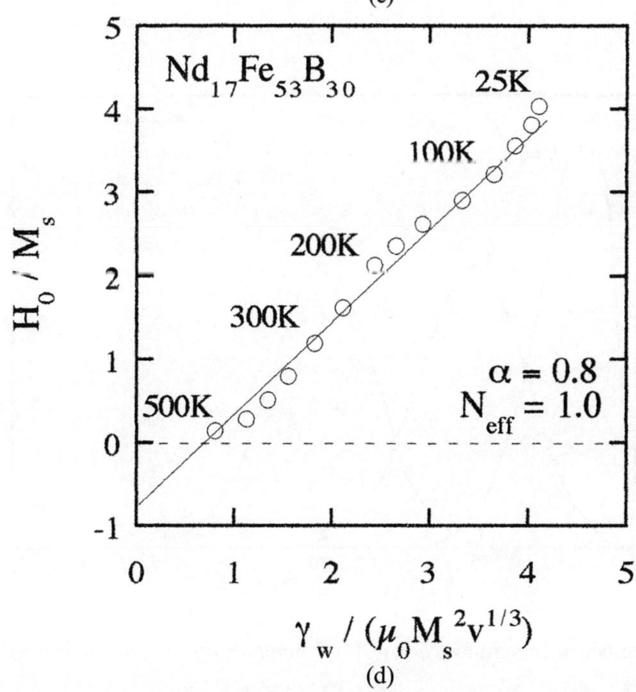

(d)

FIG. 5.31. Global approach. The plot of H_0/M_s versus $\gamma_w/\mu_0 M_s^2 v^{1/3}$ for different R–Fe–B magnets. $H_0 = H_c - 25S_v$, with $S_v = k_B T/vM_S$.
The open squares (c) are obtained for $Nd_{17}Fe_{75}B_8$ when $\beta\delta_w^3$ replacees $v\,(\beta = 1, 2, 5)$.

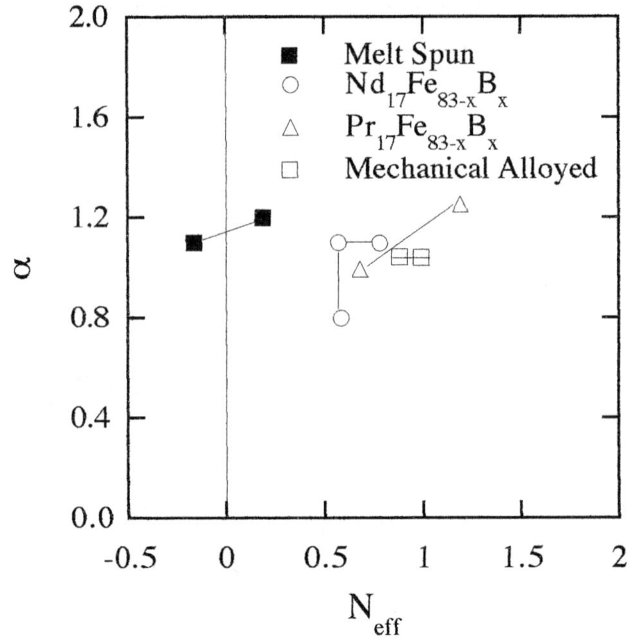

FIG. 5.32. Values of α and N_{eff} obtained within the global approach in R–Fe–B magnets of

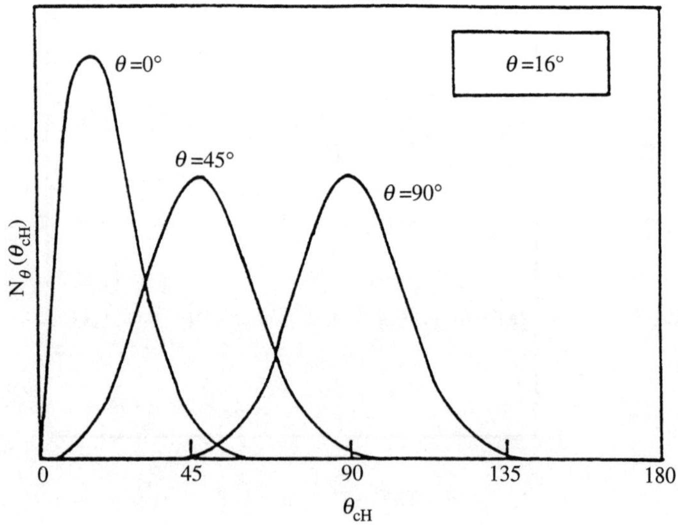

FIG. 5.33. Number of crystallites $N_\theta(\theta_{\mathrm{cH}})$ for different θ values (typical sintered magnet with $\sigma = 16°$).

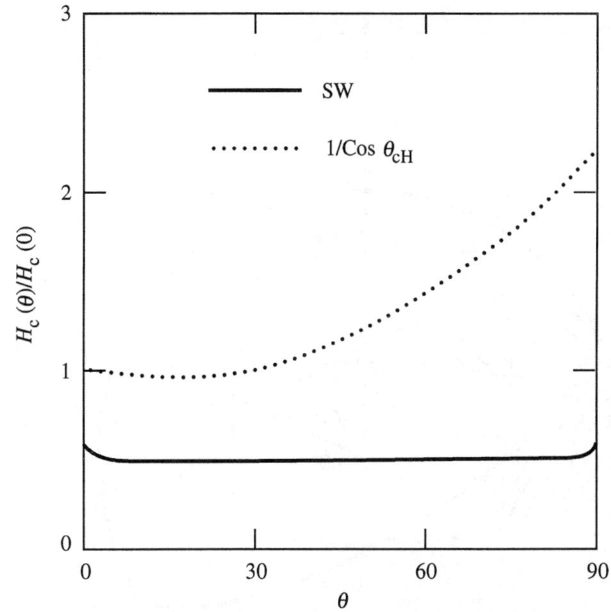

FIG. 5.34. $H_c(\theta/H_c(0)$ versus the angle θ, calculated for a magnet with grain distribution parameter $\sigma = 16°$, assuming no rotation of the moments ($1/\cos\theta_{cH}$ behaviour) [dotted curve], and for uniform collective rotation of moments [solid curve].

where $N_\theta(\theta_{cH})$ is the number of crystallites which are at θ_{cH} from H for a given θ, $h_c(\theta_{cH})$ is the angular dependence of the coercivity in the individual grains and $dh_c/d\theta_{cH}$ is its derivative. $N_\theta(\theta_{cH})$ is derived from the texture function $f(\Psi)$ and it is shown in Fig. 5.33 for different θ values for a typical sintered magnet presenting a high degree of crystallite orientation.

At each field value H, only crystallites having $h_c(\theta_{cH}) = H$ are involved in magnetization reversal. The number of different θ_{cH} values associated with a given H is small (one only for a monotonic function). The values of θ_{cH} such that χ^{irr} is maximum indicates, for a given $h_c(\theta_{cH})$ law, which crystallites are involved at the magnet coercive field H_c. The angular dependencies $H_c(\theta)$ for an assembly of grains characterized by the texture function $f(\Psi)$ can then be derived for different laws.

Applying this model to sintered magnets, where it may be assumed that the texture function is the same as that involved in the calculation of $N_\theta(\theta_{cH})$ above, $H_c(\theta)$ has thus been obtained for $h_c(\theta_{cH})$ corresponding to the Stoner–Wohlfarth (Eq. (5.5b)) and $1/\cos\theta_{cH}$ laws respectively. The calculated behaviours are shown in Fig. 5.34. In the former case, there is almost no dependence of H_c on θ, whereas in the latter case, H_c increases by approximately a factor of two as θ varies from 0 to $\pi/2$ [Elb91].

2. The alternative assumption to be considered in the derivation of $H_c(\theta)$ from $h_c(\theta_{cH})$ is that where magnetization reversal involves an avalanche pro-

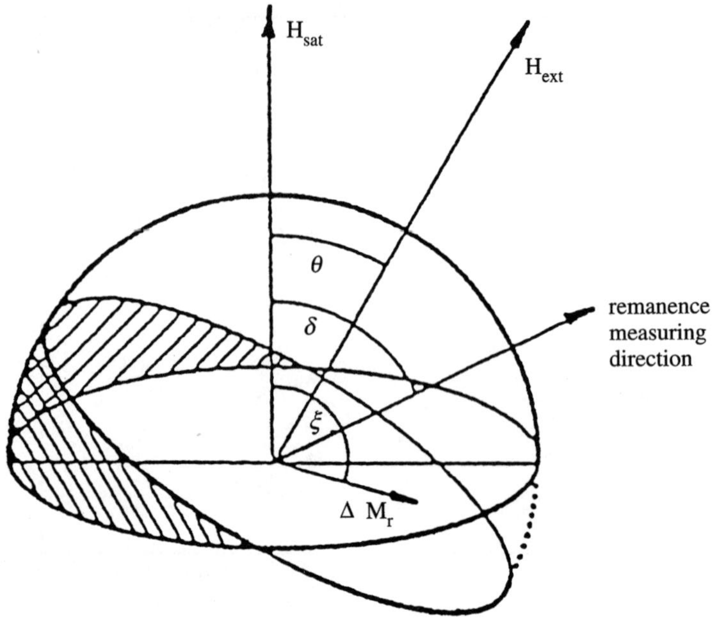

FIG. 5.35. Geometrical arrangement of the magnetic fields and the particle magnetizations.

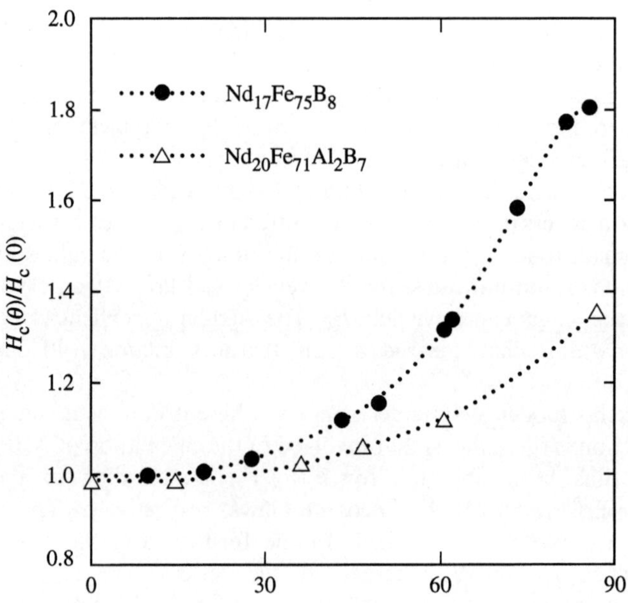

FIG. 5.36. $H_c(\theta)/H_c(0)$ for two Nd–Fe–B magnets with different values of $H_c(0)$.

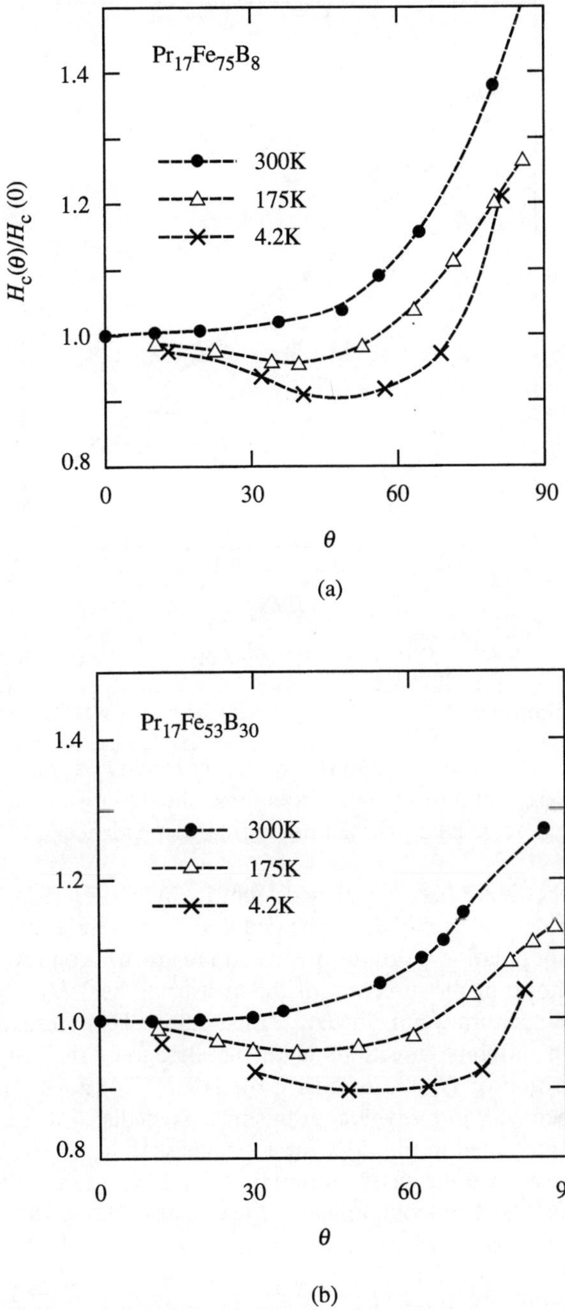

FIG. 5.37. The angular dependence of H_c for $Pr_{17}Fe_{75}B_8$ and $Pr_{17}Fe_{53}B_{30}$ at several temperatures.

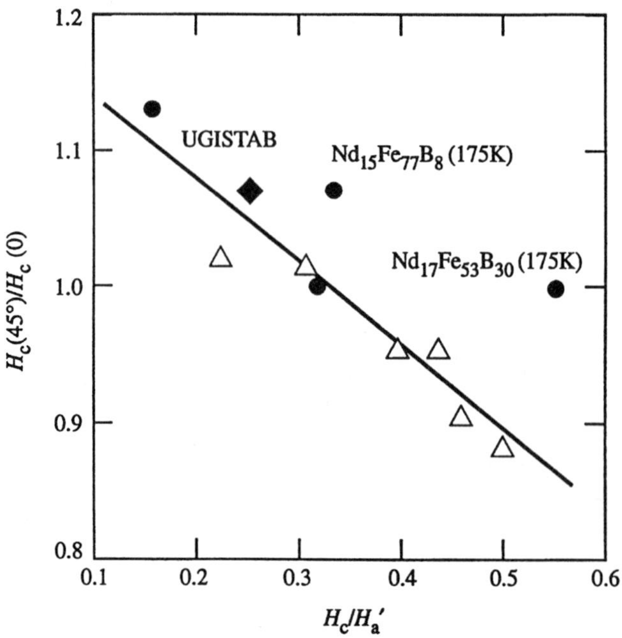

FIG. 5.38. The ratio $H_c(45°)/H_c(0°)$ as a function of H_c/H'_a, triangles are Pr magnets; circles are Nd magnets; UGISTAB is $Nd_{15}Fe_{67.5}Co_5V_4B_8Al_{0.5}$ $H'_A = \gamma_w/\mu_0 M_s \delta_w$ is a measure of bulk anisotropy.

cess. Reversal starts in crystallites where coercivity is minimum, that is, $H_c(\theta) = h_c(\theta_{cH})_{min}$, and propagates from crystallite to crystallite. It follows that the magnet's coercive field exhibits no angular dependence.

Other approaches have been developed [Sch91c, Wir92, Her92a, Ceb95] which allow the angular dependence of H_c to be deduced from a series of measurements, during which the population of grains involved in magnetization reversal is selected by appropriate choice of the direction of the measuring field H_{mes} with respect to that of the initial saturation field H_{sat} (Fig. 5.35). The measurement of the remanence in two different directions gives the change of the remanence vector ΔM_r after application of H_{mes}. From this, the law $h_c(\theta_{cH})$ governing the angular dependence of coercivity in individual grains may be deduced directly.

$H_c(\theta)/H_c(0)$ is presented in Fig. 5.36 for a magnet with $\mu_0 H_c = 0.87$ T at 300 K. H_c steadily increases as θ increases and attains, for $\theta \approx \pi/2$, a value equal to 1.8 times the $\theta = 0$ value. For comparison, $H_c(\theta)$ is also shown for an Al-doped magnet in which the room-temperature coercivity is $\mu_o H_c = 1.72$ T. The angular dependence of H_c is slower for this magnet, which has larger H_c/H_a ratio. $H_c(\theta)$ is also shown for two Pr–Fe–B magnets at several temperatures (Fig. 5.37). At 300 K, H_c (90°) is equal to 1.5 $H_c(0)$ for the first magnet and 1.3 $H_c(0)$ for the second magnet. In both magnets at low temperatures, $H_c(\theta)$ is found to decrease as θ increases from 0° to 45° and thereafter increases for larger angles.

The experimental $H_c(\theta)$ in Nd–Fe–B magnets for which $\mu_0 H_c = 0.87$ T at room temperature corresponds approximately to the one calculated above for a $1/\cos\theta_{cH}$ dependence of coercivity in individual crystallites and assuming independent reversal of magnetization from one crystallite to another (Fig. 5.36). As already discussed (Section A.1.2, p. 283), this behaviour indicates that no significant rotation of the moments takes place before magnetization reversal and thus the effect of the field is equivalent to that of its projection along the initial moment direction. For larger coercive field values, the increase in H_c with θ is reduced. This can be attributed to a nonnegligible rotation of the moments into the field direction. To illustrate this, the ratio of the coercive fields at 45° and 0° for series of Pr–Fe–B and Nd–Fe–B magnets is presented in Fig. 5.38 as a function of H_c/H_A', where $H_A' = \gamma_w/\mu_0 M_S \delta_w$ is a measure of bulk anisotropy. The larger the ratio H_c/H_A', the smaller the ratio $H_c(45°)/H_c(0°)$.

However, the decrease in H_c as θ increases, which occurs at small θ values in Pr–Fe–B magnets, cannot be interpreted within the above analysis. This has been tentatively attributed to the relative decrease, at low temperatures, of the dipolar interactions with respect to other interactions [Grö90a, Mar90].

2.2.4 *Magnetic viscosity* Whatever the mechanism determining magnetization reversal, the volume of the nucleus in which the process is initiated is an essential quantity for coercivity. The value of this volume may be evaluated from measurements of the magnetic after-effect [Woh84, Gau86] (decay of the magnetization in a fixed field, after saturation).

Magnetic viscosity can be observed in a steady reverse field near H_c. It results from the fact that the coercive energy barrier Δ may be overcome through thermal activation.

Under the assumption that the barrier crossing probability is constant, the probability $P(t)\,dt$ for Δ to be overcome between the time t and $t + dt$ is

$$P(t)\,\mathrm{d}t = (\mathrm{d}t/t)\exp(-t/\bar{t})$$

The proportion of crystallites whose magnetization is reversed between t and $t + \mathrm{d}t$ becomes

$$\mathrm{d}N/N = 1 - \exp(-t/\bar{t})$$

$\bar{t}$ is the mean waiting time required to overcome Δ. For independent processes, $\bar{t}$ is expected to follow an Arrhenius law:

$$\bar{t} = \tau_0 \exp(\Delta/k_B T) \tag{5.28}$$

where $\tau_0(\approx 10^{-11}$ s) is the characteristic time for magnetization reversal when no barrier is present [Sta60,Gau77].

The exponential time-dependence of the magnetization predicted by the above formulae is modified if it is considered that in real systems a distribution $f(\Delta)$ of energy barriers will exist. The magnetization variation associated with the

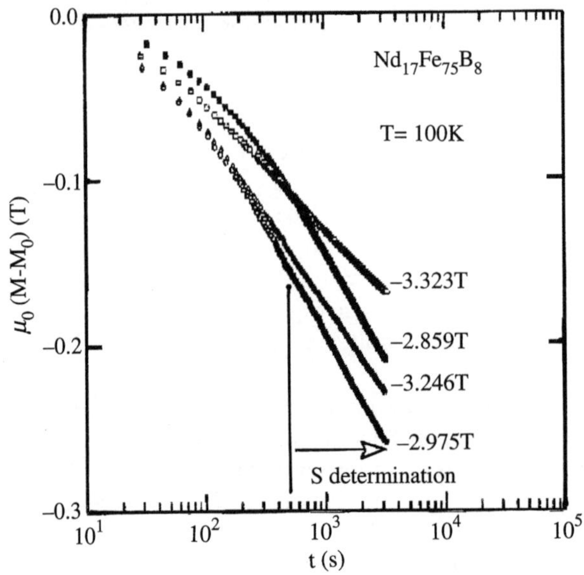

FIG. 5.39. Typical time dependence of the magnetization in sintered Nd–Fe–B. Arrows show useful region for the determination of the magnetic viscosity S when no correction of the time-origin is made. The corresponding variation of the magnetization as a function of ln t. No correction of the time-origin is made.

overcoming of the energy barriers between Δ and $\Delta + \mathrm{d}\Delta$ is

$$\mathrm{d}M = -2M_s f(\Delta)\mathrm{d}\Delta$$

Street and Wooley [Str49] and Gaunt [Gau76] have shown that over a large range of time when $f(\Delta)$ is approximately constant, a given time t can be associated with a given energy barrier Δ through

$$\Delta = k_B T \ln(t/\tau_0). \qquad (5.29)$$

Relation (5.29) is not equivalent to (5.28) because $\bar{t}$ is replaced by t. In the range of times accessible to usual experiments ($1s \leq t < 10^4$s), Δ is approximately constant, of the order of 25 $k_B T$.

The magnetic viscosity S, defined from the time dependence of the magnetization:

$$S = -\mathrm{d}M/\mathrm{d}(\ln t) = 2M_s k_B T f(\Delta)$$

can be compared to the irreversible susceptibility χ^{irr}:

$$\chi^{\mathrm{irr}} = \mathrm{d}M/\mathrm{d}H = -2M_s f(\Delta)(\partial\Delta/\partial H)$$

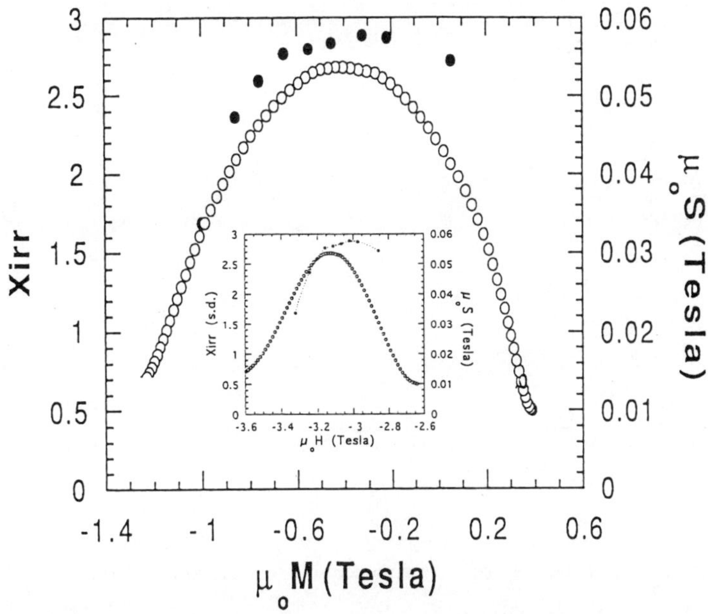

FIG. 5.40. Dependences of the magnetic viscosity μ_0S and irreversible susceptibility $\mu_0\chi^{\text{irr}}$ as a function of the irreversible magnetization. The inset shows the same variations as a function of the applied field.

Provided that S and χ^{irr} are associated with the same $f(\Delta)$ value, the magnetic viscosity coefficient S_V may be defined as

$$S_v = S/\chi^{\text{irr}} = k_B T/(\partial\Delta/\partial H) \tag{5.30}$$

To a magnetic state of the magnet corresponds a given value of $f(\Delta)$. It may be assumed that a magnetic state is uniquely determined by the value of its magnetization, and can be reached either by varying H and/or t. For Eq. (5.30) to apply, it is thus sufficient that S_v be derived from S and χ^{irr}, both measured at constant irreversible magnetization [Str89, Tay92].

The coefficient S_v has the dimension of a fluctuation field and its physical significance has been discussed by Néel [Née50, 51], Wohlfarth [Woh84] and Gaunt [Gau86]. A correspondence between time-effect and magnetic field may be derived from S_v, which leads to the equivalent thermal activation field [Giv96]:

$$H_{\text{ta}} = S_v \ln(t/\tau_0)$$

Furthermore, Wohlfarth [Woh84], Gaunt [Gau86] and Givord [Giv88a] have suggested that S_v could be expressed phenomenologically as

$$S_v = k_B T/\mu_0 v_a M_s \tag{5.31}$$

where v_a is the appropriate activation volume, defined as $\Delta m/M_s$ and where Δm

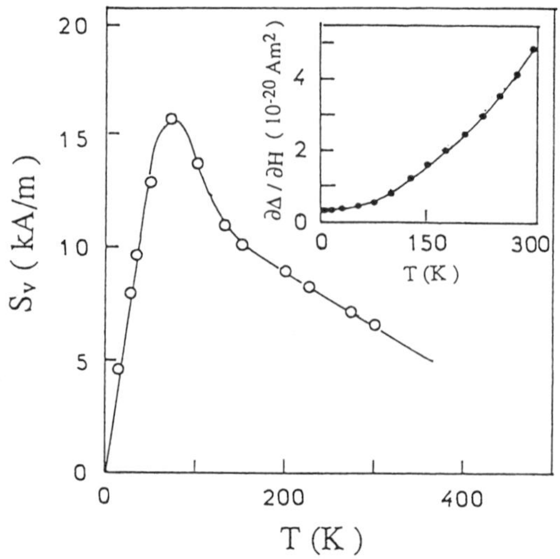

FIG. 5.41. Typical temperature dependence of the viscosity coefficient S_v in a Nd–Fe–B sintered magnet. Inset: temperature dependence of $\partial\Delta/\partial H$ deduced from S_v.

represents the magnetization variation between the initial metastable state and the state corresponding to the energy maximum [Gau86].

Note that in general $\partial\Delta/\partial H$ and v_a depend on H — see Eq. (5.7). This is, however, not always the case; for example, v_a, for pinning, is field independent to a first approximation — see Eq. (5.17). Only if $\partial\Delta/\partial H$ and v_a are independent of H will the actual process of magnetization reversal not be influenced by H.

A typical time dependence of the magnetization in several reverse fields is shown in Fig. 5.39, for a Nd–Fe–B sintered magnet, as a function of $\ln t$. The logarithmic variation obtained above 500 s indicates that the distribution f(Δ) is approximately a constant in the range of activation energies corresponding to these measurements. In this time-region, the slope of the straight line gives the viscosity $S = -dM/d\ln t$. As explained above, S should be plotted as a function of M rather than as a function of H (Fig. 5.40). S is maximum for $M \approx 0$, that is, in the vicinity of the coercive field. In Fig. 5.40, χ^{irr} appears to be approximately proportional to S, thus showing that the viscosity coefficient S_v is approximately M (and H) independent ($\mu_0 S_v = 0.025$ T). Other methods for the experimental determination of S_v are described by Street *et al.* [Str89, Est89] and Givoid *et al.* [Giv96]. The temperature dependence of S_v exhibits a maximum at about 75 K (Fig. 5.41). The field derivative of the energy barrier $\partial\Delta/\partial H$ deduced from S_v increases continuously with temperature.

The thermal variations of the activation volume thus obtained (see Eq. (5.31)) in different sintered magnets are shown in Fig. 5.42. v_a increases by more than an order of magnitude between 4.2 K and 400 K. The thermal variation of v_a is compared to that of δ_w^3 by plotting v_a/δ_w^3 (Fig. 5.43). It can be seen that the ratio is roughly constant up to room temperature and increases drastically at higher temperatures.

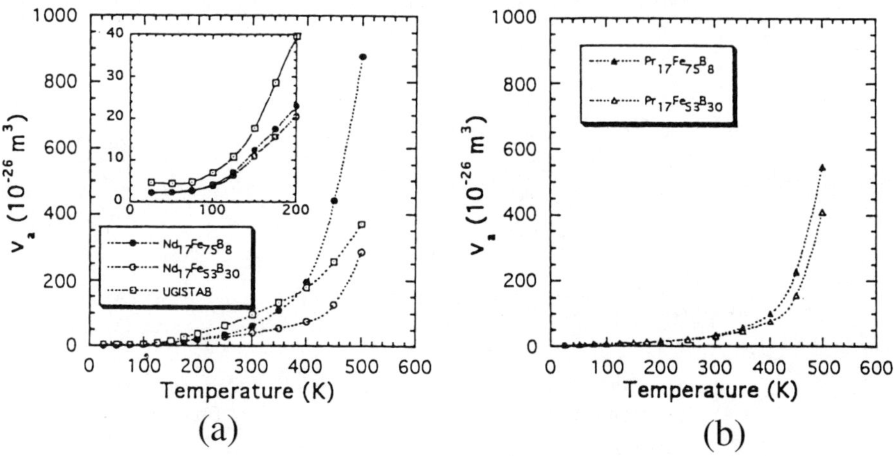

FIG. 5.42. Typical thermal variation of the activation volume v in some R–Fe–B sintered magnets.

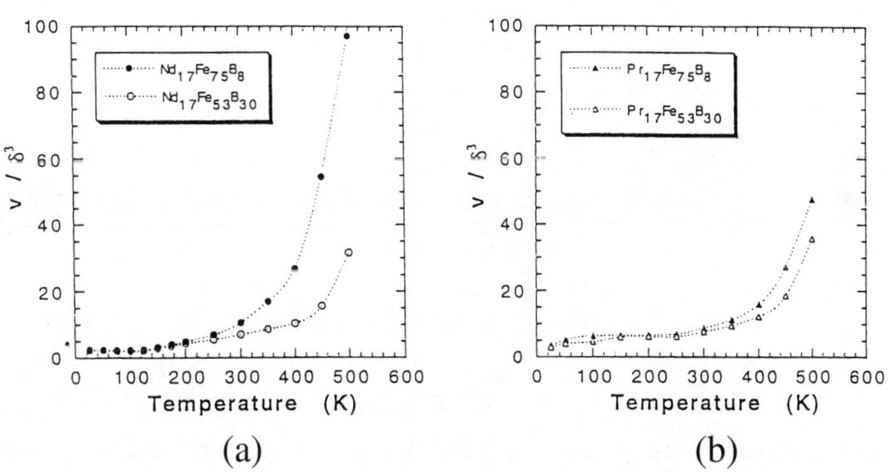

FIG. 5.43. Thermal variation of v/δ_w^3 in some R–Fe–B sintered magnets.

Considering that magnetization reversal involves nonuniform magnetization processes, the proportionality between v_a and δ_w^3 is not surprising since the domain wall width δ_w is the length scale which characterizes such processes. Note, however, that the activation volume is compared to a parameter, δ_w, characteristic of a perfect material and it is, in general, believed that the coercive process is dominated

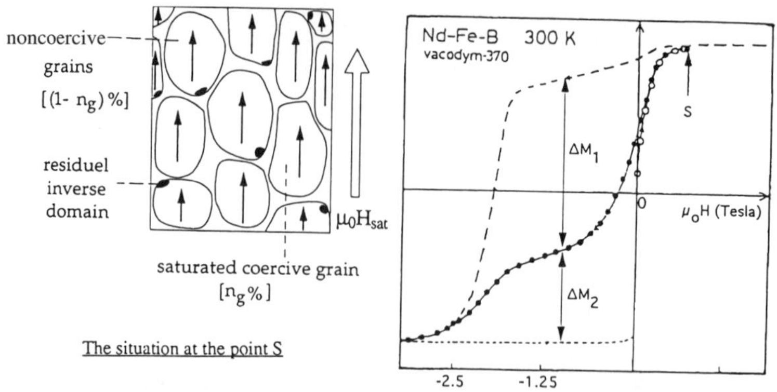

FIG. 5.44. Principle of the determination of the effective dipolar fields existing in the grains. At the point S, a proportion of the grains, n_g, have been saturated, with $n_g = \Delta M_2/(\Delta M_1 + \Delta M_2)$.

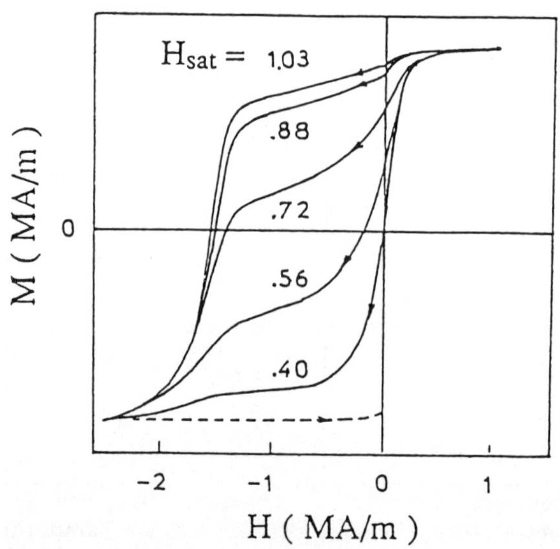

FIG. 5.45. Demagnetization curves in a Nd–Fe–B sintered magnet, for increasing values of an initial saturation field H_{sat}.

by a small volume of the total material, where properties are not exactly representative of the bulk (see also Fig. 5.31).

2.3 Direct evaluation of N_{eff}

In the case where domain wall motion within crystallites is free, Givord *et al.* [Giv91, 92a; Tay92, 94] have proposed that the maximum strength of the dipolar

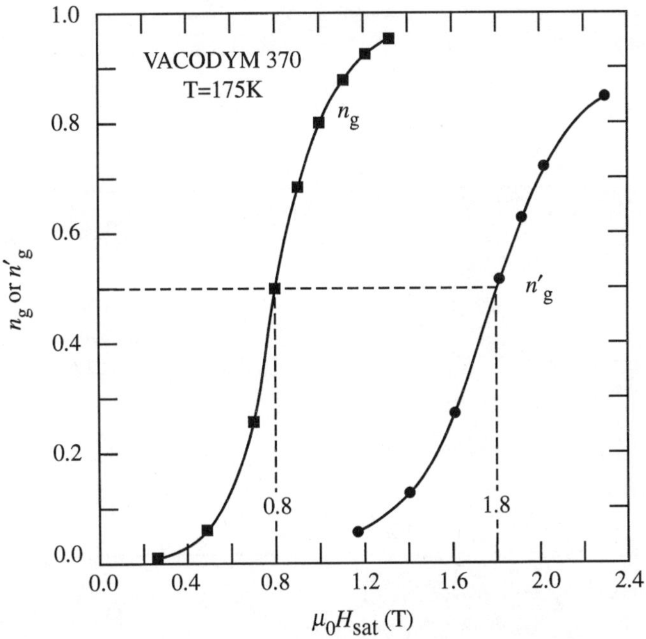

FIG. 5.46. Amount of saturated grains in two different magnetic configurations.

interactions present in each grain may be evaluated thanks to the fact that the field applied in a thermally demagnetized state works against these interactions. The value of the saturation field H_{sat} (Fig. 5.44) needed to fully eliminate domain walls in a given grain represents the maximum value of the dipolar field within this grain. If the magnetic state is similar to that existing in the vicinity of the coercive field, H_{sat} gives an estimate of the effective dipolar field acting during magnetization reversal.

Experimentally, the proportion n_g of grains which have been saturated in a given field H_{sat} (i.e. which have been made coercive by this field) may be deduced from the demagnetization curve (Fig. 5.44). Such curves observed at 300 K for increasing values of H_{sat} are shown in Fig. 5.45. The value of H_{sat} required to develop coercivity in a given proportion of grains is weakly temperature dependent between 4.2 K and 400 K, and it appears to be proportional to the spontaneous magnetization. This shows that the development of coercivity is not related to mechanisms involved in magnetization reversal, but it is actually determined by the strength of the dipolar fields in the magnet.

The above measurements were performed on needle-shaped samples with low demagnetizing factors and the values of H_{sat} obtained represent the strength of the internal dipolar fields; Fig. 5.46 shows the proportion n_g of grains which have been saturated in a given field H_{sat}. To understand the large values of H_{sat} obtained ($\mu_0 H_{sat} \approx 0.8$ T at $T = 175$ K), it must be understood that a domain wall is present

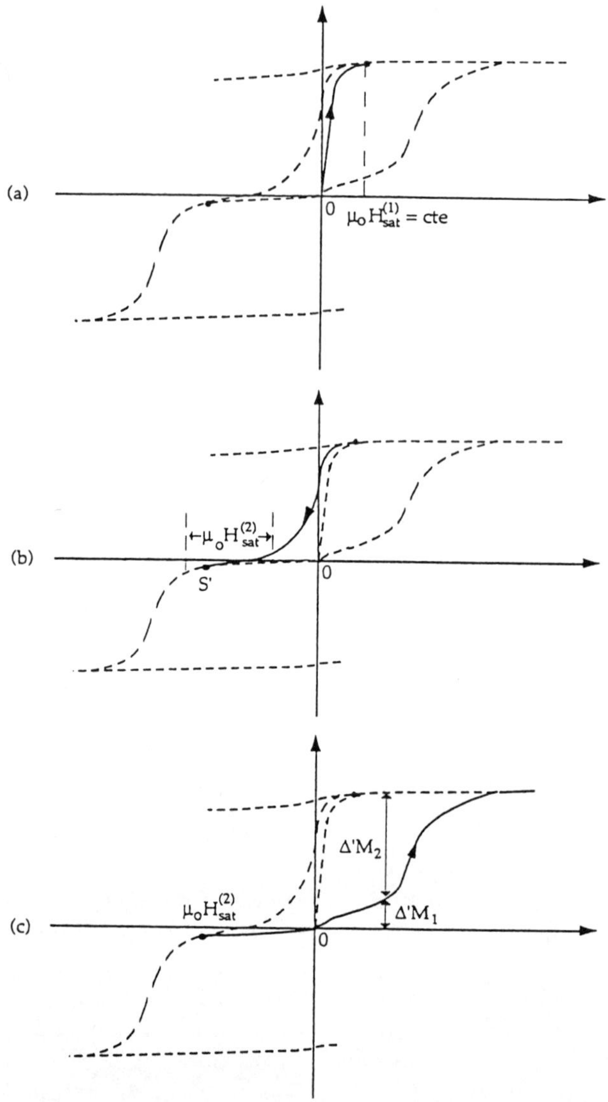

FIG. 5.47. Method for the experimental determination of $n'_g = \Delta' M_2/(\Delta' M_1 + \Delta' M_2)$ as a function of H_{sat} (configuration in which $M \approx 0$)

in a given crystallite until the applied field overcomes the dipolar field in any part of it. It is not the mean value of the dipolar field which controls magnetic saturation but, rather, its maximum value. In permanent magnets which are very heterogeneous by nature, large fluctuations in dipolar field may exist.

The dipolar fields which are revealed by the above type of measurement are, however, not the dipolar fields involved during magnetization reversal. To evaluate

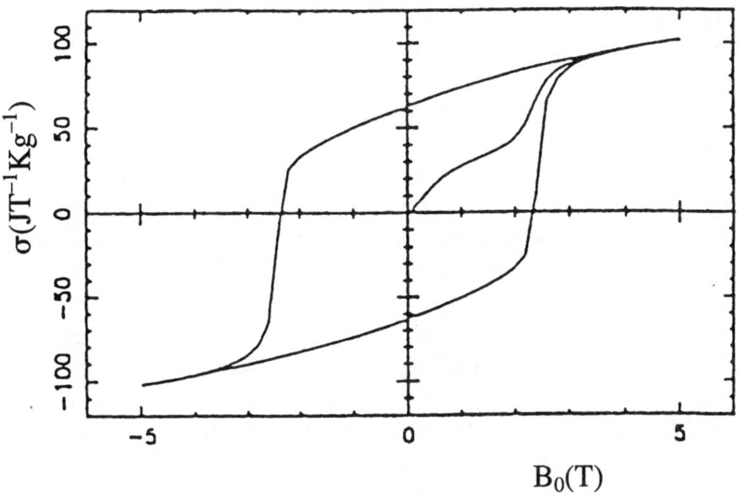

FIG. 5.48. Hysteresis loop of melt spun $Nd_{17}Fe_{72}B_{7.5}Ga_{1.5}Nb_2$.

them, a magnetization state which resembles the state in the vicinity of the coercive field should be realized experimentally. For that purpose, an initial saturation field is applied such that half the grains are fully saturated (Fig. 5.47a); these grains remain saturated in the up direction when the field direction is reversed, leading to a mean magnetization approximately equal to zero (Fig. 5.47b). The field $\mu_0 H_{sat}^{(2)}$ is now required to saturate the proportion $\Delta' M_2/(\Delta' M_1 + \Delta' M_2)$ (Fig. 5.47c) of the remaining grain population. $\Delta' M_2/(\Delta' M_1 + \Delta' M_2) = n'_g = 0.5$ corresponds to $\mu_0 H_{sat} \approx 1.8$ T (Fig. 5.46). This value is much larger than that obtained above and it may be attributed to the strongly heterogeneous character of the magnetization state where the bulk magnetization is zero but the magnetization of every crystallite is saturated.

3 R–Fe–B magnets prepared from ribbons

3.1 *General description*

Three categories of magnet may be prepared from melt-spun Nd–Fe–B ribbons [Cro84]. The ribbons are obtained by rapidly quenching an alloy of typical composition $Nd_{14}Fe_{81}B_5$ on a rotating wheel. Their microstructures are described in Chapter 6 p. 304. The remanent induction and coercive field are, respectively, $B_r = \mu_0 M_r = 0.8$ T and $\mu_0 H_c = 1.4$ T. Flakes made by grinding these ribbons can be polymer-bonded to form magnets directly. Typical hysteresis curves are shown in Fig. 5.48. The remanence B_r is reduced to 0.65 T due to the volume occupied by the polymer. Alternatively, ribbons may be consolidated to full density by hot-pressing. The magnets obtained with appropriate heat treatment have magnetic properties which are essentially similar to those of optimally quenched ribbons. Finally, crystallite alignment may be developed in a hot-pressed magnet by die-upsetting. In that case, the remanence, 1.2 T, approaches that of sintered magnets and the coercive field is of the order of $\mu_0 H_c \approx 1.2$ T. For details of processing these magnets, see Chapter 7, p. 354.

Enhanced remanence. In Nd–Fe–B ribbons containing an appropriate amount of additives such as Si [Kad86, Mat88] or prepared under controlled experimental conditions [Man91], the remanence may be larger than $\mu_0 M_s/2$, which is the maximum value expected for an isotropic system of noninteracting crystallites. This behaviour reveals the existence of exchange interactions across crystallite interfaces which are free of any nonmagnetic phase. Moments in different crystallites tend to align parallel and deviate from their local easy axis in multigrain domains. Magnets with higher-energy products may be obtained from these exchange-coupled isotropic ribbons. However, a reduction in the coercivity occurs since the effective anisotropy is reduced. Remanence enhancement has also been obtained in the Sm–Co system [Din93a]. This intergranular exchange-coupling phenomenon is also involved in the spring magnets (see Section C. 5.3, p. 280).

A close similarity may be found between these materials, in which crystallites are exchange coupled and have easy-magnetization direction distributed at random, and amorphous materials. In the HPZ model [Har73] (see p. 299), the moment configuration in the amorphous state is determined by the competition between exchange and local anisotropy. The value of the spontaneous magnetization normalized to full saturation approaches 1.0 for high exchange to anisotropy ratio and 0.5 for vanishing exchange. Regions of correlated magnetization (interaction domains) appear as a result of the different dependences on the volume of exchange and anisotropy energies [Imr75].

3.2 *Analysis*

Initial magnetization curves. The initial magnetization curve of optimally quenched ribbon is included in Fig. 5.48. The magnetization processes can be divided into two regimes. The low-field susceptibility is small, indicating that free domain wall motion does not occur. In higher fields, close to the magnet's coercive field, the irreversible susceptibility increases strongly. This behaviour shows that either the ribbons are formed of an assembly of single-domain crystallites with no domain walls, or that domain wall pinning occurs at the grain boundaries.

In ribbons which are not optimally quenched, the initial susceptibility reveals some reversible domain wall motion. Grönefeld *et al.* [Grö90b] have shown that the domain wall contribution varies with temperature. This has been interpreted by considering that the critical single domain size is temperature dependent because the wall energy varies more rapidly with temperature than the magnetization.

Similar behaviour is observed during initial magnetization of hot-pressed magnets. Some reversible domain wall motion takes place. In die-upset magnets, the large initial susceptibility is a direct consequence of the domain walls inside the crystallites.

Temperature dependence of H_c. In these categories of magnet, the coercive field decreases continuously with temperature, but the decrease above room temperature is significantly less than in sintered magnets. In melt-spun ribbons and hot-pressed

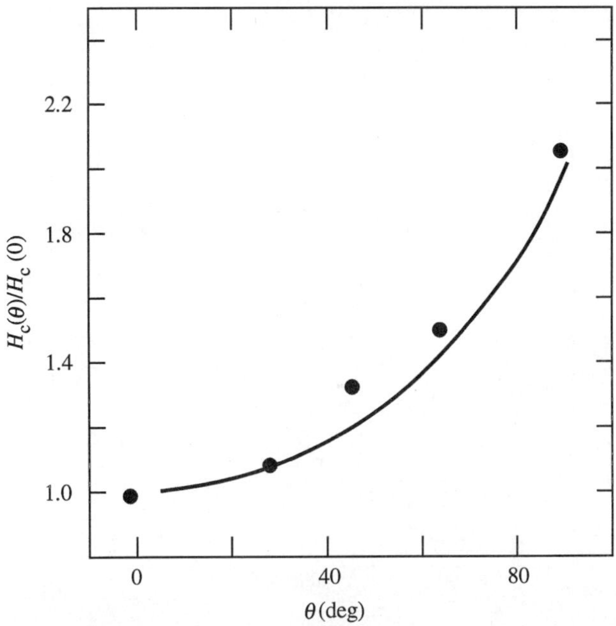

FIG. 5.49. $H_c(\theta)/H_c(0)$ for die-upset Nd–Fe–B magnet. Full line: calculation for a $1/\cos\ \theta_{cH}$ dependence of the coercive field in individual crystallites.

magnets, H_c vanishes at $T = T_c$, whereas in die-upset magnets, it vanishes approximately 20 K below T_c. The persistence of coercivity at high temperature is of considerable practical importance.

The values of α_K range from 0.75 to 1.1, those of N_{eff} range from 0.7 to 1.1 (Fig. 5.30). The former are larger than in sintered magnets, the latter are less. The reduction in N_{eff} may be attributed to the smaller size of the constituent crystallites and to their more regular shapes. Residual exchange interactions may also exist between the crystallites. This can be expected considering the evidence for exchange interactions in a remanence-enhanced system whose microstructure is not fundamentally different from that of normal ribbons.

The analysis of coercivity within the global approach, in magnets prepared from ribbons, leads to similar conclusions, that is, α is approximately the same as in sintered magnets and N_{eff} is reduced (Fig. 5.32). The value of N_{eff} obtained may even be negative, that is, interactions between crystallites reinforce coercivity. This property can hardly be attributed to dipolar interactions and would confirm the existence of exchange interactions between crystallites.

Angular dependence of H_c. The angular dependence of H_c has been examined in die-upset-magnets [Via88]. $H_c(\theta)/H_c(0)$ increases with θ in the same way as in Nd–Fe–B sintered magnets (Fig. 5.49). As discussed previously (Sections A. 1.2, p. 223, and C. 2.2.3, pp. 259–67), the results can thus be analysed in terms of a

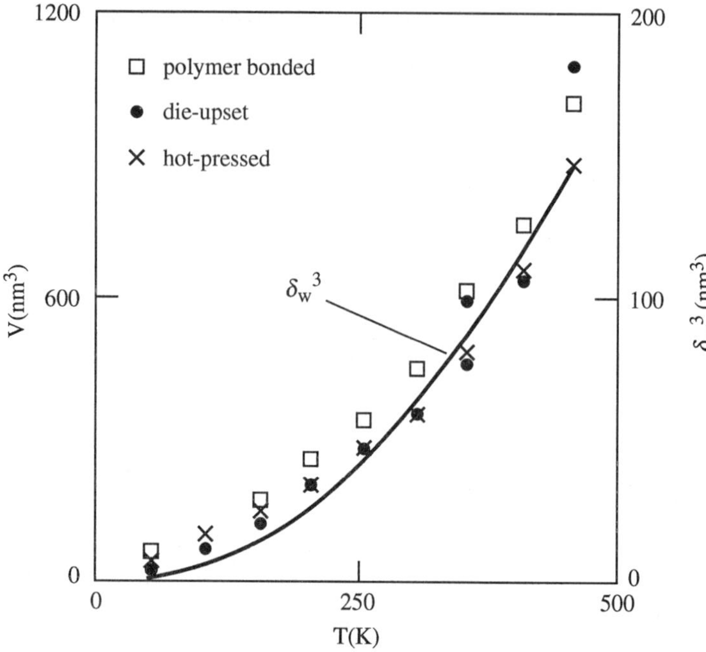

FIG. 5.50. Thermal variation of the activation volume v in Nd–Fe–B magnets prepared from ribbons: polymer-bonded, hot-pressed and die-upset. Thermal variation of δ_w^3, where δ_w is the domain wall width in the $Nd_2Fe_{14}B$ phase, is shown for comparison.

$1/\cos\theta_{cH}$ law, indicating that no significant rotation of the moments towards the field direction takes place.

Magnetic after-effect. A logarithmic dependence of the magnetization is found in melt-spun ribbons, hot-pressed magnets and die-upset-magnets. The values and temperature dependencies of the activation volume v_a are very similar for each category of magnet and resemble those characterizing sintered magnets (Fig. 5.50).

4 Some remarks about other magnets

4.1 *R–Fe–B magnets obtained by other routes*

Other types of R–Fe–B magnet may be prepared, for example by mechanical alloying [Sch91b] or by hot extrusion [Noz88, 89] (see Chapter 7). The microstructure in the former is intermediate between that of sintered magnets and that of ribbons. In the latter, the grain sizes tend to be even larger than in sintered magnets. α_K and N_{eff} parameters deduced from coercivity analysis in these magnets are shown in Figs 5.30 and 5.32.

Although very different microstructures may characterize the various categories of R–Fe–B magnet, the coercivity parameters appear to be similar in all of them.

The ratio of coercivity to anisotropy is also quite similar for different types of magnet. Magnetization reversal develops from a nucleus whose typical size is δ_w^3. N_{eff}, which represents the interactions between crystallites, may vary more significantly from one type of material to another.

4.2 *Sm–Fe–N magnetic materials*

The coercivity in the recently developed Sm–Fe–N magnets [Coe90a, 90b] has been examined by Jifan Hu *et al.* [Jif93], within the micromagnetic approach. The magnets studied were Sn-bonded in order to eliminate free-Fe, which otherwise tends to be present. These authors proposed that coercivity is governed by nucleation. A fit to the temperature dependence of H_c was obtained. The parameter α_K deduced corresponds to $r_0 = 3.3$ nm, where r_0 is the diameter of the defect determining magnetization reversal. The value of N_{eff} deduced, however, is extremely high (= 3.26). Such a large value of N_{eff} would require a very complex microstructure, which is unlikely to exist.

Coercivity of finely milled $Sm_2Fe_{17}N_3$ powder varies as $1/d$, where d is the crystallite size in the range $0.3 < d < 10$ μm (Fig. 4.11). This variation has been explained in terms of a statistical distribution of nucleation centres of different sizes in different grains [Kob95].

4.3 *Former generations of magnet*

4.3.1 *$SmCo_5$ magnets*

The magnetic properties of $SmCo_5$ magnets are similar to those of Nd–Fe–B sintered magnets. The size of the constituent crystallites is again larger than the single-domain critical size. Magnetic saturation requires the dipolar field within crystallites to be overcome. Starting from the thermally demagnetized state, $\mu_0 H_{sat} \approx 0.5$ T is obtained. The ratio of H_{sat} to the remanence is the same as in Nd–Fe–B sintered magnets. The anisotropy field is very large ($\sim$ 30 T). The angular dependence of H_c exhibits the same essential features as in sintered Nd–Fe–B magnets and it has been analysed in terms of a $1/\cos\theta_{cH}$ dependence [Bec71, Giv88b].

4.3.2 *Sm (CoFeCuZr)$_{7-8}$* magnets

The Sm $(CoFeCuZr)_{7-8}$ type of magnet was initially developed from $SmCo_5$ magnets in order to improve their remanence. Appropriate thermal treatment leads to the formation of a typical cellular microstructure at the nanometer scale (Figs 6.7 and 6.8). The boundaries of the cells consist of a high-anisotropy $SmCo_5$ type phase, whereas the core of the cells is formed of relatively less anisotropic $Sm_2(CoFe)_{17}$-type phase. Coercivity in this system arises from pinning, resulting from the existence of two magnetic phases with different anisotropies. $H_c(T)$ has been analysed within the global approach [Giv92b] and in terms of theoretical results for pinning fields of different phases [Dur88]. The angular dependence of H_c is not as expected for pinning-controlled coercivity [Giv92b].

4.3.3 *Ferrite magnets* The initial susceptibility of ferrite magnets may be high or low depending whether the size of the constituent crystallites is above or below the single-domain critical size, $d_c \sim 0.7\ \mu m$. This, however, is of no importance for the coercivity mechanism, which is nucleation controlled. However, H_c increases with increasing temperature up to 500 K. This behaviour may be directly linked to the peculiar temperature dependence of the anisotropy which increases with increasing temperature. The activation volume is of the order of $10\delta_w^3$ [Giv90].

4.4 *Spring magnets*

A new concept for hard magnetic materials has been proposed recently in which a high anisotropy magnetic phase and a low anisotropy phase are intimately intermixed and exchange coupled at the nanoscopic scale. The first such materials were iron-rich Nd–Fe–B ribbons melt-spun by Coehoorn *et al.* [Coe88, 89].

Kneller [Kne91] has discussed the properties of these systems for the case where the hard grains are oriented at random. At very small grain size, the magnetization is homogeneous in the whole sample and the mean anisotropy vanishes. There is no coercivity. At very large grain size, the magnetization in the hard crystallites follows the local easy direction but the magnetization in the soft phase follows the applied field. Nucleation can easily occur in the soft regions. The situation of interest is for crystallite sizes of the order of the exchange length in the soft and hard crystallites. The magnetization tends to align ferromagnetically in the whole sample and thus M_r may be greater than $M_s/2$, the value expected for an ideal isotropic system. In addition, the magnetization in each crystallite deviates towards the local easy axis, thus leading to the existence of some coercivity.

Skomski and Coey [Sko93] have considered a nanostructure composed of an aligned hard phase intermixed with a soft phase, where energy products up to 1 MJ/m^3 could be achieved in principle. If the crystallite size is less than a typical dimension of the order of the domain wall width of the hard phase, the material behaves as if it were homogeneous, with the anisotropy and magnetization being equal to the mean values of the constituent phases. The theoretical coercive field is equal to this mean anisotropy field. At larger crystallite size, magnetization reversal starts in the soft phase and coercivity is determined by pinning at the phase boundary. In this regime, the coercive field decreases as the crystallite size increases, until it reaches a value of the order of $H_a/4$ (see Section B. 2.1.2, p. 232).

This concept has already been applied to the development of isotropic hard magnetic materials. In the melt-spun Nd–Fe–B system [Coe88, Hir93], rapidly quenched alloys may contain a mixture of nanocrystallites of $Nd_2Fe_{14}B$ and Fe_3B or α–Fe. Coercivity of the order of $\mu_0 H_c = 0.5$ T is reached with isotropic remanence in excess of 1 T. Likewise, in mechanically alloyed Sm–Fe–N alloys a mixture of $Sm_2Fe_{17}N_{3-\delta}$ and α–Fe is present and similar coercivity and remanence have been obtained [Din93b, 94, O'Do94].

D. CONCLUSION: THE COERCIVITY MECHANISMS IN R–Fe–B MAGNETS

At the end of this chapter, it is useful to reflect on how understanding of coercivity has progressed with the evolution of experimental analysis.

Ferrite magnets, and, later, rare-earth-based magnets demonstrated that coercivity may exist in multidomain grains (i.e. grains containing domain walls in their thermally demagnetized state). The origin of coercivity in these systems was debated until recently in terms of whether coercivity is governed by nucleation or pinning, neglecting implicitly more sophisticated mechanisms. In the case of pinning, the interaction of domain walls with the microstructure is at the origin of coercivity, whereas wall motions are not involved in the case of nucleation. This is why coercivity analysis was focussed on the comparison between initial magnetization and demagnetization curves. From such analysis it is easy to distinguish whether displacement of domain walls in multidomain grains is pinned or not, but it is not possible to distinguish between other mechanisms listed on p. 224.

The experimental approaches developed more recently have undoubtedly brought about a deeper understanding of magnetization reversal processes in magnets. The present chapter has described these approaches. In summary it has been shown that:

1. Initial magnetization curves and $H_c(\theta)$ measurements may be directly related to possible coercivity mechanisms.
2. $S_v(T)$ and $n_g(H_{sat})$ are determined by the intrinsic physical properties in the critical volume but do not characterize a given mechanism.
3. The comparison of experimental $H_c(T)$ with the predictions of theoretical models and numerical simulations provides indirect insight into coercivity mechanisms.

However, the interpretation of these measurements is not entirely unambiguous.

In the case of the sintered R–Fe–B magnets, it is generally admitted that the observed free domain wall motion in the thermally demagnetized state implies that pinning is not involved. Based on this conclusion, two alternative interpretations have been proposed:

1. A first analysis retains the above assumption that, due to the absence of pinning, coercivity is dominated by nucleation. The other mechanisms are implicitly neglected. It is then further assumed, as if it were obvious, that the physical mechanism determining magnetization reversal is simply uniform rotation of the moments within the initial nucleus. This means that the Stoner–Wohlfarth model must be taken as the norm. The value 0.5 given to $\alpha_{\Psi}^{\text{eff}}$ comes from this assumption and may lead to α_K values which are larger than 1, non-physical result of the model. Difficulties also arise in the interpretation of some measurements: when the angular distribution of moment orientations is taken into account, the coercive field H_c should be almost independent of the direction

of the applied field. Actually, experimental results show that, in most cases, $H_c(\theta)$ increases monotonically with θ.

2. The alternative interpretation which we have proposed considers all the physical mechanisms as possible *a priori* and analyses to what extent each of them is compatible with experimental results. In particular, it is pointed out that $H_c(\theta)$ may follow a $1/\cos\theta$ law. This observation excludes the possibility that uniform rotation is the determinant mechanism (with $H_a^{nucleus} << H_a$), and affirms that the anisotropy field within the initial nucleus is larger than H_c and not very different from H_a. From the similarities between the characteristic dimensions of the initial nucleus, as obtained from $S_v(T)$, and calculated values of $\delta_w^3(T)$, a similar conclusion is drawn, that is, that the intrinsic physical parameters (M_s, K, A_{ex}) in the critical volume are not greatly different from those of the $Nd_2Fe_{14}B$ phase. The 'expansion mechanism', for which the activation volume (identified as the initial critical nucleus) can be entirely contained in the hard magnetic phase, appears to be essentially compatible with the results and it provides a consistent picture of coercivity. However, it must be admitted that the description of the mechanisms which have been presented is rather schematic. For example, although the passage and expansion mechanisms correspond to obviously different phenomena, it appears very likely that they occur together.

It appears at this stage that a clearer understanding of coercivity mechanisms may emerge by reconsidering the exact meaning of the term 'nucleation'. This term is not associated with a particular coercivity mechanism. It is used in the literature with different meanings which are often not clearly defined. In micromagnetic calculations, nucleation is defined as the first irreversible departure from saturation. In the micromagnetic model, nucleation is identified as coherent rotation. In the most common approach, nucleation qualitatively describes reversal that starts from a small volume, which then expands to occupy the whole grain. The different mechanisms described on p. 224, that is, true nucleation (restricted to the formation of a reverse domain), passage and expansion are aspects of the process which is commonly called nucleation. All these mechanisms, except the case where the formation of a reverse domain may be identified as uniform rotation, are compatible with a $1/\cos\theta$ law for the angular dependence of H_c. For such a law $\alpha_\Psi^{eff} = 1$ in the micromagnetic model, which yields α_K values which are smaller than 1. There is no longer a contradiction between the mechanisms included in this acceptance of the word 'nucleation' (in particular the expansion mechanism) and the results of the micromagnetic model.

As a final remark, it may be pointed out that an important element is still missing for a complete view of coercivity: due to the very small size of the critical nucleus and to the very short time required for the magnetization reversal process, the phenomena involved have not been observed directly. Coercivity can only be indirectly related to the microstructure through the micromagnetic model and recently developed micromagnetic numerical simulations. The establishment of that relationship has been one of the objectives of this chapter.

REFERENCES

[Abr60] C. Abraham and A. Aharoni, *Phys. Rev.*, **120** (1960), 1576.

[Aha58] A. Aharoni and O. Shtrikman, *Phys. Rev.*, **109** (1958), 1522.

[Aha60] A. Aharoni, *Phys. Rev.*, **119** (1960), 127.

[Aha63] A. Aharoni, *Phys. Rev.*, **131** (1963), 1478.

[Aha66] A. Aharoni, *Phys. Stat. Sol.*, **16** (1966), 3.

[Aha86] A. Aharoni, *IEEE Trans. Mag.*, **MAG-22** (1986), 478.

[Ast93a] G. Asti, F. Bolzoni and R. Cabassi, *J. Appl. Phys.*, **73**, (1993), 323.

[Ast93b] G. Asti, F. Bolzoni and R. Cabassi, Exploitation Report, Contract BREU - 150 'Analysis of Coercivity and of the Microstructure of High-Tech Hard Magnetic Materials' 1993.

[Bec71] J. J. Becker, *AIP Conf. Proc.*, **5** (1971), 1067.

[Bro45] W. F. Brown Jr, *Rev. Mod. Phys.*, **17** (1945), 15.

[Bro63] W. F. Brown Jr, *Micromagnetics*, Interscience Publishers (1963) p. 68.

[Cad88] J. M. Cadogan, J. P. Gavigan, D. Givord and H. S. Li, *J. Phys. F: Met. Phys.*, **18** (1988), 779.

[Cad92] J. M. Cadogan, D. Givord, H. S. Li, M. F. Rossignol and D. W. Taylor, *Proc. 7th. Int. Symp. on Magnetic Anisotropy and Coercivity in Rare Earth–Transition Metal Alloys* (1992), 57.

[Ceb95] F. Cebollada, M. F. Rossignol, D. Givord, V. Villas-Boas and J. M. Gonzàlez, *Phys. Rev.* B **52** (1995), 13511.

[Coe88] R. Coehoorn, D. B. de Mooij, J. P. W. B. Duchateau and K. H. J. Buschow, *J. de Physique*, **49** (1988), C8-669.

[Coe89] R. Coehoorn, D. B. de Mooij and C. de Waard, *J. Magn. Magn. Mater.*, **80**, (1989), 101

[Coe90a] J. M. D. Coey and J. Sun Hong, *Magn. Magn. Mater.*, **87** (1990), L251.

[Coe90b] J. M. D. Coey, J. Sun Hong and Y. Otani, *11th Int. Workshop on Rare-Earth Magnets and Their Applications*, Pittsburgh (1990) **2**, p. 36.

[Cro84] J. J. Croat, J. F. Herbst, R. W. Lee and F. E. Pinkerton, *J. Appl. Phys.*, **55** (1984), 2078.

[Din93a] J. Ding, Y. Liu, P. G. McCormick and R. Street, *J. Magn. Magn. Mater.*, **123** (1993), L239.

[Din93b] J. Ding, P. G. McCormick and R. Street, *J. Magn. Magn. Mater.* **124** (1993), L1.

[Din94] J. Ding, Y. Liu, R. Street and P. G. McCormick, *J. Appl. Phys.*, **75** (1994), 1032.

[Dur88] K. D. Durst, H. Kronmüller and W. Ervens, *Phys. Stat. Sol.* (a), **108**, (1988), 705.

[Eck92] D. Eckert, A. Handstein, K.-H. Müller, R. Grössinger and M. Katter, 12th Int. Workshop on RE Magnets & Their Applications, Canberra, p. 180, (1992).

[Elb91] D. Elbaz, D. Givord, H. Hirosawa, F. Missell, M. F. Rossignol and V. Villas-Boas, *J. Appl. Phys.*, **69** (1991), 5492.

[Est89] Y. Estrin, P. G. McCormick and R. Street, *J. Phys. Condens. Matter* (1989).

[Fre57] E. H. Frei, S. Shtrikman and D. Treves, *Phys. Rev.*, **106** (1957), 446.

[Fri63] J. Friedel, *Electron Microscopy and Strength of Crystals*, G. Thomas and J. Washburn (eds), Interscience, New York (1963), p. 605.

[Gau76] P. Gaunt, *Phil. Mag.*, **34** (1976), 775.

[Gau77] P. Gaunt, *J. Appl. Phys.*, **48**, (1977), 3470.

[Gau83] P. Gaunt, *Phil. Mag.*, **B48** (1983), 261.

[Gau86] P. Gaunt, *J. Appl. Phys.*, **59** (1986), 4129.

[Giv87] D. Givord, A. Liénard, P. Tenaud and T. Viadieu, *J. Magn. Magn. Mater.*, **67** (1987), L281.

[Giv88a] D. Givord, P. Tenaud and T. Viadieu, *IEEE Trans. Mag.*, **24** (1988), 1921.

[Giv88b] D. Givord, P. Tenaud and T. Viadieu, *J. Magn. Magn. Mater.*, **72** (1988), 247.

[Giv90] D. Givord, Q. Lu, M. F. Rossignol, P. Tenaud and T. Viadieu, *J. Magn. Magn. Mater.*, **83**, (1990), 183.
[Giv91a] D. Givord, Q. Lu and M. F. Rossignol, *Proc. NATO Advanced Study Institute on the Science and Technology of Nanostructured Magnetic Materials*, G. C. Hadjipanayis and G. A. Prinz (eds), Plenum Press, New York (1991), p. 635.
[Giv92a] D. Givord, Q. Lu, F. P. Missell, M. F. Rossignol, D. Taylor and V. Villas-Boas, *J. Magn. Magn. Mater*, **104–107** (1992), 1129.
[Giv92b] D. Givord, M. F. Rossignol, D. W. Taylor and A. Ray, *J. Magn. Magn. Mater.*, **104–107**, (1992), 1126.
[Giv96] D. Givord, M. F. Rossignol, V. Villas-Boas and Gonzàlez J. M., 14th International Workshop on Rare-earth Magnets and their Applications, São Paulo, Brasil, September 1996.
[Gon93a] J. M. Gonzàlez, F. Cebollada and A. Hernando, *J. Appl. Phys.*, **73** (1993), 6943.
[Gon93b] J. M. Gonzàlez and F. Cebollada, Exploitation Report, Contract BREU - 150 'Analysis of Coercivity and of the Microstructure of High-Tech Hard Magnetic Materials' 1993.
[Grö90a] M. Grönefeld, Thesis, Universität von Stuttgart (1990).
[Grö90b] M. Grönefeld and H. Kronmüller, *J. Magn. Magn. Mater*, **88** (1990), L267.
[Har73] R. Harris, M. Plischke and M. J. Zuckermann, *Phys. Rev. Lett.*, **31** (1973), 160.
[Her92a] R. Hermann and K. Elk, *J. Magn. Magn. Mater*, **117** (1992), 14.
[Her92b] A. Hernando, I. Navarro and J. M. Gonzàlez, *Europhys. Lett.*, **20** (1992), 175.
[Hil75] H. R. Hilzinger and H. Kronmüller, *Phys. Lett.*, **51A** (1975), 59.
[Hil77] H. R. Hilzinger and H. Kronmüller, *Appl. Phys.*, **12** (1977), 253.
[Hir86] S. Hirosawa, Y. Matsuura, H. Yamamoto, S. Fujimura, M. Sagawa and H. Yamauchi, *J. Appl. Phys.*, **59** (1986), 873.
[Hir89] S. Hirosawa, Y. Tsubokawa and R. Shimizu, *10th Int. Workshop on Rare-Earth Magnets and Their Applications* (The Society of Non-Traditional Technology, Tokyo) (1989), p. 465.
[Hir93] S. Hirosawa, H. Kanekiyo and M. Uehara, *J. Appl. Phys.*, **73**, (1993), 6488.
[Imr75] Y. Imry and S.-K. Ma, *Phys. Rev. Lett.*, **35** (1975), 1399.
[Jif93] Hu Jifan, I. Kleinschroth, R. Reisser, H. Kronmüller and Shouzeng Zhou, *Phys. Stat. Sol.* (a), **138** (1993), 257.
[Kad86] A. M. Kadin, R. W. McCallum, G. B. Clemente and J. E. Keen, in *Science and technology of rapidly solidified alloys*, vol. 80 of *Materials Society Symposium Proceedings*, Materials Research Society, Pittsburgh, PA (1986).
[Ker43] M. Kersten, *Phys. Z.*, **44** (1943), 63.
[Kne91] E. F. Kneller, *IEEE Trans. Mag.*, **27** (1991), 3588.
[Kob95] K. Kobayashi, R. Skomski and J. M. D. Coey, *J. Alloys Compounds*, **222** (1995), 1.
[Kou94] X. C. Kou, H. Kronmüller, D. Givord and M. F. Rossignol, *Phys. Rev.*, **B50** (1994), 3849.
[Kro78] H. Kronmüller, *J. Magn. Magn. Mater.*, **7** (1978), 341.
[Kro87a] H. Kronmüller, *Phys. Stat. Sol.* (b), **144** (1987), 385.
[Kro87b] H. Kronmüller, K. D. Durst and G. Martinek, *J. Magn. Magn. Mater.*, **69**, (1987), 149.
[Kro88] H. Kronmüller, K. D. Durst, S. Hock and G. Martinek, *J. Phys. (Paris)*, **49** (1988), C8–623.
[Kro91] H. Kronmüller, *Supermagnets, hard magnetic materials*, G. J. Long and F. Grandjean (eds), Kluwer Academic Publishers (1991), p. 461.
[Kro94] H. Kronmüller and T. Schrefl, *J. Magn. Magn. Mater.*, **129** (1994), 66.
[Li87] H. S. Li, Thesis, Université de Grenoble (1987).

[Man91] A. Manaf, R. A. Buckley, H. A. Davies and M. Leonowicz, *J. Magn. Magn. Mater.*, **101** (1991), 360.

[Mar90] G. Martinek, H. Kronmüller, and S. Hirosawa, *J. Magn. Magn. Mater.*, **89**, (1990), 369.

[Mat88] F. Matsumoto, H. Sakamoto, M. Komiya and M. Fujikura, *J. Appl. Phys.*, **63** (1988), 3507.

[Mül94] K.-H. Müller, D. Eckert, P. A. P. Wendhausen, A. Handstein, S. Wirth and M. Wolf, *EMMA '93*, Kosice (1994).

[Née50] L. Néel, *J. Phys. Rad.*, **11** (1950), 49.

[Née51] L. Néel, *J. Phys. Rad.*, **12** (1951), 339.

[Née54] L. Néel, *J. Phys. Rad.*, **15**, (1954), 225.

[Noz88] J. P. Nozières, R. Perrier de la Bathie and J. Gavinet, *J. Phys. (Paris)*, **49** (1988), C8–667.

[Noz89] J. P. Nozières, R. Perrier de la Bathie and D. Taylor, *J. Magn. Magn. Mater.*, **80** (1989), 88.

[O'Do94] K. O'Donnell, C. Kuhrt and J. M. D. Coey, *J. Appl. Phys.*, **76** (1994), 7068.

[Sag87] M. Sagawa, S. Hirosawa, H. Yamamoto, S. Fujimura and Y. Matsuura, *Jap. J. of Appl. Phys.*, **26** (1987), 785.

[Sag88] M. Sagawa and S. Hirosawa, *J. Phys. (Paris)*, **49** (1988), C8–617.

[Sch91a] H. F. Schmidts and H. Kronmüller, *J. Magn. Magn. Mater.*, **94** (1991), 220.

[Sch91b] L. Schultz, K. Schnitzke, J. Wecker, M. Katter and C. Kuhrt, *J. Appl. Phys.*, **70** (1991), 6339.

[Sch91c] F. Schumacher, *J. Appl. Phys.*, **69** (1991), 2465.

[Sch93a] T. Schrefl, H. Kronmüller and J. Fidler, *J. Magn. Magn. Mater.*, **127** (1993), L273.

[Sch93b] T. Schrefl, H. F. Schmidts, J. Fidler and H. Kronmüller, *IEEE Trans. Mag.*, **29** (1993), 2878.

[Sko93] R. Skomski and J. M. D. Coey, *Phys. Rev. B*, **48** (1993), 15812.

[Sta60] T. D. Stacey, *Aust. J. Phys.*, **13** (1960), 599.

[Sto48] E. C. Stoner and E. P. Wohlfarth, *Phil. Trans. Roy. Soc. (London)*, **240A** (1948), 599.

[Str49] R. Street and J. C. Wooley, *Proc. Phys. Soc.*, **62A** (1949), 562.

[Str89] R. Street, P. G. McCormick and Y. Estrin, *10th Int. Workshop on Rare-Earth Magnets and Their Applications*, Kyoto (1989), p. 83.

[Tay92] D. W. Taylor, Thesis, INP Grenoble (1992).

[Tay94] D. W. Taylor, V. Villas-Boas, Q. Lu, M. F. Rossignol, F. P. Missell, D. Givord and S. Hirosawa, *J. Magn. Magn. Mater.*, **130** (1994), 225.

[Ten88] P. Tenaud, Thesis, Université de Grenoble (1988).

[Via88] T. Viadieu, Thesis, Université de Grenoble (1988).

[Woh84] E. P. Wohlfarth, *J. Phys. F*, **14** (1984), L155.

[Wir92] S. Wirth and V. Christoph, *J. Appl. Phys.*, **72** (1992), 5358.

[Zij82] H. Zijlstra, in *Ferromagnetic materials*, vol. 3, E. P. Wohlfarth (ed.), North-Holland (1982), p. 37.

6

Microstructure and magnetic domains

G. C. Hadjipanayis

Hard magnetic materials are characterized by large coercive fields. The progress in developing coercivity has been remarkable [Liv80]; values have increased from 20 kA/m in magnet steels to over 3 MA/m in $SmCo_5$ [Str78, Liv81] and 18 MA/m in $SmCo_{5-x}Ni_x$ at cryogenic temperatures [Oes78]. Advances have been achieved by an interplay between theoretical models, experimental measurements and microstructure studies, all aimed at understanding the mechanisms of magnetization reversal.

In a sense, rare-earth iron magnets date from the discovery of precipitation-hardened Sm(Co, Fe, Cu, $Zr)_{7-8}$ [Oji77, Mis81] with a maximum energy product over 240 kJ/m^3, although this two-phase material contains cobalt as an essential constituent. Early cobalt-free magnets were based on amorphous or microcrystalline R–Fe binary alloys [Rhy79, Cla73, Cro84, Had80, Koo81]. However, their maximum energy product did not exceed 60 kJ/m^3. The highest coercive fields and energy products ever reported on iron-based magnets have been obtained in $Nd_2Fe_{14}B$ alloys [Had84, Cro84, Sag84], with record values exceeding 2.6 MA/m and 430 kJ/m^3, respectively. Other classes of iron-based magnet have been discovered. These include the $Nd_2Fe_{14}C$ compound [Boe88, Liu86], Sm(Fe, $Ti)_{12}$ and other compounds with $ThMn_{12}$ structure [Bus88a, Zha88, Oha87, Had87a, Li91], as well as the interstitial nitrides and carbides such as $Sm_2Fe_{17}N_3$ and $Nd(Fe_{10}Mo_2)N$ [Coe90].

The high coercive fields observed in rare-earth permanent magnets are a result of high magnetocrystalline anisotropy. However, the experimentally measured coercive fields are much smaller than the anisotropy fields because of the heterogeneous microstructure of the magnets. Nucleation of reverse domains and pinning can take place at various imperfections or inhomogeneities and this leads to lower values of coercivity (Fig. 1.26).

The metallurgical microstructure which corresponds to optimum coercivity is usually obtained after complex heat treatment. In precipitation-hardened Sm(Co, Fe, Cu, $Zr)_7$, heat treatment at $\sim$ 800 °C causes the formation of a fine cellular microstructure [Mis81] that is ideal for uniform domain wall pinning [Fid82a, Had82a]. In Nd–Fe–B magnets the heat treatment at around 600 °C does not alter the morphology inside the ϕ grains but it produces subtle changes around the grain boundaries [Kno89, Ram88] which impede the nucleation and propagation of reverse domains [Had85, Dic86].

The relationship between microstructure and hard magnetic properties of Sm(Co, Fe, Cu, $Zr)_7$, $Nd_2Fe_{14}B$, R(Fe, $Ti)_{12}$ and other Fe-based magnets will form the subject of this chapter.

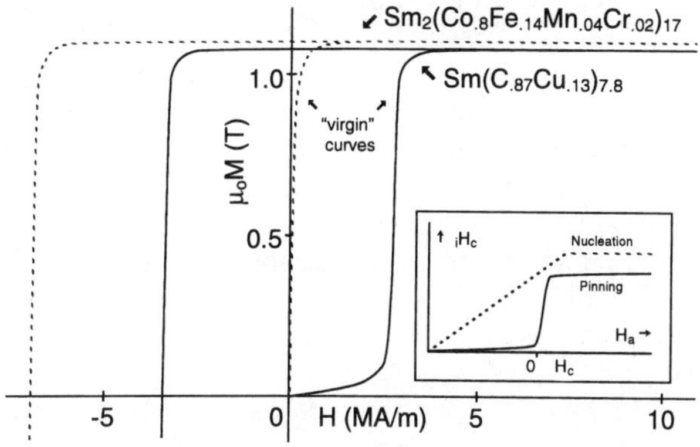

FIG. 6.1. Initial magnetization curves and field dependence of coercivity in 'nucleation' and 'pinning'-type permanent magnets [Liv80].

In Chapter 1, Section D, permanent magnets were grouped into two categories according to their virgin magnetization curve. An example is shown in Fig. 6.1. Magnets belonging in the first category initially show a steep increase in magnetization at low fields and a roughly linear increase in the coercivity H_c with applied field, reaching saturation at an internal field which is much less than H_c. Magnets of the other category initially show little change in magnetization for fields less than H_c but there is a drastic increase of magnetization for fields greater than H_c. Also, the coercivity of these magnets is almost negligible up to the critical field (H_c), above which it suddenly increases to saturation. The magnets in the first category, in which the domain walls move easily within the grains but may be pinned or nucleated near grain boundaries, are referred to as 'nucleation-type' [Kro88]. Any pins which the domain walls encounter during the initial magnetizing process are different from those encountered along the demagnetization curve. Examples of nucleation-type magnets are $SmCo_5$ and $Nd_2Fe_{14}B$. The second type of magnet, in which the domain walls are pinned inside the grains, is known as the 'pinning type' [Liv78, Had88] (pins are randomly distributed throughout the sample). In these magnets the domain walls encounter the same type of pins during the initial magnetizing process as they do during the demagnetizing process. An example of this type is the precipitation-hardened $Sm(Co, Fe, Cu, Zr)_7$ magnet.

According to the domain wall pinning models, the coercivity H_c can be determined from the following expression [Mor65] (cf. p. 235):

$$H_c = \frac{1}{2\mu_0 M_s S}\left(\frac{dU}{dx}\right)_{max} \tag{6.1}$$

where $U(x)$ is the interaction energy of the domain wall with imperfections, and it varies with position x, and S is the area of the wall pinned. Thus in order to calculate H_c, it is necessary to have an expression for the interaction energy $U(x)$.

For this, one needs to know the nature of the imperfections and how they contribute to the energy, as well as their distribution throughout the material [Gau83].

One must try to correlate the magnetic properties with the microstructure and magnetic domain structure in order to understand the origin of coercivity. Optical metallography can show the microstructure on a scale of $\geq 1\ \mu$m and domains can be revealed in polarized light using the magneto-optic Kerr effect, or by the Bitter method [Cul76a]. Scanning electron microscopy can provide information on microstructure and phase composition down to a scale of about 10 nm. But transmission electron microscope studies are unique in allowing the structure morphology, the crystal structure and the chemical composition of small regions of the samples down to the nanometer scale to be determined. Furthermore, Lorentz microscopy [Car66] allows the simultaneous observation of magnetic domains and microstructure. Knowledge of the distribution of domain walls around crystalline defects and/or phase and grain boundaries and the change of the domain pattern with the application of a magnetic field is helpful in identifying the possible pinning or nucleation mechanisms associated with high coercivities.

A. EXPERIMENTAL TECHNIQUES

1 Electron microscopy imaging and diffraction

A parallel beam of electrons accelerated by a potential V_0 is transmitted through a specimen (thin foil) and is diffracted in a number of directions by the crystal (Fig. 6.2). A typical Bragg angle for 100 keV electrons which have wavelength $\lambda = 4$pm is about 0.5°. The diffracted electron beams are brought into focus in the back focal plane of the objective lens, forming a diffraction pattern which can be magnified and studied if subsequent lenses in the microscope are arranged to focus on the rear focal plane of the objective. The diffraction pattern consists of a two-dimensional array of spots, each spot corresponding to a particular set of reflecting planes. From the diffraction pattern one can find the orientation of the foil. The objective lens also forms an image of the lower surface of the specimen in the image plane. This image can be magnified subsequently by other lenses of the instrument and is finally projected onto the fluorescent screen (Fig. 6.2).

The Hamiltonian describing the interaction of the electron beam with the crystal potential V is [Spe88]

$$\mathscr{H} = \frac{1}{2m}(\mathbf{p} + \mathrm{e}\mathbf{A})^2 - \mathrm{e}V \tag{6.2}$$

where $\mathbf{p}$ is the electron's momentum and $\mathbf{A}$ the vector potential related to the magnetic flux density, $\mathbf{B} = \nabla \times \mathbf{A}$. Schrödinger's equation for this Hamiltonian, ignoring squared terms of $\mathbf{A}$, leads to

$$\nabla^2\psi + \frac{8\pi^2 m\mathrm{e}}{h^2}(V + V_0)\psi + \frac{2\pi \mathrm{ie}}{h}\mathbf{A}\cdot\nabla\psi = 0 \tag{6.3}$$

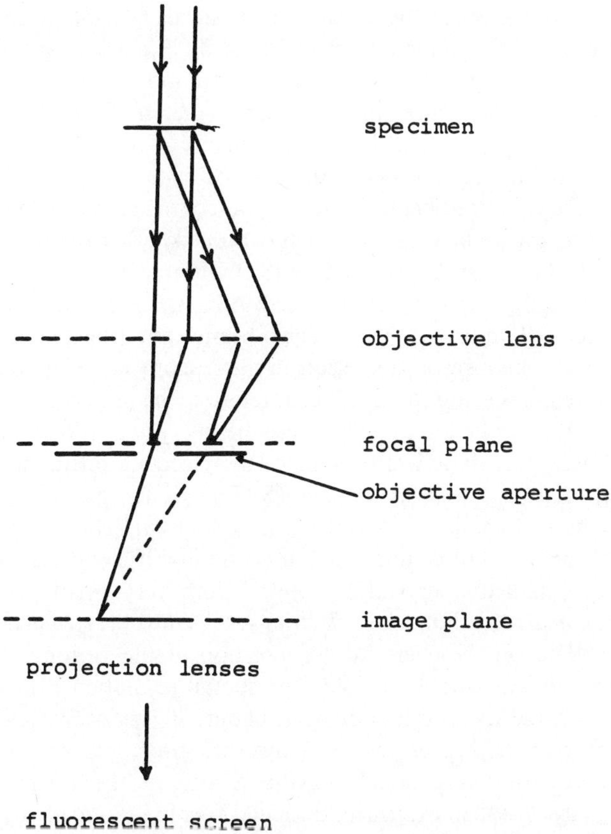

FIG. 6.2. Electron microscope image and diffraction.

where V_0 is the accelerating voltage. This equation is satisfied by the following wave function of the transmitted electrons which are actually imaged by the electron lenses in a high-resolution transmission electron microscope:

$$\psi(x, y, t) = \exp\left(\frac{i\pi}{\lambda V_0}\int_0^t V(x, y, z)\mathrm{d}z - \frac{2\pi ie}{h}\int_0^t \mathbf{A}_z(x, y, z)\mathrm{d}z\right) \qquad (6.4)$$

where λ is the electron wavelength, h is Planck's constant, and t is the thickness of the specimen. The first term in Eq. (6.4), due to the electrostatic potential in the crystal, is much larger than the second term due to magnetic flux density in the specimen. Magnetic phase shifts of up to 15 rad are common for samples of about 100 nm in thickness. The phase variation across the downstream face of the sample can be converted into intensity variation using any of the optical phase-contrast microscopy techniques. For example, in high-resolution *lattice imaging*, many diffracted beams contribute to the phase-contrast image and since the first term of Eq. (6.4) is predominant, a projection of crystal potential is observed (fine lattice periodicity). On the other hand, if a small objective aperture is used to select only the central beam, the effect of the crystal potential can be suppressed, leaving only the effect of the second, magnetic term of Eq. (6.4).

It can be seen from Fig. 6.2 that the transmitted and diffracted beams from the same area of the specimen meet in the image plane. If the diffracted beams are allowed to contribute to the final image, the quality of the image is poor, mainly because of lens aberrations. However, this can be avoided if only the transmitted beam or one of the diffracted beams is allowed to contribute to the final image. This is achieved by using a small aperture which can block either the transmitted or the diffracted rays. If the aperture is centred on the main transmitted beam blocking the diffracted rays a *bright field image* is produced. A *dark field* picture is obtained when the aperture is centred on one of the diffracted beams, thus blocking the direct beam.

The techniques of high-resolution electron microscopy [Tho79] and high-spatial-resolution microanalysis are used to characterize the microstructure at a line resolution of 0.14 nm or better, and the microchemistry at approximately the 20 nm level. Crystallographic data are obtained by electron diffraction techniques, including both *selected area diffraction* (SAD), which provides high angular resolution but limited spatial resolution, and *convergent beam electron diffraction* (CBED), which provides three-dimensional crystallographic information, accurate lattice parameter data and crystal symmetry from very small volumes. *Energy dispersive X-ray analysis* (EDS) and *electron energy loss spectrometry* (EELS) are used to obtain data on the chemical composition of the various phases and the phase boundary regions where possible. The spatial resolution is probably limited to about 30 nm at 120 kV, but is somewhat better, 20 nm, at 300 kV due to lower beam broadening at the higher energy. However, sometimes, the microstructural regions of interest are very small, having a size in the range of 1–10 nm. Quantitative analysis is then extremely difficult. Techniques involving small, bright electron sources, higher accelerating voltages and improved specimens have been developed to obtain the required microanalysis from very small regions [Lor84].

2 Magnetic domain studies

2.1 *Electron-optical techniques*

The electron-optical methods used to image magnetic domains, collectively known as Lorentz microscopy, are based on the fact that the electron beam is deflected because of stray fields outside the sample or magnetic induction inside the sample. On passing through a magnetized sample (Fig. 6.3), the electron beam is deflected by the magnetic flux intensity **B** of the sample due to the Lorentz force **F**:

$$\mathbf{F} = -e\mathbf{v} \times \mathbf{B} \tag{6.5}$$

where **v** is the velocity of electrons. This causes an acceleration in the x-direction given by

$$\ddot{x} = \frac{e\mathbf{v}\mathbf{B}}{m} \tag{6.6}$$

Since the deflection is small, one can treat v as constant and show that for the geometry of Fig. 6.3 the deflection ϕ is given by

$$\phi = \frac{\dot{x}}{v} = \frac{e\mathbf{B}t}{mv} = \frac{e\lambda t\mathbf{B}}{h} \tag{6.7}$$

These deflections are quite small. For example in a 24 nm foil of cobalt where $B_s = 1.8$ T, $\phi < 0.02°$ at 100 keV. Such a small deflection has little effect on the standard in-focus image but can lead to important contrast effects on the 'magnetic' image and on the diffraction patterns.

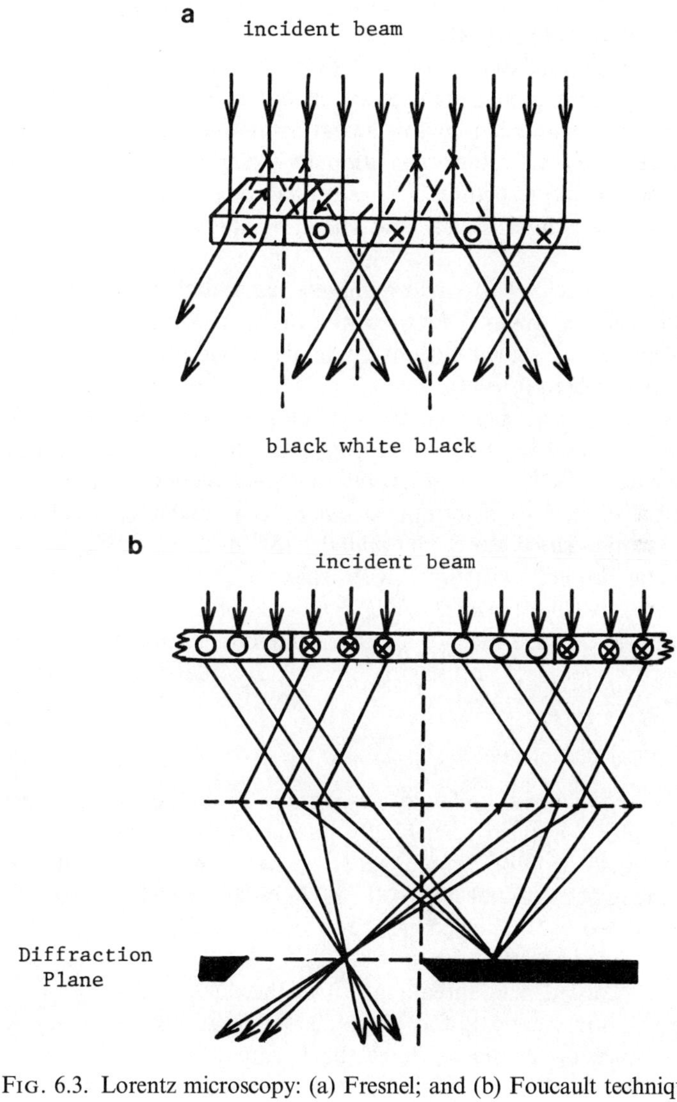

FIG. 6.3. Lorentz microscopy: (a) Fresnel; and (b) Foucault techniques.

There are two techniques used in observing the magnetic domain structure [Car66]; the *Fresnel* method where domain walls are imaged and the *Foucault* method where the magnetic domains themselves are imaged. In the Fresnel technique (Fig. 6.3a) the incident electron beam is deflected by the horizontal component of the magnetization parallel to the specimen plane. The transmitted beam will therefore diverge or converge according to the direction of the local magnetization vector. To observe the domain walls it is necessary to defocus or overfocus the objective lens so that a plane above or below the specimen is in focus. Because of the electron distribution below the specimen, the domain walls appear as alternate dark and bright lines. The difference between the underfocused and overfocused images is that the line contrast reverses as the image plane is moved from one side of focus to the other.

The Foucault technique makes use of the fact that the diffraction spots are split. The objective aperture can be used to block out one of the split central beams (Fig. 6.3b) and the magnetic domains appear as alternate light and dark bands.

The high magnetic field of the conventional electron objective lens (~ 1 T) tends to eliminate the domain structure. This effect can be minimized by:

(i) Switching off the objective lens and operating in lower magnification using the other lenses.

(ii) Raising the specimen by a few millimetres and therefore sacrificing some resolution. (Sometimes this is done by tilting the specimen.)

(iii) Using specially designed objective pole pieces which create only a small field at the specimen position [Tsu83].

Some new techniques have also been developed to image magnetic domains. *Differential phase contrast* (DPC) microscopy, together with scanning transmission electron microscopy (STEM) with double or quadrant detectors can provide the high resolution needed to determine domain wall thickness [Cha92]. *Scanning electron microscopy with polarization analysis* (SEMPA) [Ung86] can be used to image magnetic domain structures with submicron resolution. More recently, *magnetic force microscopy* (MFM), which is a variation of atomic force microscopy, has been found to provide high resolution, three-dimensional images of magnetic fields [Har94] in a non-contact imaging mode with a magnetic cantilever.

2.2 *Optical microscopy techniques*

In the *Bitter technique* a colloidal suspension of extremely fine particles of magnetite, Fe_3O_4, is applied to a highly polished and strain-free surface of the specimen [Cul72]. A drop of commercial ferrofluid is often used. The particles are attracted to the regions of high magnetic field gradient which appear near domain walls. Imagine a 180° wall intersecting the surface as in Fig. 6.4a where the spins in the wall are represented simply by the one at the center normal to the surface. This gives rise to a localized field spreading out in the directions shown in the figure. Using an optical microscope, in the 'bright field' conditions (Fig. 6.4b) the domain walls will appear as dark lines on a light background and vice versa for the 'dark field' conditions (Fig. 6.4c).

The *Kerr effect* is the rotation of polarization of a light beam when it is reflected from the surface of a magnetized sample (Fig. 6.5). The Kerr effect can be observed in three different forms depending on the direction of magnetization with respect to the plane of incidence and the reflecting surface plane (Fig. 6.5a) [Car66]. In the *polar effect* the magnetization is along the *z*-axis (as in uniaxial materials) and this produces the greatest rotation of the plane of polarization and is non-zero at normal incidence. In the *longitudinal effect*, the magnetization lies in the surface plane and in the plane of incidence *zx*. This effect is much smaller than the polar one and is found to vary with the angle of incidence, being zero at normal incidence and maximum at around 60 °C. In the *transverse effect*, the magnetization also lies in the surface plane and along the *zy*-plane, which is perpendicular to the incident plane *zx*. For this configuration there is no rotation in the plane of polarization and the effect is due to changes in the reflection coefficient of surface for light polarized along the incident plane.

A schematic diagram of a microscope for the polar Kerr effect is shown in Fig. 6.5b. The incident light is first polarized and after reflection from the magnetic specimen it passes through an analyser and then into a low-power microscope. For a specimen with two oppositely magnetized domains (Fig. 6.5c), the plane of polarization of the reflected beams is rotated in opposite senses [Cul72]. The analyser is rotated until it is crossed with one of the reflected beams, which is then extinguished and the domain appears dark, while the other beam goes through and the domain remains bright. To improve the contrast a thin layer of ZnS is

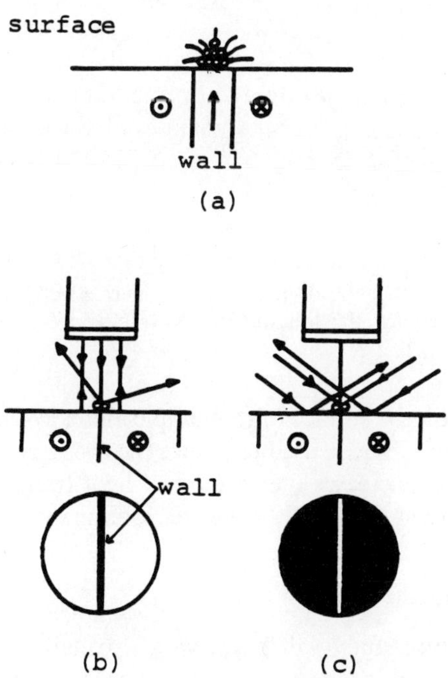

FIG. 6.4. Observation of domain walls by the Bitter method [Cul72].

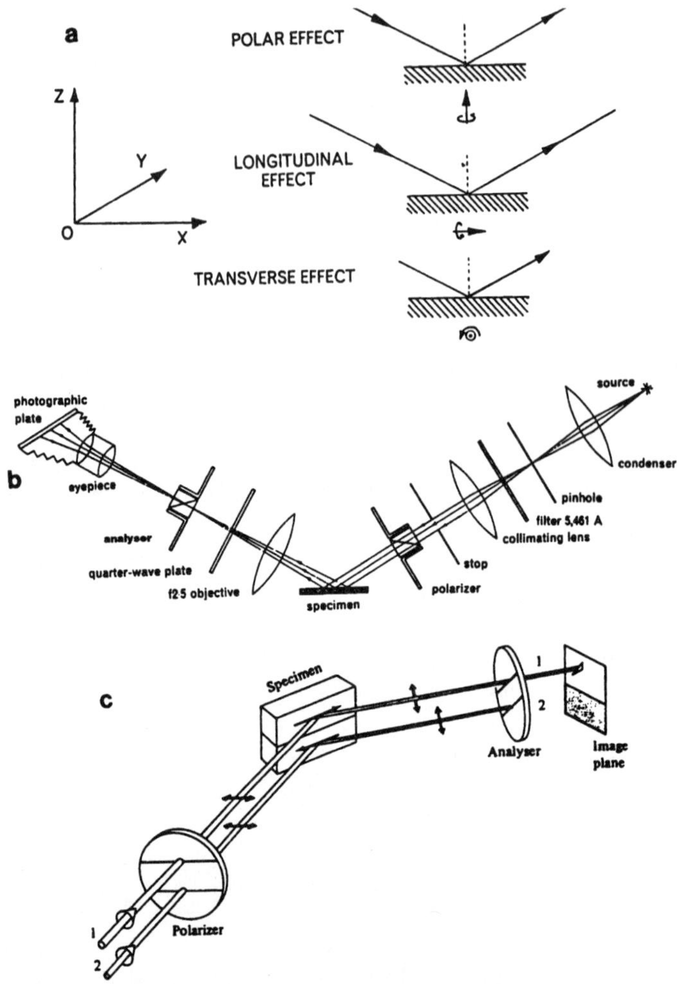

FIG. 6.5. Observation of magnetic domains by the Kerr effect [Car66]: (a) three types of Kerr effect; (b) apparatus for the longitudinal Kerr effect; (c) contrast from oppositely magnetized domains [Cul72].

sometimes deposited on the surface of the sample and a $\lambda/4$ plate is inserted in the reflected beam. Both of these are used to reduce the ordinary metallic reflection and its ellipticity. The latter increases the amount of light transmitted by the analyser and obscures the contrast due to the magnetic rotation.

3 A typical microstructure study

A typical microstructure study will begin with a metallographic examination of polished surfaces in the optical microscope. Specimens are selected and prepared as thin foils for TEM. A bright field image of the specimen under two beam conditions

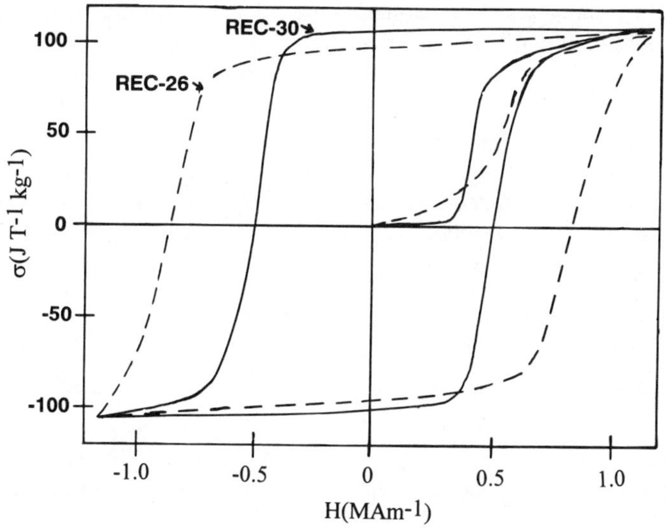

FIG. 6.6. Initial magnetization curves and hysteresis loops in precipitation-hardened magnets.

is obtained and regions or defects of interest are identified. Selected area diffraction combined with dark field microscopy provide information about the crystallographic orientation and the crystal structures of the phases present. Convergent beam electron diffraction provides high-spatial-resolution crystallographic data from regions as small as 2–5 nm. Energy dispersive X-ray analysis and electron energy loss analysis is then used to characterize the chemical composition of the various phases observed. High-resolution lattice and structural imaging of the interfaces between different phases completes the microstructural characterization. Lorentz microscopy of that region would then determine the distribution of domain walls in the area of interest. Changes in the domain wall pattern as the sample is subjected to successively larger magnetizing fields would determine the ability of the defect to impede domain wall motion and thereby determine the contribution of that defect structure to coercivity. In this fashion, material characteristics are directly related to magnetic properties which can help in evaluating theoretical models and optimizing magnetic properties.

B. STUDIES OF SPECIFIC MAGNETS

1 Precipitation-hardened rare-earth magnets

1.1 *Magnetic properties*

The hard magnetic properties of precipitation-hardened 2:17 magnets are achieved through the following sequence of heat-treatments [Oji77, Mis81]. Sintered samples are first solution heat-treated at around 1200 °C and then quenched to room temperature. The high coercivities are developed after an ageing heat-treatment at around 800 °C followed by a slow cool to 400 °C (or step ageing from 800°C to 400 °C) and then quenching to room temperature. Figure 6.6 shows the initial magnetization curves and hysteresis loops for two of these magnets. The early magnets had a lower coercivity (≈ 0.5 MA/m) and an energy product around 240 kJ/m^3 [Oji77]. Later magnets had coercivities up to 1.6 MA/m but lower energy products [Mis81, Nez79]. The later magnets were annealed at 830 °C instead of 800 °C and because of this they had a coarser cellular microstructure (see below).

1.2 *Microstructure and domains*

Transmission electron microscope studies [Fid82b, Had82b] revealed a fine cellular microstructure in the 2:17 magnets with lower coercivity (Fig. 6.7a). The cells (~ 50 nm) have 2:17 rhombohedral structure and they are surrounded by thin (4–10 nm) coherent walls with 1:5 hexagonal structure. The cell interiors consist of thin platelets which are twin-oriented (Fig. 6.7b).

In the higher coercivity 2:17 magnets the cells are apparently much coarser (~ 100 nm) (Fig. 6.8) and thin lamellae (≈ 3 nm) appear superimposed on the cellular microstructure. The lamellae have a hexagonal structure and lie along (001) planes. This structure may be of the 1:3 [Rab82] or 2:17 [Fid83] type. EDS analysis showed more Zr in the thin lamellae while more Fe was observed at cell interiors and more Cu in the cell boundaries [Had82b, Rab82, Fid83].

The cellular and lamellae microstructures develop during the 800 °C heat-treatment and do not change in the subsequent cooling. This is in contrast with the magnetic properties which change drastically during the slow cool to 400 °C. Energy dispersive X-ray analysis studies showed an increase in the Cu content of the cell boundaries during the slow cooling [Had82c]. Such an increase in Cu causes a dilution of the magnetic properties of the 1:5 phase and leads to domain wall pinning at the cell boundaries. The general metallurgical behaviour of these magnets has been explained by a model proposed by Ray [Ray84] based on the Sm–Co phase diagram and the individual interactions of Fe, Cu and Zr with $SmCo_5$ and Sm_2Co_{17}.

The magnetic domains of thermally demagnetized samples imaged by Lorentz microscopy showed a wavy nature with domain walls following the cell boundaries [Fid82b, Had82a] (Fig. 6.9). The 'wavy' nature of domain walls does not change after the application of a magnetic field, indicating that they are pinned at the cell boundaries. The domain walls are not affected by the presence of the thin lamellae. The size of magnetic domains appears to be larger in magnets with lower H_c.

1.3 *Magnetic hardening*

Correlation of the magnetic and microstructural studies of these magnets indicates pinning-type behaviour in the lower-coercivity magnets and mixed pinning and nucleation behaviour in the high-coercivity magnets [Had88b].

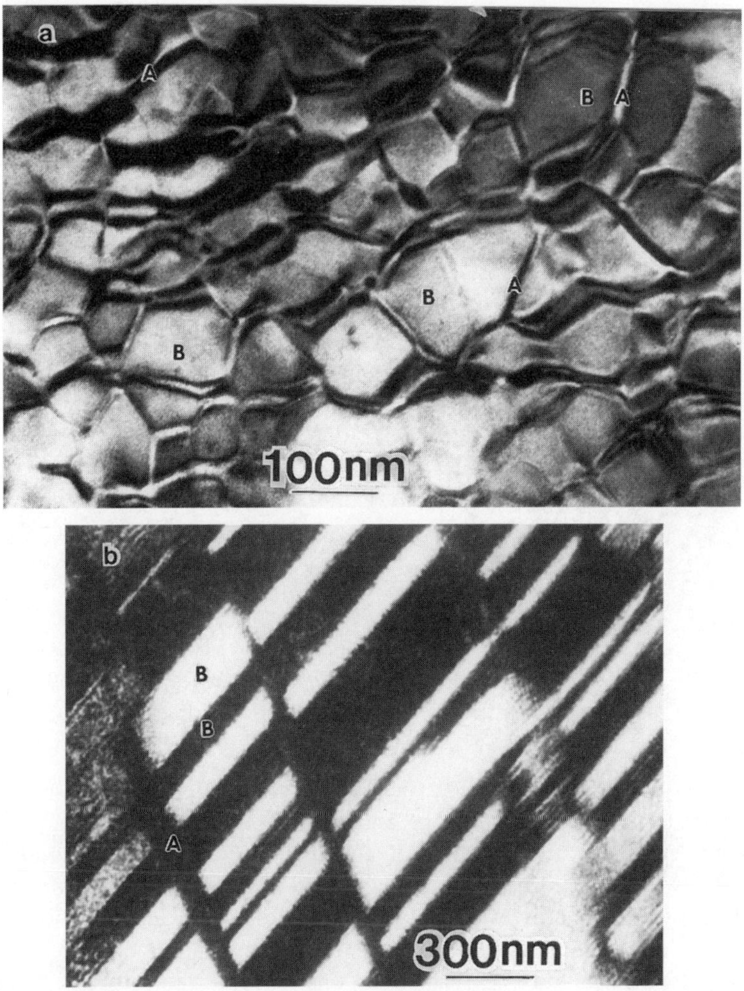

FIG. 6.7. (a) Cellular microstructure in precipitation-hardened magnets (A ≡ 1 : 5 phase, B ≡ 2 : 17 phase). (b) Twin platelets inside the 2:17 cells.

A rough estimate of the coercivity can be obtained from Eq. (6.1) considering the gradient in domain wall energy that exists across the cell boundaries using the expression [Liv77]

$$H_c = \frac{1}{2\mu_0 M_s}\left\{\frac{\Delta\gamma_w}{\delta_w}\right\} \qquad (6.8)$$

where δ_w is the domain wall thickness. However, the coercivity of these materials is found to depend strongly on the cell size [Had82b] (Fig. 6.10) and this cannot be explained by the model. Fidler *et al.* [Fid82b] took into account the magnetoelastic

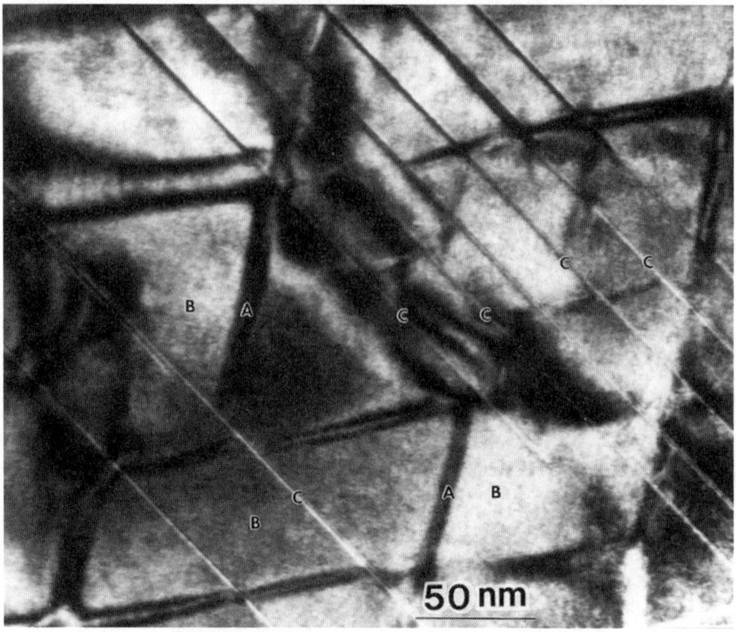

FIG. 6.8. Cellular and lamellae microstructure in 2:17 magnets with high coercivity (A ≡ 1 : 5 phase, B ≡ 2 : 17 phase, C ≡ lamellae).

coupling energy between magnetostrictive stresses of domain walls and lattice deformation strains and obtained good agreement with the experimentally obtained coercivities. More elaborate models of coercivity were proposed that take into account the 'wavy' nature of domain walls. Paul [Pau79] has treated the case of a 180° domain wall interacting with a planar defect and calculated the coercivity as a function of defect size and the dimensionless constants, $E = A_2K_2/A_1K_1$ and $F = A_2M_2/A_1M_1$, where A is the exchange constant, K the anisotropy constant and the subscripts 1 and 2 refer to the matrix and defect respectively. Using this model a correct order of magnitude of coercivity was found for many materials (with low and high coercivity) whose magnetic properties are controlled by domain wall pinning.

2 Rare-earth iron magnets

2.1 *Amorphous R–Fe alloys*

Amorphous rare-earth transition-metal alloys containing non-S-state ions show unusual magnetic hysteresis behaviour [Rhy79], which is due to the strong crystal-field interaction of the rare-earth ions in their locally distorted environments. At temperatures well below their magnetic ordering temperature, they develop a strong local magnetic anisotropy which impedes magnetic saturation and may lead to high

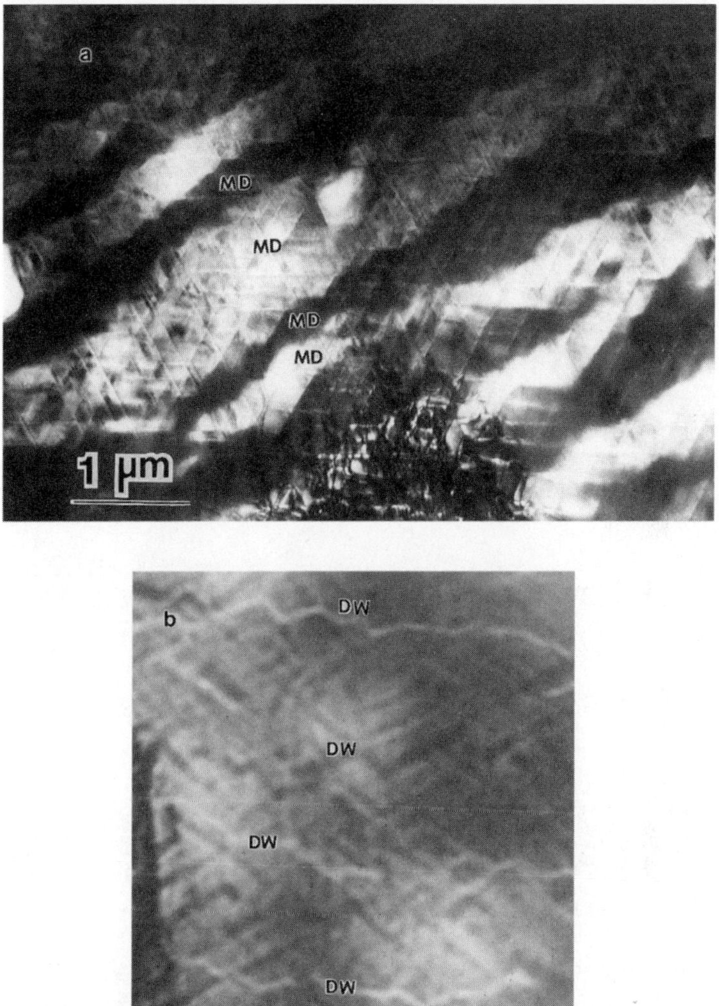

FIG. 6.9. (a) Magnetic domains (MD); and (b) domain walls (DW) in 2:17 magnets.

coercive fields, especially at cryogenic temperatures. Around the coercive field, sharp magnetization jumps are observed, together with a strong magnetic aftereffect indicating a thermal activation process.

Several approaches have been proposed to explain the magnetic properties of these alloys [Mor84]. The model Hamiltonian of Harris, Plischke and Zuckermann [Har73]:

$$\mathcal{H} = -\sum_{i>j} J\mathbf{J}_i \cdot \mathbf{J}_j - \sum_i D(\hat{\mathbf{n}}_i \cdot \mathbf{J}_i)^2 - \sum_i g\mu_B J_{iz} B_o \qquad (6.9)$$

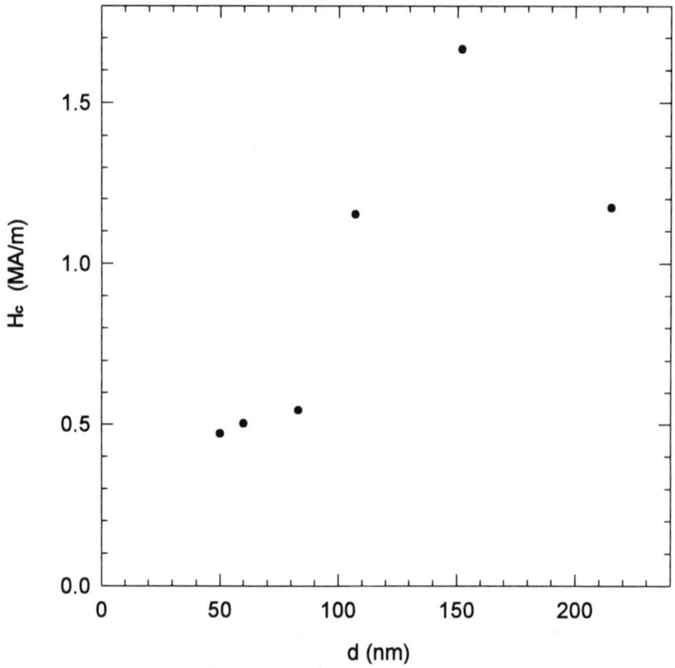

FIG. 6.10. Dependence of coercivity on cell size in 2:17 magnets.

where the sum is over all rare-earth ions provides essential insight. The first is an exchange term, with an exchange interaction J that is taken as ferromagnetic for nearest neighbours. The second term represents the random magnetic anisotropy, where $\hat{\mathbf{n}}_i$ is the local anisotropy axis. The third term is the Zeeman interaction with an external field B_e. Results are conveniently expressed in terms of α, the ratio D/J of anisotropy and exchange energy per ion. When $\alpha \sim 0$, the ground state is practically ferromagnetic and the remanence is close to the collinear saturation magnetization, but as α increases, m_r, the ratio of remanence to saturation magnetization, falls to a value of one-half as the rare-earth moments are pinned to their local easy axes (Fig. 6.11). Computer simulations where the individual spins can reorient to the position of lowest energy show a coercivity which increases abruptly when $\alpha > 0.4$ to a value of $0.2J/g\mu_B\mu_0$ [Chi77]. A molecular field treatment of Eq. (6.9) also finds a threshold for the appearance of coercivity [Cal77], but the coercivity increases linearly with α when only small reorientations of individuals spins are allowed.

In practice, the large coercivities which appear at low temperature ($H_c \geq 3$ MA/m) are consistent with model predictions. However, the magnetic domain structure of these systems is not yet well understood. Lorentz microscopy studies at room temperature show the presence of domain walls in Nd–Fe but not in Tb–Fe amorphous alloys. Anisotropy is stronger in the latter case with room-

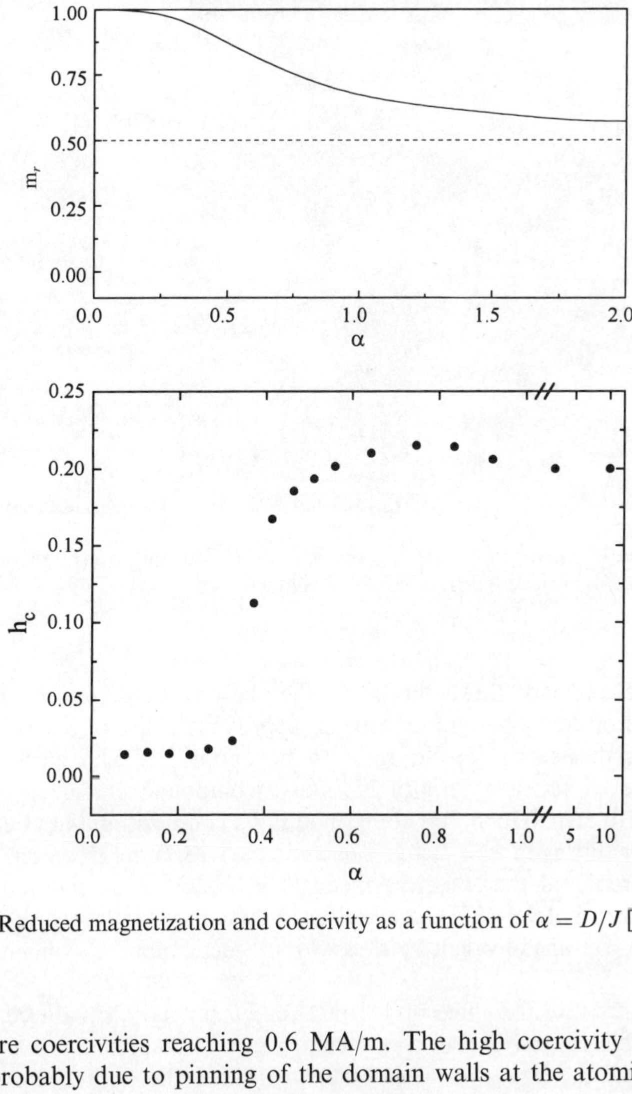

FIG. 6.11. Reduced magnetization and coercivity as a function of $\alpha = D/J$ [Moo84, Chi77].

temperature coercivities reaching 0.6 MA/m. The high coercivity in the former alloys is probably due to pinning of the domain walls at the atomic level due to fluctuations in exchange and magnetic anisotropy.

2.2 *Bulk R–Fe alloys*

Sixty years ago, Drozzina and Janus (1935) were the first to report coercivity in as-cast binary Nd-rich iron alloys. Moderately high coercivities ≈ 0.4 MA/m can be obtained in as-cast Nd–Fe and Pr–Fe alloys near the eutectic composition $R \approx 70$ at.% [Sch78, Hu88, Had88a]. The coercivity has been attributed to the presence of a highly anisotropic phase with a Curie temperature around 260 °C. The coercivity can be explained by the classical model of single-domain particles distributed in a

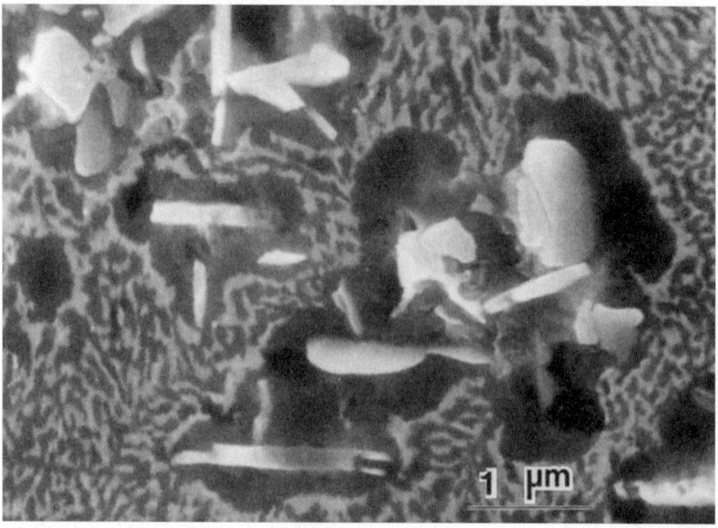

FIG. 6.12. (a) SEM micrograph showing Fe-rich Nd–Fe precipitates (bright regions) in a nonmagnetic matrix in bulk Nd-rich Nd–Fe–B alloys.

nonmagnetic Nd matrix. SEM data show the phase to be present in the form of small spherical or elongated grains with an approximate size of 5 μm (Fig. 6.12). EDS studies indicate the Fe–Nd ratio to be about 3:1. The hard phase was originally believed to be a ternary Nd–Fe–O compound stabilized by oxygen. According to Hu *et al.* [Hu88] this compound has a composition $Nd_{29}Fe_{60}O_5$ and a hexagonal structure with $a = 2.2021$ nm and $c = 1.235$ nm. However, Schneider *et al.* [Sch89] reported that this is a metastable Nd–Fe phase referred to as A_1, whose structure is not yet well characterized. Mössbauer studies [Had89a] showed the presence of four inequivalent Fe sites with magnetic moments ranging from 1.7 to 2.5 μ_B.

Similar phases are also present in melt-spun ribbons. Crystallized $Nd_{40}Fe_{60}$ samples show high coercivities at room temperature corresponding to a phase with $T_c \approx 230$ °C. In the past, Croat [Cro82] attributed the high coercivity to an unknown metastable phase, which may be the A_1 phase reported by Schneider *et al.* [Sch89a]. Stadelmaier *et al.* [Sta91] claimed that this phase is the disordered $NdFe_2$ phase which transforms to the Nd_5Fe_{17} phase [Mor90] upon annealing above 750 K, resulting in a loss of coercivity. Other studies [Str89] have shown, by thermomagnetic and transmission electron microscopy measurements, the presence of two new phases in heat-treated melt-spun samples. One of the phases has 85 at.% Fe and a hexagonal structure similar to that reported by Hu *et al.*, with $T_c \approx 250$ °C. The other phase has 70 at.% Fe and a tetragonal structure similar to $Nd_2Fe_{14}B$ with $T_c \sim 340$ °C. It has been suggested by Schneider *et al.* [Sch88] that there are two Nd–Fe–O phases which are ferromagnetic and magnetically hard.

3 Rare-earth–iron–boron magnets

3.1 *Magnetic properties*

Excellent hysteresis properties are found in Nd–Fe–B magnets made by a variety of methods. Figure 6.13 compares the initial magnetization curves and hysteresis loops for sintered and melt-spun magnets. The steep increase in magnetization at low field in sintered magnets is characteristic of a nucleation-type magnet. The behaviour of the ribbon samples is different, with the magnetization showing a slower but steady increase with field. Similar behaviour has been observed in isotropic magnets and is due both to the random orientation of grains and the small grain size. The high coercivity and remanence of Nd–Fe–B magnets are due to the tetragonal $R_2Fe_{14}B$ (ϕ) grains [Her84] with very high anisotropy field (≈ 6 MA/m). The highest coercivities in Nd–Fe–B magnets are reported in melt-spun samples [Had84, Cro84, Sag84] with values exceeding 1.2 MA/m. The values obtained in melt-spun R–Fe–B where R = Dy or Tb can be five times higher [Pin86a], but the magnetization is greatly reduced because of the ferrimagnetic structure of $R_2Fe_{14}B$ for the heavy rare-earths. The coercivity of optimally sintered Nd–Fe–B magnets is typically 1 MA/m but this can be increased by Dy or Tb substitution. Coercivity was also obtained in powdered samples [Gud86, Chr87], magnets made by liquid dynamic compaction [Chi86], thin films [Str88a, Cad86] and even in as-

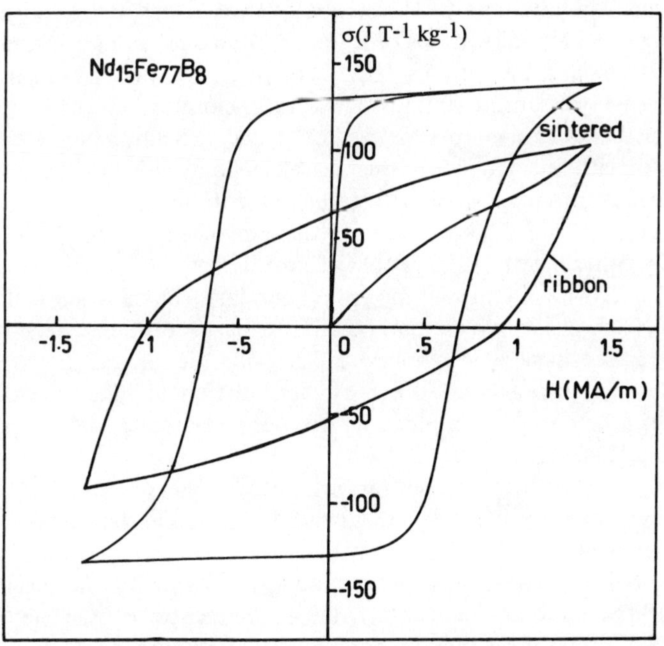

Fig. 6.13. Initial magnetization curves and hysteresis loops of melt-spun and sintered Nd–Fe–B magnets.

cast samples [Shi88, Had89b]. Record values of energy product up to 430 kJ/m^3 have been obtained in sintered Nd–Fe–B magnets. The energy product of ribbons is much lower (96 kJ/m^3) because the samples are usually isotropic.

3.2 *Microstructure studies*

Typical microstructures for sintered and melt-spun magnets are shown in Fig. 6.14. The major difference found between these magnets is the grain size [Had86a], which is much smaller in the ribbon samples where it is in the range 30–400 nm. Several phases are observed [Had85, Had87, Fid87, Fid89] in magnets with composition $R_{15}Fe_{77}B_8$ where R = Pr, Nd. The majority phase is the $R_2Fe_{14}B$ (ϕ) phase, which is usually free of any defects such as dislocations or stacking faults (Fig. 6.15). The first minority phase is the Nd-rich phase, having an fcc structure with $a = 0.52$ nm. Several Nd-rich regions (Fig. 6.16) with different Nd/Fe ratios have been observed [Fid89] including hexagonal NdO and $Nd_5Fe_2B_6O$. Other phases observed in smaller amounts are the oxide Nd_2O_3, $R_{1+\epsilon}Fe_4B_4$ (η) (Fig. 6.17) and α–Fe. The $R_{1+\epsilon}Fe_4B_4$ phase has a one-dimensional incommensurate tetragonal structure with $a = 0.71$ nm and $c = 14.46$ nm [Bez85]. Grains of this phase are characterized by a very high density of planar faults with a high density of dislocations associated with the faults. α–Fe is found in the form of small precipitates of 6 nm in size or as aggregates of larger grains about 6–10 nm in size. Other Fe-rich regions have also been observed [Fid89] with Fe/Nd ≈ 4. This phase is close to the one observed in as-cast and melt-spun Nd-rich Fe alloys [Str89]. The Nd-rich and $R_{1+\epsilon}Fe_4B_4$ phases have been observed at intersections of $R_2Fe_{14}B$ grains, but not all around the grain boundaries [Had86]. Other studies [Hir85] reported the presence of a bcc phase around the $Nd_2Fe_{14}B$ grains with an Fe:Nd ratio close to that of the ϕ phase but with a lattice parameter close to α–Fe. Magnets with low coercivity showed a microstructure with fine platelets of this phase extending from the grain boundaries into the grains. Annealing around 650 °C led to the disappearance of these platelets and enhanced coercivity. The effects of these platelets on coercivity has been challenged by other investigators [Tok86, Ram87]. Detailed microstructure studies by Ramesh *et al.* [Ram87] around the grain boundaries failed to show the presence of the bcc Fe-rich phase, which may be an artefact of ion-thinned samples. Also, amorphous regions have been reported to surround the $Nd_2Fe_{14}B$ grains in melt-spun samples by Mishra *et al.* [Mis86a], but subsequent microstructure studies showed clearly that the grain boundaries are fairly clean and free of any minority phases (Fig. 6.18).

The effect on coercivity of the various phases observed has been examined systematically in melt-spun samples [Had88a]. It was found that the Nd-rich phase plays an important rôle. In samples with stoichiometric composition $Nd_2Fe_{14}B$, little of the Nd-rich phase was present and the coercivity was approximately 0.8 MA/m. The amount of Nd-rich phase increased in samples with off-stoichiometric composition. The coercivity also increased with the amount of Nd-rich phase, from about 1.2 MA/m in $Nd_{15}Fe_{77}B_8$ to more than 2 MA/m in $Nd_{25}Fe_{67}B_8$. A greater coercivity has also been observed on B-rich melt-spun

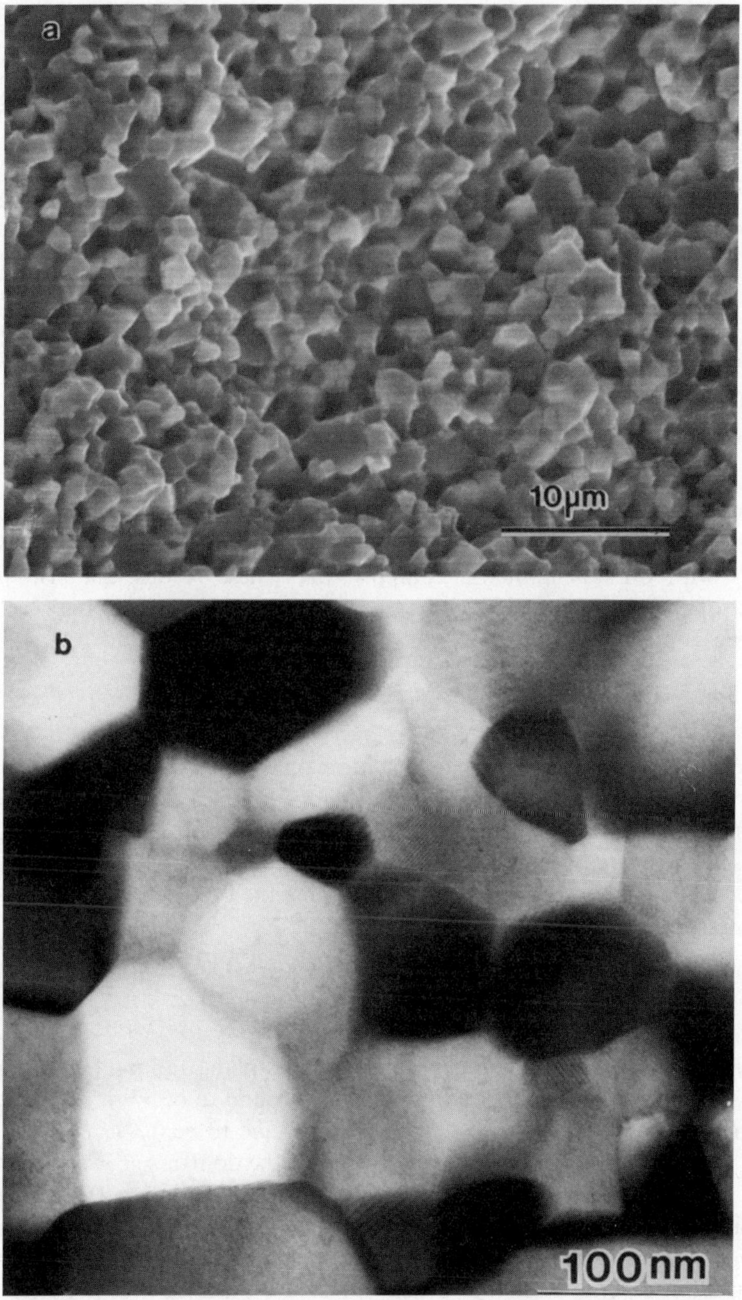

FIG. 6.14. Typical microstructures of (a) sintered; and (b) melt-spun Nd–Fe–B magnets.

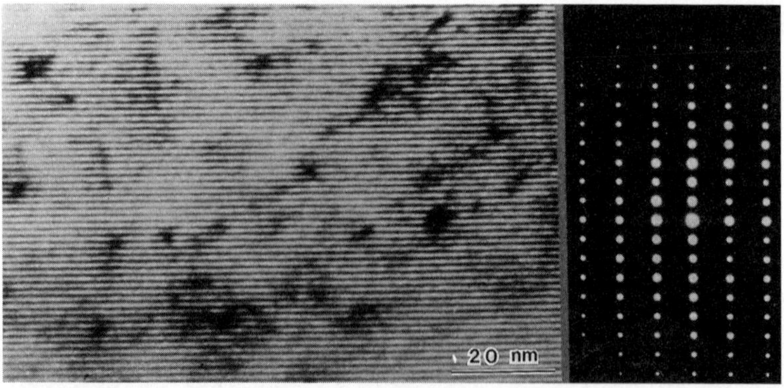

FIG. 6.15. High-resolution TEM showing lattice imaging in the $Nd_2Fe_{14}B$ phase and the corresponding electron diffraction pattern.

samples with composition $Nd_{17}B_{53}B_{30}$. However, the microstructure of these samples is quite inhomogeneous. Similar results are obtained [Sag88] on sintered magnets with off-stoichiometric compositions. These results establish that the presence of a nonmagnetic Nd-rich phase is important in obtaining high coercivities in Nd–Fe–B magnets.

3.3 *Magnetic domain studies*

3.3.1 *Kerr effect studies* The magnetic domain structure of sintered Nd–Fe–B magnets can be examined using the polar Kerr effect [Liv85; Cor89; Kro90]. Domains are observed in all the $Nd_2Fe_{14}B$ grains of thermally demagnetized samples with an average surface domain width of 0.85 μm. The Bodenberger and Hubert (1979) method can be used to determine the domain wall energy γ_w from the expression

$$\gamma_w = \frac{\mu_0 W}{\beta}\left(\frac{M_s}{4\pi}\right)^2 \tag{6.10}$$

where W is the average domain width on the basal plane and β a factor depending on the surface structure. For $Nd_2Fe_{14}B$, γ_w was found to be 35 mJ/m^2. The ‘single domain particle size’ $d_c = 18\gamma_w/\mu_0 M_s^2$ was predicted to be 0.3 μm. When a small field is applied parallel to the easy axis one set of the domains grows at the expense of the other indicating that domain walls move easily inside the grains. With increasing applied field, abrupt magnetization reversal occurs in more grains and no domain structure is observed within these grains [Sag85]. However, most of the magnetization reversal occurs at a field much less than the coercive field of bulk samples indicating that the reversal of a surface domain is easier. The nucleation of a reverse domain at the surface of sintered Nd–Fe–B is illustrated in Fig. 6.19.

In any optical or electron optical study of domains it must be borne in mind that the stray field at the surface is different from that in the bulk, so the magnitude of

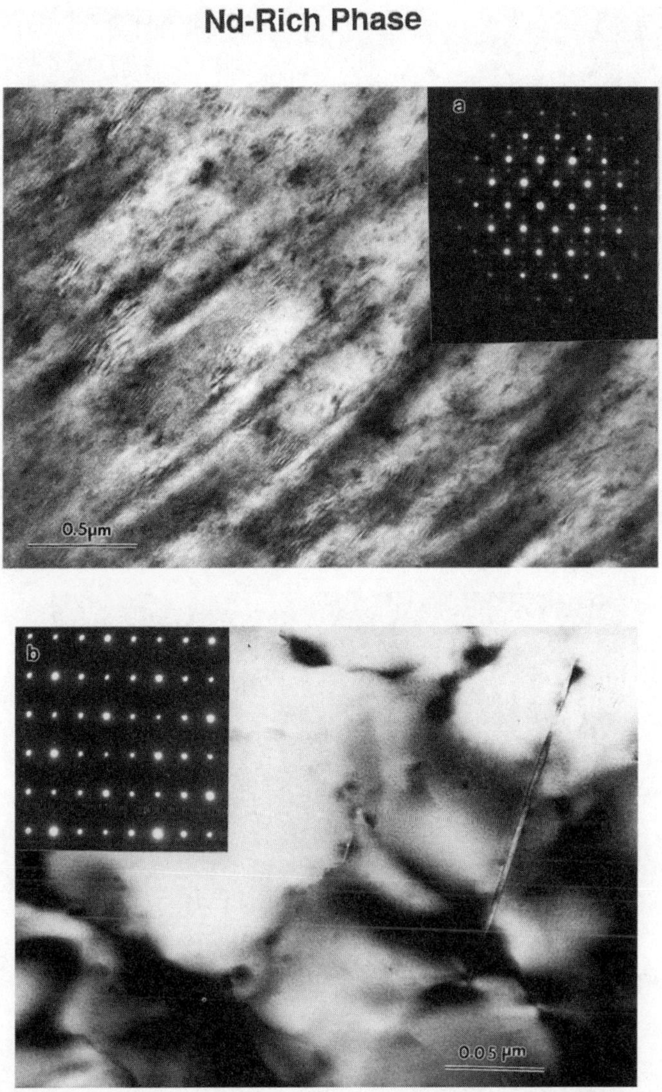

FIG. 6.16. Nd-rich phases in Nd–Fe–B magnets with corresponding electron diffraction patterns: (a) main spots fcc $a = 0.57$ nm, minor spots bcc $a = 1.15$ nm; (b) fcc $a = 0.52$ nm.

the external field required for a reversal process at the surface can be quite different from that in the bulk.

3.3.2 *Lorentz microscopy studies* Magnetic domains have been observed in as-cast [Liv84], sintered [Had87] and melt-spun magnets on both thermally demagnetized and magnetized samples [Had87c]. In sintered magnets each grain contains several

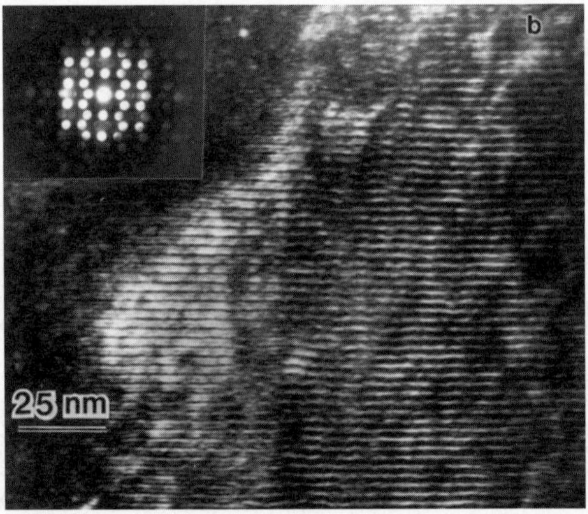

FIG. 6.17. (a) Highly faulted $Nd_{1+\epsilon}Fe_4B_4$ phase. (b) High-resolution TEM micrograph showing wavy lattice planes.

domains (Fig. 6.20a) with a size of about 0.3 μ, approximately equal to the size predicted for single domain particles [Liv85]. The domain walls in sintered magnets are found [Had85] to interact with both Nd-rich inclusions (Fig. 6.20b) and grain boundaries (Fig. 6.20c). Tilting experiments showed that the latter interaction is stronger. Studies on die-upset magnets [Mis86b] showed extended domain walls running through the grains parallel to the *c*-axis. The walls are not affected by

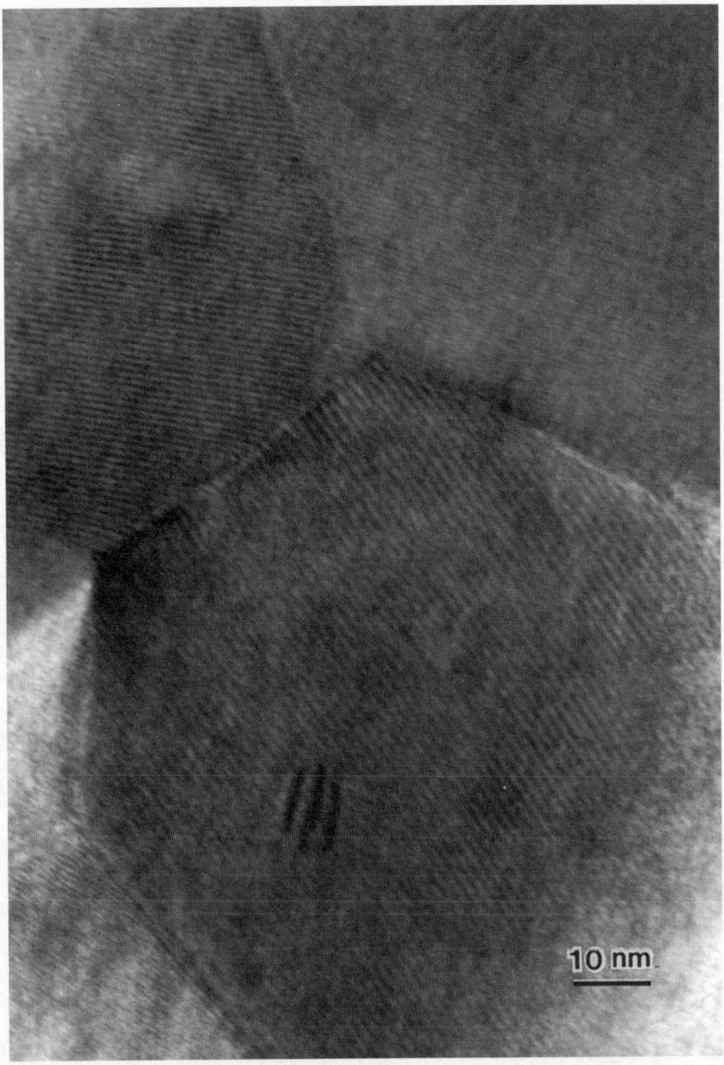

FIG. 6.18. High-resolution lattice imaging showing clear grain boundaries in melt-spun Nd–Fe–B ribbons.

crossing the grain boundaries perpendicular to the wall but they are pinned at the grain boundaries parallel to the wall.

In ribbons, the smaller grains are observed to be single domain while the larger ones may contain several domains. The domain wall contrast is shown in Fig. 6.21a. Domain walls are found around the grain boundaries indicating that they are pinned there. In addition to the conventional domains, much larger domains (Fig. 6.21b, c) containing many $R_2Fe_{14}B$ grains are sometimes observed in melt-spun samples [Had87c] with very fine grain structure (with a size below 10 nm).

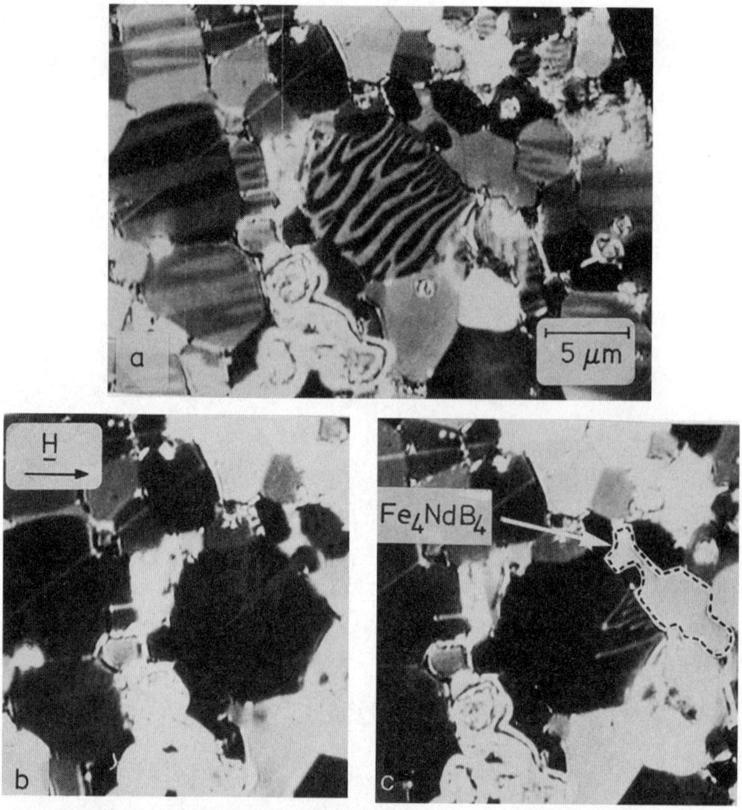

FIG. 6.19. Magnetic domains in sintered Nd–Fe–B observed by the Kerr effect [Kro90]: (a) $\mu_0 H = 0$; (b) $\mu_0 H = 1.4$ T (saturation); (c) $\mu_0 H = -0.4$ T (nucleation of reverse domains).

These domains are known are 'multigrain domains' or 'interaction domains' and are formed because of the exchange interactions (see Section 8.9) between the adjacent $R_2Fe_{14}B$ grains which are large when the surface-to-volume ratio of $R_2Fe_{14}B$ grains is large. The role of the interaction domains in the magnetization reversal mechanisms is not understood.

The results of Lorentz and Kerr microscopy studies combined with microstructure studies discussed earlier indicate that the domain walls are pinned or nucleated at grain boundaries, possibly at the disturbed layers of 2:14:1 grains neighbouring the Nd-rich grains.

3.4 *Effect of substitutions and additions on the magnetic properties and microstructure of R–Fe–B magnets*

Substitutions and additions of other elements have been made in the ternary Nd–Fe–B magnets to improve their magnetic properties. Small amounts of Co are

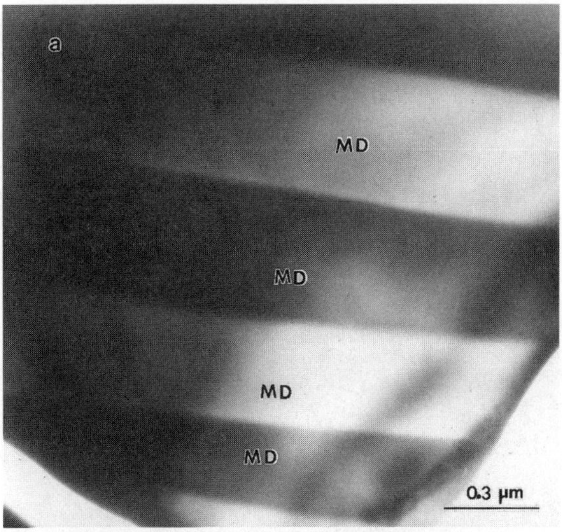

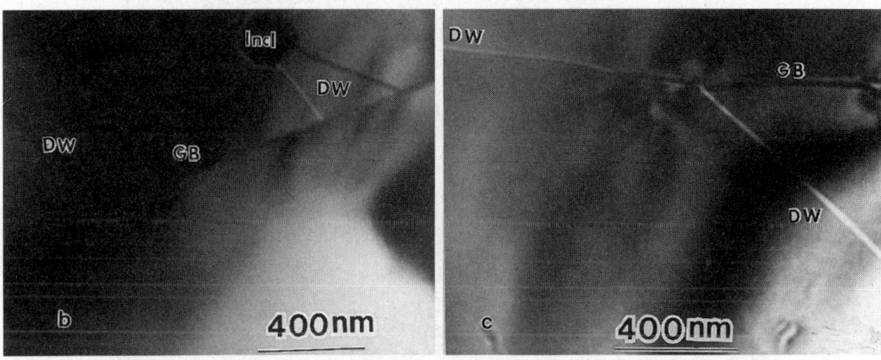

FIG. 6.20. (a) Lorentz microscopy micrographs showing (a) magnetic domains in sintered Nd–Fe–B magnets and domain walls interacting with (b) inclusions and (c) intergranular phases at grain boundaries.

used to increase the Curie temperature of the magnets. Dy and Tb substitutions are made to increase the anisotropy and coercivity of the magnets and compensate for the loss of coercivity at higher temperatures. However, because of the decrease of magnetization with Dy and Tb substitutions, other non-magnetic elements have been sought which have a similar effect on coercivity [Xia88] but without the substantial decrease in magnetization. Although their effect on H_c is not entirely understood, it is likely to be of microstructural origin because the anisotropy field of the magnets changes little [Rod88].

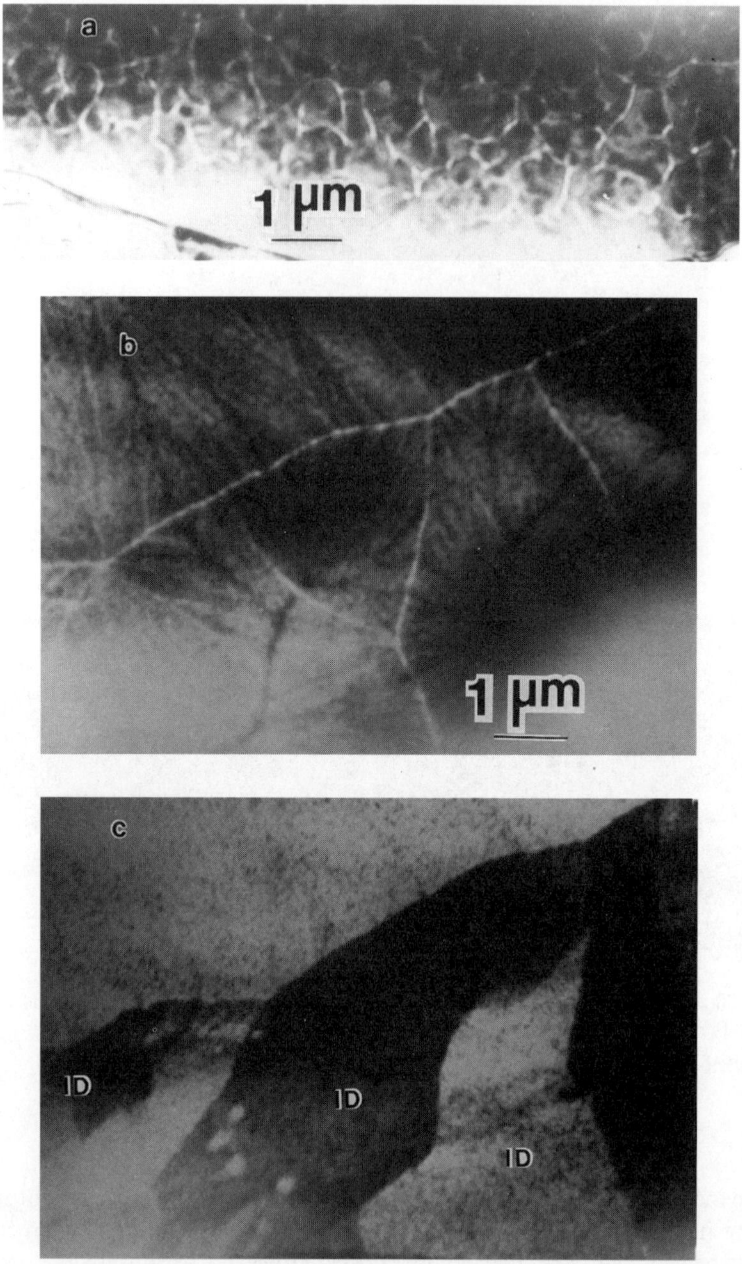

FIG. 6.21. Magnetic domains in Nd–Fe–B ribbons: (a) domain walls clustered at grain boundaries; (c) large multigrain domains (ID); and (b) corresponding domain walls.

These 'dopant' elements can be grouped into two categories [Fid92] according to their effect on microstructure. Type 1 dopants (for example, Al, Cu, Zn, Ga, Ge, Sn) which may substitute partially for Fe in the 2:14:1 phase, generally lead to a reduction in the melting point of the intergranular Nd-rich phase and to the formation of additional non-magnetic intergranular phases. Type 2 dopants (such as V, Mo, W, Nb, Ti) with a low solubility in the 2:14:1 phase are also found to form additional intergranular phases but most importantly form precipitates of binary or ternary borides within the 2:14:1 grains. In addition to enhancement in coercivity, the type 2 dopants are found to improve the corrosion resistance of Nd–Fe–B magnets. A more detailed discussion of the effects of these substitutions is given below.

3.4.1 *Co substitution*: Co substitution for Fe in R–Fe–B magnets is known [Had86c; Yam87; Gro86] to increase the Curie temperature substantially; from 310 °C in $Nd_2Fe_{14}B$ to 720 °C in $Nd_2Co_{14}B$. Because of this, the temperature coefficient of remanence is reduced [Xia88]. Even small Co substitutions extend the application of these magnets to areas where a low temperature coefficient is required. However, the coercivity of Co-substituted Nd–Fe–B samples is decreased substantially even though the anisotropy field does not change much [Yam87]. The decrease in coercivity with Co substitution has been attributed to the formation of soft magnetic phases which could serve as nucleation sites and facilitate domain wall motion. To compensate for this, small substitutions of additional elements have been used which were found to reduce the amount of such magnetic phases. According to Mizoguchi *et al.* [Miz86], a $Nd(Fe, Co)_2$ Laves phase is formed. The phase becomes nonmagnetic with Al additions and this may explain the larger coercivities in Nd–Fe(Co, Al)–B samples. However, Yamamoto *et al.* [Yam87] found that Al additions make the $Nd(Fe, Co)_2$ phase unstable, changing to Nd_3(Fe, Co, Al).

At the Co-rich end, the coercivity of Nd–Co–B samples is quite small (smaller than 50 kA/m) [Wec88], due to the rather low anisotropy field of the $Nd_2Co_{14}B$ phase (3.6 MA/m). On the contrary, the coercivity of Pr–Co–B samples is much higher ($\approx$ 1.5 MA/m) because of their much larger anisotropy field 8 MA/m, [Grö87]. Sintered Pr–Co–B magnets contain small amounts of $PrCo_5$ and $PrCo_4B$ phases in addition to the majority $Pr_2Co_{14}B$ phase [Chr89].

3.4.2 *Rare-earth substitutions*: In Chapters 1 and 2 it was shown that heavy rare-earths with negative Stevens coefficients (Tb, Dy) have a stronger crystal-field interaction than Nd but they have a ferrimagnetic structure. We noted in section D1 that melt-spun Tb–Fe–B and Dy–Fe–B exhibit enormous room temperature coercivities. It is therefore reasonable to substitute a little of the Nd by one of these elements in order to increase the anisotropy field without greatly reducing the saturation magnetization. Dy is generally used as it is the least costly of the heavy rare-earths.

The magnetic properties of $Nd_{15-x}Dy_xFe_{77}B_8$ magnets were examined by Rodewald [Rod85], who showed that the temperature coefficient of coercivity increases with the Dy concentration. A coercivity of 1.5 MA/m was obtained in $Nd_{14.3}DyFe_{76.7}B_8$, an increase of 0.5 MA/m over the Dy-free sample. Sagawa *et al.* [Sag84] also reported a coercivity of 1.6 MA/m and a saturation polarization 1.1 T in $Nd_{13.5}Dy_{1.5}Fe_{77}B_8$. Microstructural studies in Dy-substituted samples [Kii84] showed that Dy substitutes for Nd in the 2:14:1 phase and, to a minor extent, in the Nd-rich phase. Increased coercivity was also observed in sintered magnets where Dy_2O_3 was used as a sintering additive [Gha86].

In Ho-substituted Nd–Fe–B magnets the coercivity is found to increase substantially (by a factor of two) with Ho additions, even though the anisotropy field of $Ho_2Fe_{14}B$ is only slightly higher $\approx$ 6.3 MA/m) than that of $Nd_2Fe_{14}B$ (6.0 MA/m). The increase in H_c is attributed to domain wall pinning at small fcc Nd–Ho oxide inclusions which were found to exist in these magnets [Ram89].

3.4.3 *Al substitution*: Substitution of small amounts of Al for Fe in Nd–Fe–B magnets is found to decrease the magnetic moment and Curie temperature but to improve the coercivity substantially [Ma86, Hir87, Kim88]. Structural and Mössbauer studies [Yan86] showed an $8j_2$ site preference for the Al atoms. No increase in anisotropy fields was found with Al substitution, so its influence must be on the microstructure.

No definitive explanation of the effect of Al on the coercivity has yet been achieved. Fidler *et al.* [Fid89] attributed the increase of coercivity to the presence of $Nd(Fe, Al)_2$ and hcp Nd-rich phases in Al-containing magnets. Strzeszewski *et al.* [Str88b] investigated the effect of Al substitutions in $(Nd_{0.9}Dy_{0.1})_{14}Fe_{78}B_8$ and $(Nd_{0.9}Dy_{0.1})_{14}Fe_{76.6}Al_{1.4}B_8$ sintered magnets with coercivities of 0.9 and 1.4 MA/m. The anisotropy fields of both magnets, measured with the singular point detection technique, were almost the same 8.2 MA/m. Scanning transmission electron microscope studies did not show any significant differences in the structure morphology of the two samples. Nor was Al observed at grain boundaries. The only difference evident in the electron diffraction data was that the c/a ratio of the tetragonal $Nd_2Fe_{14}B$ phase was higher in the Al-containing magnets. This is consistent with preferential site occupation of the $8j_2$ sites by Al which would lead to an increase in the c lattice parameter since the atomic radius of Al is considerably higher than that of Fe. The Al substitution may then weaken the exchange interaction along the c-direction between the sheets of Fe atoms occupying all other sites (except c), thereby leading to a decrease in domain wall thickness $\delta_w = \pi\sqrt{(A/K)}$. A smaller wall thickness can lead to higher coercivities through domain wall pinning at the perturbed layers of $Nd_2Fe_{14}B$ grains close to the grain boundaries. Lorentz electron microscope studies indeed showed quite different magnetization reversals in the two samples. In the Al-containing samples the magnetization was reversed by changing the magnetization of entire grains, while for the samples without Al the magnetization was reversed by movement of domain walls inside the grains.

3.4.4 *Cu substitution*: Small Cu additions have been found to increase the coercivity of Nd–Fe–B magnets. As-cast R–Fe–B magnets do not show much coercivity; the maximum value found in bulk $Nd_{15}Fe_{77}B_8$ alloys is only 4 kA/m. Bulk $Pr_{15}Fe_{77}B_8$ alloys are similar. Shimoda *et al.* [Shi88] discovered that certain Pr–Fe–B alloys with copper additions do develop high coercivity after a heat-treatment at around 1000 °C with a maximum value of 0.6 MA/m. Similar results have been obtained in Nd–Fe–B–M alloys with M = Cu, Ag, Pb, Ga, Al [Had89]. The coercivity is further increased by Dy substitution and with an additional annealing heat-treatment at around 600 °C.

The grain structure of a typical sample is shown in Fig. 6.22. It appears that by varying the chemical composition and annealing temperature a microstructure similar to that of sintered magnets is obtained consisting of majority $R_2Fe_{14}B$ grains surrounded by a rare-earth and Cu-rich minority phase (Fig. 6.22). The addition of noble metal impurities M appears to promote the formation of an R–M-rich phase which is uniformly distributed at grain boundaries [Fid91]. The isolation of the fine 2:14:1 grains leads to an enhancement in coercivity.

The coercivity in melt-spun $Pr_{17}Fe_{78-x}Cu_xB_5$ samples increases first with Cu substitution from 1.5 MA/m in $x = 0$ to 1.8 MA/m for $x = 1.5$ and then it decreases with further Cu substitution. The increase in coercivity is believed to be related to two features observed in the microstructure studies, namely, the particle

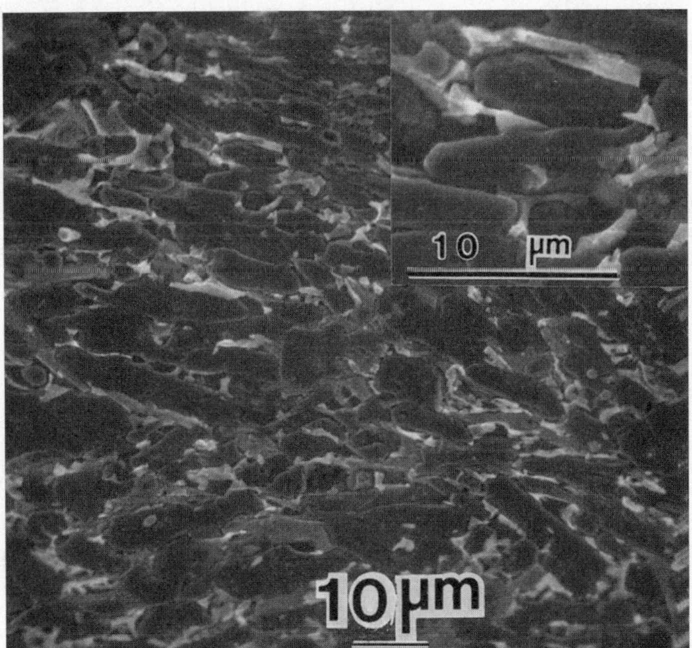

FIG. 6.22. SEM micrograph showing the grain structure of Nd–Fe–B magnets with Cu addition. Bright regions are the Nd–Cu-rich phase (insert shows the same grain structure at a higher magnification).

size and the precipitates inside the grains. Microstructure studies [Wit90a] have shown that the grain size becomes smaller with Cu additions. Also, Cu-rich precipitates were observed and the density of precipitates decreases with increasing Cu content.

3.4.5 *Si substitution*: Silicon substitution in R–Fe–B magnets was first reported by Hadjipanayis *et al.* [Had84] in a melt-spun Pr–Fe–B–Si sample. It was observed at that time that the Si-substituted samples had better squareness in their demagnetization curves. Subsequently it was found [McC87, Mat88] that small Si substitutions lead to an increased remanent magnetization and therefore a higher-energy product. For a melt-spun $Nd_{13}Fe_{79.3}Al_{0.7}B_6Si$ sample, the values of H_c and $(BH)_{max}$ reported are 1.2 MA/m and 143 kJ/m^3 respectively. This value of $(BH)_{max}$ is much higher than the normal value for isotropic samples (90 kJ/m^3) of $Nd_{13}Fe_{81}B_6$. Remanence enhancement is discussed further in Section 7. Small additions of Si area also found to increase the Curie temperature of Nd–Fe–B by 17 K [Ped86].

3.4.6 *Ga substitution*: The presence of a small amount of Ga in Nd–Fe–Co–B magnets is found to increase H_c substantially and to improve the thermal stability of the magnets considerably [End87, Tsu87]. According to Chang *et al.* [Cha89], Ga reduces the surface energy of Nd-rich liquid phase resulting in smoothed grain boundaries which affect the distribution and uniformity of the Nd-rich phase. Similar arguments have been made by Xie *et al.* [Xie88]. On the other hand, Pan *et al.* [Pan89] claimed that the high coercivity of Ga-substituted samples is due to domain wall pinning at a Ga-rich phase observed around the grain boundaries. Ga has also been found to suppress the formation of a soft Laves phase which is present in Nd–Fe–Co–B magnets [Bia89]. Similar arguments have been made by Fidler [Fid90], who found new intergranular phases containing Ga (Nd(Fe, Co, Ga) and $Nd(Fe, Co, Ga)_2$) instead of the soft ferromagnetic Co-rich Laves phase. Higher Ga substitutions (> 2 at.%) are found to decrease coercivity because of precipitation of α–Fe together with the nonmagnetic NdGa and $NdCa_2$ [Fid90] and magnetic $Nd_6Fe_{11}Ga_3$ [Li90] phases.

3.4.7 *Mo and V substitutions*: It has been found that the addition of 4 at.% V (Mo) not only increases the coercivity of R–Fe–B magnets but it also improves their corrosion behaviour substantially [Sag90, Hir90]. In particular, the coercivity of $Nd_{15}Fe_{77-x}V_xB_8$ magnets was found to increase with V up to $x = 4$ and then decrease for higher V-content. A coercivity of 1.38 MA/m was obtained in $Nd_{16}Fe_{72}V_4B_8$, which is close to that of Dy-substituted samples. The coercivity does not decrease with Co as in the case of V-free samples [Sag90]. The profound effects of V(Mo) addition on H_c lead to a significant widening of the temperature window for post-sintering heat treatment, making it easier to fabricate Nd–Dy–Fe–Co–B sintered magnets with sufficient Co content [Hir90].

It seems that the addition of V(Mo) leads to a finer grain microstructure with the $NdFe_4B_4$ phase absent. Instead, a tetragonal V_2FeB_2 or Mo_2FeB_2–boride phase is observed in addition to the $Nd_2Fe_{14}B$ and Nd-rich phases. It is believed that the V_2FeB_2 phase exists in small crystallites surrounding the $Nd_2Fe_{14}B$ grains which inhibit grain growth and lead to the observed fine microstructure.

3.4.8 *Nb, Zr substitutions*: Because of the low solubility of these elements at the sintering temperature, precipitates are formed which are rich in these elements [Pol88]. In Nb-doped magnets the $NbFe_2$ hexagonal phase ($a = 0.48$ nm; $c = 0.78$ nm) and an NbFeB phase have been reported [Par87]. Xiao *et al.* [Xia88] reported the existence of a 2:17 phase and an additional new phase at higher Nb concentrations. However, the interaction of domain walls with these precipitates is not strong enough to explain the coercivities [Ish89]. In fact the drop in coercivity at higher Nb content is attributed to these phases which presumably are magnetically soft. Small amounts of Nb and Zr are also used as grain growth inhibitors to form a fine grain microstructure that can lead to high coercivity.

3.5 *As-cast Nd-rich Nd–Fe–B alloys*

The microstructure of as-cast Nd-rich Nd–Fe–B alloys is completely different from that of alloys of approximate composition of $Nd_{16}Fe_{76}B_8$, where ϕ is the major phase [Sch89b]. A hard magnetic phase with a Curie temperature around 260 °C and a coercivity of 0.42 MA/m has been observed in an arc-melted $Nd_{80}Fe_{15}B_5$ sample. The high coercivity has been attributed to the metastable A_1 phase observed in as-cast Nd-rich Nd–Fe alloys [Sch89a]. Annealing at higher temperatures leads to a transformation of this phase to $Nd_2Fe_{14}B$, which is present in the form of small needle-shape crystallites randomly distributed in a nonmagnetic Nd-rich matrix, resulting in coercivities greater than 1.1 MA/m.

3.6 *Conclusions on magnetic hardening in Nd–Fe–B magnets*

The steep increase in the demagnetization curve of sintered magnets implies easy domain wall motion without significant pinning inside the ϕ grains. The high coercivity of the magnets can then be explained by either the nucleation of reverse domains or by domain wall pinning at grain boundaries. This interpretation is consistent with the microstructure studies which show the presence of several phases with defect-free 2:14:1 as the majority phase and the Nd-rich phase as the principal minority phase usually occupying the interstices between grains. The amount of Nd-rich intergranular phase increases with the Nd content and it is found to lead to higher coercivity. It is for this reason that the Nd–Fe–B sintered magnets are based on the composition $Nd_{15}Fe_{77}B_8$ rather than the stoichiometric $Nd_{12}Fe_{82}B_6$. Dopants are usually found to influence the wetting of the Nd-rich liquid phase and the surface smoothness of the ϕ grains. A lower dihedral angle between the ϕ grains allows more penetration of the liquid phase into the grain boundaries which then isolates the grains.

According to the nucleation model, the presence of nonmagnetic phases at the grain boundaries serves to increase the nucleation field and promotes magnetic decoupling of adjacent grains, both of which increase the coercivity. The hypothesis of decoupling of ϕ grains in Nd-rich Nd–Fe–B magnets is not in agreement with the remanence curve relationship, $M_r(H) = M_r(\infty) - 2M_d(H)$, which is expected for an assembly of noninteracting single-domain particles [Woh58] and for uniform domain wall pinning materials [McC64]. ($M_r(H)$ is the initial remanence, $M_r(\infty)$ the saturation remanence and $M_d(H)$ the demagnetization remanence.) The smallest deviations from the above relationship have been found for magnets with stoichiometric composition and the largest in Nd-richer Nd–Fe–B magnets. For the latter, this implies either strong interactions between adjacent ϕ grains or localized domain wall pinning at grain boundaries. Pinning of domain walls at grain boundaries (phases) is required to prevent complete reversal of the entire magnet from a single weak nucleation site.

The assumption in the domain wall pinning model is that domain walls are not completely removed in the available magnetic fields but, rather, they are pushed to grain boundaries, where they are pinned. It is extremely difficult to observe these domain walls because of the small width $\delta_w = 4$ nm and the strong contrast at grain boundaries. In such a case, the trapped domain walls behave like 'nuclei' and nucleation and unpinning effects are almost the same.

Domain wall pinning has been suggested [Str88] to take place at an 'epitaxial shell' of a hard magnetic phase at the surface of the ϕ grains. From the temperature dependence of coercivity and ac thermomagnetic data, the Curie temperature of this phase was estimated to be 50–80 °C lower than the Curie temperature of the $Nd_2Fe_{14}B$ phase. Later studies suggested that this phase may be an oxygen stabilized metallic Nd–Fe–O [Sta89].

The fine and nonuniform microstructure of ribbons makes it difficult to distinguish between nucleation and pinning models. Larger grains contain domain walls and their behaviour is similar to that of nucleation-type magnets. Grains below 150 nm consist of a single magnetic domain with domain walls separating adjacent grains. Magnetization and remanence curve measurements [Pin86b] showed that magnetization reversal in the latter grains occurs by flipping the moment of a few grains at a time. The hysteresis behaviour of these grains is characteristic of interacting single-domain particles where the interactions among the grains induce domain wall propagation. The magnetization reversal processes in the fine-grained magnets can be explained better using the micromagnetic models discussed in Chapter 5. The interactions are much stronger in smaller grains (below 20 nm) and this leads to the 'interaction domains'. The coercivity in the latter is probably controlled by the motion of domain walls in the 'interaction domains'.

4 Rare-earth–iron carbides

Studies have shown that carbon can partly or totally replace boron in the 2:14:1 structure [Gue87, Boe88, Coe88]. The $R_2Fe_{14}C$ phase is not present in the as-cast samples but is obtained by suitable heat treatment (Chapter 3.4.3). The $R_2Fe_{14}C$ compounds are more difficult to prepare with light rare-earths than with heavy rare-earths [Coe88] because the $R_2Fe_{14}C$ phase becomes stable at lower temperatures at the beginning of the rare-earth series. For $Nd_2Fe_{14}C$, very long annealing times are required below 890 °C. At higher temperatures a 2:17 carbide phase $R_2Fe_{17}C_x$ is stable [Bus88b].

4.1 *As-cast R–Fe–C*

High coercivities have been achieved by annealing as-cast Dy–Fe–C and Dy(Nd)–Fe–B(C) alloys [Liu86, Had88b, Eis91]. In the latter alloys small substitutions of B have been found to facilitate the formation of the $R_2Fe_{14}C_{1-x}B_x$ phase, even with annealing at $T \geq 800$ °C, leading to a substantial coercivity (1.2 MA/m) and a remanence of 0.6 T [Eis91]. In Dy–Fe–C the as-cast samples consist mostly of the $Dy_2Fe_{17}C_x$ phase and they are magnetically soft. A coercivity greater than 0.95 MA/m is obtained [van88] in samples heat-treated for over 24 hours at 900 °C. A steep increase in magnetization is observed at low fields, similar to that found in anisotropic sintered Nd–Fe–B magnets, indicating the presence of rather large grains with a preferred orientation. Thermomagnetic analysis shows the presence of two phases: the majority $Dy_2Fe_{14}C$ phase with T_c around 300 °C and a minority Dy–Fe–C phase with a T_c around 40 K. The microstructure consists of a continuous matrix phase identified as $Dy_2Fe_{14}C$ with some embedded islands of the minority phase Dy–Fe–C. This type of microstructure can explain the magnetic hysteresis behaviour. The domain walls move easily inside the matrix phase, resulting in a steep increase in magnetization at low fields. Domain wall pinning possibly takes place at the interface between the $Dy_2Fe_{14}C$ and Dy–Fe–C phases, resulting in high coercivities. Similar results have been obtained in Nd–Fe–C alloys but the coercivity is much lower because of the presence of α–Fe inclusions within the $Nd_2Fe_{14}C$ phase. The iron acts as a nucleation centre for reverse domains.

4.2 *Melt-spun Nd–Fe–C*

Melt-spun Nd–Fe–C ribbons showed large coercivities (≈ 0.8 MA/m) when annealed at 720 °C [Coe88]. The as-spun samples had a mixture of an amorphous phase and the $Nd_2Fe_{17}C_x$ phase which were transformed into a fine-grained $Nd_2Fe_{14}C_x$ after the heat-treatment. Cobalt substitution in $Nd_{14}Fe_{78-x}Co_xC_8$ ribbons initially increased the coercivity to reach the highest value of 3.1 MA/m at 32 at.% Co. After this point the coercivity dropped rapidly with increasing Co content due to a transition from the 2:14:1 phase to the 1:5 phase at higher Co concentrations.

5 Alloys with $ThMn_{12}$-type structure

The Sm(Fe, M)$_{12}$ alloys with M = Ti, Mo, V, ... have a Curie temperature similar to that of $Nd_2Fe_{17}B$ and sufficiently high magnetization and magnetocrystalline anisotropy to make them worth considering for permanent-magent development [Li91]. The alloys for most of the rare-earths have an easy axis parallel to the *c*-axis of the tetragonal 1:12 cell (Fig. 1.21) but only the Sm-containing compounds generally have a high anisotropy. For Sm(Fe, Mo)$_{12}$ the anisotropy field is about 7.2 MA/m. Originally, the coercivities obtained in mechanically alloyed Sm–Fe–Mo alloys [Sch88] and melt-spun Sm–Fe–Ti alloys [Had87a] were relatively low 0.4 MA/m. However, the coercivities were found to increase substantially to (0.6 MA/m) when small amounts of B and Si were added and melt-spun, overquenched ribbons were crystallized at higher temperatures (750 – 1000 °C) [Sin89].

The crystallization temperatures of the melt-spun 1:12 alloys were in the range of 650–680 °C, about 100 °C higher than those for 2:14:1 alloys. The samples had to be annealed at temperatures much higher than the crystallization temperature for optimum magnetic hardening. This is because the grains of the anisotropic phase (1:12 and 2:14:1) must be of the right size for maximum coercivity. The addition of B and Si into the Sm(Fe–T)$_{12}$ alloys not only improves the mechanical properties of the ribbons but also leads to a better crystallized microstructure with a smaller amount of α–Fe. Vanadium has also been found to increase the coercivity of melt-spun Sm–Fe–Ti-V ribbons [Oka89, Pin89]. Coercivities exceeding 0.8 MA/m have been obtained in melt-spun Sm–Fe–Ti-V magnets, with the highest value of 0.85 MA/m observed in $Sm_8Fe_{75.5}Ga_{0.5}Ti_8V_8$ [Wan89].

Transmission electron microscope studies on melt-spun samples revealed large areas of different microstructure consisting of a majority 1:12 phase with adjacent regions of α–Fe and Fe–Ti. The microstructure of the 1:12 phase (Fig. 6.23) shows irregular grains of 20–50 nm in size. They are much smaller than the calculated single-domain size $d_c \simeq 0.4\ \mu$m. Lattice imaging in the majority grains is consistent with the lattice constants of the 1:12 phase ($a = 0.845$ nm, $c = 0.485$ nm).

The coercivity of such an array of randomly oriented interacting nanocrystals can be modelled by analogy with Eq. (6.9). The index i now refers to the nanocrystals, $g\mu_B J$ is their moment and D and J are replaced by effective single-particle anisotropy and interparticle exchange interactions. As their ratio decreases, the remanence is enhanced over 0.5, but the coercivity falls (Fig. 6.11). The calculated value of coercivity for melt-spun Sm(Fe_{11}Ti) is 1.2 MA/m, in good agreement with the measured value at low temperature (1 MA/m) [Sun90].

The presence of α–Fe may be critical in limiting the coercivities to about 0.8 MA/m. Since the anisotropy field of these 1:12 compounds is greater than that of $Nd_2Fe_{14}B$, the coercivities are expected to increase if the amount of α–Fe can be minimized. Sintered Sm–Fe–Ti magnets show a low coercivity which is attributed to their nonuniform microstructure [Wan89]. The grain structure of the sintered magnets is not well developed, but they contain at least four phases; the 1:12 phase, α–Fe, a Sm-rich Fe–Ti phase, and another Sm-rich phase (possibly a samarium

FIG. 6.23. Fine grain microstructure in melt-spun Sm–Fe–Ti alloys with the 1:12 structure and corresponding electron diffraction pattern.

oxide). More promising for magnet development are the nitrides and carbides based on the 1:12 structure which are discussed in the next section.

Mechanically alloyed $Sm_{20}Fe_{70}Ti_{10}$ samples [Sch90] have shown room-temperature coercivities exceeding 4 MA/m. These enormous values are attributed to the presence of a new phase which has a tetragonal structure with $a = 0.8$ nm, $c = 1.24$ nm, and a T_c of 380 °C, which may be related to the A_2-type phase Nd_5Fe_{17}. Similar behaviour is observed [Kam88] in sputtered $Sm_{19}Fe_{73}Ti_8$ films and in melt-spun samples annealed at 800 °C or below [Sch91]. At higher annealing temperatures the 1:12 phase is the majority phase. The high coercivity phase is unsuitable for permanent magnets because of its low remanence ($B_r \sim 0.3$ T).

6 Interstitial rare-earth iron nitrides and carbides

6.1 *2:17 Nitrides*

R_2Fe_{17} compounds with light rare-earths have a large saturation magnetization but cannot be used for permanent magnets because of their planar anisotropy and low Curie temperature [Wal85]. However, it was found by Coey and Sun [Coe90] that the introduction of interstitial nitrogen atoms into the 2:17 structure leads to a dramatic improvement in magnetic properties. The $Sm_2Fe_{17}N_3$ compound develops a uniaxial anisotropy with an anisotropy field exceeding 11 MA/m and has a much higher Curie temperature (about 470 °C). The intrinsic magnetic properties of this compound were discussed in Chapter 2 and the principles of preparing it from Sm_2Fe_{17} by gas–solid reaction were presented in Chapter 4.

The nitrogenation kinetics in $Sm_2Fe_{17}N_x$ have been investigated by Mukai *et al.* [92] using magneto-optic Kerr microscopy. These studies showed that a nitride layer of $Sm_2Fe_{17}N_x$ is first formed at the surface of the Sm_2Fe_{17} particle (Fig. 6.24) and its thickness grows upon further nitrogenation until the whole particle is completely nitrogenated. The same studies showed that when a partially nitrided particle is annealed its nitrogen content becomes uniform throughout the particle through atomic diffusion from the surface layer to the particle interior. The estimated domain wall energy for a fully nitrogenated compound $Sm_2Fe_{17}N_{2.6}$ is 28 mJ/m^2 and the single domain size is 0.3 μm. The nitrogenated powders have a small coercivity $\sim$ 100 kA/m despite their large anisotropy field. In imperfectly nitrogenated particles, the centre remains soft and acts as a built-in nucleation centre for reverse domains. Generally, the problem is solved by using quite fine powders for nitrogenation and milling afterwards to yield particle sizes of a few

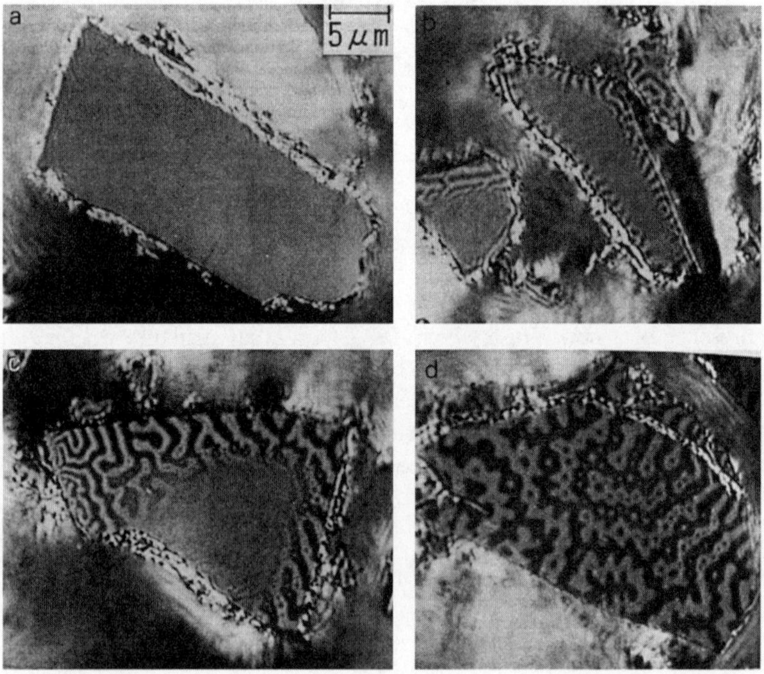

FIG. 6.24. Kerr effect studies in $Sm_2Fe_{17}N_x$ compounds nitrogenated at 500 °C showing the magnetic domain structure at different stages of nitrogenation: (a) as-milled; (b) 1 minute; (c) 8 hours; (d) homogenized [Muk92].

microns, as described in Chapter 4. Optimum oriented powder has a square loop with $H_c = 1$ MA/m and $(BH)_m \sim 240$ kJ/m^3. Much higher coercivities 2.4 MA/m have been obtained in mechanically-alloyed [Sch90] and melt-spun [Kat91, Pin92] samples which were subsequently nitrogenated. Both of these particles have a fine-grained microstructure with grain sizes around 100 nm, but the microstructure is isotropic and the remanence is only ~ 0.7 T.

Several studies have been made on bonded $Sm_2Fe_{17}N_x$ magnets using low-melting-point soft metals (Al, Bi, Sn, Zn, ...) [Ota91, Kuh92, Hen93]. It was found that in the Zn-bonded magnets a paramagnetic Fe_3Zn_7 phase is formed at the grain boundaries which eliminates α–Fe and decouples the $Sm_2Fe_{17}N_x$ grains, leading to significantly enhanced coercivities. Energy products around 160 kJ/m^3 have been obtained in these compounds. Similar properties have also been obtained in epoxy-bonded magnets using ball milled (in *n*-hexane) $Sm_2Fe_{17}N_3$ powders [Taj95]. For both the above magnets very high compaction pressures are needed to obtain the best properties. Zinc-bonded mechanically alloyed $Sm_2Fe_{17}N_3$ shows a coercivity as high as 3.5 MA/m [Kuh92].

It has been reported by Suzuki *et al.* [Suz94] that useful coercivities can be obtained in coarse $Sm_2(Fe, V)_{17}N_x$ powders, thus allowing the production of bonded magnets without the need for very fine milling and high compaction pressures. Microstructure studies on the $Sm_2Fe_{15.5}V_{1.5}N_x$ particles showed laminated structures with submicron spacers which may lead to domain wall pinning, as indicated from the shape of initial magnetization curves.

6.2 *1:12 Nitrides*

The intrinsic magnetic properties of $R(Fe, M)_{12}$ compounds (M = Mo, Ti, V) also change upon nitrogenation (Chapter 2C, 4.2), with resulting anisotropy fields exceeding 6.4 MA/m, making these compounds potential candidates for permanent magnets.

The best hard magnetic properties have been obtained in mechanically alloyed $Nd(Fe, Mo)_{12}N_x$ compounds with coercivity values in the range 0.4–0.6 MA/m [End92, Gon92]. Al-bonded magnets made with mechanically alloyed powders showed an increase in coercivity by 0.2 MA/m [Gon93]. Melt-spun samples also showed coercivities with a value of 0.4 MA/m in $NdFe_{10}V_2N_x$ [Sch91] and 0.2 MA/m in $NdFe_{11}TiN_x$ [Hir92]. The latter compounds, however, had a much higher remanence (~ 0.9 T). The lower coercivities in $NdFe_{11}TiN_x$ compounds have been attributed to their highly disordered microstructure and the presence of α–Fe.

6.3 *2:17 and 1:12 Carbides*

6.3.1 *Carbides made by arc-melting*: Investigation of the conditions under which the tetragonal $R_2Fe_{14}C$ phase is formed led to the identification of an interstitial phase with composition $R_2Fe_{17}C_x$ with $x < 1.5$ for heavy rare-earths and $x = 1$ for light rare-earths [de M88, Gub89, Dire89]. Crystal structure studies have shown that the ternary carbides are interstitial compounds with structures based on a rhombohedral Th_2Zn_{17} or hexagonal Th_2Ni_{17}-type structure. The Curie temperature increases strongly with x, rising from $T_C = 130$ °C for $x = 0$ to 247 °C for $x = 1.5$ for R = Sm [Gub89]. The introduction of interstital carbon atoms into the 2:17 structure also leads to a dramatic change in the magnetic anisotropy of the $Sm_2Fe_{17}C_x$ compound [Dir89]. The value of K_1 increases with increasing x and a transition from an 'easy-plane' to an 'easy-axis' anisotropy takes place at about $x = 0.1$ [Wan90].

The arc-melted carbides are quite stable, retaining the 2:17 structure even at temperatures above 100 °C. Substitution of A1 and Ga for Fe in $Sm_2Fe_{17}C_x$ helps the formation of high-carbon rare-earth compounds with 2:17-type structure. Shen *et al.* [She93] have prepared single-phase $R_2Fe_{17-x}Ga_xC_2$ compounds by arc-melting and found that the Curie temperature increases initially and then decreases with Ga substitution while M_s decreases monotonically. These compounds have anisotropy fields exceeding 11 MA/m. Cheng [Che94] found similar behaviour in $Sm_2Fe_{17-x}Al_xC_z$, where the $x = 2$ compound has an anisotropy field of 13.6 MA/m. The increase in Curie temperature with A1, Ga and C has been attributed to the expansion of Fe–Fe bonds which compensates more for the dilution of the Fe sublattice. The stability of the arc-melted carbides indicates that carbon is more strongly bonded than nitrogen in the gas–solid reaction samples.

Melt-spun alloys with composition $Sm_2Fe_{17-x}M_xC_y$, M = Al, Ga [Zhe94] have been found to have coercivities exceeding 0.9 MA/m at room temperature. The high coercivity obtained may be attributed to the addition of A1 and Ga which allows a higher solid solubility of carbon in the samples with a corresponding increase in the anisotropy filed. Microstructure studies revealed the coexistence of an α–Fe and a 2:17-type phase. In samples annealed at high temperatures, the distribution of these phases is inhomogeneous, with a coarse grain size leading to a shoulder in the demagnetization curve. However, in samples annealed at lower temperatures (700 °C) the grains are finer and uniform, leading to a smooth demagnetization curve with a lower coercivity but a higher reduced remanence, characteristic of nanocomposite exchange-coupled magnets (see Section 7).

6.3.2 *Carbides made by gas–solid and solid–state reactions*: Larger amounts of carbon ($x \sim 2.8$) can be introduced in the $Sm_2Fe_{17}C_x$ compounds by solid–gas reaction using hydrocarbon gases (methane, acetylene or butane) [Coe91, Lia92], resulting in better magnetic properties. The solid–gas reaction in $RFe_{10}SiC_x$ alloys has been found [Sin94] to induce a phase transformation from the 2:17:C_x structure (below 600 °C) to the $BaCd_{11}$-type structure (above 600 °C). These novel compounds have a lower Curie temperature than those of the 2:17:C_x but they are stable up to high temperatures. The $Sm_2Fe_{17}C_x$ compounds have also been prepared by solid-state reaction (see Chapter 4), by blending, for example, nanocrystalline Sm_2Fe_{17} powders with amorphous carbon and sintering in the temperature range 500–550 °C [Kuh94]. The magnetic properties of these samples, however, are inferior to those prepared by gas–solid reaction, with a maximum coercivity of only 0.15 MA/m and a remanence of 0.4 T. Microstructure studies showed an incomplete and nonuniform carbon distribution throughout the particle.

7 Exchange hardened magnets

The characteristic feature of these magnets is a nanocrystalline structure with clean grain boundaries which leads to a strong exchange interaction between the magnetic moments of adjacent grains. This interaction pulls the magnetic moments off the *c*-axis, resulting in large interaction domains and enhanced remanence compared with isotropic microcrystalline magnets. This behaviour has been observed in single-phase and two-phase systems.

7.1 *Single-phase materials* ($R_2Fe_{14}B$)

Small Si and Al substitutions in melt-spun isotropic Nd–Fe–B magnets were found to result in values of reduced remanence $m_r = M_r/M_s$, greater than that expected for an isotropic distribution of uniaxial grains [Cle88]. For an array of randomly oriented, non-interacting, single-domain particles, the Stoner–Wohlfarth model (Chapter 5, p. 228) predicts a remanence m_R of 0.5 and a coercivity $H_c \simeq K_1/\mu_o M_s$. The enhanced remanence in Si-substituted samples has been investigated by correlating the magnetic properties with the structure and microstructure [Had88c]. The effect was initially observed only in a small number of flakes. X-ray diffraction and transmission electron microscope data did not show any significant preferred orientation to explain the large value of remanence observed. Indeed, electron diffraction data suggest that the $Nd_2Fe_{14}B$-type grains are randomly oriented. The size of these grains (12–15 nm) is much smaller than that of ordinary melt-spun Nd–Fe–B samples (40–80 nm). The highest values of m_r for $Nd_{13}Fe_{79.3}Al_{0.7}B_6Si$ and $Pr_{15}Fe_{77}B_6Si_2$ are 0.70 and 0.64, respectively.

Subsequent studies by Manaf *et al.* [Man93] showed that the Si addition is not necessary to obtain an enhanced remanence and that this behaviour is not a property of a few flakes but, rather, of all the melt-spun sample. In fact, through careful control of roll speed, ejection nozzle size and ribbon thickness a nanocrystal-like structure was obtained in Nd–Fe–B samples with a mean grain size < 30 nm and with a remanence exceeding 0.8 T.

Larger values of remanence and smaller values of coercivity than those predicted by the Stoner–Wohlfarth theory can be explained by considering the interactions among single-domain particles. Because of the very fine size of grains (large surface to volume ratio) and the clean boundaries, a large exchange interaction is expected between the neighbouring magnetic moments of each grain. According to the model proposed by Callen *et al.* [Cal77], when the interaction is increased, m_r is increased from 0.5 to 1, while the coercivity is reduced from $K_1/\mu_o M_s$ to zero. This behaviour is similar to that illustrated for decreasing α in Fig. 6.11. From the observed values of remanence the interaction parameters can be found and the coercivity predicted by the model is about 1.6 MA/m. This value is close to the experimental one of 1.2 MA/m. A micromagnetic approach is used in recent models to explain the magnetic hysteresis behaviour of these magnets (Section H.2.2).

7.2 *Two-phase materials/nanocomposite magnets*

High values of remanence and reduced remanence with relatively high coercivities and reversible demagnetization curves (second quadrant) can also be achieved in nanocomposite magnets consisting of a fine mixture of hard and soft phases. Such systems include $R_4Fe_{77}B_{19}$, Fe-rich Nd–Fe–B and Sm–Fe–B and Sm–Fe–N, and Co-rich Sm–Co alloys. Kneller and Hawig [Kne91] proposed a one-dimensional model to explain the behaviour of this type of magnet and showed that for optimum coercivity the soft regions should have dimensions of the order of the domain wall width of the hard phase δ_w, which is 3–4 nm, similar to the exchange length l_{ex} in both hard and soft phases (Table 1.8). They were also able to explain qualitatively the observed reversible demagnetization curves and hence they named these magnets 'exchange-spring' magnets.

Analytical calculations by Skomski and Coey [Sko93] and micromagnetic simulations by Schrefl *et al.* [Sch94] agree with the result of Kneller and Hawig on the critical particle dimensions. The dependence of coercivity on phase composition and grain size was discussed explicity for the 2:14:1/α–Fe system by Kronmuller *et al.* [Kro94]. Furthermore, these theoretical investigations showed that both the coercivity and remanence increase with decreasing particle size, unlike the homogeneous 2:14:1 magnets in which high remanence is achieved at the expense of coercivity.

7.2.1 $R_4Fe_{77}B_{19}$: Heat treated melt-spun ribbons with an approximate composition $Nd_4Fe_{77}B_{19}$ have been found to have a coercivity of 0.3 MA/m [Coe 89]. This coercivity is attributed to dispersion of the hard magnetic phase $Nd_2Fe_{14}B$ in a soft Fe_3B matrix.

DSC and X-ray diffraction studies [Wit90b] show that as-spun $R_4Fe_{77}B_{19}$ samples are amorphous and have a crystallization temperature in the range 570–600 °C. When crystallized, the samples are magnetically hard and have a high saturation magnetization. The highest magnetization is in samples containing Pr ($\sigma = 155$ JT^{-1} kg^{-1}). The highest coercivity was obtained in a sample with $R \equiv Nd_2Dy_2$, but for $R \equiv Dy_4$, the coercivity was only 0.1 MA/m. Thermomagnetic data showed the presence of two phases, a majority Fe_3B phase with $T_C \sim 500$ °C and a minority $R_2Fe_{14}B$ phase with $T_C = 300$ °C. Transmission electron microscope studies on heat-treated melt-spun ribbons revealed a fine-grained, complex multiphase nanostructure consisting of $R_2Fe_{14}B$, Fe_3B and α–Fe phases with a grain size in the range 5–50 nm (Fig. 6.25), which approaches that necessary for exchange hardening.

7.2.2 *Fe-rich Nd–Fe–B*: Enhanced remanences have been observed in melt-spun Nd–Fe–B alloys over a range of compositions [Man92]. Low Nd content alloys (8–10 at.%) showed a remanence of 1.1 T but with decreased coercivities (350–500 kA/m) [Man93]. The microstructure of ribbons with optimum coercivity was found to consist of a fine mixture of 30 nm $Nd_2Fe_{14}B$ grains and 10 nm α–Fe grains. The high remanence is attributed to both exchange coupling and the presence of α–Fe.

FIG. 6.25. Transmission electron micrograph and selected area diffraction pattern in a melt-spun $Nd_6Fe_{77}B_{19}$ sample showing fine precipitates of $Nd_2Fe_{14}B$, Fe_3B and α–Fe.

Much higher remanences up to 1.36 T with reduced values ranging from 0.6–0.7 have been obtained [Wit94] in lower R-content alloys with composition $Nd_{3.5}(Fe, Nb)_{93}B_{3.5}$ and $R_6(Fe, Nb)_{88}B_6$ with R = Dy, Nd, Pr, Tb. The as-spun $R_6(Fe, Nb)_{88}B_6$ ribbons crystallize in two steps; at first the $Y_3Fe_{62}B_{14}$ and α–Fe phases are formed when R = Dy, Nd and Pr and subsequently they transform to 2:14:1 and α–Fe upon heating above 700 °C (Fig. 6.26). The intermediate phase in the case of $Tb_6(Fe, Nb)_{88}B_6$ is of the $TbCu_7$ type.

7.2.3 *Sm–Fe–N and Sm–Co alloys*: Remanence enhancement has been obtained [Din93a, Odo94] in mechanically alloyed Fe-rich Sm_7Fe_{93} powders consisting of fine soft α–Fe and hard $Sm_2Fe_{17}N_x$ grains with an average size of 20 nm. The highest remanence obtained was 1.38 T with $m_r = 0.8$, $H_c \sim 0.3$ MA/m and an energy product of 210 kJ/m^3. Similar behaviour has been observed in mechanically alloyed Sm_xCo_{100-x}, $x = 11$–15, powders consisting of a mixture of hard $SmCo_5$ and soft Sm_2Co_{17} [Din93a]. The nanoscale microstructure was developed after annealing the amorphous mechanically alloyed powders at temperatures around 700 °C. The largest $(BH)_m$ obtained in these alloys was 143 kJ/m^3.

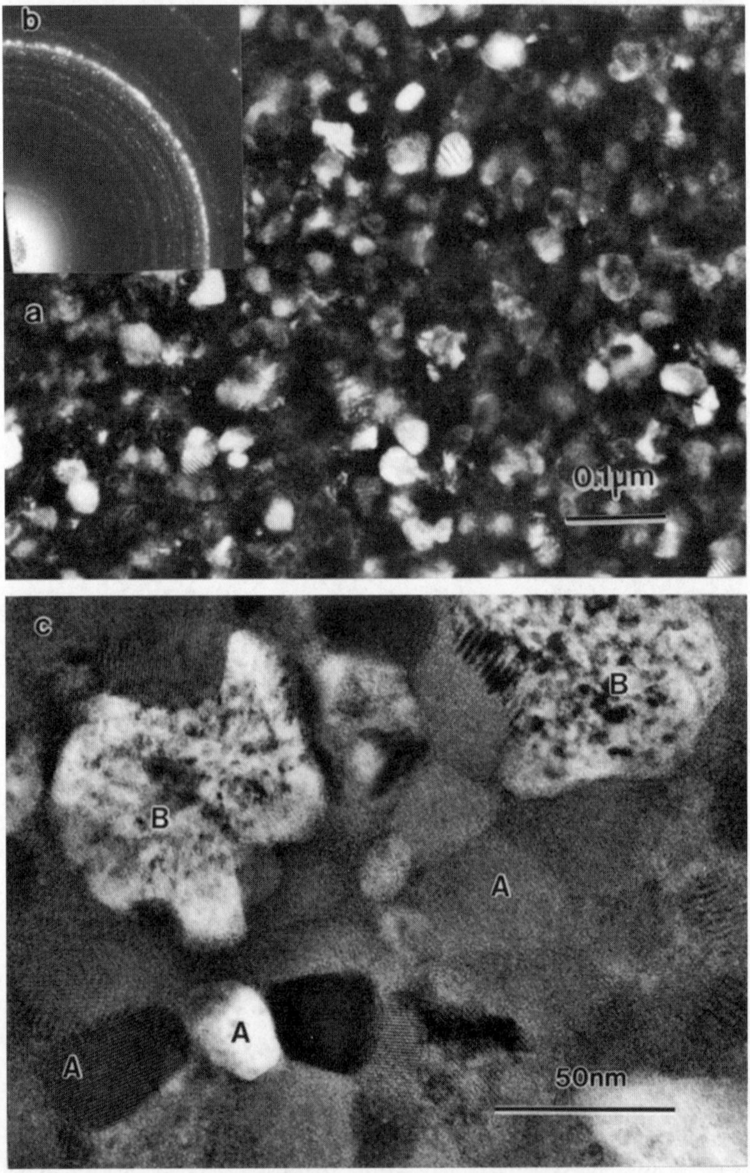

FIG. 6.26. (a) Microstructure in nanocomposite melt-spun $Nd_6Fe_{87}Nb_1B_6$. (b) Corresponding electron diffraction pattern. (c) Higher-resolution micrograph showing (A) 2:14:1 grains, and (B) α–Fe grains.

REFERENCES

[Bez85] A. Bezinge, H. F. Braun, J. Muller and K. Yvon, *Sol. Stat. Comm.*, **55** (1985), 131.

[Bia89] W. Biao, C. P. Xin and H. Fang, *10th Int. Workshop on Rare-Earth Magnets and Their Applications*, Kyoto, Japan (1989), 491.

[Bod79] R. Bodenberger and A. Hubert, *Phys. Stat. Solidi*, A**56**, (1979), 637.

[Bus88a] K. H. J. Buschow, *J. Appl. Phys.*, **63** (1988a), 3130.

[Bus88b] K. H. J. Buschow, C. J. M. Denissen and B. D. deMooij, *J. de Physique*, C**8** (1988b), 593.

[Cad86] F. J. Cadieu, T. D. Cheung and L. Wickramasekara, *J. Magn. Magn. Mater.*, **54–57** (1986), 535.

[Cal77] E. Callen, Y. J. Lin and J. R. Cullen, *Phys. Rev.*, B**16** (1977), 263.

[Car66] R. Carey and E. D. Isaac, *Magnetic domains and techniques for their observation*, Academic Press Inc., New York (1966).

[Cha86] M. H. Chandehari, *Intermag Conf.*, Phoenix, Arizona (1986).

[Cha89] W. C. Chang, C. H. Lin, K. D. Lin, Y. M. Chien and S. K. Chen, *10th Int. Workshop on Rare-Earth Magnets and their Applications*, Kyoto, Japan (1989), p. 509.

[Cha78] J. N. Chapman, P. E. Batson, E. M. Waddell and R. P. Ferrier, *Ultramicroscopy*, **3** (1978), 203.

[Cha92] J. N. Chapman, R. Ploessl and D. M. Dounet, *Ultramicroscopy*, **47** (1992), 331.

[Che94] Z. Cheng, B. Shen, J. Zhang, F. Wang, H. Gong, W. Zhan and J. Zhao, *J. Appl. Phys.*, **76** (1994), 6734.

[Chi76] M. C. Chi and R. Alben, *AIP Conf. Proc.* **34** (1976), 316.

[Chi86] T. S. Chin, Y. Hara, E. J. Lavernia, R. C. O'Handley and N. J. Grant, *J. Appl. Phys.*, **59** (1986), 1297.

[Chr87] C. Christodoulou, J. Schlup and G. C. Hadjipanayis, *J. Appl. Phys.*, **61** (1987), 3760.

[Cla73] A. E. Clark, *Appl. Phys. Lett.*, **23** (1973), 624.

[Cle88] G. B. Clemente, J. Ekenard, J. P. Bradley, *J. Appl. Phys.*, **64** (1988), 5299.

[Coe88] R. Coehoorn, D. B. de Mooij, J. P. W. B. Duchateau and K. H. J. Buschow, *J. de Physique*, C**8** (1988), 669.

[Coe89] R. Coehoorn, D. B. de Mooij and C. de Waard, *J. Magn. Magn. Mater.*, **80** (1989), 101.

[Coe90] J. M. D. Coey and H. Sun, *J. Magn. Magn. Mater.*, **87** (1990), L251.

[Coe91] J. M. D. Coey, H. Sun, Y. Otani and D. P. F. Hurley, *J. Magn. Magn. Mater.*, **98** (1991), 76.

[Cor89] W. D. Corer and M. J. Hawton, *Concerted European action on magnets*, I. V. Mitchell *et al.* (eds), Elsevier, London (1989), p. 424.

[Cro81] J. J. Croat, *J. Magn. Magn. Mater.*, **24** (1981), 125.

[Cro82] J. J. Croat, *J. Appl. Phys.*, **53** (1982), 3161.

[Cro84] J. J. Croat, J. F. Herbst, R. W. Lee and F. E. Pinkerton, *J. Appl. Phys.*, **55** (1984), 2078.

[Cul72] B. D. Cullity, *Introduction to magnetic materials*, Addison-Wesley, London (1972).

[deM87] D. B. de Mooij and K. H. J. Buschow, *Phillips J. Research*, **42** (1987), 246.

[deM88] D. B. de Mooij and K. H. J. Buschow, *J. Less-Common Met.*, **142** (1988), 349.

[deM88] F. R. de Boer, R. Verhoef and Zhi-dong Zhang, *J. Magn. Magn. Mater.*, **73** (1988), 263.

[Dic86] R. C. Dickerson, K. R. Lawless and G. C. Hadjipanayis, *Mater. Research Soc. Symp. Proc.*, **58** (1986), 199.

[Din93a] J. Ding, P. G. McCormick and R. Street, *J. Magn. Magn. Mater.*, **124** (1993a), 1.

[Din93b] J. Ding, Y. Liu, P. G. McCormick and R. Street, *J. Magn. Magn. Mater.*, **123** (1993b), L239.

[Dir89] M. W. Dirken, R. C. Thiel, L. J. de Jourge, T. H. Jacobs and K. H. J. Buschow, *J. Less Common Met.*, **155** (1989), 339.

[Dro35] V. Drozzina and R. Janus, *Nature*, **135** (1935), 36.

[Eis91] J. Eisses, D. B. de Mooij, K. H. J. Buschow and G. Martinek, *J. Less Common Met.*, **171** (1991), 17.

[End87] M. Endoh, M. Tokunaga and H. Harada, *IEEE Trans. Mag.*, **MAG-23** (1987), 2290.

[End92] M. Endoh, K. Nakamura and H. Mikami, *IEEE Trans. Mag.*, **MAG-28** (1992), 2560.

[Fid82a] J. Fidler and P. Skalicky, *J. Magn. Magn. Mater.*, **27** (1982a), 127.

[Fid82b] J. Fidler and P. Skalicky, *Proc. 3rd Int. Symp. on Magn. Anisotropy and Coercivity in RE-Transition Metal Alloys*, J. Fidler (ed.), Baden, Austria (1982b), p. 585.

[Fid83] J. Fidler, *Phil. Mag.*, **A47** (1983), L19.

[Fid87] J. Fidler, *IEEE Trans. Magn.*, **MAG-23** (1987), 2106.

[Fid89a] J. Fidler, K. G. Knoch, H. Kronmuller and G. Scheider, *J. Mater. Res.*, **4** (1989a), 806.

[Fid89b] J. Fidler and K. G. Knoch, *J. Magn. Magn. Mater.*, **80** (1989b), 48.

[Fid91] J. Fidler and J. Bernardi, *J. Appl. Phys.*, **70** (1991), 6456.

[Fid92] J. Fidler, *Proc. 6th Int. Symp. on Magn. Anisotropy and Coercivity in RE-Transition Metal Alloys*, Canberra, Australia (1992), p. 11.

[Fue88] C. D. Fuerst and J. F. Herbst, *J. Appl. Phys.*, **63** (1988), 3324.

[Gau83] P. Gaunt, *Phil. Mag.*, **48** (1983), 261.

[Gon92] W. Gong and G. C. Hadjipanayis, *IEEE Trans. Mag.*, **MAG-29** (1992), 2563.

[Gon93] W. Gong and G. C. Hadjipanayis, *J. Appl. Phys.*, **73** (1993), 245.

[Grö86] R. Grössinger, R. Krewenka, X. K. Sun, R. Eibler, H. R. Kirchmayr and K. Buschow, *J. Less Common Met.*, **124** (1986), 165.

[Grö87] R. Grössinger, *9th Int. Workshop on Rare-Earth Magnets and their Applications*, Part II (1987), p. 301.

[Gub90] P. C. M. Gubbens, A. M. van de Kraan, T. H. Jacobs and K. H. J. Buschow, *J. Magn. Magn. Mater.*, **80** (1990), 165.

[Gud86] K. Gudimetta, C. N. Christodoulou and G. C. Hadjipanayis, *Appl. Phys. Lett.*, **48** (1986), 670.

[Gue87] M. Gueramin, A. Bezinge, K. Yvon and J. Muller, *Sol. State Commun.*, **64** (1987), 639.

[Had80] G. C. Hadjipanayis, S. G. Cornelison and D. J. Sellmyer, *J. Magn. Magn. Mater.*, **21** (1980), 101.

[Had82a] G. C. Hadjipanayis, R. C. Hazelton, K. R. Lawless and L. S. Horton, *IEEE Trans. Mag.*, **MAG-18** (1982a), 1480.

[Had82b] G. C. Hadjipanayis, *Proc. 3th Int. Symp. on Magn. Anisotropy and Coercivity in RE-Transition Metal Alloys*, Baden, Austria (1982b), p. 609.

[Had82c] G. C. Hadjipanayis, R. C. Hazelton, S. H. Wollins, A. Wysierkierski and K. R. Lawless, *Proc. 3th Int. Symp. on Magn. Anisotropy and Coercivity in RE-Transition Metal Alloys*, Baden, Austria (1982c), p. 667.

[Had83] G. C. Hadjipanayis, R. C. Hazelton and K. R. Lawless, *Appl. Phys. Lett.*, **43** (1983), 797.

[Had84] G. C. Hadjipanayis, R. C. Hazelton and K. R. Lawless, *J. Appl. Phys.*, **55** (1984), 2073.

[Had85] G. C. Hadjipanayis, Y. F. Tao and K. R. Lawless, *8th Int. Workshop on Rare Earth Magnets and their Applications*, Dayton, Ohio (1985), p. 657.

[Had86a] G. C. Hadjipanayis, *ASM Materials Week, Proc. on Soft and Hard Magn. Mater.*, Orlando, Florida (1986a), p. 89.
[Had86c] G. C. Hadjipanayis, *Mater. Res. Soc. Symp. Proc.*, **58** (1986c), 187.
[Had87a] G. C. Hadjipanayis, S. H. Aly and S. F. Cheng, *Appl. Phys. Lett.*, **51** (1987a), 2048.
[Had87b] G. C. Hadjipanayis, *Can. J. Phys.*, **65** (1987b), 1200.
[Had87c] G. C. Hadjipanayis and W. Gong, *J. Magn. Magn. Mater.*, **66** (1987c), 390.
[Had88a] G. C. Hadjipanayis and A. Kim, *J. Appl. Phys.*, **63** (1988a), 3310.
[Had88b] G. C. Hadjipanayis, N. Venkateswaran and J. Strzeszewski, *J. de Phys. Colloq.*, C**8** (1988b), 639.
[Had89a] G. C. Hadjipanayis, A. Tsoukatos, J. Strzeszewski, G. J. Long and O. A. Pringle, *J. Magn. Magn. Mater.*, **78** (1989a), L1.
[Had88c] G. C. Hadjipanayis and W. Gong, *J. Appl. Phys.*, **64** (1988c), 5559.
[Had89b] G. C. Hadjipanayis, M. Zhang and C. Gao, *Appl. Phys. Lett.*, **54** (1989b), 1812.
[Had90] G. C. Hadjipanayis and L. Withanawasam, *Proc. 6th Int. Symp. on Magn. Anisotropy and Coercivity in RE-Transition Metal Alloys*, Pittsburgh, Pennsylvania (1990), p. 530.
[Har73] R. Harris, M. Plischke and M. J. Zuckermann, *Phys. Rev. Lett.*, **31** (1973), 160.
[Har94] U. Hartmann, *Advances in Electronic and Electron Physics*, **87** (1994), 49.
[Hen93] P. A. P. Hendhausen, D. Eckert, A. Handestein, J. H. Muller, G. Leither and R. Skomski, *J. Appl. Phys.*, **73** (1993), 6044.
[Her84] J. F. Herbst, J. J. Croat, J. D. Pinkerton and W. Yelon, *Phys. Rev.*, B **29** (1984), 4176.
[Hir85] K. Hiraga, M. Hirabayashi, M. Sagawa and Y. Matsuura, *Japan J. Appl. Phys.*, **24** (1985), L30.
[Hir87] S. Hirosawa, Y. Miguchi, K. Tokunaga, H. Yamamoto, S. Fujimura and M. Sagawa, *IEEE Trans. Mag.*, **MAG-23** (1987), 2120.
[Hir90a] S. Hirosawa, T. Tomizawa, S. Mino and A. Hamamura, *IEEE Trans. Mag.*, **MAG-26** (1990a), 1960.
[Hir90b] S. Hirosawa, A. Hanaki, H. Tomizawa, S. Mino and A. Hamamura, *IEEE Trans. Mag.*, **MAG-26** (1990b), 1960.
[Hir92] S. Hirosawa, K. Makita, T. Ikegami and M. Umemoto, *Proc. 7th Int. Symp. on Magnetic Anisotropy and Coercivity in RE-Transition Metal Alloys*, Canberra, Australia (1992), p. 389.
[Hu88] B. P. Hu, J. M. D. Coey, C. J. Cardin, E. J. Devlin and I. R. Harris, *J. Less Common Met.*, **144** (1988), L29.
[Hua91] M. A. Huang, L. Y. Zhang, B. M. Ma, Y. Zheng, J. M. Elbicki, W. E. Wallace and S. G. Sankar, *J. Appl. Phys.*, **70** (1991), 6028.
[Iri92] T. Iriyama, K. Kobayashi, N. Imaoka, T. Fukuda, H. Kato, Y. Nakagawa, *IEEE Trans. Mag.*, **MAG-28** (1992), 2326.
[Ish89] T. Ishikawa, Y. Hamada and K. Ohmori, *10th Int. Workshop on Rare-Earth Magnets and their Applications*, Kyoto, Japan (1989), p. 421.
[Kam88] N. Kamprath, N. C. Lin, H. Hedge and F. J. Cadieu, *J. Appl. Phys.*, **64** (1988), 5720.
[Kan94] Y. Kaneko and N. Ishigaki, *J. Magn. Eng. Performance*, **3** (1994), 228.
[Kat90] M. Katter, J. Wecker and L. Schultz, *J. Magn. Magn. Mater.*, **92** (1990), L14.
[Kat91] M. Katter, J. Wecker and L. Schultz, *J. Appl. Phys.*, **70** (1991), 3188.
[Kii90] A. A. Kiiski, J. A. Raustela, M. J. Talritie and V. K. Lindroos, *10th Int. Workshop on Rare-Earth Magnets and their Applications*, Kyoto, Japan (1990), p. 501.
[Kim88] A. S. Kim, *J. Appl. Phys.*, **63** (1988), 3519.

[Kne91] E. F. Kneller and R. Hawig, *IEEE Trans. Mag.*, **MAG-27** (1991), 3588.
[Kno89] K. G. Knoch, G. Schneider, J. Fidler, E. T. Hening and H. Kronmuller, *IEEE Trans. Mag.*, **MAG-25** (1989), 3426.
[Koo81] N. C. Koon and B. N. Das, *Appl. Phys. Lett.*, **39** (1981), 840.
[Kro88] H. Kronmuller, K. D. Durst and M. Sagawa, *J. Magn. Magn. Mater.*, **74** (1988), 291.
[Kro90] H. Kronmuller in *Science and technology of Nanostructured materials*, G. C. Hadjipanayis and G. Prinz (eds), Plenum Press, New York (1990), p. 657.
[Kro94] H. Kronmuller, T. Schrefl, R. Fischer and J. Fidler, *Proc. 8th Int. Symp. on Magn. Anisotropy and Coercivity in RE-Transition Metal Alloys*, Birmingham, UK (1994), 1.
[Kuh92] K. Kuhrt, K. O'Donnell, M. Katter, J. Wecker, K. Schnitzke and L. Shultz, *Appl. Phys. Lett.*, **60** (1992), 3316.
[Kuh94] C. Kuhrt and J. Wecker, *13th Int. Workshop on RE Magnets and their applications*, Birmingham, UK (1994), 773.
[Li90] H. S. Li, B. P. Hu, J. M. Cadogan, J. M. D. Coey and J. P. Carigan, *J. Appl. Phys.*, **67** (1990), 4841.
[Li92] H. S. Li and J. M. D. Coey in *Handbook of ferromagnetic materials*, K. H. J. Buschow (ed.), North Holland, Amsterdam, Vol. 6 (1992), p. 1.
[Lia92] L. X. Liao, X. Chen, Z. Altounian and D. H. Ryan, *Appl. Phys. Lett.*, **60** (1992), 129.
[Liu86] N. C. Liu and H. H. Stadelmaier, *Mater. Lett.*, **4** (1986), 377.
[Liv78] J. D. Livingston, *IEEE Trans. Mag.*, **MAG-14** (1978), 668.
[Liv80] J. D. Livingston, General Electric, Report No. 80CRD139 (1980).
[Liv81] J. D. Livingston, *J. Appl. Phys.*, **52** (1981), 2544.
[Liv84] J. D. Livingston, General Electric, Report No. 84CRD281 (1984).
[Liv85a] J. D. Livingston, *8th Int. Workshop on RE-Magnets and their applications*, Dayton, Ohio (1985a), 423.
[Liv85b] J. D. Livingston, *J. Appl. Phys.*, **57** (1985b), 4137.
[Lor84] M. H. Loretto, *Electron beam analysis of materials*, Chapman and Hall, New York (1984).
[Ma86] B. M. Ma and K. S. V. L. Narasimhan, *IEEE Trans. Mag.*, **MAG-22** (1986), 916.
[Man92] A. Manaf, M. Leonowicz, H. A. Davies and R. A. Buckley, *Mat. Letters*, **13** (1992), 194.
[Man93] A. Manaf, P. Z. Zhang, I. Ahmad, H. A. Davies and R. A. Buckley, *IEEE Trans. Mag.*, **MAG-29** (1993), 2866.
[Mat88] F. Matsumoto, H. Sakamoto, M. Komiya and M. Fujikura, *J. Appl. Phys.*, **63** (1988), 5307.
[McC64] R. A. McCurrie and P. Gaunt, *Proc. Int. Conf. on Magnetism* (1964), 780.
[McG87] R. W. McGallum, A. M. Kadin, G. B. Clemente and J. E. Keem, *J. Appl. Phys.*, **61** (1987), 3577.
[Men78] A. Menth, H. Nagel and R. S. Perkins, *Ann. Rev. Mater. Sci.*, **8** (1978), 21.
[Mis81] R. K. Mishra, G. Thomas, T. Yoneyama, A. Fukuno and T. Ojima, *J. Appl. Phys.*, **52** (1981), 2517.
[Mis86a] R. K. Mishra, *J. Magn. Magn. Mater.*, **54–57** (1986a), 450.
[Mis86b] R. K. Mishra and R. W. Lee, *Appl. Phys. Lett.*, **48** (1986b), 733.
[Miz86] T. Mizoguchi, I. Sakai, H. Liu and K. Inomata, *IEEE Trans. Mag.*, **MAG-22** (1986), 919.
[Moo84] K. Moorjani and J. M. D. Coey, *Magnetic glasses*, Elsevier, Amsterdam (1984).
[Mor90] J. M. Moreau, L. Paccard, J. P. Nozières, E. P. Missell, G. Schneider and V. Villas-Boas, *J. Less Comm. Met.*, **163** (1990), 245.

[Mor65] A. Morrish, *Physical principles of magnetism*, John Wiley, New York (1965).
[Muk92] T. Mukai and T. Fujimoto, *J. Magn. Magn. Mater.*, **103** (1992), 165.
[Oes78] H. Oesterreicher, *J. Appl. Phys.*, **15** (1978), 341.
[Oha87] K. Ohashi, T. Yokoyama, R. Osugi and Y. Tawara, *IEEE Trans. Mag.*, **MAG-23** (1987), 3101.
[Oji77] T. Ojima, S. Tomizawa, T. Yoneyama and T. Hori, *IEEE Trans. Mag.*, **MAG-13** (1977), 1317.
[Oka89] M. Okada, K. Yamagishi and M. Homma, *Mater. Trans.*, JIM **30** (1989), 374.
[Ota91] Y. Otani, A. Moukarika, H. Sun and J. M. D. Coey, *J. Appl. Phys.*, **69** (1991), 6735.
[Pan89] S. Pan and M. Ruahang, *10th Int. Workshop on Rare-Earth Magnets and their Applications*, Kyoto, Japan (1989), p. 537.
[Par87] S. H. F. Parker, P. Grundy and J. Fidler, *J. Magn. Magn. Mater.*, **66** (1987), 74.
[Pau79] D. I. Paul, *J. Appl. Phys.*, **50** (1979), 2128.
[Ped86] A. T. Pedziwiatr and W. E. Wallace, *Sol. Stat. Comm.*, **60** (1986), 653.
[Pin92] F. E. Pinkerton and C. D. Fuerst, *Appl. Phys. Lett.*, **90** (1992), 2558.
[Pin86a] F. E. Pinkerton, *J. Magn. Magn. Mat.*, **54–57** (1986a), 579.
[Pin86b] F. E. Pinkerton and J. D. Van Wingerden, *J. Appl. Phys.*, **64** (1986b), 5556.
[Pin89] F. E. Pinkerton and D. J. Van Wingerden, *IEEE Trans. Mag.*, **MAG-25** (1989), 3306.
[Pol88] R. J. Pollard, P. J. Grundy, S. F. H. Parker and D. G. Lord, *IEEE Trans. Mag.*, **MAG-24** (1988), 1626.
[Pro64] M. Pruton, *Thin ferromagnetic films*, Butterworths, London (1964).
[Rab82] L. Rabenberg, R. K. Mishra and G. Thomas, *J. Appl. Phys.*, **53** (1982), 2389.
[Ram87] R. Ramesh, J. K. Chen and G. Thomas, *J. Appl. Phys.*, **61** (1987), 2993.
[Ram88] R. Ramesh and G. Thomas, *Acta Metall.*, **36** (1988), 3137.
[Ram89] R. Ramesh, G. Thomas and B. M. Ma, *Acta Metall.*, **37** (1989), 1421.
[Ray84] A. E. Ray, *J. Appl. Phys.*, **55** (1984), 2094.
[Rhy79] J. J. Rhyne, *Handbook of physics and chemistry of rare-earths*, K. A. Gschneider and L. Eyving (eds), North Holland, Amsterdam (1979), p. 259–294.
[Rad88] W. Rodewald and W. Feinengel, *IEEE Trans. Mag.*, **MAG-24** (1988), 1683.
[Sag85] M. Sagawa, S. Fujimura, Y. Yamamoto, Y. Matsuura, S. Hirosawa and K. Hiraga, *Proc. 4th Int. Symp. on the Coercivity and Anisotropy of RE-Transition Metal Alloys*, Dayton, Ohio (1985), p. 587.
[Sag86] M. Sagawa, S. Fujimura, H. Yamamoto, Y. Matsuura, S. Fujimura, H. Tokuhara and K. Hiraga, *IEEE Trans. Mag.*, **MAG-22** (1986), 910.
[Sag88] M. Sagawa and S. Hirosawa, *J. Mater. Res.*, **3** (1988), 45.
[Sag90] M. Sagawa, P. Tenaud, F. Vial and K. Hiraga, *IEEE Trans. Mag.*, **MAG-26** (1990), 1957.
[Sch87] G. Schneider, E. Henig, G. Petzow and H. H. Stadelmaier, *Z. Metallkde*, **78** (1987), 697.
[Sch88] G. Schneider, G. Martinez, H. H. Stadelmaier and G. Petzow, *Mater. Lett.*, **7** (1988), 215.
[Sch89a] G. Schneider, F. I. G. Landgraf and F. P. Missel, *J. Less Common Met.*, **153** (1989a), 169.
[Sch89b] G. Schneider, F. J. P. Landgraf, V. Villas-Boas and F. P. Missel, *10th Int. Workshop on Rare-Earth Magnets and their Applications*, Kyoto, Japan (1989b), p. 63.
[Sch90a] K. Schnitzke, L. Schultz, J. Wecker and M. Katti, *Appl. Phys. Lett.*, **57** (1990a), 2853.
[Sch90b] K. Schnitzke, L. Schultz, J. Wecker and M. Katter, *Appl. Phys. Lett.*, **56** (1990b), 587.

[Sch94] T. Schrefl, J. Fidler and H. Kronmuller, *Phys. Rev.*, B **49** (1994), 6100.
[Sch88] L. Schultz and J. Wecker, *J. Appl.*, **64** (1988), 5711.
[Sch91] L. Schultz, K. Schnitzke, J. Wecker, M. Katter and C. Kuhrt, *J. Appl. Phys.*, **70** (1991), 639.
[She93] B.-G. Shen, L.-S. Kong, F.-W. Wang and L. Cao, *Appl. Phys. Lett.*, **63** (1993), 2288.
[Shi79] T. Shimada, K. Kasai and K. Teraishi, *4th Int. Workshop on R-Co Magnets*, Hakoue, Japan (1979), p. 335.
[Shi88] T. Shimoda, K. Akioka, O. Kobayashi and T. Yamagami, *J. Appl. Phys.*, **64** (1988), 5290.
[Sin89] E. W. Singleton, J. Strzeszewski and G. C. Hadjipanayis, *Appl. Phys. Lett.*, **54** (1989), 1934.
[Sin94] E. W. Singleton, G. C. Hadjipanayis, V. Papaefthymiou, Z. Sun and W. B. Yelon, *J. Appl. Phys.*, **75** (1994), 6000.
[Sko93] R. Skomski and J. M. D. Coey, *Phys. Rev.*, B **48** (1993), 15812.
[Spe90] J. C. H. Spence, *Science and technologies of nanostructured magnetic materials*, G. C. Hadjipanayis and G. A. Prinz (eds), Plenum, New York (1990), p. 279.
[Sta89] H. H. Stadelmaier and K. J. Strnat, *10th Int. Workshop on Rare-Earth Magnets and their Applications*, Kyoto, Japan (1989), p. 53.
[Sta91] H. H. Stadelmaier, G. Schneider, E. Th. Henig and M. Ellner, *Mat. Lett.*, **10** (1991), 303.
[Sto48] E. C. Stoner and E. P. Wohlfarth, *Phil. Trans. R. Soc. London Ser.*, A**240** (1948), 599.
[Str78] K. J. Strnat, *J. Magn. Magn. Mater.*, **7** (1978), 351.
[Str88] K. J. Strnat, H. F. Mildrum, M. Tokunaga and H. Harada, *J. Appl. Phys.*, **63** (1988), 3321.
[Str88a] J. Strzeszewski, G. C. Hadjipanayis, K. Aylesworth and D. J. Sellmyer, *Mater. Sci. Eng.*, **89** (1988a).
[Str88b] J. Strzeszewski, G. C. Hadjipanayis and A. Kim, *J. Appl. Phys.*, **64** (1988b), 5568.
[Str90] J. Strzeszewski, A. Tsoukatos and G. C. Hadjipanayis, *J. Appl. Phys.*, **67** (1990), 4966.
[Sun90] H. Sun, Y. Otani and J. M. D. Coey, *J. Appl. Phys.*, **61** (1990).
[Suz94] S. Suzuki, S. Suzuki and M. Kawasaki, *J. Appl. Phys.*, **76** (1994), 6708.
[Taj95] S. Tajima, T. Hatton and Y. Kato, *IEEE Trans. Mag.*, **MAG-31** (1995).
[Ten90] P. Tenaud, F. Vial and M. Sagawa, *IEEE Trans. Mag.*, **MAG-26** (1990), 1930.
[Tho79] G. Thomas and M. J. Goringe, *Transmission electron microscopy of materials*, John Wiley, New York (1979).
[Tok86] M. Tokunaga, M. Tobise, N. Meguro and H. Harada, *IEEE Trans. Mag.*, **MAG-22** (1986), 904.
[Tsu83] K. Tsuno and T. Taoka, *Japan J. Appl. Phys.*, **22** (1983), 1041.
[Tsu87] T. Tsutai, I. Sakai, T. Mizoguchi and K. Inomata, *Symp. on Rare Earth, Japan* (1987), p. 12.
[Ung86] J. Unguris, G. G. Hembree, R. J. Celotta and D. T. Pierce, *J. Magn. Magn. Mater.*, **54** (1986), 1629.
[van88] R. van Mens, D. B. de Mooij and K. H. J. Buschow, *J. Appl. Phys.*, **64** (1988), 5708.
[Wal73] W. E. Wallace *Rare-earth intermetallics*, Academic Press, New York (1973).
[Wal85] W. E. Wallace, *Prog. Solid State Chem.*, **16** (1985), 127.
[Wan90a] Y. Z. Wang, G. C. Hadjipanayis, A. Kim, N. C. Lin and D. J. Sellmyer, *J. Appl. Phys.*, **67** (1990a), 4954.
[Wan90b] Y. Z. Wang, G. C. Hadjipanayis and D. J. Sellmyer, *Proc. 6th Int. Symp. on Magn. Anisotropy and Coercivity in RE-Transition Metal Alloys*, Pittsburgh, Pennsylvania (1990b), p. 181.

[Wan90c] Y. Z. Wang and G. C. Hadjipanayis, *J. Magn. Magn. Mater.*, **87** (1990c), 375.

[Wec88] J. Wecker and L. Schultz, *J. de Physique*, **C8** (1988), 611.

[Wit90] L. Withanawasam, Y. Zhang and G. C. Hadjipanayis, *Proc. 6th Int. Symp. on Magn. Anisotropy and Coercivity in RE-Transition Metal Alloys*, Pittsburgh, Pennsylvania (1990), p. 240.

[Wit94] L. Withanawasam, G. C. Hajipanayis and R. F. Krause, *J. Appl. Phys.*, **75** (1994), 6646.

[Woh58] E. P. Wohlfarth, *J. Appl. Phys.*, **29** (1958), 595.

[Xia88] Y. Xiao, S. Liu, H. F. Mildrum, K. S. Strnat and A. E. Ray, *J. Appl. Phys.*, **63** (1988), 3516.

[Xie88] J. Q. Xie, C. H. Wu, Y. C. Chuang and F. M. Yang, *J. Magn. Magn. Mater.*, **75** (1988).

[Yam87] H. Yamamota, S. Hirosawa, S. Fujimura, K. Tokuhara, H. Nagata and M. Sagawa, *IEEE Trans. Mag.*, **MAG-23** (1987), 2100.

[Yan86] Y. C. Yang, W. J. James, X. D. Li, H. Y. Chen and L. G. Xu, *IEEE Trans. Mag.*, **MAG-22** (1986), 757.

[Zha88] Z. R. Zhao, Y. G. Ren, D. J. Sellmyer, J. Strzeszewski and G. C. Hadjipanayis, *J. Appl. Phys.*, **63** (1988), 3699.

[Zhe95] Y. H. Zheng, F. M. Yang and G. C. Hadjipanayis, *J. Magn. Magn. Mater.*, **140–144** (1995).

7
Magnet processing

I. R. Harris

A. Nd–Fe–B

Unlike their predecessors based on Sm–Co, the Nd–Fe–B alloys can be processed into permanent magnets by many different routes. The most important group of processing routes consists of variants of powder metallurgy, where monocrystalline powder with grains a few microns in size is oriented and sintered to yield a uniform, fine grain structure with a high degree of *c*-axis orientation, where each grain is isolated from its neighbours by Nd-rich grain boundary phases. Such a microstructure is necessary to achieve the greatest energy product for a particular magnet composition. A second group of processes yield polycrystalline powder or flakes composed of many submicron crystallites which are isotropic, or at best imperfectly oriented. Melt-spinning is used to make this material industrially, and it is particularly suitable for bonded magnets. A third type of process avoids the use of powders by hot-working cast ingots to achieve an appropriate microstructure with a high degree of *c*-axis orientation in relatively simple shapes such as arc segments.

It should be emphasized that the majority of commercial Nd–Fe–B magnets are of the sintered-type. Nevertheless, magnets based on melt-spun materials are of growing importance, and other processes which are currently at an early stage of development may eventually rival the conventional powder route.

1 Alloy production

The starting point is the basic alloy in ingot or powder form. Some commonly used compositions are given in Table 7.1. The alloys are usually produced either by

Table 7.1 Some commonly used Nd–Fe–B ingot compositions

Composition (at %)	Comments
$Nd_{15}Fe_{77}B_8$	Original Sumitomo composition
$Nd_{16}Fe_{76}B_8$	Additional Nd to allow for oxidation during processing
$Nd_{13.5}Dy_{1.5}Fe_{77}B_8$	Dy added to enhance the coercivity
$Nd_{13.5}Dy_{1.5}Fe_{76}Nb_1B_8$	Nb added to enhance loop squareness and coercivity
$Nd_{15}Fe_{76}Al_1B_8$	Al added to enhance the coercivity
$Nd_{15}Fe_{68}Co_5V_4B_8$	V and Co added to enhance coercivity, Curie point, temperature coefficient of H_{cj} and corrosion resistance

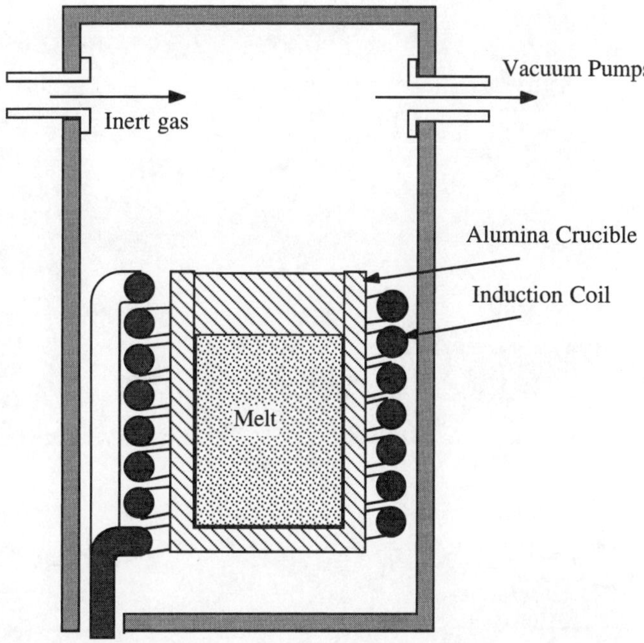

FIG. 7.1. A schematic representation of an induction furnace.

Table 7.2 Typical analysis (wt%) of neodymium, ferro-boron and $Nd_{15}Fe_{77}B_8$ alloy used in the production of sintered magnets

Neodymium	97.3%Nd, 1.4%Pr, 0.3%La, <0.1%Sm, 0.2%Ca, <0.05%Mg, <0.01%C, 0.02%O
Ferro-boron	17–20%B, 1.5%Al, 0.5%Si, balance Fe
$Nd_{15}Fe_{77}B_8$	0.1–0.15%Al, <0.05%Ca, <0.05%Mg, 0.015%O

induction melting the metallic constituents or by a process of coreduction of oxides. Each method will be outlined.

Induction melting. Figure 7.1 shows a schematic representation of an induction furnace. The constituents are melted in a crucible which is usually made of alumina. It is unnecessary to use high-purity metal. Typical analyses are shown in Table 7.2. Neodymium is normally around 98% pure, with the balance being mainly other rare-earth metals. Electrolytic iron (> 99.9% Fe) is employed and boron is added normally in the form of ferro-boron, which can contain a significant quantity of aluminium. This could create difficulties of composition control as aluminium is

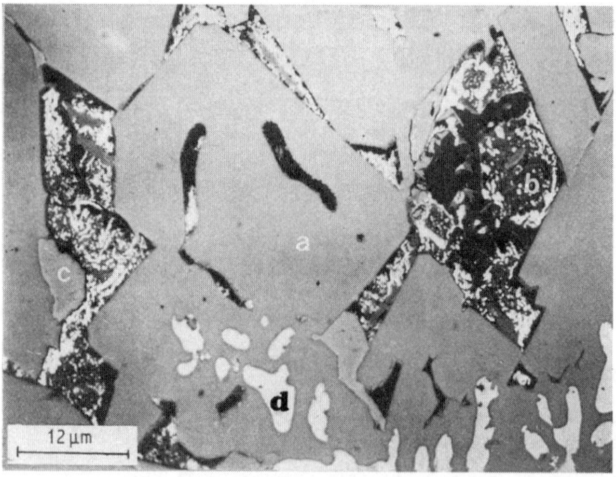

(a)

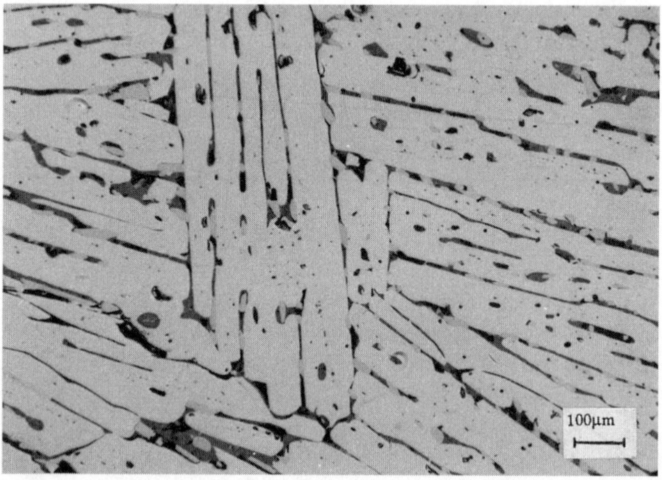

(b)

FIG. 7.2. Typical as-cast structure of $Nd_{16}Fe_{76}B_8$ ingot. (a) a: $Nd_2Fe_{14}B$ phase ϕ, b: Nd-rich material, c: $NdFe_4B_4$ phase η, d: α–Fe dendrites. (b) Lower magnification picture showing slab-like nature of the $Nd_2Fe_{14}B$ grains.

often added to improve the coercivity (see Table 7.1). Typically, the system is evacuated after loading the charge and pure argon is admitted. The mixture is then melted by induction heating and the molten alloy is poured into a mould. Cylindrical castings or rectangular slabs can be produced and a typical 'as-cast' structure is shown in Fig. 7.2. Reasonably rapid cooling is advantageous in

FIG. 7.3. A variety of Nd–Fe–B sintered magnets.

avoiding the formation of free iron dendrites in the casting which toughen the material, thus making it harder to crush prior to fine milling.

Co-reduction process. The co-reduction process was first developed for $SmCo_5$ [Her75, Dom76] and subsequently adapted for $Sm_2(Co, Fe, Cu, Zr)_{17}$ [Li81] and Nd–Fe–B [Her85, Zho87]. Possible synthesis pathways are

$$8Nd_2O_3 + 64Fe + \underset{\text{(ferro-boron)}}{20Fe_{0.6}B_{0.4}} + 24Ca \rightarrow Nd_{16}Fe_{76}B_8 + 24CaO \tag{7.1}$$

$$8Nd_2O_3 + 4B_2O_3 + 5Fe_2O_3 + 66Fe + 51Ca \rightarrow Nd_{16}Fe_{76}B_8 + 51CaO \tag{7.2}$$

The advantages of these reactions are that the reduction of rare-earth oxide is part of the alloy production process and the product is in the form of relatively fine powder (~ 20 μm average diameter) which can be readily milled to the final size.

Reaction (7.2) is preferred over reaction (7.1) because of the difficulty of controlling the impurity content, especially Al and C, in the ferro-boron. The reaction is carried out at 1100 °C in vacuum. Calcium oxide and excess calcium are separated from the Nd–Fe–B powder by leaching in water. The coreduced material has a significantly higher oxygen content than that of the induction melted alloy.

2 Monocrystalline powder processing

A major advantage of powder metallurgy is that a wide variety of sintered shapes (Fig. 7.3) can be produced which are machined to their final tolerances. The major disadvantage as far as Nd–Fe–B is concerned is that rare-earth transition-metal powders are very reactive and oxidize readily, particularly in damp environments. This means that the powders have to be produced and handled in a controlled environment and this increases the processing costs significantly.

2.1 Conventional powder route

The conventional route indicated in Fig. 7.4 [Sag84, Orm84, 85, Nar85] is virtually identical to that employed in the production of $SmCo_5$ [Das69, Benz70]. It was adapted for Nd–Fe–B by Sagawa and coworkers at Sumitomo. The process begins with jaw crushing and then coarse milling the induction-melted alloy prior to the production of fine powder by attritor or jet milling. The problem associated with the presence of α–Fe (seen in Fig. 7.2) is that it toughens the alloy significantly so that it becomes much less brittle than $SmCo_5$ or $Sm_2(Co, Fe, Cu, Zr)_{17}$, and is therefore difficult to crush. This problem can be overcome very effectively by the use of the hydrogen decrepitation process or by starting from coreduced or atomized Nd–Fe–B powder.

Fine milling. Fine powder with an average size of order 3 μm is needed for the subsequent stages of magnet fabrication. Great care has to be exercised when milling Nd–Fe–B. An attritor mill uses hardened steel balls and a solvent such as cyclohexane to aid dispersal and to limit oxidation. In the early days when the alloy was milled under freon, explosions and even fatalities occurred due to the violent reaction of the rare-earth with free radicals. No problems appear to have been experienced with cyclohexane, but a disadvantage of using a solvent is that the milling process is slower compared with jet milling and the solvent has to be removed from the powder by an additional drying operation. There is also the possibility of carbon pick-up and hence the neutralization of carbide-forming constituents such as niobium. The problem of oxygen pick-up during attritor milling has been considered by Ormerod [Orm84, 85], who showed that there is a progressive build-up of oxygen content with milling time, and that the Nd–Fe–B powder is significantly more reactive than Sm–Co powders.

An alternative to the attritor is the jet miller, shown schematically in Fig. 7.5. The Nd–Fe–B particulate material is loaded via a hopper into the milling chamber and high-pressure (5–10 bars) gas jets result in the collision of the particles which break up into progressively finer powder, the finest of which falls from the milling chamber via a central hole into collecting bins. High-purity nitrogen can be used and the gas is recycled after use. Jet milling is the preferred route since it is continuous, whereas attritor milling is a batch process. In addition, it avoids the environmental problems associated with the use of solvents such as cyclohexane.

Production of green compacts. The fine powder is aligned magnetically and pressed to make 'green compacts' ready for sintering which typically are around 60% dense. The pressing load is sufficient to make compacts having enough

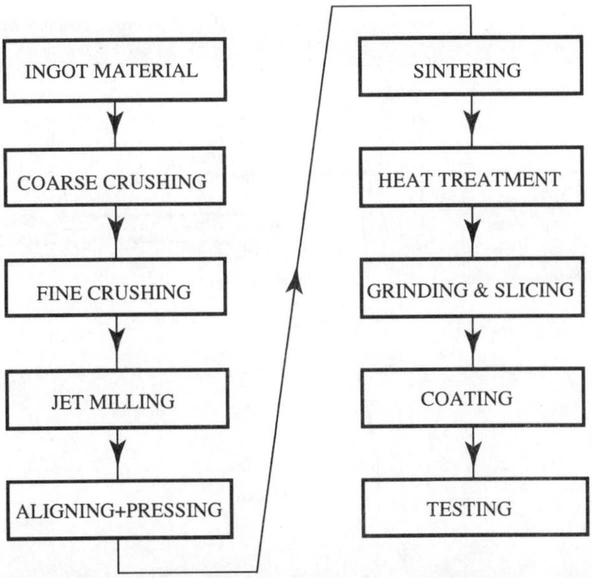

a)

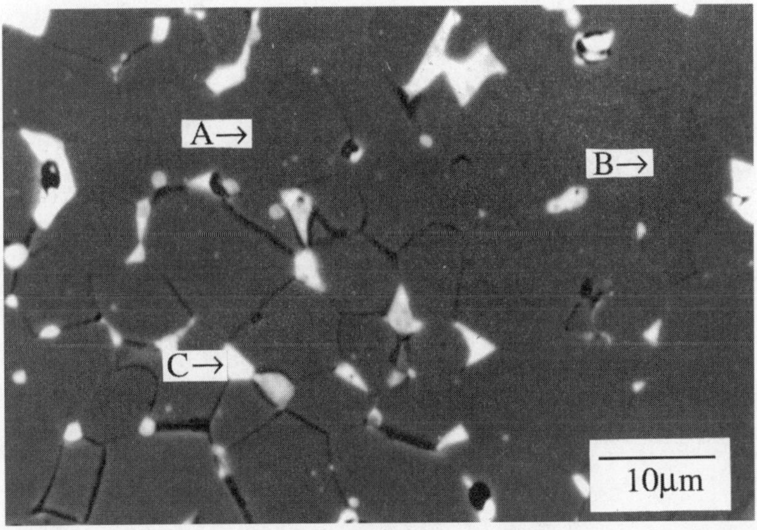

b)

FIG. 7.4. (a) The conventional powder route for the production of sintered magnets. (b) Back-scattered electron micrograph of a sintered $Nd_{15}Fe_{77}B_8$ magnet showing (A) $Nd_2Fe_{14}B$ (ϕ phase), (B) $Nd_{1.1}Fe_4B_4$ (η phase), and (C) Nd-rich phase (Photo, courtesy D.G.R. Jones).

strength to be handled but without significant misorientation of the crystallites. Additives, such as wax, may be added to the powder to improve its fluidity during the pressing operation, and these are driven off by the subsequent heat treatment.

Aligned green compacts can be produced by (a) aligning and pressing along the same axis; (b) aligning perpendicular to the pressing axis; and (c) isostatic pressing

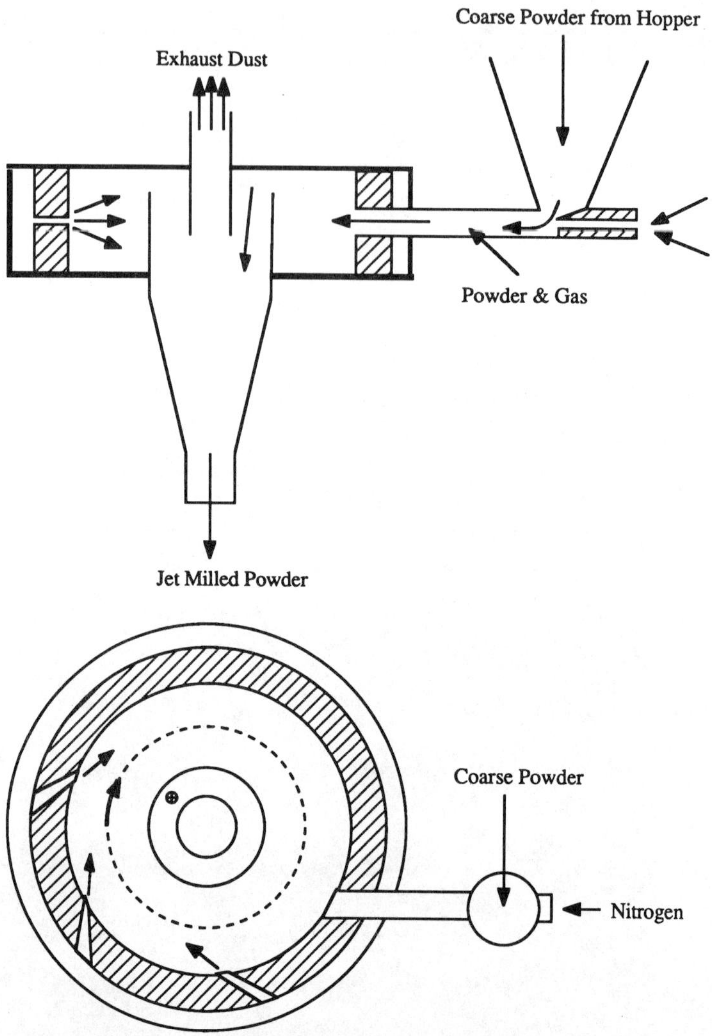

FIG. 7.5. A schematic representation of a jet miller.

of prealigned powder (Fig. 7.6). The most commonly used of these methods is (a) but the degree of alignment, and hence the quality of the magnet, improves progressively on going from (a) to (c). In the case of methods (a) and (b), alignment fields of around 1 T are provided by electromagnets, whereas in method (c), alignment is achieved by means of a pulsed magnetic field of around 6 T using a coil powered by a bank of capacitors.

Method (a) is favoured if the output is in the form of large numbers of cylindrical or ring-type magnets. If, on the other hand, rectangular or cubic shapes are required, then method (b) can be employed to produce moderate-sized blocks.

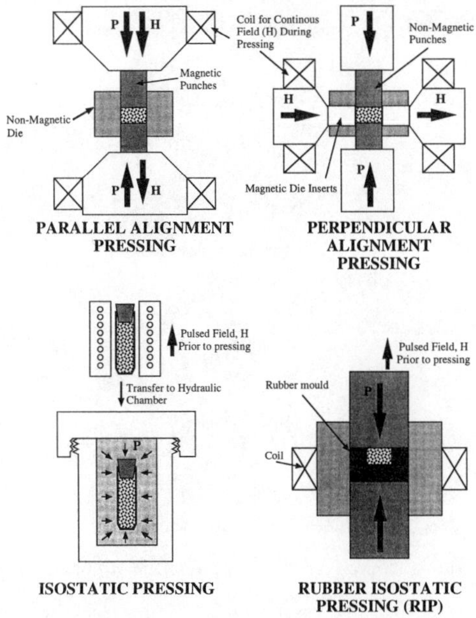

FIG. 7.6. A schematic representation of the various routes for powder alignment and pressing (including rubber isostatic pressing).

Large blocks which can then be sliced after the sintering operation are produced by method (c).

A cross between the uniaxial and isostatic pressing is the so-called 'rubber isostatic pressing (RIP)' method developed by Sagawa and coworkers [Sag93, 94]. The RIP arrangement is shown schematically in Fig. 7.6. The magnet powder is subjected to a strong pulsed field (4 T) just prior to compaction in the rubber mould. This ensures that the powder is very well oriented. The powder is then pressed isostatically while the orientation is maintained.

Sintering operation. Densification is achieved by heating the aligned green compact in a sintering furnace (Fig. 7.7) under high-purity argon. The compacts are normally held at the appropriate sintering temperature T_s for one hour before cooling to room temperature. The temperature, T_s, used for the $Nd_{15}Fe_{77}B_8$ composition is around 1080 °C [Sag84], but somewhat higher temperatures are required for dysprosium-containing magnets [War89]. Liquid-phase sintering and densification occur with the melting of the grain boundary material at and above 630 °C. The neodymium-rich material is very prone to oxidation [Orm84, 85] but when care is exercised to avoid oxidation both prior to and during the sintering process, then very good sintered densities ($\sim$ 99%) can be achieved.

There is a trend towards large-capacity plants ($>$ 20 t/m) with continuous heating furnaces [Nag94] and these offer three main advantages compared with the vacuum

FIG. 7.7. A sintering furnace. The green compacts are stacked in stainless steel trays (a) which are approximately 25 cm square.

chamber furnace. These are, (1) each magnet is subjected to the same temperature–time cycle; (2) quality control can be performed on line; and (3) a continuous heating furnace may be linked directly with the presses and with the subsequent machining steps, thus allowing a steady material flow resulting in shorter overall processing times and faster feedback for production control.

Magnetic properties of the sintered magnets depend critically on their subsequent thermal history [Sag84, Orm84, 85, Bai85, Nar85, Oka85, Tok85, Zho86, Hsu87, McG89a, Wit90]. Quenching from T_s results in severely degraded properties compared with those obtained on slow cooling, as shown in Fig. 7.8 [Wit90]. An anneal of one hour at 650 °C gives the best properties irrespective of the initial cooling condition. The quenched magnets differ from the slowly cooled magnets in that if they are annealed at temperatures around 400 °C their coercivity becomes progressively lower with ageing time, whereas no such changes are observed in the slowly cooled magnets. Optimum values can be regained in the aged magnets by annealing at 650 °C [Wit89]. The microstructure of the sintered magnets with oriented grains of $Nd_2Fe_{14}B$ separated by Nd-rich grain boundary regions is discussed in Chapter 6. Magnets with $(BH)_{max}$ values 400 kJ/m^3 and $H_c = 1040$ kA/m have been produced [End94] by the hydrogen decrepitation (HD) route (see later) with the aid of material design and by careful control of the oxygen content.

A variation on the normal sinter route is to blend single-phase $Nd_2Fe_{14}B$ powder with powder of an appropriate liquid-phase sintering agent. Such a method has enabled magnets with very high $(BH)_{max}$ values (> 400 kJ/m^3) to be produced [Ots90]. Because $Nd_2Fe_{14}B$ forms by a peritectic reaction it contains a large amount

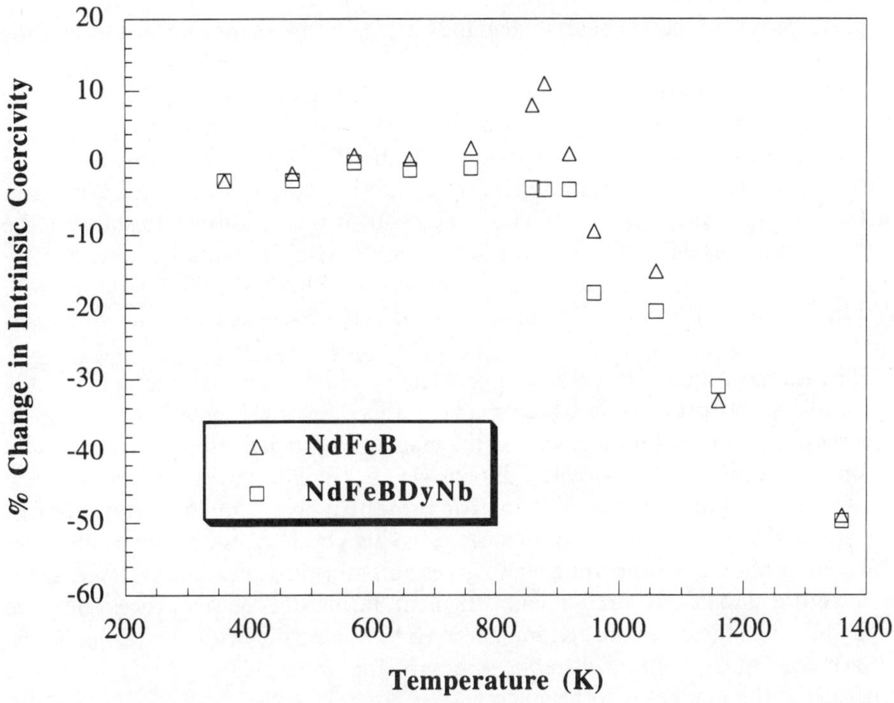

FIG. 7.8. The effects of quenching from different temperatures on the intrinsic coercivities of Nd–Fe–B and Nd–Dy–Fe–Nb–B sintered magnets. Each magnet was held for one hour at a given temperature and then quenched to room temperature.

of free iron in the cast condition and hence a long, high temperature homogenization treatment is required to produce single-phase material prior to the milling operation. The homogenization process can be accelerated by small additions of Nb to the stoichiometric alloy [Ahm94]. An alternative is to produce the stoichiometric alloy by means of the direct reduction process [Ste92] and to mill this powder with a Nd-rich sintering agent [Ala94].

Machining. Sintered magnets can be machined when the dimensions are critical, using a centreless grinding machine and nonreactive fluid. Alternative methods include electroerosion, which can be used to produce complex shapes with a wire saw, and annular sawing to produce slices from large sintered blocks. The Nd–Fe–B magnets are less brittle than magnets based on Sm–Co and better yields are achieved due to less chipping. Some deterioration in magnetic properties may appear after machining [Sta86, Ima87, Mül87, Nis90] but the magnets are not normally subject to any further annealing treatments at this stage. Because of the high raw material cost, it is obvious that losses on machining should be kept to a minimum since an effective means of recycling the waste has not yet been developed.

Coating. The final step in the manufacturing procedure is coating the sintered magnets, provided the application demands it. The poor corrosion resistance of the alloy which is associated with the reactivity of the rare-earth and multiphase nature of the material is one of the two principal difficulties associated with Nd–Fe–B, the other being the rapid decrease in coercivity with increasing temperature.

Studies of corrosion of Nd–Fe–B-type magnets have been concerned with the electrochemical behaviour of the alloys [Chi88, War89, Bal90, Blo90], laser surface amorphization [Arm89], alloy composition and additives to stabilize the grain boundaries [Kim87, 90, Ham89, Ten90, Hir90], metallic coatings by electroless, electrolysis and ion vapour deposition [Blo90, Mit90, Rib90, Tor90], chemical etchants [Kim89, Dic90] and the effects of selected organic chemicals [Dic90].

The ideal solution to the corrosion problem would be to develop an intrinsically stable alloy, and progress has been made in this direction [Ten90, Hir90, Shi94]. Otherwise, the only solution is to coat the magnet to provide a barrier to corrosion agents, primarily water vapour. Bloch *et al.* [Blo90] have summarized the requirements for an effective coating. The protective layer should be thin but free of pores and cracks, with a smooth surface having good adhesion to the substrate. The coating should be impermeable to gases and moisture and exhibit the required temperature stability. Hydrogen embrittlement during the coating process must be negligible. In some cases the coating needs to be electrically insulating. Finally, the surface layer must often be glueable, or compatible with the special metallic layers necessary if the magnet is to be soldered.

Proper preparation of the surface of the magnet is usually critical. It is also important to avoid introducing significant quantities of hydrogen into the alloy because this can cause bursting of the coating, and failure of the magnet itself. The coating must be thin enough to avoid undue increase in the air gaps but thick enough to provide long-term stability.

Imaizumi *et al.* [Ima87] have demonstrated that improved humid corrosion properties can be obtained by means of a 0.3-μm-thick oxidation layer, formed by a second post-sintering heat treatment at 600–1000 °C in a low-pressure oxygen atmosphere (10^{-7} bar). Other ways of producing a passivated layer include exposing powder or magnets to a nitrogen or carbon dioxide atmosphere [Bog91, 93] or immersion in an acid bath [Cav92]. Edgley and coworkers [Edg93, 94] have shown that an adherent oxide layer is formed by annealing Nd–Fe–B-type magnets in air at around 400 °C. Electron microscope studies [Edg93] revealed that this layer consists of aligned fine crystals of iron interspersed with particles of Nd_2O_3. In humid environments the oxidized magnets exhibited enhanced corrosion resistance compared with that of the nonoxidized samples.

Effective resin coatings can be achieved by means of a technique called barrel painting, which has been developed by Sagawa and coworkers [Sag94]. This technique is illustrated in Fig. 7.9 and consists of the following steps:

1. Formation of uncured resin layer where the magnets are dipped in an uncured resin diluted with a solvent and then dried.

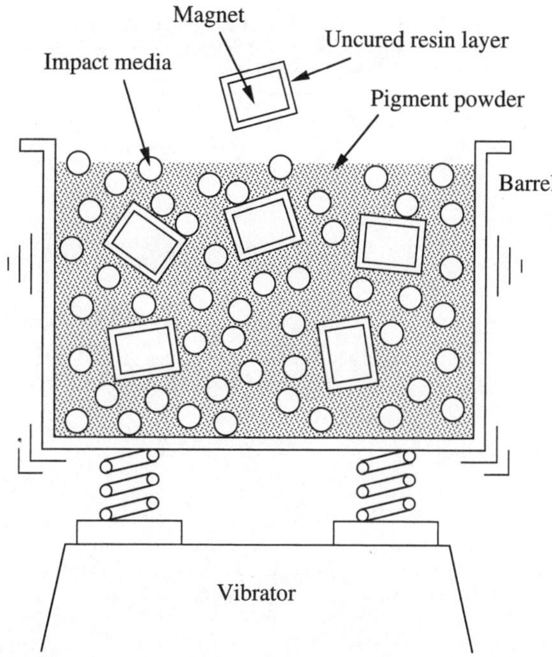

FIG. 7.9. The barrel painting method for resin coating of magnets.

2. Vibration in a barrel of pigment powders and steel or ceramic balls to prepare a well-dispersed mixture.
3. The magnets are put into the barrel and the pigment is embedded into the resin layer by the impact media.
4. Curing the resin, thus enhancing the adhesion of the coating.

Various coating techniques are summarized in Table 7.3. The best overall results have been obtained with electrolytic nickel coatings, and the coating procedure is summarized in Fig. 7.10.

2.2 *Alternative powder routes*

Some alternative means of producing powdered material for sintering are now considered.

2.2.1 *Atomization* This is a way of producing powder directly from the molten alloy [Nar86a]. Atomization is a misnomer as the powder is relatively coarse ($\sim 20\ \mu$m) and milling is still required to reduce it to the size required to make sintered magnets. The experimental arrangement is shown in Fig. 7.11. Molten alloy is poured through an orifice and 'atomized' by an argon gas stream to produce the powder. A small amount of oxygen can be added to the atmosphere in the chamber as it has been found [Nar86b] that the resultant powder is stabilized by

Table 7.3 Coating procedures for Nd–Fe–B sintered magnets

Coating materials	Technique
Aluminium	Ion vapour deposition
Electropaint	Cathodic
Titanium nitride	Plasma assisted physical vapour deposition
Silicon oxide	Chemical vapour deposition
Nickel/zinc	Electroplating or electroless plating
Zinc	Dacromet process*
Resin coating	Air spray of bulk material, barrel painting

* Coating in an aqueous dispersion containing zinc flakes, chromium and proprietary organics. This is followed by furnace drying to remove water and final baking at 300 °C.

passivation of the particle surface and produces sintered magnets with squarer loop shapes compared with the nonoxygen-treated powder. Magnets made from atomized powder are found to have a finer, more uniform grain size compared with those of a number of other types, and they exhibit very square demagnetization loops [War89].

A variation of atomization which has been used to produce Nd–Fe–B layers is liquid dynamic compaction (LDC) [Chi86, Tan86]. A stream of molten alloy converted into a spray of fine particles by a jet of high-velocity gas is collected on a substrate and a solid layer of Nd–Fe–B is built up. Disadvantages of the technique are that the layer is isotropic, with a large variation in the grain size with depth.

Atomized powder has been encapsulated in a steel container and then extruded at high temperatures [Cha89]. The extruded rods exhibited a radial texture and were subject to an optimizing heat treatment at 900–950 °C, where slight grain growth improved the loop squareness. A field of 2.8 MA/m was required to magnetize the material to an optimum condition. The radial texture is most suitable for permanent magnet motors for electrical machines, as shown in Fig. 7.12. A variant of the process is back extrusion, where the powder is pressed down a central hole and forced back into the space between two coaxial cylinders, where it forms a radially oriented ring magnet suitable for cutting into arc segments. This process has also been applied to melt-spun material (see later).

2.2.2 *Hydrogen decrepitation (HD)* The use of hydrogen in the processing of rare-earth magnets was first proposed by Harris and coworkers in 1979 [Har79]. Early work concerned hydrogen decrepitation of $SmCo_5$, but the process was later

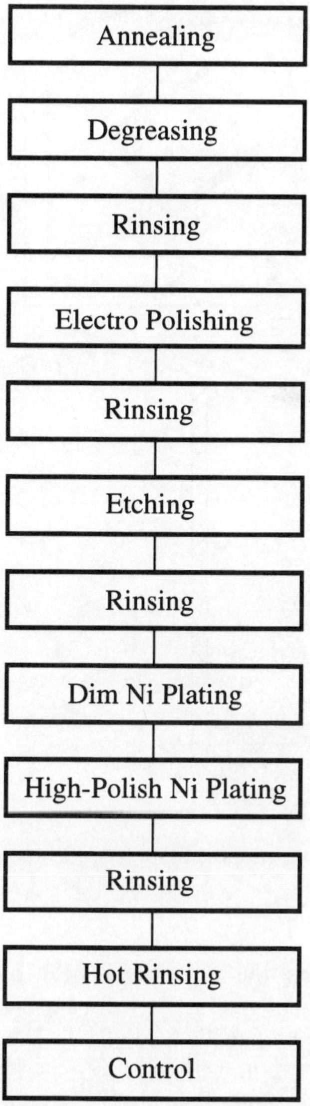

FIG. 7.10. A schematic representation of the nickel coating procedure.

extended to $Sm_2(Co, Fe, Cu, Zr)_{17}$-type alloys [Kia84, 85]. Soon after the discovery of $Nd_2Fe_{14}B$ magnets, studies showed that alloys such as $Nd_{16}Fe_{76}B_8$ reacted readily with hydrogen at moderate pressures in a strongly exothermic reaction which led to decrepitation of the ingot [L'He84, Oes84, Har85, 89, Cad86, Wie87, Fru89]. The process has been exploited [McG86, 89b] to produce sintered magnets with excellent properties and the overall process is shown in Fig. 7.13.

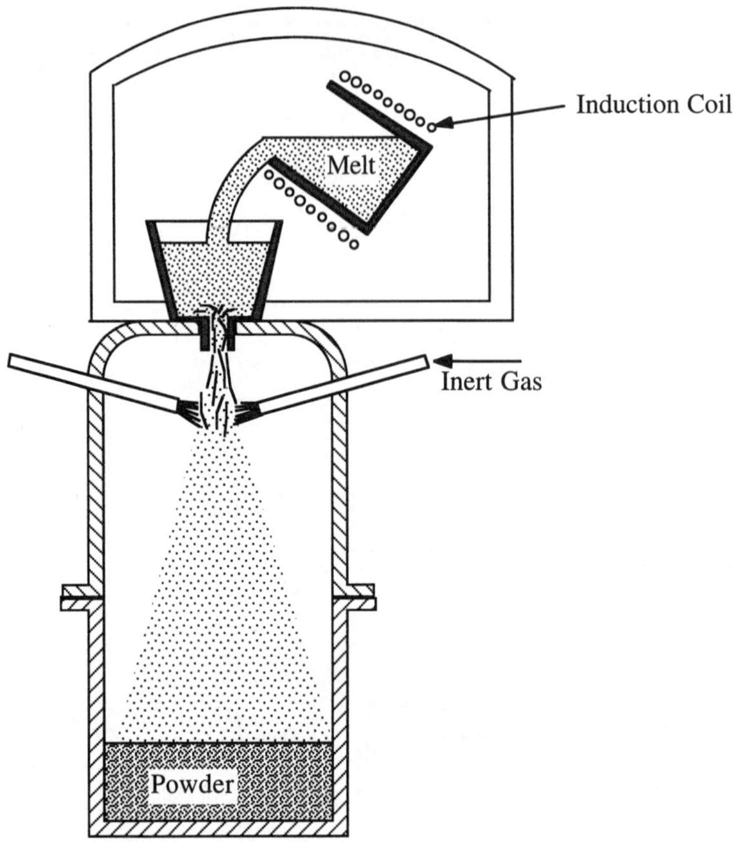

FIG. 7.11. A schematic representation of the atomization process.

Although hydrogen lowers the anisotropy field of $Nd_2Fe_{14}B$, [Oes84] the hydrided powder is still sufficiently anisotropic for the *c*-axes of the crystallites to be aligned effectively by a modest magnetic field of 1 T. The hydrogen is then removed from the alloy during a vacuum anneal prior to sintering. Figure 7.14 shows that hydrogen is desorbed in three stages [Wil90, 91, Nak90], first from the $Nd_2Fe_{14}B$ matrix phase and then in two steps from the Nd-rich phases. Benefits of hydrogen decrepitation are numerous. The major advantage is that the normally tough induction-melted ingots can easily be reduced to a relatively fine powder which is extremely friable. This means that the jaw crushing and coarse milling operations, which are difficult and expensive in energy, can be eliminated, and the time for jet milling can thus be reduced considerably (Fig. 7.15). Intergranular failure produces single-crystal particles with very clean surfaces which are ideal for sintering, and it is possible to control the particle size and shape distribution by varying the HD conditions. Oxygen pick-up during processing is reduced as desorption of hydrogen

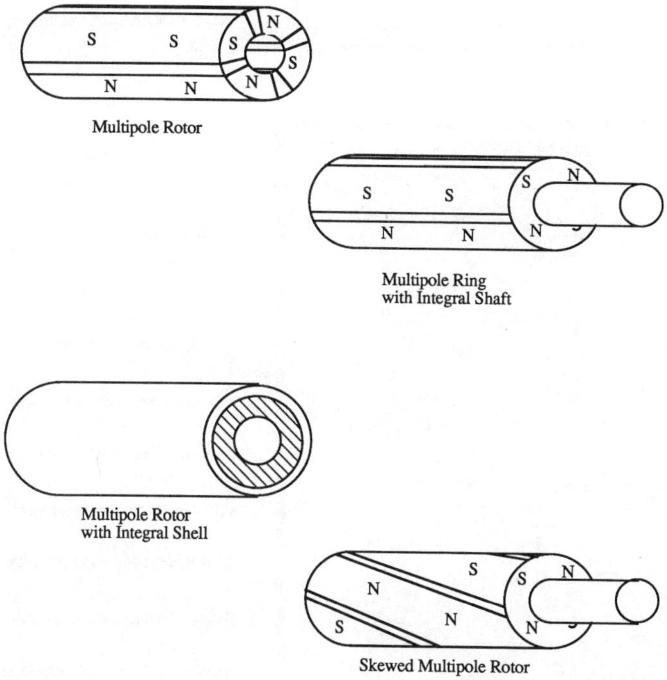

FIG. 7.12. Various forms of extruded rods using atomized powder for electric motor applications.

on heating the green compacts creates a nonoxidizing environment. The near-zero remanence of the compacts makes them easy to handle.

The morphology of the hydrided material [McG90a, 90b], together with a typical particle-size distribution, are shown in Fig. 7.16. Flake-like powder is often produced by an 'onion skin' decrepitation mechanism. Thus, hydrogen diffuses into the bulk to a particular depth below the surface where a critical strain resulting from the volume difference between the hydrided and nonhydrided material causes fracture at the interface and hence flaking. This fracture behaviour is characteristic of tough material such as ingot-containing dendritic α–Fe (Fig. 7.2). Homogenized ingots exhibit different hydrogen decrepitation behaviour [McG90c], with a faster reaction rate and evidence of intergranular and transgranular failure more typical of the HD-behaviour of 2:17 alloys [Kia85].

The hydrogen decrepitation process can be applied in the manufacture of the whole family of Nd–Fe–B alloys, and it has been adopted by magnet manufacturers throughout the world. Their magnets tend to have finer grain size and hence greater mechanical strength than the magnets produced by the conventional powder process. The HD process has also been employed to produce coercive powder from hot-pressed, die-upset and hot-rolled material, and these processes will be described later in the chapter.

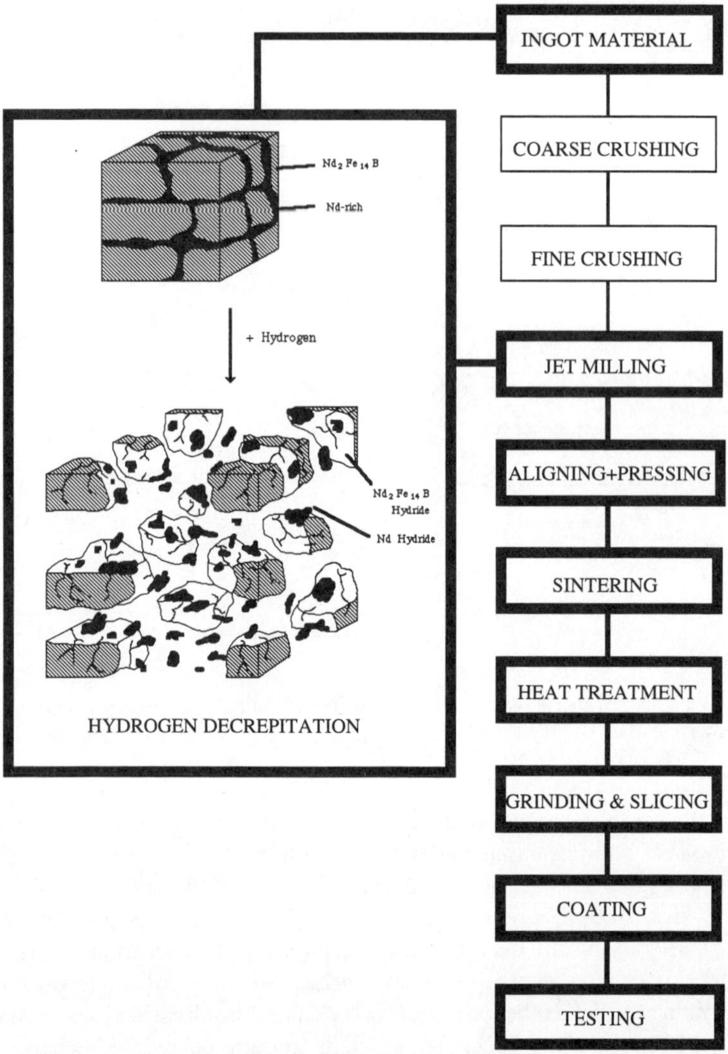

FIG. 7.13. The hydrogen decrepitation (HD) process.

Hydrogen has also been employed to characterize Nd–Fe–B-type ingots, powders and sintered magnets. Thus the hydrogen uptake [Cad86, Har87a, b, c, 89, Scho87] enthalphy of hydrogen absorption [Har87a, b, c, Scho87, McG89c] and the thermal desorption behaviour [Har87a, b, c, 90, Scho87, Rup88, McG89c, Nak90, Wil90, 91] can be used to determine the proportion of Nd-rich material in a particular sample and the extent of oxidation of this material.

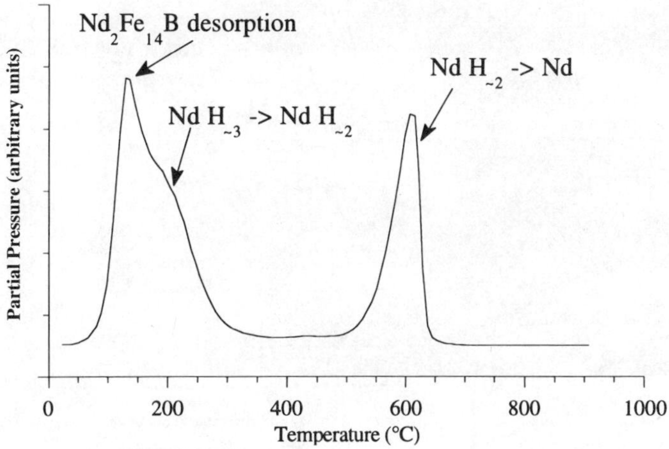

FIG. 7.14. Mass spectrometer trace of the hydrogen desorption behaviour of hydrided $Nd_{16}Fe_{76}B_8$ powder.

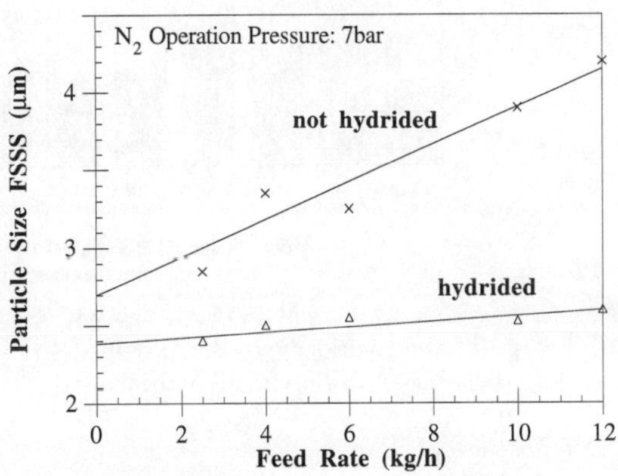

FIG. 7.15. Particle size of jet milled powder (determined by Fisher Sub Sieve Sizer, FSSS) vs feed rate for hydrided and nonhydrided feedstock.

3 Polycrystalline powder processing

There are three main independent methods of processing Nd–Fe–B into coercive particles where each particle is constituted by many submicron grains which may be contiguous or separated by a grain-boundary phase. This material can be mixed with epoxy resin or rubber and moulded into appropriate shapes, or combined with a thermoplastic polymer such as nylon and used as a feed stock for injection moulding of complex net-shape components. Many polymer-loaded magnets are

(a)

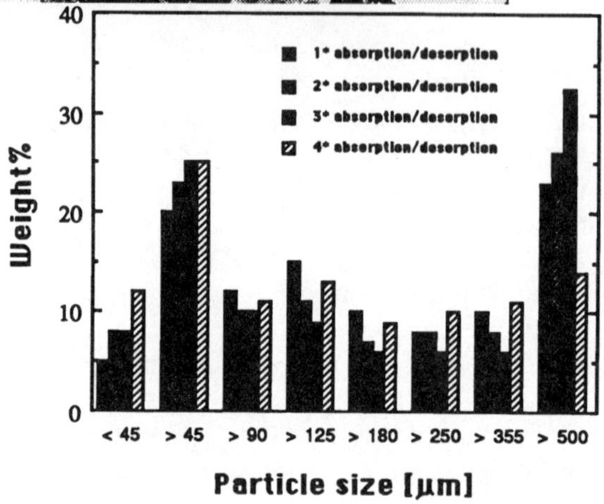

(b)

FIG. 7.16. (a) Morphology of the hydrided powder obtained from the cast material of $Nd_{16}Fe_{76}B_8$ alloy. (b) Particle-size distribution of hydrided powder after various cycles [Scho87].

made with strontium ferrite ($SrFe_{12}O_{19}$), as magnetic filler, but the availability of coercive Nd–Fe–B powder greatly expands the potential of these types of magnet. Alternatively, the particulate material can be hot-pressed to produce fully dense isotropic or anisotropic magnets.

3.1 *Melt-spinning*

Melt-spinning is a direct production route which had no precedent for any permanent magnet before Nd–Fe–B. The development of melt-spun magnets by Croat and Herbst of General Motors [Cro84, 88, 90, 94; Her91, Pan89] ran parallel with that of sintered magnets by Sagawa at Sumitomo. Because of the very rapid cooling rate and the consequent avoidance of free iron, the compositions of the

Table 7.4 Some commonly used alloy compositions (at.%) for melt-spun magnets

Nd	Fe	B	Co	Ga
13.9	80.6	5.5	–	–
14.0	63.6	5.5	16.9	–
12.2	77.0	5.4	5.4	–
12.5	65.6	5.4	16.5	–
13.9	77.8	5.5	2.8	–
13.9	72.4	5.9	7.2	0.6

melt-spun alloys are closer to stoichiometry than the sintered magnets. Some typical compositions are summarized in Table 7.4.

In the melt-spinning process, a jet of molten alloy is fired at a rapidly rotating, water cooled copper wheel where the liquid solidifies at a cooling rate of up to 10^6 K/s and is thrown off the wheel in the form of the fine ribbon. In the case of Nd–Fe–B, the brittle nature of the alloy results in the ribbon breaking into short lengths or flakes. The physical state of the ribbon depends upon the quenching rate and therefore upon the wheel speed, ejection conditions and melt temperature. There is a strong dependence of the coercivity on the wheel speed [Cro84] and this is shown in Fig. 7.17. The usual manufacturing procedure is to overquench the ribbon and then to anneal it as a subsequent processing step in order to produce the required grain size and coercivity. The ribbons are around 30 μm thick and a typical average grain size for the annealed ribbon is $\sim$ 50 nm, a hundred times smaller than in the sintered magnets.

The melt-spun ribbon is magnetically isotropic and it forms the basis of a range of commercial magnets. The various production routes are shown schematically in Fig. 7.18. The simplest process involves light milling of the ribbon material, a short anneal at 700 °C to develop the required coercivity, mixing with a resin with a loading of around 80 vol. % and compression moulding. The powder can also be blended with polymer to make injection-moulded polymer-bonded magnets and this allows the rapid production of net-shape magnets with loadings of around 60 vol. %. These isotropic polymer-bonded magnets are used extensively in applications where temperatures do not exceed approximately 100 °C. A problem is their propensity to corrode under hot, humid conditions because of the permeability of the polymer matrix to oxygen and water. The problem of corrosion was critical when resin-bonded magnets were employed in vehicle starter motors, where they were found to be particularly susceptible to salt-spray corrosion. This difficulty was overcome by encapsulating the magnets in a metal skin inside the motor. Polymer-bonded magnets for use in less-demanding environments can also be protected from corrosion by coating procedures similar to those described earlier for sintered magnets. The surfaces can be coated with ion-vapour deposited aluminium or with electropaint, for example. Methods have been developed for

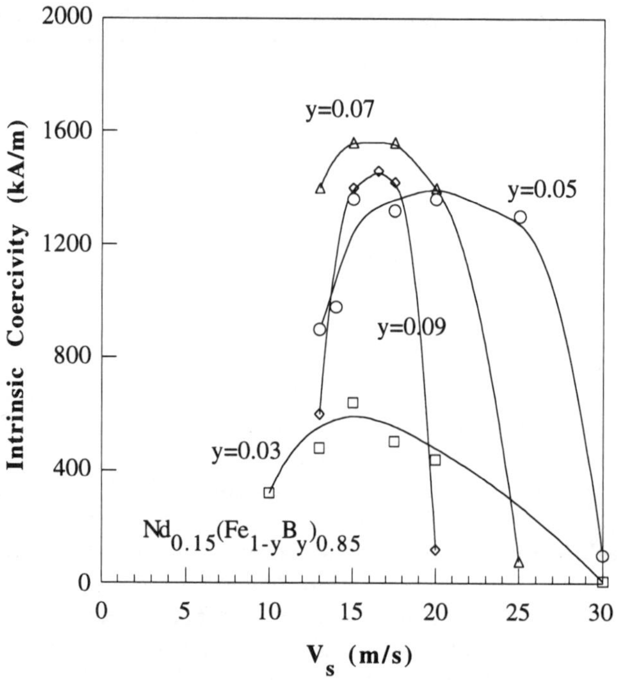

FIG. 7.17. Coercivity of melt-spun ribbon vs wheel speed [Cro84].

coating the individual particles which give improved corrosion resistance and compaction of the final product [Kas89].

High-density, isotropic magnets can be made with a metal matrix by rotary forging the melt-spun powder at room temperature with a soft metal powder such as aluminium [Row89, 90, Bret90]. A very high loading on the powder compact can be achieved by limiting the contact area of the upper tool and applying a differential rotary action of the tool and the powder sample. With a loading of 90 vol. % of flake and a PTFE binder, energy products of around 84 kJ/m^3 can be obtained [Tat94a, 94b]. A disadvantage of the isotropic bonded magnets is their relatively low-energy product due to the lower remanence because of the limited volume fraction of magnetically active material. Mechanical strength might also be a limiting factor for certain applications. Advantages are that an orientating magnetic field is unnecesary, which allows for faster cycle time and greater flexibility in the size and shape of the bonded parts which can be produced. The temperature coefficient of intrinsic coercivity of the isotropic magnets is significantly better than that for the oriented magnets made by the conventional powder route and the coefficient of the remanence is improved by small additions of cobalt [Bur90]. In electric motors the high electrical resistance is an advantage with regard to the suppression of eddy currents.

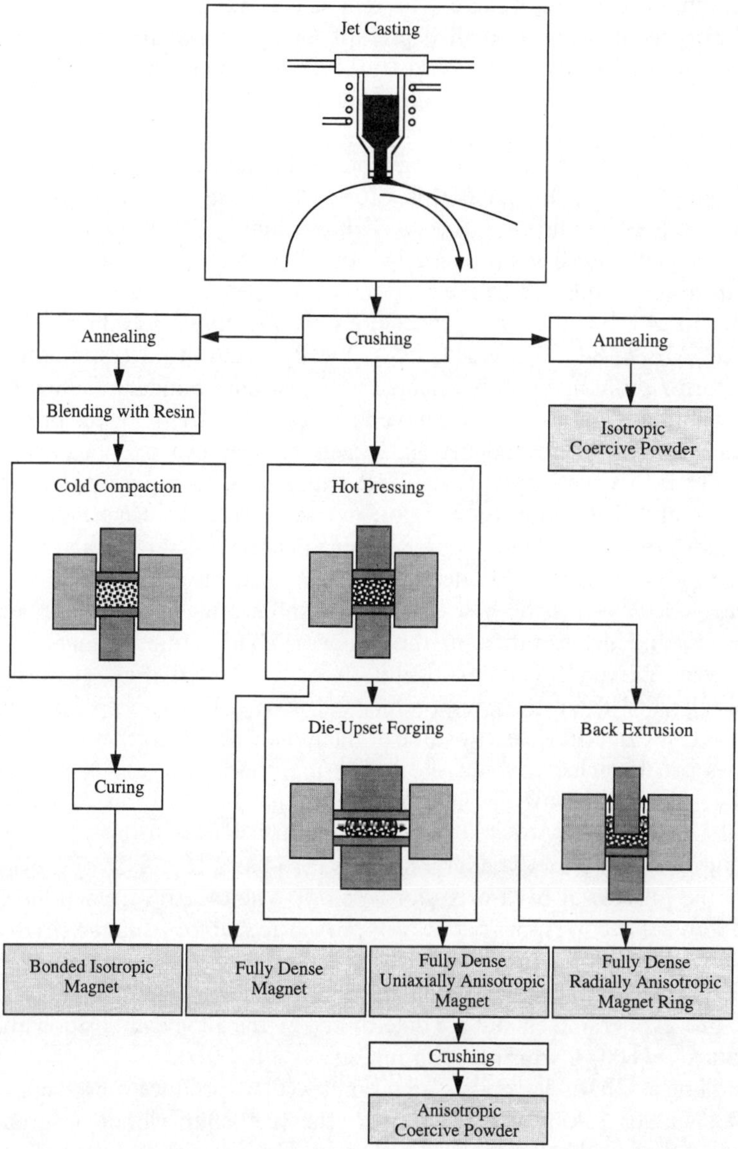

FIG. 7.18. A schematic representation of various production routes for production of magnets from melt-spun ribbon.

The polymer or metal matrix can be eliminated by hot-pressing the melt-spun flake at 750 °C to form a fully dense compact. A cross-section of such a compact reveals the boundaries of the bonded ribbon particles which are typically 30 μm thick. It should be noted that the individual grains are too fine ($\sim$ 40 nm) to be seen by standard metallography. The compacts are pressed between graphite or special alloy dies in an inert atmosphere. The remanence values (~ 0.8 T) of this second type of magnet indicate a small degree of enhancement in the properties. The compacts are mechanically very hard and difficult to machine. They exhibit much better corrosion resistance than the resin-bonded product but are often coated to give improved corrosion stability.

A third type of magnet is produced from melt-spun flake by the process of die-upset forging, that is, the material is allowed to increases its dimensions at right angles to the pressing direction [Lee85, Tok89, Shin92, Cro94, Leo94]. Fully dense isotropic magnets are first produced by hot-pressing and these are then die-upset forged to about a quarter of their original thickness. The process results in the development of a strong *c*-axis texture along the pressing direction which has been attributed [Mis86, 88, 90, Iwas89, Li90, Xie90, Chu91] to a combination of the shear deformaion in favourably oriented grains, grain boundary sliding, boundary migration and diffusion slip. The detailed process parameters for the production of die-upset forged Nd–Fe–B magnets have been reported by Tokunaga *et al.* [Tok89, Shin92]. These workers found that the $Nd_{14}Fe_{80}B_6$ composition has the best workability and the compression ratio (inverse of reduction in height) should be larger than four to achieve fully aligned material. Grain coarsening during die-upsetting resulted in a decrease in H_c. It was found that gallium ($Nd_{14}Fe_{79.25}B_6Ga_{0.75}$) is the best element for enhancing H_c without unacceptable losses in B_r and deterioration in the hot workability. Improvement in thermal stability was achieved by cobalt-substitution for the Fe ($Nd_{14}Fe_{71.75}Co_{7.5}B_6Ga_{0.75}$), thus raising T_c without a corresponding fall in H_c. The preferred orientation is clearly evident from the excellent value of the remanence. A further improvement in properties can be achieved by a short (five minute) anneal at 800 °C but longer anneals result in excessive grain growth and the deterioration of the magnetic properties.

Rings with radial *c*-axis orientation can be produced from isotropic cylinders or rings by the process of back extrusion [Yos94]. The die arrangement is shown in Fig. 7.18 and this corresponds to a well-proven technology for the production of very thin walled metallic tubes, such as those used in the manufacture of shells. As with the die-upset process, the speed (strain rate) and the degree of deformation are not free-process parameters but are determined by the alloy composition and by the temperature (~ 800 °C) during deformation.

Hot-rolling methods have also been employed to produce anisotropic powder [Muk90, Wad90, Sak91]. In this process the melt-spun ribbon is ground into powder and then packed into steel containers under vacuum and hot-rolled to 82% thickness reduction at 700 °C. The slabs produced by this method are then pulverized to form anisotropic powder with a particle size of around 250 μm. This powder can then be aligned in a magnetic field (up to 1.5 T) and consolidated by hot-pressing using a form of electro-sintering already employed to produce isotropic compacts [Wad90]. Intricate magnet parts such as rings and segments have been fabricated in net-shape using this technique.

The fully dense anisotropic compact can be ground to form stable anisotropic powder which can then be used to make anisotropic bonded magnets [Tok89,

(a) (b)

FIG. 7.19. Typical morphologies of lightly crushed melt-spun ribbon (a) and of crushed anisotropic powder (b).

Cro90]. The HD process has been employed [Dos91] to produce anisotropic powder and in this case the hydrogen has to be removed from the powder by an appropriate degassing treatment. The anisotropic powder exhibits a more granular shape (compared with the flake-like isotropic ribbon material) with a nonequiaxed morphology, and the two types of powder are shown in Fig. 7.19. The melt-spun flake and the anisotropic powder can be stored in air for a considerable time without any deterioration in the magnetic properties, in contrast to milled Nd–Fe–B powder and the crushed sintered Nd–Fe–B powder, both of which exhibit a rapid deterioration in magnetic properties on exposure to air [Esh90]. This can be attributed to the different behaviour of single-grain and multigrain particles and possibly to the different nature of the grain boundary material.

A typical $(BH)_{max}$ value for an anisotropic bonded magnet is around 120 kJ/m^3. However, the powders are less stable than powders of the isotropic melt-spun material and, because of their shape, they are difficult to align magnetically.

Aluminium-bonded, rotary forged magnets [Row89] have been produced from the die-upset anisotropic powder with $(BH)_{max}$ values of 147 kJ/m^3. These exhibited better mechanical properties compared with those of corresponding magnets made using isotropic flake [Bret90] and this was attributed to the less angular particle shape of the anisotropic powder and better wetting by the aluminium.

Much effort has been directed towards producing anisotropic ribbon straight from the wheel, as this would have obvious advantages over the methods outlined above. Unfortunately, this has not proved possible, although some encouraging results were reported by Coehoorn and Duchateau [Coe89], who showed that, for a $Nd_{13.5}Fe_{79.6}B_{6.9}$ alloy, for instance, melt-spun flakes exhibited a pronounced *c*-axis texture normal to the surface of the flakes at the free side but a much weaker texture on the wheel side. For optimum composition and quench rate, $(BH)_{max}$ values 10–15% better than in the case of the isotropic melt-spun flakes were obtained; these are well below the fully anisotropic value.

Substantially enhanced magnetic properties in cast melt-spun ribbons have been reported for alloys containing small concentrations of silicon [Ber86, Dav89] (e.g. $Nd_{13.2}Fe_{79.6}B_6Si_{1.2}$). The ribbon material was magnetically isotropic, even though $B_r > 0.5B_s$. Enhanced properties have also been reported for melt-spun alloys containing small amounts of vanadium [Yam89] or zirconium [Yaj88]. Exchange coupling between randomly oriented nanocrystals can explain these results [Kne91, Sko93, Kro94] but as the remanence is enhanced the coercivity declines. Davies and coworkers [Dav94] have shown that, by very careful control of the grain size, enhanced remanences can be obtained in melt-spun material, without the necessity of minor alloy additions. It has also been shown [Man93a, 93b, Dav94] that it is possible to achieve enhanced magnetic properties in Fe-rich (e.g. $Nd_9Fe_{86}B_5$) melt-spun alloys. The microstructure of these ribbons was shown to consist of nanoscale (~ 15 nm) grains of α–Fe phase dispersed within 30 nm grains of $Nd_2Fe_{14}B$. These alloys might provide a means of enhancing the $(BH)_{max}$ value of bonded melt-spun magnets without the necessity of developing anisotropy in the ribbon.

Another type of nanocrystalline melt-spun alloy containing $Nd_2Fe_{14}B$ and with the composition $Nd_{4.8}Fe_{77}B_{19.2}$ has been reported by Coehoorn *et al.* [Coe88]. This was produced by melt-spinning to the amorphous state, followed by annealing (at 670 °C) to form a crystalline, nanophase structure of Fe_3B, $Nd_2Fe_{14}B$ and α–Fe. The magnetic properties of this material ($B_r = 1.2T, H_c = 240$ kA/m) were ascribed to the ferromagnetic exchange coupling between the nanocrystalline grains, and the very fine microstructure is an essential feature of these magnets. Thus it can be resin bonded to form magnets but the problem with hot-pressing to form fully dense magnets is the coarsening of the microstructure at elevated temperatures. With regard to the rather low initial coercivity [Coe88], the performance of the magnets has been improved considerably by minor alloy additions such as Ga [Hir93, Kan93] and Cr [Hir94]. Bonded magnets based on these materials occupy a useful range of properties between those of fully dense, hard ferrites and conventional bonded rare-earth magnets.

The exchange-coupled nanocrystalline magnets are often referred to as 'exchange spring' magnets (see Chapter 5, p. 280)

3.2 *Mechanical alloying*

Mechanical alloying is well established in the field of superalloys [Ben70] and has now been applied to the experimental production of Nd–Fe–B magnets by Schultz and coworkers [Sch87, 88a, 88b, 88c, 92]. The technique is still in a research and development stage and has yet to be applied on an industrial scale. Nevertheless, it should be considered here as a possible alternative to melt-spinning.

Stadelmaier *et al.* [Sta85] prepared Nd–Fe–B isotropic magnets by reaction of Nd, Fe and Fe_2B powders. Mechanical alloying enables the intimate mixing of such a mixture prior to reaction. Mechanical alloying of Nd–Fe–B powder has been performed by planetary ball milling powders of the constituent elements under argon. This treatment first produces powder with a layered microstructure of Fe and Nd with the submicron B particles remaining underformed and embedded in the Fe/Nd interfaces. Atomic mixing does not occur and the hard magnetic phase $Nd_2Fe_{14}B$ is only formed by a subsequent heat treatment. The reaction begins at relatively low temperatures and takes place with short reaction times at higher temperatures because of the extremely fine distribution of the reactants in the milled material. Optimum coercivities are obtained after an annealing at 700 °C for 15–30 min, which yields a grain size of around 50 nm. Resin-bonded magnets exhibiting isotropic properties can be made from this material and compacted magnets can be made by hot uniaxial pressing. Magnetically anisotropic magnets can be prepared by die-upset forging at high temperatures with process conditions similar to those employed for the melt-spun ribbon.

3.3 *The hydrogenation, disproportionation, desorption and recombination (HDDR) process*

Coercive powder can be produced from the cast and homogenized Nd–Fe–B-type ingots by the HDDR-process, and this is illustrated in Fig. 7.20. When Nd–Fe–B alloys are heated in hydrogen to above 650 °C, the $Nd_2Fe_{14}B$ matrix phase disproportionates into iron, neodymium hydride and ferroboron [L'Hé84, Har85, Oes85, Cad86] and in 1985, Oesterreicher [Oes85] proposed that it should be possible to recombine such a disproportionated mixture by heating under vacuum, thus producing a fine grained mirostructure, as observed previously for hydrogen storage alloys.

In 1989, Takeshita and Nakayama [Tak89] reported that coercive powder could be produced by heating Nd–Fe–B alloys under hydrogen and then vacuum annealing. McGuiness and coworkers [McG90a, 90b] interpreted this behaviour in terms of what they termed a hydrogenation, disproportionation, desorption and recombination (HDDR) reaction.

The disproportionation reaction can be represented as:

$$Nd_2Fe_{14}B + (2 \pm x)\,H_2 \rightleftharpoons 2NdH_{2\pm x} + 12Fe + Fe_2B + \Delta H \qquad (7.3)$$

ΔH is the exothermic (absorption) or endothermic (desorption) heat of reaction. The particular value of x depends on the temperature and hydrogen pressure.

The disproportionated mixture (see Fig. 7.20) consists of a very finely divided mixture of the components [Xia92, Gut94a, 94b, Manw94, Mar94] and on removal of the hydrogen by vacuum annealing, the Nd, Fe_2B and Fe recombine to form $Nd_2Fe_{14}B$ with a submicron ($\sim 0.3\ \mu m$) grain size. Thus the HDDR process converts the initial, coarse grained ingot material (Fig. 7.2) into very fine grained powder (Fig. 7.20), which, if optimally processed, exhibits appreciable intrinsic coercivity. The best magnetic properties for the HDDR powder are achieved by an annealing treatment which leads to an Fe-free, uniformly fine grained ($\sim 0.3\ \mu m$) microstructure. Too low an annealing temperature results in the persistence of α–Fe, whereas too high a temperature results in the rapid growth of large faceted grains [Har92, McG92, Jon94]. Both these features are inimical to the development of coercivity. These microstructures are illustrated in Fig. 7.21.

Takeshita and coworkers [Tak90, 92, Nak93] have shown that small additions of certain elements such as Zr, Hf or Ga (e.g. $Nd_{12.5}Fe_{69.9}Co_{11.5}B_{6.0}Zr_{0.1}$) have the remarkable effect of producing anisotropic HDDR powder from which resin-bonded magnets with $(BH)_{max}$ values around 140 kJ/m^3 have been produced [Tak90, 92, Nak93, 94]. According to Nakayama *et al.* [Nak94], it was necessary to homogenize the alloys at 1120 °C for 20 h in 1 bar argon. They were then crushed into blocks of size less than $20 \times 20 \times 20$ mm^3, then heated in hydrogen at 1 bar to

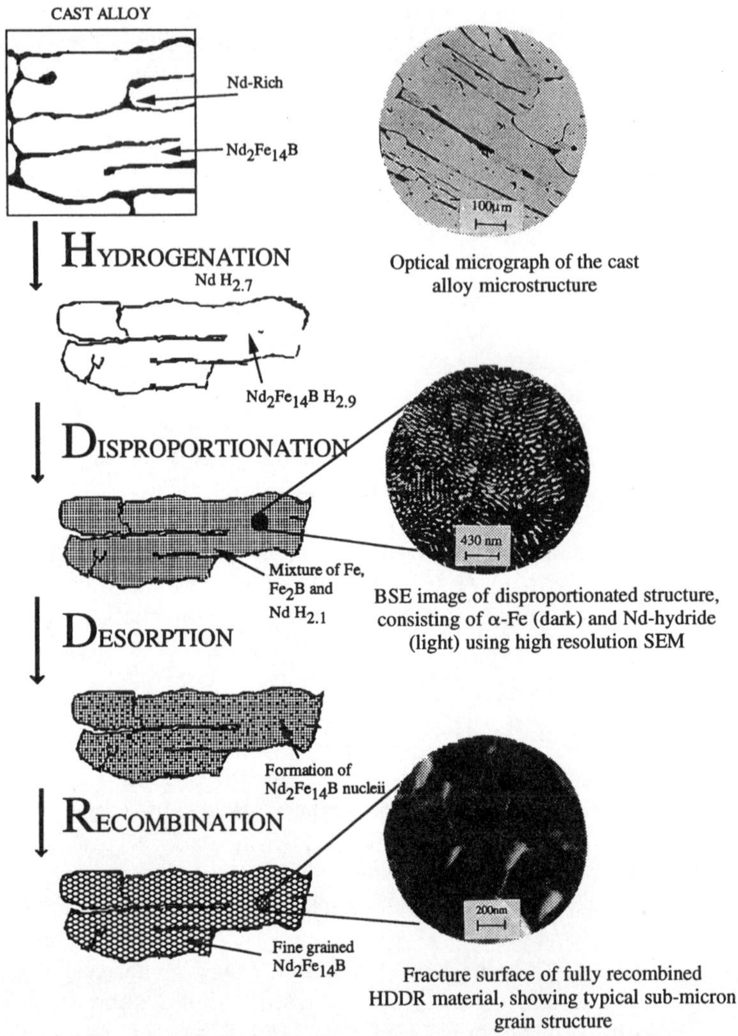

FIG. 7.20. A schematic representation of the HDDR process with typical miscrostructures.

830 °C and kept at this temperature for 10 min. The system was then evacuated to a high vacuum ($\sim 10^{-8}$ bar) for about 40 min and the material was subsequently quenched in argon gas. The anisotropic bonded magnets were prepared by mixing the HDDR powder (crushed to a particle size of < 420 μm) with epoxy resin and moulding in an alignment field of 2.5 T. The magnets were then set by heating at 100 °C for 2 h.

The HDDR powder has also been found to be very suitable for the production of hot-pressed magnets [Har92, McG92, Tak92, Manw94]. The hot-pressed magnets reported by McGuiness *et al.* [McG92] and by Manwaring and coworkers [Man94]

(a)

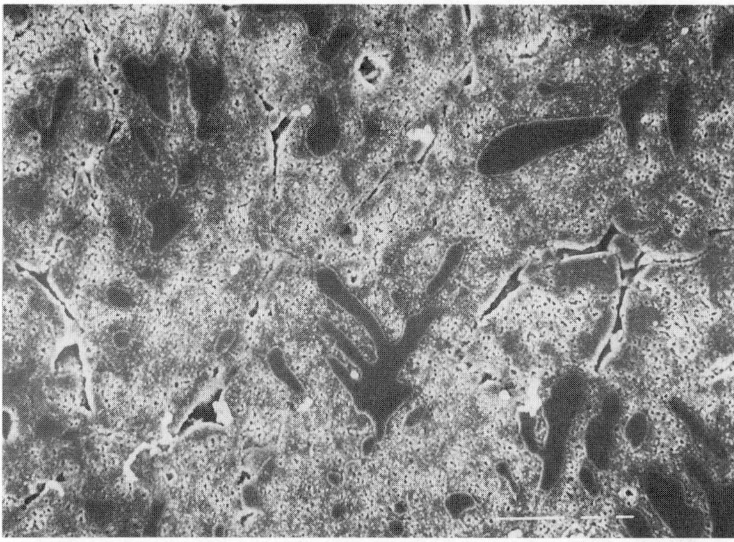

(b)

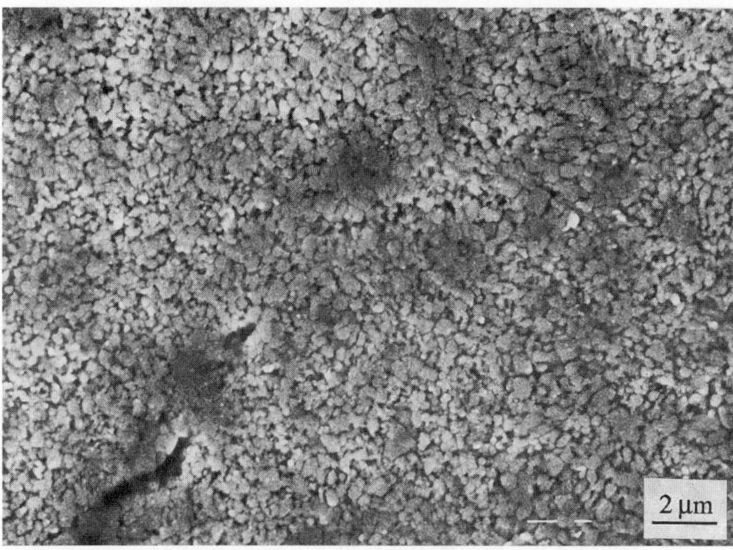

(c)

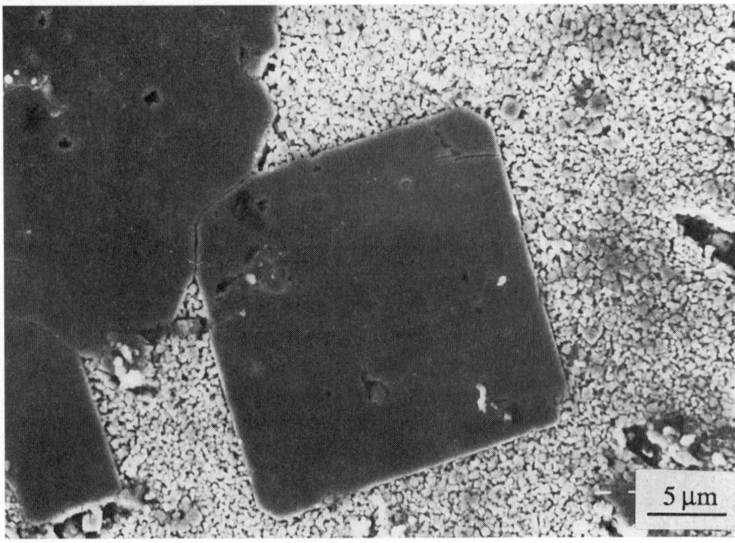

FIG. 7.21. The microstructure of an HDDR particle (a) below optimum processing temperature showing the presence of α–Fe dendrites; (b) at optimum temperature showing uniform fine grain size; (c) above optimum temperature showing the presence of large facetted grains coexisting with very fine grains.

were produced by making a green compact of $\sim$ 7 gm using coarse ground isotropic HDDR powder. The green compacts were then pressed by 790 °C using a load of 2 tonnes/cm^2 until full density had been obtained. Hot-pressed magnets based on the composition $Nd_{16}Fe_{76}B_8$ exhibited $B_r \approx 0.8$ T, $H_c \approx 1160$ kA/m and $(BH)_{max} \approx 108$ kJ/m^3 and the enhanced value of B_r for this composition indicated a degree of upsetting during the hot-pressing operation. Electron microscope observations [Est94] of these magnets indicated an average grain size of $\sim 0.4\ \mu$m, which means that the enhanced remanence could not be ascribed to exchange coupling. Die-upset forging experiments by McGuiness *et al.* [McG92] indicated that the hot-pressed HDDR compacts ($Nd_{16}Fe_{76}B_8$ exhibited a rate of increase of B_r with an upset ratio very similar to that observed by Tokunaga *et al.* [Tok89] for melt-spun compacts based on ($Nd_{14}Fe_{80}B_6$). The coercivities of the HDDR magnets, however, exhibited a much smaller decrease with upset ratio than those of the melt-spun magnets.

It is possible to avoid decrepitation of the alloy during the HDDR processing by introducing the hydrogen at or above the minimum disproportionation temperature. This has been called the S(solid)–HDDR process [Gut94a, 94b]. Rapid heating of a near-stoichiometric alloy ($Nd_{2.1}Fe_{14}B$) in a hydrogen atmosphere also avoids decrepitation [Zha91]. A useful consequence of the HDDR process is that it is a very effective means of homogenizing near-stoichiometric compositions which contain α–Fe dendrites in the cast state [Har92]. The α–Fe can be removed quickly compared with standard homogenization treatment by annealing the alloy in the disproportionated state at a relatively modest temperature ($\approx$ 750 °C).

4 Casting and hot-working

One of the most cost-effective means of producing anisotropic, fully dense Nd–Fe–B magnets would be to cast them into slabs with a fine grain size and with the *c*-axis of the grains aligned perpendicular to the surface of the slab. In the case of Pr–Fe–B–Cu alloys, appreciable coercivities (up to 1070 kA/m) can be obtained in the cast condition [Shi88a, Far91, Myc93]. Unfortunately, the orientation is such that there is a more or less random *c*-axis orientation in the plane at right angles to the predominant cooling direction, so that it is not possible to produce highly aligned anisotropic magnets from the cast material without subsequent hot deformation. Isotropic cast magnets based on $Nd_2Fe_{14}C$ can be produced [Bus88, Gri91, Fri94] simply by casting, for example, a Nd–Fe–Cu–C–B alloy and then subsequently annealing between 900 °C and 950 °C for 5–6 days, during which time fine grain $Nd_2Fe_{14}C$ is formed by a very slow solid-state reaction. By this means it is possible to produce magnets with $B_r \approx 0.61$ T, $H_c \approx 600$ kA/m and $(BH)_{max} \approx 61$ kJ/m^3. To obtain magnets from cast material with much higher $(BH)_{max}$ values it is necessary to subject the alloys to a hot deformation process. Because of the necessary scale of the operation to make them economical, these processes are usually hot-extrusion or hot-rolling. In the Nd–Fe–B (and Pr–Fe–B) alloys containing excess rare-earth, the processes are much aided by the presence of an intergranular liquid phase at the deformation temperature.

The first report of a direct mechanical method of making Nd–Fe–B magnets from cast material was by Ervens [Erv89], who extruded rods at 800 °C which exhibited a radial magnetic orientation. The magnetic properties were found to depend upon the extrusion conditions such that higher remanence with a lower coercivity appeared at lower extrusion velocities. Unfortunately, these magnets were unstable and there was an irreversible loss in the coercivity even on standing at room temperature.

The extrusion process was much improved by Nozières and coworkers [Noz89a, 89b, 90], who encapsulated the Nd–Fe–B alloy in an iron sheath which was capped and sealed by welding in an inert atmosphere in order to prevent oxidation and loss of the liquid phase during extrusion. The bar was then extruded at temperatures in the range 750–850 °C and a high deformation ratio (>10) was required to achieve a pronounced radial texture. A final anneal was needed to optimize the magnetic properties. Metallographic studies of the extruded bar showed that the mechanical working resulted in a decrease in the average grain size to around 0.5–10 μm, with the $NdFe_4B_4$ phase strung out along the hot-working direction. The radially

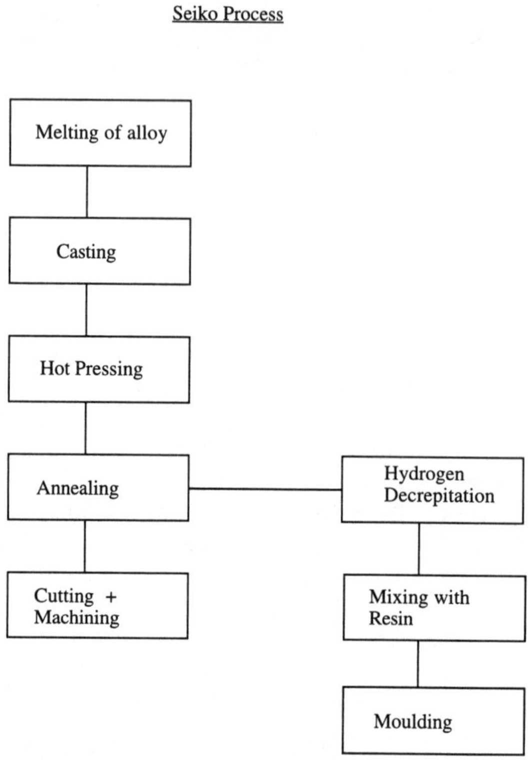

FIG. 7.22. A schematic representation of the Seiko, hot-pressing production route.

oriented magnets can be used as arc segments and, for some applications, it may be unnecessary to remove the Fe-sheath by machining.

Uniaxial texture can be imparted by hot-pressing which results in higher remanences but lower coercivities compared with those of the extruded bars. An important advance in hot-working came with the discovery of Shimoda and coworkers [Shi88a, 88b] at Seiko, that Pr–Fe–B alloys exhibited much higher coercivities in the cast state compared with those of the corresponding Nd–Fe–B alloys and much better magnetic properties after hot-pressing. These workers also found that hot-pressing are greatly facilitated by small additions of Cu to the alloy, for example $Pr_{17}Fe_{76.5}B_5Cu_{1.5}$. The Seiko manufacturing route is illustrated schematically in Fig. 7.22 and the Pr alloys are hot-pressed at around 1000 °C to produce good *c*-axis alignment parallel to the pressing direction. At this temperature the grain boundary phase is in the liquid state and this allows the alloy to deform and to change the alignment of the grains. The copper addition results in the formation of a Pr–PrCu eutectic mixture at the grain boundaries which has a significantly lower melting point than that of the non-Cu-containing eutectic. An additional heat treatment of 950 °C for 6 h and 475 °C for 2 h was

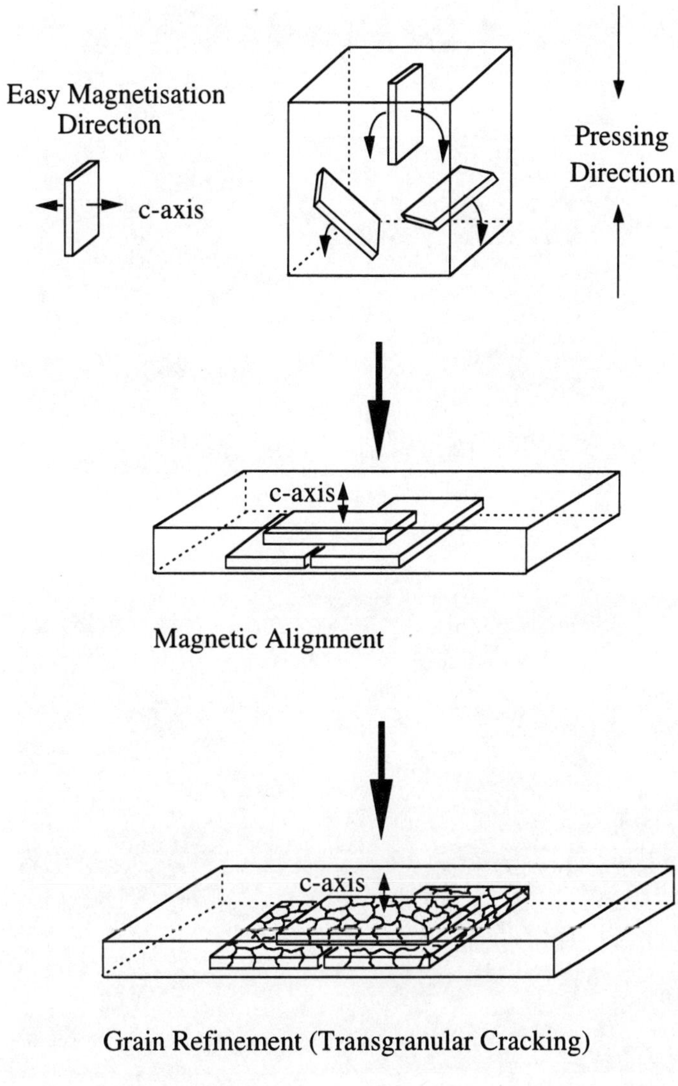

FIG. 7.23. Proposed model for the hot-pressing behaviour of a $Pr_{20}Fe_{74}B_4Cu_2$ alloy [Kwo92a, 92b].

found to result in an improvement in the magnetic properties, and a $(BH)_{max}$ value of 240 kJ/m^3 and a H_c of 870 kA/m have been reported for these magnets.

Kwon and Harris [Kwo92a, 92b] have examined the hot-pressing behaviour of a $Pr_{20}Fe_{74}B_4Cu_2$ alloy and obtained magnetic properties comparable with those of standard sintered $Nd_{16}Fe_{76}B_8$-type magnets. The Pr-alloy exhibited plate-like grains with the *c*-axis perpendicular to the plate surface and it was proposed that alignment occurred during the hot-pressing process at ~ 950 °C, by rotation and

(a)

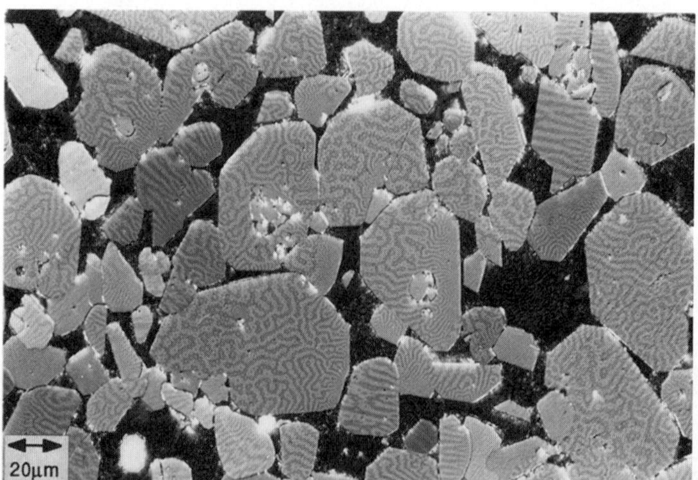

(b)

FIG. 7.24. Kerr-effect pictures of partially aligned (65% thickness reduction) $Pr_{20}Fe_{74}B_4Cu_2$ alloy, hot-pressed at 900 °C: (a) parallel; and (b) perpendicular to the upset forging direction.

subsequent fracture of the grains, as shown schematically in Fig. 7.23. The microstructure of the partially aligned, hot-pressed material is shown in Fig. 7.24 and the orientation of the grains is revealed by the Kerr-effect contrast. The alignment process is only possible because of the large proportion of intergranular liquid phase in this alloy at the hot-pressing temperature. The cast ingot contains an extensive amount of free iron and this was thought to be necessary to produce the initial fine grain size [Del94]. The free iron was eliminated after the hot-pressing operation, and this was ascribed to the acceleration of the reaction between the iron and the Pr-rich liquid phase as a result of the cracks in the matrix grains which are terminated at the Fe dendrites. The coercivity was improved significantly by annealing at $\sim 500\ °C$ and the amount of the intergranular δ-phase $Pr_6Fe_{13}Cu$ increased as a result of this treatment.

The origin of the texture formation induced by hot-pressing in Nd–Fe–B alloys and the mechanical properties at elevated temperatures have been investigated by Luo and Zhang [Luo87, 89, 92], and a mechanism involving the rotation of grains and slip in the basal plane of the tetragonal $Nd_2Fe_{14}B$ phase has been proposed.

Hot-rolling has also been developed as a means of producing magnets from Pr–Fe–B–Cu alloys [Shi89, 90, Ohk89, 93, Ara94]. The main advantage of this technique, compared to hot-pressing, is that it allows large amounts of material to be processed in a single operation. The ingot is wrapped in a steel sheath and hot-rolled at $\sim 1000\ °C$ to give a deformation of $> 70\%$. For best results it is necessary to obtain an ingot with a fine, well-developed dendritic structure (achieved by having a low B-content and hence free iron in the cast ingot) and then applying a load perpendicular to the growth direction of the grains in order to reduce the thickness. During the rolling operation the *c*-axes are aligned with the direction of the thickness reduction. An additional heat treatment of 950 °C for 6 h and 475 °C for 2 h was found to result in an improvement in the magnetic properties. $(BH)_{max}$ values of $\sim 240\ kJ/m^3$ and H_c of ~ 870 kA/m have been reported for these magnets and, unlike the hot-pressing method, good magnetic properties have also been obtained for hot-rolled Nd–Fe–B–Cu alloys, but in this case the post-annealing treatment must be altered slightly to optimize the properties [Shi90, Iwa92]. However, the coercivities of these materials were significantly less than those of the corresponding Pr-alloys.

Yuri and Ohki [Yur94] have studied the hot-rolling behaviour of $Pr_{20}Fe_{76}B_4$ and have observed plate-like grains in this alloy very similar to those reported by Kwon and Harris [Kwo92a, 92b) for the alloy $Pr_{20}Fe_{76}B_4Cu_2$. Yuri and Ohki proposed a model very similar to that of Kwon and Harris to explain the hot-rolling behaviour of their alloy. They also proposed a preferential cracking of the grains in a plane parallel to the *c*-axis and annealing subsequent to rolling was thought to involve the preferential growth of the *c*-axis-aligned grains.

Fruchart and coworkers [Fru94] have shown that fine grain size Nd–Fe–B–Cu alloys (hot-rolled or hot-pressed) can be powdered by the hydrogen decrepitation (HD) process and the initial remanence and coercivity can be largely restored by suitable hydrogenation and annealing treatments, thus providing powder suitable for anisotropic bonded magnets. This could improve the economics of the hot-working processes if excess material could be used for the production of anisotropic bonded magnets.

The obvious advantage of the hot-working processes is that they do not involve the handling of fine powdered material with all the attendant contamination problems. Thus, significantly lower oxygen and carbon contents are reported for the hot-worked material compared with those of the standard Nd–Fe–B sintered magnets. Disadvantages include the need for high-temperature deformation processes, the need to encapsulate the material, less orientation control compared with that of magnetic alignment and more limited shapes.

B. Sm–Fe–N

The possibility of converting the Sm_2Fe_{17} phase, which has a lower Curie temperature (≈ 113 °C) and basal-plane anisotropy, into $Sm_2Fe_{17}N_3$, which has a much higher Curie temperature (480 °C) and *c*-axis anisotropy with anisotropy field ($\mu_0 H_a$) of 14 T, was signalled by Coey and Sun [Coe90] and developed independently by Iriyama, Kobayashi and coworkers [Iri92]. The principles of gas-phase nitrogenation were discussed in Chapter 4.

The preparation of $Sm_2Fe_{17}N_x$ magnets can be divided into the following steps, (i) preparation of the Sm_2Fe_{17} alloy; (ii) production of powder (if necessary) prior to nitrogenation; (iii) nitrogenation; (iv) pulverization to fine powder (preferably jet milling); and (v) mix powder with resin or soft metal; (vi) align (for anisotropic magnets) and press. Successful laboratory-scale production routes are summarized in Fig. 7.25:

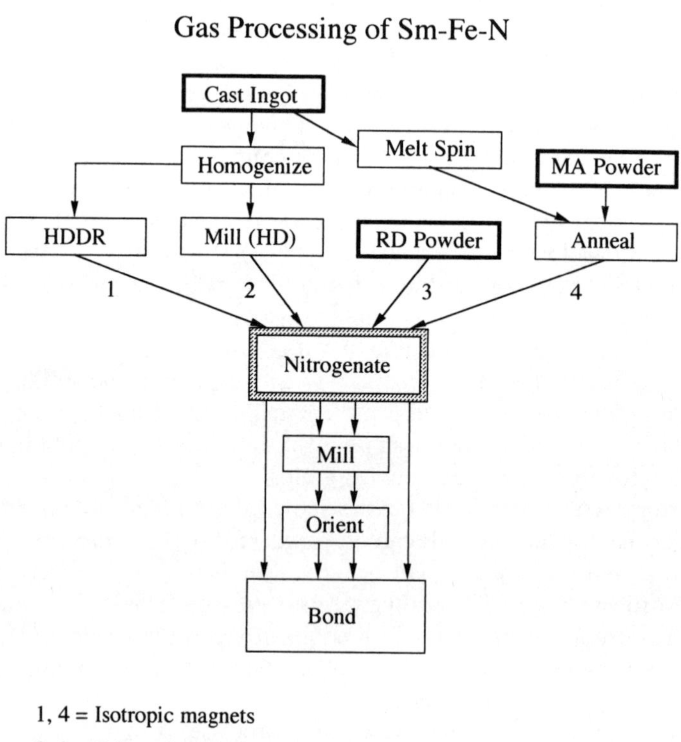

FIG. 7.25. Gas-processing routes for producing Sm–Fe–N magnets.

Processing routes to turn Sm–Fe–N into useful magnets are constrained by the metastability of the $Sm_2Fe_{17}N_x$ phase, which excludes any heat treatment steps in the process during or after nitriding. The product is pre-eminently suited for making bonded magnets using coercive powder.

For the production of bonded anisotropic magnets, monocrystalline particles are required and these can be obtained from induction-melted or calciothermic material. Fine grain, coercive polycrystalline powder can be used for the production of isotropic bonded magnets and these materials can be prepared by melt-spinning, mechanical alloying or HDDR.

It can be seen from the Sm–Fe phase diagram (Fig. 3.3) that Sm_2Fe_{17} is produced by a peritectic reaction of liquid and α–Fe with wide separation of the liquidus and peritectic temperature. This means that, in the cast state, the alloy $Sm_{10.5}Fe_{89.5}$ (Sm_2Fe_{17} composition) contains large amounts of free iron in the form of dendrites together with some Sm-rich phases. Free iron (in this form) and the Sm-rich phases have a deleterious effect on the coercivity of the nitrided powder. Thus, long homogenization treatments (e.g. 48 h at 1000 °C) are required to make the alloy predominantly single-phase and, unless the alloy is sealed in, for instance, a silica tube, there is a problem of samarium loss during annealing. Such a long homogenization treatment would have an adverse effect on the economics of production of magnets based on cast $Sm_2Fe_{17}N_x$. The only practical means of producing large amounts of cast material is by induction melting but this will lead to relatively large iron dendrites in the alloy, thus requiring a long homogenization treatment to produce single-phase material. Formation of α–Fe can be suppressed by minor additions of niobium [Pla92] or tantalum alloy [Saj94]. These effects are attributed [Rei94] to narrowing of the field of primary α–Fe crystallization such that Sm_2Fe_{17} rather than α–Fe becomes the primary crystallization product at Nb contents >4–5 at.% (i.e. $Sm_{10.5}Fe_{84.5}Nb_5$). In these alloys there is a dilution of the ferromagnetic Sm_2Fe_{17} phase by the paramagnetic Laves phase $NbFe_2$ but these additions obviate the need for a long homogenization treatment.

Once the ingot material is in a suitable form it can be milled, nitrogenated and then milled again to produce single-crystal particles of $Sm_2Fe_{17}N_x$ suitable for the production of anisotropic bonded magnets, as explained in Chapter 4.

An alternative means of producing single-phase monocrystalline Sm_2Fe_{17} is by means of the calciothermic route using Sm_2O_3, Ca and Fe. This material is produced in powder form and can be nitrogenated and milled to yield single-crystal particles suitable for magnet production [Bre94]. This material will contain significantly more oxygen than powder obtained from the ingot.

Other means available to produce Sm_2Fe_{17} in polycrystalline form are rapid solidification [Kat91, Chr92, Yan93], mechanical alloying [Sch90, Liu92, Schu92] and HDDR [Sug92] but as these form very fine grain coercive material, it is only possible to use them to make isotropic bonded magnets. Very high coercivities ($\approx$ 2.4 MA/m) have been achieved with mechanically alloyed powder after post-annealing and nitrogenation. The very fine grain sizes have enabled two-phase $Sm_2Fe_{17}N_3/\alpha Fe$ exchange spring magnets to be produced from the mechanically alloyed powder of the alloy Sm_7Fe_{93} [Din93, O'Don94]. The enhanced remanence of these magnets compensates, to some extent, for their isotropic character, and mechanical alloying could represent the best method of producing these types of magnet provided adequate coercivity can be achieved.

Fine grain ($\sim 0.3\ \mu$m) powder of $Sm_2Fe_{17}N_x$ can be produced from the induction-melted or calciothermic material by the HDDR process [Sug92, Zho92, Oka94, Wen94] in a similar manner to that employed for $Nd_2Fe_{14}B$. The hybride disproportionation reaction can be represented as:

$$Sm_2Fe_{17} + (2 \pm x)H_2 \rightleftharpoons 2SmH_{2\pm x} + 17Fe + \Delta H \qquad (7.4)$$

ΔH is the exothermic (absorption) or endothermic (desorption) heat of reaction. The particular value of x depends on the temperature and hydrogen pressure.

The nature of the disproportionated mixture has been revealed by SEM [Zho92] and TEM [Oka94] studies and the latter have shown that it consists of a very fine mixture of rods of $SmH_{2\pm x}$ in an Fe matrix which is a very similar microstructure to that of disproportionated $Nd_2Fe_{14}B$ [Gut94b]. The recombined microstructure consists initially of fine grains ($\sim 0.3\ \mu$m) and the coercivity depends strongly on the temperature of recombination [Wen94]. The HDDR powder can be used to produce isotropic bonded magnets. It might be possible to produce anisotropic HDDR powder by modification of the process, by appropriate alloying additions.

The nitrogenation reaction must be carefully controlled for optimum results. It has been reported [Kob94] that the use of nitrogen at high pressure, pre-treatment with hydrogen and nitrogenation with ammonia or ammonia/hydrogen mixture can all help to promote the nitrogenation reaction by allowing the nitrogenation treatments to be carried out at lower temperatures for shorter times, thereby suppressing the harmful disproportionation reaction.

Because of the disproportionation of the $Sm_2Fe_{17}N_x$ phase at temperatures $\gtrsim 600$ °C, fully dense sintered or hot-pressed magnets based on $Sm_2Fe_{17}N_x$ have not been produced. Unless the phase can be stabilized by appropriate alloying its use will be confined to bonded magnets.

In the case of anisotropic magnets, some useful polymer-bonded magnets have been produced [Suz92]. Much work has also been carried out [Ota91, Hua91, Suz92, Mül92, Wen92, Rod92] on Zn- (and Sn-) bonded magnets where a low-temperature reaction with the zinc at about 420 °C results in an appreciable improvement in the coercivity. This is probably related to the presence of a disproportionated SmN/Fe layer on the particles which reacts with the Zn to form Fe_3Zn_7. A problem with these magnets is that the reaction leads to porosity and hence to a mediocre density of the compact. Metal-bonded magnets do not appear to be a realistic commercial proposition, and the future of $Sm_2Fe_{17}N_x$ rests on the cost-effective production of highly coercive, monocrystalline powder which can be used to produce anisotropic compression or injection-moulded magnets.

Conclusion

This chapter has shown the remarkable versatility of the Nd–Fe–B alloys, in that they can be processed effectively into permanent magnets by many routes.

Processing is the key to economic magnet production since processing costs far outweigh the costs of the raw materials, and it is by the exploitation of large-scale production that high-performance rare-earth iron magnets will find their way to mass markets such as electric motors in automobiles. Currently, the leading method is the powder/sinter route, which is ideal for the mass production of small components with optimal magnetic properties. Disadvantages of the process are the difficulty and expense of handling reactive powder and the large number of steps involved. The melt-spun magnets satisfy a different market because bonded magnets are not, at present, made from ingot material. Compression-moulded and injection-moulded isotropic magnets made from the melt-spun material suit lower-cost markets where high-energy products and elevated-temperature performance are not required. Anisotropic magnetics can be produced from die-upset forged ribbon material, and bonded magnets with energy products comparable with those of sintered $SmCo_5$ can be made from the crushed compacts. In the future, the isotropic and anisotropic coercive powder produced by the HDDR process from ingot material might be suitable for both bonded and hot-pressed magnets. The hot-working routes present a possible long-term alternative to both the powder and melt-spun routes, and might provide material for fully dense and bonded magnets.

Anisotropic bonded magnets made from Sm–Fe–N may also become available.

REFERENCES

[Ahm94] F. M. Ahmed, D. S. Edgley and I. R. Harris, *13th Int. Workshop on RE Magnets and their Applications*, Birmingham, UK, Sept. (1994), p. 463.

[Ala94] T. K. R. Alander, M. J. Talvitie, K. T. J. Aittoniemi and V. K. Lindroos, *13th Int. Workshop on RE Magnets and their Applications*, Birmingham, UK, Sept. (1994), p. 301.

[Ara94] A. Arai, O. Kobayashi, F. Takagi, K. Akioka and T. Shimoda, *J. Appl. Phys.*, **75** (1994), 6631.

[Arm89] J. V. Armstrong, J. M. D. Coey, J. G. Lunney and M. E. G. Lyons, *Hyperfine Interaction*, **46** (1989), 467.

[Bai85] T. Bailey and I. R. Harris, *J. Mat. Sci. Lett.*, **4** (1985), 151.

[Bal90] H. Bala, S. Szymura and J. J. Wyslocki, *J. Mat. Sci.*, **25** (1990), 571.

[Ben70] J. S. Benjamin, *Met. Trans.*, **1** (1970), 29.

[Benz70] M. G. Benz and D. L. Martin, *Appl. Phys. Lett.*, **17** (1970), 176

[Ber86] R. Bergeron, K. Canavan, W. R. McCallum and J. Keem, European Patent Application 0195219 (1986).

[Blo90] W. Bloch, K. Grendel and H.Staubach, *11th Int. Workshop on RE Magnets and their Applications*, Pittsburgh, Pa. Oct. (1990), p. 108.

[Bog91] Y. Bogatin, M. Robinson and J. Ormerod, *J. Appl. Phys.*, **70** (1991), 6594.

[Bog93] Y. Bogatin, D. Li and P. Behan, *J. Mat. Eng. Perf.*, **2** (2) (1993), 173.

[Bre94] S. Brennan, K. Kobayashi and J. M. D. Coey, *13th Int. Workshop on RE Magnets and their Applications*, Birmingham, UK, Sept. (1994), p. 795.

[Bret90] R. L. Brett, N. Rowlinson, M. M. Ashraf, I. R. Harris and P. Bowen, *J. Appl. Phys.*, **67** (1990), 4622.

[Bur90] E. Burzo and H. R. Kirchmayr, *Handbook of rare earth metals and alloys* (1990), p. 72.

[Bus88] K. H. J. Buschow, D. B. de Mooij and C. J. M. Denissen, *J. Less-Common Met.*, **141**, (1988), L15.

[Cad86] J. M. Cadogan and J. M. D. Coey, *Appl. Phys. Lett.*, **48** (1986), 442.

[Cav92] P. Cavolloti, B. Bozzini, R. Cecchini, G. F. Bava, H. A. Davies and C. Hoggarth, *J Magn. Magn. Mat.*, **104–107** (1993), 1216.

[Cha89] V. K. Chandhok and Bao-Min Ma, US Patent No. 4881984 (1989).

[Chi86] T. S. Chin, Y. Hara, E. J. Lavernia, R. C. O'Handley and N. J. Grant., *J. Appl. Phys*, **59** (1986), 1297.

[Chi88] T. S. Chin, R. T. Chang, W. T. Tsai and M. P. Hung, *IEEE Trans. Mag.*, **MAG-24** (1988), 1927.

[Chr92] C. N. Christodoulou and T. Takeshita, *Rep. Magnetics working committee, Mag. Conf.*, Tokyo; IEEE MAG-92-245, Tokyo (1992), 53.

[Coe88] R. Coehoorn, D. B. Mooij, J. P. W. B. Duchateau and K. H. J. Buschow, *J. de Phys.*, **C8** (1988), 669.

[Coe89] R. Coehoorn and J. P. W. B. Duchateau, *CEAM Report*, Elsevier Appl. Science, (1989), 558.

[Coe90] J. M. D. Coey and H. Sun, *J. Magn. Magn. Mater.*, **87** (1990) L251.

[Coe90a] J. M. D. Coey, Hong Sun and Y. Otani, *Proc. 6th Int. Symp. on Mag. Anisotropy and Coercivity in RE Trans. Metal Alloys*, Pittsburgh, Pa., Oct. (1990), p. 36.

[Cro84] J. J. Croat, J. F. Herbst, R. W. Lee and F. E. Pinkerton, *Appl. Phys. Lett.*, **44** (1984), 148.

[Cro88] J. J. Croat, *J. Mat. Eng.*, **10** (1) (1988), 7.

[Cro89] J. J. Croat, *J. Less-Common Met.*, **148** (1989), 7.

[Cro90] J. J. Croat, *11th Int. Workshop on RE Magnets and their Applications*, Pittsburg, Pa., Oct. (1990), p. 7.

[Cro94] J. J. Croat, *13th Int. Workshop on RE Magnets and their Applications*, Birmingham, UK, Sept. (1994), p. 65.

[Chu91] T. Y. Chu, L. Rabenberg and R. K. Mishra, *J. Appl. Phys.*, **69** (1991), 6046.

[Das69] D. K. Das, *IEEE Trans. Mag.*, **MAG-5** (1969), 214.

[Dav89] H. A. Davies, K. J. A. Mawella, R. A. Buckley, G. E. Carr, A. Manaf and A. Jha, *CEAM Report*, Elsevier Appl. Science (1989), p. 543.

[Dav94] H. A. Davies, *Proc. 8th Int. Symp. on Mag. Anisotropy and Coercivity in RE Trans. Metal Alloys*, Birmingham, UK, Sept. (1994), p. 465.

[Del94] G. W. Delamore, G. P. Hatch, P. J. McGuiness, G. J. Mycock and I. R. Harris, *Proc. 8th Int. Symp. on Mag. Anisotropy and Coercivity in RE Trans. Metal Alloys*, Birmingham, UK, Sept. (1994), p. 253.

[Dic90] E. D. Dickens and A. M. Mazany, *J. Appl. Phys.*, **67** (1990), 4613.

[Din93] J. Ding, P. G. McCormick and R. Street, *J. Magn. Magn. Mater.*, **124** (1983), L1.

[Dom76] H. G. Domazer and K. J. Strnat *2nd Int. Workshop on RE-Co Magnets and their Applications*, Dayton, Ohio, May (1976), p. 348.

[Dos91] M. Doser, V. Panchanathan and R. K. Mishra, *J. Appl. Phys.*, **70** (1991), 6603.

[Edg93] D. S. Edgley, J. M. LeBreton, D. Lemarchand, I. R. Harris and J. Teillet, *J. Magn. Magn. Mater.*, **128** (1993), L1.

[Edg94] D. S. Edgley, F. M. Ahmed, I. R. Harris, J. S. Taylor, K. A. Watts and J. M. LeBreton, *13th Int. Workshop on RE Magnets and their Applications*, Birmingham, UK, Sept. (1994), p. 283.

[End94] M. Endoh and M. Shindo, *13th Workshop on RE Magnets and their Applications*, Birmingham, UK, Sept. (1994), p. 397.

[Erv89] W. Ervens, *CEAM Report*, Elsevier Appl. Science, (1989), p. 655.

[Esh90] L. J. Eshelman, K. A. Young, V. Panchanathan and J. J. Croat, *J. Appl. Phys.*, **64** (1988), 5295.

[Est94] E. Estevez, J. Fidler, C. Short and I. R. Harris, *Proc. 8th Int. Symp. on Mag. Anisotropy and Coercivity in RE Trans. Metal Alloys*, Birmingham, UK, Sept. (1994), p. 235.

[Far91] R. N. Faria, J. S. Abell and I. R. Harris, *J. Appl. Phys.*, **70** (1991), 6104.

[Fri94] K. Fritz, B. Reinsch and G. Petzow, *13th Int. Workshop on RE Magnets and their Applications*, Birmingham, UK, Sept. (1994), p. 493.

[Fru89] R. Fruchart, P. L'Heretier, D. Fruchart, P. Wolfers and A. Yaouanc, *CEAM Report*, Elsevier Appl. Science, (1989), p. 230.

[Fru94] D. Fruchart, S. Miraglia, L. Pontonnier, P. De Rango, J. L. Soubeyroux, R. Fruchart, P. Mollard and R. Perrier de La Bâthie, *13th Int. Workshop on RE Magnets and their Applications*, Birmingham, UK, Sept. (1994), p. 511.

[Gri91] B. Grieb, K. Fritz and E. T. Henig, *J. Appl. Phys.*, **70** (1991), 6447.

[Gut94a] O. Gutfleisch, N. Martinez, M. Verdier and I. R. Harris, *J. Alloys and Compounds*, **215** (1994), 227.

[Gut94b] O. Gutfleisch, N. Martinez, I. R. Harris, M. Matzinger and J. Fidler, *Proc. 8th Int. Symp. on Mag. Anisotropy and Coercivity in RE Trans. Metal Alloys*, Birmingham, UK, Sept. (1994), p. 243.

[Ham89] T. Hamada, T. Hayakawa, Y. Matsuura, US Patent No. 4837114 (1989).

[Har79] I. R. Harris, J. Evans and P. Nyholm, British Patent No. 1554384 (1979).

[Har85] I. R. Harris, C. Noble and T. Bailey, *J. Less-Common Met.*, **106** (1985), L1.

[Har87a] I. R. Harris, P. J. McGuiness, D. G. R. Jones and J. S. Abell, *Phys. Scripta.* **T19b** (1987), 435.

[Har87b] I. R. Harris, *9th Int. Workshop on RE Magnets and their Applications*, Bad Soden, (1987), p. 249.

[Har87c] I. R. Harris, *J. Less-Common Met.*, **131** (1987), 245.

[Har89] I. R. Harris, J. S. Abell, P. J. McGuiness, E. J. Devlin and P. A. Withey, *CEAM Report*, Elsevier Appl. Science, (1989), p. 489.

[Har90] I. R. Harris and P. J. McGuiness, *11th Int. Workshop on RE Magnets and their Applications*, Pittsburgh, Pa., Oct. (1990), p. 29.

[Har92] I. R. Harris, *12th Int. Workshop on RE Magnets and their Applications*, Canberra, Australia, July (1992), p. 347.

[Her75] C. S. Herget and H. G. Domazer, *Goldschmidt Informiert* 4/**75** (35) (1975), 3.

[Her85] C. S. Herget, *8th Int. Workshop on RE Magnets and their Applications*, Dayton, Ohio, May (1985), p. 407.

[Herb91] J. f. Herbst, *Supermagnets, hard magnetic materials*, G. J. Long and F. Grandjean (eds), NATO ASI Series, Kulwer Acad. Pub. (1991), p. 69.

[Hir90] S. Hirosawa, S. Mino, H. Tomizawa and K. Tokuhara, *11th Int. Workshop on RE Magnets and their Applications*, Pittsburgh, Pa., Oct. (1990), p. 136.

[Hir93] S. Hirosawa and H. Kanekiyo and M. Uehara, *J. Appl. Phys.*, **73** (1993), 6488.

[Hir94] S. Hirosawa and H. Kanekiyo, *13th Int. Workshop on RE Magnets and their Applications*, Birmingham, UK, Sept. (1994), p. 87.

[Hsu87] S. E. Hsu, K. L. Wang and L. C. Su, *IEEE Trans. Mag.*, **MAG-23** (1987), 2515.

[Hua91] M. Q. Huang, Y. Zheng, K. Miller, J. M. Elbicki, S. G. Sankar, W. E. Wallace and R. Obermyer, *J. Magn. Magn. Mater.*, **102** (1991), 91.

[Ima87] N. Imaizumi, N. Inoue and K. Takahashi, *IEEE Trans. Mag.*, **MAG-23** (1987), 3610.

[Iri92] T. Iriyama, K. Kobayashi, N. Imaokot, T. Fukuda, H. Kato and Y. Nakagawa, *IEEE Trans. Magn.*, **MAG-28** (1992), 2326.

[Iwa92] E. Iwamura, T. Yuri, A. Hanaki, H. Mitani and K. Itayama, *12th Int. Workshop on RE Magnets and their Applications*, Canberra, Australia, July (1992), p. 705.

[Iwas89] K. Iwasaki, Y. Nozawa, S. Tanigawa, M. Tokunaga and H. Harada, *J. Mat. Eng.*, **11** (3) (1989), 209.

[Jon94] D. G. R. Jones, N. Martinez, D. Pere, D. Lavielle and I. R. Harris, *13th Int. Workshop on RE Magnets and their Applications*, Birmingham, UK, Sept. (1994), p. 413.

[Kan93] H. Kanekiyo, M. Uehara and S. Hirosawa, *IEEE Trans. Mag.*, **MAG-29** (1993), 2863.

[Kas89] A. Kass, US Patent No. 4832891 (1989).

[Kat91] M. Katter, J. Wecker, L. Schultz, *J. Appl. Phys.*, **70** (1991), 15.

[Kia84] A. Kianvash and I. R. Harris, *J. Mat. Sci.*, **19** (1984), 353.

[Kia85] A. Kianvash and I. R. Harris, *J. Mat. Sci.*, **20** (1985), 682.

[Kim87] A. S. Kim and J. M. Jacobson, *IEEE Trans. Mag.*, **MAG-23** (1987), 2509.

[Kim89] A. S. Kim, *J. Mat. Eng.*, **11** (1) (1989), 95.

[Kne91] E. F. Kneller and R. Hawig, *IEEE Trans. Mag.*, **MAG-27** (1991), 3588.

[Kob94] K. Kobayashi, *13th Int. Workshop on RE Magnets and their Applications*, Birmingham, UK, Sept. (1994), p. 717.

[Kro94] H. Kronmüller, T. Shrefl. R. Fisher and J. Fidler, *Proc. 8th Int. Symp. on Mag. Anisotropy and Coercivity in RE Trans. Metal Alloys*, Birmingham, UK, Sept. (1994), p. 1.

[Kub82] O. Kubaschewski, *Iron-binary phase diagrams*, Springer-Verlag, (1982), p. 2853.

[Kwo92a] H. W. Kwon, P. Bowen and I. R. Harris, *12th Int. Workshop on RE Magnets and their Applications*, Canberra, Australia, July (1992), p. 705.

[Kwo92b] H. W. Kwon, P. Bowen and I. R. Harris, *J. Alloys and Compounds*, **189** (1992), 131.

[Lee85] R. W. Lee, *Appl. Phys. Lett.*, **46** (1985), 790.

[Leo94] M. Leonowicz and H. A. Davies, *13th Int. Workshop on RE Magnets and their Applications*, Birmingham, UK, Sept. (1994), p. 623.

[L'He84] P. L'Heritier, P. Chaudouet, R. Madar, A. Rouault, P. Senateur and R. Fruchart, *CR Acad. Sci.*, **299** (13) (1984), 849.

[Li81] D. Li, J. Liu, E. Xu, G. Hua and X. Su, *5th Int. Workshop on RE–Co Magnets and their applications*, Roanoke, VA. (1981), 571.

[Li90] L. Li and C. D. Graham Jr, *J. Appl. Phys.* **67** (1990), 4756.

[Liu92] Y. Liu, M. P. Dallimore, T. Alonso and P. G. McCormick, *12th Int. Workshop on RE Magnets and their Applications*, Canberra, Australia, July (1992), p. 24.

[Luo87] Y. Luo, N. Zhang, and C. D. Graham Jr, *J. Appl. Phys.*, **61** (1987), 3442.

[Luo89] Y. Luo and N. Zhang, *10th Int. Workshop on RE Magnets and their Applications*, Kyoto, Japan, May, Vol. 2 (1989), p. 275.

[Luo92] Y. Luo and T.-S. Xia, *12th Int. Workshop on RE Magnets and their Applications*, Canberra, Australia, July, (1992), p. 473.

[McG86] P. J. McGuiness, I. R. Harris, E. Rozendaal, J. Ormerod and M. Ward, *J. Mat. Sci.*, **21** (1986), 4107.

[McG89a] P. J. McGuiness, A. J. Williams, I. R. Harris, E. Roendaal and J. Ormerod, *IEEE Trans. Mag.*, **MAG-25** (1989), 3773.

[McG89b] P. J. McGuiness, E. Devlin, I. R. Harris, E. Rozendaal and J. Ormerod, *J. Mat. Sci.*, **24** (1989), 2541.

[McG89c] P. J. McGuiness, I. R. Harris, U. D. Scholz and H. Nagel, *Zeitchrift für Physikalische Chemie Neue Folge*, **163** (1989), 687.

[McG90a] P. J. McGuiness, X. J. Zhang, X. J. Yin and I. R. Harris, *J. Less-Common Met.*, **158** (1990), 359.

[McG90b] P. J. McGuiness, X. J. Zhang, H. Forsyth and I. R. Harris, *J. Less-Common Met.*, **162** (1990), 379.

[McG90c] P. J. McGuiness, A. Ahmed, D. G. R. Jones, I. R. Harris, S. Burns and E. Rozendaal, *J. Appl. Phys.*, **67** (1990), 4626.

[McG92] P. J. McGuiness, C. Short, A. F. Wilson and I. R. Harris, *J. Alloys and Compounds*, **184** (1992), 243.

[Man93a] A. Manaf, R. A. Buckley and H. A. Davies, *J. Magn. Magn. Mater.*, **128** (1993), 302.

[Man93b] A. Manaf, M. A. Al-Khafaji, P. Z. Zhang, H. A. Davies, R. A. Buckley and W. M. Rainforth, *J. Magn. Magn. Mater.*, **128** (1993), 128.

[Man94] C. A. F. Manwaring, G. Keegan, H. Nagel, W. Krönert, J. H. Vincent, S. P. S. Sangaha and I. R. Harris, *13th Int. Workshop on RE Magnets and their Applications*, Birmingham, UK, Sept. (1994), p. 553.

[Mar94] N. Martinez, D. G. R. Jones, O. Gutfleisch, D. Lavielle, D. Peré and I. R. Harris, *J. Appl. Phys.*, **76** (1994), 6825.

[Mis86] R. K. Mishra and R. W. Lee, *Appl. Phys. Lett.*, **48** (1986), 733.

[Mis88] R. K. Mishra and E. G. Brewer and R. W. Lee, *Appl. Phys.*, **63** (1988), 3528.

[Mis90] R. K. Mishra, T. Y. Chu and L. K. Rabenburg, *J. Magn. Magn. Mater.*, **84** (1990), 88.

[Mit90] P. Mitchell, *IEEE Trans. Mag.*, **MAG-26** (1990), 1933.

[Muk89] T. Mukai and H. Sakamoto, *Appl. Phys. lett.*, **54** (1989), 1597.

[Muk90] T. Mukai, Y. Okazaki, H. Sakamoto, M. Fujikura and T. Inaguma, *11th Int. Workshop on RE Magnets and their Applications*, Pittsburgh, Pa. Oct. (1990), p. 72.

[Mül87] K.-H. Müller, A. Handstein, D. Eckert and J. Schniedner, *Phys. Stat. Sol.* (A), **99** (1987), K61.

[Mül92] K.-H. Müller, P. A. P. Wendhausen, D. Eckert and A. Handstein, *7th Int. Symp. on Mag. Anisotropy and Coercivity in RE Trans. Metal Alloys*, Canberra, Australia, July (1992), p. 34.

[Myc93] G. J. Mycock, R. N. J. Faria and I. R. Harris, *J. Alloys and Compounds*, **201** (1993), 23.

[Nag94] H. Nagel and W. E. Krönert, *13th Int. Workshop on RE Magnets and their Applications*, Birmingham, UK, Sept. (1994), p. 391.

[Nak90] J. Nakagawa and O. Onitsuka, *11th Int. Workshop on RE Magnets and their Applications*, Pittsburgh, Pa. Oct. (1990), p. 341.

[Nak93] R. Nakayama and T. Takeshita, *J. Appl. Phys.*, **74** (1993), 2719.

[Nak94] R. Nakayama, T. Takeshita, M. Itakura, N. Kuwano and K. Oki, *J. Appl. Phys.*, **76** (1994), 412.

[Nar85] K. S. V. L. Narasimhan, *J. Appl. Phys.*, **57** (1985), 4081.

[Nar86a] K. S. V. L. Narasimhan and E. J. Dulis, US Patent No. 4585473 (1986).

[Nar86b] K. S. V. L. Narasimhan, C. Willman and E. J. Dulis, US Patent No. 4588349 (1986).

[Nis90] N. Nishio, H. Yamamoto, M. Nagakura and M. Uehara, *IEEE Trans. Mag.*, **MAG-26** (1990), 257.

[Noz89a] J. P. Noziéres, R. Perrier de la Bâthie and D. W. Taylor, *J. Magn. Magn. Mater.*, **80** (1989), 88.

[Noz89b] J. P. Noziéres and R. Perrier de la Bâthie *CEAM Report*, Elsevier Appl. Science, (1989), p. 659.

[Noz90] J. P. Noziéres, PhD Thesis, Laboratoire de Magnétisme Louis Néel, Grenoble (1990).

[O'Don94] K. O'Donnell, C. Kuhrt and J. M. D. Coey, *13th Int. Workshop on RE Magnets and their Applications*, Birmingham, UK, Sept. (1994), p. 851.

[Oes84] K. Oesterreicher and H. Oesterreicher, *Phys. Stat. Sol.* (A), **85** (1984), K61.

[Oes85] H. Oesterreicher, *Proc. 4th Int. Symp. on Magn. Anisotropy and Coercivity in RE Trans. Metal Alloys*, Dayton, Ohio, May (1985), p. 705.

[Ohk89] T. Ohki, T. Yuri, M. Miyagawa, Y. Takahashi, C. Yoshida, S. Kambe, M. Higashi and K. Itayama, *10th Int. Workshop on RE Magnets and their Applications*, Kyoto, Japan, May (1989), p. 399.

[Ohk93] T. Ohki and T. Yuri, *IEEE Trans. Mag.*, **MAG-29** (1993), 2800.

[Oka85] M. Okada, S. Sugimoto, C. Ishizaka, T. Tanaka and M. Homma, *J. Appl. Phys.*, **57** (1985), 4146.

[Oka94] M. Okada, K. Saito, H. Nakamura, S. Sugimoto and M. Homma, *13th Int. Workshop on RE Magnets and their Applications*, Birmingham, UK, Sept. (1994), p. 763.

[Orm84] J. Ormerod, *Nd–Fe Permanent magnets - their present and future applications*, I. V. Mitchell (ed.) (CEC), Brussels, (1984), p. 69.

[Orm85] J. Ormerod, *J. Less-Common Met.*, **111** (1985), 49.

[Ota91] Y. Otani, A. Moukarika, H. Sun, J. M. D. Coey, E. Devlin and I. R. Harris, *J. Appl. Phys.*, **69** (1991), 6735.

[Ots90] E. Otsuki, T. Otsuka and T. Imai, *11th Int. Workshop on RE Magnets and their Applications*, Pittsburg, Pa. Oct. (1990), p. 328.

[Pan89] V. Panchanathan, *Mat. and Manufacturing Processes*, **4** (4) (1989), 505.

[Pla92] A. Platts, I. R. Harris, J. M. D. Coey, *J. Alloys and Compounds*, **185** (1992), 251.

[Rei94] B. Reinsch, H. H. Stadelmaier and G. Petzow, *13th Int. Workshop on RE Magnets and their Applications*, Birmingham, UK, Sept. (1994), p. 733.

[Rib90] R. W. Ribitch, *11th Int. Workshop on RE Magnets and their Applications*, Pittsburgh, Pa., Oct. (1990), p. 93.

[Rod92] W. Rodewald, M. Velicescu, B. Wall and G. W. Reppel, *12th Int. Workshop on RE Magnets and their Applications*, Canberra, Australia, July (1992), p. 191.

[Row89] N. Rowlinson, M. M. Ashraf and I. R. Harris, *J. Magn. Magn. Mater.*, **80** (1989), 93.

[Row90] N. Rowlinson, M. M. Ashraf and I. R. Harris, *J. Magn. Magn. Mater.*, **87** (1990), 93.

[Roz87] E. Rozendaal, J. Ormerod, P. J. McGuiness and I. R. Harris, *9th Int. Workshop on RE Magnets and their Applications*, Bad Soden, Aug. (1987), p. 275.

[Rup88] B. Rupp, A. Resnik, D. Shaltiel and P. Rogl, *J. Mat. Sci.*, **23** (1988), 2133.

[Sag84] M. Sagawa, S. Fujimura, N. Togawa, H. Yamamoto and Y. Matsuura, *J. Appl. Phys.*, **55** (1984), 2083.

[Sag93] M. Sagawa and H. Nagata, *IEEE Trans. Mag.*, **MAG-29** (1993), 2747.

[Sag94] M. Sagawa, H. Nagata, O. Itatani and T. Watanabe, *13th Int. Workshop on RE Magnets and their Applications*, Birmingham, UK, Sept. Suppl. (1994), p. 13.

[Saj94] B. Saje, A. E. Platts, S. Kobe Besenicar, I. R. Harris and D. Kolar, *IEEE Trans. Mag.*, **MAG-30** (1994), 690.

[Sak91] H. Sakamoto, M. Fujikura and T. Mukai, *J. Appl. Phys.*, **69** (1991), 5832.

[Sch90] K. Schnitzke, L. Schultz, J. Wecker and M. Katter, *J. Appl. Phys.*, **57** (1990), 2853.

[Scho87] U. D. Scholz, W. E. Krönert and H. Nagel, *9th Int. Workshop on RE Magnets and their Applications*, Bad Soden, Aug. (1987), p. 267.

[Schu87] L. Schultz, J. Wecker and E. Hellstern, *J. Appl. Phys.*, **61** (1987), 3583.

[Schu88a] L. Schultz, *Proc. DGM Conf. on New Materials by Mech. Alloying Techniques*, Calw/Hirsau, Oct., E. Arzt and L. Schutz (eds) (1988), p. 53.

[Schu88b] L. Schultz and J. Wecker, *Mat. Sci. Eng.*, **99** (1988), 127.

[Schu88c] L. Schultz, K. Schnitzke and J. Wecker, *J. Appl. Phys.*, **64** (1988), 5302.

[Schu92] L. Schultz, *Matr. Science Forum*, **89–90** (1992), 687.

[Shi88a] T. Shimoda, K. Akioka, O. Kobayashi and T. Yamagani, *J. Appl. Phys.*, **64** (1988), 5290.

[Shi88b] T. Shimoda, K. Akioka, O. Kobayashi and T. Yamagani, *J. de Phys.*, C**8** (1988), 631.

[Shi89] T. Shimoda, K. Akioka, O. Kobayashi and T. Yamagani, *10th Int. Workshop on RE Magnets and their Applications*, Kyoto, Japan, May (1989), p. 389.

[Shi90] T. Shimoda, K. Akioka, O. Kobayashi, T. Yamagani and A. Arai, *11th Int. Workshop on RE Magnets and their Applications*, Pittsburgh, Pa., Oct. (1990), p. 17.

[Shi94] M. Shimotomai, *13th Int. Workshop on RE Magnets and their Applications*, Birmingham, UK, Sept. (1994), p. 267.

[Shi92] M. Shinoda, K. Iwasaki, S. Tanigawa and M. Tokunaga, *12th Int. Workshop on RE Magnets and their Applications*, Canberra, Australia, July, (1992), p. 12.

[Sko93] R. Skomski and J. M. D. Coey, *Phys. Rev.*, **48** (1993) 15812.

[Sta85] H. H. Stadelmaier, N. A. Elmasry and S. R. Stallard, *J. Appl. Phys.*, **57** (1985), 4149.

[Sta86] H. H. Stadelmaier and N. C. Liu, *Mat. Lett.*, **4** (1986), 304.

[Ste92] M. Steinhorst, *12th Inst. Workshop on RE Magnets and their Applications*, Australia, Canberra, July (1992), p. 652.

[Sug92] S. Sugimoto, H. Nakamura, M. Okada and M. Homma, *12th Int. Workshop on RE Magnets and their Applications*, Canberra, Australia, July (1992), p. 372.

[Suz92] S. Suzuki and T. Miura, *IEEE Trans. Mag.*, **MAG-28** (1992), 994.

[Tak89] T. Takeshita and R. Nakayama, *10th Int. Workshop on RE Magnets and their Applications*, Kyoto, Japan, May (1989), p. 551.

[Tak90] T. Takeshita and R. Nakayama, *11th Int. Workshop on RE Magnets and their Applications*, Pittsburgh, Pa., Oct. (1990), p. 49.

[Tak92] T. Takeshita and R. Nakayama, *12th Int. Workshop on RE Magnets and their Applications*, Canberra, Australia, July (1992), p. 670.

[Tan86] S. I. Tanigawa, E. J. Hara, E. J. Lavernia, T. S. Chin, R. C. O'Handley and N. J. Grant, *IEEE Trans. Mag.*, **MAG-22** (1986), 746.

[Tat94a] C. Tattam, A. J. Williams, J. N. Hay, I. R. Harris, S. F. Tedstone and M. M. Ashraf, *13th Int. Workshop on RE Magnets and their Applications*, Birmingham, UK, Sept. (1994), p. 573.

[Tat94b] C. Tattam, A. J. Williams, J. N. Hay, I. R. Harris, S. F. Tedstone and M. M. Ashraf, *J. Appl. Phys.*, **76** (1994), p. 6831.

[Ten90] P. Tenaud, F. Vial, A. Barzasi, A. Duchêne and M. Sagawa, *11th Int. Workshop on RE Magnets and Their Applications*, Pittsburgh, Pa., Oct. (1990), p. 123.

[Tok85] M. Tokunaga, N. Meguro, M. Endoh, S. Tanigawa and H. Harada, *IEEE Trans. Mag.*, **MAG-21** (1985), 1964.

[Tok89] M. Tokunaga, Y. Nozawa, K. Iwasaki, S. Tanigawa and H. Harada, *J. Magn. Magn. Mat.*, **80** (1989), 80.

[Tor90] S. Tori and G. Bava, *CEAM Report*, Elsevier Appl. Science, (1990), p. 628.

[Wad90] M. Wada and F. Yamashita, *IEEE Trans. Mag.*, **MAG-26** (1990), 2601.

[War89] M. Ward and J. S. Taylor, *CEAM Report*, Elsevier Appl. Science, (1989), p. 600.

[Wen92] P. A. P. Wendhausen, D. Eckert, A. Handstein and K.-H. Muller, *Phys. Stat. Sol.* (A), **129** (1992), K45.

[Wen94] P. A. P. Wendhausen, B. Gebel, N. M. Dempsey K.-H. Müller and J. M. D. Coey, *13th Int. Workshop on RE Magnets and their Applications*, Birmingham, UK, Sept. (1994), p. 831.

[Wie87] G. Wiesinger, G. Hilscher and R. Grossinger, *J. Less-Common Met.*, **131** (1987), 409.

[Wil90] A. J. Williams, P. J. McGuiness and I. R. Harris, *IEEE Trans. mag.*, **MAG-26** (1990), 1945.

[Wil91] A. J. Williams, P. J. McGuiness and I. R. Harris, *J. Less-Common Met.*, **171** (1991), 149.

[Wit89] P. A. Withey, E. J. Devlin, J. S. Abell and I. R. Harris, *J. Magn. Magn. Mater.*, **80** (1989), 67.

[Wit90] P. A. Withey, H. M. Kennett, P. Bowen and I. R. Harris, *IEEE Trans. Mag.*, **MAG-26** (1990), 2619.

[Xia92] Y. Xiao, J. Liu, B. Qui and M. Liu, *12th Int. Workshop on RE Magnets and their Applications*, Canberra, Australia, July (1992), p. 258.

[Xie90] J. Q. Xie and C. D. Graham Jr, *Proc. 6th Int. Symp. on Mag. Anisotropy and Coercivity in RE Trans. Metal Alloys*, Pittsburgh, Pa., Oct. (1990), p. 308.

[Yaj88] K. Yajima, H. Nakamura, O. Kohmoto and T. Yoneyama, *J. Appl. Phys.*, **64** (1988), 5528.

[Yam89] H. Yamamoto, M. Nagakura, Y. Ozawa and T. Katsuno, *IEEE Trans. Mag.*, **MAG-25** (1989), 4123.

[Yan93] C. J. Yang, W.-Y. Lee and H.-S. Shin, *J. Appl. Phys.*, **74** (1993), 6824.

[Yos94] N. Yoshikawa, H. Yamada, Y. Iwasaki, K. Nagata and Y. Kasai, *13th Int. Workshop on RE Magnets and their Applications*, Birmingham, UK, Sept. (1994), p. 635.

[Yur94] T. Yuri and T. Ohki, *13th Int. Workshop on RE Magnets and their Applications*, Birmingham, UK, Sept. (1994), p. 645.

[Zha91] X. J. Zhang, P. J. McGuiness and I. R. Harris, *J. Appl. Phys.*, **69** (1991), 5838.

[Zho86] S. Z. Zhou, L. Li, L. D. Zhang and Q. Hu, *J. Magn. Magn. Mater.*, **54** (1986), p. 54.

[Zho87] S. X. Zhou, D. Li and R. J. Shuai, *9th Int Workshop on RE Magnets and their Applications*, Bad Soden, Aug. (1987), p. 339.

[Zho92] S. Zhou, J. Yang, M. Zhang, D. Ma, F. Li and R. Wang, *12th Int. Workshop on RE Magnets and their Applications*, Canberra, Australia, July (1992), p. 44.

8

Static applications

H. A. Leupold

The advent of rare-earth permanent magnets (REPM) has brought about magnetic structures that were previously impracticable [Rot78, Tau85]. So different are these remarkable materials from the earlier magnets that conventional design wisdom is inadequate to fully exploit their unique characteristics. Indeed, the conventional approach can lead to error or to the employment of quite inappropriate design procedures. The salutary characteristics stem from two basic attributes of the rare-earth materials: (i) large intrinsic moments per unit volume; (ii) extraordinarily high resistance to demagnetization by external or internal demagnetizing fields.

The magnetization offers high flux density; the coercivity enables the magnets to maintain this flux density in the face of very high demagnetizing fields engendered by flat aspect ratios. Thus Sm–Co and Nd–Fe–B can be fashioned, with impunity, into shapes that would cause demagnetization of Alnico materials. This point is illustrated in Fig. 8.1, which shows the second quadrant of the hysteresis loop of a typical REPM. Note that the constancy of magnetization in reverse fields as high as 1.5 T results in a linear B versus $\mu_0 H$ curve of unit slope throughout the entire quadrant. This linearity of the magnetization curve makes possible magnetic structures in which the magnet is exposed to demagnetizing fields in excess of the remanence. The importance of this property to design simplification will become obvious from the following discussions, where $\mu_0 H$ and $\mu_0 M$ will be written B_0 and J, respectively, for convenience in curve plotting and calculations. This procedure has the advantage that in the relationship between field, flux density and magnetization $\mathbf{B} = \mu_0(\mathbf{H}+\mathbf{M})$, written $\mathbf{B} = \mathbf{B}_0 + \mathbf{J}$, all three of the principal quantities involved are of the same kind and are measured in teslas.

Magnetic circuit design with the older magnetic materials, such as Alnico, was largely an intuitive hit-or-miss affair, as is clear from the B vs B_0 curve for Alnico in Fig. 8.1. Flux densities are low except for geometries with the very highest B to B_0 ratios, that is, those with the very lowest demagnetizing factors. Because the state of magnetization of each magnet depends upon the geometry of the structure in which it is placed, exact analytic solution for all but the simplest configurations is difficult or impossible. Reliance is then placed on numerous approximations, physical models, electrical analogues or rule of thumb.

Because of the analytical simplification afforded by the magnetic rigidity of the REPMs, the range of applicability of traditional approaches to magnetic design has been greatly extended. These approaches fall into four broad types: (i) analogy of magnetic configurations to electric circuits; (ii) analytical solutions through Maxwell's equations; (iii) reduction of permanent magnet arrays to distributions of equivalent pole densities or current sheets and the insertion of those distributions into Coulomb's law or the Biot–Savart law; and (iv) finite-element computer

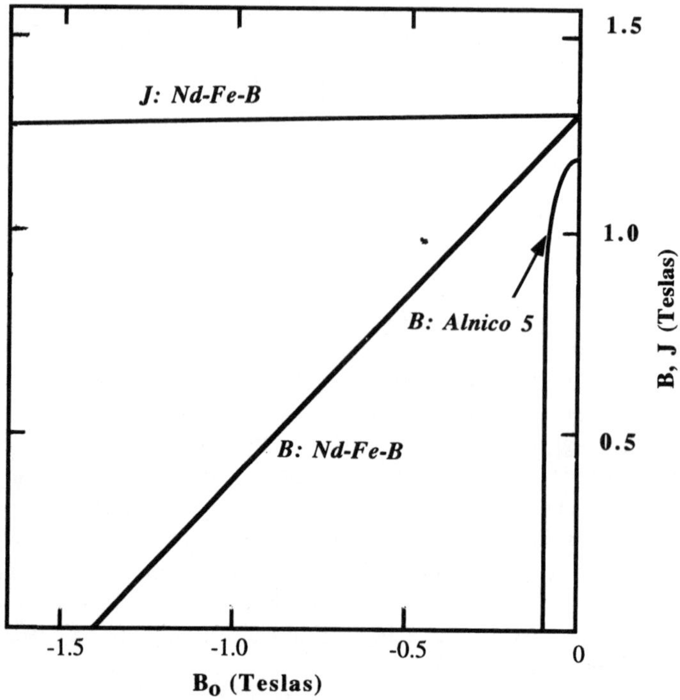

FIG. 8.1. Demagnetization curves for Nd–Fe–B and Alnico 5 magnets.

solution of a tentative design, the plausibility of which is previously established by one of the other approaches, usually the first. In this chapter we will discuss these approaches and employ them to solve illustrative problems.

A. MAGNETIC CIRCUIT DESIGN

1 Magnetic analogue of Ohm's law

In the past, elementary courses in general physics often included a brief treatment of magnetic circuits that featured a magnetic analogue of Ohm's law, introduced in Chapter 1 (p. 50). In this scheme, the rôle of electric current I is played by the magnetic flux Φ, that of the electromotive force V by the magnetomotive force (mmf) F, that of the electric conductance G by the magnetic permeance P and that of the resistance R by the reluctance R, so that the magnetic Ohm's law reads:

$$\Phi = PF = F/R \tag{8.1}$$

in respective correspondence to

$$I = GV = V/R$$

Permanent magnets can then be regarded formally as magnetic 'batteries', the sources of mmf, and materials of high permeability, such as soft iron or permalloy, as near-perfect flux conductors, by analogy to the near-perfect current conductor, copper, in electrical circuits. Air gaps and materials of low permeability play the rôle of magnetic resistors. The analogy is completed by the identification of electrical solenoids wound about permeable circuit members as flux 'generators', which can be either AC or DC, depending upon the currents sent through them.

Although this approach, like most analogies, is philosophically and mnemonically gratifying, it is rarely used in practice as outlined, and has now been dropped from elementary courses. The barrier to more general usefulness was essentially twofold. In the first place, unlike electric currents, magnetic fluxes are not confined to neat, analytically tractable paths like wires, but fill virtually all of space. This difficulty is not too serious, because, in many cases, the space around a magnetic circuit can be divided into flux paths of which the boundaries are planes, cylindrical arcs or segments of spherical surfaces that emanate normally from the surfaces of the circuit to connect points of different magnetic potential. The permeances of these simplified paths can be calculated approximately through standard formulae (Fig. 8.2), and if the division is made judiciously, surprisingly good approximations to the true fluxes can often be obtained.

The second, and more serious, difficulty that has prevented a simple, direct application is that magnetic materials such as Alnicos, generate no unique magnetomotive force (mmf), because the mmfs depend upon the circuit into which they are inserted. The great value of REPMs in the simplification of circuit design arises because these materials exhibit the same total circuital mmf regardless of the nature of the circuit in which they find themselves. Although this advantage has been discussed previously [Rot78, Tau85, Leu81], it is still not as widely appreciated as are the high-energy products and coercivities of the REPMs; consequently, circuits employing these materials are still often analysed by unnecessarily cumbersome procedures.

To illuminate the origins of these difficulties and to highlight the contrasts between the rare-earth magnets and their predecessors, it is useful to consider the derivation of the magnetic Ohm's law. We begin by writing the circuital form of Ampère's law.

$$\oint \mathbf{H} \cdot \mathrm{d}\mathbf{s} = \pm I \tag{8.2}$$

where the magnetic field H is in A/m, the current I, linked by the loop integral, is in ampères, and the sign depends upon whether the current direction in the coils produces an mmf aiding or opposing that of the magnet. Assuming H_{m} is the average field over the length of the path in the magnet L_{m}, we can write

$$H_{\mathrm{m}} L_{\mathrm{m}} + \int_A^B \mathbf{H}_{\mathrm{L}} \cdot \mathbf{d}s = \pm I \tag{8.3}$$

where A and B are at the boundaries of the magnet, the line integral is along a flux line in the region outside the magnet, including the airgap, and H_{L} is the field along

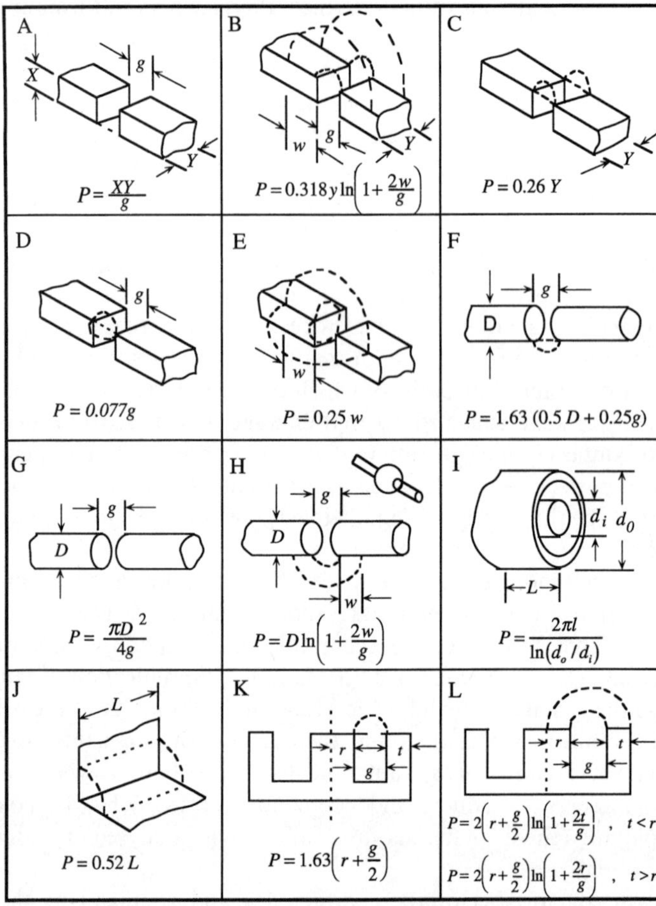

FIG. 8.2. Standard permeances.

that line. Outside the magnet, $\mu_L H_L = B_L$, therefore

$$H_m L_m + \int_A^B \frac{\mathbf{B}_L \cdot \mathrm{d}\mathbf{s}}{\mu_L} = \pm I \tag{8.4}$$

but $B_L = \Delta\Phi/\Delta A$, where ΔA is a differential element of area normal to a flux line of differential length $\mathrm{d}s$ and $\Delta\Phi$ is the element of flux passing through it. Consequently,

$$H_m L_m + \int_A^B \frac{\Delta\Phi}{\mu_L \Delta A} \mathrm{d}s = H_m L_m + \Delta\Phi \int_A^B \frac{\mathrm{d}s}{\mu_L \Delta A} = \pm I \tag{8.5}$$

Of course, the area ΔA and permeability μ_L associated with flux $\Delta\Phi$ may vary with progression along the flux line, that is, they are functions of s. We recognize that the expression in the integral is just the differential reluctance $dR = ds/(\mu_L \Delta A)$, so that

$$\int \frac{ds}{\mu_L \Delta A} = \Delta R_t^E = \frac{1}{\Delta P_t^E} \tag{8.6}$$

where ΔP_t^E is the permeance of the tube enclosing flux $\Delta\Phi$ exclusive of the permeance of its path within the magnet.

Upon integration over s, Eq. (8.5) becomes

$$\frac{\Delta\Phi}{\Delta P_t^E} = \pm I - H_m L_m \tag{8.7}$$

Since these tubes are all in parallel and fill all of space, their summation and that of their fluxes yield $\Phi_m/P_t^E = \pm I - H_m L_m$, where Φ_m is the total flux emanating from the magnet; therefore, using $\Phi_m = B_m A_m$, where B_m is the average flux density within the magnet, we obtain

$$B_m = P_t^E[\pm I - H_m L_m]/A_m \tag{8.8}$$

For Nd–Fe–B, the demagnetization curve is reversible, linear and of the form

$$B_m = \mu_r H_m + B_r \tag{8.9}$$

where μ_r is the recoil permeability and B_r is the remanent flux density, $\mu_0 M_r$. Solving for H_m, we obtain

$$H_m = (B_m - B_r)/\mu_r = \Phi_m/\mu_r A_m - B_r/\mu_r \tag{8.10}$$

Multiplication of Eq. (8.8) by A_m and combination of the results with Eq. (8.10) yields

$$\Phi_m = P_t^E[-\Phi_m L_m/\mu_r A_m + B_r L_m/\mu_r \pm I]$$

$$\Phi_m = \frac{(B_r L_m/\mu_r \pm I)}{(R_t^E + R_m)} \tag{8.11}$$

This is the magnetic form of Ohm's law where the total Φ_m is analogous to the total current and the reluctances $R_t^E = 1/P_t^E$ and $R_m = L_m/\mu_r A_m$ correspond to the total circuit resistance external to the battery and the battery internal resistance respectively. Corresponding to the battery emf is $L_m B_r/\mu_r$ or $-L_m H_c$ and $\pm I$ is the analogue of the emf from a zero impedance, constant voltage source. This simple and useful formula is applicable to rare-earth magnets because of the linear relationship between H_m and B_m expressed by Eq. (8.9), which, in turn, is a consequence of the constancy of magnetization in all parts of the magnet, regardless of stimuli arising from the rest of the circuit or the magnet's own shape. In contrast, Alnicos do not have reversible demagnetization curves that are expressible in simple analytical form. Therefore, in dealing with circuits containing such magnets, one must proceed less directly and with recourse to B–B_0 graphs such as those shown in

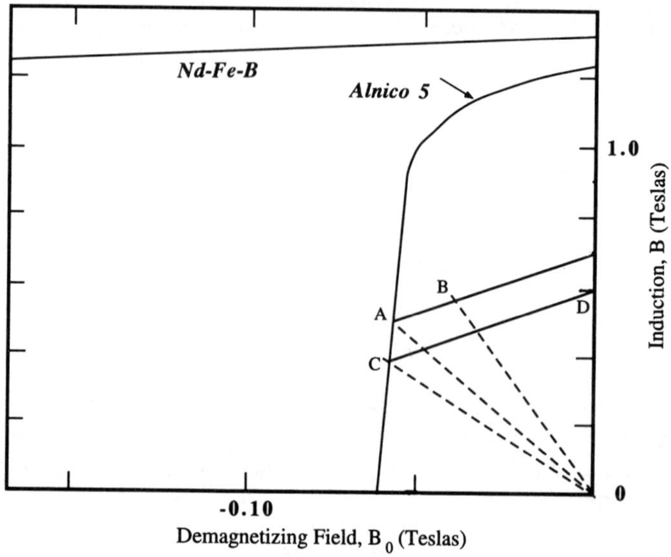

FIG. 8.3. Determination of B_m in an Alnico 5 magnet. The load line OA is determined from the total external permeance as described in the text. B_m is the value of B at intersection A. A decrease in P_t^E would lower the operating point to C, the base of a new reversible minor loop CD. The reversible linear demagnetization curve for Nd–Fe–B magnets of 1.3 T remanence is shown for comparison.

Fig. 8.3 [Leu82]. For simplicity, we consider as an example a circuit with no electromagnetic sources of flux. In such a case, Eq. (8.8) can be written

$$B_m/H_m = L_m P_t^E/A_m$$

Then, one must calculate P_t^E, which, together with the use of magnet length and the cross-sectional area in (8.11), yields B_m/H_m. Next, a load line with slope $B_m/\mu_0 H_m = B_m/B_{0m}$ is drawn through the origin and its intersection with the second quadrant B–B_0 curve determines B_m. The product $B_m A_m$ then yields Φ_m, which is the total flux output of the magnet; Φ_m is distributed among the various flux paths in the circuit in accordance with Ohm's and Kirchoff's laws.

If the demagnetizing field of the Alnico magnet is lessened by an increase in P_t^E resulting from the narrowing of an air gap within the circuit, the operating point of the magnet moves along an approximately linear minor loop to point B in Fig. 8.3, in accordance with the new value of B_m/B_{0m}. When changes in the circuit are such as to move the operating point from left to right on the minor loop, we can use an expression such as (8.9) with the intercept of the minor loop with the B axis, B_i in place of B_r, and the slope of the minor loop μ_i/μ_0 in place of μ_r/μ_0. Consequently, one may define a magnet mmf of the same form as that of the rare-earth magnets, that is

$$F = B_i L_m/\mu_i \qquad (8.12)$$

However, should the demagnetizing field increase due to widening of an airgap, temporary removal of the magnetic from the circuit or application of an external demagnetizing field, the operating point would move to the left along the demagnetization curve to *C*, the base of a new minor loop *CD*. The magnet would also have a new mmf given by the new line constants B_i' and μ_i'/μ_0

$$F = B_i' L_m / \mu_i' \tag{8.13}$$

Hence, we see that no unique mmf can be assigned to a traditional permanent magnet, and if the useful magnetic field is to be modulated by variation of gap length or electrically generated mmf's, the magnet mmf will always be that corresponding to the lowest point on the demagnetization curve reached over the course of the modulation cycles. An additional complication in determining the mmf of an alnico magnet is that the lengths L_m appearing in Eqs. (8.12) and (8.13) are effective rather than actual lengths. Because of demagnetizing fields, the magnetization tends to be nonuniform, so that different parts of the magnet lie on different load lines, and therefore produce different mmf's and inhomogeneous fields. This causes effective lengths to be shorter than actual ones and all the other germane quantities, such as H_m and B_m, to have only a rough average significance. This problem exists for rare-earth magnets as well but to a lesser extent because only H can vary from point to point while for Alnicos, both H and M vary. For some configurations, use of geometrical lengths and average field values can cause significant errors, and estimations of effective lengths must be made [Rot78]. Magnetic quantities and their electric analogues are summarized in Table 8.1.

2 Example of magnetic gap field calculations

In Fig. 8.4(a) is pictured a simple magnetic circuit with the space around it divided into permeance paths. The flux and field of interest are those in the magnet gap P_g. To calculate these quantities we must first find the permeances associated with the 14 external flux paths. All are of standard form and can be calculated by means of formulae found in Fig. 8.2 and in [Leu81, Rot40]. To aid in the visualization of the forms of these paths, brief descriptions of each are summarized in Table 8.2. The 14 permeances are in parallel with each other and in series with the internal permeance P_m of the magnet. (See Fig. 8.4(b).). Almost rigid magnets ($B_r/B_{0c} = 1.05$) of $B_r = 1.32$ T are employed. From the values of the quantities shown, the total circuit reluctance is calculated to be $R_t = R_m + R_t^E = 3.333 + 0.345 = 3.678\ \mathrm{m}^{-1}$. The mmf: $F = L_m B_{0c} = L_m B_r / 1.05 = 0.4\ B_r/1.05 = 0.381\ B_r = 0.503$ T/m. Here, as previously stated, we use B_0 rather than H so that $F = L_m B_0$ is really a quasipotential measured in T/m. There is no difficulty involved when this value is used to obtain a flux via the formula $\Phi = PF$ because

$$F(\text{true}) = B_0 L/\mu_0 = F(\text{pseudo})/\mu_0$$

and

$$P(\text{true}) = \mu_0 \mu_p A/L = \mu_0 P(\text{pseudo}),$$

Table 8.1 Magnetic circuit quantities and their electric counterparts

	Magnetic		Electric		
Magnetomotive force	ΔU, F	amperes	Electromotive Force	ΔV, ε	volts
Magnetic potential	U	amperes	Electric potential	V	volts
Magnetic flux	Φ	webers	Electric current	I	amperes
Permeance	P	webers/ampere	Conductance	G	siemens, amperes/volt
Reluctance	R	amperes/weber	Resistance	R	ohms, volts/ampere
Magnetic field	H	amperes/m	Electric field	E	volts/m
Flux density	B	teslas, webers/m^2	Current density	J	amperes/m^2
Permeability	μ	tesla-m/ampere	Conductivity	σ	siemens/m
Magnetic field energy	W	joules	Electric power	P	watts
MMF of a magnet	$L_m H_c$	amperes	EMF of a battery	V_b	volts

Table 8.2 Magnetic flux paths of Fig. 8.4a

Path description	Paths	Formulae
Gaps with opposing parallel faces	P_{MG}, P_G, P_M	$P = xy/g$
Paths between parallel linear edges of opposing surfaces	P^1_{GE}, P^2_{GE}, P^3_{GE}	$P = 0.26y$
Annular cylindrical sections extending between coplanar surfaces	P^1_{ML}, P^2_{ML}, P^1_{GL}, P^2_{GL}	$P = 0.318y\ln(1 + 2w/y)$
Spherical quadrants ('orange slices') connecting opposing corners of opposing parallel surfaces	P_{GC}	$P = 0.077g$
Semicylindrical sections backstreaming from the sides of the magnet	P^1_{MS}, P^2_{MS}	$P = 0.318y$
Quadrants of spherical shells ('melon slices')	P_{GB}, P_{MB}	$P = 0.25w$
'Orange-slice' sections backstreaming from longitudinal edges of magnet	P_{MC}	$P = y/16$

In the formulae: x and y are dimensions of the rectangular gap faces perpendicular to the flux lines and g refers to gap lengths parallel to the flux lines respectively. w refers to thicknesses of cylindrical and spherical shells. The multiplying factors given below indicate the number of flux paths of that particular shape.

Calculated values of permeances in metres

$P_{MG} = 0.200$	$P^1_{GL} = 0.349$	$2 \times P_{GC} = 0.031$	$P^2_{MS} = 0.318$
$P_G = 1.000$	$2 \times P^2_{GL} = 0.140$	$2 \times P_{GB} = 0.100$	$2 \times P_{MC} = 0.050$
$P^1_{GE} = 0.260$	$2 \times P^1_{ML} = 0.052$	$2 \times P_{MB} = 0.050$	$P_M = 0.300$
$2 \times P^2_{GE} = 0.104$	$P^2_{ML} = 0.129$	$2 \times P^1_{MS} = 0.076$	$P^E_t = 2.901$
$2 \times P^3_{GE} = 0.042$			

$P_t = 0.2719$

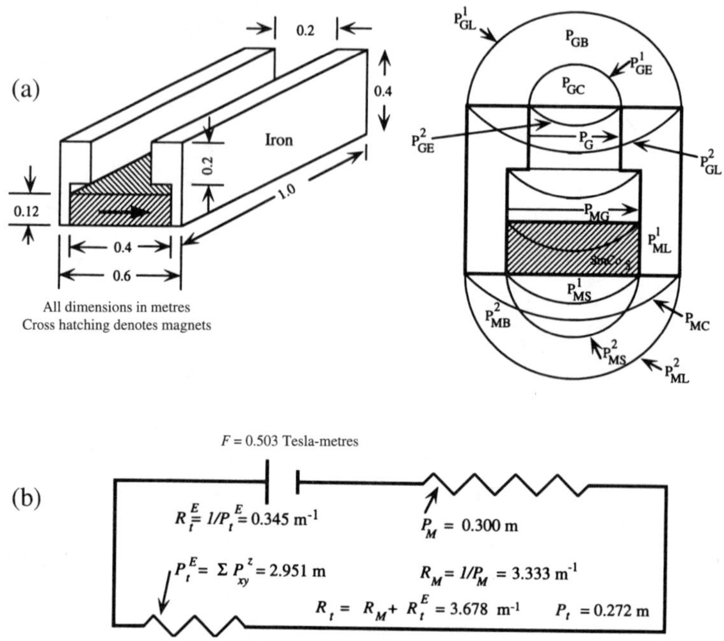

FIG. 8.4. (a) Division of space around magnetic circuit into approximate flux paths. (b) Schematic of equivalent electric circuit.

where μ_p is the permeability relative to that of empty space. Hence,

$$\Phi = P(\text{true})\, F(\text{true}) = [F(\text{pseudo})/\mu_0][\mu_0 P(\text{pseudo})] = F(\text{pseudo})P(\text{pseudo})$$

so that when B_0L is used for the potential F, and the relative permeability, $\mu_p A/L$, for P, the same results are obtained as from the true values of F and P. In most problems μ_p is either equal to 1.0, as in air gaps and in rigid permanent magnets, or to ∞, as in iron yokes and pole pieces. Rarely do paramagnetic substances occur in design problems but they present no special difficulties, as their μ_ps are between unity and infinity in the permeance formulae. When passive materials such as iron are operated close to or above saturation the approximation $\mu_p = \infty$ does not hold and the method of permeance estimation is not readily practicable.

Accordingly, we find the total flux emanating from the magnet

$$\Phi_m = FP_t = (0.503)(0.272) = 0.137 \text{ Wb}$$

Since all permeances in P^E_t are in parallel, the gap flux is given by

$$\Phi_g = P_g \Phi_m / P^E_t = (1)(0.137)/(2.90) = 0.047 \text{ Wb}$$

and the average field in the gap

$$B_g = \Phi_g / A_g = 0.047/0.2 = 0.236 \text{ T}$$

If the calculation were to be made for the same configuration with Nd–Fe–B replaced by Alnico 5, one would proceed as follows:

1. Calculate the total external permeance $P_t^E = 2.901\,\text{m}$ and total circuit permeance $P_t = 0.2719$ m.
2. Find the ratio B_m/B_{0m} via the formula $B_m/B_{0m} = -L_m P_t^E/A_m = -(0.4)(2.901)/(0.2) = -9.67$.
3. Draw the line $B_m/B_{0m} = -9.67$ together with the second quadrant of the Alnico demagnetization curve (Fig. 8.3).
4. Find the magnet operating point at intersection (A) and the value of B_m at that point, that is, $B_m = 0.514$ T, as in Fig. 8.3.
5. Calculate total flux $\Phi_m = B_m A_m = 0.062$ W.
6. Find $\Phi_g = P_g \Phi_m / P_t^E = 0.021$ W.
7. Find $B_g = \Phi_g / A_g = 0.107$ T, less than half the field obtained when Nd–Fe–B is used.

We see that the first and last two steps in this calculation must also be completed for Nd–Fe–B and Φ_m obtained from $\Phi_m = F/R_t$ in lieu of step (5). But the time-consuming steps, (2)–(4), are unnecessary when REPMs are used, and no external aids such as graphs are needed.

Magnetic circuit design by means of the electric circuit analogue, although seemingly crude, yields valuable qualitative guides to design adjustment and, in many cases, good quantitative results as well. The latter are often surprisingly accurate, especially when the calculated quantities of interest involve averages over large areas, as in the case of magnetic flux calculation. If a local field or flux density calculation is made, agreement of calculated values with actual values is not as good. These points are illustrated in Fig. 8.5[Leu73]. Note the remarkable agreement in geometric detail between the rough permeance calculations and the computer plot of Fig. 8.5. Point A is the point which, according to the calculations, is the demarcation between the flux lines which go from the limb to the iron plate and those which return to the other end of the magnet. Point B is the calculated effective equator of the magnet. Note how it is the approximate symmetry centre for the P_0 flux lines. Point C is the calculated point that marks the switching of the centre post flux lines from the permeable plate to the ring magnet. Point D is the predicted switch-over point from P_r paths to P_b paths. Point E is the demarcation between fluxes emanating from the ring magnet to the iron and to the centre post. The Φ obtained from these calculations was 2.12×10^{-5} W, compared with 2.18×10^{-5} W yielded by our longhand permeance calculations, agreement to better than 2%. In contrast, calculations of the flux density at Point G yield 0.39 T for the analogue method and 0.30 T for the computer, an approximately 30% disagreement.

3 Flux confinement by cladding

Experience in magnetic design teaches us that, in most magnetic circuits, much of the flux generated by the permanent magnets is wasted by distribution among

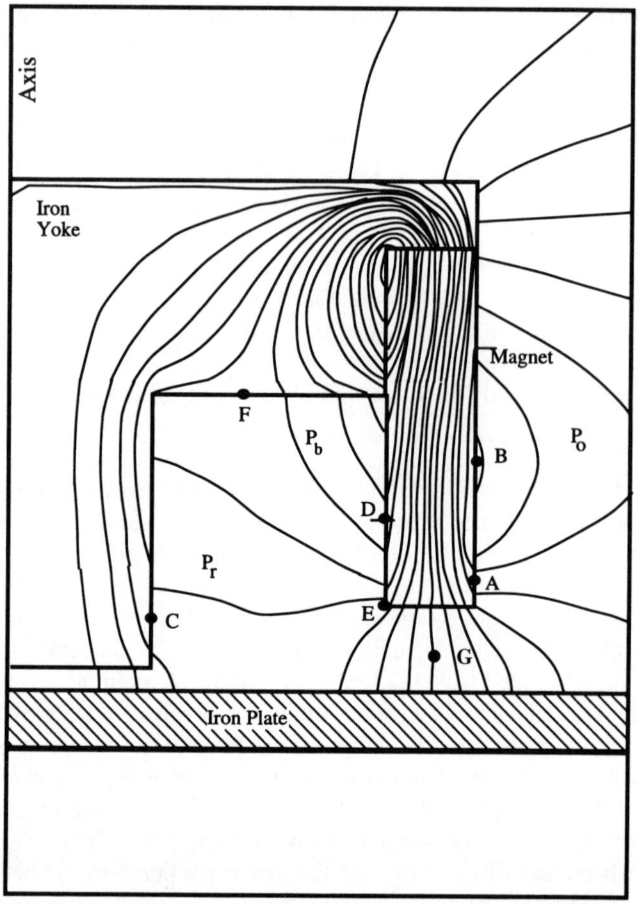

FIG. 8.5. Computer plot of flux lines produced by a cylindrically symmetric speaker magnet of E-shaped cross-section in proximity with an iron plate.

unwanted flux paths exterior to the work space of interest. The electric analogue method suggests a solution to this problem. In an electrical circuit, current can be kept from flowing along extraneous paths either by insulation or by employment of compensating or bucking potentials, as in a potentiometer. Magnetic insulation is impracticable pending the arrival of room-temperature superconductors with sufficiently high lower critical fields (H_{cl}). However, with rigid rare-earth permanent magnets such as Sm–Co or Nd–Fe–B, the potentiometric method has proved to be very fruitful in magnetic design.

In illustration, let us consider the simplest of magnetic circuits, the horseshoe-like configuration of Fig. 8.6. Clearly, much of the generated flux leaks into regions P_L. The equivalent current in the electric analogue would flow through wires G_L. The

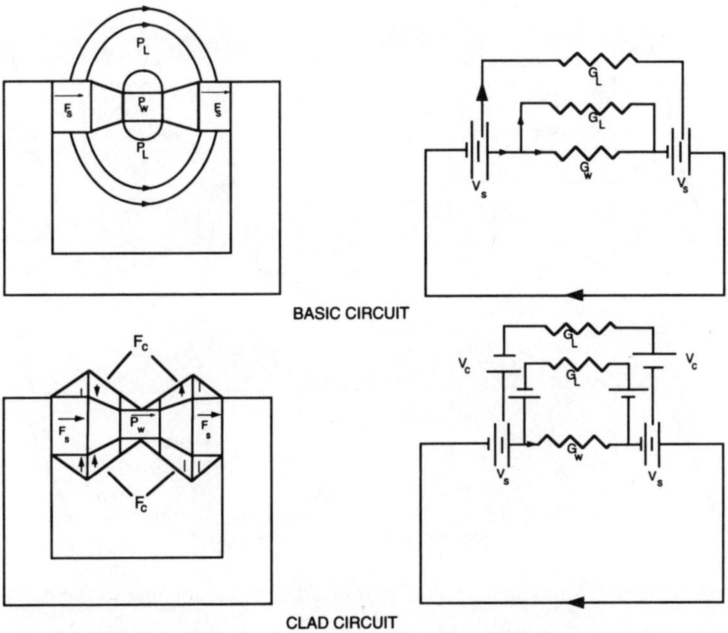

FIG. 8.6. Gap field intensification through cladding. The magnetic circuits are shown with their electrical analogues.

analogy is not exact, since the G_L are discrete, while the P_L are continuous. However, this slight discrepancy is irrelevant to illustrating the principle of cladding. In the electric circuit, batteries V_c of the appropriate EMF are placed in G_L, so that no current flows there. To effect the analogous suppression of flux in P_L of the corresponding magnetic circuit, it is clear that permanent magnets F_c analogous to the batteries V_c must be placed there. We begin by assuming that this goal has been accomplished and then work backwards to find the appropriate magnet configuration. By hypothesis, the total flux furnished by the magnets flows exclusively within the cross-section of the circuit and is given by the magnetic Ohm's law (Eq. 8.1)

$$\Phi = F/R_t = -2L_m B_{0c}/(R_m + R_g) = 2L_m B_r/(R_m + R_g) \tag{8.14}$$

where L_m is the length of a single magnet, B_r is the magnet remanence, R_m and R_g are the reluctances of the magnets and gap, respectively, and B_{0c} is the coercivity, which for our perfectly rigid magnets is equal to $-B_r$. The reluctivity of the magnets is assumed to be unity, and that of the yoke zero. For total flux confinement to be effected, every point on the surface of, or exterior to, the final configuration must be at the same potential. This condition is fulfilled automatically for points on the equipotential surface of the iron yoke. Therefore, cladding must be placed around the magnets and the gap so that its outer surface potential is everywhere equal to

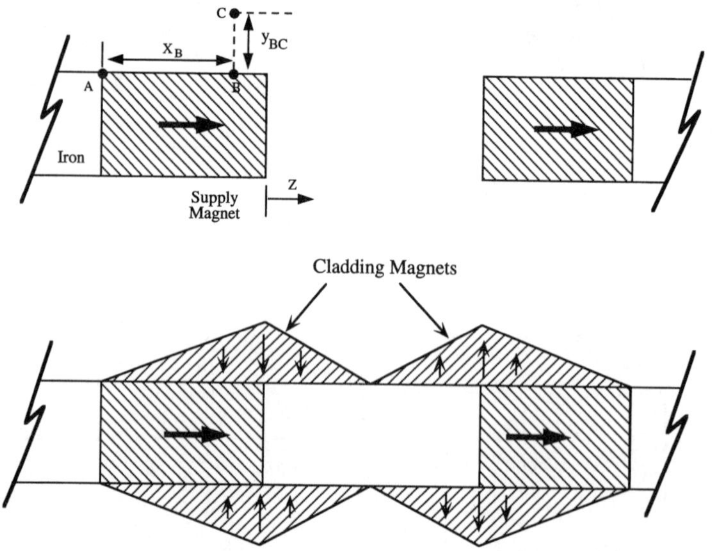

FIG. 8.7. Determination of cladding thickness of a horseshoe magnet in the gap region.

that at the yoke surface. Therefore, the potential difference between any point C on the cladding surface and point A on the yoke must be equal to zero and given by the line integral along any path connecting A and C. We consider the path A B C shown in Fig. 8.7:

$$F_{ABC} = F_{AB} + F_{BC} = 0 \tag{8.15}$$

Now $F_{AB} = F_m x_B / 2L_m$, where F_m is twice the mmf across one of the magnets and $F_{BC} = B_{0d} y_{BC}$ where B_{0d} is the radial magnetic field in the cladding and y_{BC} is the cladding thickness at point B. Since $F_m = \Phi R_m$, Eq. (8.15) becomes

$$0 = \Phi R_m x_B / 2L_m + B_{0d} y_{BC} \tag{8.16}$$

Substitution of (8.14) in (8.16) yields

$$y_{BC} = F_{BC}/B_{0d} = -F_{AB}/B_{0d} = -\Phi R_m x_B / 2L_m B_{0d} \tag{8.17}$$

Since, by hypothesis, no flux flows to the exterior, the radial flux density in the cladding magnet must also be zero. From the demagnetization curve of a magnetically rigid material, we know that at zero flux density, the field is equal to the coercivity $B_{0c} = -B_r$. Substitution of this value for B_{0d} in Eq. (8.17) yields

$$y_{BC} = 2L_m B_r R_m x_B / 2L_m B_{0c}(R_m + R_g) = \frac{R_m x_B}{(R_m + R_g)} \tag{8.18}$$

and the outer surface of the cladding about the supply magnet is a truncated cone, whose half-angle is $\tan^{-1}[R_m/(R_m + R_g)]$. The cladding thickness y reaches maximum at the end of the magnet, which is at the gap edge. The thickness declines beyond the edge because the field in the gap B_{0g} is in the direction opposite to that of B_{0m} and is given by

$$B_{0g} = B_g = B_m = \Phi/A_m$$

where A_m is the cross-sectional area of the gap. The potential declines with distance, z from the gap edge as

$$\Delta F = -zB_{0g} = zB_m = z\Phi/A_m$$

Inserting (8.14) we obtain for Φ

$$\Delta F = -2zL_m B_r/A_m[R_m + R_g] \qquad (8.19)$$

At the midgap ΔF is equal to $-F$ at the gap edge so that F is zero there and takes on increasingly negative values beyond. The negative potentials are balanced by cladding of opposite polarity, which increase in thickness up to the other gap edge, after which is declines linearly to zero. The resulting antisymmetric structure is shown in Fig. 8.7. In practice, the gap cladding is sometimes omitted for easier access. This results in a reduction in gap field and some bowing of the field lines, which is often tolerable but becomes more serious as the gap length, g, increases.

The efficacy of cladding is demonstrated by consideration of the unclad structure and its identical clad counterpart in Fig. 8.6. The latter has a gap field of 2.2 T, as compared with only 0.5 T for the unclad structure. Mass efficiency of the cladding is demonstrated by the 7 kg mass of such an assembly compared with the 55 kg required to produce only 1.6 T in an unclad configuration. The 2.2 T delivered by the clad version also compares favorably with that usually delivered by electromagnets that are many times as large, and is generated without power supply or current source. Cladding is especially effective when the leakage permeances are a large fraction of the total external permeance. When the same horseshoe-like structure has no tapered pole pieces as in Fig. 8.7, most of the external permeance is in the gap and cladding raises the gap field from 0.5 T to only 0.8 T.

4 Clad permanent-magnet solenoids

Traditionally, uniform magnetic fields extending over distances that are long, compared to the diameter of a cylindrical working space, are produced by electrical solenoids. These are cumbersome, requiring power supplies which often entail the expenditure of considerable energy. The rare-earth materials provide solenoidal fields with permanent magnet structures. A particularly useful and ingenious design for such a configuration was conceived by Neugebauer and Branch [Neu73] to focus the electron beam in a microwave klystron (Fig. 8.8). The required flux density was to have a magnitude of 0.15 T in a direction parallel to the axis of a cylindrical space 21.6 cm long and 7.0 cm in diameter. The REPM magnet used to supply the flux is in the form of an axially magnetized annular shell with μ_r approximately equal to unity forming the perimeter of the work space. Abutting against its ends are disks of a passive ferromagnet of high permeability and

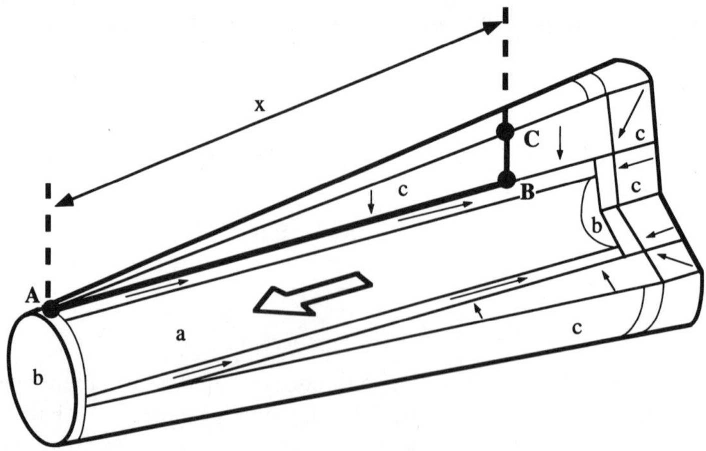

FIG. 8.8. Permanent magnet solenoid of the Neugebauer–Branch type. a = flux supply magnet; b = pole pieces; c = cladding magnets.

saturation magnetization, such as iron or permalloy. These disks serve as pole pieces to guide the flux produced by the magnet into the cylindrical work space. Again, we require that all the flux generated by the magnet passes through the work space. To accomplish this, the magnet must have a cross-sectional area A_m dictated by conservation of flux:

$$B_m A_m = B_w A_w = B_{0w} A_w \tag{8.20}$$

where B_m is the flux density in the magnet and B_w is that in the work cavity, and where A_w is the cross-sectional area of the cavity. If, as we assume, flux confinement has been accomplished, and since the end disks are equipotential surfaces, the field in the magnet B_{0m} must equal B_w. B_m can then be determined from (8.20) and from the equation for the demagnetization curve of a rigid magnet:

$$B_m = B_{0m} + B_r = B_w + B_r.$$

Substitution of this expression into (8.20) yields A_m:

$$A_m = B_w A_w/(B_w + B_r) = A_w/(1 + B_r/B_w)$$

To find the cladding thickness necessary for flux confinement, we take our zero potential to be at the left pole piece in Fig. 8.8. For any point C on the outer surface of the cladding to be at zero potential, the line integrals of H from A to B and from B to C must be equal and opposite, that is,

$$B_w x = -B_{0c} t_B \tag{8.21}$$

Since no flux flows radially through the cladding, B_{mc} is zero; and the rigid-magnet demagnetization curve shows that B_{0c} is equal to the coercivity B_{0c}, which in turn is equal to $-B_r$. Therefore, from (8.21) we see that the thickness, t, anywhere must be

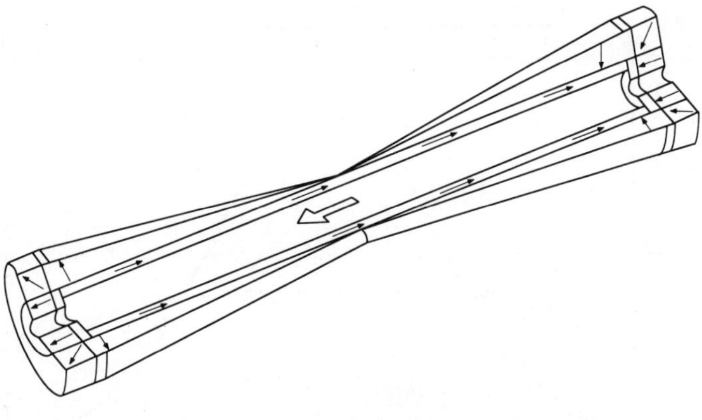

FIG. 8.9. Because the zero potential reference has been moved to the middle from the end, the structure pictured here has less than half the mass of the Neugebauer design of Fig. 8.8. Solenoidal fields up to 0.5 T are easily produced in this manner, in a structure much lighter than an equivalent electric solenoid and its power supply [Neu72, Cla86, Leu86].

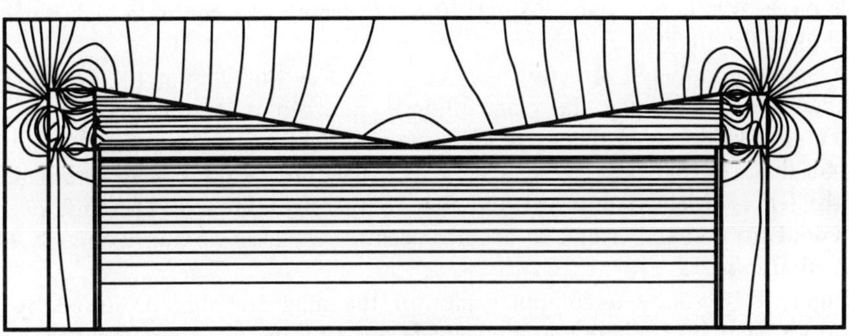

FIG. 8.10. Flux plot of magnetic field generated by the structure of Fig. 8.9. Apparent flux 'crowding' towards the periphery is because each line represents a unit of flux in an annular ring of given thickness.

$$t = B_{\mathrm{w}} x / B_{\mathrm{r}} \tag{8.22}$$

Again, t is linear in x and a conical surface results, as shown in Fig. 8.8. Because it is at the same potential as the right end of the supply magnet, the pole piece on the right must also be encased in cladding of maximum thickness.

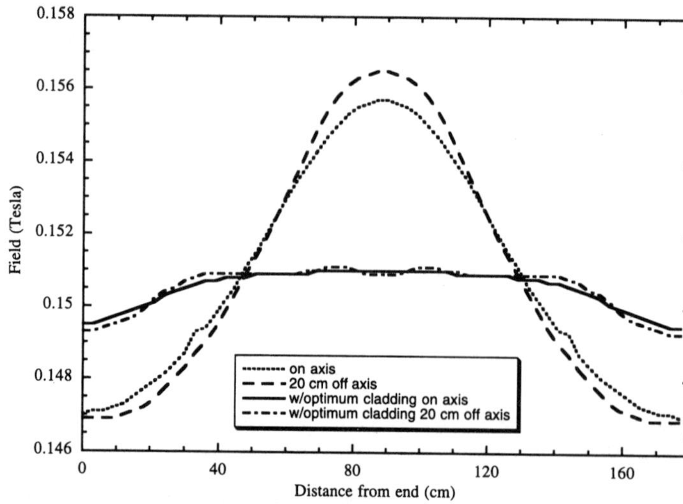

FIG. 8.11. Axial magnetic-field profile of the structure of Fig. 8.10.

The actual structure of Neugebauer and Branch employed two tandem chambers with oppositely directed fields, so that cladding of the bases of the high potential pole pieces was unnecessary; a single iron disc served as a common high-potential pole piece for the two chambers.

A single-chambered structure can be made less bulky than the Neugebauer configuration by taking the plane midway between pole pieces to be at zero potential [Cla86,Leu86]. Then, the cladding is in the form of two conical structures of half the base thickness of the Neugebauer structure. (See Fig. 8.9.) The cladding to the left is oriented outward and that to the right inward. The shift of zero potential from chamber end to chamber centre results in an approximately 60% reduction in mass.

Figure 8.10 shows a computer plot of the magnetic flux produced by the improved structure. There is some leakage due to imperfect cladding at the ends. Figure 8.11 shows the axial field as a function of distance from a chamber end and the much improved field uniformity obtained after a simple correction to the cladding slope is made. The correction consists of an increase of cladding thickness Δt according to the formula

$$\Delta t = (H_{\mathrm{d}} - H_{\mathrm{a}})t/H_{\mathrm{a}}$$

where H_{d} is the desired field and H_{a} is the actual field.

Figure 8.12 illustrates the impracticality of Alnico for permanent magnet solenoids. Since the cladding thickness is inversely proportional to the coercivity, it must be ten times as thick for Alnico. This translates to an approximately hundred-fold increase in mass.

There are many useful variants of the permanent magnet solenoid. For example, the internal field need not be constant but can be varied along the axis by

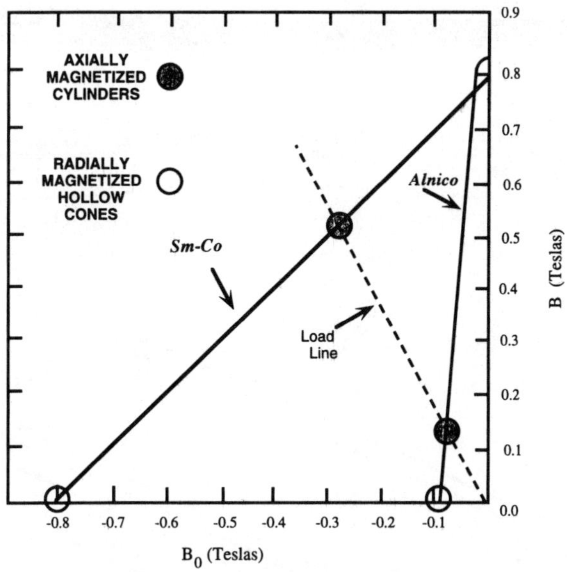

FIG. 8.12. Demagnetization curves and operating points for the magnets in a solenoidal field source.

appropriate adjustments in the supply magnet and cladding thickness [Cla86, Leu84]. As an example, if the field is to increase linearly with progression along the axis, the cross-sectional area of the supply magnet must also increase linearly while the cladding thickness is parabolic. In such a structure, the most efficient placement of zero potential is at the magnetic centre C of the configuration, that is, where the line integral of the field along the axis over the length of the chamber attains half its value, that is, $\int_0^c \mathbf{H} \cdot d\mathbf{l} = \int_c^L \mathbf{H} \cdot d\mathbf{l}$ where 0 and L are the ends of the working space. Clearly, C lies on the high field side of the geometric centre and the cladding is concave on the high field side and convex on the low field side.

The structural variations necessary to obtain the desired fields can be effected parametrically as well as geometrically, that is, material properties of the magnets can be varied rather than their dimensions [Cla86,Leu84,Leu86]. This procedure is not as efficient with regard to the weight and bulk of the structure because the uniform cladding thickness must be the maximum thickness in the geometric structure. On the other hand, the ring magnets are all of the same dimensions and are easier to manufacture.

The principles outlined can be applied to other field configurations such as in the cylindrical structure of Fig. 8.13 in which the field is confined to an annular ring by the placement of cladding on both the inner and outer boundaries of the working space [Cla87]. In this arrangement, the supply magnets can be located at either the inner or outer boundary of the working space, or both. Of course, the specified requirements must be such that the inner hollow is spacious enough to

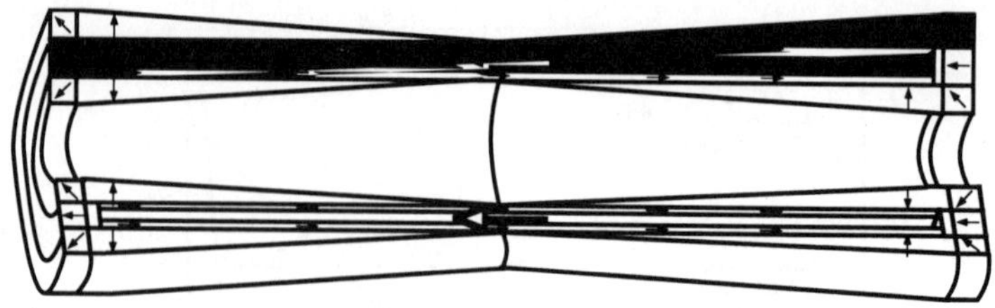

FIG. 8.13. Permanent magnet annular field source.

(a)

(b)

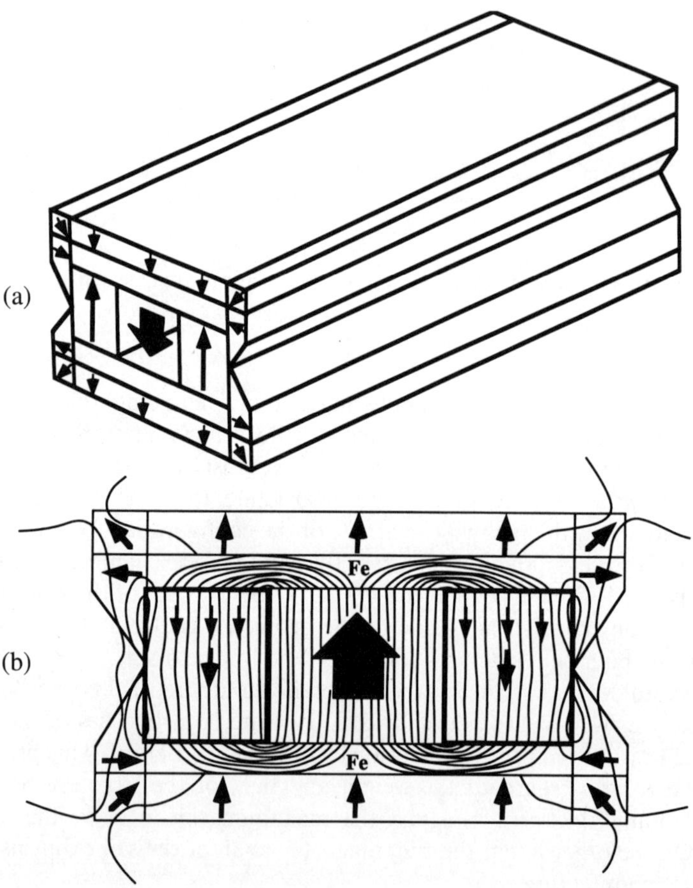

FIG. 8.14. Transverse field source for a rectangular cavity (a) and its sectional flux plot (b) in the cavity. The large arrow indicates the field in the rectangular central cavity, the medium-sized arrows indicate the orientations of the permanent magnets and the small arrows, the directions of the flux lines. The supply magnets are marked by thicker lines. All other magnets in the flux plot are cladding. Note the efficiency of the cladding in the flux confinement as evidenced by the negligible leakage at the corners only.

accommodate both the inner cladding and all the supply magnet. Clad rectangular structures with fields normal to the axis can be produced as in Fig. 8.14 [Leu86]. Such configurations are useful for magnetic resonance imaging (MRI). These applications will be discussed more fully in the next section.

All clad structures lose some of their field uniformity if access holes are drilled into their interior cavities. Such holes can often be made to be axially symmetric so that the field distortions are also axially symmetric and can be at least partially compensated for by methods similar to that employed in the compensation illustrated in Fig. 8.6.

If only one hole is needed in a structure that has its zero of potential at an end, it is usually better to drill the hole at that end since then no penetration of a cladding magnet is necessary and there is no field reversal in transit.

B. STRUCTURES COMPOSED ENTIRELY OF RIGID PERMANENT MAGNETS

1 Linear structures

Iron flux guides or yokes to shape and intensify the field in a working space can be used with the Alnicos but for many working-space geometries, it would be better to arrange magnetic poles so that maximum field is produced where it is wanted. Of course, there are no free magnetic poles; they are always accompanied by a like number of opposite poles of equal strength. It is the goal of yokeless magnet design to place the component magnets so that their poles are as effectively placed as possible to produce the desired field where needed while the opposite poles are more remotely placed or negated. Because the optimal arrays often entail very strong demagnetizing fields, the art of yokeless magnet design has very limited applicability with magnets of low coercive force.

Simple all-magnet structures that produce extraordinarily high fields can be obtained through considerations of symmetry and a very useful theorem [Hal81,Hal85]. The theorem states that if the magnetization of an infinite line source oriented perpendicular to its axis is rotated about that axis, the field it produces remains everywhere constant in magnitude and is everywhere rotated by the same angle in the opposite sense. This is easily demonstrated. For an infinite line dipole the tangential and radial field components are

$$H_\theta = \lambda \sin\theta / 2\pi r^2; \quad H_r = \lambda \cos\theta / 2\pi r^2 \qquad (8.23)$$

Therefore the magnitude of **H** is

$$|\, \mathbf{H} \,| = (H_\theta^2 + H_r^2)^{1/2} = \lambda / 2\pi r^2$$

and is independent of dipolar orientation.* λ is the moment per unit line length and θ is measured from the dipolar axis. Equation (8.23) indicates that the field orientation

* Note that a point dipole m produces a field $H_\theta = m \sin\theta / 4\pi r^3$; $H_r = 2m \cos\theta / 4\pi r^3$ which does not have this property.

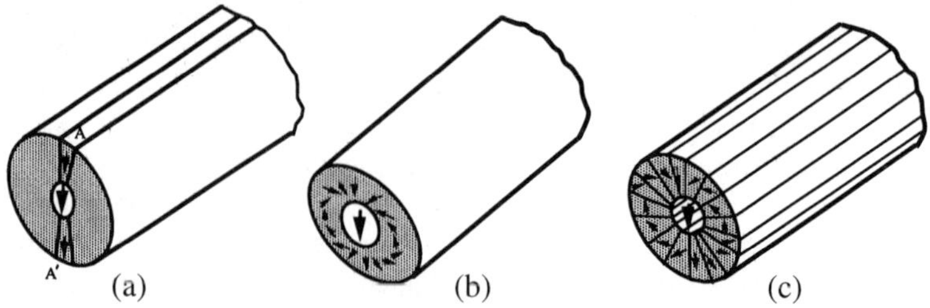

FIG. 8.15. (a) If a cylindrical dipolar source is to produce a uniform field in a cylindrical cavity as shown above, it is clear from symmetry that the infinitesimal elements A and A must be oriented as shown. The large arrow indicates the field in the inner cavity. (b) Ideal structure resulting from rotation theorem. (c) Segmented structure approximating (b).

angle α with respect to the axis is twice θ. If we then widen θ by a rotation of the magnetization, α must change by the same amount in the opposite direction to preserve the relationship $\alpha = 2\theta$. This principle has been used extensively by Halbach in his design of various particle beam devices [Hal80, 81a, 81b, 85].

As an example of this principle's application, we consider a cylindrical cavity of infinite length in which it is desired to generate a uniform field transverse to the principal axis. If this is to be accomplished with permanent-magnet material in the form of a circumscribed cylindrical shell, it is clear that the orientation of magnetization must have reflection symmetry, as in Fig. 8.15. Also by symmetry, the magnetization of the infinitesimal sector, **A**, must be radially oriented in the direction of the desired field.

If all the other sectors of the annular magnet ring were also pointed radially inward, each would contribute a field, $\mathrm{d}B_\mathrm{w}$, at the centre of the cavity equal to that of **A** in magnitude, but oriented at an angle of θ to that of **A**, where θ is the azimuthal angle. Of course, this would result in mutual cancellation of all the $\mathrm{d}B_\mathrm{w}$. According to our theorem, if the orientation of each segment is now changed by $-\theta$, the fields of the individual segments would all be in the same direction everywhere and hence would add without cancellation. Therefore, the prescription of the theorem is that the orientation of magnetization be in the direction γ, where

$$\gamma = 2\theta$$

The magnetic field in the cavity can then be found either by determination of the surface, σ_s, and volume, σ_v, pole densities arising from the polarization **J** and insertion of the results into Coulomb's law, or by application of Maxwell's equations to the boundaries of the magnetic material [Zij85]. The pole densities are given by $\sigma_\mathrm{s} = \hat{\mathbf{n}} \cdot \mathbf{J}$ and $\sigma_\mathrm{v} = -\nabla \cdot \mathbf{J}$ respectively, where $\hat{\mathbf{n}}$ is the unit vector normal to the surface under consideration. The resulting field in the cavity (w) is uniform and given by

$$B_\mathrm{w} = B_\mathrm{r} \ln(R_2/R_1) \tag{8.24}$$

where R_1 and R_2 are the inner and outer radii and B_r is the remanence of the magnet. Thus it can be seen that there is no limit in principle to the magnetic field that can be obtained in such a structure, provided only that the outer radius R_2 be made large enough. Due to the logarithmic dependence of B_w on R_2, material bulk quickly becomes prohibitively large with relatively small field increments. Nevertheless, fields of twice the material remanence should be practicable, namely 2.0–2.5 T for the rare-earth materials with the highest remanence and coercivity. If, for example, sintered Nd–Fe–B with $B_r = 1.2$ T is used, the working space is 2.5 cm in diameter and the outer structural diameter is 15 cm, the internal field will be given by $B_w = 1.2 \ln(15/2.5) = 2.1$ T, which is an impressive field to be generated in so large a working space by a structure that can be packed into a cylinder 15 cm in diameter and which generates no stray field.

The materials used in these REPM circuits need a high coercivity to withstand the antiparallel demagnetizing fields acting on some segments, and a high anisotropy (rigidity) to resist the transverse fields acting on others.

Manufacture can be facilitated by approximating the cylindrical boundaries with circumscribed polygons so that the structure is made of trapezoidal segments as in Fig. 8.15c. If this is done, the expression for the field becomes

$$B_w = B_r \left[\frac{\sin(2\pi/N)}{2\pi/N} \right] \ln(R_2/R_1) \tag{8.25}$$

where N is the number of segments. Polygons with as few as eight sides still generate a field 90% of that obtained with ideal circular cylindrical surfaces and with continuously varying magnetization.

Structures of this type are useful for any application in which high transverse fields are required in elongated cavities as in nuclear magnetic resonance imagers, biasing fields in electronic filters and general utility laboratory magnets. Cross-sectional slices of the structures can be used as segments of wigglers and undulators of free electron lasers to provide higher fields with less bulk and mass than is possible with more conventional structures (Fig. 8.16), [Hal80,81a,b,82,85]. Use of such structures and of their quadrupolar forms in particle beam guidance and MRI has been pioneered by Halbach [Hal80,81a,b,85], Holzinger [Hol82] and Zijlstra [Zij85].

Their cylindrical structures can also be made to be mechanically adjustable [Leu88]. This is done by fabricating the magnet in two rings with one nested inside the other, as in Fig. 8.17. Then the field within the central cavity will be the vector sum of the fields produced by the individual rings. If the rings are dimensioned so that each produces a field of equal magnitude and they are rotated with respect to each other by an angle β, the magnitude of the field within the cavity is just

$$B_w(\beta) = B_w(0)\cos(\beta/2) \tag{8.26}$$

where $B_w(0)$ is the maximum field, which occurs $\beta = 0$. Thus any field in the range $\pm B_w(0)$ is easily available. For field equality of the individual rings, the radius R_3

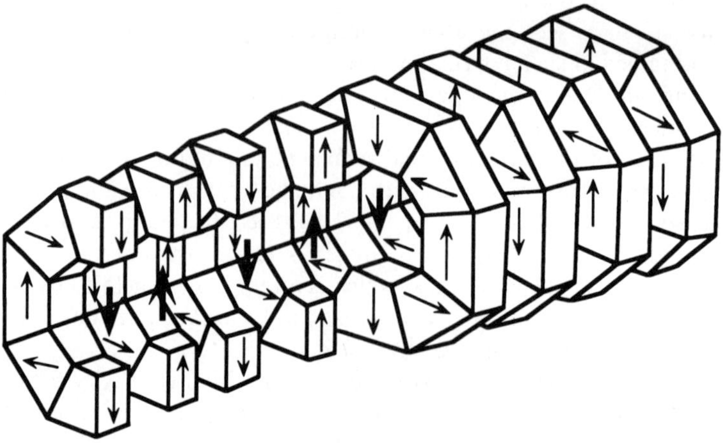

FIG. 8.16. Wiggler made from slices of octagonal cylindrical transverse field sources.

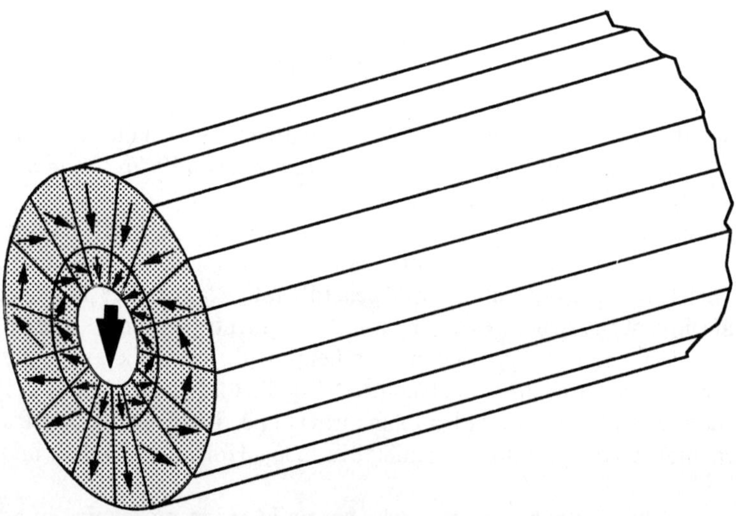

FIG. 8.17. Adjustable circular–cylindrical transverse field source. The inner and outer rings are rotated in opposite senses to vary the field in the inner bore.

of the boundary between them must be the geometric mean of the outer and inner radius of the device $R_3 = \sqrt{(R_1 R_2)}$. Variable flux sources of the type shown in Fig. 8.17 can be approximated with only eight segments in each ring. They are very useful in laboratory instrumentation where a variable field is required [Coe94]. For example a compact vibrating-sample magnetometer has been described where a

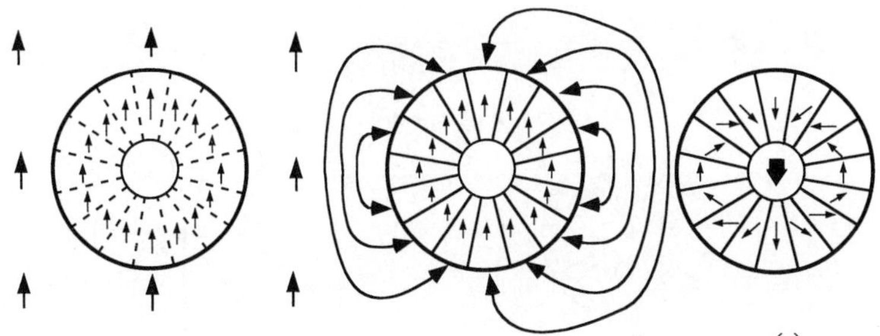

FIG. 8.18. Construction of circular–cylindrical transverse field source. (a) Structure is magnetically aligned (small arrows) in a uniform magnetic field (large arrows) and sliced into segments (dotted lines). (b) Pieces are interchanged as shown and bonded to form the finished structure in (c).

variable field of 0–1.2 T is generated in a 26 mm bore by a 7 kg Nd–Fe–B structure [Cug94a].

Fields of higher multipole value than dipole can also be generated by an appropriately faster change in variation of orientation, γ, with θ, so that

$$\gamma = ((n/2) + 1)\theta \tag{8.27}$$

where n is 2 for a dipole, 4 for a quadrupole, and so on [Hal85].

To make a ring of this type it is not necessary to make the many pieces with different orientations individually [Leu87]. For example, in the manufacture of a dipolar source, it is only necessary to orient a ring in a single unidirectional field, as in Fig. 8.18a. The ring is then cut into as many sectors as the desired approximation to a perfect ring demands. The pieces are interchanged according to the prescription $\theta \rightleftarrows -\theta$, as shown in Fig. 8.18b, and glued together to form the desired structure in Fig. 8.18c. This is possible because every possible orientation of magnetization with respect to the local radius is contained in the structure of Fig. 8.18a. They need only be reassembled to yield the dipolar source of Fig. 8.18c. The same result can also be obtained, by a 180° rotation of each segment about its radial axis. This may be achieved with a flexible sheet of bonded isotropic material by rolling it into a cylinder, magnetizing in a uniform transverse field and then rerolling it up in the opposite sense [Bel93].

A similar procedure can be used to make higher pole sources, but then **n** structures like that of Fig. 8.18a are required to yield the **n** of the desired n-poled arrays. The sequence of assembly of a quadrupolar source is shown in Fig. 8.19.

2 Polygonal cavities

It is often desirable to produce a strong transverse magnetic field in a cylindrical cavity of equilateral polygonal cross-section. It is sometimes further required that the magnetic field be completely confined to within the outer boundaries of the

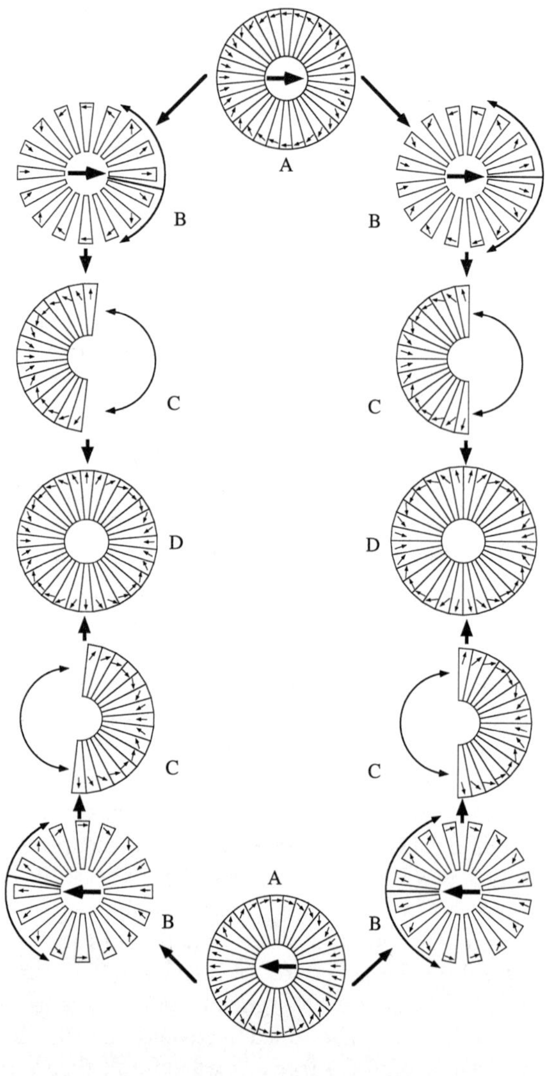

FIG. 8.19. Formation of two quadrupole sources from two dipole sources. (1). Start with two dipolar structures **A**. (2) Separate every other segment from adjacent segments to form two new structures **B** for each original structure. (3) Compress structures **B** in the direction of the circular arrows to form structures **C**. (4) Assemble structures **C** to form structures **D**.

structure with the boundaries being parallel and similar to the cavity boundaries. The latter requirement is to ensure the possibility of successive nesting of similar structures within each other to take advantage of the additive property of such fields to produce a large interior field equal to the sum of the components. Because

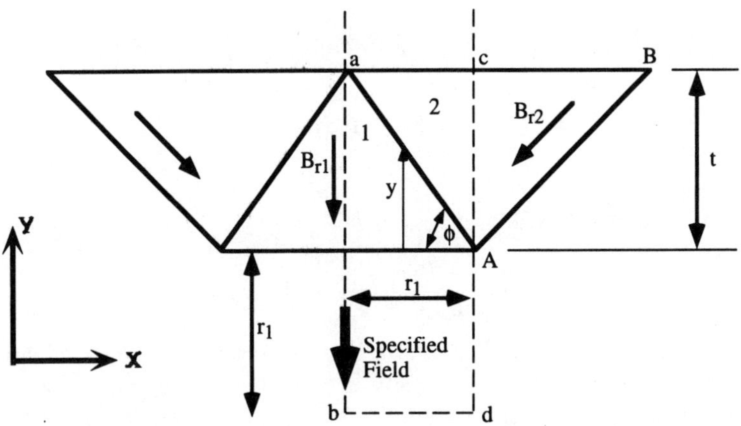

FIG. 8.20. Determination of the configuration of the upper segment of a square dipolar field source.

of its usefulness in MRI [Abe88, Leu88], a square cross-section due to Abele will be considered as an example of the method of determination of the structure.

We begin with a consideration of the top segment of the square magnet. It is of trapezoidal cross-section with 45° base angles. Because of square symmetry, only one half of it need be considered. For simplicity, it is desirable to have fields, magnetization and flux density constant over any of the segments into which the half trapezoid is to be divided. It is not possible to accomplish this together with flux confinement and generation of the desired field with a single-piece. Accordingly, to attempt to fulfil all the requirements, we divide the segment into two triangular pieces 1 and 2, as shown in Fig. 8.20. The polarization $J = B_r$ is the same for both. Since it is clear that the orientation of the magnetization in segment 1 should be in the direction of the desired flux density B_w, we need only determine the thickness t and the orientation of the magnetization in segment 2 that will fulfil the requirements:

1. That there be no flux exterior to the structure.
2. That the specified uniform flux density, B_w, be generated in the interior cavity.
3. That the exterior boundary of the structure be a square.

The first condition applied to the circuital form of Ampère's law $\oint B_0 \mathrm{d}l = 0$ dictates that

$$B_{01}y + B_{02y}(t - y) + B_w r_1 = 0$$

or if paths of integration **ab** or **cd** of Fig. 8.20 are chosen:

$$B_{01}t = B_{02y}t = -B_w r_1 \tag{8.28}$$

so that

$$B_{01} = B_{02y}$$

The second condition applied to Gauss' law at the cavity boundary yields $B_1 = B_w$ and since $B_1 = B_{01} + J = B_{01} + B_r$, it follows that

$$B_{01} = B_w - B_r \tag{8.29}$$

Insertion of (8.29) into (8.28) determines t:

$$t = \frac{-B_w r_1}{B_w - B_r} = r r_1/(1 - r) \tag{8.30}$$

where $r = B_w/B_r$, that is, the fraction of the remanence flux density of the magnet that is manifested as useful flux density in the cavity. The fraction r cannot be chosen arbitrarily if the outer boundary is to be a square. To demonstrate this, we apply Ampère's and Gauss's laws to the outer surface. For **B** and **H** to be zero everywhere exterior to the outer boundary, these laws require that $B_{02x} = 0$ and $B_{2y} = 0$ so that B_{02} and B_2 must be mutually perpendicular and normal and parallel, respectively, to the said boundary and since $\mathbf{B}_{02} + \mathbf{B}_r = \mathbf{B}_2$

$$(B_{02})^2 + (B_2)^2 = B_r^2 \tag{8.31}$$

Since the outer boundary is a square with its sides parallel to those of the inner boundary, application of Gauss's law at the boundary (1,2) yields

$$B_2 \sin\phi = B_1 \cos\phi \tag{8.32}$$

$$B_2 = B_1 \cot\phi \tag{8.33}$$

where ϕ is the angle between (1,2) and the inner boundary.

If we insert (8.33) into (8.31), we obtain

$$B_1^2 \cot^2\phi + B_{02}^2 = B_r^2 \tag{8.34}$$

but

$$B_{01} = B_{02y} = B_{02}, \quad \cot\phi = r_1/t, \quad B_1 = B_w \text{ and } B_0^1 = B_w r_1/t \tag{8.35}$$

so that (8.34) becomes

$$2B_w^2(r_1/t)^2 = B_r^2 \tag{8.36}$$

Hence

$$2(r r_1/t)^2 = 1 \tag{8.37}$$

From (8.30) and (8.37) we obtain

$$r = \frac{\sqrt{2} - 1}{\sqrt{2}} = 0.293 \tag{8.38}$$

Inserting this value into Eq. (8.37) we obtain the ratio of r_1 to t:

$$\frac{r_1}{t} = \frac{1}{\sqrt{2} - 1} = 2.41$$

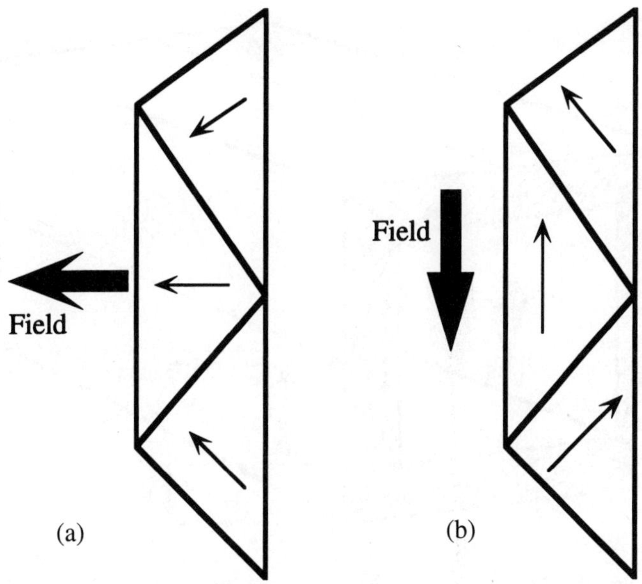

FIG. 8.21. (a) Side segment of square if relative magnetic orientation were that of Fig. 8.20. (b) After −90° rotation of all magnetizations.

Because from (8.31), (8.35) and (8.36) we have that $B_{02} = B_2$ and the angle α between $\mathbf{B}_{r2}$ and $\mathbf{B}_2$ is

$$\alpha = \tan^{-1}\frac{B_{02}}{B_2} = 45^\circ$$

so that the direction of $\mathbf{B}_{r2}$ is parallel to the boundary AB of Fig. 8.20. By symmetry, the directions of magnetization in the side forming the bottom of the configuration are just the inverse mirror images of those just formed for the top side. The magnetization orientations in the other two sides are found by the dipolar rotation theorem. If the side shown in Fig. 8.21 had its magnetization vectors arranged relative to its sides as in the top side, the resulting field would point to the left, as in Fig. 8.21(a). For the field to reinforce that of the top side it must be rotated 90° to make it point straight down as in Fig. 8.21(b) and therefore the magnetizations must all be rotated −90°. The complete resulting configuration is shown in Fig. 8.22. A permanent magnet material of remanence 1.0 T produces a flux density of $B_w = 0.293$ T in the square cavity and there is no limit in principle to the intensity that can be attained by successive circumscription of similar structures so that the total flux density is $0.293n$ T, where n is the number of layers used.

Such square structures are particularly useful as MRI magnets because they can easily be compensated for small random defects incurred in manufacture and

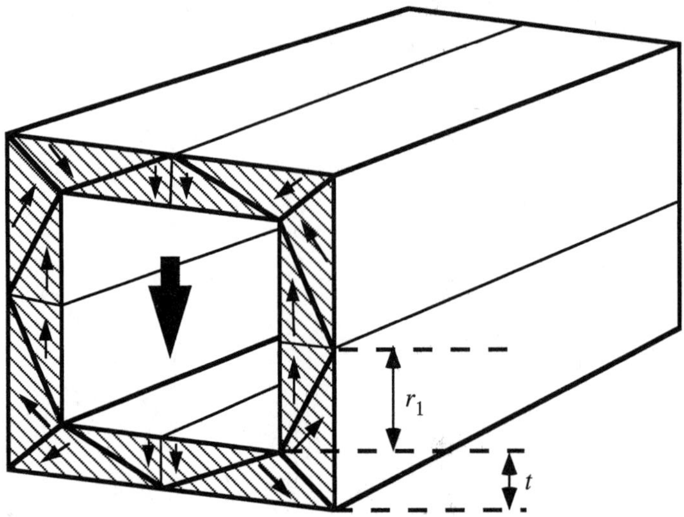

FIG. 8.22. Square dipolar field source. Large arrow shows field in cavity. Small arrows show orientations of constituent magnets.

assembly [Abe89]. This is accomplished by placing small dipoles at the inner or outer periphery of the structure, usually in the inner corners. The magnet is constructed from sectional slices that are compensated individually. The field that an ideal slice should generate is calculated and then compared with the measured actual field. The deviation field is then Fourier analysed to determine the magnitudes and directions of the compensating dipoles needed. Fortunately, satisfactory compensation can usually be achieved with terms no higher than dipolar. See [Abe89, 90] for details.

The same procedure can be used to find a triangular structure such as that of Abele [Abe89] which fills the same requirements as those specified for the square structure (Fig. 8.23). Table 8.3 compares the triangular, square, octagonal and circular configurations with regard to structural parameters. The triangular structure produces a higher field with a single layer than does the square. The circular structure produces a higher field than either with the same amount of magnetic material but requires more pieces for a near approximation (99%) to its theoretical performance. Therefore, if simplicity of manufacture is the prime consideration, the triangular section is best, but it is the worst in field strength produced for a given quantity of magnetic material. The circle is best in the latter regard but, as noted, it is the most complex. The square is intermediate and can be easily dipole-compensated for structural defects. The octagonal approximation to the circle is simple and produces a strong field but does not afford total flux confinement.

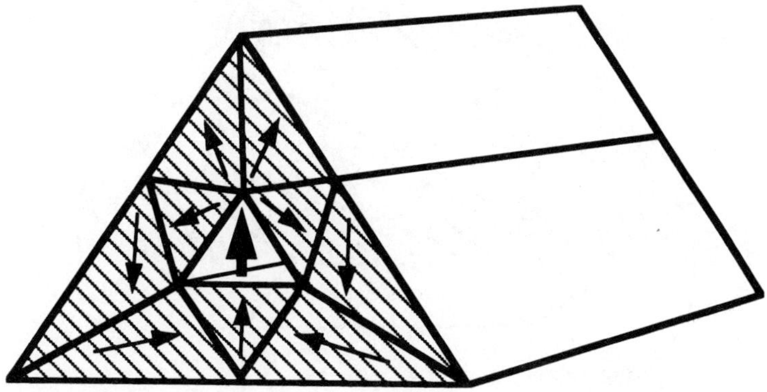

FIG. 8.23. Triangular dipolar field source. Large arrow shows field in cavity. Small arrows show orientations of constituent magnets.

Table 8.3 Comparison of cylindrical confined flux sources with similar inner and outer boundaries. B_w is the field produced in a 1 cm cavity, $A_t(\text{cm})^2$ is the total cross-sectional area of the structure, $A_m(\text{cm}^2)$ is the cross-sectional area occupied by material. The remanence B_r is taken to be 1 T in the last two columns

Structure	B_w/B_r	A_t	A_m	B_w/A_t	B_w/A_m
Triangular	0.50	5.2	3.9	0.96	1.3
Square	0.29	2.0	1.0	1.5	2.9
Octagonal	0.32	1.7	0.83	1.9	3.9
Circular	0.35	1.6	0.78	2.2	4.5

Uniform fields can be provided in cylindrical cavities of any cross-section. The shape need not be symmetrical or even externally convex. The details of the generation of such fields can be found in references by Abele *et al.* [Abe88,89a–c,90] and Leupold *et al.* [Leu92].

If the required fields are less than about half the remanence, open compact sources for fields in axially finite cavities can be made from such sections and the open faces then clad according to the same principles described in the discussion of permanent magnet solenoids. A clad section of an octagonal source is illustrated in Fig. 8.24.

End effects arise in unclad structures of finite length [Abe93]. This is illustrated in Fig. 8.25 for octagonal cylinders of different lengths, l. The field is essentially that of an infinite cylinder except where an end is approached to within a distance approximately equal to the structural diameter, d, where they begin to decline, falling to about one-half maximum at the end.

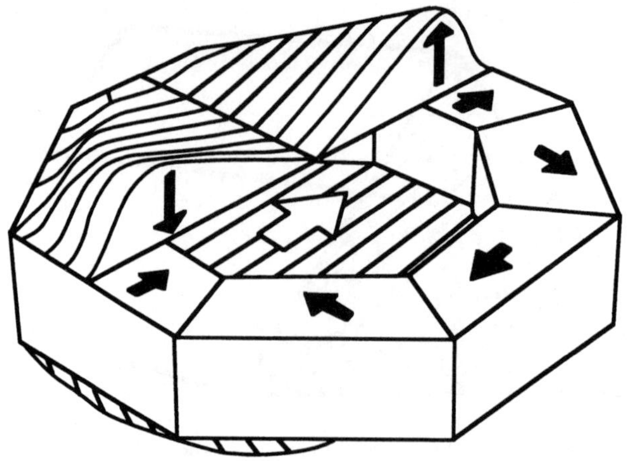

FIG. 8.24. Section of an octagonal pyx. The source of the field is the octagonal ring. The rest of the magnets are cladding to keep the field in the interior (large arrow) confined and uniform.

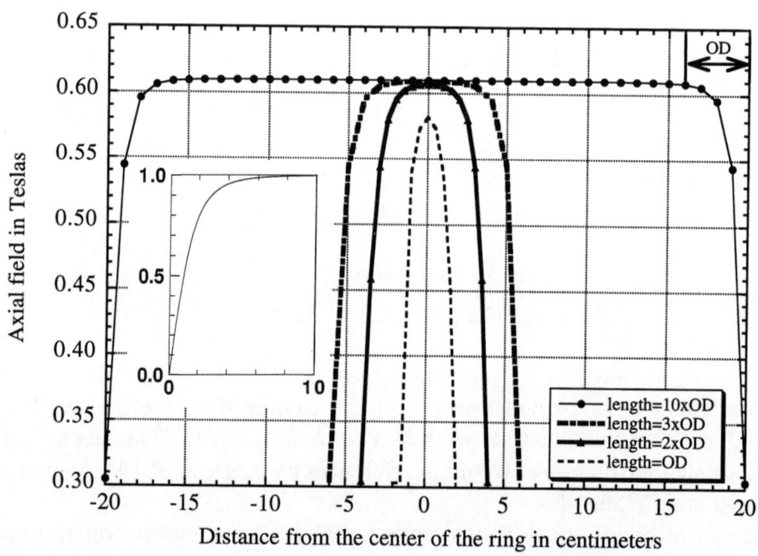

FIG. 8.25. Transverse field along the axis of an octagonal approximation to a magic ring, with ID = 2 cm, OD = 4 cm, $B_r = 1T$. The insert shows the field at the centre, in reduced units, as a function of length, in cm.

For $l = 3d$, the constant portion of the curve is still essentially at full field. For $l = 2d$ there is no region of constant field but the maximum is nearly as great as in the longer structure. For $l = d$ the maximum field is smaller. Similar end effects are present in transverse field structures of different cross-sectional geometry.

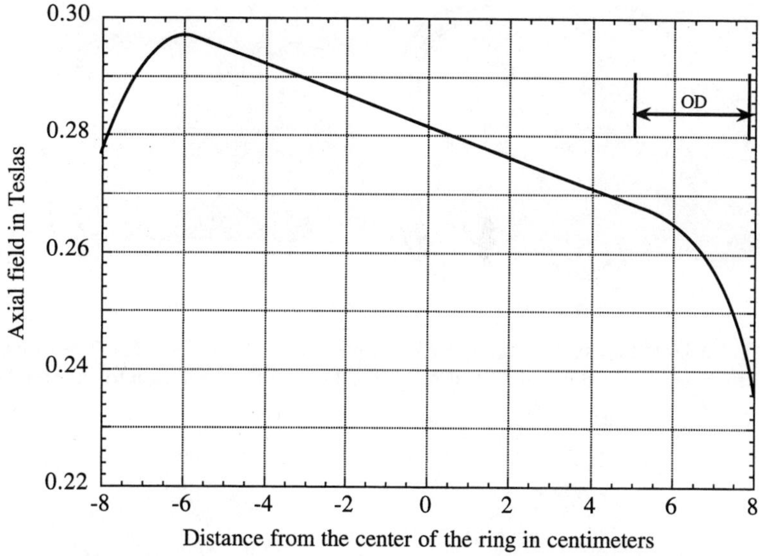

FIG. 8.26. Transverse field along the axis of an octagonal approximation to a magic ring with a tapered field.

Since permanent magnet MRI imagers are open ended, it would seem that their axial lengths would have to be prohibitive to provide a sufficiently large region of space with the required magnetic field uniformity. Abele [Abe 93] has provided an ingenious solution wherein by a judicious removal from the magnet cross-section of slices of the proper thickness and location, the field harmonics that constitute the end effects are largely removed. A typical case cited is for a magnet with a length of twice its bore in which the field inhomogeneity over a central sphere of one-quarter bore diameter is reduced by two orders of magnitude by the removal of five slices. Other examples and details of the procedure can be found in [Abe93], page 336.

In finite structures, the rotation theorem, Eq. (8.23) holds only approximately, because of the change in effective dimensionality. As a result, in structures of squat aspect ratio, the formula $\gamma = 2\theta$ does not yield the most efficient configuration [Zij85], but the approximation is a good one for $l/d > 3$.

The fields in all such arrays can be made to taper in magnitude with progression along the axis. The field plot in Fig. 8.26 is for one in which the field has a linear taper which is imposed by a like taper in the remanence of the magnet. The graph shows end effects similar to those of the taper-free structures. Tapered fields of this type are useful in channelizers.

3 Rod ensembles

Particularly useful but simple configurations can be assembled from transversely magnetized rods arranged in circular, cylindrical arrays [Coe94, Cug94b]. Figure

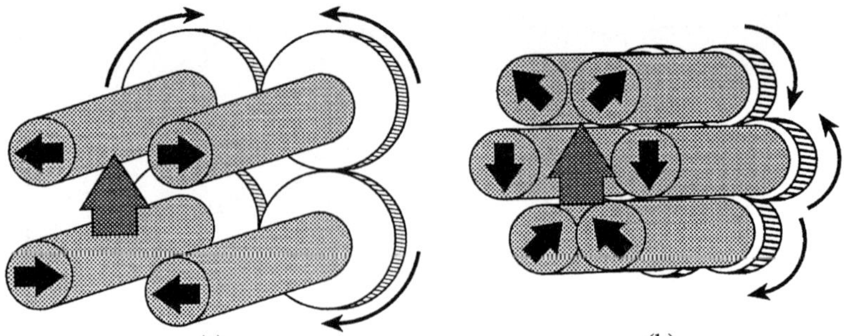

(a) (b)

FIG. 8.27. Variable flux sources composed of four and six rotatable rods. The orientations shown correspond to the maximum field positions. Directions of rotation to vary the field are indicated.

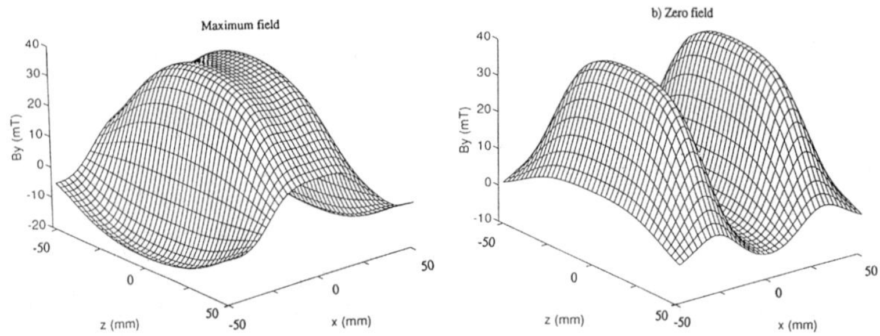

FIG. 8.28. Flux plots for the structure of Fig. 8.27a with the rods oriented for maximum and zero field at the centre.

8.27 shows two such configurations of four and six rods. The field can be varied by rotation of the individual rods and the field profiles of Fig. 8.28 are for the maximum and minimum fields obtainable for the four-rod configuration known as the 'magic mangle'. The mangle has been used by Coey *et al.* [Coe94] to apply variable magnetic fields to specimens examined on optical microscopes and to create sinusoidially time-varying fields for beam line experiments. Prototype models produce variable fields ranging from $\pm$ 35 mT to $\pm$ 0.8 T, with a uniformity of 10^{-3} over a volume of radius 3 mm.

The more rods in such an ensemble, the more closely the field approximates that of a magic cylinder. For example the resemblance of the axial field of the six-rod 'magic colt' of Fig. 8.27 to that of the magic cylinder of Fig. 8.15b is obvious. Rotating rods of this type have also been used in periodic arrrays to produce travelling magnetic fields [Yon95a,b].

4 Three-dimensional structures

If a thin section of any cylindrical transverse flux source is rotated 180° about its polar axis its locus forms a closed chamber in which, of course, there are no end effects as in a finite cylinder and in which the fields are greater than in the

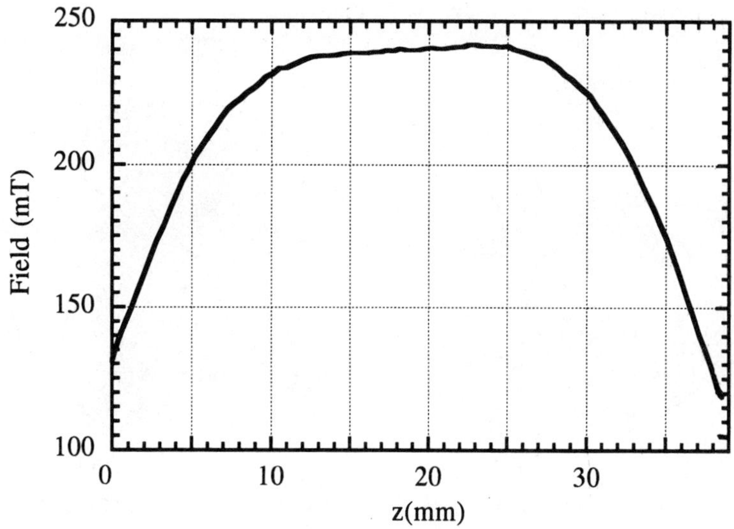

FIG. 8.29. Field profile on the axis of the six-rod magnet of Fig. 8.27 at maximum field.

corresponding cylinders. For example, the spherical source pictured in Fig. 8.30 produces a uniform field in its cavity that is four-thirds as great as that in a cylinder with the same inner and outer radius [Zij85, Leu87]. If B_r is 1.2 T and a flux density of 2.0 T is desired in a cavity 2.5 cm in diameter, the outer diameter of the sphere need be only 10 cm.

The fields produced in the spherical cavities are not only high but remarkably uniform along the axis, even when axial tunnels are bored through the poles. Figure 8.31 shows the computer-calculated field profiles for a sphere with $r_2/r_1 = 2$ and axial tunnels of varying diameter. It is clear that for tunnels up to one-fourth the internal cavity diameter, the internal field profile is negligibly affected in amplitude and only slightly in uniformity. This suggests great utility for short Faraday rotators and possibly in-tandem periodic structures for particle beam focus. The latter is attractive because the field in the tunnels can be made to be nearly the reverse of those in the cavities. Because the square wave field configurations have higher axial field amplitudes than are possible in conventional periodic magnet stacks and because the squareness results in an average field value close to maximum, better focussing should result than from the conventional sinusoidal configuration.

A drawback of the spherical structure is that it does not confine magnetic flux completely [Leu88], but generates a small dipolar field outside its outer surface. If this field is troublesome for a particular application, it can easily be cancelled by means of a uniformly magnetized shell on the outer surface of the spherical source. The shell is mounted with its magnetic orientation in opposition to the external field produced by the spherical source and is given just sufficient magnetic moment to cancel it. Since a uniformly magnetized spherical shell produces no field in its interior, it does not alter the specified field produced by the spherical source in the working space.

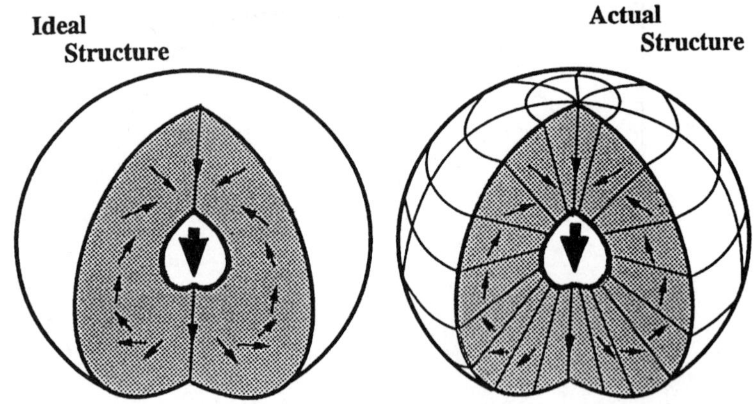

FIG. 8.30. Spherical magnet structure. The field in the central cavity is given by $B_0 = (4B_r/3)\ln(r_2/r_1)$, where B_r is the remanence and r_2 and r_1 are the outer and inner radii respectively.

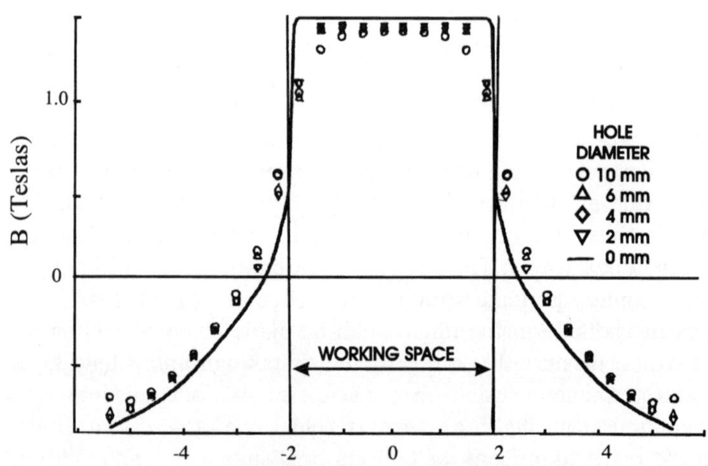

FIG. 8.31. Axial field profile of the magnet structure of Fig. 8.30. Note the constancy of the field in the cavity and also the relative insensitivity to the diameter of the access holes.

As in the permanent magnet solenoid, longitudinal field taper can be built into both magic spheres and magic cylinders [Leu92]. Because of the very high fields and

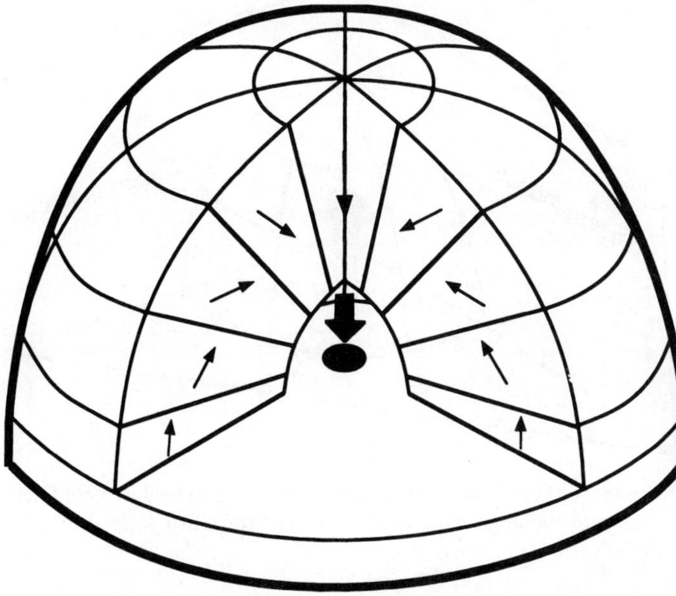

FIG. 8.32. Section of a hemisphere of a spherical field source like that of Fig. 8.30 rests on an iron slab which takes the place of the missing hemisphere.

field gradients produced by these structures, they seem promising for use in ore and slurry separators where magnetic species are to be extracted. The figure of merit of such devices is the product of field strength with its gradient. The former induces large moments m in paramagnetic particles and the later provides the force $-m\nabla B_0$ that acts on those moments to pull the particles through the medium.

The gradients may be provided parametrically, with the remanence of the cylinder or sphere varying directly with the polar angle θ. Such a remanence provides a corresponding linear field gradient in the direction of the polar axis in the sphere and the polar plane in the cylinder. Variable magnetic field gradients are very conveniently achieved using the rotatable rod ensembles [Cug94b].

Conventional permanent-magnet separators use a rotating multipole drum and moving belt configuration. Some designs exploit eddy currents to separate metallic and nonmetallic material [Ger94].

4.1 *Structural simplification with magnetic mirrors*

All these dipolar structures have antimirror symmetry in the equatorial plane perpendicular to the polar axis. For this reason, any of them can be cut at the equatorial plane and either half placed on a high-permeability, high-saturation slab of material such as iron or permalloy to produce the same field in the interior that is produced in the cavity of an entire structure [Leu88]. This works because the

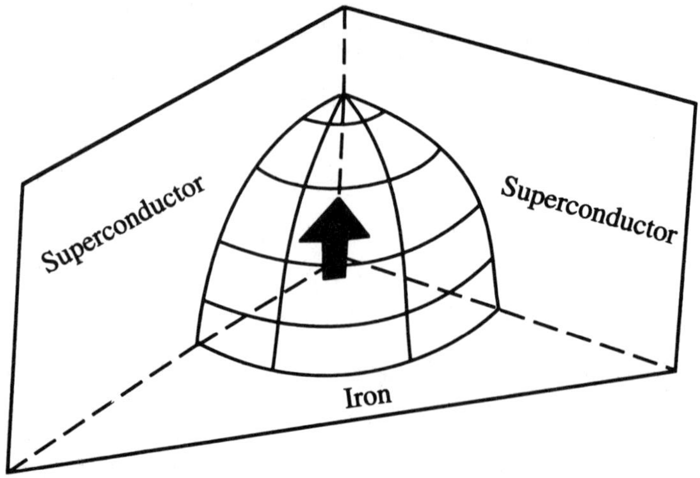

FIG. 8.33. An eighth-sphere structure that produces the same field as a complete sphere with magnetic mirrors (superconductors) and antimirrors (iron). For simplicity, only the exterior and direction of polarity are shown.

antimirror image of the half-structure formed in the slab produces the same field in the hemispherical cavity that would be produced by the missing half. Such a procedure affords significant structural simplification since only slightly more than half as many different magnet segments of complex form need be manufactured for the semistructure. The working volume is reduced by half, but if the full volume of the working space of the whole structure is needed, it may pay to use a larger half-structure to furnish it. Although more magnetic material would be needed, this drawback may be compensated by the increased simplicity afforded by the use of fewer pieces.

An igloo-like structure made of half a sphere and a slab of iron is pictured in Fig. 8.32. Such a source made of Nd–Fe–B would provide a working flux density of 2 T in a 2.5 cm cavity with an outer diameter of only 9 cm and a weight of 3 kg of permanent-magnet material of 1.4 T remanence. This is an easily portable source and is potentially a very convenient general-use laboratory magnet. It also has the advantage of an easily accessible interior as lead holes can be drilled through the expendable iron slab rather than through expensive magnet material, as in the sphere. This advantage affords greater flexibility with regard to modes of operation, as iron slabs with different numbers, positions and sizes of holes would be easily interchangeable. Another advantage of the igloo is that transit through the iron is effected without encounter of field reversals. This is important for some electron-beam and Faraday rotator applications.

If room-temperature superconductors of sufficiently high lower critical field H_{cl} were ever to become available, it would be practical to simplify such structures further by generating mirror rather than antimirror images in an axial plane [Leu88]. With various combinations of perfectly diamagnetic ($B = 0$) superconducting mirrors combined with perfectly paramagnetic ($B_0 = 0$) iron antimirrors, many configurations would be possible. The octant structure is a particularly impressive example as it reduces the practical minimum of 120 segments needed to approximate a spherical source to only fifteen, as shown in Fig. 8.33.

5 Periodic permanent-magnet stacks (PPMs)

5.1 *Travelling wave tubes*

Travelling wave tubes (TWTs) are sources of microwave/millimetre wave radiation that employ an electron beam which is focussed and guided by an axial magnetic field. Sometimes the field is furnished by an electric solenoid but solenoids tend to be bulky and rely on cumbersome power supplies. In military devices, especially airborne and ballistic ones, mass and bulk must be minimized. To obtain the necessary high field intensities in small-bore cylindrical structures, the fields are usually made periodic and are supplied by appropriate permanent magnet stacks. The simplest configuration is of the form shown in Fig. 8.34a, where axially oriented magnets supply magnetic flux to the interior by way of interspersed iron pole pieces. Unfortunately, most of the flux leaks to the exterior rather than passing through the bore, where it is wanted. The pole pieces can be indented or tapered at their external boundaries to increase the reluctance of the external flux paths to reduce the flux there, as in Fig. 8.34b and c respectively. This is what is usually done [Leu90]; however the rigid permanent magnets make alternatives to this arrangement possible. An obvious stratagem is to use magnets that are magnetized radially and are stacked to alternate between inward and outward orientations, as in Fig. 8.34d. Such a stack is compared with the conventional array with regard to field and size in Fig. 8.35. The comparison shows a significant advantage in favour of the radial magnet stack, especially in the technologically interesting region between 0.25 and 0.50 T, where a conventional stack must be three times as massive to attain equal field with the same bore and period. An even better structure may be achievable by the use of the principle employed in the transverse circular cylindrical magnets already discussed. There, large fields are obtained through gradual linear variation of the magnetization γ orientation with the coordinate angle θ. If the change in orientation of the magnetization with distance x along the axis of the radial magnet stack is made to be more gradual rather than in abrupt 180° jumps, the much improved field–bulk curve of Fig. 8.35 results. Here, the size advantage over the conventional stack at the fields of interest is of an order of magnitude and more, while at the higher fields it rises by two orders of magnitude. However, gradual variation of magnetization is not technologically feasible. Fortunately, as in the case of the rings, rather coarse orientation increments of 90° between adjacent segments, as in Fig. 8.34e, produce a field 94% of that produced by the continuous variation. It is this insensitivity that changes the concept of gradual variation of orientation from an interesting speculative concept to a powerful technological tool.

Although the hybrid TWT magnets have impressive mass and field advantages over the conventional designs, they are more difficult to build for two reasons.

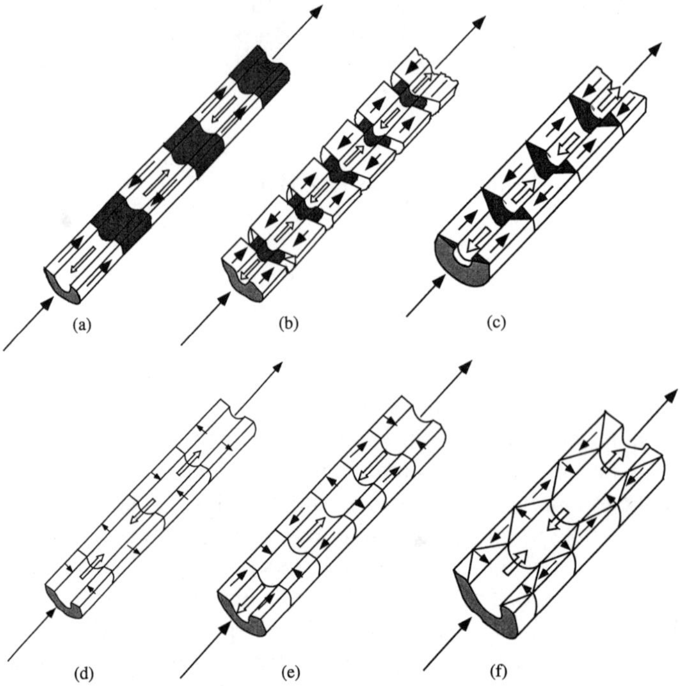

FIG. 8.34. (a) A simple PPM stack. (b) Stack with indented pole pieces. (c) Stack with tapered pole pieces. (d) Radially oriented magnets. (e) Hybrid PPM stacks. (f) Version of **e** with triangular magnet cross-sections. Small arrows show magnet orientations. Large open arrows show the working fields in the internal cavities. The long arrows at the ends show the paths of the electron beams. Darkly shaded areas are iron pole pieces.

First, it is hard to manufacture small-bore radially oriented ring magnets because of the difficulty of guiding sufficient flux through their bores to radially align and magnetize the high coercivity materials. Second, because there are no iron pole pieces through which small amounts of perturbing flux can be conveniently led to compensate for small field irregularities incurred in manufacture and assembly, all-magnet stacks are much harder to balance.

A simple and very effective alternative is to make pole pieces of triangular toroidal cross-section as in Fig. 8.34c [Leu87, Leu90]. This increases the magnetic permeance of the bore relative to that of the exterior so that more flux flows through the former. It also increases the average effective length of the axially oriented magnets, thereby increasing the magnetomotive force that drives the magnetic circuits. Both effects increase the field amplitude in the bore.

While it does not afford the spectacular mass and bulk advantages of the hybrid stack, the triangular poled array is still smaller by a factor of two than that of a conventional stack with indented poles at a bore field amplitude of 0.5 T. Moreover, it has none of the manufacturing and balancing difficulties associated with the hybrid stack. It is probably the best all-around practical alternative at present.

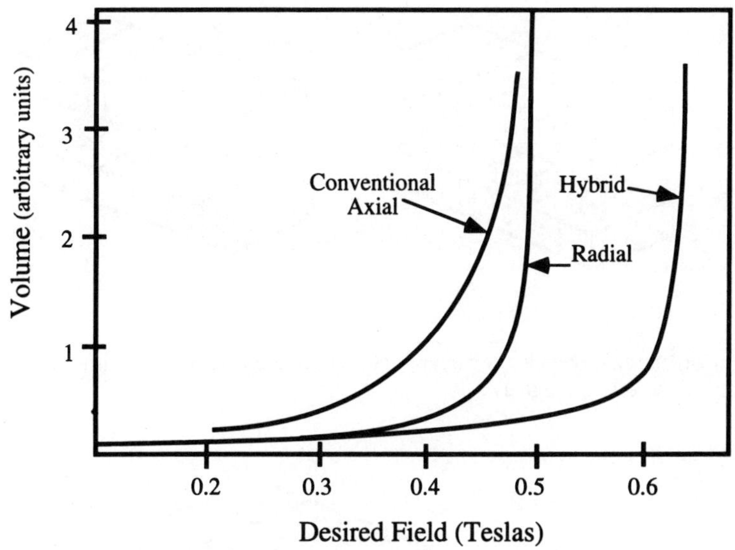

FIG. 8.35. Comparison of the volumes of the structures of Fig. 8.34.

5.2 *Wigglers and undulators*

Periodic transverse field sources are called wigglers and are used to provide lateral oscillatory motion to electron beams in free-electron lasers and synchrotron sources. It is this lateral acceleration that causes the electron to radiate. If the product of the flux density in teslas and the wiggler period in centimetres is of the order of one or less, the beam radiates coherently and laser action results. When this occurs, the wiggler is called an undulator [Hal85, Mar85].

The resonance condition is easily derived. The cyclotron frequency of an electron is given by

$$f_c = eB/2\pi m$$

and for the usual relativistic electron beam, the frequency of the field oscillation is

$$f_v \approx c/l$$

where l is the period of the wiggler. For resonance to occur these frequencies must be equal: hence $Bl = 2\pi mc/e = 0.0107$ Tm ~ 1 Tcm, an easy-to-remember condition.

As usual, when magnetic fields with a rapid spatial variation are required, permanent magnets offer the best solution. The simplest permanent magnet wiggler is just a series of bar magnets arranged as in Fig. 8.36a. More efficient would be planar arrays of magnets of alternately longitudinal and transverse orientation (see Fig. 8.36b). The principle used here is that of the hybrid TWT structure and, as in that structure, a high field-to-mass ratio results. But, like the TWT hybrid, it is

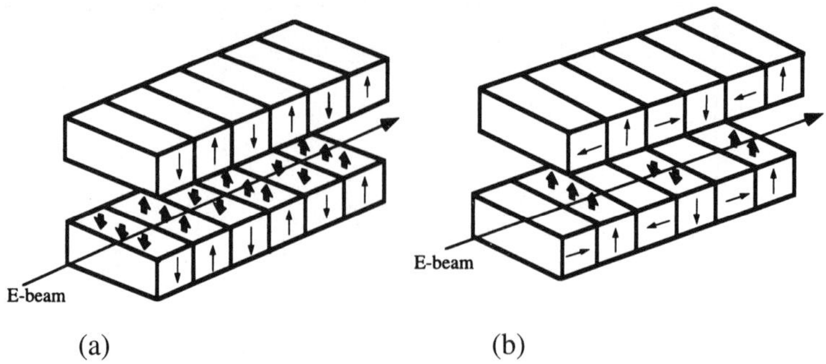

FIG. 8.36. (a) Simple wiggler structure. (b) High field structure with 90° change in orientation of successive magnets.

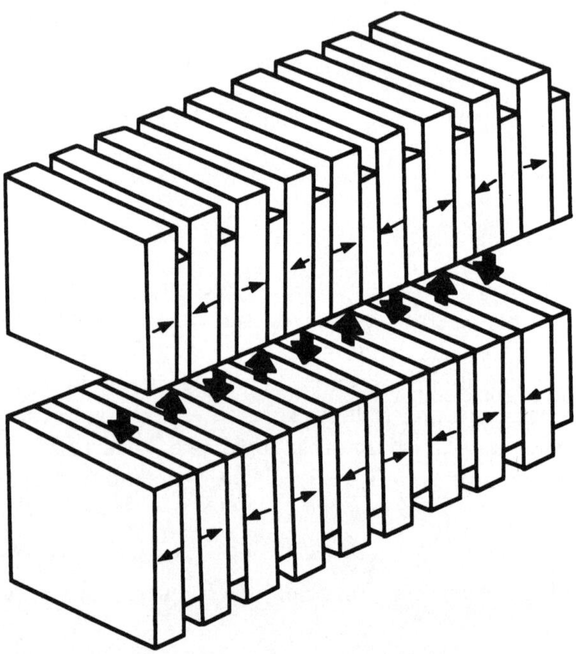

FIG. 8.37. Stack with alternately oriented longitudinal magnet interspersed with indented iron pole pieces.

somewhat difficult to adjust because of its lack of iron pole pieces to which one can attach compensating shims. For this reason, the less bulk-efficient but more tractable arrangement of Fig. 8.37 is often used [Hal85]. It can produce very high fields, is simple, relatively easy to adjust and probably the most practical choice

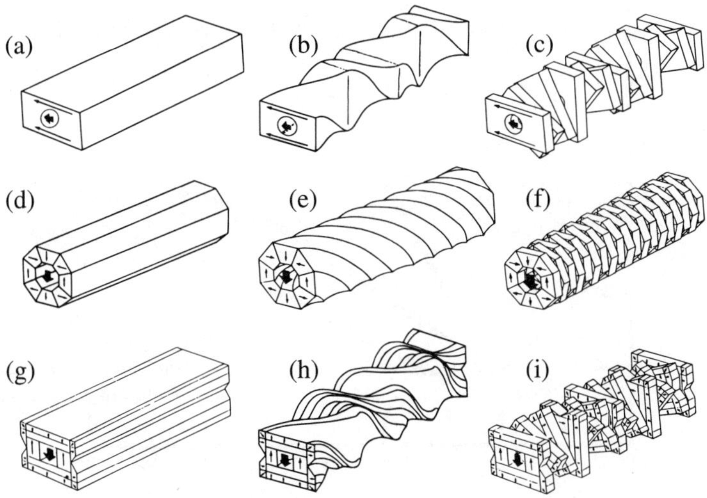

FIG. 8.38. Three twister structures: (a–c) twister bases; (d–f) ideal structures derived from the bases; (g–i) practical approximations for ideal structures.

where minimum mass and bulk are not considerations. As in the corresponding TWT, its indented or tapered pole pieces allow less flux leakage than flush ones.

Axial arrays of cross-sectional slices of any of the cylindrical transverse structures already discussed have the advantage of high-field capability and minimal flux leakage to the structural exterior, but they also suffer from relative complexity and cannot readily be adjusted by shims.

The radiation from wigglers and undulators is polarized in the plane of the oscillatory motion of the electron beam. Sometimes, it is desirable that the radiation be circularly polarized, in which case a different magnetic array is needed to impart the necessary helical motion to the electron beam. This can be accomplished by either a bifilar helical electric solenoid or a stack of wiggler elements arranged so that successive elements differ in orientation by a small angle θ. This results in a field that is constant in magnitude but whose orientation changes continuously with progression along the beam axis. Figure 8.38 shows such arrays together with those with untwisted cylinders from which they are derived and the stacks of finite segments with which they are approximated in practice. The structures shown in Fig. 8.38b and c will be recognized as the ones having already been discussed in cylindrical form. The structure in Fig. 8.38a is a simple, transversely magnetized bar magnet with a hole bored along its axes to accommodate the electron beam. The magnetic field in the hole is generated by the dipolar distribution on the inner surface that arises from the constant magnetization of the material. This field is slightly reduced by the presence of the more remote poles that the magnetization forms on the rectangular ends. The field is also progressively reduced with rate of twist and coarseness of twist increments. Fortunately, the effect of the latter is small for reasonable increments. For example, the 36° change between successive

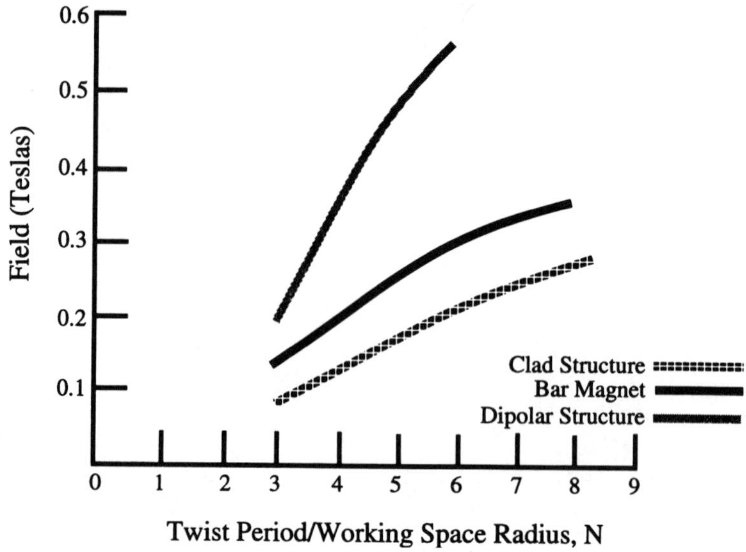

FIG. 8.39. Maximum field attainable in three twister structures as a function of twist period to working space ratio.

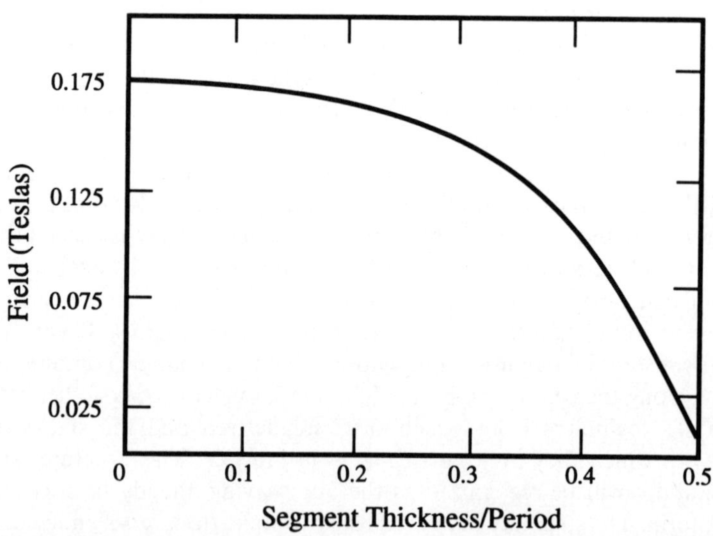

FIG. 8.40. Maximum field attainable as a function of segment thickness/period for the octagonal structure of Fig. 8.38b.

segments used in a prototype results in a field diminution of the order of only 1%. Figures 8.39 and 8.40 compare the properties of the three structures with regard to parameters of interest. Although not quite so desirable as the structure of

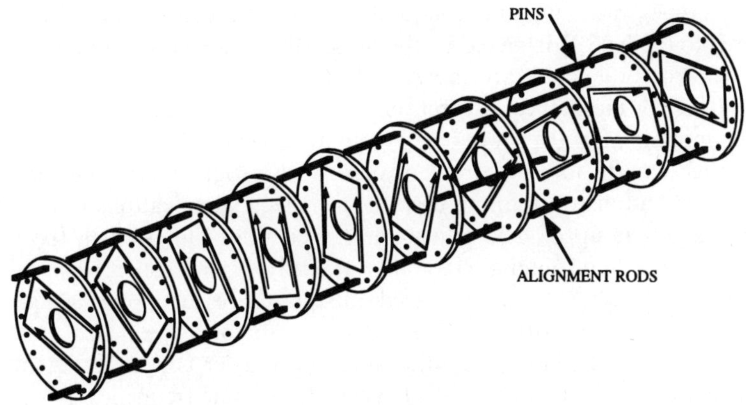

FIG. 8.41. Exploded view of a 'twister' prototype.

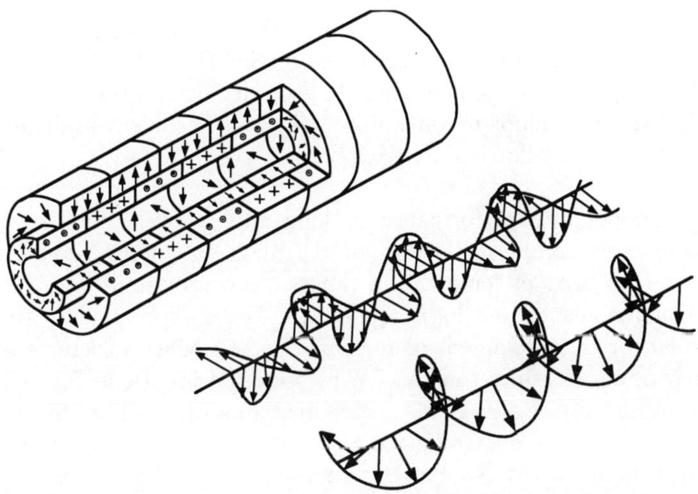

FIG. 8.42. Addition of fields of two nested twisters to form a 'wiggler' field.

Fig. 8.38b, the simple bar magnet of Fig. 8.38a was chosen for a prototype design for a high-powered radar source which is still far superior to the solenoid it will replace and less expensive than the other structures. The only advantage of the configuration in Fig. 8.38c is that the iron equipotential it forms tends to smooth aberrations in the material and structure to give a more uniform field. The twister design was chosen over the wiggler because circularly polarized radiation affords better discrimination on some military targets. Also, in a twister the magnetic field is at a maximum throughout its length while a wiggler produces an average field of approximately 0.7 maximum.

A prototype was built and found to deliver the theoretical field of 0.13 T compared to the 0.06 T delivered by the presently employed solenoid with a current of 200 A. Structural details are shown in Fig. 8.41. The magnet slices are mounted in brass discs for mechanical protection. Rods through holes near the periphery of the brass mounts offer the proper alignment and provide flexibility in the twist rate so that pitch can be adjusted or even varied with progression along the axis. The field strength and distribution agree with the design specifications to about 1%.

A wiggler made up of field adjustable nested rings discussed in the preceding section affords an intriguing possibility [Leu88]. If the inner stack of such a structure is rotated by 90° and displaced one-quarter period along the axis, its field will add vectorially with that of the outer ring to form the helical pattern of a twister (see Fig. 8.42). Thus, the adjustable ring wiggler is not only quantitatively adjustable, but can be operated as a twister of the same frequency as well.

Perspectives

While much progress has been made in rare-earth permanent-magnet flux source technology over the past two decades based on Sm–Co and Nd–Fe–B magnets, the demand for improvements is perennial. As projected devices become more complex, the technological demands become more stringent. There is considerable room for improvement in the following areas.

1. Better temperature performance of temperature-compensated magnets. This includes temperature coefficients of the magnets' energy products and their extension to broader temperature ranges. Also important is further improvement of remanence and loop squareness. These developments are absolutely essential for the attainment of higher fields in smaller structures operating at greater power densities such as TWTs projected for use in future lightweight, airborne devices. Raising the coercivities and squaring of the loops of the currently best Sm_2Co_{17} temperature-compensated magnets might be a solution. Compensation of the $Nd_2Fe_{14}B$ compounds might be an even better one as their energy products are generally much higher than those of Sm_2Co_{17}, but they suffer from low Curie temperatures.
2. Improved magnet alignment in novel magnet geometries by obtaining sufficient flux where it is needed during the fabrication and magnetization processes. Sometimes, this is very difficult in the case of conical shell magnets aligned normal to the cone surfaces, which are needed for the cladding of pole pieces in horseshoe-type magnetic circuits. Also difficult is the radial magnetization of toroids with small holes and broad annuli.
3. Improvement of fabrication and assembly techniques through the development and use of better cements and dies.
4. A better variety of square-looped material with low-energy products is needed for widespread implementation of parametric field shaping. This might be achieved by the development of very high coercivity ferrites, or the dilution of oriented REPM powder in a metal or epoxy matrix.

Investigation of deposition of REPM materials in the form of films on surfaces of different geometries should be pursued [Cad84]. Energy products must be raised, hysteresis loops squared and mechanical and thermal ruggedness assured. Especially promising is the possibility of depositon of films of in-surface and normal-to-surface orientations upon each other. Success in such a venture could lead to a whole generation of microdevices that are unattainable at present, including miniature field sources for microelectronics, micromotors and miniature wigglers.

In conclusion, the REPMs are very much more than just very powerful Alnicos. They afford not only the possibility of performing conventional tasks better, but offer a cornucopia of design options that were impossible before their advent. As mindsets acquired in the course of design experience with conventional magnets are discarded, the technological riches offered by rare-earth permanent magnets can be fully and speedily realized.

Bibliography

Structures of permanent magnets, M. G. Abele, John Wiley, New York (1993) — a systematic and mathematically rigorous study of the design of cylindrical magnetic shells that confine uniform transverse fields to their inner chambers. The cylinders are of arbitrary cross-sectional shape. Indispensable work for efficient REPM design.

Permanent magnets and their applications, R. J. Parker and R. J. Studders, John Wiley, New York (1962) — introductory work predating REPM technology. It discusses some elementary theory, manufacturing procedures, magnet properties, design techniques and applications, with some useful graphs, tables and charts summarizing material properties and magnetic circuits.

Advances in permanent magnetism, R. J. Parker, John Wiley, New York (1992) — partly an update of Parker and Studders above, but with the inclusion of REPMs, more circuit theory and an expanded section on magnetic instrumentation and measurement.

Electromagnetic devices, H. C. Roters, John Wiley, New York; Chapman and Hall, London (1940) — covers practical aspects of magnet design in the pre-computer, pre-REPM era. It includes an extensive treatment of the estimation of the permeance method of magnet design. Copious illustrative examples.

Philips J. Res., **40** (1985), 259–304, H. Zjilstra, — a useful summary of transverse field structures for cylindrical cavities which also includes some discussion of spherical structures. The consequences of finite length are discussed in detail.

REFERENCES

[Abe89] M. G. Abele, Technical Report, TR-15, New York University School of Medicine, 1 March (1989).

[Abe89] M. G. Abele, Technical Report, TR-20, New York University School of Medicine, 1 March (1989).

[Abe93] M. G. Abele *Structures of permanent magnets*, John Wiley, New York (1993).

[Abe89] M. G. Abele, R. Chandra, H. Rusinek, H. A. Leupold and E. Potenziani II, *IEEE Trans. Magn.*, **MAG-25** (1989), 3904.

[Abe88] M. G. Abele and H. A. Leupold, *J. Appl. Phys.*, **64** (1988), 5988.

[Abe90] M. G. Abele and H. Rusinek, *J. Appl. Phys.*, **67** (1990), 4644.

[Bel93] J. Bell, UK Patent.

[Cad84] F. J. Cadieu, T. D. Cheung, L. Wickramasekara and S. H. Aly, *J. Appl. Phys.*, **55** (1984), 2622.

[Cla86] J. P. Clarke and H. A. Leupold, *IEEE Trans. Magn.*, **MAG-22** (1986), 1063.

[Cla87] J. P. Clarke, E. Potenziani II and H. A. Leupold, *J. Appl. Phys.*, **61** (1987), 3468.

[Coe94] J. M. D. Coey and O. Cugat, *13th Int. Workshop on Rare Earth Permanent Magnets and their Applications*, Birmingham, UK, 11–14 September (1994), p. 41.

[Cug94a] O. Cugat, R. Byrne, J. McCaulay and J. M. D. Coey, *Rev. Sci. Instrum.*, **65** (1994), 3570.

[Cug94b] O. Cugat, P. Hansson and J. M. D. Coey, *IEEE Trans. Magn.*, **MAG-30** (1994), 4602.

[Ger94] R. Gerber, in *Applied magnetism* R. Gerber *et al.* (ed.) Kluwer, Dordrecht (1994), p. 165.

[Hal80] K. Halbach, *Nucl. Instr. and Meth.*, **169** (1980), 1.

[Hal81] K. Halbach, *Nucl. Instr. and Meth.*, **187** (1981), 109.

[Hal81a] K. Halbach, *5th Int. Workshop on Rare Earth–Cobalt Permanent Magnets and their Applications*, Roanoke, Virginia USA, 7–10 June (1981), p. 73.

[Hal85] K. Halbach, *8th Int. Workshop on Rare Earth–Cobalt Permanent Magnets and their Applications*, Dayton, Ohio USA, 5–9 May (1985), p. 103.

[Hol82] R. F. Holzinger, *6th Int. Workshop on Rare Earth Permanent Magnets and their Applications*, J. Fidler (ed.) Tech. U. of Vienna, (1982), p. 147.

[Leu81] H. A. Leupold, *5th Int. Workshop on Rare Earth–Cobalt Permanent Magnets and their Applications*, Roanoke, Virginia USA, 7–10 June (1981), p. 270.

[Leu86] H. A. Leupold and E. Potenziani II, *IEEE Trans. Mag.*, **MAG-22** (1986), 1078.

[Leu88] H. A. Leupold and E. Potenziani II, *J. Appl. Phys.*, **63** (1988), 3487.

[Leu88] H. A. Leupold, E. Potenziani II and M. G. Abele, *J. Appl. Phys.*, **64** (1988), 5994.

[Leu84] H. A. Leupold, E. Potenziani II and J. P. Clarke, *Shaping of Cylindrically Symmetric Magnetic Fields with Permanent Magnets*, US Army Technical Report, DELET-TR-84-12, December (1984), Fort Monmouth, New Jersey USA.

[Leu87] H. A. Leupold and J. P. Clarke, *IEEE Trans. on EDS*, Part II, Vacuum electron devices, **ED-34** (1987), 1868.

[Leu87] H. A. Leupold, E. Potenziani II, J. P. Clarke and D. Basarab, *Material Research Society Bulletin*, **96** (1987), 279.

[Leu90] H. A. Leupold, E. Potenziani II and A. Tauber, *J. Appl. Phys.*, **67** (1990), 4656.

[Lue92] H. A. Leupold, E. Potenziani II and A. S. Tilak, *IEEE Trans. Magn.*, **MAG-28** (1992), 3045.

[Leu73] H. A. Leupold, F. Rothwarf, C. G. Campagnoulo, H. Lee and J. E. Fine, *Magnetic Circuit Design Studies for an Inductive Sensor*, US Army Technical Report, TR-ECOM-4158, October (1973) Fort Monmouth, New Jersey USA.

[Mar85] T. C. Marshall, *Free electron lasers*, Macmillan Publishing Co., New York, NY (1985).

[Mor86] A. B. C. Morcos, H. A. Leupold and E. Potenziani II, *IEEE Trans. Magn.*, **MAG-22** (1986), 1066.

[Neu72] W. Neugebauer and E. M. Branch, *Applications of Cobalt–Samarium Magnets to Microwave Tubes*, Technical Report, Microwave Tube Operations, General Electric, 15 March (1972), Schenectady, New York USA.

[Par62] R. J. Parker and R. J. Studders, *Permanent magnets and their applications*, John Wiley, New York (1962).

[Rot40] H. C. Roters, *Electromagnetic devices*, John Wiley, New York and Chapman and Hall, London (1940).

[Rot78] F. Rothwarf, L. J. Jasper and H. A. Leupold, *3rd Int. Workshop on Rare Earth-Cobalt Permanent Magnets and their Applications*, University of California, San Diego, 27–30 June (1978), p. 255; Proceedings book by University of Dayton, School of Engineering, Dayton, Ohio, USA, 45469.

[Tau85] A. Tauber, H. A. Leupold and F. Rothwarf, *8th Int. Workshop on Rare Earth–Cobalt Permanent Magnets and their Applications*, Dayton, Ohio USA, 5–9 May (1985), p. 103.

[Yon95a] J.-P. Yonnet and A. Picard, *IEEE Trans. Magn.*, **26** (1990), 290.

[Yon95b] J.-P. Yonnet and A. Picard, *J. Mag. Magn. Mater.*, **83** (1990), 273.

[Zij85] H. Zijlstra, *Philips J. Research*, **40** (1985), 259.

9

Magnetomechanical devices

J.–P. Yonnet

The forces acting between permanent magnets are exploited in magnetic bearings, magnetic couplings and holding devices. For all these applications, the effects (force, torque) depend quadratically on the polarization of the magnet J. If a coupling is designed with $SmCo_5$ magnets ($J = 0.9$ T), then replacing the magnets by Nd–Fe–B ($J = 1.3$ T), the torque is more than doubled. Magnetomechanical devices need high-polarization magnets. Moreover, these devices very often use repulsion forces between the magnets. The polarization must remain constant, even in a large reverse field. The coercive field must be high. Rare-earth iron permanent magnets are frequently the most suitable.

Magnetomechanical devices utilize the forces exerted between magnets. The forces can be calculated easily by adding all the contributions of elementary magnets. In the first part of this chapter, we give expressions for the field and the forces. Next, we present two examples of magnetomechanical devices: magnetic bearings and synchronous couplings.

A. FORCES EXERTED BETWEEN PERMANENT MAGNETS

The interaction between two magnets can be calculated directly or approximated by an analytical expression. Modern rare-earth magnets exhibit a rigid and constant magnetization. The magnet can be represented by charged surfaces with magnetic 'poles' or by a distribution of surface currents. The following results have been obtained by calculating the forces exerted between magnetic charges whose density is:

$$\sigma = \mathbf{J} \cdot \hat{\mathbf{n}}$$

where $\hat{\mathbf{n}}$ is a unit vector normal to the magnet surface. A cuboidal magnet, for example, can be represented by two charged surfaces (Fig. 9.1).

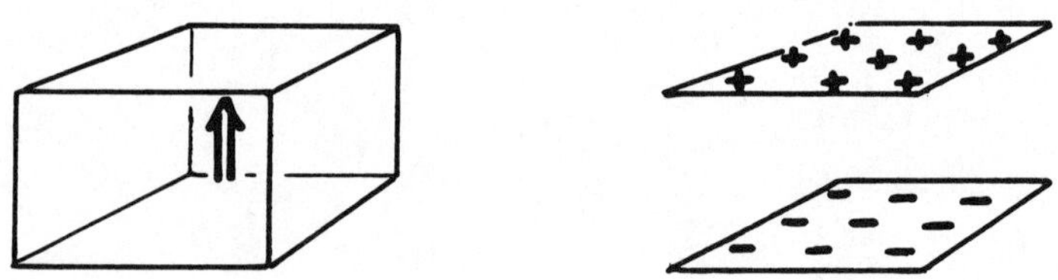

FIG. 9.1. Equivalence between the polarization $\mathbf{J}$ and the surface charge density σ.

In addition to this Coulombian model, the dipolar calculation will be presented which can give accurate results when the distance between the magnets is large in

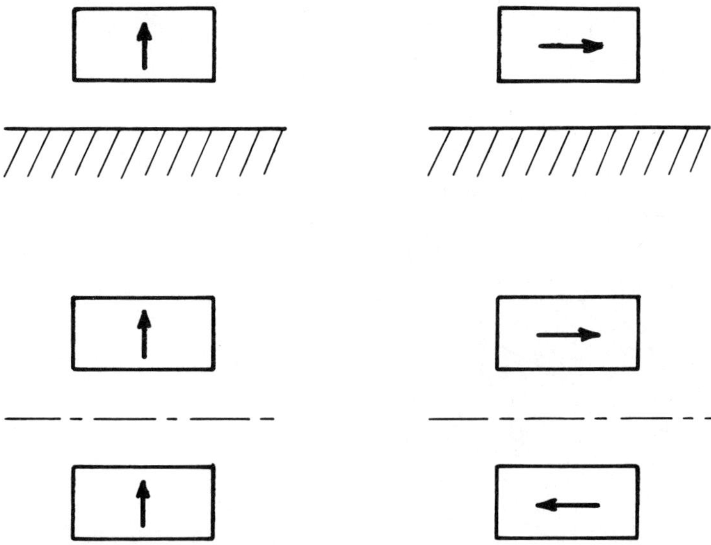

FIG. 9.2. Equivalence of two different magnet-iron systems (above) and the same magnets with their images (below).

comparison with the magnet dimensions. These models are useful provided the geometrical shape of the system is not too complicated. Otherwise, it is necessary to resort to numerical computation, by a finite-element method, for example. But, for most applications, the analytical models presented are sufficient and they allow the shape to be easily optimized. The force expressions are developed for a pair of interacting magnets. For entire systems, these elementary forces must be added following the superposition principle. When soft iron pieces are used, their surfaces are magnetic equipotentials. If the size of the soft magnetic piece is sufficient, the magnet–iron system can be replaced by the magnet and its image (Fig. 9.2). The attractive force is identical to that calculated for a pair of magnets.

1 Calculation by the Coulombian method

1.1 *Field created by a long bar-shaped magnet*

Let us consider an infinitely long magnet in the z-direction (Fig. 9.3), transversely magnetized in the y-direction, of cross-section $2a \times 2b$. The magnetic field $\mathbf{H}$ created by the magnet is

$$\begin{cases} H_x = \dfrac{+\sigma}{4\pi\mu_0} \ln \dfrac{[(a-x)^2+(b+y)^2][(a+x)^2+(b-y)^2]}{[(a+x)^2+(b+y)^2][(a-x)^2+(b-y)^2]} \\ H_y = \dfrac{-\sigma}{2\pi\mu_0}\left[\tan^{-1}\dfrac{x-a}{y-b} - \tan^{-1}\dfrac{x+a}{y-b} - \tan^{-1}\dfrac{x-a}{y+b} + \tan^{-1}\dfrac{x+a}{y+b}\right] \end{cases} \tag{9.1}$$

with $\sigma = |\,\mathbf{J}\,|$.

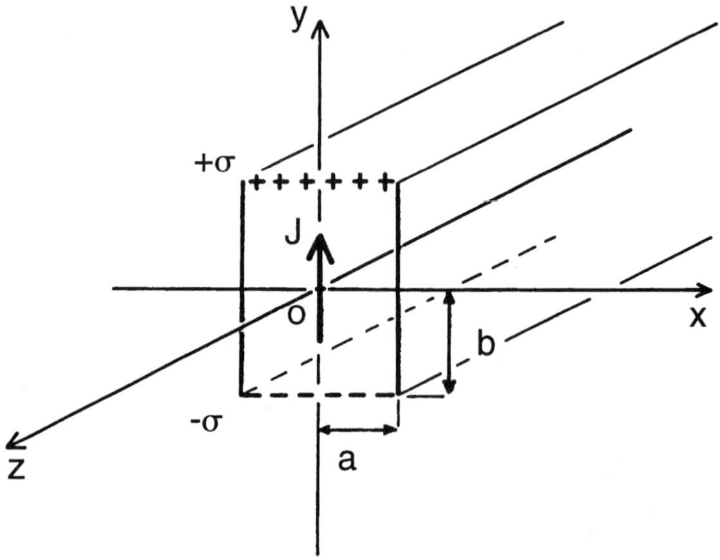

FIG. 9.3. Long transversely magnetized bar.

An equivalent form of Eq. (9.1) is

$$H_i = \frac{-\sigma}{4\pi\mu_0} \sum_{i=0,1} \sum_{j=0,1} (-1)^{i+j} \psi_i$$

where, for H_x:

$$\psi_x = \ln[(a + (-1)^i x)^2 + (b + (-1)^j y)^2]$$

and, for H_y:

$$\psi_y = 2\tan^{-1} \frac{a + (-1)^i x}{b + (-1)^j y}$$

These expressions allow us to calculate the magnetic field not only outside the magnet but also inside it, where the **H** field is called the demagnetizing field.

1.2 *Forces and stiffnesses between two magnets*

The force on a magnet can be calculated from the surface charge and the scalar potential. Here we consider the case of two long bar-shaped magnets magnetized in

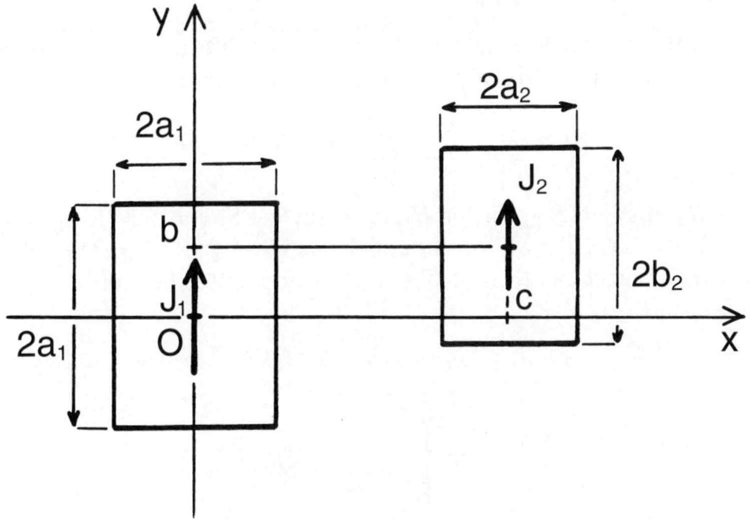

FIG. 9.4. Two long parallel magnets in interaction.

parallel directions, as shown in Fig. 9.4. The force expression contains 32 terms [Yon 81b]. The force per unit length is

$$\frac{F}{L}=\frac{\sigma_1\sigma_2}{4\pi\mu_0}\sum_{i=0,1}\sum_{j=0,1}\sum_{k=0,1}\sum_{l=0,1}(-1)^{i+j+k+l}\phi(u_{ij},v_{kl}) \tag{9.2}$$

where for F_x,

$$\phi_x=-u\ln(u^2+v^2)-2v\tan^{-1}\frac{u}{v}$$

and for F_y,

$$\phi_y=-v\ln(u^2+v^2)-2u\tan^{-1}\frac{u}{v}$$

with

$$u_{ij}=c+(-1)^i a_1+(-1)^j a_2 \quad \text{and} \quad v_{kl}=d+(-1)^k b_1+(-1)^1 b_2$$

With the same parameters, the stiffnesses are given by

$$\frac{K_x}{L}=-\frac{K_y}{L}=\frac{+\sigma_1\sigma_2}{4\pi\mu_0}\sum_{i=0,1}\sum_{j=0,1}\sum_{k=0,1}\sum_{l=0,1}(-1)^{i+j+k+l}\ln(u_{ij}^2+v_{kl}^2) \tag{9.3}$$

The stiffness is defined as the force variation for a small displacement. It is a very important parameter for magnetic bearings. The stiffness definition and the relation between the different stiffnesses will be developed in the section on magnetic bearings.

1.3 *Field created by a cuboidal magnet*

The analytical three-dimensional calculation is more difficult but remains possible. The magnetic field created by a charged rectangular surface (Fig. 9.5) is given by

$$H = \frac{}{4\pi\mu_0} \sum_{i=0,1} \sum_{j=0,1} (-1)^{i+j} \varepsilon(S_i, T_j, R) \tag{9.4}$$

where for H_x, $\varepsilon_x = \ln(R - T)$, for H_y, $\varepsilon_y = \ln(R - S)$ and for H_z, $\varepsilon_z = \tan^{-1}\frac{ST}{RZ}$, with $S_i = X - (-1)^i a$; $T_j = Y - (-1)^j b$ and $R = (S_i^2 + T_j^2 + Z^2)^{\frac{1}{2}}$. The field created by a cuboidal magnet can then be obtained by adding the field of its two charged surfaces $+\sigma$ and $-\sigma$.

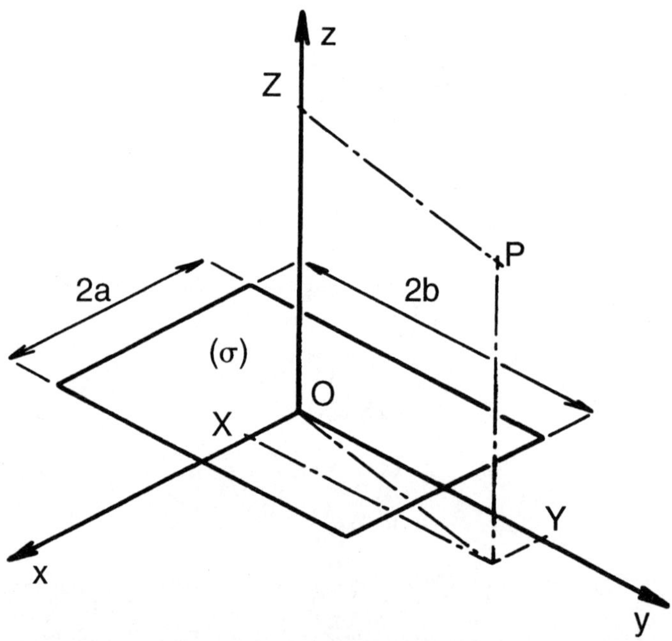

FIG. 9.5. Calculation of the magnetic field created by a rectangular surface $2a \times 2b$, bearing a uniform pole density σ.

1.4 *Forces exerted between two cuboidal magnets*

Consider the two cuboidal magnets with parallel faces having polarization J and J', centred at O and O' with dimensions as shown in Fig. 9.6. The vector OO' is (α, β, γ). The analytical expressions are relatively complicated. Akoun [Ako84] has shown that the forces between the magnets are given by

$$F = \frac{\sigma\sigma'}{4\pi\mu_0} \sum_{i=0,1} \sum_{j=0,1} \sum_{k=0,1} \sum_{l=0,1} \sum_{p=0,1} \sum_{q=0,1} (-1)^{i+j+k+l+p+q} \phi(u_{ij}, v_{kl}, w_{pq}, r) \tag{9.5}$$

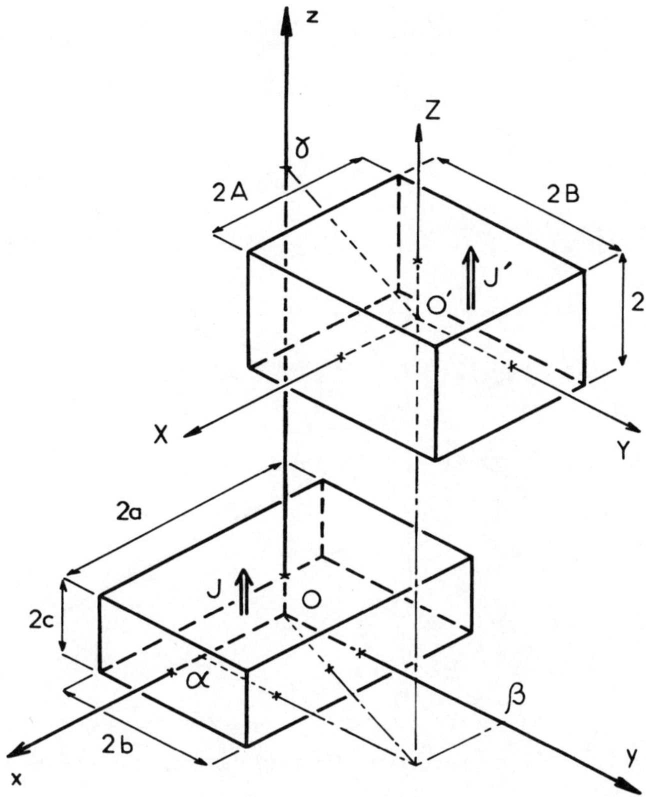

FIG. 9.6. A pair of cuboidal magnets.

where for F_x,

$$\phi_x = \frac{1}{2}(v^2 - w^2)\ln(r-u) + uv\ln(r-v) + vw\tan^{-1}\frac{uv}{rw} + \frac{1}{2}ru$$

for F_y,

$$\phi_y = \frac{1}{2}(u^2 - w^2)\ln(r-v) + uv\ln(r-u) + uw\tan^{-1}\frac{uv}{rw} + \frac{1}{2}rv$$

and for F_z,

$$\phi_z = -uw\ln(r-u) - vw\ln(r-v) + uv\tan^{-1}\frac{uv}{rw} - rw$$

with $u_{ij} = \alpha + (-1)^j A - (-1)^i a$; $v_{kl} = \beta + (-1)^1 B - (-1)^k b$; $w_{pq} = \gamma + (-1)^q C - (-1)^p c$; and $r = (u_{ij}^2 + v_{kl}^2 + w_{pq}^2)^{\frac{1}{2}}$. These expressions can easily be incorporated in a small program to calculate many devices, such as couplings or holding magnet systems.

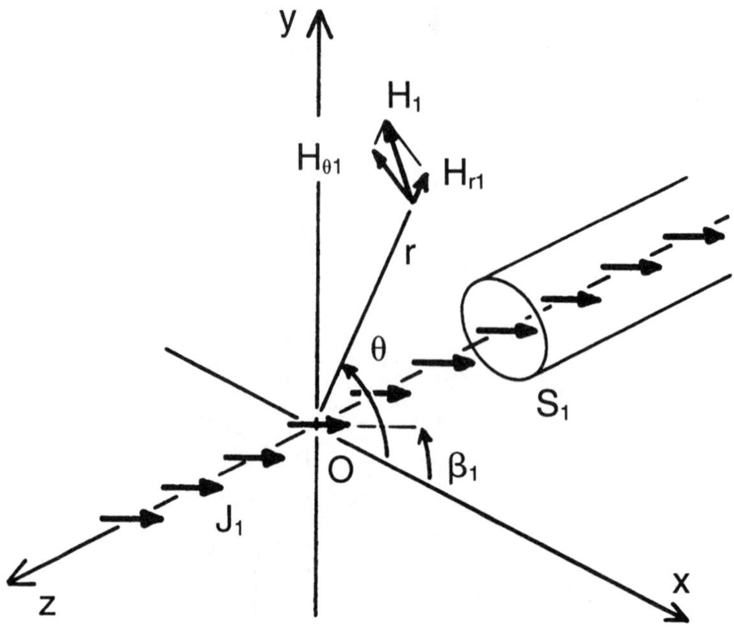

(a)

FIG. 9.7. A long line dipole (a) and the corresponding flux pattern in a perpendicular plane ▶ (b).

2 Calculation with the dipolar model

At distances greater than the dimensions of the magnet, the magnetic moment can be considered as concentrated at the centre [Yon81a]. Then the calculation of the field, forces and stiffnesses is very easy. It is based on the expressions for the field due to a point or line dipole.

2.1 *Field created by a long bar-shaped magnet*

The field $\mathbf{H}_1$ created by an infinitely long dipole was discussed in Chapter 8 (p. 401). For the dipole of Fig. 9.7, the field lies in the plane xOy, with components H_r and H_θ:

$$\begin{cases} H_{r1} = \dfrac{1}{2\pi}\dfrac{\lambda_1}{r^2}\cos(\theta - \beta_1) \\ H_{\theta 1} = \dfrac{1}{2\pi}\dfrac{\lambda_1}{r^2}\sin(\theta - \beta_1) \end{cases} \tag{9.6}$$

where $\lambda_1 = J_1 S_1/\mu_0$ is the magnetic moment per unit length.

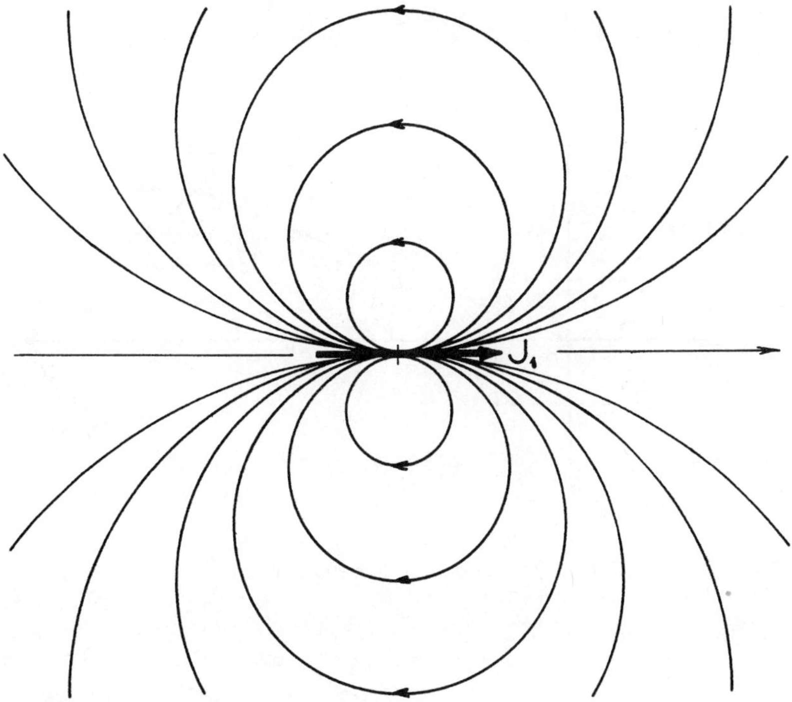

(b)

The modulus is $|\mathbf{H}_1| = (1/2\pi)\,\lambda_1/r^2$ and its phase is 2θ. The flux lines are circles, tangent to the $\mathbf{J}_1$ vector (Fig. 9.7b).

Equation (9.6) is equivalent to Eq. (9.1) when $r > a, b$, and it can be used not only for a rectangular section but also for any section S_1.

2.2 *Forces and stiffnesses between two magnets*

We now consider the interaction between two long magnetic dipoles of length L in the coordinate system shown in Fig. 9.8. The forces F_x and F_y are given by

$$\begin{aligned} \frac{F_x}{L} &= \frac{-\mu_0}{\pi}\,\frac{\lambda_1\lambda_2}{r_{12}^3}\cos(\beta_1+\beta_2-3\theta) \\ \frac{F_y}{L} &= \frac{\mu_0}{\pi}\,\frac{\lambda_1\lambda_2}{r_{12}^3}\sin(\beta_1+\beta_2-3\theta) \end{aligned} \tag{9.7}$$

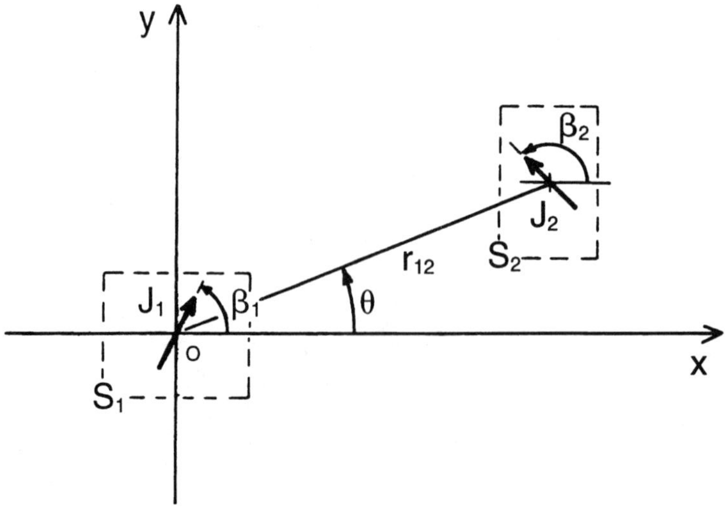

FIG. 9.8. Two long magnets in interaction.

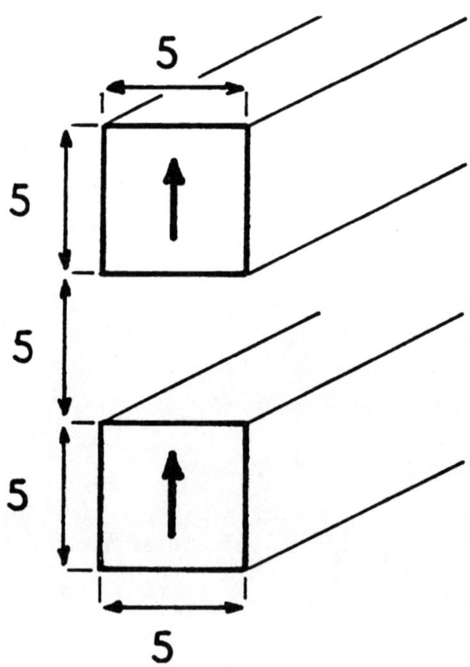

FIG. 9.9. Example of two square magnets in attraction with $\lambda = 20$ A m used in the comparison of different methods of calculation. Dimensions are in millimetres.

The stiffness expressions are

$$\frac{K_x}{L} = \frac{-K_y}{L} = \frac{-3J_1J_2}{\pi\mu_0}\frac{S_1S_2}{r^4}\cos(\beta_1+\beta_2-4\theta)$$
$$= \frac{-3\mu_0}{\pi}\frac{\lambda_1\lambda_2}{r^4}\cos(\beta_1+\beta_2-4\theta) \tag{9.8}$$

Finally, we compare the results obtained from the complete analytical expressions with those based on the dipolar model for a simple example. If we consider the attractive forces exerted between the two parallel square rods shown in Fig. 9.9, we obtain the following results using the analytical expressions: $(F_y/L) = 158\ \text{N m}^{-1}$ from Eq. (9.7) and $(K_y/L) = 47.5 \times 10^3\ \text{N m}^{-2}$ from Eq. (9.8), whereas by using the complete analytical expressions (Eqs (9.2) and (9.3) F_y is found to be 3% smaller and K_y is 7% smaller. In this example, an airgap equal to the dimensions of the magnet sections marks the limit between the two methods. For a lower airgap, it is necesary to use the complete analytical expressions. For a larger airgap the dipolar model is quite sufficient.

2.3 *Three-dimensional dipolar field*

The field created by a point dipole of moment M_1 has axial symmetry about the dipole axis (Fig. 9.10). It is given by

$$\begin{cases} H_{r1} = \dfrac{1}{4\pi}\dfrac{2M_1}{r^3}\cos\alpha \\ H_{\alpha 1} = \dfrac{1}{4\pi}\dfrac{M_1}{r^3}\sin\alpha \end{cases} \tag{9.9}$$

with $M_1 = J_1 v_1/\mu_0$.

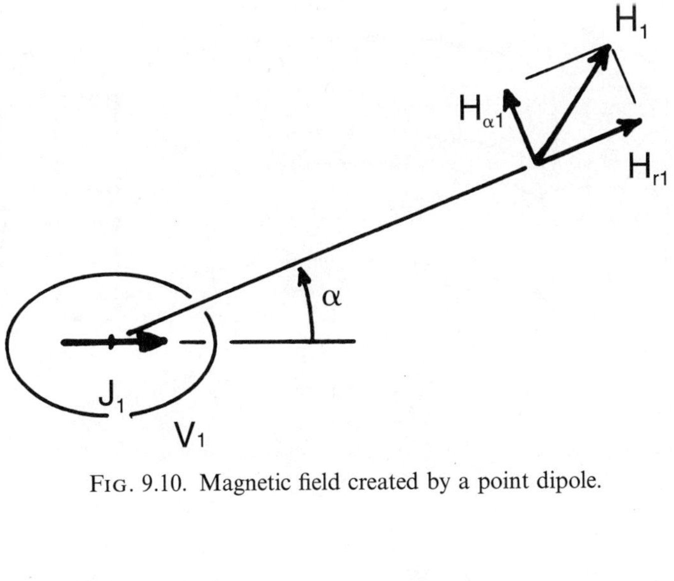

FIG. 9.10. Magnetic field created by a point dipole.

Let us consider another dipole $\mathbf{M}_2$ situated in this field $\mathbf{H}_1$. The interaction energy is: $W_{12} = -\mu_0 \mathbf{M}_2 \mathbf{H}_1$ with $\mathbf{M}_2 = \mathbf{J}_2 v_2/\mu_0$ and the forces are the derivatives of this energy $\mathbf{F} = -\nabla(W_{12})$.

B. PERMANENT-MAGNET BEARINGS

1 Bearing operation principle

The simplest radial bearing is made with two concentric ring-shaped magnets in repulsion (Fig. 9.11a). In the centred position, this system does not exert any force. Whenever one magnet is radially displaced, the radial force tries to restore the

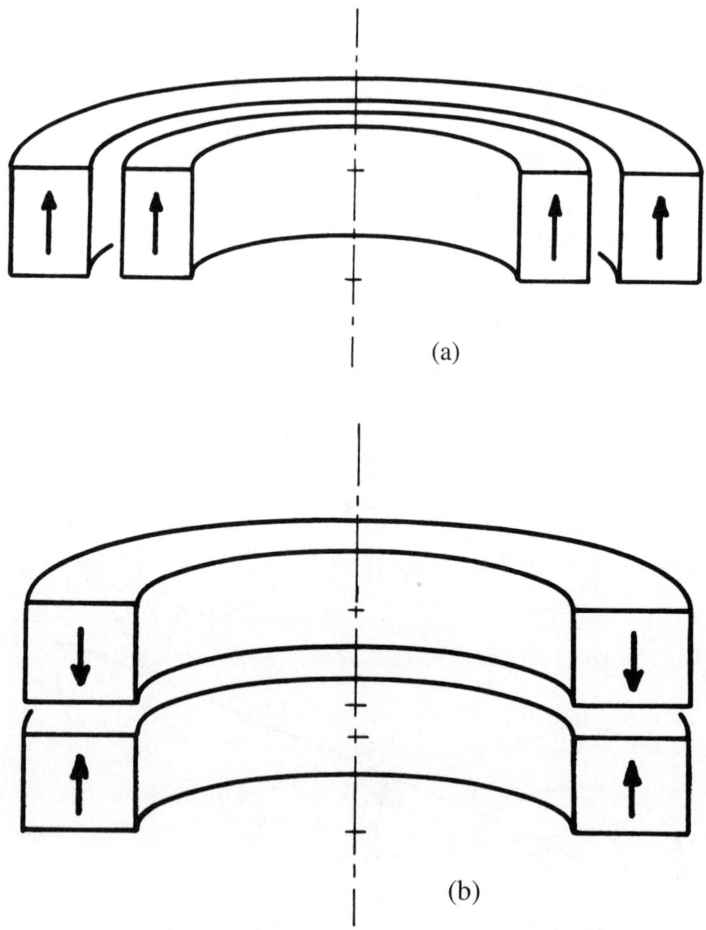

FIG. 9.11. Radial magnetic bearing with axially magnetized magnets (a); and axial magnetic bearing with axially magnetized magnets (b).

centred position. The system is radially stable, but the axial and angular instabilities are quite severe: the two rings do not want to stay in the same plane. With this example, it is easy to understand how a permanent-magnet bearing operates. A radial bearing is a fully passive system used for centring an axis. There also exist axial bearings which are used to control the position along an axis, as shown in Fig. 9.11b.

These bearings are very simple and cheap. They need no power. They are extremely reliable, with a long lifetime. They are used either in partially magnetic suspensions, in combination with a mechanical bearing, for example, or in fully magnetic suspensions, but in the last case the stability must be achieved by an active bearing. Complete stability is impossible using passive magnetic bearings alone.

2 The stability of permanent-magnet suspensions

2.1 *Force, stiffness and equilibrium*

The forces exerted between the two parts of a magnetic bearing can be split into two components F_z (axial force) and F_r (radial force) (Fig. 9.12). In the centred position, the radial force is nil. For a small displacement dr, the restoring force is dF_r.

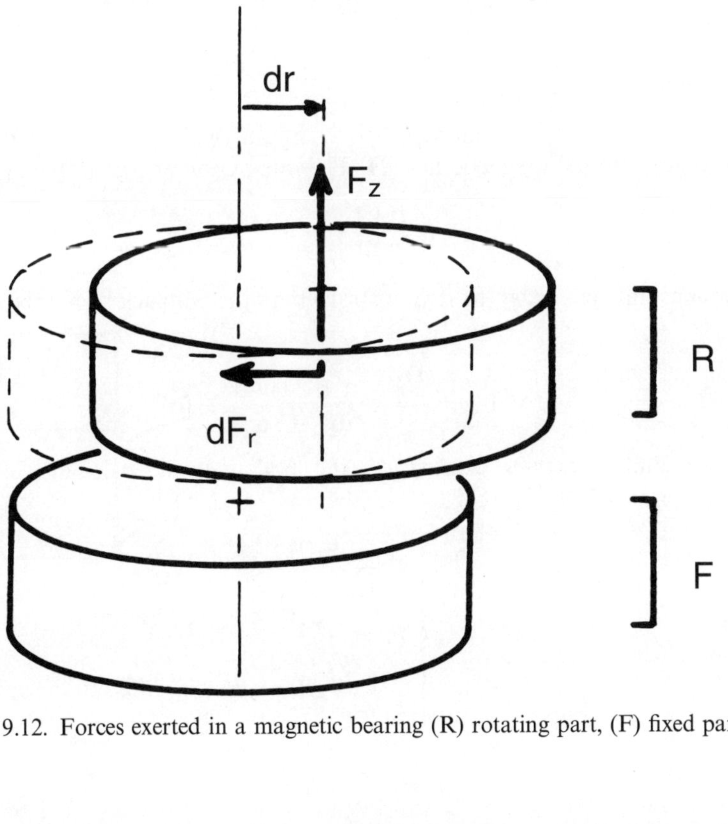

FIG. 9.12. Forces exerted in a magnetic bearing (R) rotating part, (F) fixed part.

Around the centred position, we can define the stiffness K_r by

$$K_r = -\frac{dF_r}{dr}$$

The axial stiffness is similarly defined by

$$K_z = -\frac{dF_z}{dz}$$

If $K_r > 0$, F_r is a restoring force. The system is stable in the r-direction. To obtain stable equilibrium of a body, it must be submitted to no net forces:

$$F_x = F_y = F_z = 0 \qquad \text{in Cartesian coordinates}$$

or

$$F_r = F_z = 0 \qquad \text{in cylindrical coordinates}$$

The equilibrium stability can be reached only if all the displacements create restoring force, that is,

$$K_x > 0, K_y > 0, K_z > 0 \qquad \text{for a Cartesian system} \tag{9.10.a}$$

or

$$K_r > 0, \quad K_z > 0 \qquad \text{for a cylindrical system} \tag{9.10.b}$$

2.2 *Earnshaw's theorem*

Let us consider a body of volume V, possessing a rigid magnetic polarization **J**, situated in an external magnetic field **H**. The interaction energy in the system is

$$W = -\iiint_v \mathbf{J} \cdot \mathbf{H} \, dv$$

As **J** is constant, it is easy to demonstrate that the Laplacian of this energy is zero:

$$\nabla^2 W = \frac{d^2 W}{dx^2} + \frac{d^2 W}{dy^2} + \frac{d^2 W}{dz^2} = 0$$

The forces and stiffnesses exerted on the body can be calculated from the interaction energy $\mathbf{F} = -\nabla W$:

$$\begin{cases} F_x = -\dfrac{dW}{dx} \\ F_y = -\dfrac{dW}{dy} \\ F_z = -\dfrac{dW}{dz} \end{cases}$$

$$K_x = \frac{-\mathrm{d}F_x}{\mathrm{d}x} \begin{cases} K_x = \dfrac{\mathrm{d}^2 W}{\mathrm{d}x^2} \\ K_y = \dfrac{\mathrm{d}^2 W}{\mathrm{d}y^2} \\ K_z = \dfrac{\mathrm{d}^2 W}{\mathrm{d}z^2} \end{cases}$$

By combining the above equations, we obtain:

$$K_x + K_y + K_z = 0 \tag{9.11.a}$$

For a cylindrical system: $K_x = K_y = K_r$, the sum of the stiffnesses becomes

$$K_z + 2K_r = 0 \tag{9.11.b}$$

These results demonstrate Earnshaw's theorem. Earnshaw [Ear1839] originally proved the instability of an electric charge in an electrostatic field. The result is extended to the instability of a magnetic moment in a magnetostatic field since Eq. (9.11) is incompatible with the equilibrium stability conditions, Eq. (9.10). To achieve stability for a magnetic bearing, the three translational stiffnesses (K_x, K_r and K_y) must be positive. The sum of the stiffnesses cannot be zero. Consequently, if the system is stable in two perpendicular directions, it will be unstable in the third one.

Consequences of Earnshaw's theorem for rotating magnetic bearing systems are:

1. If K_r is positive, K_z is negative. A radial bearing is unstable along the z-axis.
2. If the K_z is positive, K_r is negative. An axial bearing is always radially unstable.

Earnshaw's theorem refers only to translational stiffnesses. The three angular stiffnesses are independant. For example, special couplings can transmit a torque when there is no translational force or stiffness between the two halves. As far as the angular stability is concerned, the basic configurations are generally unstable for radial bearings, and stable for axial bearings. For whole suspensions, built with a combination of several bearings, Earnshaw's theorem remains valid. Unfortunately, it is not possible to compensate the axial instability of a passive bearing by another passive bearing.

3 Bearing configurations

3.1 *Rotating bearings*

Two elementary configurations of radial bearings are shown in Fig. 9.13 (1A and 1B). The first one is axially magnetized, the second one is radially magnetized. In most cases, the curvature can be neglected. Consequently all the forces and stiffnesses exerted by the two systems are identical. Yonnet [Yon78] has demonstrated that they have the same interaction energy. More generally, if the magnetization direction of the first magnet is rotated through an angle α, and the second one through $-\alpha$, the interaction energy is conserved. Since axially

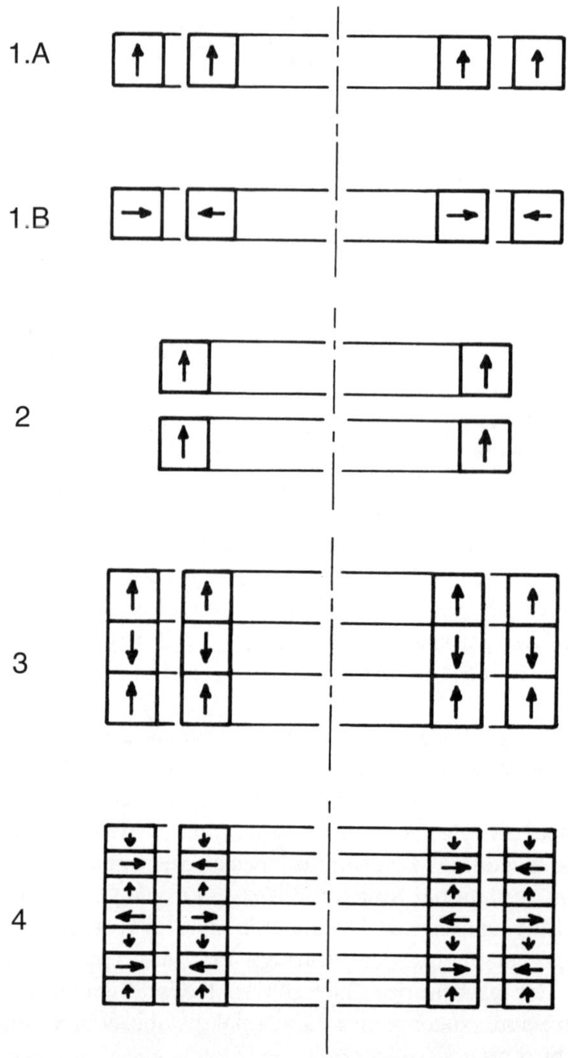

FIG. 9.13. Some configurations of permanent-magnet radial bearings: 1A and 1B are the elementary designs which create no axial force in the centred position, A = axial magnetization, B = radial magnetization. Type 2 is a radial bearing working by attraction. Type 3 and 4 are stacked structures used to increase the stiffness.

magnetized rings are easier to produce than radial ones, type A must be preferred over type B.

Many other configurations of radial bearings can be used. Some of the most common are shown in Fig. 9.13. Type 2 exerts a large axial force which must be cancelled either by a symmetrical arrangement or by an active system. By using stacked structures, it is possible to increase the stiffness. With type 3, using n elementary bearings, the stiffness is multiplied approximately by $2n$–1. Type 4 is

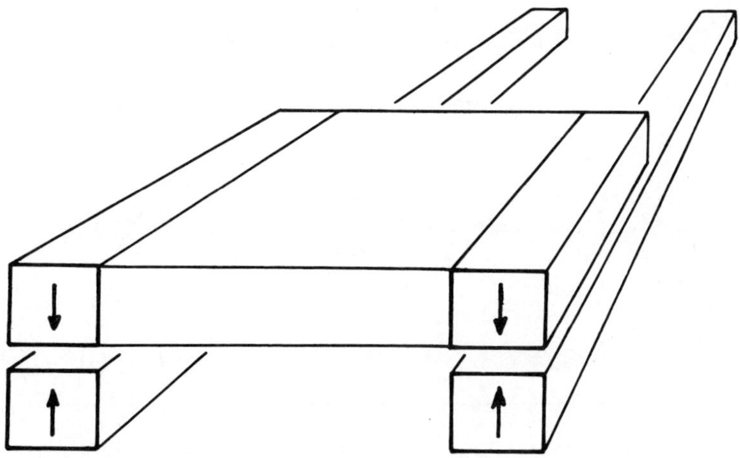

FIG. 9.14. Elementary configuration of a permanent-magnet linear bearing.

made with axially and radially magnetized rings. Compared with type 3 using the same magnet volume, the stiffness is increased by about 1.8.

For axial bearings, the different configurations can be obtained from the radial ones by reversing the magnetization direction in one-half of the system.

3.2 *Sliding bearings*

Although rotating bearings are the most often used, linear bearings also exist. They ensure translational guidance. For example, a train can be supported by magnetic levitation (Maglev). Its movement along the track is controlled by a linear motor. Its stable position is governed by the magnetic suspension, which controls five degrees of freedom.

Configurations having magnets on the vehicle and on the track have been proposed. The magnet arrangements can be deduced directly from those for rotating bearings. For example, the elementary system shown in Fig. 9.14 provides a high vertical force and axial stability, but the transverse stability must be ensured by an active system. This configuration is the linear version of the rotating bearing shown in Fig. 9.11b.

The use of sliding permanent-magnet bearings is limited to special applications. For a train, the cost of magnets all along the track is prohibitive. Other solutions are preferable, [Jay81] for example, an iron track attracted by magnets fixed on the vehicle or eddy-current repulsion.

4 Bearing calculations

4.1 *Effect of curvature*

For most radial permanent-magnet bearings, the gap between the two magnets, and all the cross-section dimensions are small in comparison with the average radius. If

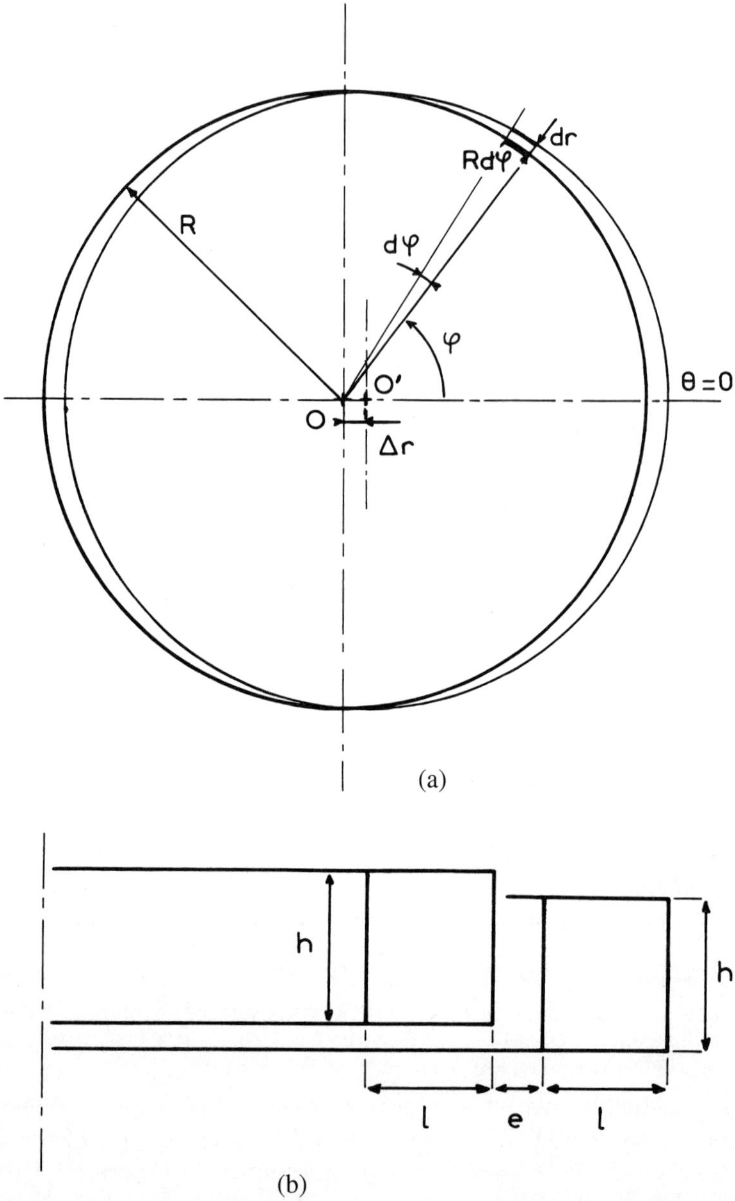

FIG. 9.15. (a) A system with cylindrical symmetry. Displacement Δr along the direction $\varphi = 0$ produces a relative displacement $\mathrm{d}r = \Delta r \cos\varphi$ along the circumference. (b) An elementary magnetic bearing realized with two rings of identical cross-section.

this condition is realized, the magnetic field around the magnets is only slightly influenced by the curvature. Then the magnetic field and its variations are the same as for a linear system of infinite length and it is possible to neglect the curvature without impairing the accuracy of the calculation.

By integrating the effects over the φ angle (Fig. 9.15a) it follows that

$$F_z = 2\pi R\left(\frac{F_z}{L}\right)$$

$$F_r = \pi R\left(\frac{F_x}{L}\right)$$

$$K_z = -2K_r = 2\pi R\left(\frac{K_z}{L}\right)$$

where (F/L) are the forces per unit length calculated for linear bar-shaped magnets.

4.2 *Stiffness calculation*

Very often, the sections of the ring magnets are identical (Fig. 9.15b). In this case, the expression for the radial stiffness can be simplified to

$$K_r = \frac{-J^2}{4\pi\mu_0} p. \ln \frac{(2l+e)^2 e^2 \left[(l+e)^2 + h^2\right]^2}{(l+e)^4 \left[(2l+e)^2 + h^2\right](e^2 + h^2)}$$

where p is the average circumference of the bearing. For example, an elementary bearing with square section ring ($l = h = 5$ mm, $e = 1$ mm, $J = 1$ T, $p = 160$ mm) has a radial stiffness of 24 N/mm. It can rotate at 100 000 rpm with a belted rotating magnet.

5 Uses of permanent-magnet bearings

Permanent-magnet bearings are the key components of magnetic suspensions. A fully magnetic suspension can be built with two radial passive bearings and an active axial system. The active element typically involves an electromagnetic servosystem with a positive sensor. In any event, as a consequence of Earnshaw's theorem, at least one axis must be active (or controlled by a superconducting system). The suspension cannot be built entirely out of permanent-magnet bearings.

As explained at the beginning, the magnets for these bearings should have as high a polarization as possible because the forces and stiffnesses depend quadratically on this parameter. Moreover, the magnets may be submitted to a high demagnetizing field. Many configurations work in repulsion. The magnets are subject to their own demagnetizing field as well as to the external field created by the opposite magnet. The magnet must possess a high coercive field. Rare-earth magnets are very well adapted for these applications. In addition, the magnets must be homogeneous to avoid the creation of eddy currents in the opposite magnets and the surrounding metal parts.

Typically one of the two magnets is rotating at high speed. Belting techniques can be used, but the magnets need to have good mechanical properties.

Permanent-magnet bearings are mainly used in high-rotation-speed systems such as momentum wheels for satellites [Pou78] and other flywheel energy storage systems, turbomolecular pumps [Kab88] and ultracentrifuges. They are also components of systems where very low friction is necessary, such as watt-hour metres.

C. PERMANENT-MAGNET COUPLINGS

1 Coupling configurations

Permanent-magnet couplings are invaluable when it is necessary to transmit motion into a closed container, for example under vacuum or in an agressive environment.

Two types of synchronous coupling are mainly used (Fig. 9.16):

1. the coaxial type, operating with a cylindrical airgap,
2. the face-type coupling with a planar airgap.

In a face-type coupling, the equally sized rings are alternately magnetized up and down. A large attractive force is created in the axial direction. In a coaxial-type coupling, the two halves of the coupling are arranged concentrically and the rings are alternately magnetized inwards and outwards. The radial instability is troublesome, but in the centred position the axial and radial forces are nil.

When is it necessary to modify the maximum transmitted torque, face-type couplings are used with an adjustable airgap. Nevertheless, the axial force must be

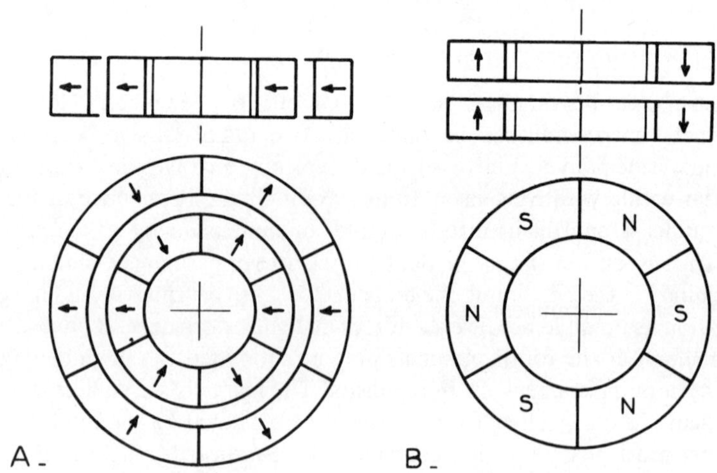

FIG. 9.16. The two classical types of permanent-magnet coupling: A coaxial coupling; B face-type coupling.

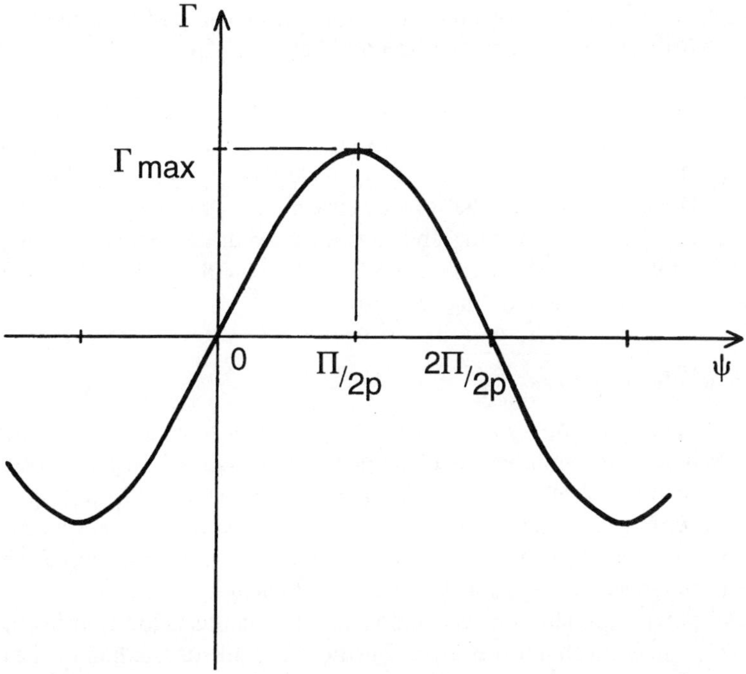

FIG. 9.17. Torque Γ versus the angular position ψ_{2p} is the pole number.

absorbed by the bearings, which limits the coupling size. For transmission of large torques, coaxial couplings are generally used. Other types of permanent magnet coupling exist, but their use is limited to special applications: these include linear couplings for the control of linear motions, metre couplings which are used for the transmission of angular information when the torque is very small and permanent magnet gears, which consist of two noncoaxially mounted rings, each with different number of poles.

The main characteristic of the coupling is its curve $\Gamma(\alpha)$, whose shape often resembles a sine wave (Fig. 9.17). For example, a 12-pole coaxial coupling (Fig. 9.16a) made with $SmCo_5$ magnets (cross-section 10 × 10 mm, airgap 3mm, average radius 40 mm) is able to transmit a maximum torque of 8.5 Nm.

2 Calculation of permanent-magnet couplings

The calculation of the forces exerted in a permanent-magnet coupling is relatively difficult because it is a real three-dimensional problem. Two calculation methods can be used

1. numerical simulation, using a finite-element method, for example, which allows the complex geometry and the nonlinearities of the materials to be taken into account,

2. analytical calculation, assuming that the materials have linear characteristics and that the real geometry can be simplified.

Each type of calculation has its advantages. By means of numerical simulation, it is possible to optimize a given coupling. On the other hand, analytical calculation is better adapted to general study and to provide the approximate dimensions of a coupling. The expressions of the forces between cuboidal magnets can be used to calculate the transmitted torque and give very accurate results. As an example, Yonnet [Yon93] and Krist [Kri93] have published analytical calculations for a coaxial- and a face-type coupling respectively.

3 Magnets for couplings

When the maximum torque is exceeded, the driver part slips from the driven part and the magnets are submitted to a high demagnetizing field. In some positions, the magnets are directly in repulsion. The magnets must be able to support these conditions without demagnetization, so the coercive force must be very high. Since the torque is directly proportional to the square of the polarization J, the use of rare-earth magnets allows realization of very efficient systems.

Several other considerations are important. The magnets are rotating, and must be subject to high mechanical stresses. Furthermore, the driven half of the coupling is very often situated inside a container, and is submitted to particular conditions of temperature, pressure and corrosion from which it may be necessary to protect it by coating or encapsulation.

In conclusion, rare-earth magnets are very well adapted to magnetomechanical applications because of their high polarization and coercive force. For the applications discussed, magnetic bearings and magnetic couplings, these magnets are the best choice because they can tolerate a demagnetizing field comparable to the magnetization, and the desired effects (the stiffness or the torque) depend quadratically on the magnets' polarization. There is a market for the better-quality sintered rare-earth iron magnets, and also for cheaper bonded isotropic magnets with good magnetic characteristics. The magnets must also possess very good mechanical and physical properties. For high-speed magnetic bearings in particular, the magnetization homogeneity must be excellent.

Magnetomechanical applications generally require simple magnet shapes. The performances of other devices can easily be calculated from the expressions of the forces between elementary magnets in two or three dimensions. The analytical expressions developed here for the bearings, and cited for the couplings, can be used in many other applications.

REFERENCES

[Ako84] G. Akoun and J-P. Yonnet, *IEEE Trans. on Mag.*, **MAG 20**, (1984), 1962–1964.
[Ear39] S. Earnshaw, *Trans. Cambridge Philosophical Society*, **7** (1839), 97–112.

[Jay81] R. V. Jayawant, *Electromagnetic levitation and suspension techniques* Edward Arnold, London 1981.

[Kab88] H. P. Kabelitz and K. J. Fremerey, *Vacuum*, **38** (1988), 673–676.

[Kri93] P. Krist, R. Gerber, J. Boehm and S. Withley, *IEEE Trans. Magn.*, **MAG 29** (1993), 2935–2937.

[Pou78] P. C. Poubeau, *2nd Int. Workshop on RE–Co PM and their Applications*, Dayton, Ohio (1978), pp. 157–178.

[Yon78] J-P. Yonnet, *IEEE Trans. Magn.*, **MAG 14** (1978), 803–805.

[Yon81a] J-P. Yonnet, *IEEE Trans. Magn.*, **MAG 17** (1981), 1169–1173.

[Yon81b] J-P. Yonnet, *5th Int Workshop on Rare-Earth Cobalt Permanent Magnets and their Applications*, Roanoke, Virginia (1981).

[Yon93] J-P. Yonnet, S. Hemmerlin, E. Rullière and G. Lemarquand, *IEEE Trans. Magn.*, **MAG 29** (1993), 2932–2934.

10

Permanent-magnet motors

R. Hanitsch

New electromechanical drive systems endowed with a combination of desirable characteristics are a considerable impetus to technological progress. Taking a robot axis drive as an example, it should combine a low moment of inertia and low torque pulsations with high power/weight ratio, high torque overload capability and high efficiency. The advent of rare-earth permanent magnets has opened up the prospect of significantly improved designs for such devices, as well as for a range of other industrial drives, small electrical machines and micromotors.

The essential feature of all electromagnetic devices based on permanent magnets is the flux density **B** generated by the permanent-magnet array. The Lorentz force $q\mathbf{v} \times \mathbf{B}$ on moving charges (currents) in the solid conductors that form a part of electric motors, loudspeakers or measuring devices is proportional to B. Likewise, the induced emf in generators, eddy-current brakes or moving-coil microphones, which is given by Faraday's law, is also proportional to B.

Devices equipped with Nd–Fe–B benefit from the high remanence and near-ideal loop shape of the magnets. Consequent advantages include a higher torque for a given frame size, improved efficiency, better dynamic response and reduced weight and volume. However, there are also some difficulties: handling and assembly of magnetized parts of industrial drives demand special procedures; the temperature sensitivity of the coercivity requires that particular attention be paid to the thermal aspects of any new design and special computer aided design techniques have had to be developed. Detailed design studies have been made on both brushed and brushless DC motors, where Nd–Fe–B magnets have been considered for radial and axial field designs employing either slotted or slotless configurations [Hon80, Han86, How88]. These machines exemplify the trend towards high-speed, high-flux-density drives which is basically motivated by the quest for the most efficient use of energy.

In this chapter, the physical properties of interest to the design engineer are first reviewed, and then the designs of some representative permanent-magnet machines are discussed.

A. MATERIAL PROPERTIES

Permanent magnets based on Nd–Fe–B come in either the sintered or bonded forms from commercial suppliers. Anisotropic sintered Nd–Fe–B is produced by power metallurgy. Different grades, with different microstructure, some including additions such as dysprosium or cobalt, are available from many companies. These magnets offer excellent magnetic properties [Kum88, Kun85], including $(BH)_{max}$ values that are more than 30% greater than those of Sm–Co. materials (Fig. 10.1). Furthermore, sintered $Nd_2Fe_{14}B$ alloys are significantly stronger than $SmCo_5$

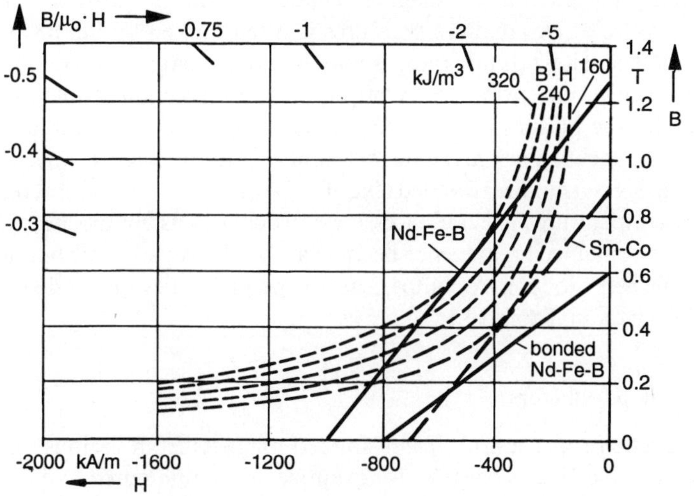

FIG. 10.1. Comparison of *B*(*H*) characteristics of sintered and polymer bonded Nd–Fe–B

Table 10.1 Properties of selected commercial sintered and bonded Nd–Fe–B magnets

Quantity	Unit	Cast/sintered magnets Nd–Fe–B	Bonded magnets Nd–Fe–B
$(BH)_{max}$	kJ/m^3	80–350	35–65
B_r	T	0.7–1.4	0.45–0.65
H_c	kA/m	450–1600	310–420
$(1/B_r)(\partial B_r/\partial T)$	%/°C	–(0.07–0.16)	–0.2
$(1/H_c)(\partial H_c/\partial T)$	%/°C	–(0.37–0.9)	–0.4
μ_{rec}	–	1.04–1.3	1.15
Magnetizing field	kA/m	1600–2800	1600–2800
Resistivity	μΩm	1.4–1.6	180
Density	kg/m^3	7200–7500	5000–6000
Bending strength	MPa	240–270	50

alloys, allowing the design engineer more flexibility in his or her use of the magnet material in new designs. There are limitations on the shapes of magnets produced by sintering, and they often require machining on silicon carbide or diamond grinding wheels to produce the required surface finish and dimensions.

Isotropic bonded magnets are also quite widely available. They incorporate flakes of melt-spun Nd–Fe–B in a matrix of resin, plastic, rubber or metal, but their energy product is only about 20% of that of the sintered magnets on account of the reduced packing fraction of magnetic material and the absence of crystalline orientation. Use at elevated temperatures is often limited by the properties of the polymer matrix. These disadvantages may be offset against the possibility of near-net-shape fabrication using the compression or injection-moulding techniques developed by the plastics industry for loaded polymers. For example, a magnetic rotor for a synchronous machine or stepper motor can be injection-moulded directly onto its spindle. Bonded Nd–Fe–B magnets are of interest wherever operation temperatures are not excessive, particularly as composites with oriented crystallites or enhanced remanence become available. Anisotropic bonded Sm_2Co_{17} is an established product. A comparison of the properties of bonded and sintered magnets is given in Table 10.1.

1 Magnetic parameters

Some magnetic properties of typical sintered magnets from various manufacturers are listed in Table 10.2. Koerdym, for example, is manufactured in different grades, each with a reversible permeability of about 1.05. The Nd–Fe–B magnets are obviously superior to Sm–Co, at least at room temperature. As temperature increases, the magnetic values change and losses occur. One distinguishes reversible and irreversible losses, the latter being irrecoverable on cooling back to room temperature. Reversible losses are often quoted as temperature coefficients

Table 10.2 Magnetic properties at room temperature for selected grades of sintered Nd–Fe–B magnets

Grade	$(BH)_{max}$ (kJ/m^3)	B_r (T)	$_BH_c$ (kA/m)	H_c (kA/m)	$(1/B_r)(\partial B_r/\partial T)$ (%/K)
Neomax-35	286	1.2	836	915	–0.12
RES 270	216	1.1	750	835	–0.14
NERONIT 270	270	1.2	700	800	–0.10
Crumax 261	206	1.04	790	1580	–0.09
Koerdym 260	240–290	1.1–1.25	700–900	750–1100	–0.10
Vacodym 335HR	260–335	1.2–1.3	700–1000	1000	–0.11

Table 10.3 Typical temperature coefficients, operating temperatures and Curie temperatures

Magnet material	$(1/B_r)(\partial B_r/\partial T)$ (%/K)	$(1/H_c)(\partial H_c/\partial T)$ (%/K)	Max. operation temp. (°C)	Curie temp. (°C)
$Nd_2Fe_{14}B$	–0.13	–0.55	150	310
$SmCo_5$	–0.045	–0.25	250	720
Alnico	–0.02	+0.03	500	800

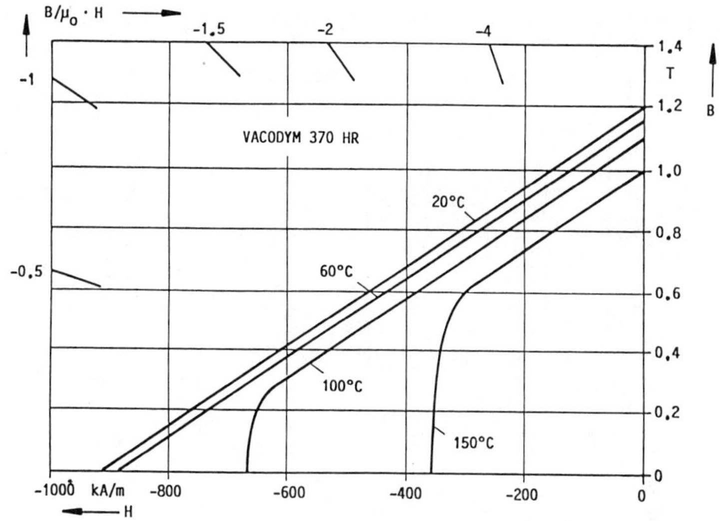

FIG. 10.2. Influence of temperature on the $B(H)$ characteristic of Nd–Fe–B.

$(1/X)(\partial X/\partial T)$, where X stands for the remanence or the coercivity. Table 10.3 gives typical values for various magnets. In this comparison, Nd–Fe–B is the least attractive. It may be better to use Sm–Co, or even Alnico, when elevated operating temperatures are unavoidable. Figure 10.2 shows demagnetization curves at several temperatures for one grade of Nd–Fe–B. The rapid decline above 100 °C is obvious. Some improvement in the high-temperature performance is achieved in grades which incorporate a little dysprosium, to increase coercivity (2 wt% of dysprosium will almost double H_c to 1600 kA/m), and cobalt, to increase the Curie point. However, there is a price to pay in terms of lower remanence and energy product, and greater difficulty in magnetizing the material. After an initial heat treatment to stabilize the magnet, the maximum operating temperature of grades with no additives having a remanence of 1.25 T is about 100 °C, whereas other grades, containing dysprosium, have a remanence of only 1 T, but may be used up to 180 °C. At operating points above the knee in the 120 °C demagnetization curve, the changes in magnetic values are essentially reversible on cooling. Irreversible losses occur under increased shearing [För90]. To cope with this problem, a material with large coercivity should be chosen, and it is necessary to make the magnet thicker and the permeance factor larger.

The irreversible losses at elevated temperature are related to local spontaneous demagnetization, which depends on the coercivity and operating point. Some typical data are shown in Fig. 10.3. Additional losses may occur because of oxidation or corrosion. The characteristics indicate outstanding magnetic properties at room temperature. At elevated temperatures, the high-temperature coefficients of coercivity and remanence must be taken into consideration, and the operating point must not fall below the knee if the magnet is to remain

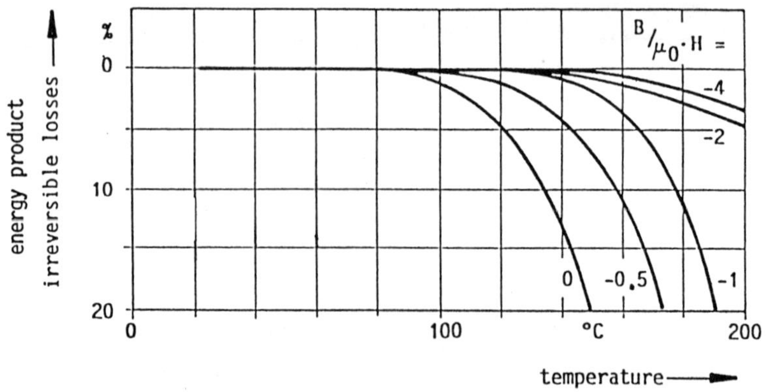

FIG. 10.3. Irreversible flux losses for different load lines versus temperature.

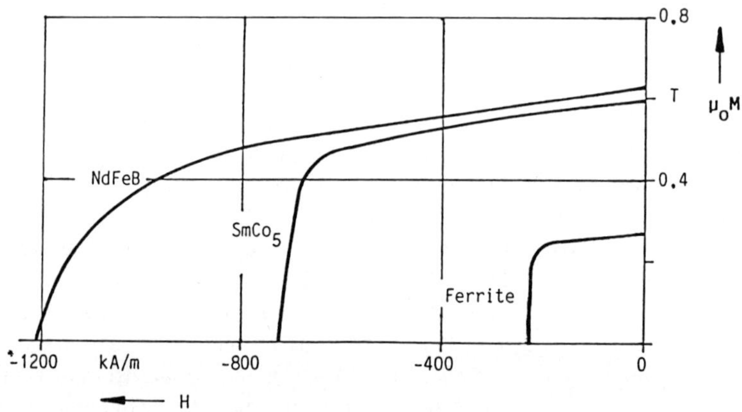

FIG. 10.4. Comparison of polymer-bonded magnets.

undamaged. These features of Nd–Fe–B magnets motivate the engineer to design circuits that ensure proper cooling of the magnets. The hurdle of poor thermal stability can be overcome through innovative design.

Compared with fully consolidated Nd–Fe–B, the density of bonded magnets is much lower and the magnetic properties are also reduced. Nevertheless, the magnetic parameters are adequate for many applications in devices operating at ambient temperatures not exceeding 50 °C. The energy product is around 50 kJ/m^3. Machine designers have several sets of conditions to meet. The available magnetic field is certainly important, but the total system must perform cost-effectively. There are some quite significant advantages of bonded magnets, including almost unlimited shape versatility with good dimensional tolerances, low weight and the possibility of incorporating shafts, contacts, gears, etc., during moulding. Flexibility of design is one of the main attractions of bonded magnets.

A comparison of isotropic Nd–Fe–B with other bonded magnets is shown in Fig. 10.4. Oriented bonded ferrite and Sm–Co–magnets have energy products which are about 50% of their sintered counterparts. Bonded Nd–Fe–B, due to its isotropic nature, has a value only 20% of the sintered equivalent. The typical applications for bonded magnets are:

1. replacement for economic reasons of a sintered product by a redesigned system with bonded magnets;
2. electromagnetic systems where bonded magnets are the only viable choice, for example because of intricate shape.

The example of a magnetic rotor for a synchronous machine or stepper motor which can be injection-moulded directly onto the spindle has already been mentioned. Another application is the use of a polymer bonded one-off moulding instead of a multipole magnet assembly. Bonded Nd–Fe–B certainly has a market niche and interest is growing [How88, Hop87, Shi87].

2 Non-magnetic parameters

Not only magnetic properties, but mechanical and thermal properties of permanent magnets are also relevant to machine design. The comparison of Nd–Fe–B and Sm–Co sintered magnets in Table 10.4 shows that the two materials are rather similar with respect to resistivity, thermal conductivity, specific heat and hardness. Their thermal conductivity is much smaller than that of mild steel (50 W/(m K)), stainless steel (15 W/(m K)) or Alnico (25 W/(m K)). Rare-earth sintered magnets are generally quite hard and brittle, although Nd–Fe–B is somewhat tougher than Sm–Co, and less susceptible to breakage and chipping. Sm–Co is about 15% denser than Nd–Fe–B, but the most significant differences are found in the thermal expansion and mechanical strength.

Table 10.4 Nonmagnetic properties of sintered Sm–Co and Nd–Fe–B-magnets

	Unit	Sm–Co	Nd–Fe–B
μ_{rec}		~1.05	~1.05
Density	kg/m^3	8200–8400	7200–7500
Specific resistivity	$10^{-6}\Omega$m	0.86	1.40–1.60
Vickers hardness		500–550	550–600
Specific heat	J/kg K	370	405
Thermal conductivity	W/m K	10	7
Bending strength	MPa	150	250
Compressive strength	MPa	820	1100
Tensile strength	MPa	36	75
Thermal expansion coefficient			
$C_{\parallel}(0\ldots100\ ^\circ C)$	$10^{-6}/^\circ$C	8	5.2
$C_{\perp}(0\ldots100\ ^\circ C)$	$10^{-6}/^\circ$C	11	–0.8

Nd–Fe–B is approximately twice as strong as Sm–Co, although the strength of any generic permanent magnet will vary significantly from one grade to another. The transverse rupture strength, for example, varies from 220 to 440 MPa and it is correlated with the density of the sintered alloy.

Nd–Fe–B has a negative coefficient of thermal expansion normal to the direction of magnetization, whereas other magnetic and constructional materials have positive coefficients. Values, in units of $10^{-6}/°C$, are 12 for steel, 17 for copper and 24 for aluminium. Ferrites and Sm–Co are in the range 10–15, but the mean value for Nd–Fe–B is 1. The anomalous thermal expansion (invar effect) is intrinsically related to the 3d magnetism of iron. It causes difficulties when assembling machines by glueing magnets to other materials. This mismatch can be alleviated, if necessary, by introducing some cobalt into the alloy.

3 Corrosion

On account of the neodymium-rich material at the grain boundaries and the high iron content of Nd–Fe–B magnets, precautions against corrosive attack may be necessary. The corrosion rate of Nd–Fe–B magnets is considerably higher than that of iron, and metallurgical investigations reveal that the phenomenon is closely related to the presence of multiple phases [Kum88]. Whereas Nd–Fe–B is fairly stable in dry conditions, a humid environment will lead to appreciable corrosion and consequent degradation of the magnetic properties. After 3000 h in distilled water at room temperature, the magnetization decreased by more than 40%.

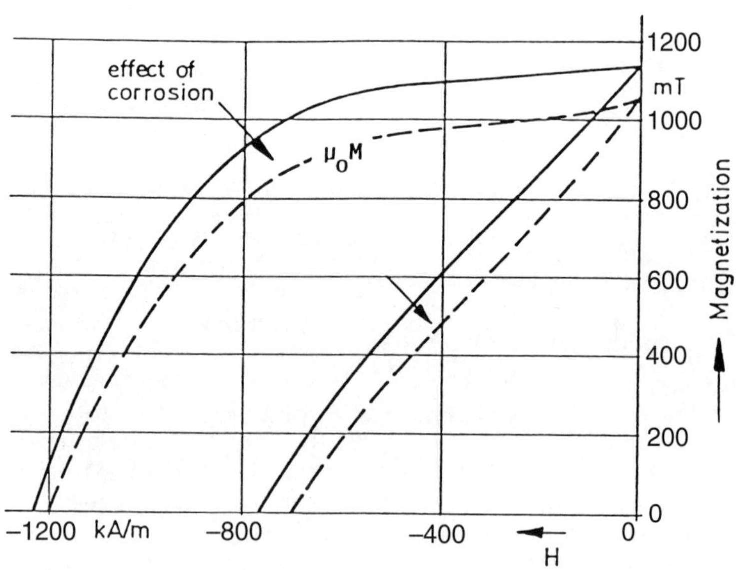

FIG. 10.5. Effect of humid corrosion on demagnetization characteristics: $M(H)$ and $B(H)$ for sintered material.

Table 10.5 Types of coatings and flux losses

Group	Type of coating	Result
1	uncoated	flux loss 4%
2	electrodeposited Ni 30 ± 5 μm	flux loss 1%
3	aluminium-chromate 17 ± 8 μm	flux loss 0.5%

Corrosion in acidic conditions is very rapid. Hydrogen-rich environments pose problems because of the tendency towards absorption and decrepitation [Kun85, Tok87, Hop87, Shi87]. The results of some corrosion tests are summarized in Fig. 10.5.

The obvious solution is to metallize or plastic-coat the magnets. In bonded magnets, the individual grains or flakes are already coated so they remain stable as long as the corrosive medium does not penetrate the polymer matrix. Aluminium, zinc or nickel coatings are used for sintered magnets. Table 10.5 shows the effects of a 300 h exposure in air at 60 °C at a relative humidity of 90%, following different coating procedures.

B. ELECTRICAL MACHINES

1 DC machines

Historically, the earliest electrical machines were axial-field designs (Faraday's disc motor), but these were soon replaced by the radial-field machines familiar in numerous everyday applications. Now it appears that, due to the high-energy product of Nd–Fe–B and the possibility of manufacturing magnets in the form of thin discs, the axial air gap is again a desirable configuration [Cha88, Mil90].

The classical, radial air gap DC motor has a field winding on the stator and a rotating armature winding that is separately excited via a commutator with brushes. Permanent magnets may replace the field winding in a brushed DC motor. Alternatively, permanent magnets may replace the windings on the rotor, with the armature winding now on the stator. This is the brushless DC motor, and the current in the field windings must now be commutated electronically. The three classes of DC motor are illustrated in Fig. 10.6. A permanent magnet DC motor can often be designed that is smaller, lighter and more efficient than the equivalent machines with its copper windings. The actual size reduction achievable is determined by the choice of magnetic material. Having decided on the type of magnet to be used, detailed design work is necessary to determine the best magnet configuration to achieve the desired performance. The magnet dimensions, together with the effective air-gap length and flux leakage factors, determine the open circuit

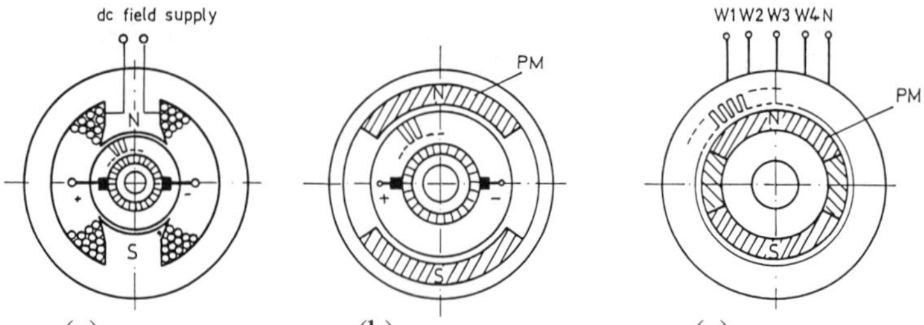

FIG. 10.6. Classes of DC motor: (a) wound field DC commutator; (b) permanent-magnet commutator; (c) permanent-magnet brushless DC motor.

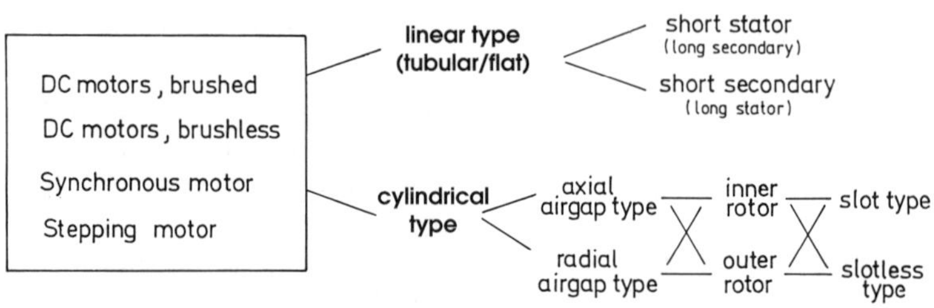

FIG. 10.7. Classification of motors.

magnetic flux [Moh83, Rah88, Weh86]. The magnet dimensions and layout of the magnetic circuit are critical in determining the magnet's ability to withstand the demagnetizing fields produced by the armature reaction, particularly in the stalled conditions. The standard method of 'load line' calculations is quite helpful as a preliminary design procedure, but the use of finite-element methods is advantageous for detailed design work. This method, together with an interactive computer aided design (CAD) system will give the design engineer the chance to exploit the materials in the optimum way. Numerical optimization may be used to find the best design for a given set of constraints. A combination of random search methods, evolution strategy and simulated annealing is a powerful design tool [Ham92, 94]. A three-dimensional approach is often necessary which takes proper account of the magnetic, thermal and mechanical characteristics needed for a thoughtful design. A further classification of permanent-magnet motors by their geometry, is given in Fig. 10.7.

Windings are buried in slots in a soft iron former in the slotted types, whereas the slotless types have a winding that is in the air gap. The designs shown in Fig. 10.6, for example, are all cylindrical slotted motors with an inner rotor and a radial air gap. The outstanding characteristic of DC machines is their versatility. By means of various combinations of armature winding and the permanent-magnet system, a wide variety of speed–torque characteristics can be achieved. In this chapter, brushed DC machines will be analyzed first, and then brushless DC motors will be examined.

1.1 *Brushed type*

Direct current motors with a wound rotating armature and permanent magnets in place of the stator field winding are used in many applications in the power range up to a few kW. The permanent-magnet DC motor is widely accepted for both general purpose drives and for servo drives [Han83, Pfa84, Hen86]. Servomotors have a linear speed–torque characteristics and high torque-to-inertia ratio. Their principal application is in automatic position control systems requiring fast response and a high power-to-weight ratio. The latter characteristic is especially important in airborne control systems. Brushed permanent magnet DC motors are widely used in hand-held tools.

The equations for a DC machine can be derived from either the magnetic-field or the coupled-circuit viewpoint. The electromechanical coupling terms are the magnetic torque T and the generated voltage U_i (often called induced voltage) in the rotor winding.

$$T = K\Phi I_\mathrm{a} \tag{10.1}$$

$$U_\mathrm{i} = K\Phi\omega = 2\pi K\Phi n \tag{10.2}$$

where

$$K = \frac{pz}{2\pi a} \quad \text{and} \quad \omega = 2\pi n$$

I_a = armature current, z = total number of conductors in the armature winding, a = number of parallel paths through the armature winding, ω = angular velocity of rotor (armature), Φ = total flux per pole, p = number of pole-pairs, n = frequency.

If the resistance of all windings in the armature circuit and of the brushes is represented by the resistor R, the voltage equation for static motor operation can be derived:

$$U = U_i + I_\mathrm{a} R \tag{10.3}$$

For any motor, the dynamic equation for the mechanical system is

$$T = T_\mathrm{L} + J\frac{\mathrm{d}\omega}{\mathrm{d}t} \tag{10.4}$$

where J is the moment of inertia and T_L is the load torque opposing rotation. The differential equation of motion of the mechanical system, the volt–ampere equation for the armature circuit and the magnetization curve, describe the system performance of permanent-magnet machines.

It is evident from Eqns (10.1) and (10.2) that with increased flux due to improved magnets the machine's performance will improve. The effects of magnetic saturation of the soft iron parts of the magnetic circuit, however, have to be examined more thoroughly in machines with Nd–Fe–B magnets.

Under load conditions the armature reaction mmf modifies the magnetic field produced by the permanent magnet. This is known as the demagnetizing effect of cross-magnetizing armature reaction due to saturation [Cha88, Mil90].

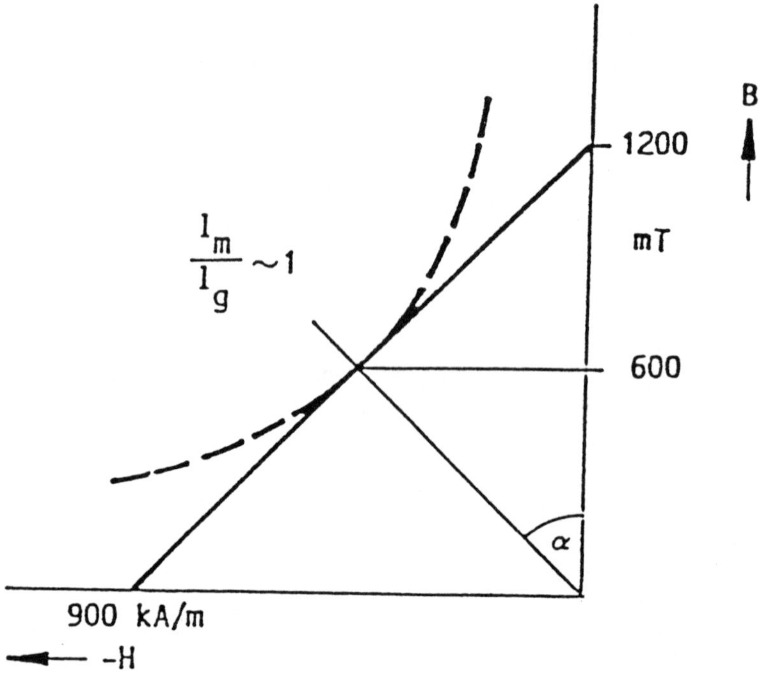

FIG. 10.8. Load line for magnet system at $(BH)_{max}$-point.

The effect on the flux distribution is of interest because the limits of successful commutation are influenced directly. The flux magnitude is important because it is directly proportional to the voltage generated and the torque per unit armature current [Nür87].

The flux distortion due to cross-magnetization is a factor limiting the short-time overload of a brushed DC machine. The unpleasant effects of cross-magnetizing armature reaction may be minimized in the design and construction of the machine. One way of reducing the flux density in the air gap is by increasing the operation angle α in Fig. 10.8. At the $(BH)_{max}$ point, the flux density is about half of the remanent induction. In this case the ratio 'thickness of magnet/thickness of air gap' is close to unity: $l_m/l_g \sim 1$. Motors with nonferrous rotors or air-gap windings have large air gaps. Slotless motor types will benefit from this design approach due to the high remanence and coercivity of the Nd–Fe–B material.

Figure 10.9 shows the typical second quadrant demagnetization curve, where the intersection with the B-axis represents the remanence B_r and those with the H-axis, the coercivities H_c and ${}_BH_c$. The static working point of the magnet is given by the intersection of the demagnetization curve with the load line (point A_1). The magnetic circuit of the electric machine under load always presents external demagnetizing fields which cause the magnet to function in dynamic conditions between several working points. For example, point A_3 is the working point for

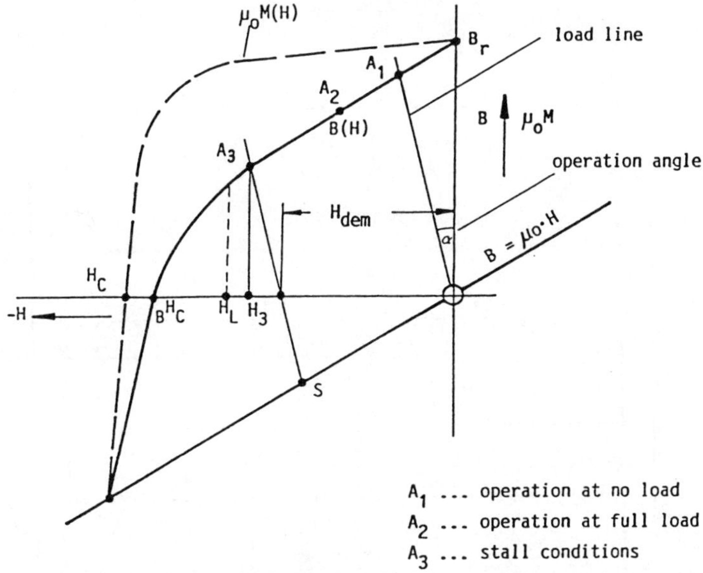

FIG. 10.9. Functional behaviour of permanent-magnet motor.

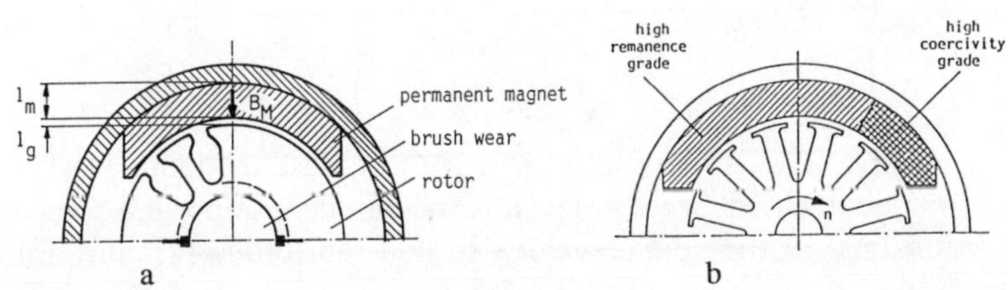

FIG. 10.10. Brushed DC motor designs: (a) standard permanent-magnet stator; (b) split segment stator to avoid demagnetization.

stall condition and A_2 the full load point in Fig. 10.9. To prevent any irreversible demagnetization at the trailing edge of the magnet, the following relation must be satisfied: $H_3 < H_L$, where H_L is the limiting field strength at the knee of the demagnetization curve. In order to avoid irreversible demagnetization of the permanent magnet due to cross-magnetization a grade with very high coercivity might be used. But the cost of this material is high. Another approach is to split the magnet segment into two sections. Depending on the direction of rotation (Fig. 10.10), for the larger section material with high remanence is used, while for the trailing edge material with high coercivity is applied. Because of the high coercivity of Nd–Fe–B and the knee in the curve close to it, these permanent magnets can be dimensioned to the minimum volume and can work on the $(BH)_{max}$ point [Moh83, How87, Ham91].

Table 10.6 Simplified design procedure

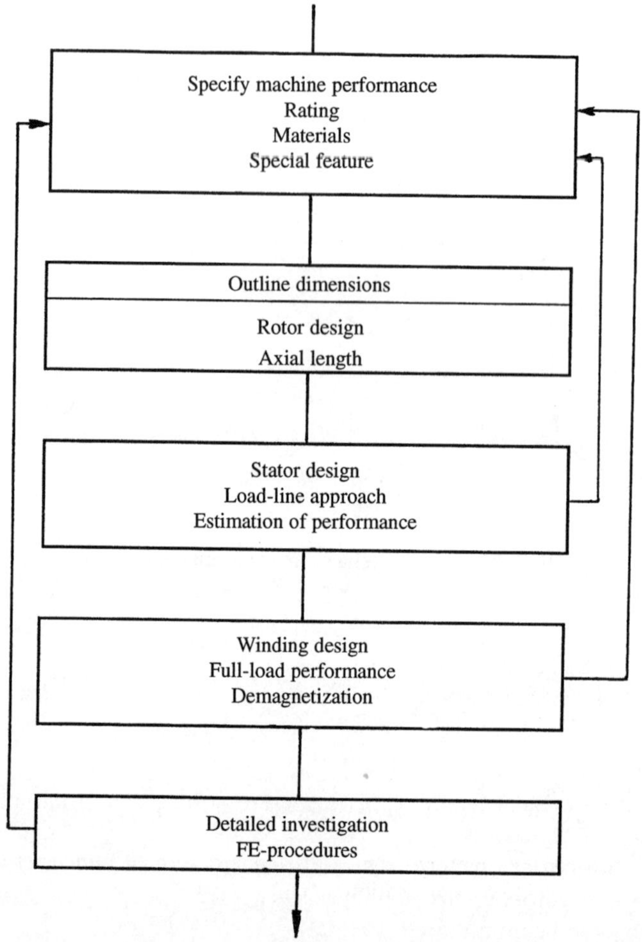

Table 10.6 outlines the design procedure for a motor. Sometimes it emerges from the pre-design for a Nd–Fe–B motor that the magnet should be rather thin and expensive to produce. A complete redesign is then necessary. Cup-type and disc-type DC motors are alternate designs with large airgaps. They operate as servomotors because of their fast response due to low moment of inertia and small electrical time constant. Figure 10.11 shows the principle of the design. Because of the high-energy product offered by Nd–Fe–B magnets, they serve the quest for miniaturization, since the minimum magnet volume required to establish a specific flux density in an air gap is inversely proportional to $(BH)_{max}$. Demonstrating the advantages of Nd–Fe–B over ferrite, various comparative design studies for brushed d.c. motors have shown that the volume V_m of the magnet can be reduced by a factor of 4–6 [Mil88, Eva87].

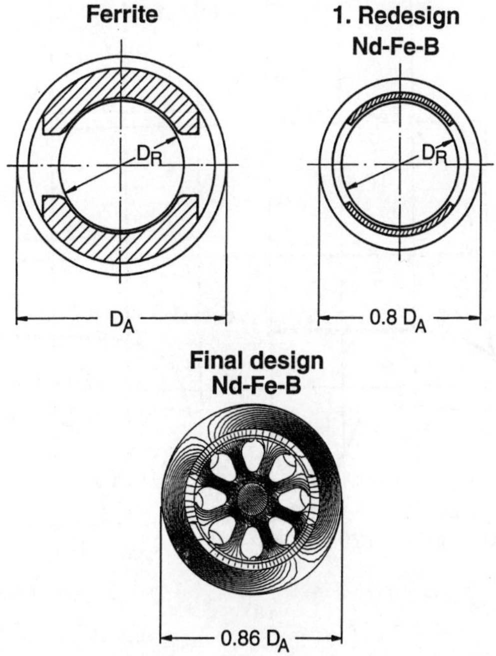

FIG. 10.11. Typical brushed DC motor redesign. Frame size is reduced when ferrite is replaced by Nd–Fe–B. In the final design the integration with the load is taken into account.

The reduced magnet thickness permits a larger diameter armature and therefore a reduction in the active length of the motor. No magnet overhang is necessary for the Nd–Fe–B motor; it had to be designed, however, to a higher tooth flux density. Of course, the motors compared had a similar quality of commutation.

Attaining good commutation is still, to some extent, an empirical art rather than a quantitative science. An obstacle to quantitative analysis lies in the complicated behaviour of the carbon–copper contact [Nür87].

A special feature of the permanent-magnet excited d.c. motor is the cogging torque. The number of rotor slots and the arc of the magnet segment have a strong influence on the cogging torque. For the classical two-pole design with segments magnetized in a radial direction, the cogging torque was calculated for four, eight and twelve slots. From Fig. 10.12 it is seen that the cogging torque can be minimized by selecting the pole arc in the right way. For example, in the case of the rotor with twelve slots the cogging torque can be minimized by using a pole arc of 123° or 153°. Another approach to reduce the cogging torque is to use magnet segments with diametral magnetization instead of radial magnetization [Itt91, Mar88].

Brushed DC motors with permanent magnets are typically available in the power range from 0.2 to 5000 W. The speed range is from 500 rpm to 20 000 rpm.

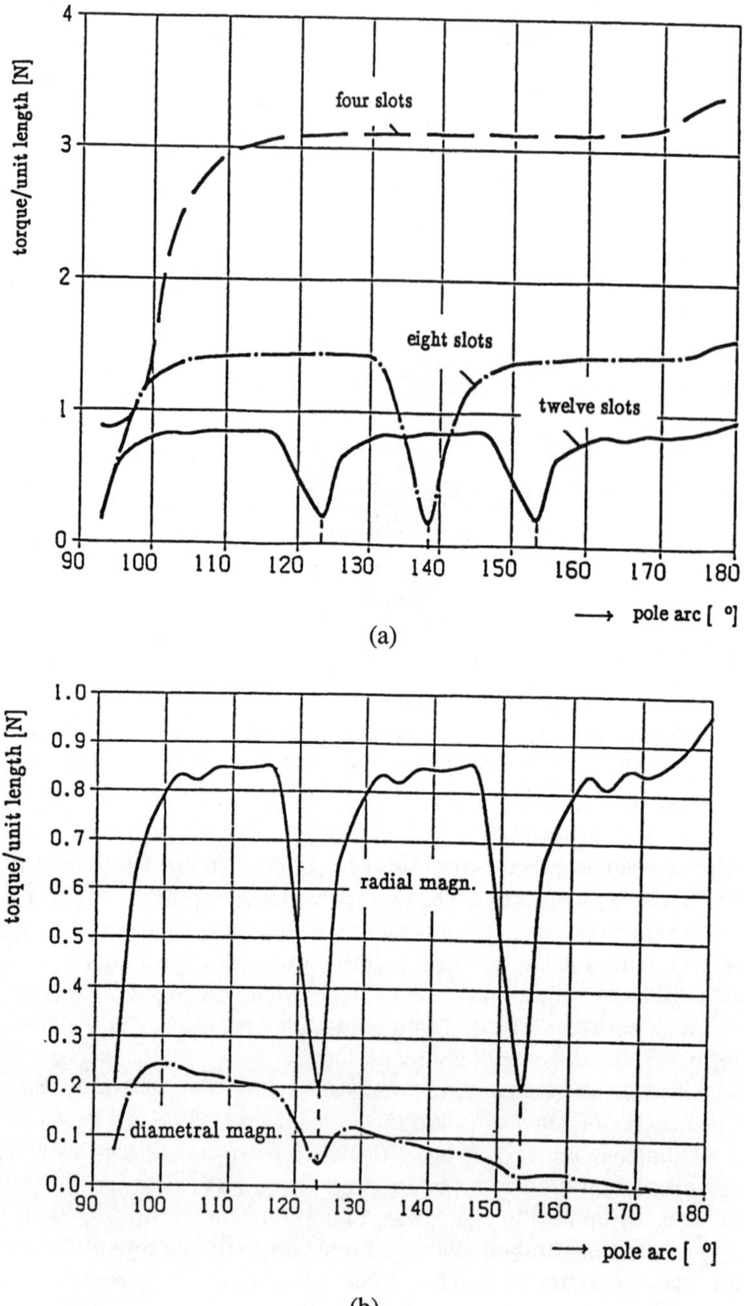

FIG. 10.12. Cogging torque versus magnet pole-arc for a two-pole motor: (a) influence of number of slots; (b) influence of type of magnetization.

The ratio of starting torque to rated torque is between 4 and 10. The efficiency for operation at rated power is 0.4 to 0.8.

1.2 *Brushless type*

The brushed DC motor with its mechanical commutator is widely used on account of its good dynamic and static behaviour. But at speeds well above 10 000 rpm it is necessary to replace the mechanical commutator by an electronic one. Of course, a very-high-speed motor may require air-cushion bearings or a magnetic suspension as discussed in Chapter 9B. A DC motor with electronic commutation offers the additional advantages of long lifetime, low maintenance, low noise, wide speed range, easy speed control, high starting torque, high efficiency and compatibility for explosive atmospheres [Dot90, Mil90].

In a brushless d.c. motor, a multiphase armature winding is located on the stator and the rotor is a rotating permanent-magnet assembly. Normally, three or four phases are used for economic reasons. Electronic commutation is realized by the use of electronic switches: transistors, FETs, etc. These switches control the flow of current in the multiphase winding according to the rotor position in order to give maximum torque. Therefore, rotor position sensors are a vital part of a brushless DC motor [Mil90].

Figure 10.13a shows the basic circuit diagram of a two-pole, four-phase, star connected brushless d.c. motor. In this case, the current will flow in each phase for a period of about 90° for every rotor revolution. During the time in which the current flows through one phase of the stator windings, the phase displacement angle between the rotor flux vector Φ_R and the stator flux vector Φ_s changes from 45° to 135°, as shown in Fig. 10.13b. The average angle is 90°; this device therefore displays the typical behaviour of a DC machine. For the simplified four-phase arrangement considered here, the torque varies due to the sinusoidal flux distribution and the assumed constant current. An increase in the number of stator phases will provide only a slight increase in the average torque, but it requires more power switches for the electronic commutation circuit. A way of minimizing the torque pulsations will be discussed later [Mil90, Sit87, Zim83]

Sensors for detecting the rotor position may be Hall generators, magnetoresistance sensors search coils or photodiodes. The proper choice of position sensor is an important step in the design of a brushless DC motor. A typical circuit diagram is shown in Fig. 10.14. The Hall generators 13 and 14 control the switching of the armature windings 1, 2 and 3, 4 respectively. The angular displacement between these two Hall generators is 90°.

For reasons of simplicity, we focus on the arrangement with four phases, arranged as an air-gap winding and a sinusoidal flux distribution produced by the cylindrical permanent-magnet rotor. Figure 10.15 shows the equivalent circuit for one winding. The angle α in Fig. 10.15a gives the actual position of the rotor and the angular velocity is defined: $\omega = \mathrm{d}\alpha/\mathrm{d}t$. Each winding is characterized by: w = number of turns, d = average diameter of winding, l = active length, f_W = winding parameter ($f_W \leq 1$).

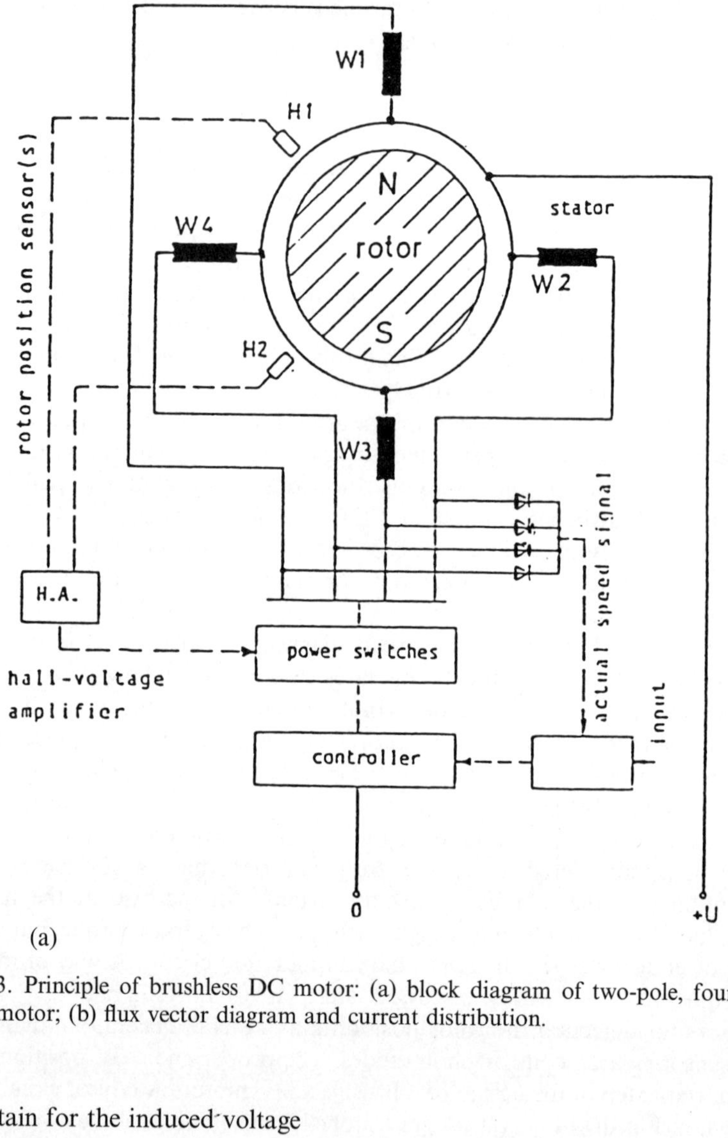

(a)

FIG. 10.13. Principle of brushless DC motor: (a) block diagram of two-pole, four-phase brushless motor; (b) flux vector diagram and current distribution. ▶

We obtain for the induced voltage

$$u_i = k_B \omega \sin \alpha \tag{10.5}$$

where

$$k_B = w \, dl \, f_W \hat{B} \tag{10.6}$$

During $\alpha_{on} \leq \alpha \leq \alpha_{off}$ we energize the winding from the DC source with a voltage U. Here $\hat{B} = \sqrt{2} B_{rms}$. For the electronic switch a voltage drop of u_s is assumed:

$$U = k_B \cdot \omega \sin \alpha + iR + L \frac{di}{dt} + u_S; \quad \alpha_{on} \leq \alpha \leq \alpha_{off} \tag{10.7}$$

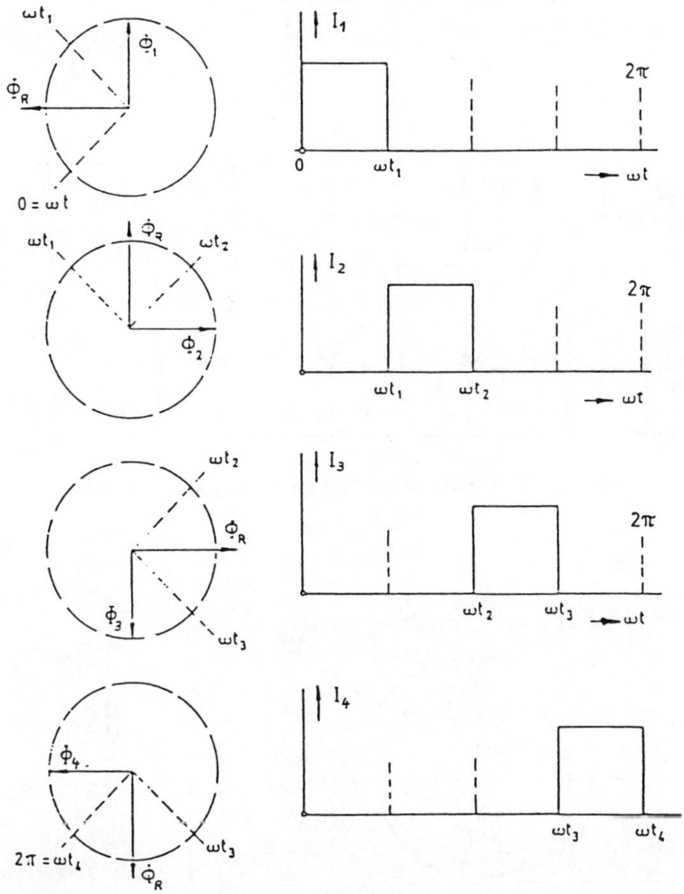

(b)

The instantaneous torque is given by the equation

$$t_{el} = k_B i \cdot \sin\alpha = T_L + J d\omega/dt \tag{10.8}$$

In order to produce a continuous torque the phases have to be energized in sequence. With four phases and a star connection of the windings, this will give

$$\alpha_{off} - \alpha_{on} = 90°.$$

Because of the number of phases $N = 4$ we can write: $\omega \approx$ const. $= \Omega$. For steady-state conditions and for $L = 0$ we obtain for the current $i\,(\alpha)$ if we neglect the very small voltage drop U_s across the electronic switch

$$i(\alpha) = \frac{U}{R} - \frac{k_B\Omega}{R}\sin\alpha \quad \text{for } \alpha_{on} \leq \alpha \leq \alpha_{off} \tag{10.9}$$

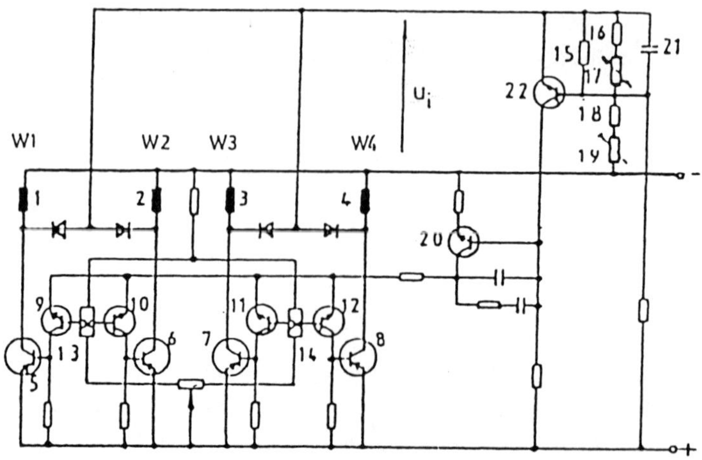

FIG. 10.14. Circuit diagram with speed control features.

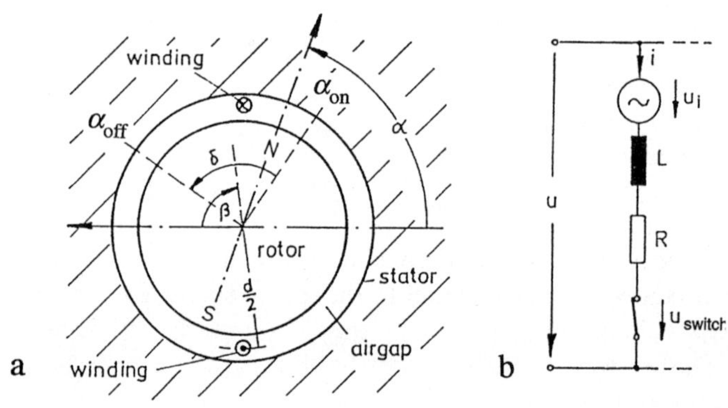

FIG. 10.15. Circuit for one phase of brushless DC motor: (a) schematic arrangement; (b) circuit representation.

For additional calculations we introduce

average current: $$I = \frac{1}{2\pi}\int_0^{2\pi} \sum_{j=1}^{N} i_j(\alpha)\mathrm{d}\alpha \tag{10.10}$$

average torque: $$T_{el} = \frac{Nk_B}{2\pi}\int_0^{2\pi} i(\alpha)\sin(\alpha)\,\mathrm{d}\alpha \tag{10.11}$$

current flow angle: $$\delta = \alpha_{off} - \alpha_{on} \tag{10.12}$$

current phase angle: $$\beta = \pi - (\alpha_{on} + \alpha_{off})/2 \tag{10.13}$$

We now obtain for the torque and voltage

$$T_{\mathrm{el}} = \frac{a_1 k_{\mathrm{B}}}{\delta} I - \frac{(\delta a_2 - 2a_1^2) k_{\mathrm{B}}^2 N}{4\pi R \delta} \Omega$$
$$U = \frac{a_1 k_{\mathrm{B}}}{\delta} \Omega + \frac{2\pi R}{\delta N} I \tag{10.14}$$

with

$$a_1 = 2 \sin \beta \sin(\delta/2) \tag{10.15}$$
$$a_2 = \delta - \sin \delta \cos 2\beta \tag{10.16}$$

The control parameters δ and β govern these equations.

The second term in Eq. (10.14) is typical for the brushless DC motor. Because of the fact that the induced voltage varies with position, a small reduction of the torque can be observed. The standard four-phase electronic commutation will work with $\delta = \beta = 90°$. However, with advanced control strategies we can change the current flow angle and use this parameter for speed control.

Figure 10.16 shows the voltage across the idealized power switch and the current through one phase. The current flow angle is equal to the current phase angle $\beta = 90°$. Due to the sinusoidal flux distribution which induces a sinusoidal voltage, there will be a torque ripple. These fluctuations can be reduced when the magnetic circuit is changed to a trapezoidal flux distribution (Figs 10.17; 10.18). More details are given by [Han86, Han87].

In some applications, the drive motor must rotate at a constant speed independent of the load torque and battery voltage. An electronic speed regulator must than be used. However, a tachogenerator is not required for speed measurement. Because of the induced emf in the armature windings and the diodes in the circuit, U_{i} is proportional to the motor speed. By comparison of the reference speed voltage with the actual speed voltage U_{i} transistor 22 can be controlled. Together with transistor 20 the output signal of transistor 22 affects (after amplification) the emitter voltage of the transistors 9–12, and then the power transistors 5–8 (Fig. 10.14).

The design of the electronic commutation system has a strong influence on the performance characteristics of an 'electronic motor'. The motor discussed employed analog processing of data from the rotor position sensors. However, a significant advantage is derived from the use of digital components [Han83].

Figure 10.19 shows a block diagram of the principle of such a control circuit. The armature emf is used for the sensing process of the rotor position, its instantaneous value being passed through diodes to the voltage-controlled oscillator (VCO). This produces a pulse train at a voltage-controlled frequency. The main components of the digital control are the shift registers and the main logic circuit. During start-up, the motor behaves like a stepper motor.

With the adoption of solid state components and microelectronics in complex control schemes, reliability becomes an important consideration. Reliability prediction enables the system engineer to evaluate alternative designs [Mil88, She80].

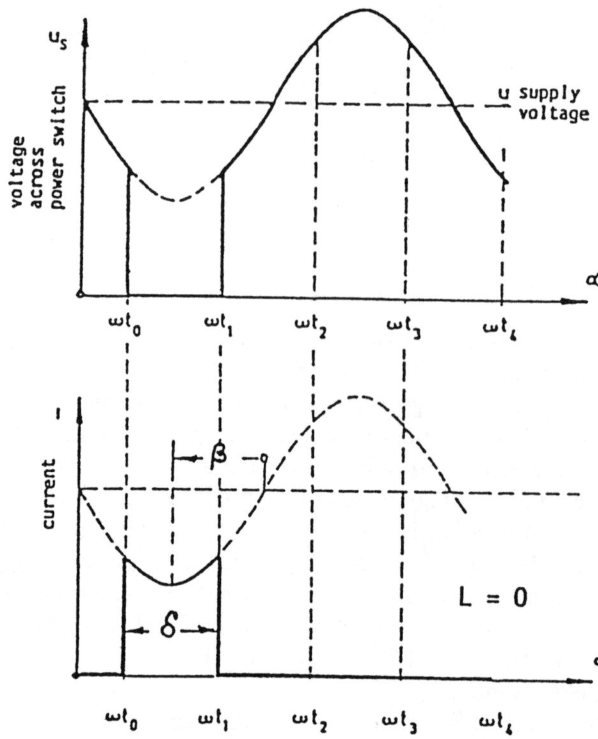

FIG. 10.16. Voltages and current vs position (ideal representation for sinusoidal flux distribution).

Beside the classical brushless DC motor, some special designs are in use due to recent advances in the magnet technology and semiconductor devices. These are

1. brushless linear DC motors;
2. brushless disc-type motors;
3. brushless cup-type motors;
4. brushless segment-type motors.

Again in these types of motor, the armature windings are located on the stator and they form a layer (1), disc (2), cup (3) or segment (4). The exciting field windings are replaced by an arrangement of permanent magnets which move linearly (1) or rotate (2, 3, 4). Figure 10.20 indicates how changing the control unit can cause the brushless motor hardware to behave in different ways [Han83]. Figure 10.21 illustrates some brushless DC motor designs.

(a) Linear brushless DC motors in comparison with the AC linear motor, have the advantage of easier speed control. As already pointed out, the number of

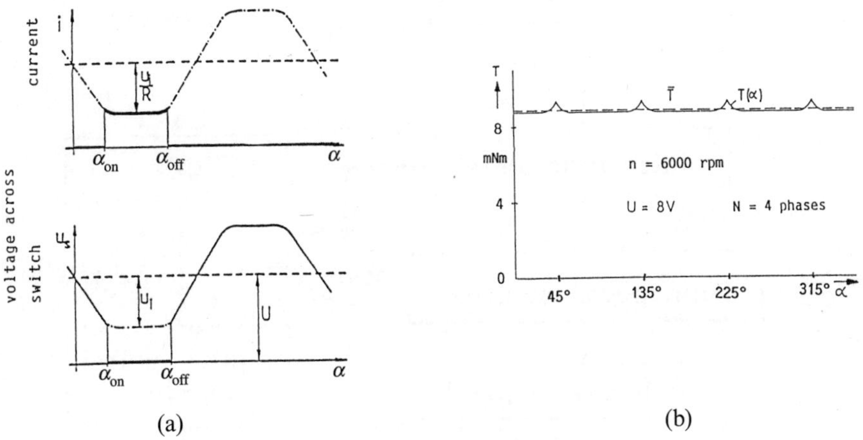

FIG. 10.17. Current voltage and torque vs position for trapezoidal flux distribution: (a) current and voltage across the switch; (b) torque for four-phase motor.

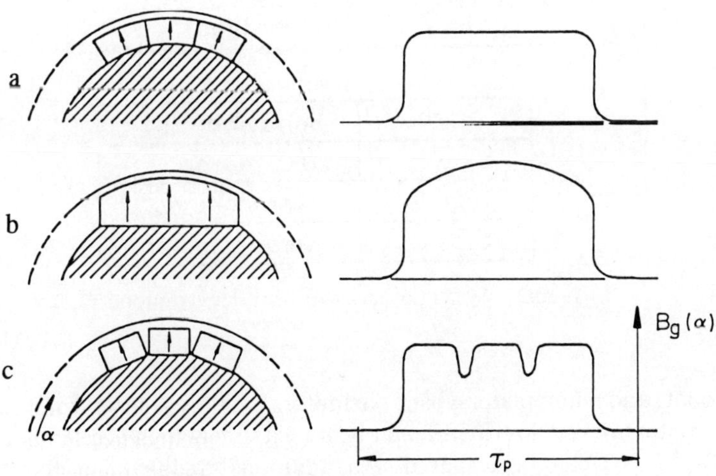

FIG. 10.18. Magnetic circuit designs and airgap flux density.

armature phases N denotes the number of switches per pole of the stator. Examples of linear stepping motors with Nd–Fe–B magnets are given in various papers [She80, Han87a].

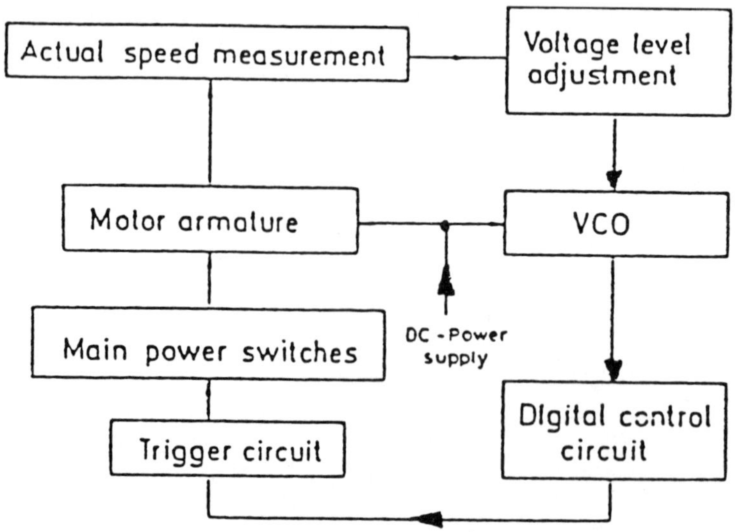

FIG. 10.19. Block diagram of brushless DC motor with voltage controlled oscillator.

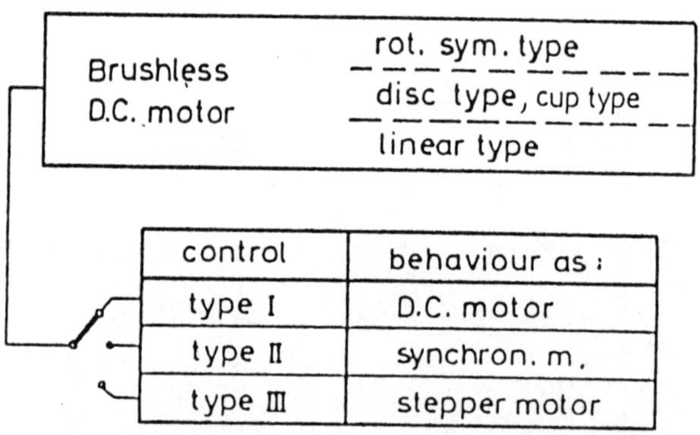

FIG. 10.20. System overview for brushless motors.

(b) For robots and manipulators, pancake motors with disc-shaped rotors are used. These motors have a low inertia and hence extremely high accelerations (up to 260 000 rad/s^2) can be achieved. An additional requirement for drives in automatic industrial production is a very high accuracy of positioning, which means that the motor should have a very low torque pulsation [Hen86, Han87b]. Figure 10.21c shows the principle of the disc-type motor, which is derived from the typical brushless d.c. motor. The air-gap winding is fixed to the stator construction while the permanent magnets are mounted on a disc-type rotor. A photodiode was used as a position sensor for the rotor in a prototype design [Han87b]. The electronic commutator operates with $\delta = \beta = 90°$, which

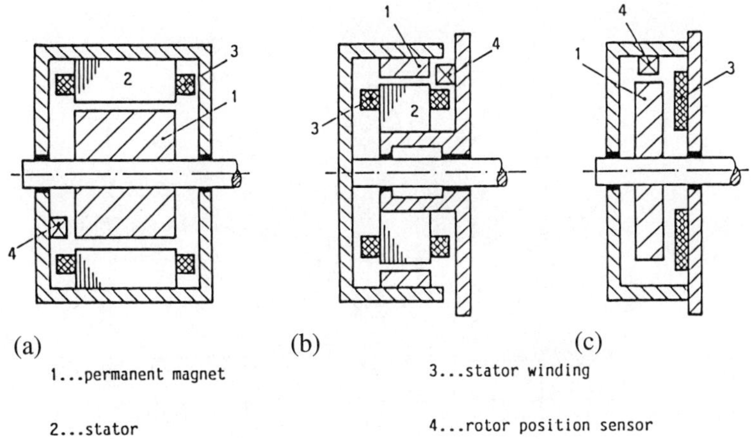

FIG. 10.21. Schematic representations of different brushless DC motors: (a) classical cylindrical design, inner rotor with slots; (b) cup-type cylindrical design, outer rotor with slots; (c) disc-type design, without slots.

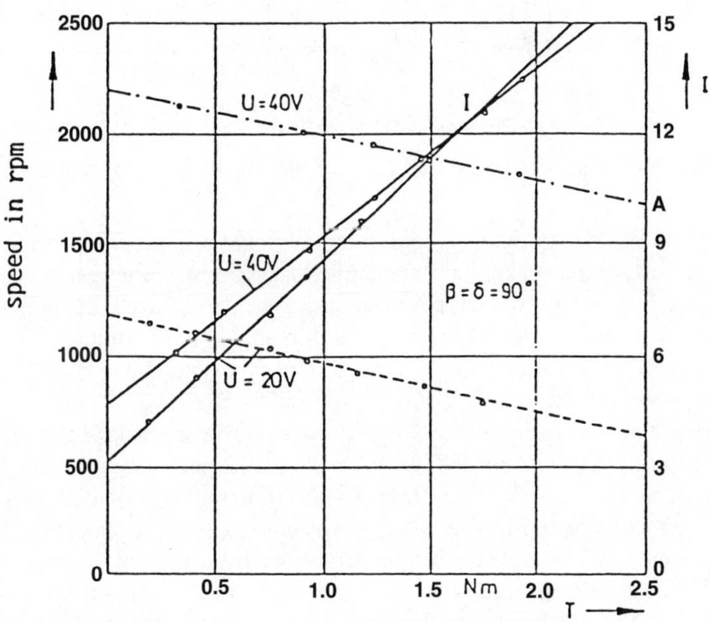

FIG. 10.22. Speed and current vs torque for brushless DC disc-type motor.

means that the current flow angle δ and the phase reference angle β are equal. Results obtained from an experimental six-pole motor are given in Fig. 10.22. The linear speed–torque characteristic shows that by variation of the applied voltage, very wide speed ranges may be obtained. The rotor construction was a

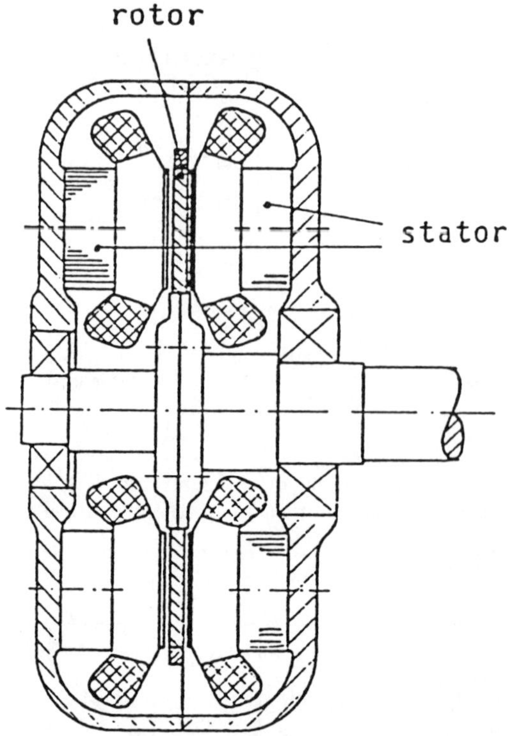

FIG. 10.23. Symmetric brushless low-inertia disc-type motor (slotted type)

compromise of conflicting requirements, since optimum performance characteristics require, on the one hand, the largest possible magnetic flux. On the other hand, however, the requirement of a low rotor moment of inertia demands a small rotor diameter and thin magnets. Torque fluctuations can again be avoided by appropriate design steps of the magnetic circuit that lead to a machine with trapezoidal emf (Fig. 10.18). Because of the fixed air-gap winding, the heat transfer from the armature to the housing is not ideal. Often, forced ventilation is needed to control the motor temperature. For these reasons, machines with a central permanent magnet rotor disc and windings arranged as indicated in Fig. 10.23 are preferred because they have the lowest rotor inertia and the heat transfer from the windings to the housing is very good. The stator windings are sometimes embedded in slots as in conventional machines. For particular applications, where a low noise level is required, air-gap windings are recommended. The mechanical fixture of the air-gap winding can cause problems at high temperatures.

Due to the fact that the end windings require some space, the inner diameter of the stator cannot be too small. The permanent-magnet rotor discs experience high mechanical stresses under impulse currents and eddy current losses in the rotor disc have to be reduced by design measures such as subdivision of the

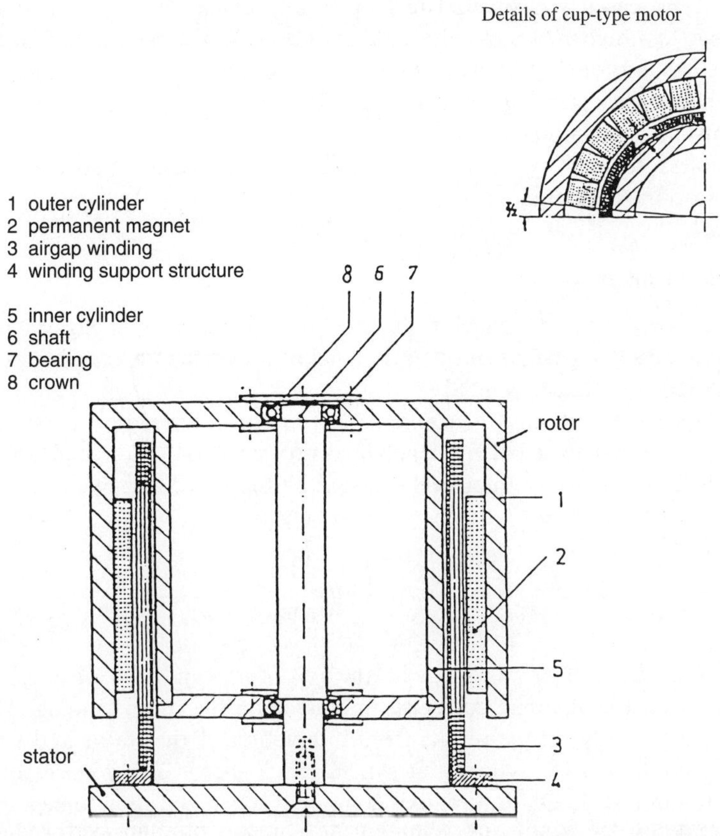

FIG. 10.24. Cup-type motor with air-gap winding.

magnet disc or the use of bonded magnets. These examples demonstrate that the design must accommodate all the factors that stress the magnet, not only electromagnetic but thermal and mechanical as well.

(c) For cup-type motors, Figs 10.21 and 10.24 illustrate the design principle. The armature is constructed by nesting preformed wire-wound coils to form a cylindrically shaped air-gap winding. The whole armature assembly is encapsulated in epoxy resin to provide rigidity and strength. As there is no iron in the armature, the armature reaction flux is small [Cha83]. The undesirable phenomenon of cogging caused by the interaction of poles and discrete teeth is eliminated because of the air-gap winding and a second ferromagnetic cylinder which is mechanically coupled to the outside cylinder with its permanent magnets. Using this construction, the iron losses are reduced [Eva87]. In Fig. 10.24, prefabricated magnets are glued to a steel ring in order to reduce magnet cost and losses.

The increased size of electrical loads on vehicles is expected to lead to a change to a higher system voltage, for example 48 V. This will favour the use of brushless motors for a wide range of new applications including electrical power-assisted steering, electrical antilock brakes, electrically driven water pumps and a combined starter-generator.

(d) Brushless segment-type motors have been constructed, but the number of applications is limited.

2 Synchronous machines

Here again we describe the machine in terms of relatively simple physical reasoning. Detailed analysis of permanent-magnet synchronous machines can be found in the literature [Bin79, Mel88, Rah88].

The classical synchronous machine, be it a motor or generator, is one where the stator is connected to a balanced polyphase (often two- or three-phase) electrical system, its travelling-wave mmf being excited by the phase currents in combination. The stator field rotates at synchronous speed (angular velocity) for a $2p$-pole machine:

$$\omega_s = \omega/p \text{ rad/s}$$
$$n_s = \frac{120f}{2p} \text{ rpm}$$

The rotor carries a single winding with the axis of its mmf fixed with respect to the synchronous rotor. A direct current is impressed on the rotor winding. The ability to produce a steady torque results from interaction of the stator and rotor fields when the rotor is also revolving at synchronous speed. The concept of dividing mmfs into direct- and quadrature-axis components will be found to be a great help.

The losses in the rotor winding due to DC currents disappear when the winding is replaced by permanent magnets and, consequently, the efficiency of the synchronous machine improves. Figure 10.25 compares the idealized classical three-phase synchronous machine with permanent-magnet designs.

2.1 *Line start motors*

The power-to-weight ratio of induction-start permanent-magnet synchronous motors is greater than for comparable asynchronous motors. Their efficiency is also higher as a result of the absence of rotor winding losses. Although synchronous motors tend to be more expensive than induction motors, they require a lower volt–ampere input for a given load torque. This makes them logical candidates for use in AC variable-speed drives where the inverter is normally the most expensive system component.

The magnetic circuit of the synchronous machine stator with permanent magnets on the rotor is of the same form as a conventional polyphase machine. The magnetic circuit of the rotor, however, can be designed in different ways. In line start motors, high remanence and high coercivity are equally important. During start-up, the rotor-magnet revolutions lag the rotating stator field which causes the

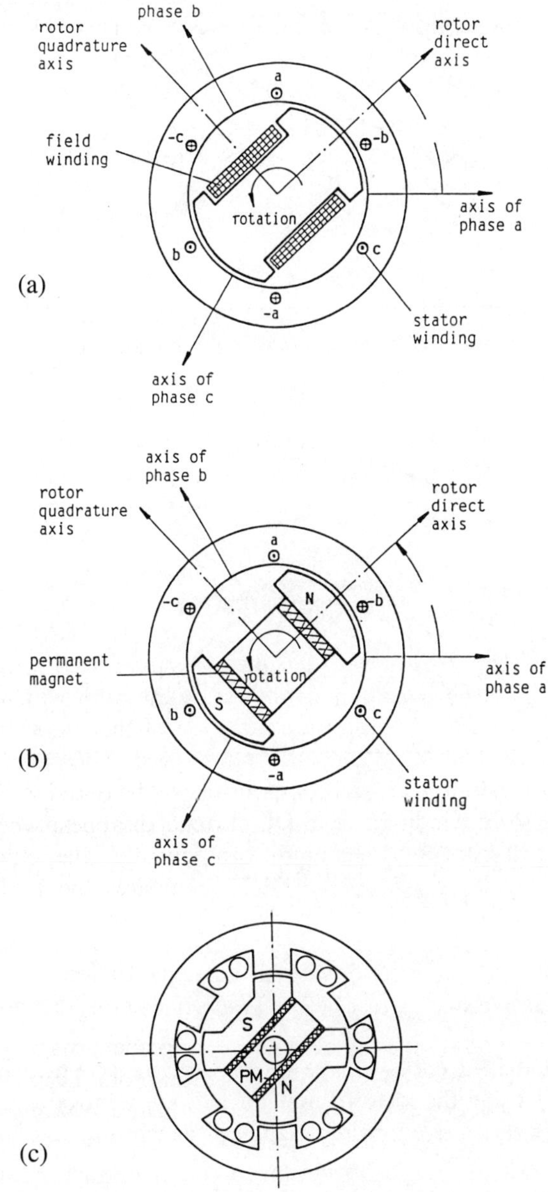

FIG. 10.25. Synchronous machines: (a) idealized conventional machine; (b) idealized synchronous machine with permanent magnets; (c) schematic of a doubly salient permanent-magnet motor.

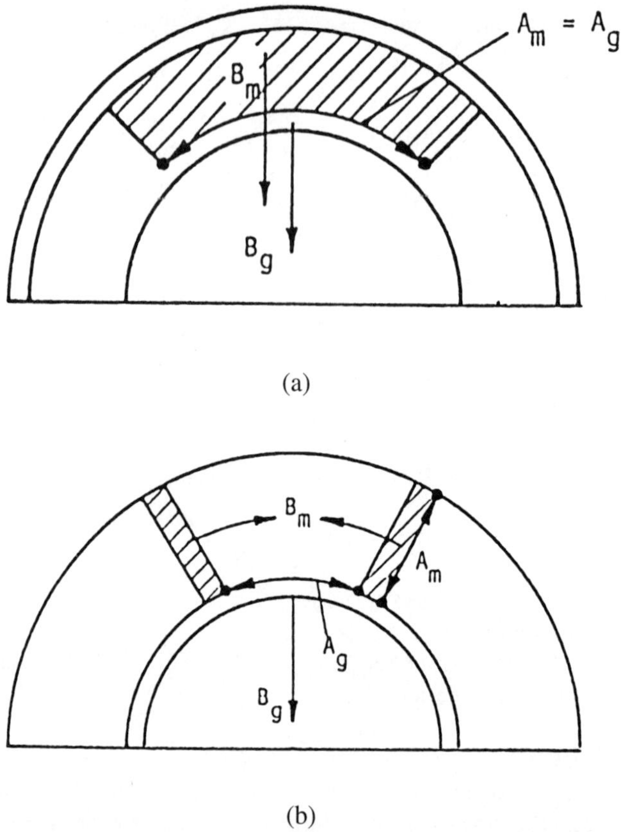

FIG. 10.26. Principle of flux concentration. With radial magnetization (a) $B_g = B_m$, but with tangential magnetization (b) it is possible that $B_g > B_m$.

working point to move up and down the demagnetization characteristic. The highest demagnetization occurs just before synchronization; therefore, a sufficient coercivity is needed.

In designs where the current layer is restricted, a high flux density is needed which can be achieved by using the ideas of flux concentration introduced in Chapter 8. From Fig. 10.26 we can derive for the air-gap flux density B_g assuming that there is no stray flux:

(a) $B_g = B_m$, when the magnet faces the air-gap,

(b) $B_g = 2(A_m/A_g) \cdot B_m$, when two magnets contribute to the air-gap flux.

In Fig. 10.27 we see embedded magnets in flux concentration arrangements.

For line start motors, an outer cage winding is needed to facilitate line starting. Under steady-state conditions, however, the cage is inoperative. During run-up, the

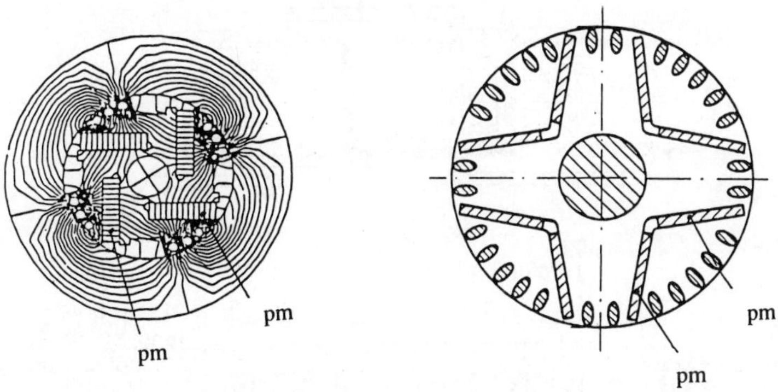

FIG. 10.27. Embedded magnets in rotors of permanent-magnet synchronous machines.

effective asynchronous torque is the sum of the cage induction torque and a braking torque due to the motion of the permanent magnet [Mec91, Cha92]. Because the magnets will be exposed to demagnetizing fields during run-up a careful field analysis is necessary.

In recent years, much work has been done on the modelling and design of line start motors, with emphasis being placed on the best choice for the machine parameters [Hon80, Seb86, Bun87, May88]. From this work we know that the torque is dependent upon the following two values:

1. Magnet saliency $ms = X_d/X_q$. The quantities X_d and X_q are called the direct-axis synchronous reactance and quadrature-axis synchronous reactance [Schi87, Mil84].
2. Ratio between the open circuit voltage due to the magnets and the applied terminal voltage ($r_{o,v}$).

Permanent-magnet machines should be designed with $0.5 < ms < 0.9$ and $0.4 < r_{o,v} < 0.6$. Stator resistance and leakage reactance have a less significant effect, but both are useful in limiting the currents during run-up. Decreasing the number of poles will result in decreased motor performance, with the exception of efficiency and power factor.

The high coercivity of Nd–Fe–B magnets poses fewer problems for the design engineer but a careful thermal analysis is needed on account of the temperature sensitivity of the material. For example, in a synchronous motor for inverter-fed application, the torque of the $Nd_2Fe_{14}B$-type machine at 20 °C is about 18% higher than that of the $SmCo_5$-type machine. At an elevated temperature of 120 °C, however, the torque for both types of magnet is almost identical. The temperature increase from 20 °C to 120 °C results in a torque decrease from 2.8 kNm down to 2.0 kNm at the same load angle of 20°.

The application of a special linear synchronous motor with Nd–Fe–B permanent magnets in a Maglev system for urban transportation is another example of advanced drive technology. The 'M-Bahn' in Berlin was run by an electromagnetic

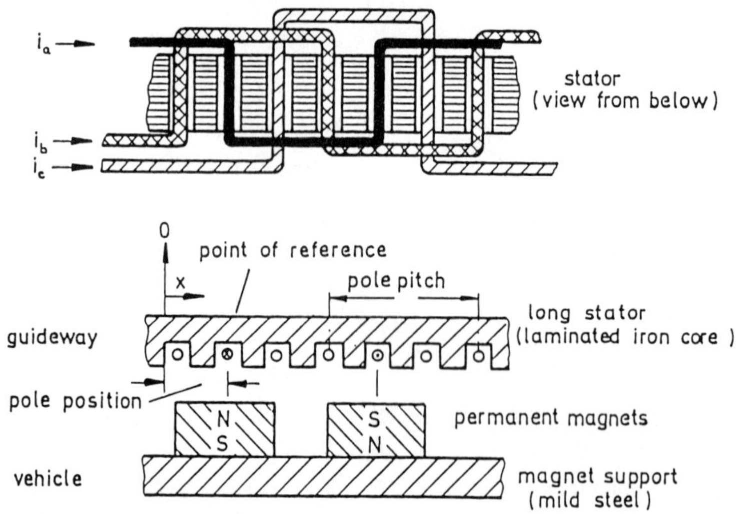

FIG. 10.28. Principle of linear synchronous motor for 'M-Bahn'.

travelling field that moves along the guideway. This field is generated by a three-phase current fed in through cable windings that have been laid inside the guideway. Figure 10.28 shows the principle of the linear synchronous motor. The magnetic field pulls the vehicles along the guideway through interaction with the high-energy permanent magnets on the vehicle. Compared with Sm–Co magnets used at first, the Nd–Fe–B material offers lower magnet weight due to its lower density. Because of the improved remanence and coercivity it was possible to shorten the bogies in which the magnets are mounted and to increase the air gap by about 30%. A larger air gap permits higher guideway tolerances and shorter bogies result in weight savings.

2.2 *Generators*

An a.c. generator without exciter, slip rings, brushes or field winding is obviously advantageous. The removal of these components makes the permanent-magnet generator very attractive. The advent of new magnetic materials such as Nd–Fe–B created an opportunity to improve generator performance without change in size [Mil84, Jab87, Tan81]. Flywheel igniters and bicycle dynamos are the types of permanent magnet generator produced in large volumes. These mass products could have been equipped with Nd–Fe–B magnets, but for the emphasis on economy. On the other hand, there are examples of automobile generators and wind energy converters which can be improved considerably with Nd–Fe–B magnets, but problems arise here from temperature sensitivity and corrosion of the magnet material [Seb86, Bun87].

Provided that the load-line of the magnetic circuit intersects the linear section of the magnet characteristic, the temperature-induced change flux will be reversible.

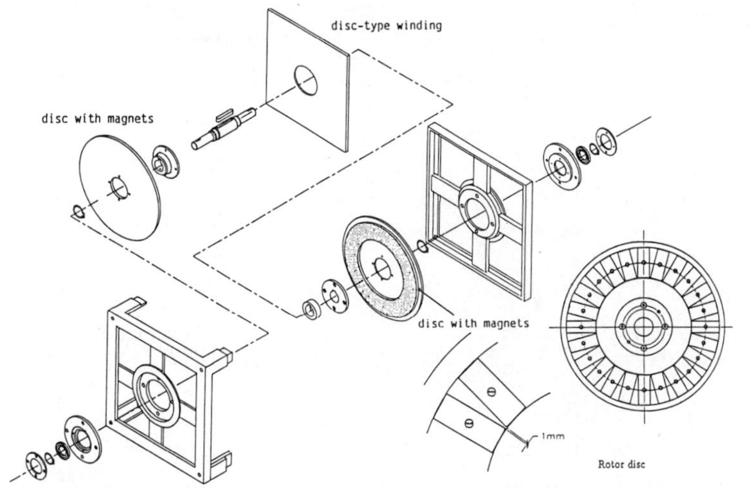

FIG. 10.29. Exploded view of disc-type synchronous generator for wind energy converters showing details of the rotor disc.

It should therefore be possible to accommodate the reversible effects on the polarization and the temperature coefficient for copper and aluminium in most applications [Bou80, Weh86, May88].

Because of the growing pressure for reduced pollution, the interest in electric vehicles, especially hybrid electric cars, has increased. In this context engineers are developing a high-speed alternator which would be suitable for use in turbine-powered hybrid vehicles. The gas turbine's efficient combustion process will result in improved fuel economy and cleaner exhaust. The alternator will be one of the permanent-magnet type, making use of Nd–Fe–B, and will rotate at speeds up to 100 000 rpm. The high speed will reduce the weight and volume of the alternator, so that a 30 kW alternator is expected to weigh about 10 kg. Prototypes with liquid cooling have been built with specific power of almost 10 kW/kg.

In general, significant savings in generator volume and cost can be obtained by improving cooling (fluid cooling, heat pipes, improved heat sinks) and increasing machine speed. An additional advantage of the permanent-magnet rotor is that fluid cooling can be applied to the stator. Nd–Fe–B magnets offer the potential for significant reduction in size and weight. For wind energy converters the permanent-magnet excited generator is widely accepted. Figure 10.29 shows an exploded view of a disc-type generator with 24 poles. The three-phase winding is centred between the two discs on which the magnets are mounted. Details of the assembled magnets on the rotor disc are shown.

The total losses in a well-designed electrical machine are usually a small fraction of the power-conversion level. The power needed to provide the excitation for d.c. machines is from 0.5% to 5% of the rated power. In the case of the synchronous

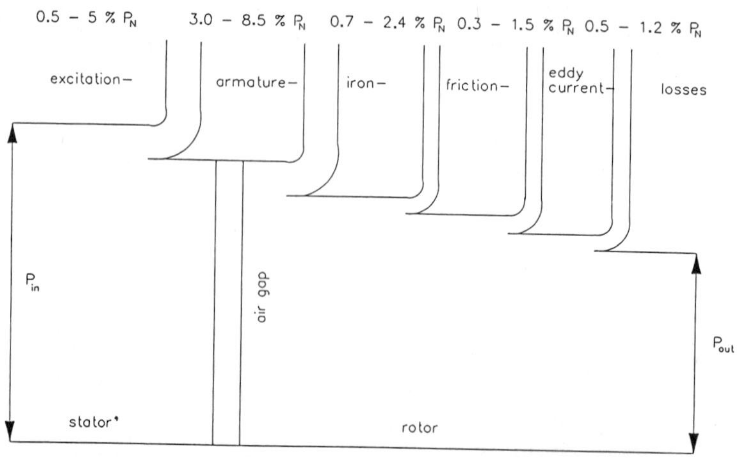

FIG. 10.30. Power flow in a synchronous machine.

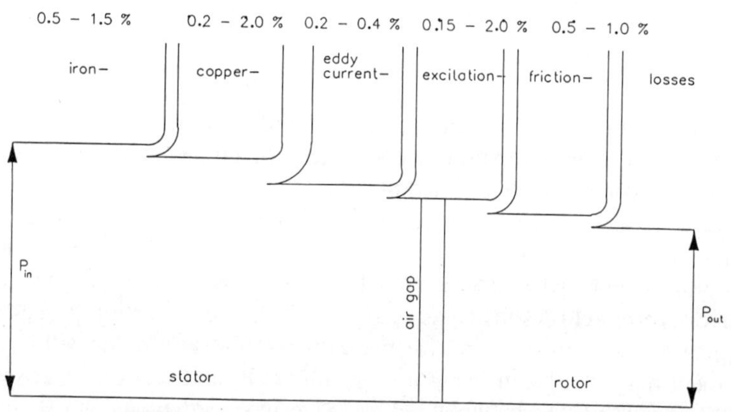

FIG. 10.31. Power flow in a DC machine.

machine, about 0.2–2% of the rated power is used for excitation [Han86, Schi87]. From the power flows indicated in Figs. 10.30 and 10.31 it can be seen that in synchronous and DC machines a fair fraction of the total power losses are due to excitation. When the excitation winding is replaced by permanent magnets, these losses disappear and the efficiency is improved. Due to the electrical conductivity of the permanent-magnet material, however, we have to pay attention to the eddy current losses in this material. It is better to use small segments or blocks in the axial and circumferential directions rather than large magnet blocks. Nonsynchronous harmonics in the machine will then produce only small losses. We then have a rotor with almost no losses without the usual reduction in the power capacity of the machine. The arrangement of the magnets, the magnet volume and the lay-out of the PWM (pulse width modulation) inverter have a considerable influence on the system characteristics. The reduction of losses for the total machine is acknowledged as a specific advantage of permanent-magnet machines. Their efficiency is higher as a result of the lack of winding losses from excitation.

Optimally efficient operation for permanent-magnet synchronous machines is achieved through the best trade-off between copper losses and core losses. The size and variation of the core losses are important features of a permanent-magnet machine. In general, the core loss in a permanent-magnet machine will be higher than in an induction machine because of the harmonic fields associated with the magnets and the armature reaction harmonics in salient pole designs [Zes86, Jia88]. Nonsalient pole designs such as full pitch magnets and protruding magnets show little load-dependent core loss. The embedded (inset) magnet machine core loss can double as the machine goes from no load to full load operation.

Three-phase synchronous generators with self-start capabilities are available for synchronous speeds $\leq$ 3600 rpm depending on frequency (50 or 60 Hz) and number of pole pairs. The lowest power is around 200 W. Prototypes with permanent magnets are on the market with output power up to 15 kW. A ratio of starting torque/rated torque $\leq$ 3 can be achieved.

2.3 *Stepping motors*

A large variety of stepping motors is in use; they can be classified into several types according to construction and principle of operation [Ken84]:

1. variable reluctance stepping motors (VR motor),
2. permanent-magnet stepping motors (PM motors),
3. hybrid stepping motors.

The VR motor works in such a way that the magnetic reluctance becomes a minimum. The air gap between the stator teeth and rotor teeth must be small as possible to produce a high torque and to attain high accuracy in positioning.

A stepping motor using a permanent magnet in the rotor is called a PM motor. The principle of a PM motor is indicated in Fig. 10.32. A cylindrical permanent magnet is employed as the rotor, and the stator has four teeth. Each tooth carries a coil which may also be considered as a phase. If the phases are excited in a proper sequence (Phases 1 $\rightarrow$ 2 $\rightarrow$ 3 $\rightarrow$ 4) the rotor will move clockwise. The step angle is 90° in this case. In order to decrease the step angle in a PM motor, the number of magnetic poles and stator teeth must be increased. Disadvantages of the PM motor are the large costly magnet volume and the limitation of maximum flux density level by the remanence of the magnet.

An alternative to the PM motor is the hybrid stepping motor with its small step angle. The term 'hybrid' derives from the fact that the motor is operated under the combined principles of the PM motor and the VR motor [Jen87, Lis76]. The important feature is the structure of the rotor. A cylindrically shaped magnet lies in

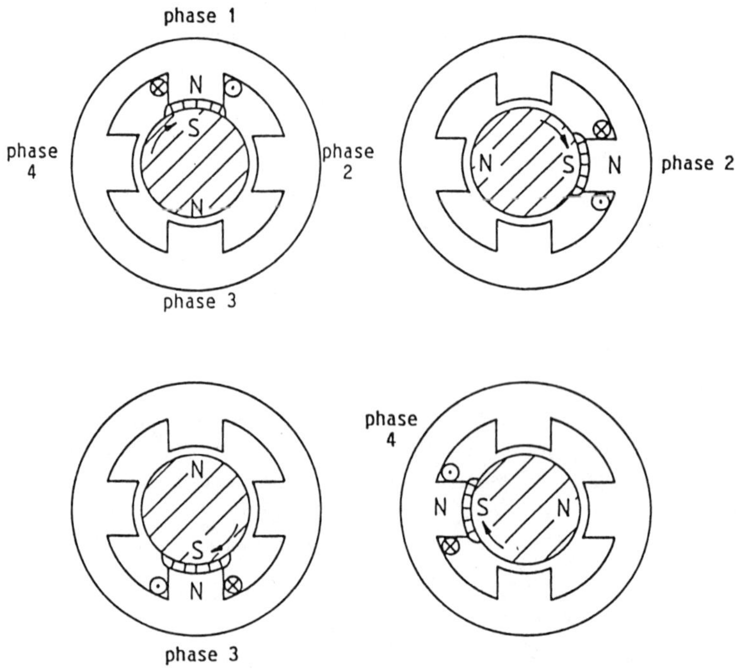

FIG. 10.32. Steps in a four-phase permanent-magnet stepping motor.

the core of the rotor, as indicated in Fig. 10.33. Magnetized in the axial direction, a unidirectional field is produced. The two poles of the magnet are covered with toothed sections made of laminated silicon steel or sintered steel. The teeth on the two pole sections are misaligned with each other by half a tooth pitch.

The winding arrangement and coil connections are different from those in the VR motor. A four-phase hybrid motor is also chosen for comparison. Coils of two different phases are wound on the same pole in bifilar arrangement so that different magnetic polarities can be achieved upon excitation. The magnetic field produced by this type of energized winding is a heteropolar field. The interaction of these two fields creates the torque. The four-phase 200-step hybrid motor with a step angle of 1.8° is in widespread use. In order to increase the torque, multistack hybrid motors are used. Nd–Fe–B magnets are advantageous for motor designers of unipolar stepper motors. The construction on the left in Fig. 10.33a has a magnet made of Alnico. The length of the magnet depends on its coercivity. The Sm–Co magnet shown in the middle of the diagram has nearly the same remanence, but a much higher coercive field strength. Therefore, the magnet can be much shorter. The system on the right built with Nd–Fe–B magnets has almost the same magnet size but 50% more flux. The torque of the motor will increase accordingly and there is a small inertia increase. As a result, the dynamics of the motor with Nd–Fe–B magnets are better. The cross-section of the hybrid motor, Fig. 10.33b, shows the toothed structures in stator and rotor designed to realize a small step angle.

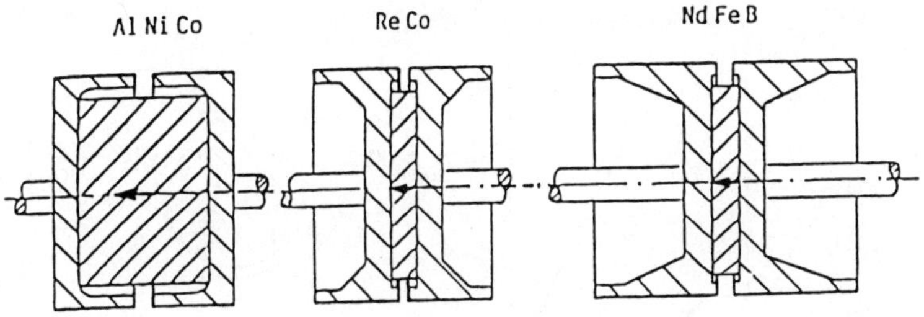

(a)

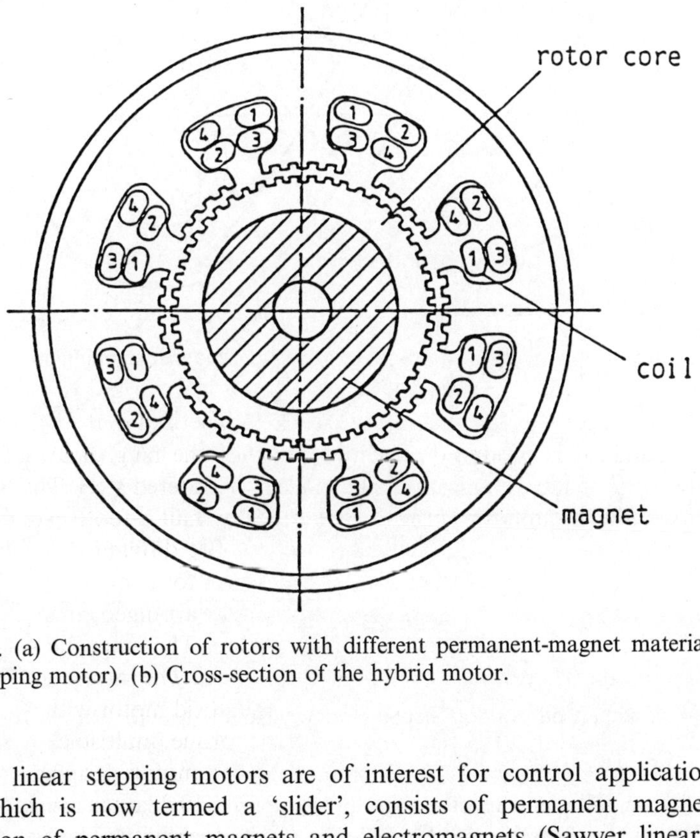

(b)

FIG. 10.33. (a) Construction of rotors with different permanent-magnet materials (for a hybrid stepping motor). (b) Cross-section of the hybrid motor.

Finally, linear stepping motors are of interest for control applications. The 'rotor', which is now termed a 'slider', consists of permanent magnets or a combination of permanent magnets and electromagnets (Sawyer linear motor) [Che88, Hin74]. Figure 10.34 shows a comparison of the performance of linear stepping motors made of Nd–Fe–B or Sm–Co [Han87].

In some applications, the high-torque ripple inherent in some stepper motor designs creates noise. These problems are reduced by dividing each full step into half- or quarter-steps. Microstepping is done by increasing the current in one phase in small increments and at the same time decreasing the current in the other phase. This results in a displacement of the torque function by increments which are

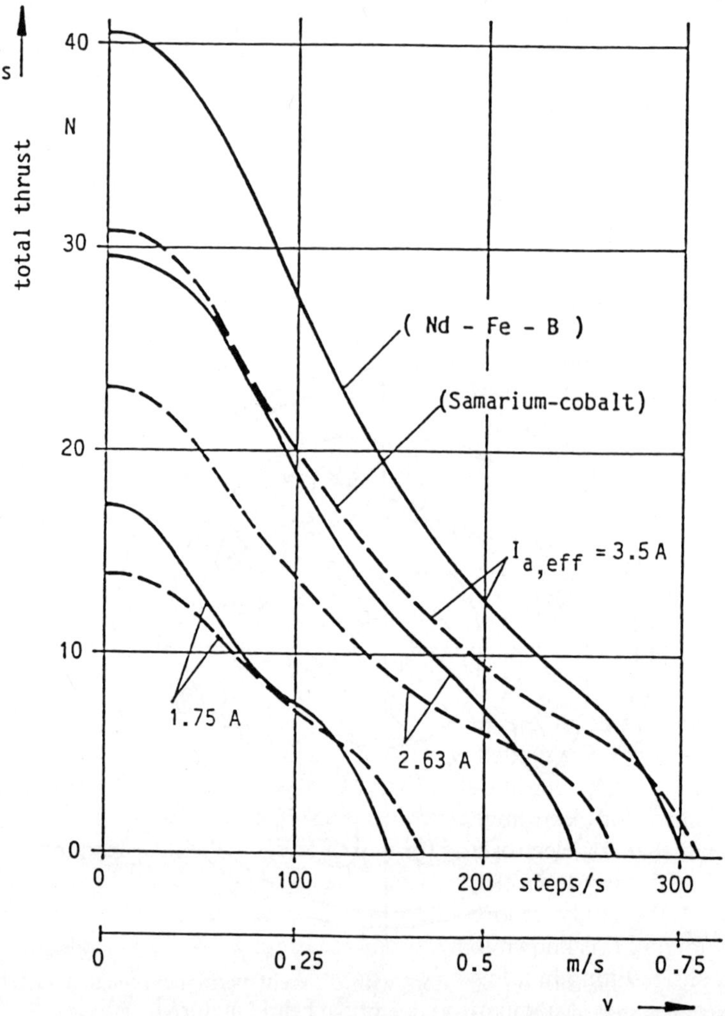

FIG. 10.34. Comparison of linear stepping motor characteristic for Nd–Fe–B and Sm–Co.

smaller than a full step. Special disc-type stepper motors were developed to cope best with microstepping [Pra88, Lan79]. The rotor of one motor design consists of a thin disc of Nd–Fe–B material axially magnetized with many poles of alternating polarity. In comparison with hybrid stepping motors, the air gap is relatively long which leads to linearization of the magnetic circuit. As a result, the disc-type stepping motor shows very low cogging torque and good linearity between current and torque [Bai60].

All the stepping motors described so far are of the polyphase type. Stepping motors with single-phase supply are widely used in watches, timers and counters.

3 Micromotors

A micromachine is defined as 'a very small apparatus for applying mechanical power and having extremely miniaturized parts, each with a definite purpose'. In the case of micromotors, microgenerators or microactuators the dimensions of the electromechanical energy converter are in the millimetre and submillimetre range.

Before performing any calculation it is important to consider which type of drive motor is best suited for miniaturization. We can look at different principles and different types of field distribution. We will focus on forces based on field-energy changes. The electromagnetic energy converter and the electrostatic devices belong to this group.

In both types of energy converter, most of the energy is stored in the gap. Therefore, this parameter is used to compare the electromagnetic with the electrostatic devices. The energy density in the gap of an electromagnetic device is

$$W_{\mathrm{m}} = \frac{1}{2\mu_0} B^2$$

With $B = 1.5$ T, limited by the saturation, the energy density is close to $90 \cdot 10^4$ J/m^3. The energy density of an electrostatic motor is given by

$$W_{\mathrm{e}} = \frac{1}{2} \varepsilon_0 E^2$$

E is the magnitude of the electric field in the air gap. Because of the electrical breakdown of air, E is limited in classical machines to $3 \cdot 10^6$ V/m. With this figure we obtain an energy density of 40 J/m^3. For macroscopic dimensions, the stored energy density for electromagnetic devices is therefore four orders of magnitude larger than that for electrostatic devices. However, Paschen reported a dramatic increase in the breakdown electric field as the product of gap thickness and pressure decreases. One study reports a sustained electric field of $3.2 \cdot 10^7$ V/m in a 12.5 μm airgap. It is known that any surface roughness will concentrate the electric field locally, resulting in a lower average breakdown field.

For the purpose of comparison, an electric field of $3 \cdot 10^7$ V/m is assumed. This results in an energy density of $4 \cdot 10^3$ J/m^3. A breakdown strength as high as $200 \cdot 10^6$ V/m is obtaned in SiO_2 thin films. In this case the energy density of the electric field reaches $700 \cdot 10^3$ J/m^3. For tiny dimensions, the torque density of electrostatic devices is obviously competitive with that of electromagnetic devices. However, the problem is the high voltage, required to drive the motor. For medical applications electromagnetic devices are therefore more suitable [Han94].

As pointed out, in the millimetre range electromagnetic devices are good candidates to produce torque or thrust. In order to reduce assembly problems, flat axial flux micromotors with etched planar windings of the brushless type have been built and put into operation. The rotor is an assembly of flat magnets or a thin disc made of high-energy product permanent magnets. While for larger motors, rotor position sensing devices have to be used, for micromotors the position sensor can be avoided. Just by

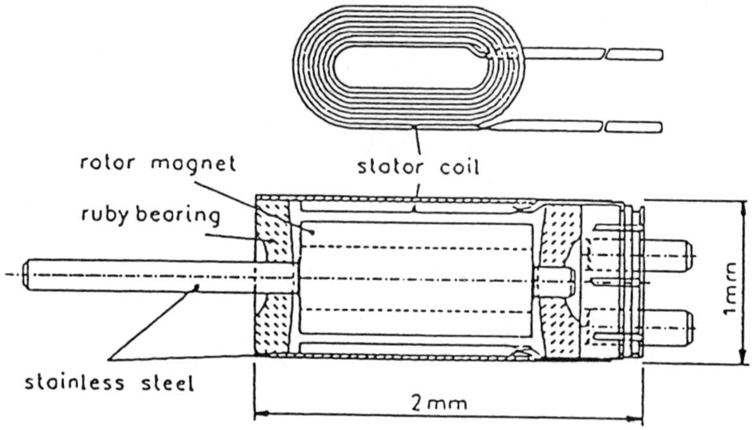

(a)

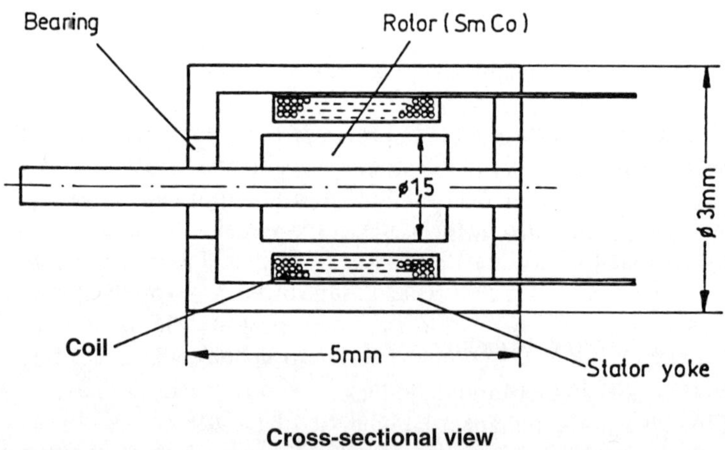

(b)

FIG. 10.35. Micromotors: (a) synchronous motor; (b) brushless motor.

handling the information from measured currents and voltages, the position of the rotor for miniature stepping motors can be estimated quite accurately.

To realize a high current layer in the micromotor, a multilayer etched winding is desirable. Four-layer etched windings have been investigated where the conducting material for the prototypes is gold with some addition of palladium. Extremely high accuracy is necessary in the course of processing multilayer etched windings using thick film technology. As substrate, different ceramic materials or glass is used. The cross-section of the conductors can range from 3750 μm^2 and 7500 μm^2 and the width between conducting elements varies between 150 μm and 200 μm. The current density may vary from 10^3 A/mm^2 to 10^4 A/mm^2. Thin ferromagnetic layers can also be realized using a ferrofluid with a permeability of approximately 10. This approach helps to reduce the effective airgap [Han94]. With the advent of high-temperature superconductors, further novel miniaturized actuators and motors are feasible. This material is ideal for planar winding structures in the millimetre range and therefore the combination with permanent magnets may create a new cluster of micromotors.

The brushless designs lead, depending on the electronic control, to brushless d.c. motors or synchronous motors. The two-pole or four-pole, two-phase designs dominate for micromotors with radial magnetic fields. Two examples of these motors are shown in Fig. 10.35. The torque constant, depending on the design features for the synchronous motor ranges from $2 \cdot 10^{-7}$ Nm/A to $7 \cdot 10^{-7}$ Nm/A. The four racetrack stator coils, having eight turns of flattened 35 μm copper wire, are arranged symmetrically in the stator with ruby bearings. Assembly of such microdevices is rather complicated.

The range of applications for microdevices is extremely wide, starting from microconnectors, slide bearings, actuators and vibration sensors to micromotors. These micromechanical devices are useful for the adjustment of glass-fibers, laser mirror adjustment, medical- and bio-engineering applications. Microactuators are used in crosspoint switching arrays, electrostatic microvalves for wafer-scale gas chromatography and piezoelectrically pumped cavities for ink-jet printers. A large field of applications will be microsurgery, endoscopic surgery and demand cellular drug delivery. Much greater precision in surgery will be possible using microforceps or knives and high-speed microdrills in combination with special catheters. By inserting the micromotor intravascularly, surgery will be done without large openings of vessels.

4 Magnetization and assembly

The direction of magnetization should coincide with the anisotropy axis for the optimal value for the remanent induction to be achieved. The various grades of Nd–Fe–B alloy have a range of coercive field strengths and remanences, but magnetic fields from 1500 to 2000 kA/m (1.9 T to 2.5 T) are needed to saturate the coercivity of the magnets (Fig. 10.36).

In comparison, for the full development of the coercivity in $SmCo_5$, fields from 2500 to 3000 kA/m, are required, but for low-coercivity Sm_2Co_{17} magnets a field strength of about 1000 kA/m is sufficient, but complete magnetization of a high-coercivity 2:17 magnet requires field strengths from 4000 to 5000 kA/m. Alnico has a low coercivity, which makes it possible to magnetize the magnets in the assembled machine with fields between 100 and 400 kA/m which can be generated by built-in magnetizing windings. Nd–Fe–B and Sm–Co, in most applications, have to be magnetized before assembly, which raises the problem of handling magnetized parts.

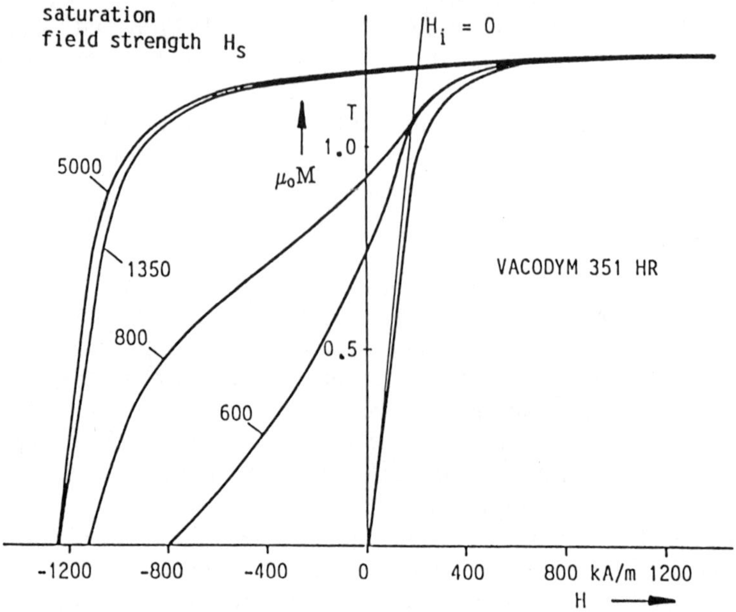

FIG. 10.36. Influence of magnetizing field strength on characteristic $M = f(H)$.

Table 10.7 Typical magnet geometries

Type of motor	Magnet geometry
linear motor	strip
axial field	disc
radial field	ring
stepper motor	disc; ring

Multipole magnets are widely used. There is a trend towards using one-piece magnets in a ring, disc or strip configuration. Discs less than 1 mm thick can be manufactured from Nd–Fe–B if necessary. Table 10.7 gives an overview of the shapes used for different types of permanent-magnet motor [Ric87, How87].

In order to magnetize these discs or strips properly, iron-cored or air-cored magnetizing tools must be developed. These provide a magnetizing field in excess of 1500 kA/m. Special fixtures have been designed to impart multipole patterns onto the permanent magnet components.

The big advantages of an air-cored magnetizer are that magnets can easily be removed from the device after magnetization. Iron-cored magnetizing tools are limited to a flux density of 2.2 T unless special pole pieces are used to concentrate

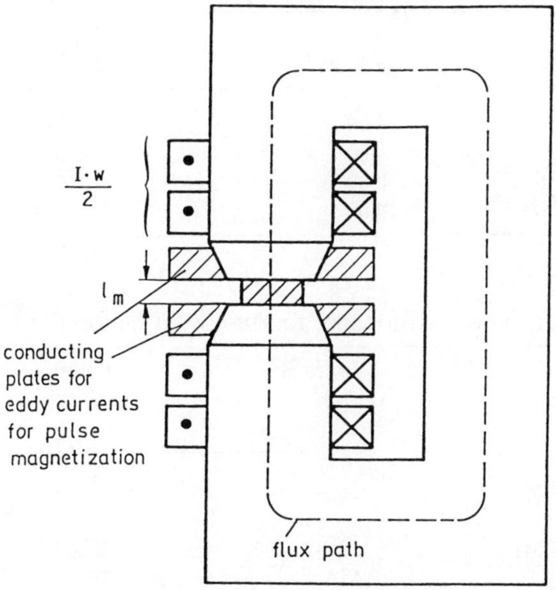

FIG. 10.37. Iron-cored magnetizer.

the flux (Fig. 10.37). Furthermore, the resulting magnets and the tool attract each other, and a special assembly jig may be required.

It can be difficult to magnetize short pole pitches with high accuracy. A good magnetizer is characterized by solid mechanical strength with wires well embedded in synthetic material to ensure electric insulation. A capacitor discharge circuit is used to generate a high-current pulse in the magnetizing coil. A short pulse time is necessary to prevent overheating, but eddy currents induced in the magnet material hinder fast penetration of the field into the magnet material. When designing a magnetizer for a new application, the interactions between pulse shape, pulse duration, magnetizer geometry, magnet geometry and material properties must all be considered [Ric87]. The penetration of the field into the magnet is reduced if its conductivity or its cross-section is increased. Therefore, short current pulses can be chosen for ferrites and magnets with a small cross-section, but magnet pieces with a larger cross-section require longer pulses. Magnetization up to saturation is produced by a magnetic field of strength H_s. The magnetic energy W_m is given by

$$W_m = \frac{\mu_0}{2} H_s^2 \cdot V_m$$

where V_m is the volume of the magnet. This equation will give approximate values because the losses due to electric resistances and eddy currents are not taken into

Table 10.8 Energy necessary for magnetization

Magnet type	H (kA/m)	W_m/V_m (J/m^3)
Alnico	400	$\sim 1.10^5$
Ferrite	800	$\sim 4.10^5$
Sm–Co	4000	$\sim 100.10^5$
Nd–Fe–B	2000	$\sim 25.10^5$

Table 10.9 Comparison of performance for air-cooled motors

Power (W)	Type	Efficiency (%)	Power/weight (W/kg)
15	DC (PM)	36–40	10–15
100	DC (PM)	55–82	30–60
	AC	57	35
3000	DC (PM)	85	160
	AC	80	75

DC (PM): permanent-magnet DC motor.
AC: induction motor.

account. In addition the volume of the magnetizing fixture might contribute to the losses (Table 10.8). About 10–100 times more energy is needed for the magnetization of rare-earth alloys than for the magnetization of Alnico or ferrites.

Conclusions

In addition to its superior magnetic characteristics, Nd–Fe–B has the advantages of greater strength and lower density compared with Sm–Co. The problems in using Nd–Fe–B in industrial drives, automobiles, static applications and actuators are sometimes its temperature limitations (usable only up to about 150 °C) and its tendency to corrode. With coatings such as aluminium chromate or zinc chromate, reasonable corrosion protection can be achieved. The new permanent-magnet motors offer greater efficiency and a better power-to-weight ratio than other types of motor. Table 10.9 gives a comparison of performance for industrial drives at rated speed of about 3000 rpm.

Prototypes have been developed with Nd–Fe–B magnets having power-to-weight ratio of up to 1000 W/kg and claims of much higher ratios under special cooling conditions have been published [Mil88].

Overall, this chapter illustrates the many good opportunities that exist for electrical machine designers arising from the excellent performance characteristics of Nd–Fe–B permanent magnets.

Bibliography

Electric motors and their controls, an introduction, T. Kenjo, Oxford Science Publishers, 1991 — excellent for beginners.

Electric motor handbook, B. J. Chalmers, Butterworths, London, 1988 — very useful for industrial applications.

Brushless servomotors — Fundamentals and applications, Y. Dale and S. Kinoshita, Clarendon Press, Oxford, 1990 — good overview of servomotors.

Magnetische Werkstoffe, H. Förster and R. Hanitsch, Verlag TÜV Rheinland, 1990 — information about magnetic materials for design engineers.

Stepping motors and their microprocessor controls, T. Kenjo, Clarendon Press, Oxford, 1984 — introduction to stepping motor control.

Computer-aided design in magnetics, D. A. Lowther and P. P. Sylvester, Springer Verlag, Berlin, 1986 — introduction to CAD of magnetic circuits.

Brushless permanent-magnet and reluctance motor drives, T. J. E. Miller, Clarendon Press, Oxford, 1990 — overview of brushless permanent-magnet motors.

Die Prüfung elektrischer Maschinen, W. Nürnberg and R. Hanitsch, Springer Verlag, Berlin 1987 — testing procedures for all types of electric motor.

Finite elements for electrical engineers, P. P. Sylvester and R. L. Ferrari, Cambridge University Press, Cambridge, 1986 — standard text on finite-element techniques.

REFERENCES

[Bai60] S. J. Bailey, *Control Engineering*, Dec. (1960), 97.

[Bin79] K. J. Binns, M. A. Jabbar and G. E. Parry, *Proc. IEE*, **126** (1979), 741.

[Bol81] R. Boll and L. Borek, *Siemens Forsch.- und Entwickl. Ber.*, **10** (1981), 83.

[Bou80] N. Boules, W. R. Canders and H. Weh *Archiv f. Elektrotechnik*, **62** (1980), 283.

[Bun87] E. Bunzel *Elektrie* (1987), 2, 57.

[Cha92] B. J. Chalmers, *Intl Aegean Conf. on Electrical Machines and Power Electronics*, Vol. 2/2 (1992), p. 399.

[Cha83] J. C. Chang, *VDI-Berichte*, **482** (1983), 17.

[Che88] Y. Cheng-Huan and S. A. Nasar, *Electric Machines and Power Systems*, **15** (1988), 381.

[Eva87] P. D. Evans, D. Brown and J. F. Eastham, *Proc. of Beijing International Conference on Electrical Machines* (1987), Beijing, China Academic Publisher, Book II, p. 729.

[Ham91] K. Hameyer and R. Hanitsch, *J. Magn. Magn. Mater.*, **101** (1991), 286.

[Ham92] K. Hameyer, *Beitrag zum automatischen optimalen Entwurf von elektromechanischen Wandlern am Beispiel eines Stellantriebs*, Reihe Elektrotechnik, Verlag Shaker, Aachen (1992).

[Ham94] K. Hameyer and R. Hanitsch, *Proc. Compumag*, Miami (1993).

[Han83] R. Hanitsch, *Bull. SEV/VSE*, **74** (1983), 23, 1344.

[Han86a] R. Hanitsch and J. C. Chang, *Proc. of 21st. UPEC*, London (1986), p. 371.

[Han86b] R. Hanitsch, U. Lorenz and D. Petzold, *Elektrische Energietechnik*, Springer, Berlin (1986).

[Han87a] R. Hanitsch and L. Neubauer, *Proc. of Beijing International Conference on Electrical Machines* (1987), Beijing, China Academic Publisher, Book II, p. 746.

[Han87b] R. Hanitsch, IEE Conf. Publn, **282** (1987), p. 255.
[Han89] R. Hanitsch, C. S. Park and K. Hameyer, *J. Magn. and Magn. Mater.*, **80** (1989), 131.
[Han91] R. Hanitsch, L. Walkhoff, R. Belmans and W. Geysen, *Fifth Intl Conf. on Electrical Machines and Drives*, IEE Conf. Publ. No. 341 (1991), p. 126.
[Han92] R. Hanitsch, *Intl Aegean Conf. on Electrical Machines and Power Electronics*, Vol, 2/2 (1992), p. 415.
[Han93] R. Hanitsch and D.-S Choi, *Proc. of PCIM* (1993), Nürnberg, p. 456.
[Han94a] R. Hanitsch and N. Parspour, *Proc. IEEE IECON '94*, (1994), 1310.
[Han94b] R. Hanitsch, *RGE*, No. 8 (1994), p. 9.
[Hen86] G. Henneberger, *ICEM '86, München, Proc. Part 3*, p. 1079.
[Hin74] W. E. Hinds and B. Nocito, *Proc. Third Annual Symposium on Incremental Motor Control Systems and Devices*, University of Illinois (1974), p. W1–W10.
[Hos80a] V. B. Honsinger, *IEEE Trans. on Power Apparatus and Systems*, **PAS-99**, No. 4 (1980), p. 1503.
[Hos80b] V. B. Honsinger, *IEEE Trans. on Power Apparatus and Systems*, **PAS-99**, No. 4 (1980), p. 1510.
[Hop87] D. M. Hopstock, *Proc. of 9th Intl. Workshop on Rare-Earth Magnets*, Bad Soden (1987), p. 667.
[How87] D. Howe and T. S. Birch, *Proc. of 9th Intl. Workshop on Rare-Earth Magnets*, Bad Soden (1987), p. 65.
[How88] D. Howe, *Proc. of Actuator 88, VDE*, Berlin (1988), p. 135.
[Itt91] H. G. Ittstein, Dissertation, Technical University of Berlin, D83 (1991).
[Jab87] M. A. Jabbar, *Proc. of Electric Energy Conf.*, Adelaide (1987), p. 272.
[Jen87] M. K. Jenkins, T. S. Birch and D. Howe, IEE Conf. Publn 282 (1987), p. 270.
[Jia88] J. Jiang, H. May and H. Weh, *Etz-Archiv*, **10** (1988), 249.
[Ken94] T. Kenjo, *Stepping motors and their microprocessor controls*, Clarendon Press, Oxford (1994).
[Kum88] K. Kumar, *J. Appl. Phys.*, **63** (1988), R13–R57.
[Kun85] K. Kuntze, D. Kohake, R. Beranek, S. Stieler and C. Heiden, *Journal de Physique*, **46**, (1985), C6–253.
[Lan79] T. Langley and H. K. Kidd, *Proc. Intl Conf. on Stepping Motors and Systems*, Leeds (1979), p. 32.
[Lia94] Y. Liao and T. A. Lipo, *Electric Machines and Power Systems*, **22** (2)(1994), 259.
[Lis76] M. Liska and B. Ulrich, *Conf. on Small Electrical Machines*, IEE, London (1976), p. 15.
[Mar88] M. Marinescu, *Etz-Archiv*, **10** (1988), 3, 83.
[May88] R. Mayer, B. Orlik and J. Poschadel, *Etz-Archiv*, **10** (1988), 3, 73.
[Mec91] B. C. Mecrow and A. G. Jack, *Fifth Intl Conf. on Electrical Machines and Drives*, IEE Conf. Publ No. 341 (1991), p. 26.
[Mel88] P. H. Mellor, K. J. Binns and M. Al-Taee, IEE Conf. Publn **291** (1988), p. 265.
[Mil84] T. J. E. Miller, *IEEE Trans. on Power Apparatus and Systems*, **PAS-103** (1984), 1822.
[Mil88] T. Miller, *Power Eng. Journal*, **2** (1988), No. 1, p. 55.
[Moh83] A. Mohr, *Etz-Archiv*, **5** (1983), 385.
[Moh83] A. Mohr, *Etz-Archiv*, **5** (1983), 8. 249.
[Ort87] E. S. Ortoli, D. Pavlik and J. R. Repp, *Proc. of 9th Intl Workshop on Rare-Earth Magnets*, Bad Soden (1987), p. 41.
[Pfa84] G. Pfaff and L. Sach, *Tz für Metallbearbeitung*, **78** (1984), 15.
[Pra88] F. Prautzsch, *Proc. of Actuator 88, VDE*, Berlin (1988), p. 191.
[Rah88] M. A. Rahman and A. M. Osheiba, *Electr. Machines and Power Systems*, **14** (1988), 195.

[Ric87] K. M. Richardson and E. Spooner, IEE Conf. Publn 282 (1987), p. 250.

[Sch87] R. F. Schiferl, R. S. Colby and D. W. Novotny, *Proc. of Electric Energy Conf.*, Adelaide (1987), p. 286.

[Seb86] T. Sebastian, G. R. Slemon and M. A. Rahman, *IEEE Trans. Magn.*, **MAG-22** (1986), 1069.

[She80] M. A. Shehata and R. Hanitsch, *Electrical Review*, **206** (1980), 45.

[Shi87] T. Shimoda, *Proc. of 9th Intl Workshop on Rare-Earth Magnets*, Bad Soden (1987), p. 651.

[Sit87] A. M. Sitzia, B. J. Chalmers, IEE Conf. Publn **282** (1987), p. 260.

[Tan81] G. Tan, D. Howe and T. S. Birch, IEE Conf. Publn **202** (1981) p. 125.

[Tok87] M. Tokunaga, M. Endoh and H. Harada, *9th Intl Workshop on Rare-Earth Magnets*, Bad Soden (1987), p. 477.

[Tyr88] C. Tyren, *Proc. of Actuator 88, VDE*, Berlin (1988), p. 500.

[Weg76] K. Wegener, Dissertation, Technical University of Berlin (1976).

[Weh86] H. Weh and H. May, *Proc. ICEM '86, München*, Part 3, 1107.

[Zes86] D. Zesheng, I. Boldea and S. A. Nasar, *IEEE Trans. Magn.*, **MAG-22**, (1986), p. 107.

[Zim83] P. Zimmermann, *Wt-Zeitschr.ind. Fert.*, **73** (1983), 629.

11

Actuators

D. Howe

The emergence of high-energy rare-earth permanent magnets has facilitated the development of new and improved forms of limited motion actuators, both linear and rotary. Applications span all market sectors, from disk head drives and industrial robots to pneumatic pumps and automatic car door locks. Thus, requirements regarding power and displacement, speed of response, positional accuracy and duty cycle are very wide. In general, however, all actuators provide a force acting directly to move a load, and thus tend to be simpler electromagnetic structures than rotary motors with gears.

Although there are many configurations, to match the motion and space envelope requirements of the application, the basic classes of actuator are moving-coil, moving-iron and moving-magnet. Within these the air-gap length may either remain constant with travel, the useful magnetic force being parallel to the air-gap and perpendicular to both the direction of current flow and the field of the permanent magnet, or vary with travel, the useful magnetic force being exerted normally across the air-gap. Linear actuator strokes typically range from a few microns to several centimetres, with equivalent displacements for rotary devices. Acceleration rates can exceed 20 g. Operation can be broadly divided into latching, oscillatory, open-loop and closed-loop servo-controlled, when the positional accuracy, resolution and repeatability are determined by the associated position sensor and servo-amplifier. Examples of three practical actuation devices, shown in Fig. 11.1, are the following:

1. Moving-coil (constant air-gap) actuator — often referred to as a voice-coil actuator since such actuators were originally used for loudspeakers and similar vibration transducers. However, they are now employed for a wide variety of applications, such as read/write head positioning on computer disk drives which currently represents the largest application for Nd–Fe–B magnets, and beam focusing in optical systems [Kim87].

   In general, moving-coil actuators possess a low inertia, but they require flying lead connections. The active length of the moving-coil can either be longer than that of the magnet (or the working air-gap in the case of a flux focusing design) in order to improve the force capability, or shorter, so as to reduce the moving mass and the electrical time constant. In both cases, the force can be enhanced by employing flux concentration, as illustrated in Fig. 11.1a. The force may vary with the coil displacement, depending on the location of the magnets and the combined effect of the armature reaction field resulting from current in the coil and the reluctance force which results from a variation in the coil inductance. There is no preferred detent position (i.e. alignment position when the coil is

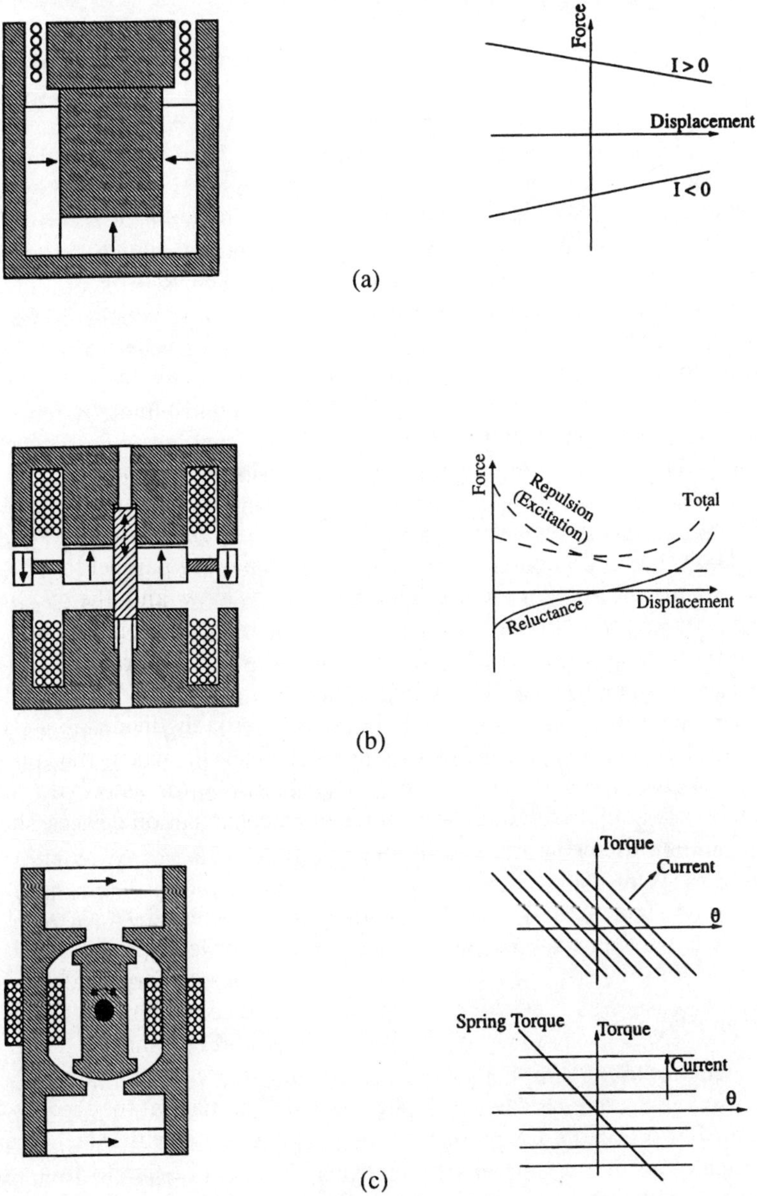

FIG. 11.1. Alternative types of permanent-magnet limited-motion actuator and typical static performance characteristics: (a) flux-focusing linear moving-coil actuator (constant air-gap); (b) bistable linear moving-magnet actuator (variable air-gap); (c) rotary moving-iron actuator (constant air-gap).

unexcited), and no force ripple (because the actuator is slotless), and moving-coil actuators do not suffer significantly from the effects of hysteresis.

2. Moving-magnet (variable-air-gap) actuator — which overcomes the major weakness of the conventional solenoid moving-iron actuator, in which the developed force is weakest at the beginning of motion when the air-gap is at its largest, which in turn limits the achievable speed and stroke. When the coils in the bistable configuration shown in Fig. 11.1b are unexcited the moving-magnet assembly is held in a stable position at the limit of its travel.

   Application of a current of the right polarity to the appropriate coil results in a repulsive force being exerted on the magnet, which is a maximum at the beginning of its motion and which is augmented by a reluctance force as the magnet passes through its mid-travel point in being attracted to the opposite pole structure. Thus, fast motion can be achieved over relatively long strokes, without the need for flying leads [Leq90]. However, relative motion between the magnet and the iron circuit can introduce significant hysteresis, which can ultimately limit the dynamic performance of moving-magnet devices.

3. Moving-iron (constant-air-gap) actuator — of which the topology shown in Fig. 11.1c exhibits the highest torque per volume capability [Bol90]. Its design has usually been orientated towards the production of a family of torque–displacement characteristics which possess considerable stiffness (i.e. $dT/d\theta > 0$) in order that it can be operated in open-loop mode. However, low stiffness designs for use in open-loop/return spring and closed-loop servo-controlled systems are also possible. The permanent magnets produce a polarizing field which accounts for the majority of the torque-producing flux, the control windings effectively serving to redistribute the flux and displace the rotor, in accordance with the minimum reluctance alignment principle.

The foregoing form only a representative selection of the many actuator types and topologies. Whilst static performance characteristics reflect their specific force or torque producing capability, it is generally the dynamic performance of an actuator which is of most significance.

1 Basic principles

Despite their apparent simplicity, the design and analysis of limited motion permanent-magnet actuators is extremely challenging. It is complicated by the spatial distribution of the magnetostatic field, nonlinearities in the magnetization characteristic of soft magnetic materials, field variations caused by the excitation current, eddy current effects caused by time-varying fields and motion, inductance variations with displacement and coil resistance changes due to heating. Furthermore, given the wide variety of actuators and performance requirements, it is difficult to generalize regarding design strategies and analysis techniques.

All permanent-magnet actuators exploit one or possibly two fundamental force production mechanisms, namely, excitation force — which is essential for energy

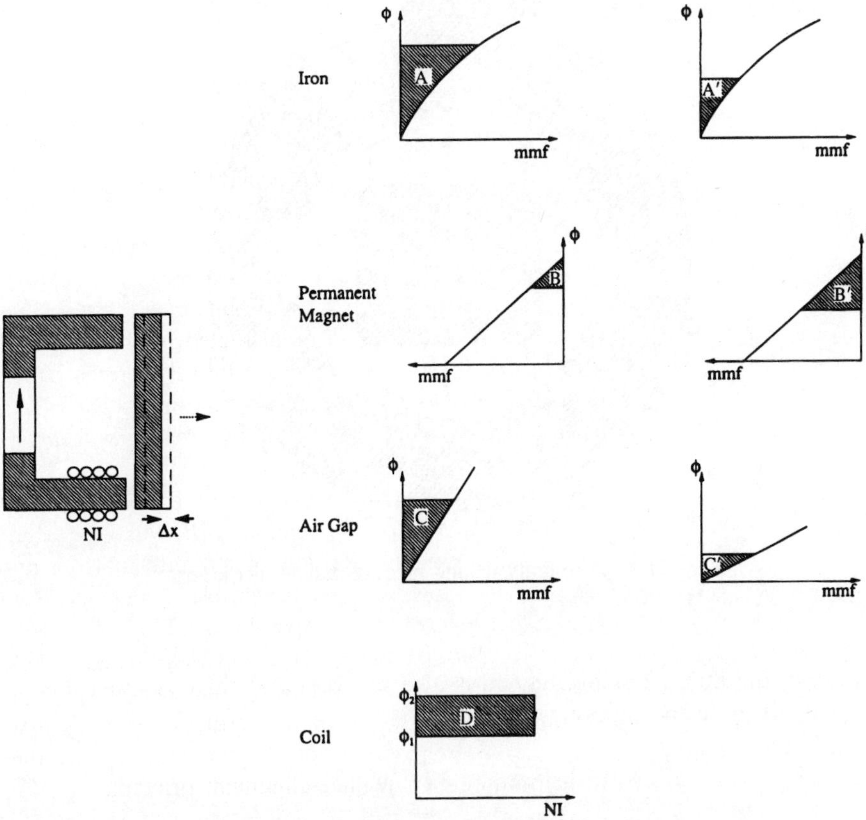

FIG. 11.2. Flux-mmf diagrams for calculation of energy changes in nonlinear magnetic circuit.

conversion, and reluctance force — which may be present either as an undesirable component or as an additional mechanism but which cannot, alone, facilitate controlled motion. Moreover, force is essentially a static effect, and energy conversion cannot occur until there is some movement. The most appropriate methods for calculating the net force and the dynamic performance depend on the actuator geometry and the extent of nonlinearities at both actuator and system levels, as well as the mode of operation.

Energy conversion results in an interchange between electrical and mechanical energy through the coupling magnetic field, and the most general method for calculating forces and torques is from changes in energy. With reference to Fig. 11.2, the flux/mmf diagrams show the stored magnetic energy associated with the various sections of a simple moving-iron actuator before and after the air-gap length has been increased by Δx. The net change in stored energy is given by the difference in the shaded areas, i.e. $(A+B+C)-(A'+B'+C')$. As a result of the

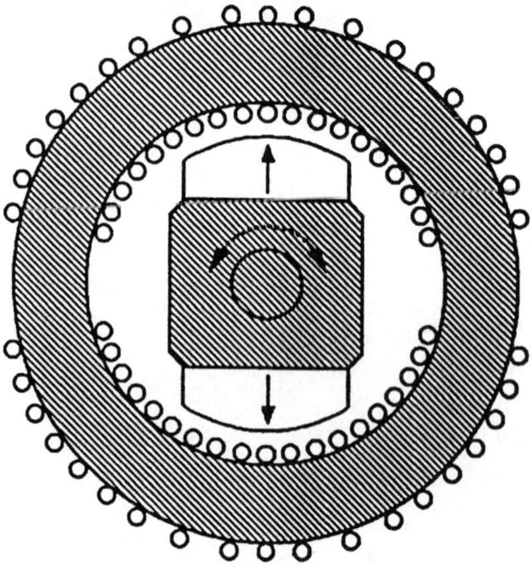

FIG. 11.3. Toroidally wound moving-magnet actuator.

movement, the flux φ linking the coil is also reduced, and there is a net flow of electrical energy to the supply given by:

$$\int_{t_1}^{t_2} vi\mathrm{d}t = \int_{\varphi_1}^{\varphi_2} Ni\mathrm{d}\varphi$$

which is shown by the shaded area D on the coil flux/mmf diagram. For conservation of energy, the mechanical energy $\Delta E = F\Delta x$ must equal the change in energy stored in the magnetic field minus the change in electrical energy. Hence the average force is given by:

$$F = \frac{(A+B+C)-(A'+B'+C')-D}{\Delta x} = \frac{\Delta E}{\Delta x}$$

Alternative expressions for the force may be developed from the same considerations in terms of either the change in co-energy with displacement or from electric circuit concepts. Co-energy E' is defined as $\int_{\mathrm{v}}\int B\mathrm{d}H\mathrm{dv}$, so the difference between the change in stored energy in the field and the electrical work done is equal to the change in co-energy, and the force can be calculated from:

$$F = \frac{\Delta E'}{\Delta x}$$

where the total co-energy is equal to the sum of components evaluated for all parts of the magnetic circuit. Alternatively, the magnet in the rotary moving-magnet actuator of Fig. 11.3 could be represented as a coil, with an mmf equal to the product of its coercivity and active length, in series with a permeance, having the same area and length as the magnet but with its demagnetization characteristic displaced to the first-quadrant. The force can then be expressed as:

$$F = \frac{1}{2}I^2\frac{dL}{d\theta} + \frac{1}{2}I_m^2\frac{dL_m}{d\theta} + II_m\frac{dM}{d\theta}$$

The first two terms are reluctance force components — which are negligible in this particular case, and the third term is the excitation force, which can be shown to be equivalent to the Lorentz force equation:

$$F = BIL$$

where L is the active length of the conductors which carry the current I normal to the field produced by the magnets.

In order to predict the dynamic behaviour of a given actuation system, the system parameters, including resistances and inductances, and mass, stiffness and friction, must be evaluated, and the electrical and mechanical equations of motion must be solved. The choice of models and simplifying assumptions will depend greatly on the system and its application, as well as on the degree of refinement required in the solution. Some of the considerations will be highlighted in the following application studies.

2 Application studies

2.1 Voice-coil motor for disk drive

Numerous designs of linear and rotary voice-coil actuator have been developed over the years as disk drives have shrunk in size, increased in storage capacity and accessed data at ever faster rates. Future advances in thin-film disk technology and read/write head devices will necessitate even faster spindle motors and voice-coil positioning actuators. Hence, high-energy magnets are a key enabling technology. Indeed, the main impetus behind the development of grades of Nd–Fe–B, with an energy product of 400 kJ/m^3 has been the need to make very thin drives for portable computers.

Figure 11.4a shows a typical cylindrical voice-coil motor. By controlling the current, any desired track can be accessed. The required velocity profile will vary depending on the distance which the head is required to travel, but it is typically triangular, which requires a constant current for uniform acceleration. The faster the current increases, the shorter will be the access time. However, since the self-inductance of the coil would limit the rate of rise of current significantly, disk drive VCMs feature a copper sleeve (shorted-turn), concentric to the moving-coil but fixed to the pole structure. Its function is to reduce the coil time constant and thereby the access time, since when a voltage is applied to the coil, eddy currents are induced in the sleeve by changes in the coil current, and these produce a magnetic field which opposes the inducing field. Thus, the effective inductance is reduced [Wag92].

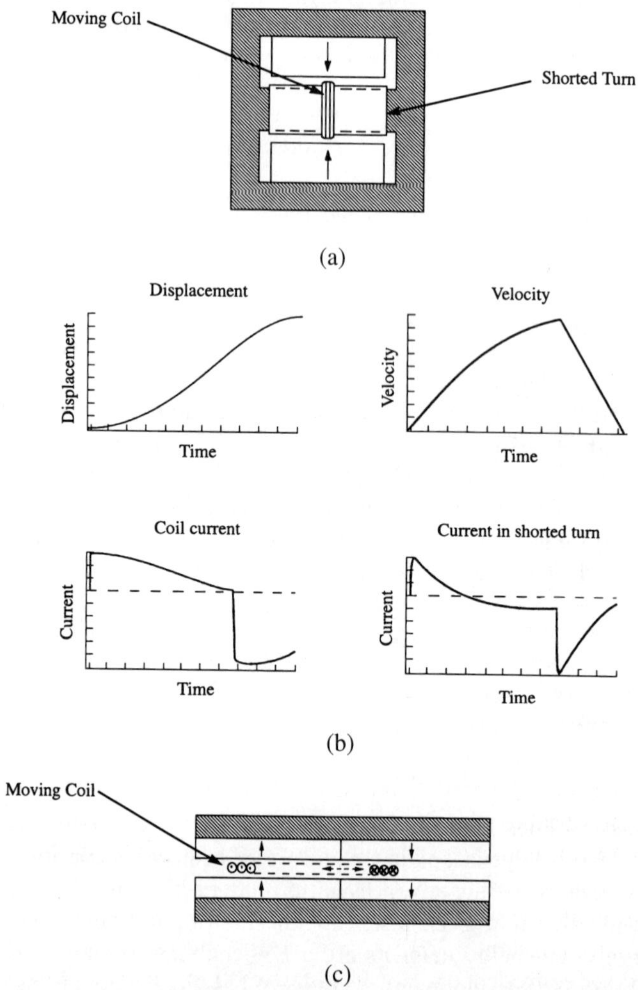

FIG. 11.4. Cylindrical voice-coil motor with shorted turn (a), and typical dynamic waveforms (b). An alternative flat coil design (c).

The transient dynamic performance of the actuator can be obtained by solving the differential equations derived from a transformer equivalent circuit model and the equation of motion, namely,

$$Ki_{\mathrm{e}} = m\frac{\mathrm{d}v}{\mathrm{d}t} + k_{\mathrm{d}}v + F_{\mathrm{f}}$$

$$V = R_{\mathrm{c}}i_{\mathrm{c}} + L_{\mathrm{c}}\frac{\mathrm{d}i_{\mathrm{c}}}{\mathrm{d}t} - M\frac{\mathrm{d}i_{\mathrm{s}}}{\mathrm{d}t} + K_{\mathrm{e}}v$$

$$O = R_{\mathrm{s}}i_{\mathrm{s}} - M\frac{\mathrm{d}i_{\mathrm{c}}}{\mathrm{d}t} + L_{\mathrm{s}}\frac{\mathrm{d}i_{\mathrm{s}}}{\mathrm{d}t}$$

where K, K_d and K_e are the actuator force constant, the damping constant and the back-emf constant respectively; m and F_f are the moving mass and the friction force respectively; R_c, L_c and R_s, L_s are the resistance and inductance of the moving coil and shorted-turn respectively; and M is the mutual inductance between the coil and shorted turn. Figure 11.4b shows typical dynamic profiles.

An alternative VCM design uses a flat square coil attached to a lever. The coil then moves in an arc in the air-gap formed by two adjacent pairs of thin magnets, Fig. 11.4c. Due to the large effective air-gap the coil has an inherently low inductance. This is the design commonly used in personal computer disk drives.

2.2 *Reciprocating moving-magnet actuators*

A wide range of electromechanical devices, ranging from diaphragm air-compressors and Stirling cycle cryo-coolers to fuel pumps and artificial hearts, employ linear reciprocating actuators [Cla95]. The normal mode of operation is to drive the actuator at the mechanical resonant frequency so that no work is done in accelerating/decelerating the moving mass and optimum efficiency and displacement are obtained. The mechanical resonant frequency is a function of both the inertia and compliance of the system. Typically, the required stroke varies between 4 mm and 8 mm, at frequencies between 20 Hz and 60 Hz.

The most suitable actuator topology is the moving-magnet type, since it has a high specific force capability, no flying leads and good thermal dissipation, and for which the high-energy product to density ratio of Nd–Fe–B magnets is highly advantageous. Figure 11.5a shows an axi-symmetric 2-pole configuration, for which the design methodology must take account of the electromechanical system and its inherent nonlinearities, predominantly the stiffening spring characteristic of the load. A full state-variable system simulation would be very lengthy, so it is more convenient to use an equivalent circuit representation of the mechanical system during electromagnetic design synthesis optimization.

In order to estimate the system performance over a range of operating frequencies, the derived actuator constants, electrical parameters and moving mass are coupled to a mass–capacitance equivalent circuit which represents the mechanical load, as shown in Fig. 11.5b and Table 11.1. In most reciprocating actuation systems, the load exhibits a highly nonlinear stiffening spring characteristic. For example, in the case of

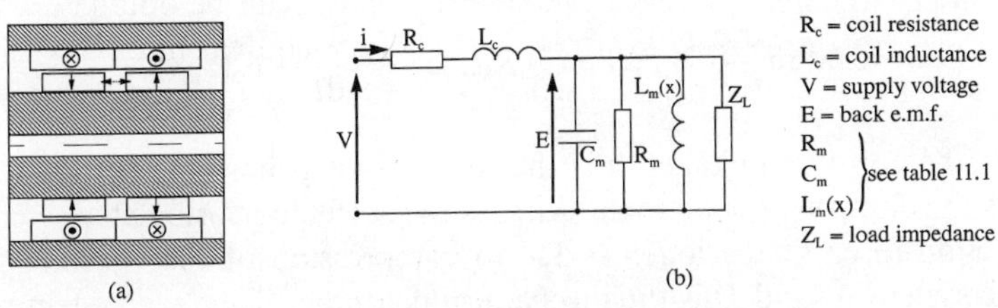

FIG. 11.5. Axi-symmetric reciprocating moving-magnet linear actuator (a), and electrical analogue of system (b).

Table 11.1 Mass–capacitance analogues for electrical equivalent circuit

Electrical	Mechanical	
Current	Force, F	
Voltage	Velocity, V	
Inductance, $L_m(z)$	Spring rate, $K_s(z)$	$L_m(z) = K^2/K_s(z)$
Capacitance, C_m	Mass, m	$C_m = m/K^2$
Conductance, I/R_m	Damping, D	$R_m = K^2/D$
Impedance, Z_L	Pneumatic load/losses	

K is the (force/back-emf) constant of the actuator.

a diaphragm air-compressor this is due to the deformation of the pumping membrane and its associated elasticity, the suspension springs — which offer a high transverse stiffness but a nonlinear axial spring rate, and the associated gas spring rate. In general, the spring rate has the form:

$$K_s(z) = a + bz^2$$

where z is the axial displacement, and the constants a and b can be derived from measured force–displacement characteristics.

A typical design goal is to minimize the copper loss and overall device mass, whilst also minimizing the volume of the magnet and meeting the thrust and stroke requirements.

Although the equivalent circuit approach can provide reasonable estimates of the displacement, input current, copper loss and system efficiency, a more accurate prediction of the dynamic performance can be obtained from a nonlinear state-variable model, the differential equations of which are:

$$v(t) = \mathrm{i}(t)R + e(t)$$

$$\text{where } e(t) = \frac{\delta\lambda}{\delta z}\frac{\mathrm{d}z}{\mathrm{d}t} + \frac{\delta\lambda}{\delta i}\frac{\mathrm{d}i}{\mathrm{d}t}$$

$$\text{and } m\frac{\mathrm{d}^2 z}{\mathrm{d}t^2} = F_e(i, z) - K_s(z)z - D\frac{\mathrm{d}z}{\mathrm{d}t} - F_c(z, p)$$

where λ is the coil flux-linkage, m is the total moving mass of the armature and load, F_e is the electromagnetic force, $K_s(z)$ is the stiffening spring characteristic, F_c is the reaction force of the load — due to compression of the working gas in a pneumatic system — and D is the mechanical damping. Since the system resonant frequency of a compressor is a function of the load pressure, due to the gas spring rate, the state-variable simulation must include a model of the compression system.

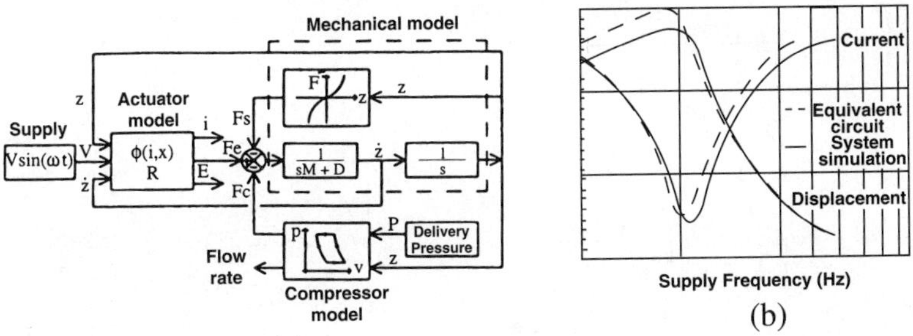

(a) (b)

FIG. 11.6. State-variable simulation block diagram of reciprocating diaphragm air-compressor, and predicted frequency responses.

For example, for a positive displacement air-compressor, a reversible polytropic compression cycle could be used, that is,

$$p \cdot v^n = \text{constant}$$

where p is the gas pressure, v is the compression space volume, and n is the compression index. Thus, for a given load delivery pressure, the output flow rate and the reaction force F_c on the delivery membrane can be calculated.

A typical simulation block diagram of the system is shown in Fig. 11.6a. Time-domain solutions of the differential equations which govern the electromagnetic, mechanical and pneumatic models and span a range of frequencies enables the displacement and current frequency response characteristics of Fig. 11.6b to be obtained.

2.3 *Moving-iron actuator for optical scanning system*

Most optical scanning systems use a moveable mirror and focusing optics to scan the object plane and focus the incoming light onto a sensor, as shown in Fig. 11.7a. However, this imposes severe demands on the actuator and drive which are used to position the mirror, since, typically, a raster scanning mode of operation may require an output shaft to follow the position profile shown in Fig. 11.7b, with a

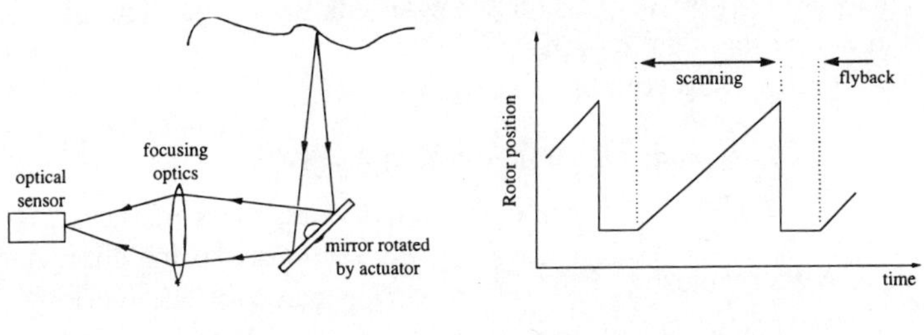

(a) (b)

FIG. 11.7. Optical scanning system, with typical actuator position–time waveform.

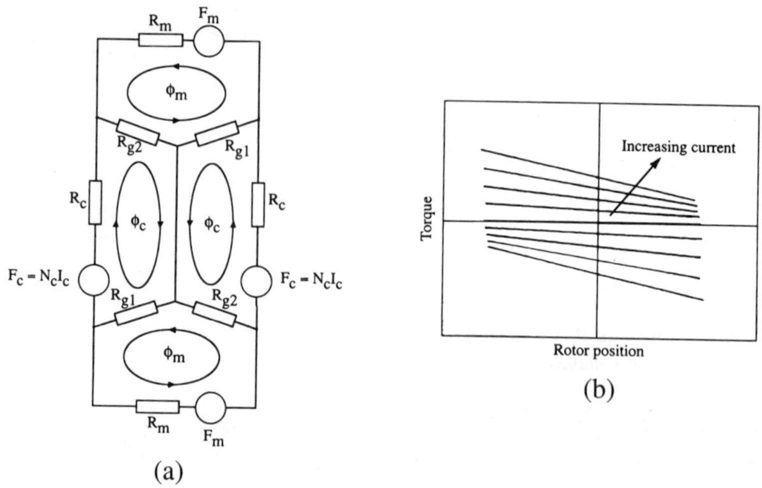

FIG. 11.8. Simple magnetic equivalent circuit of moving-iron actuator (a), and typical 'idealized' torque–displacement characteristics (b).

period of 20 ms and a flyback time of around 2–3 ms, whilst maintaining a linearity during the scan of about 0.1%.

Since the moving-iron actuator topology of Fig. 11.1c is capable of a very high torque-to-inertia ratio it is used extensively in such scanning systems. The form of its torque–displacement characteristics can be deduced from the approximate magnetic equivalent circuit shown in Fig. 11.8a, in which the magnet is represented as a source of mmf $F_m = H_{cn}l_m$ in series with a reluctance $R_m = \dfrac{l_m}{\mu_0\mu_r A_m}$, where A_m and l_m are the cross-sectional area and length of the magnets, and the rotor iron reluctance is assumed to be zero (although the analysis can be refined to account for a nonzero reluctance). The air-gap reluctances are:

$$R_{g_1} = \frac{g}{\mu_0 r(\alpha - \theta)w} \text{ and } R_{g_2} = \frac{g}{\mu_0 r(\alpha + \theta)w}$$

where α is the angle of overlap between the salient stator and rotor poles at zero displacement, θ is the rotor displacement angle, g is the air-gap length, r is the air-gap radius and w is the axial length.

The corresponding loop equations are simply:

$$H_{cn}l_m = \varphi_m(R_m + R_{g1} + R_{g2}) + \varphi_c(R_{g2} - R_{g1})$$
$$N_cI_c = \varphi_m(R_{g2} - R_{g1}) + \varphi_c(R_c + R_{g1} + R_{g2})$$

the solution of which yields:

$$\varphi_m = \frac{2aRN_cI_c + [2R + R_c(1 - a^2)]H_{cn}l_m}{4R^2 + 2R(R_c + R_m) + R_cR_m(1 - a^2)}$$

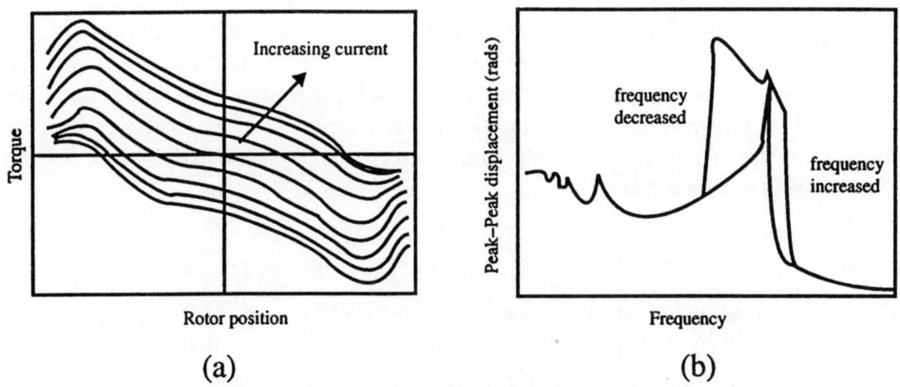

FIG. 11.9. Typical practical torque–displacement characteristics of polarized moving-iron actuator (a), and resulting dynamic frequency response (b).

$$\varphi_c = \frac{2aRN_{cn}I_m + [2R + R_m(1 - a^2)]N_c I_c}{4R^2 + 2R(R_c + R_m) + R_c R_m(1 - a^2)}$$

where $R = \dfrac{g}{\mu_0 r\alpha w}$ and $a = \dfrac{\theta}{\alpha}$

Thus, $B_{g1} = \dfrac{(\varphi_m - \varphi_c)}{r(\alpha - \theta)w}$; $B_{g2} = \dfrac{(\varphi_m + \varphi_c)}{r(\alpha + \theta)w}$; $B_m = \dfrac{\phi_m}{A_m}$

from which the co-energies associated with the air-gaps and magnets can be calculated. For example, if $I_c = 0$ and $R_c = 0$, so that φ_m is constant, the total co-energy is associated only with the four air-gaps (and also equals the stored magnetic energy), and is:

$$E' = 2\frac{gwr}{2\mu_0}[B_{g1}^2(\alpha - \theta) + B_{g2}^2(\alpha + \theta)]$$
$$= \frac{g}{\mu_0 wr}\left[\frac{(\varphi_m - \varphi_c)^2}{(\alpha - \theta)} + \frac{(\varphi_m + \varphi_c)^2}{(\alpha + \theta)}\right]$$

Thus,

$$F = \frac{\Delta E'}{\Delta\theta} = 0$$

Similarly, the lumped magnetic circuit approach can be used to deduce performance characteristics when the coils are excited, for both linear and nonlinear operation, Fig. 11.8b. However, it is adequate only for relatively low levels of saturation. In practice, finite element analysis is required to account for the effects of saturation, flux leakage and fringing, which can result in significant nonlinearities in the inductance and emf constant, Fig. 11.9a. In turn, these can severely affect the overall system dynamic performance. For example, Fig. 11.9b shows a predicted frequency response characteristic, assuming a low level of mechanical damping so as to highlight some of the operational problems associated with the dynamic operation of nonlinear systems. These include jump resonance phenomena, when the actuator may resonate at two entirely different amplitudes depending on its immediate operating history, and also resonances which occur due to the generation of harmonic torques.

2.4 *Bistable polarized moving-iron actuator for proportional control valve*

There are numerous possible topologies of bistable actuator, the configuration shown in Fig. 11.10a being a variant on the one shown in Fig. 11.1b for operating a hydraulic proportional control valve. Since many such applications demand a fast response, a transient time-domain analysis of the complete electromechanical system is necessary in order to accurately account for the effects of induced eddy currents, which cause the electromagnetic force on the moving plunger to be a function not only of its position and the excitation current but also their derivatives [Smi94]. This is achieved by coupling a transient finite element model of the actuator to other functional models representing the electrical and mechanical elements of the system, and simultaneously solving the governing differential equations, namely,

$$v(t) = i(t)R + \frac{\delta\lambda}{\delta z}\frac{\mathrm{d}z}{\mathrm{d}t} + \frac{\delta\lambda}{\delta i}\frac{\mathrm{d}i}{\mathrm{d}t}$$

$$m\frac{\mathrm{d}^2 z}{\mathrm{d}t^2} = F_{\mathrm{e}}\left(i, \frac{\mathrm{d}i}{\mathrm{d}t}, z, \frac{\mathrm{d}z}{\mathrm{d}t}\right) - F_{\mathrm{p}} - D\frac{\mathrm{d}z}{\mathrm{d}t}$$

where F_{p} is a static pre-load, and

$$\frac{\delta}{\delta r}\left(\frac{1}{\mu}\frac{\delta A_\theta}{\delta z}\right) + \frac{\delta}{\delta z}\frac{1}{\mu}\left(\frac{\delta A_\theta}{\delta r} + \frac{A_\theta}{r}\right) = -J_{\mathrm{s}} + \frac{\delta M_r}{\delta z} - \frac{\delta M_z}{\delta r} + \sigma\frac{\delta A_\theta}{\delta t}$$

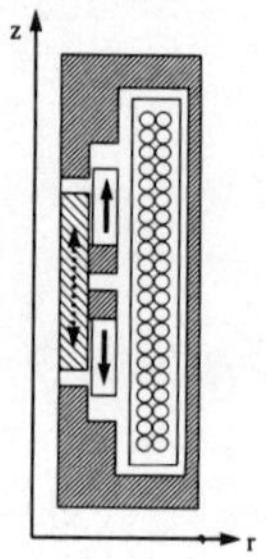

(a)

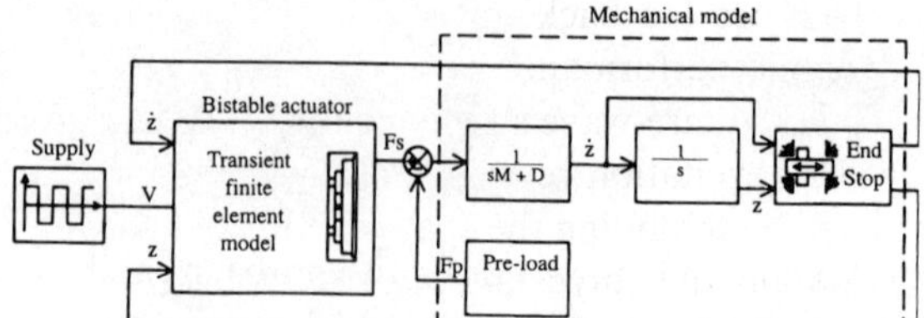

(b)

FIG. 11.10. Polarized moving-iron actuator (a), and dynamic simulation model (b).

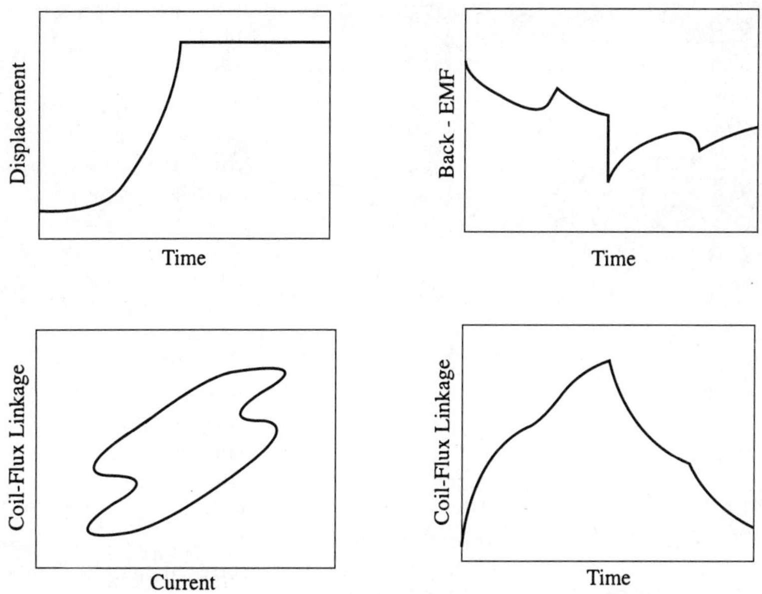

FIG. 11.11. Dynamic performance characteristics of polarized moving-iron actuation system.

where A_θ is the circumferential component of the vector magnetic potential (assuming an axi-symmetric device), J_s is the excitation current density and M_r and M_z are the magnetization components, the source term $\sigma(\delta A_\theta/\delta t)$ being associated with the induced eddy currents arising from motion as well as a time-varying excitation source.

Figure 11.10b shows a typical simulation block diagram of the electromechanical system, in which the functional block labelled 'bistable actuator' solves for the nonlinear transient electromagnetic field corresponding to the instantaneous position and velocity of the plunger, and calculates the resulting electromagnetic force. This is then summed with a spring pre-load force to give the net force acting on the plunger, from which its instantaneous velocity and position are calculated, both of which are fed back to the 'bistable actuator' block. Figure 11.11 shows various dynamic performance characteristics when the actuator is driven from a bi-polar, 10 Hz square-wave supply voltage. The response time is $\cong$ 30 ms from the instant the coil excitation reverses polarity, the travel time being $\cong$ 5 ms. Induced eddy currents, by inhibiting the change of flux, effectively increase the damping and help to maintain the force holding the plunger at its initial position. The area enclosed by the coil flux-linkage versus current trajectory represents the electromechanical energy conversion and eddy current loss per cycle. The extent of the induced eddy currents is evident in the instantaneous field plots shown in Fig. 11.12.

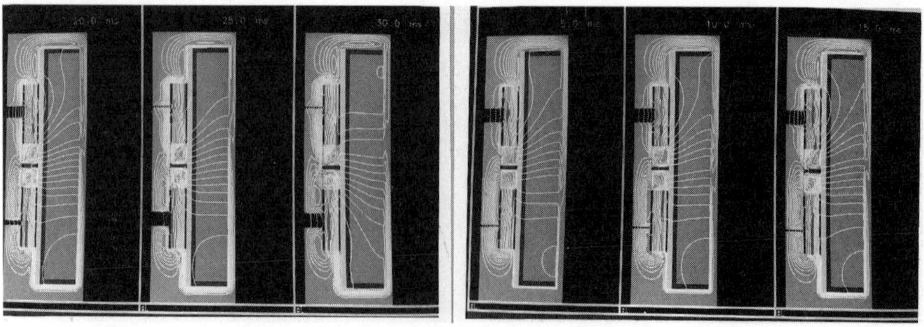

FIG. 11.12. Transient field distributions during dynamic operation of moving-magnet actuation system.

Conclusions

Due to the fact that they can offer short response times, the use of limited motion actuators for positioning applications is increasing in all market sectors, and, in addition to single-degree-of-freedom devices of the type which have been described, multiple-degree-of-freedom actuators are also emerging. However, despite the relative simplicity of direct-drive limited motion actuators, their design and dynamic performance analysis tends to be somewhat more demanding than for continuous motion electric motors. This chapter has described a few of the large number of possible actuator topologies, and outlined some of the application considerations. For many applications, the required specific force capability, and hence size, weight and dynamic response requirements, can be realized only throught the use of Nd–Fe–B magnets.

REFERENCES

[Bol90] H. R. Bolton and Y. Shakweh, *IEE Proc.*, **137B**(1) (1990), 1–13.

[Cla95] R. E. Clarke, D. S. Smith, D. Howe and P. H. Mellor, *Proc. 1st Int. Symp. on Linear Drives For Industry Applications* (1995), 417–420.

[Kim87] J. Kimble, *Machine Design* (1987), 117–121.

[Leq90] B. Lequesne, *IEEE Trans. on Industrial Applications*, **26** (1990), 401–407.

[Smi94] D. S. Smith, PhD Thesis, University of Sheffield (1994).

[Wag92] J. A. Wagner, *Conf. Record IEEE Industrial Applications Annual Meeting*, **1** (1992), 148–155.

Formula index

Subject index